S. M. Stone.

New Perspectives in Sponge Biology

New Perspectives in Sponge Biology

Edited by Klaus Rützler

in cooperation with
Venka V. Macintyre and Kathleen P. Smith

Papers contributed to the
Third International Conference on the Biology of Sponges
convened by
Willard D. Hartman and Klaus Rützler
Woods Hole, Massachusetts, 17–23 November 1985

Smithsonian Institution Press
Washington, D.C.   London

Production Editor: Nancy P. Dutro.

Library of Congress Cataloging-in-Publication Data

International Conference on the Biology of Sponges (3rd : 1985 : Woods
Hole, Mass.)
New perspectives in sponge biology : Third International Conference on the
Biology of Sponges, convened by Klaus Rützler and Willard D. Hartman,
Woods Hole, Massachusetts, 17–23 November 1985 / edited by Klaus
Rützler in cooperation with Venka Victoria Macintyre and Kathleen T.
Smith.
p. cm.
Includes bibliographical references and index.
ISBN 0–87474–784–8
1. Sponges—Congresses. 2. Sponges, Fossil—Congresses.
I. Rützler, Klaus, II. Macintyre, Venka Victoria.   III. Smith,
Kathleen T.   IV. Title.
QL371.I57   1985
593.4—dc20   90–9996

British Library Cataloguing-in-Publication Data is available.
Manufactured in the United States of America.
97 96 95 94 93 92 91 90 5 4 3 2 1

∞ The paper used in this publication meets the minimum requirements of
the American National Standard for Permanence of Paper for Printed
Library Materials Z39.48—1984.

For permission to reproduce illustrations appearing in this book, please
correspond directly with the owners of the works, as listed in the individual
captions. The Smithsonian Institution Press does not retain reproduction
rights for these illustrations individually, or maintain a file of addresses for
photo sources.

Foreword

At a time in the history of the biological sciences when sophisticated new technological advances have turned the attention of many biologists to the study of the molecular and cellular bases of life and a concentration on a relatively few organisms that can readily be raised under laboratory conditions, this volume presents work concentrating on a specific group of organisms, the sponges. Biologists who orient their studies around specific groups of organisms apply all the sophisticated techniques of modern biology to these organisms, as well as many of the traditional approaches. In many ways the approach of the organismal biologist is broader than that of the reductionist since his or her interests might run the gamut from molecular biology to population biology or from developmental biology to paleobiology in an attempt to understand the general biological processes of the organisms involved.

Organismal biologists regard their approach to the study of biology as vital and timely. It is important for man to have a full knowledge of the organisms with which he shares this planet, in part to give man a sense of his place in nature and in part to make him aware of the usefulness to him of these organisms. Sponges, to take the case in question, are ubiquitous in aquatic environments, both marine and freshwater. Over that part of the earth's surface covered by coral reefs the total biomass of sponges may equal that of the corals that are responsible for building the frameworks of the reefs. The calcareous skeletons of sclerosponges help to fill in vacant spaces in the reef framework and are also responsible for framework construction below the depths where reef-building scleractinian corals thrive. Sponges are also important in the productivity of temperate and cold-water seas and of fresh-water environments. The more direct uses of sponges to man include delicate cleansing and medicinal uses. Since antiquity, sponges with a fine, netlike skeleton of protein fibers have found favor in human bathing activities, and the same water-holding capacity combined with fine, soft texture renders them useful in certain industrial applications even today in the face of synthetic substitutes. More recently, certain sponges have been found to harbor antibiotic and anti-cancer substances which have become of considerable interest to the pharmaceutical industry.

The scope of the papers in this volume attest to the breadth of interest of those organismal biologists who concern themselves with sponges.

WILLARD D. HARTMAN
Professor of Biology and Acting Director
Peabody Museum of Natural History

Contents

Introduction

The biology of the phylum Porifera is a rapidly growing field. Its growth can be measured by the increasing frequency of national and international symposia, workshops and conferences.

Until the late 1960s sponge workers had never gathered for formal discussions of progress and problems in their work. Part of the reasons was certainly that these specialists were few and scattered, that their discipline was considered eccentric at best by other biologists and therefore difficult to fund, and that travel was tedious and rather expensive.

Since 1968, when the first international conference on sponge biology convened in London (Fry, 1970), similar events took place at ever-increasing rate. In 1975 a symposium in Albany, New York (Harrison and Cowden, 1976) brought together most North American investigators of sponge biology. Paris, in 1978, was the location of the second international conference (Lévi and Boury-Esnault, 1979). In 1983 a group of European sponge taxonomists met at Sherkin Island Marine Station, Ireland (Jones, 1987). Woods Hole, Massachusetts, accomodated the third international conference in 1985 (this volume). Workshops to create data bases and develop new methods and concepts in taxonomy of Northeastern Atlantic (including Mediterranean) sponges were held in Marseille in 1986 (Vacelet and Boury-Esnault, 1987), Geneva in 1987, and Genova in 1988. And at the time of this writing (September 1988) another international conference on fossil and Recent sponges is being staged in Berlin (Keupp and Reitner, 1988).

It was sad to learn that one participant of the Woods Hole conference did not live to see his contribution in print. Takaharu Hoshino was one of the most productive new sponge taxonomists of his native Japan. We were also informed that Gerardo Green, a contributor to the conference proceedings, recently died at a young age. He worked on systematics and chemical defense of Mexican sponges.

Before introducing this volume another important recent event should be noted. The library, papers, photographs, and microscope-slide preparations of Max Walker de Laubenfels, eminent although controversial American sponge worker, have been turned over to the scientific public after almost 30 years in private possession. The

1, 2, M. W. de Laubenfels, 1934; 3, A. Hyatt, ca. 1880; 4, P. S. Galtsoff, 1928; 5, H. V. Wilson, 1928; 6, W. J. Sollas, ca. 1880; 7, J. S. Bowerbank, ca. 1880; 8, S. O. Ridley, ca. 1885; 9, R. Kirkpatrick, ca. 1900; 10, A. Dendy, ca. 1900; 11, G. P. Bidder, ca. 1910; 12, M. Burton, ca. 1925; 13, F. Ferrer, ca. 1920; 14, E. Topsent, ca. 1930;

IS TERRARVM.

15, J. Cotte, 1929; 16, W. Arndt, ca. 1930; 17, J. Thiele, ca. 1900; 18, F. E. Schulze, ca. 1900; 19, K. Schröder, ca. 1930; 20, L. Breitfuss, ca. 1910; 21, W. Weltner, ca. 1910; 22, G. C. J. Vosmaer, ca. 1890; 23, H. van Trigt, ca. 1920; 24, N. Miklucho–Maclay, ca. 1880; 25, S. Hozawa, ca. 1900; 26, I. Ijima, ca. 1900; 27, Y. Okada, 1932.

entire collection was donated to the School of Marine Sciences of the University of Miami, Florida. Books and reprints are now incorporated in the library of that institution. Other documents and, above all, the original microscope slide preparations on which all of de Laubenfels's taxonomic descriptions are based have been transferred (through the generosity of Gil Voss, Miami) to the Porifera section of the National Museum of Natural History, Smithsonian Institution.

Photographs of important sponge workers of the past shown on these pages are part of the de Laubenfels legacy, courtesy of the School of Marine Sciences, University of Miami. It would have been tempting to supplement these pictures with a more complete selection from other sources, such as the collections of the British Museum (Natural History) (Shirley Stone, pers. comm.) but time and space constraints did not permit this. However, several illustrated biographies and obituaries are published and can be consulted by readers interested in the following sponge scientists: L. M. Lambe (Kindle, 1920); E. A. Minchin (Woodcock, 1925; Corliss, 1979a); C. Keller, E. Haeckel (Keller, 1928); A. Wierzejski (Siedlecki, 1935) H. V. P. Wilson (Costello, 1961); R. von Lendenfeld, F. Urban (Kraus, 1967); W. E. Ankel (Steiner, 1967); O. F. Müller, J. B. P. A. Lamarck, W. S. Kent, E. Haeckel, J. Leidy (Corliss, 1979b); A. F. Norman (Mills, 1980); W. Arndt (Kühlmann, 1985); E. O. Schmidt (Desqueyroux-Faundez and Stone, in prep.).

New Perspectives in Sponge Biology is the result of the third international conference on the topic held at the Swope Center of the Marine Biological Laboratory in Woods Hole, Massachusetts, during the week of 17–23 November 1985. The conference was cosponsored by the National Museum of Natural History, Smithsonian Institution, and the Peabody Museum of Natural History, Yale University.

It is crucial to document again the state of the science of sponge biology, a field considered by many the most enigmatic and least studied in organismic biology. New theoretical ground is broken in most of the disciplines, many papers cross disciplinary boundaries, bodies of disparate data are synthesized—due primarily to personal interaction during the conference—and new methods of analysis are described and demonstrated. Highlights of important new developments include the following:

In evolutionary biology and paleobiology studies continue on the relationship of Recent sclerosponges to enigmatic fossil groups, with evidence accumulating that stromatoporoids, chaetetids, and possibly also favositids belong to the Phylum Porifera.

Biochemical investigations explore the diversity of natural products in sponges. The early work on sterols and fatty acids of sponges has been extended using the more efficient methodology of today. Work on terpenoids and carotenoid pigments is proceeding at this time with new application to taxonomic problems.

Work on comparative ultrastructure of sponges proceeds apace as biologists explore the diversity of cell types and attempt to work out their evolution. There have been extensive studies of intra- and interspecific grafting among sponges and a search for the immunological basis for rejection or acceptance reactions. Studies of hexactinellid sponges confirm that these forms have mostly syncytial organization.

The developmental potential of the cells of larval sponges continues to be of interest as we seek to learn more about totipotency and cell transformation in these animals. The classical problem of reaggregation of dissociated sponge cells continues to interest biologists, now largely at the molecular level.

Micromorphological studies with the aid of scanning electron microscopy are revealing much about the internal structure and function of sponges. An extensive comparative study of the choanocyte chambers of sponges is providing evidence of considerable diversity in the functional significance of the material of which sponge skeletons are constructed and of the influence of the environment on skeletal form.

The description of new species of sponges proceeds with no seeming end to the discovery of new forms. Novel characters for sponge taxonomy are examined and new systematic techniques are employed. Some attention is given to biogeographic exploration.

The ecology of coral reef sponges is an active field of research at present. The discovery of the widespread occurrence of symbiotic bacteria, cyanobacteria, and zooxanthellae in reef sponges has led to the demonstration of the importance of the metabolites of these monerans and protists to the nutrition of their sponge hosts. Distribution studies and studies of species interactions are also being pursued. The difficult question of determining the contribution of autotrophic nutrition to the livelihood of sponges possessing photosynthetic symbionts is being studied in freshwater sponges.

Acknowledgments

The Third International Conference on the Biology of Sponges was made possible by several grants aiding travel of participants. We wish to acknowledge support from The National Museum of Natural History, the Office of Fellowships and Grants, the Walcott Fund, and the Directorate of International Activities, Smithsonian Institution. We thank Richard Fiske, James Tyler, Mary Tanner, Marsha Sitnik, Roberta Rubinoff, Catherine Harris, Francine Berkowitz, and Saundra Thomas. Additional support came from the Peabody Museum of Natural History, Yale University, and SeaPharm, Harbor Branch Foundation.

We are grateful to all who helped with the organization of the conference, particularly Kathleen Smith, Marty Joynt, Elaine Sarkkinen, Britt Wheeler, Carol Ailes, and Judith Petroski.

Publication support for this volume was provided by the National Museum of Natural History and the Peabody Museum of Natural History. We are grateful to Robert Hoffmann, James Tyler, Roger Cressey, and Willard Hartman. Administrative help was received from Mary Tanner and Marty Joynt. We thank Cornelia Holley and Lois Edwards for typing and editorial assistance. Yoshiki Masuda provided the photographs of conference participants. For help with design, illustrations, and photography we acknowledge Jo Moore, Molly Ryan, Mike Carpenter, and Victor Krantz. The quality of many papers was improved by comments of numerous reviewers, particularly Ralph Chapman, Clive Evans, Willard Hartman, Clifford Jones, Dave Pawson, Rob van Soest, and Jean Vacelet.

References Cited

Corliss, J. O. 1979a. A Salute to Fifty-four Great Microscopists of the Past: A Pictorial Footnote to the History of Protozoology. Part II. *Transactions, American Microscopical Society*, 98:26–58.

———. 1979b. *The Ciliated Protozoa*. Second Edition. Pergamon Press: New York. 455 pp.

Costello, D. P. 1961. Henry Van Peters Wilson. *Biographical Memoirs, National Academy of Sciences*, 35:351–383.

Desqueyroux-Faundez, R., and S. M. Stone. In prep. The Sponges of E. O. Schmidt, An Illustrated Catalogue.

Fry, W. G., ed. 1970. *The Biology of the Porifera*. Symposia of the Zoological Society of London 25. London: Academic Press. 512 pp.

Harrison, F. W., and R. R. Cowden, eds. 1976. *Aspects of Sponge Biology*. New York: Academic Press. 354 pp.

Jones, W. C., ed. 1987. European Contributions to the Taxonomy of Sponges. *Publications of the Sherkin Island Marine Station*, 1:1–140.

Keller, C. 1928. *Lebenserinnerungen eines Schweizerischen Naturforschers*. Zürich: Orell Füssli. 162 pp.

Keupp, H., and J. Reitner, eds. 1988. Fossil and Recent Sponges (Abstracts). Berlin: Dietrich Reimer. 57 pp.

Kindle, E. M. 1920. Memorial of Lawrence M. Lambe. *Bulletin, Geological Society of America*, 31:88–97.

Kraus, O. 1967. Die Poriferen-Bibliothek Lendenfeld/Urban im Senckenberg-Museum. *Natur und Museum*, 97:186–188.

Kühlmann, D. H. H. 1985. Professor Dr. Walther Arndt, Wissenschaftler und Antifaschist, Kustos am Museum für Naturkunde Berlin 1921–1944. *Mitteilungen, Zoologisches Museum Berlin*, 61:287–334.

Lévi, C., and N. Boury-Esnault, eds. 1979. *Biologie des Spongiaires*. Colloques Internationaux du C.N.R.S. 291. Paris: Centre National de la Recherche Scientifique. 533 pp.

Mills, E. L. 1980. One "Different Kind of Gentleman": Alfred Merle Norman (1831–1918), Invertebrate Zoologist. *Zoological Journal of the Linnean Society*, 68:69–98.

Siedlecki, M. 1935. Prof. Dr. Antoni Wierzejski. *Mémoirs, Academie Polonaise des Sciences Mathématiques et Naturelles* (series B), 3(9):i–viii.

Steiner, G. 1967. W. E. Ankel, 70th Birthday. *Zeitschrift für Morphologie und Ökologie der Tiere*, 60:1–4.

Vacelet, J., and N. Boury-Esnault, eds. 1987. *Taxonomy of Porifera*. Berlin: Springer. 332 pp.

Woodcock, H. M. 1925. E. A. Minchin. *Parasitology*, 17:157.

Paleobiology

THEO M. G. VAN KEMPEN*
Geologisch Instituut
Universiteit van Amsterdam
Nieuwe Prinsengracht 130
NL 1018VZ, Amsterdam, The Netherlands

On the Oldest Tetraxon Megascleres

Abstract

A revision is given of an assemblage of Australian quadriradiate sponge spicules of Middle Cambrian age, formerly attributed to dialytine calcareous sponges. The present assignment of these spicules to soft-bodied demosponges is based primarily on morphological comparison with fossil and recent spicule types and on a review of the relevant literature. The triaene, as a basic type of demosponge tetraxon megasclere, is found to have already existed in the Middle Cambrian.

In 1978 I reported on an assemblage of isolated, recrystallized sponge spicules found in a core sample from a water borehole—Gidyea 1, at a horizon of approximately 37 m—drilled in the central part of the Georgina Basin, Northern Territory, Australia (van Kempen, 1978). The well-preserved opaquely translucent, chalcedonic spicule pseudomorphs (index of refraction between 1.530 and 1.544) come from beds known as the Ranken Limestone, and have a definite early Middle Cambrian (Ordian) age based, among other factors, on the occurrence of some diagnostic species of trilobites (Gatehouse, 1967; Kruse, 1983). The sample consisted of more than 200 spicules from lithistids, soft-bodied demosponges, hexactinellids, heteractinellids, and occasional monaxon spicules of undetermined origin. The triaene-shaped four-rayed spicule forms among them were formerly assigned to the dialytine Calcarea for reasons given below. Recent questions about the validity of this assignment have prompted a reconsideration of the true nature of these Cambrian spicules.

Furthermore, SEM-micrographs of smooth tetraxon spicules from the Baltic area demonstrate that representatives of the Tetractinellida already existed in the Early Paleozoic. The Ordovician orthotriaene megasclere figured here is preceded by similar spicule forms that were already profusely present in the Australian spicule sam-

*Present address: Vrije Universiteit, Instituut voor Aardwetenschappen, Postbus 7161, The Netherlands.

ple; orthotriaenes probably represent the primary form of the triaene megasclere.

Former Assignment

The Australian spicules under discussion were formerly assigned to dialytine Calcarea on the following grounds:

1. Grossly seen, owing to their T-shaped form, the assemblage of predominantly four-rayed spicules was considered suggestive of spicules belonging to dialytine Calcarea.

2. Differences between the general characteristics of the Australian spicules and those of recent dialytine sponges were presumed to be the possible result of evolutionary changes.

3. Until recently, it was still widely assumed that the smooth tetraxon megasclere, which is a common constituent both in the skeletons of choristid (astrophorid and spirophorid) and homosclerophorid (carnosid) soft sponges and in many fossil and extant lithistid genera, did not appear until the Early Carboniferous (Finks, 1967, 1970; Finks and Hill, 1967; Bergquist, 1978; Wiedenmayer, 1980).

Morphology of the Spicules

All the spicules in the Australian sample were grouped according to the kind of sponges they represent and then were photographed. Approximately 45 were four-rayed forms having an unpaired ray that was varyingly longer than the three grossly equidimensional paired rays and thus had the shape of triaenes (Figures 1, 2). Two types of triaenes could be recognized, namely, orthotriaenes and plagiotriaenes (as defined by Wiedenmayer, 1977:42; and orthotriaenes as also defined by Reid, 1968:35) together with somewhat intermediate forms. The orthotriaenes are more numerous than the plagiotriaenes. The latter have long shafts and comparatively short, straight spinelike clads that form angles between 125 and 130 degrees with the shaft (Figures 1, 2a). The clads of the orthotriaenes are either straight (Figures 1, 2b), or somewhat convexly curved upward (Figures 1, 2c); the latter form, which is generally stouter than the former, resembles the clads of orthotriaenes as in Lendenfeld's (1907:plate XXIX:22, 24) recent species *Stelletta sigmatriaena* or in Lebwohl's (1913:plate II:49–51) *Stelletta pilula*. The clads of the orthotriaenes grossly form angles of 90–110 degrees with the shaft; the latter tend to become transitional to plagiotriaenes. Von Lendenfeld (1907:plate 32, 37, 38) mentions plagioorthotriaenes in *Isops gallica*. The sporadic three-rayed spicules in the Australian sample (pd in Figure 1) are probably modifications of the triaenes and represent plagiodiaenes; similar modifications are known from extant choristid species, for example, *Thenea megaspina*, *Isops texoteuches* (von Lendenfeld, 1907:plate XXI:

20,21; plate XXXVI:13,14), and *Calthropella simplex* (Sollas, 1888:plate X:21–26).

Dimensions of spicules shown on Figure 2 are listed in Table 1. The total length of a spicule is taken as the distance from a point halfway between the tips of clads to the tip of the shaft. The length of the shaft is measured from the tip up to the base of the cladome. The shaft is thickest in diameter at a point just beneath the base of the clads, where the sides of the shaft become parallel. If present, only complete rays of individual clads of the spicules were measured; otherwise the best preserved broken clad was used. Because the three clads of the spicule in Figure 2c1 are complete, a range is given for their individual length as well as the distance between the tips of their rays. All measurements are based on actual spicules.

Arguments for a Demosponge Provenance

ARGUMENT 1. In a recent paper, van Soest (1984:216) expressed serious doubt concerning the dialytine nature of the Australian spicules under investigation. He noted, for example, the presence of numerous conspicuous, long-shafted triaenes that are not typical of modern dialytine (or other spiculate) Calcarea, and the absolute preponderance of quadriradiate spicules over triradiate individuals, which is not characteristic of the skeleton of any extant dialytine representative.

Lending further support to van Soest's view is the fact that the Late Paleozoic (Lower Carboniferous) dialytine spicule assemblage reported by Hinde (1887:176)—which is still the oldest known occurrence of loose spicules from a calcareous sponge—closely matches the spicule types that constitute the skeleton in a modern dialytine sponge, in spite of the great time span involved. Dialytine sponges seem to be very conservative organisms as far as their spicular morphology and diversity are concerned.

ARGUMENT 2. Triaenes (and true calthrops, which were lacking in the sample) are generally considered to represent a basic type of the tetraxon demosponge megasclere. Triaenes are common and widely dispersed among recent choristids and lithistids; they are also known to abound in

Table 1. Dimensions (μm) of the spicules represented in Figure 2 (for method of measurement see text)

Dimension	Plagiotriaene	Orthotriaene	
	2a,a1	2b,b1	2c,c1
Total length	1038	625	550
Length of shaft	930	570	425
	(lacking tip)		(incomplete)
Diameter of shaft	49	35	68
Length of clads (rays)	200	125	337–350
		(incomplete)	
Distance between tips of rays	335	250	618–662

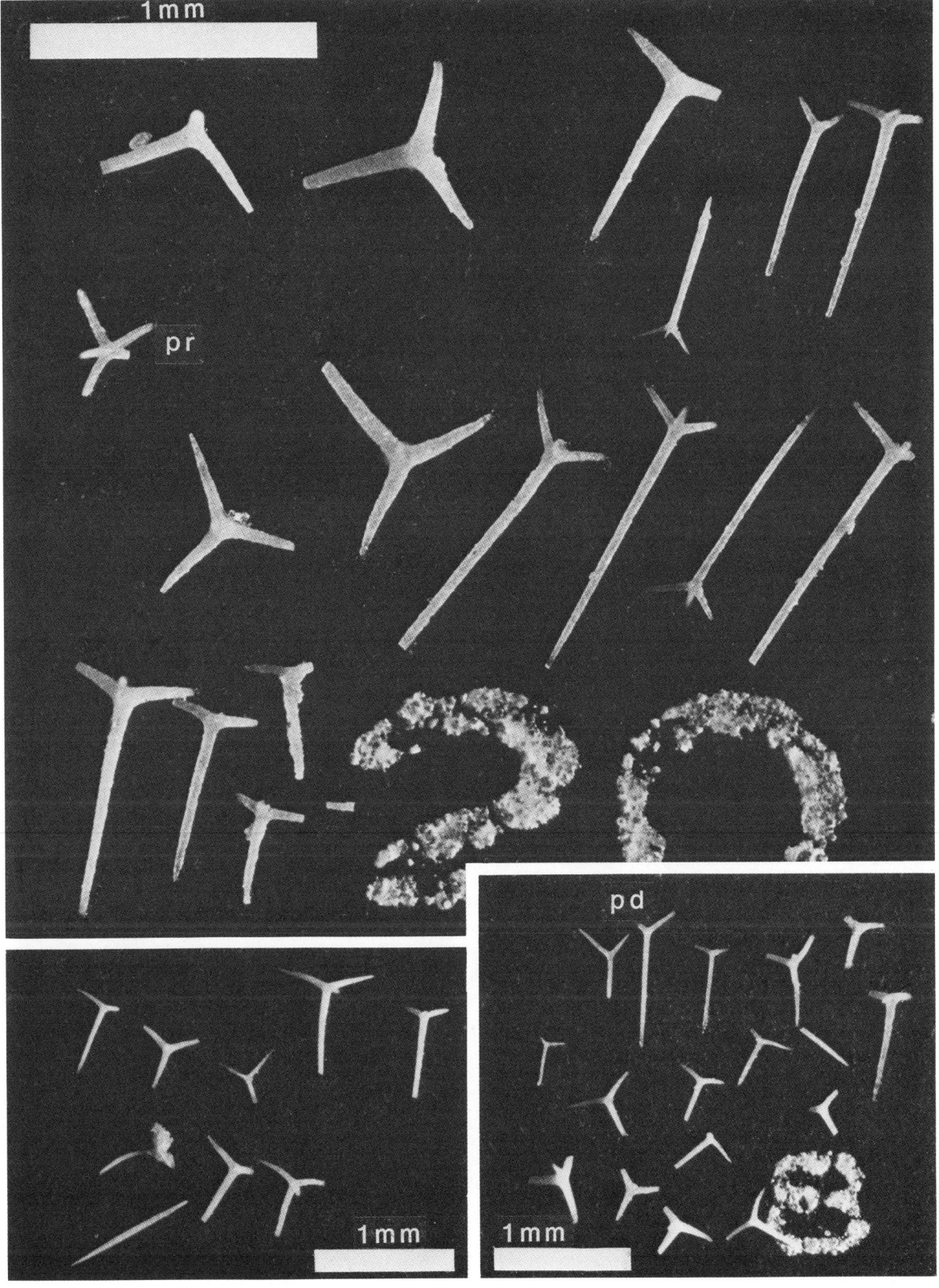

Figure 1. Isolated orthotriaenes and plagiotriaenes from the early Middle Cambrian Ranken Limestone, Georgina Basin, Australia: three-rayed forms (*pd*, lower right) apparently represent plagiodiaenes; irregular pentaradiate spicule (*pr*, upper left) is of unknown origin.

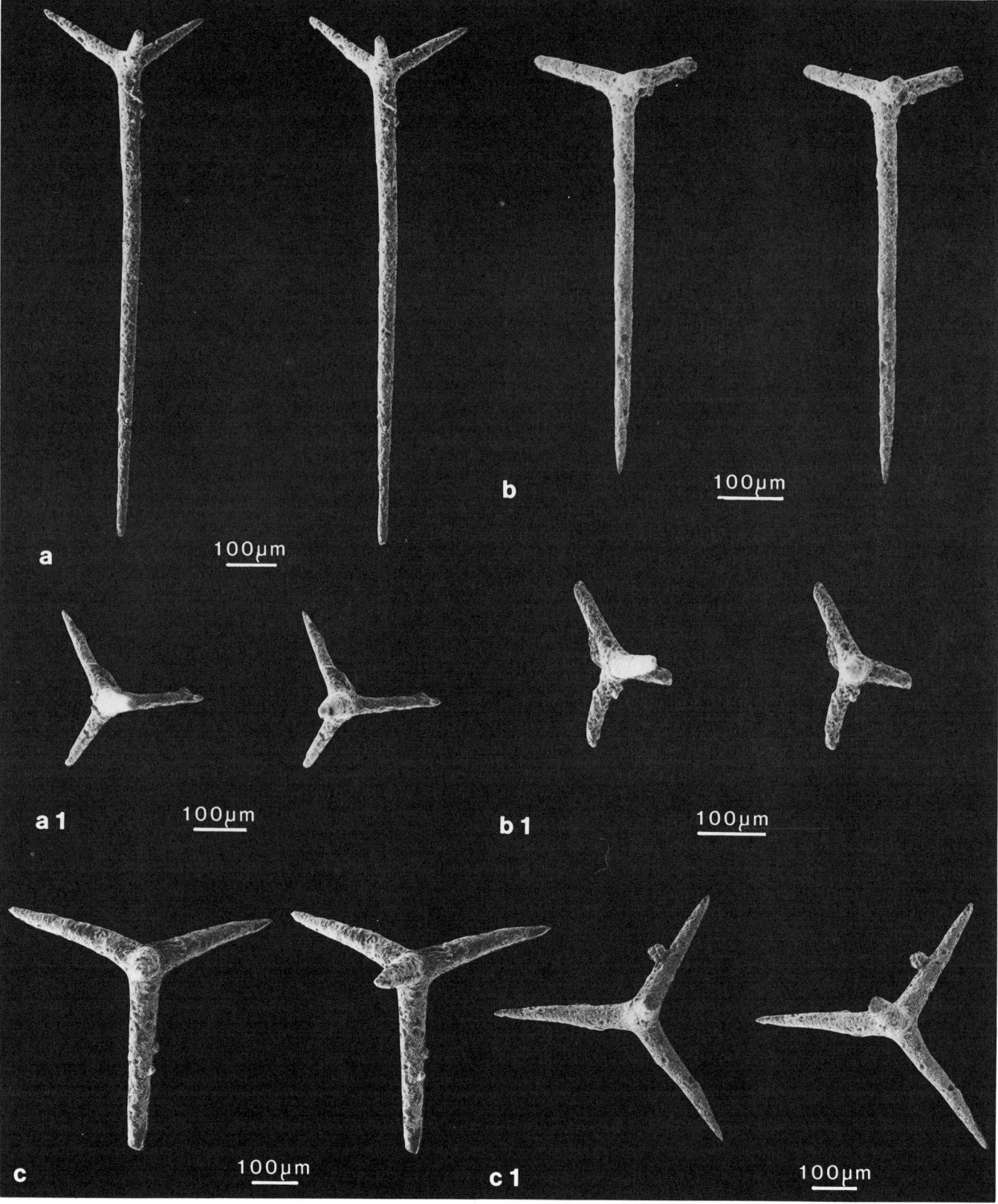

Figure 2. SEM stereopairs of smooth tetractinellid spicules from the early Middle Cambrian Ranken Limestone: *a*, intact pla-giotriaene; *b*, orthotriaene with straight clads, all broken; *c*, orthotriaene with upwardly curved clads and shaft lacking part of tip. Each spicule shown in side view *(a, b, c)* and from underside of cladome *(a1, b1, c1)*.

Mesozoic spicule assemblages and are represented in assemblages from older strata (Hinde, 1880, Upper Cretaceous; Schrammen, 1924, Upper Cretaceous; Hinde, 1885, Lower Cretaceous; Schönlaub, 1973, Lower-Upper Cretaceous; Moczydlowska and Paruch-Kulczycka, 1978, Jurassic-Upper Cretaceous; Schrammen, 1936, Upper Jurassic; Geyer, 1958, Upper Jurassic; Mostler, 1976, Triassic; Alexandrowicz, 1978, Lower Carboniferous; and

others). Of the triaenes, orthotriaenes and plagiotriaenes are the most common types of spicules (Wiedenmayer, 1977:50). They probably represent the basic types of triaene spicules, and they are the forms profusely represented in the Australian spicule sample. Note that one of the spicule types, shown in Figure 3a–a4, is an orthotriaene from the Late Ordovician, which, in its general features, closely resembles the Cambrian forms.

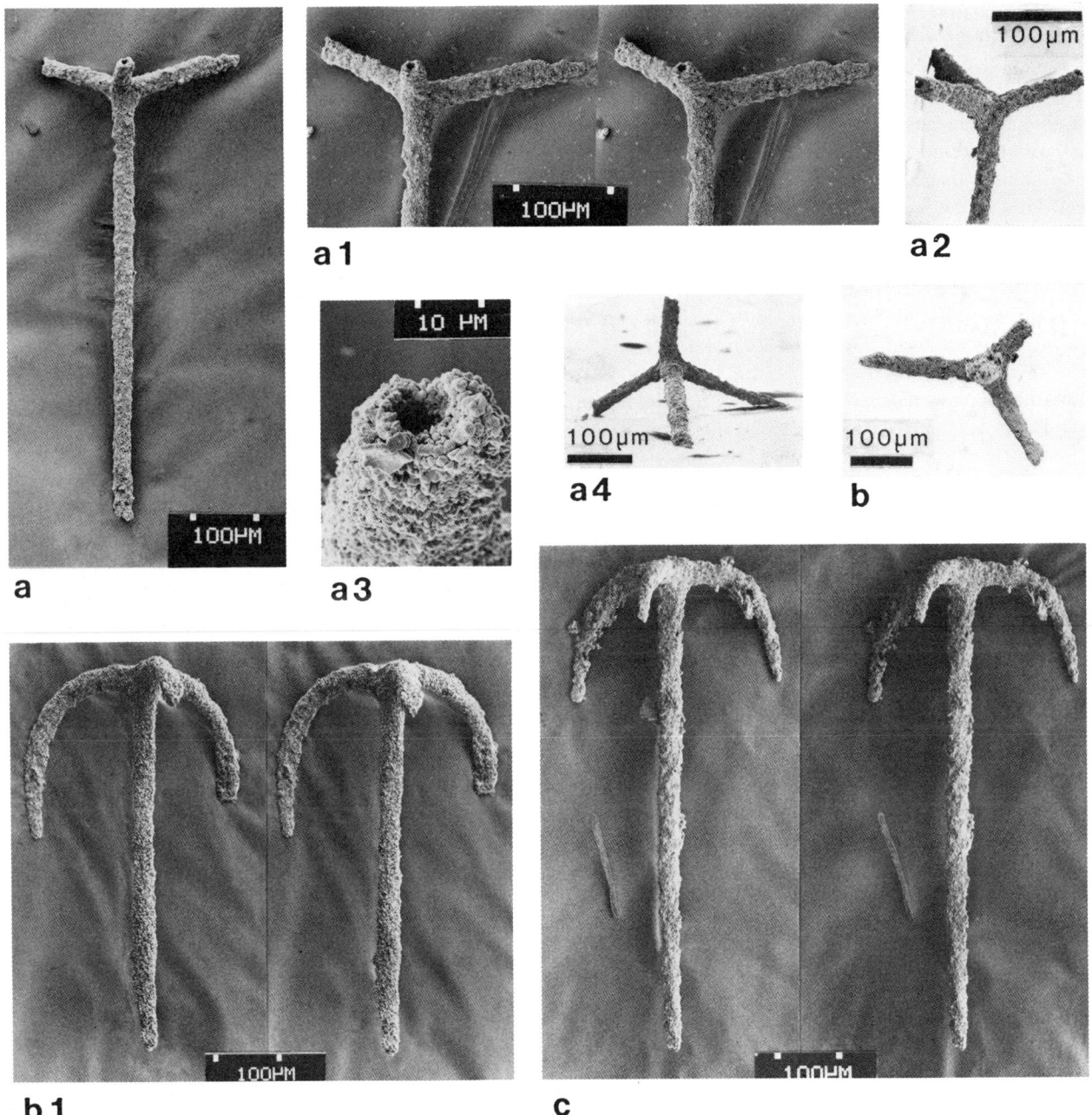

Figure 3. SEM micrographs of smooth tetractinellid spicules from the Baltic Late Ordovician: *a*, orthotriaene, side view (*a1*, close-up stereopair of clads; *a2*, base of cladome, ray to left showing axial hollow that corresponds with ray pointing to left in *a1*; *a3*, close-up of aperture; *a4*, same view from underside of cladome); *b*, anatriaene, seen from underside of cladome (*b1*, side view, stereopair); *c*, sideview, stereopair anatriaene.

ARGUMENT 3. The literature on Early Paleozoic tetraxon sponge spicules shows a steady increase of their presence in Ordovician and Cambrian times. Eisenack (1968:86) and Reif (1968:735), for example, reported the occurrence of some fragments of "tetractine" spicules of Late Ordovician age originating from Baltic areas; Reif speculated that these were from choristid or carnosid sponges. Reid (1968:24; 1970:72) pointed out that the choristid tetraxon spicule forms known from the Lower Carboniferous of Northern Ireland and Scotland are far too diverse to represent the true first choristids. Subsequently, Mostler and Mosleh-Yazdi (1976:17,19) describe and figure spicules of Upper Cambrian age from the Alborz Mountains, Iran, which they identified as protetraenes and calthrops. However, the protetraenes are quite aberrant from the spicule types usually found in fossil assemblages or in modern tetractinellid sponges and therefore are probably not from this group of sponges, unless they belong to an unknown choristidlike representative. The irregular calthrops they figure may have come from an ancestral representative of the homosclerophorids. Eisenack (1978:647) reports finding a fragmentary glauconite mold of a lumpy protriaenelike spicule from the Llanvirnian (late Early Ordovician) of Öland, Sweden. Rigby (1983:18) noted that "choristid sponges may have had a long, but unknown, Early Paleozoic history." Kruse (1983) mentions triacts and orthotriaenes in a spicule assemblage obtained from the same beds in the Georgina Basin (Ranken Limestone) as the spicules under discussion here.

The Baltic Spicules

The Baltic spicules figured in the SEM-micrographs (Figure 3) are from the residues of Öjlemyr-flints treated with hydrofluoric acid. These flints were found as erratic boulders either in the kaolin sands on the island of Sylt, N.W. Germany, or at Wielen, W. Germany, near the Dutch border. They are of Late Ordovician, Ashgillian age, which is equivalent to the stages F_{1c}–F_2 of the Estonian chronostratigraphical classification (Schallreuter, 1984).

Since the spicules could not be examined directly, the dimensions given below had to be inferred from the scale bars in the SEM-micrographs. As far as possible, they were measured in the same way as the Australian spicules shown in Figure 2.

ORTHOTRIAENE (FIGURE 3A–A4)

PROVENANCE: Öjlemyr-flint (type Wielen; see Schallreuter, 1984:10), erratic from Wielen, W. Germany. Collection U. von Hacht, Hamburg.
AGE: F_{1c}–F_2 (Ashgill).

In its general characteristics, the eroded, long-shafted spicule closely resembles the corresponding Australian counterparts from the Middle Cambrian (Figures 1, 2b). Small pieces of the tip of the shaft and individual clads are missing, but the parts of the clads that remain are long enough to be recognized as the cladome of a dichotriaene spicule. The overall length of the spicule is 800 μm. In the complete state, the length of the three clads may each have been about 210 μm; the incomplete clad shown at the right-hand side of Figure 3a and 3a1 is about 175 μm. Two of the broken clads clearly show the aperture of an axial canal (Figure 3a–a3). The shaft is 700 μm long, and maximum thickness is about 35 μm. The orthotriaene is a form known from extant families such as the Stellettidae, Geodiidae, and Pachastrellidae, and it also resembles the orthotriaenes in the Cretaceous choristid sponge *Theneopsis steinmanni* Zittel (see Schrammen, 1910–1912:55, Figure I:5a; however, it is unlike the specimen figured in de Laubenfels 1955, Figure 21:7).

ANATRIAENES (FIGURE 3B, B1, C)

PROVENANCE: Öjlemyr-flint (type Braderup; see Schallreuter, 1984:9), erratic from the island of Sylt, N.W. Germany. Sy 159, collection U. von Hacht, Hamburg.
AGE: F_{1c}–F_2 (Ashgill).

Two specimens of the same type of tetraxon spicule are represented. The comparatively large downward curving clads form a wide, grapnel-like arch. The clads gently taper into pointed tips and the space formed between the clads and shaft is almost rectangular. The distance between the tips of the two rays, which in both micrographs are in the same plane, is about 330 and 355 μm. Both spicules, which lack the acute tip of their shaft, are 670 and 830 μm long. Maximum thickness of the shafts is roughly 40 μm for both spicules. In their overall form both spicules resemble certain hexactinellid anchoring spicules (see Hinde and Holmes, 1892:251, plate 15:22) which, however, have cladomes with four arms forming angles of 90 degrees. Fossil anatriaenes with wide-angled clads that closely resemble the Baltic spicules are described and figured by Mostler (1976:6–7) from the alpine Middle Triassic (Ladinian). In extant choristid species wide-angled anatriaene clads are seen for example in the genus *Tetilla* (*Tethya* in von Lendenfeld, 1907, plates XV:3, 4, XVI:28; Sollas, 1888, plate V:6), *Cinachyra* (von Lendenfeld, 1907, plate XV:40–42, 50), and *Thenea* (Sollas, 1888, plate VI:5), but these spicules are invariably smaller, although they may have shafts as long as 20 mm(!).

Discussion

The question of interest here is which group of demosponges do both the triaenes in the Australian sample and the Baltic representatives belong to? Lithistids most probably were not the source group since early Paleozoic lithistids, as far as can be traced in the fossil record, developed exclusively smooth megascleres of the monaxonid type, supplementary to their desma skeleton. Accordingly, there can be little doubt that the two types of triaenes in the Australian and Baltic sample are of choristid provenance. Probably all of the geologically younger (i.e., post-Cambrian) lithistids, excluding a few extant species, ever since they initiated the development of smooth tetraxon megascleres, produced only certain kinds of smooth tetraxon spicules, namely, dicho-, phyllo-, or discotriaenes and intergrading forms. These spicule forms, in comparison with the ortho-, ana-, and plagiotriaenes in the samples at hand, are likely to be more specialized kinds of triaenes adapted to form an outer or cortical skeleton and appearing at a later stage in the evolutionary development of the triaene spicule (see also Dendy, 1921: 100). The monaxons in the early lithistid stock seem to have had primarily a supporting function—such as the reinforcement of skeletal trabeculae in many anthaspidellids. In contrast, the dichotriaenes and allied forms in geologically younger lithistids functioned as a blanket or protective device, whether or not in combination with modified, often densely packed small desmas.

The first appearance in the fossil record of the group dicho-, phyllo-, and discotriaenes, including their derivatives, is not yet known. As far as I have been able to ascertain, the oldest known examples date from the Middle Triassic (Pelsonian; see Mostler, 1976). However, the triaene forms represented from that time are already so diverse that the ancestral form(s) probably originated in the Paleozoic.

Conclusions

Siliceous orthotriaenes and plagiotriaenes were already present in the early Middle Cambrian (Ordian), in addition to siliceous anatriaenes, which were present in the Baltic Late Ordovician (Ashgill).

However, from the rich fossil record known to date it appears, firstly, that early Paleozoic lithistids developed only smooth monaxonid megascleres, apart from their principal and cortical skeleton of desmas, and, secondly, that geologically younger lithistids apparently developed no triaene forms other than dicho-, phyllo-, or discotriaenes, and their intermediate forms.

Consequently both the orthotriaenes and plagiotriaenes in the Australian Middle Cambrian sample examined here and the orthotriaene and anatriaenes from the Late Ordovician Öjlemyr-flint, are considered to have originated from choristid, or choristidlike, soft-bodied sponges (Tetractinellida).

Acknowledgments

Thanks go to U. von Hacht, Hamburg, for permission to reproduce the SEM-micrographs from the Geological-Palaeontological Institute of Hamburg. R. W. M. van Soest reviewed the manuscript. C. W. Mulder-Blanken provided the SEM-micrographs of the Australian material at the Laboratory for Electron Microscopy of the University of Amsterdam, which is thanked for allowing me to use the facilities. L. H. Gonggryp and J. J. Wiersma provided practical assistance, and J. H. Baker helped to improve the English text.

Literature Cited

Alexandrowicz, S. W. 1978. Sponge Spicules from the Lower Carboniferous of the Olkusz Area. *Bulletin de l'Academie Polonaise des Sciences*, 26(2):87–94.

Bergquist, P. R. 1978. *Sponges*. Berkeley: University of California Press. 268 pp.

Dendy, A. 1921. The Tetraxonid Sponge Spicule—a Study in Evolution. *Acta Zoologica*, 2:95–152.

Eisenack, A. 1968. Mikrofossilien eines Geschiebes der Borkholmer Stufe, Baltisches Ordovizium, F2. *Mitteilungen aus dem Geologischen Staatsinstitut in Hamburg*, 37:81–94.

――――. 1978. Beitrag zur Glaukoniet-Forschung. A Contribution to Glauconite Research. *Neues Jahrbuch für Geologie und Paläontologie, Monatshefte*, 11:641–656.

Finks, R. M. 1967. The structure of *Saccospongia laxata* Bassler (Ordovician) and the Phylogeny of the Demospongia. *Journal of Paleontology*, 41(5):1137–1149.

――――. 1970. The Evolution and Ecologic History of Sponges during Palaeozoic Times. Pages 3–22 in *The Biology of the Porifera*, edited by W. G. Fry. Symposia of the Zoological Society of London, 25, London: Academic Press.

Finks, R. M., and D. Hill. 1967. Porifera and Archaeocyatha. Pages 333–345 in *The Fossil Record, a Symposium with Documentation*, edited by W. B. Harland et al. London: Geological Society of London.

Gatehouse, C. G. 1967. First Record of Lithistid Sponges in the Cambrian of Australia. *Bulletin, Bureau of Mineral Resources, Geology and Geophysics (Canberra, Australia)*, 92:57–67.

Geyer, O. F. 1958. Ueber Schwammnadeln aus dem Weissen Jura g von Würgau (Oberfranken). *Naturforschungs Gesellschaft Bamberg, Bericht*, 36:9–14.

Hinde, G. J. 1880. Fossil Sponge Spicules from the Upper Chalk found in the Interior of a Single Flint-stone from Horstead in Norfolk. Thesis, Munich. 83 pp.

――――. 1885. On Beds of Sponge-Remains in the Lower and Upper Greensand of the South of England. *Philosophical Transactions, Royal Society of London*, 174(2):403–453.

――――. 1887. Sponges of Palaeozoic and Jurassic Strata. Pages 93–188 in *A Monograph of the British Fossil Sponges*, 1. London: Palaeontographical Society of London.

Hinde, G. J., and W. M. Holmes. 1892. On the Sponge-Remains in the Lower Tertiary Strata near Oamaru, Otago, New Zealand. *Journal, Linnean Society of London, Zoology*, 24(151):177–255.

Kempen, T. M. G. van. 1978. Anthaspidellid Sponges from the Early Paleozoic of Europe and Australia. *Neues Jahrbuch für Geologie und Paläontologie, Abhandlungen*, 156:305–337.

Kruse, P. D. 1983. Middle Cambrian 'Archaeocyathus' from the

Georgina Basin Is an Anthaspidellid Sponge. *Alcheringa*, 7:49–58.

Laubenfels, M. W. de. 1955. Porifera. Pages E21–E122 in *Treatise on Invertebrate Paleontology, Part E, Archaeocyatha and Porifera*, edited by R. C. Moore. Lawrence, Kansas: Geological Society of America and University of Kansas Press.

Lebwohl, F. 1913. Japanische Tetraxonida 1. Sigmatophora und 2. Astrophora metastrosa. *Journal of the College of Science (Tokyo Imperial University)*, 35, (2), 1–116.

Lendenfeld, R. von. 1907. Die Tetraxonia. Pages 57–373 in *Wissenschaftliche Ergebnisse der Deutschen Tiefsee-Expedition auf dem Dampfer "Valdivia" 1898–1899*, edited by Carl Chun. Jena: Gustav Fischer.

Moczydlowska, M., and J. Paruch-Kulczycka. 1978. An Analysis of Siliceous Sponge Spicules from the Oxfordian of Wrzosowa and Zawodzic and the Campanian of Bonarka. *Kwartalnik Geologiczny*, 22(1):83–103. (In Polish with English summary).

Mostler, H. 1976. Poriferenspiculae der Alpinen Trias. *Geologisch-Paläontologische Mitteilungen Innsbruck*, 6(5):1–42.

Mostler, H., and A. Mosleh-Yazdi. 1976. Neue Poriferen aus Oberkambrischen Gesteinen der Milaformation im Elburzgebirge (Iran). *Geologisch-Paläontologische Mitteilungen Innsbruck*, 5(1):1–36.

Reid, R. E. H. 1968. Microscleres in Demosponge classification. *University of Kansas Paleontological Contributions*, Paper 35:1–37.

———. 1970. Tetraxons and Demosponge Phylogeny. Pages 63–89 in *The Biology of Porifera*, edited by W. G. Fry. Symposia of the Zoological Society of London, 25, London: Academic Press.

Reif, W. E. 1968. Schwammreste aus dem Oberen Ordovizium von Estland und Schweden. *Neues Jahrbuch für Geologie und Paläontologie, Monatshefte*, 12:733–744.

Rigby, J. K. 1983. Fossil Demospongia. Pages 12–39 in *Sponges and Spongiomorphs*. Notes for a short course organized by J. K. Rigby and C. W. Stearn. Knoxville: University of Tennessee Studies in Geology 7.

Schallreuter, R. 1984. Geschiebe-Ostrakoden 1. Ostracodes from Erratic Boulders 1. *Neues Jahrbuch für Geologie und Paläontologie, Abhandlungen*, 169(1):1–40.

Schönlaub, H. P. 1973. Schwamm-Spiculae aus dem Rechnitzer Schiefergebirge und ihr Stratigrafischer Wert. *Jahrbuch, Geologische Bundesanstalt, Austria*, 116:35–49.

Schrammen, A. 1910–12. Die Kieselspongien der oberen Kreide von Nordwestdeutschland. *Palaeontographica*, Supplement, 5. 385 pp.

———. 1924. Die Kieselspongien der oberen Kriede von Nordwestdeutschland. III. und letzter Theil. Mit Beitragen zur Stammesgeschichte. *Monographien zur Geologie und Paläontologie*, Series 1, 2:1–159.

———. 1936. Die Kieselspongien des oberen Jura von Süddeutschland. *Palaeontographica*, 84 and 85 (A), 1–114.

Soest, R. W. M. van. 1984. Deficient *Merlia normani* Kirkpatrick, 1908, from the Curaçao Reefs, with a Discussion on the Phylogenetic Interpretation of Sclerosponges. *Bijdragen tot de Dierkunde*, 54(2):211–219.

Sollas, W. J. 1888. Report on the Tetractinellida Collected by H.M.S. "Challenger" during the years 1873–1876. Report on the Scientific Results of the Voyage of H.M.S. Challenger, 25. clxvi + 458 pp.

Wiedenmayer, F. 1977. *Shallow-water Sponges of the Western Bahamas*. Basel and Stuttgart: Birkhäuser. 336 pp.

———. 1980. Siliceous Sponges, Development through Time. Pages 55–85 in *Living and Fossil Sponges*. Notes for a short course, edited by W. D. Hartman, J. W. Wendt, and F. Wiedenmayer. Miami: University of Miami.

ROBERT M. FINKS
Department of Geology
Queens College
City University of New York
Flushing, New York 11367

Late Paleozoic Pharetronid Radiation in the Texas Region

Abstract

Calcareous sponges are abundant in the Pennsylvanian and Permian marine deposits of Texas and neighboring states. They build reefs in the Permian. All seem interrelated because they are built of a nonspicular skeleton composed of spherulitic calcium carbonate, probably originally aragonite. The skeleton consists either of internal trabeculae that outline anastomosing tubular spaces, probably originally filled with sponge tissues, or a cortical, sheetlike covering, pierced by pores and frequently outlining modular, hollow structures (chambers), probably also originally filled with soft parts. Combinations of the two also occur. Six types of skeleton are recognized, based on various degrees of development of the trabeculae or the cortical skeletons, types of pores, and sizes of spherulites. They are considered evolutionary lineages.

Most of these types, even individual genera, persist into the Triassic. In some Triassic genera, occasional calcareous spicules are embedded in the spherulitic calcareous skeleton. In others the spherulites become flattened and flakelike. In the Jurassic and Cretaceous, most trabecular types have large calcareous spicules coring the trabeculae, coated by what may be flaky spherulites. It is proposed that spicules were gradually added to a spherulitic skeleton, and that most of these sponges are related. The anastomosing habit of the soft parts and the presence of flaky spherulites in the living genera *Murrayona* and *Paramurrayona* suggest that these sponges are Calcinea.

Texas and its neighboring states were the site of the most persistent epicontinental seas in North America during the Late Paleozoic. This persistence, together with a position near the paleoequator, provided a very favorable site for the development of the apparently tropical group of sponges (see map in Finks, 1970, fig. 12) with a skeleton containing nonspicular calcium carbonate. These sponges may conveniently be called "pharetronids," as proposed by Zittel (1878). Steinmann (1882) recognized two subcategories among the forms Zittel included: the Inozoa, those in which the nonspicular skeleton has a fibrous structure; and the Sphinctozoa, those in which the nonspicular skeleton is sheetlike and outlines hollow chambers. It is apparent from other contributions to this volume that the pharetrones in this broad sense are a polyphyletic group. However, there is considerable evidence that the Late Paleozoic forms are largely interrelated. There are intergradations between them, their nonspicular skeleton is spherulitic, and the skeletal parts tend to outline anastomosing tubular spaces.

As I have indicated elsewhere (Finks, 1983), in my view the anastomosing tubular spaces (cf. Hartman, 1958) and the spherulitic skeleton (cf. *Murrayona* Kirkpatrick, 1910, and *Paramurrayona* Vacelet, 1967) point to a calcinean affinity for these Late Paleozoic forms, and for their probable descendants of Triassic and later times.

Structural Types of Pharetronids

The nonspicular calcareous skeleton may be divided into two structural categories: (1) an internal one arranged as fibers or trabeculae and (2) an external (cortical or peripheral) one in the form of a porous or nonporous sheet. The two may coexist in the same sponge, but one or the other is usually dominant. Those forms in which the internal skeleton dominates are the typical Inozoa. Those in which the external skeleton dominates are the typical Sphinctozoa. In this second group, the external skeleton outlines chambers that appear to reflect an episodic manner of growth. Each chamber represents a single bout of growth, and the external skeleton appears to have formed in toto, after the new unit of soft parts attained its full dimensions (no incompletely formed chambers have ever been found among fossils). It should be noted that the chambers were probably not hollow in life, but rather were filled with complexly organized soft parts, for a trabecular internal skeleton may exist within the chambers of some species. Sphinctozoa also have imperforate vesicles or diaphragms (rather analogous to rugose-coral dissepiments), that appear to mark the upward (or distal) withdrawal of living tissue. Each is continuous with a secondary lining of the part of the chamber distal to it (see Figure 1). Anatomical terms useful for understanding sphinctozoan skeletons are shown in Figure 2. They are more fully defined in Finks (1983).

An episodic manner of growth is also found in some Inozoa (for example, in some species of *Fissispongia* of the Late Paleozoic) and produces an external appearance similar to that of typical Sphinctozoa. In the case of *Fissispongia*, it is clear that the modular and nonmodular growth forms are closely related (undescribed species; Finks, in prep.). In a general way the modular growth correlates with a more strongly developed cortical layer.

The internal fibrous or trabecular skeleton of the Inozoa outlines anastomosing tubular spaces which are mainly oriented longitudinally and radially, and provide continuous passageways between the exterior surface and

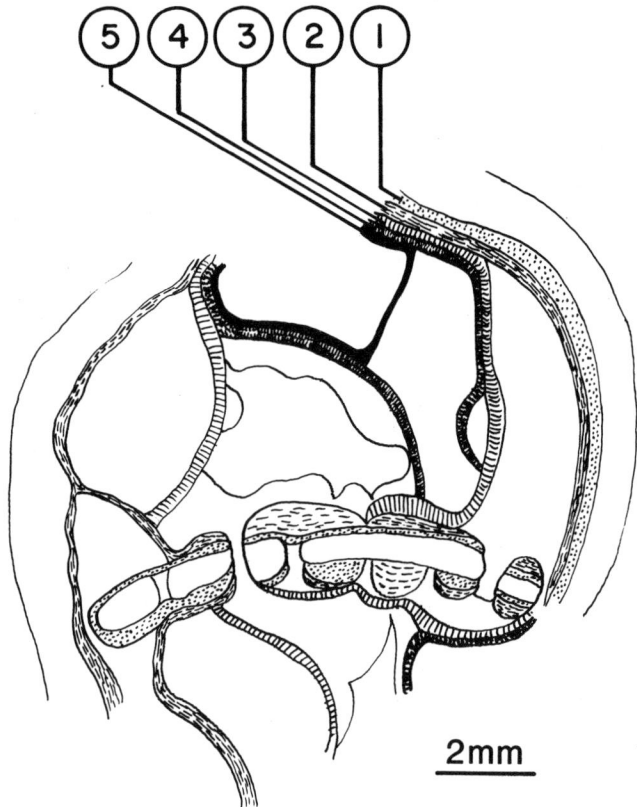

Figure 1. Drawing of a longitudinal thin section (after a photograph in an unpublished manuscript of N. D. Newell) of a specimen of *Girtyocoelia typica* King (Pennsylvanian, Missourian, Plattsburg Formation, SE Kansas) showing successive layers (numbered sequentially 1–5) of secondary vesicular skeletal tissue formed during progressive distal withdrawal of sponge soft tissue from chamber. At stage 3, sponge tissue in lower chamber still communicated through a pore in interwall with sponge tissue in upper chamber, being confined in both chambers to axial region bounded by layer 3. By stage 5, soft tissue remained only at distal (upper) end of upper chamber.

the cloaca and (or) upper surface. These tubular spaces are the only places in which the sponge tissue could have been located; thus they are not canals, although a canal could have occupied the center of each tube. The tendency to form anastomosing tubes is characteristic of Calcinea (Hartman, 1958). An example from the Texas Permian is illustrated here (Figure 3).

Although pharetronids from the Mesozoic and later may contain spicules embedded in their nonspicular skeleton, no spicules have been found thus in the Pennsylvanian and Permian pharetronids of the Texas region; their skeleton appears to be completely nonspicular. At the microscopic level, it is composed of spherulites that are spheroidal rather than flake-shaped (flaky spherulites occur in some Triassic and later pharetronids, and will be discussed below).

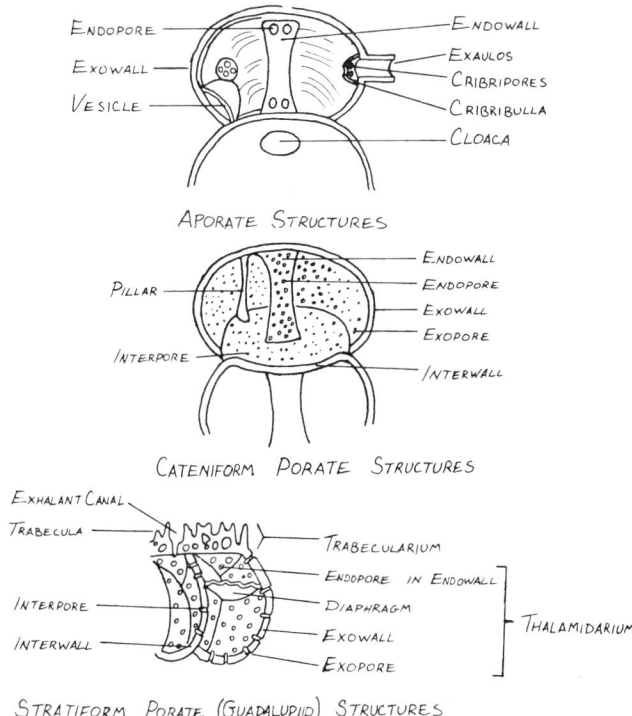

Figure 2. Morphological features of Sphinctozoa. Aporate Structures = type 1; Cateniform Porate Structures = type 2; Stratiform Porate (Guadalupiid) Structures = type 5. (From Finks, 1983, Figure 1.)

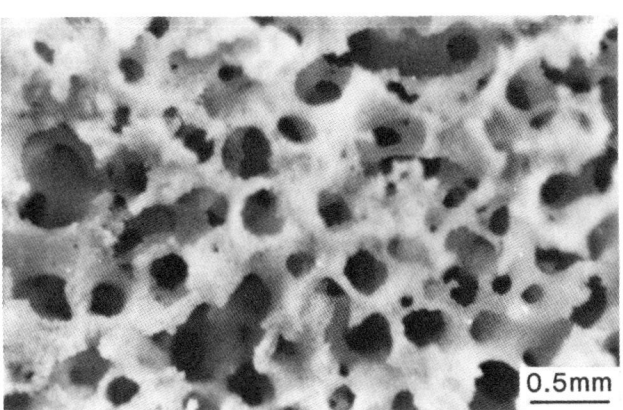

Figure 3. Inozoan internal trabecular net outlining anastomosing tubular spaces, *"Virgola"* sp., USNM 127729, USNM loc. 703, Permian, Road Canyon Formation, Old Word Ranch, Glass Mountains, Texas.

The structural types present in the Pennsylvanian and Permian beds of the Texas region are as follows:

TYPE 1. Skeleton exclusively peripheral and "aporate," with openings in it usually limited to inhalant cribribullae provided with exauli (see Figures 2, 4*a*) and to exhalant endopores or cribribullae leading to a narrow central cloaca. The chambers are spheroidal and in linear series

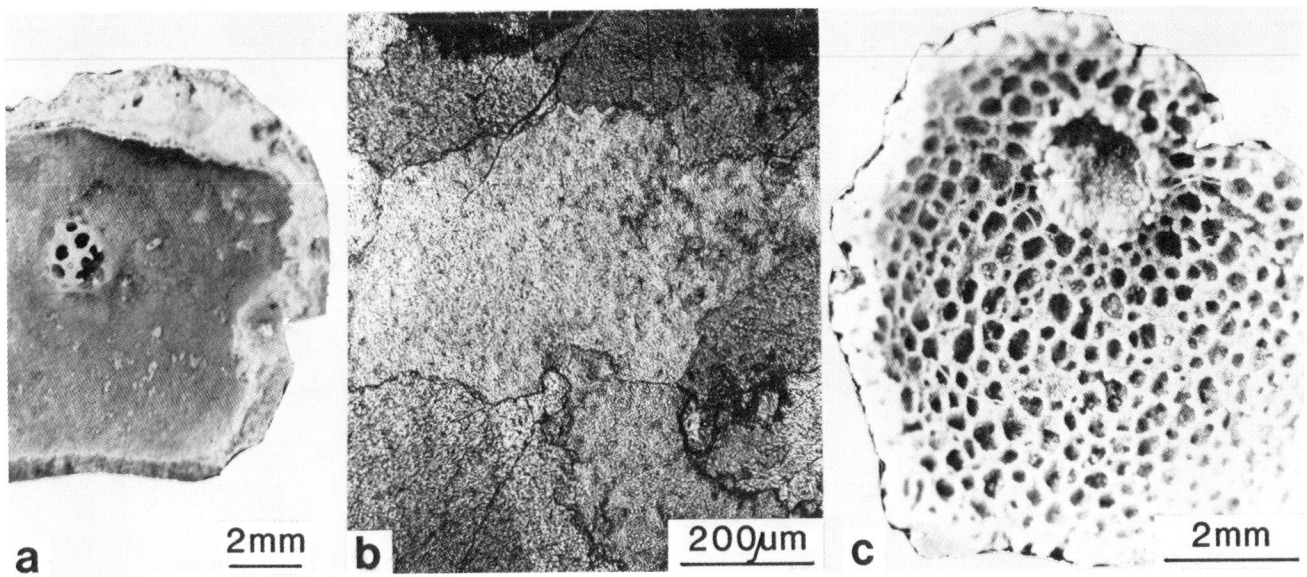

Figure 4. Pharetronid structural type 1, *Girtyocoelia dunbari* King: *a*, cribribulla, viewed from within chamber, USNM 127728, Permian, Glass Mountains, Texas; *b*, thin section showing spherulites, USNM 127730, USNM loc. 707q, Permian, Cathedral Mountain Formation, near Clay Slide, Glass Mountains, Texas, crossed nicols; *c*, view of interior of exowall showing unusually preserved trabecular net; broken portion of endowall at top; note also indications of exopores in this normally aporate genus; USNM 127734, Permian, Glass Mountains, Texas.

(cateniform). Habit ramose with subparallel branches. Large vesicles are present in earlier chambers. Juveniles (protocysts) are adnate and lack a cloaca. Spherulite diameter is 20–60 μm. Example: *Girtyocoelia* Cossmann, 1909 (Figure 4*b*).

TYPE 2. Skeleton exclusively peripheral, penetrated by closely spaced exopores, which are canal-like and sometimes anastomose within the wall; rare, very large exopores in one genus have lips. A broad central calcified cloaca (endowall) penetrated by endopores is present in one genus; otherwise either a terminal oscule is found in each chamber or the chambers are connected only by interpores. The chambers are in linear series, toroidal if a cloaca is present, hemispheroidal if not. Habit cylindroidal. Spherulite diameters 20–50 μm. Examples: *Amblysiphonella* Steinmann, 1882 (with cloaca); *Girtycoelia* King, 1933 (no cloaca) (Figure 5*a*).

TYPE 3. Skeleton dominantly internal and trabecular. A partial or complete peripheral skeleton may be added, in which case growth may be modular and sphinctozoanlike. Small short exauloslike structures may be present. Habit ramose with subparallel branches. The commonest genus (*Fissispongia* King, 1938) has a double cloaca. Spherulite

diameters very small (15–20 μm). Externally similar forms with a single cloaca, or with a ring of canals parallel to a central cloaca, also occur (new genera; Finks, in prep.) (Figure 5*b*).

TYPE 4. Skeleton both internal (trabecular) and external, with a gradational time series showing decrease of former and increase of latter. Earliest forms resemble type 3 somewhat but are larger and coarser. Broad single central cloaca (a Permian new genus has multiple cloacae; Finks, in prep.). External skeleton with characteristic lipped pores (labripores). Later forms, in which external skeleton dominates, are sphinctozoanlike and convergent with type 2 (above). Habit ramose with subparallel branches. Spherulite diameters large (40–130 μm). Examples: *Maeandrostia* Girty, 1908; *Stylopegma* King, 1943 (Figure 5*c*).

TYPE 5. Skeleton exclusively peripheral, outlining hemicylindrical chambers arranged in a sheet (thalamidarium). Walls pierced by small pores; chambers containing diaphragms. Exhalant surface of sheet covered with a layer of anastomosing trabeculae (trabecularium) in which radial (astrorhizalike) exhalant canal systems are sunk, each convergent on an oscule. Sheet may have the

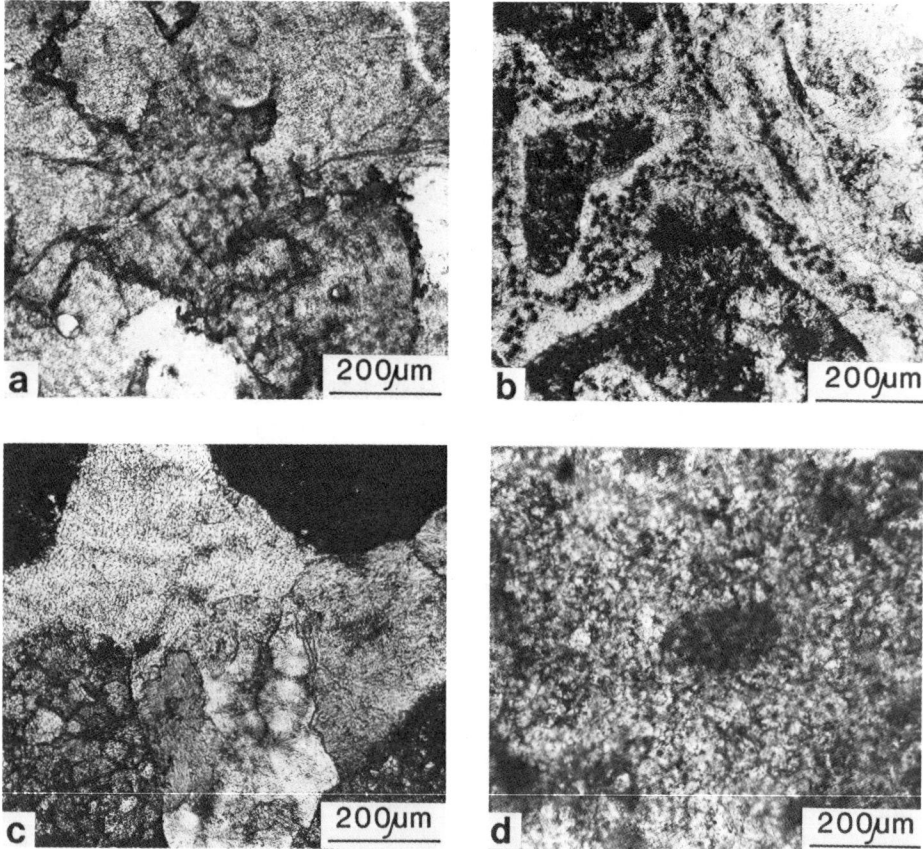

Figure 5. Examples of pharetronid structural types: *a*, type 2, thin section showing spherulites, *Girtycoelia typica* King, USNM 127731, USNM loc. 518o, Pennsylvanian, Stanton (?) Formation, Neodesha, Kansas; *b*, type 3, *Fissispongia jacksboroensis* King, thin section showing spherulites, USNM 127732, Pennsylvanian, Gonzales Formation, 4.5 miles north of Finis, Texas, plane-polarized light; *c*, type 4, *Stylopegma* sp., thin section showing spherulites, USNM 127733, USNM loc. 702h, Permian, Neal Ranch Formation (shale just above Bed 12), Center of Wolfcamp Hills, Texas, crossed nicols; *d*, type 5, *Guadalupia cylindrica* Girty, holotype, thin section showing spherulites, USNM 118136, USGS loc. 2926 green, Permian, Capitan Formation, Capitan Peak, Guadalupe Mountains, Texas, crossed nicols.

form of cups, branching straps, partly rolled-up tubes with lateral openings, or complete tubes. Trabecularium is on the inner or upper surface. Spherulite diameters 20–50 μm. Example: *Guadalupia* Girty, 1909 (Figure 5*d*); *Cystothalamia* Girty, 1909; *Cystauletes* King, 1943.

TYPE 6. Skeleton dominantly or wholly internal trabecular. Sponge large, conical-fungiform or ramose with complex internal canals in addition to the anastomosing tubes outlined by the trabeculae. External cortex partial or absent. Spherulites large (50–400 μm). Examples: "Pharetronid A" (Finks, 1960) (Figure 6*a*); "Pharetronid B" (Finks, 1960); "*Virgola*" spp. (Finks, 1960).

Relationships and Radiation

The six types of structure noted above show certain interrelationships. Types 2 and 5 both show a dominantly peripheral skeleton pierced by numerous small, subequal pores and both have imperforate diaphragms or vesicles within the chambers. The spherulites are small in both cases (20–50 μm), but type 2 has cateniform organization and type 5 has stratiform organization. The latter also possess a unique trabecular layer on the exhalant surface.

Type 1 is distinctive because of an apparent lack of pores, but otherwise resembles type 2 in its exclusively peripheral skeleton and cateniform organization. The

Figure 6. Examples of pharetronid structural type 6: *a*, "Pharetronid A" of Finks, 1960, thin section showing spherulites, USNM 127737, USNM loc. 703, Permian, Road Canyon Formation, Old Word Ranch, Glass Mountains, Texas; *b*, "*Stellispongia*" *variabilis* Muenster, thin section showing trabecula with isodiametric spherulites (specimen used for drawing in Finks, 1983, fig. 3), USNM 127736, Triassic, St. Cassian Beds, Seelandalp (Carbonin), Italy; *c*, *Sestrostomella robusta* Zittel, thin section showing flaky spherulites lining a tubular space, USNM 127735, Triassic, St. Cassian Beds, Seelandalp (Carbonin), Italy; *d*, *Sestrostomella robusta* Zittel, thin section of same specimen as in *c*, showing general view of trabecular net with flaky spherulites parallel to trabecular surfaces (specimen used for drawing in Finks, 1983, fig. 2, but a different area); *e*, *Myrmecidium hemisphericum* (Goldfuss) (specimen identified by Zittel as *Myrmecium hemisphericum* Goldfuss), thin section showing trabeculae composed of flaky spherulites or "sinuous spicules," AMNH 1281/1, Jurassic, Hossingen, Württemberg, Germany; *f*, *Blastinia costata* Goldfuss (specimen identified by Zittel), thin section showing trabeculae cored by one or two triactins coated by flaky spherulites or "sinuous spicules," AMNH 1218/1, Upper Jurassic, Sontheim, Württemberg, Germany.

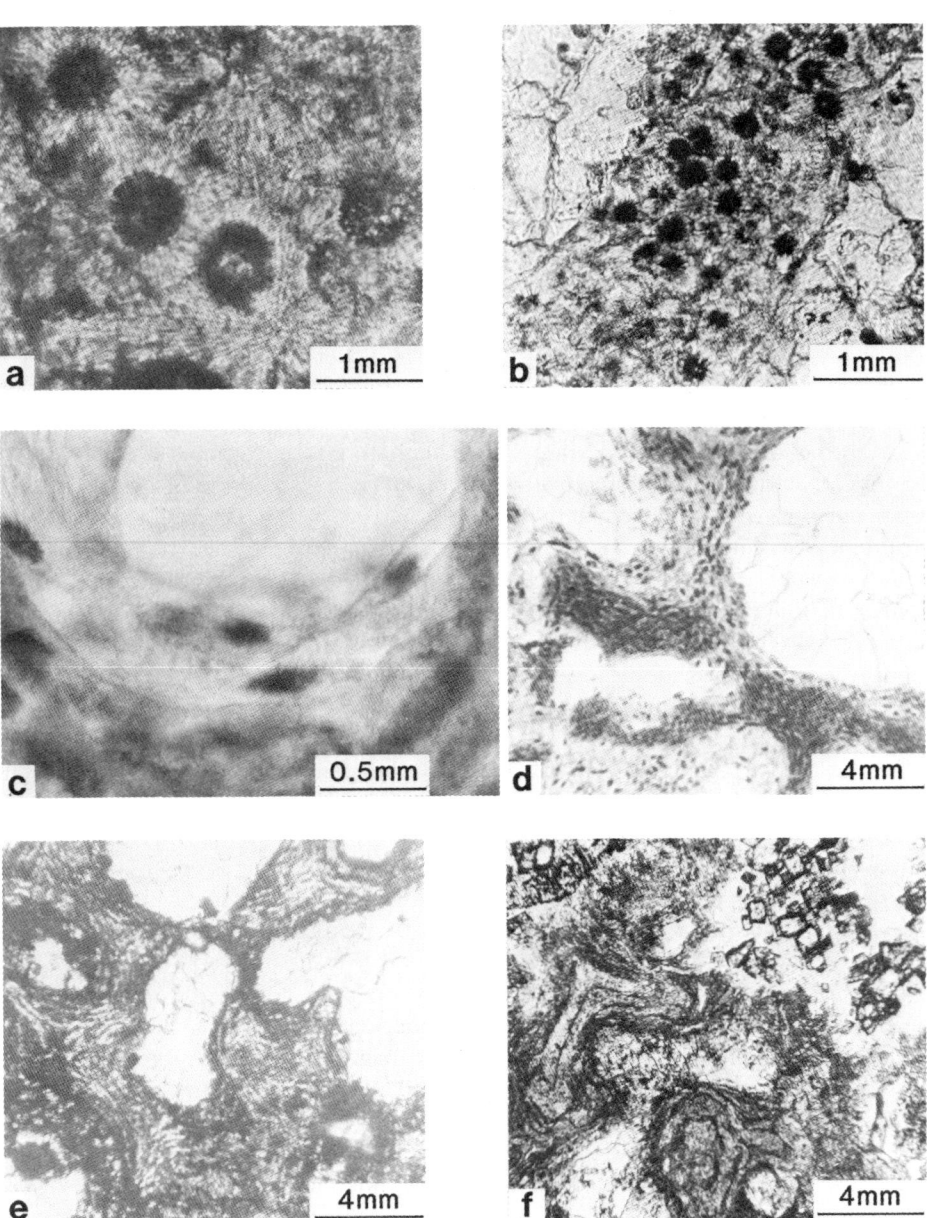

spherulites are the same size (20–60 μm) and vesicles also
occur in the chambers. However, traces of fine pores, and
even of trabecular tissue, are preserved on the inner surfaces of chambers in a few specimens of type 1 (Figure 4c).
It may be that type 1 represents a more heavily calcified
branch of the same stock. The presence of such special
structures as exauli and cribribullae indicates that we are
not dealing merely with heavily calcified individuals of
type 2 species.

Types 3, 4, and 6 also have certain features in common.
In this group the skeleton is dominantly internal and trabecular, with the trabeculae outlining anastomosing tubular spaces. A peripheral skeleton is often present also,
and is characterized by lipped pores (labripores) that are
more variable in size and more widely spaced, than the
pores of types 2 and 5. In type 4, there is a development
toward reduction of the internal trabecular skeleton and
enlargement of the cortical skeleton, with concomitant
development of chambers, so that it comes to resemble the
members of type 2. Nevertheless these convergent forms
(*Stylopegma* spp.) show their relationship to their own
group in the presence of labripores of variable size, as well
as in the larger size of their spherulites (70–90 μm vs. 20–
50 μm). Spherulite size in this group is roughly correlated
with the coarseness of the trabecular net. In types 4 and 6
it is large, never being smaller than 40 μm and ranging up
to 400 μm, although most genera are in the 50–130 μm
range. Type 3 stands apart in having very small spherulites (15–20 μm), and a finer trabecular net, although
one that is morphologically similar to type 4.

At the beginning of the appearance of the pharetronids
in the Texas region (Des Moinesian or mid-Pennsylva-
nian, equal to Westphalian C), representatives of types 3,
4 and 5 are present in the form of *Fissispongia*, *Maeandrostia*,
and *Cystauletes*, respectively. It is possible that types 3 and
4 are endemic forms. They have not been reported from
other parts of the world this early, and the genera involved
appear to be confined to the Texas region, although their
respective structural types attain a more worldwide distribution in the Permian and Triassic. Type 5 (*Cystauletes*),
however, was already present in the Moscovian of Spain
(Van de Graaf, 1969).

These types are joined in the Texas region during the
succeeding Missourian epoch by representatives of type 1
(*Girtyocoelia*) and type 2 (*Girtycoelia*). The latter two types
certainly did not originate in the Texas region for repre-
sentatives of type 1 (*Sollasia*, Steinmann, 1882) and type 2
(*Amblysiphonella*) are known from the Moscovian of Spain
(Van de Graaf, 1969; Steinmann, 1882) and even earlier
(Rigby, this volume). *Amblysiphonella* itself appears in the
Texas region in the latest Pennsylvanian (Virgilian).

Types 3 and 4 show the greatest diversity during the
Pennsylvanian, with several species each of *Fissispongia*
and *Maeandrostia* (Finks, in prep.). This lends additional
support to an endemic (or near-endemic) origin for these

two groups. They show their greatest diversity in the succeeding earliest Permian (Neal Ranch Formation) patch
reefs, where they are among the chief reefbuilders. Several
new genera of type 3 appear, along with several species of
the persistent *Fissispongia*. Although *Maeandrostia* of type 4
disappears, it is replaced by new forms in which the internal skeleton is reduced, the peripheral skeleton is becoming more dominant, and a "sphinctozoan" chambered
organization is beginning.

At the start of the Permian, type 2 forms (*Amblysiphonella*, *Girtycoelia*) disappear and are replaced by
similar, apparently convergent, species of the type 4 lineage (various new species of *Stylopegma* King, 1943 and
related new genera, Finks, in prep.). These forms attain a
peak of diversity in the mid-Permian patch reefs (Road
Canyon and Word formations), but persist into the latest
Permian Capitan barrier reef. The radiation of this group
would appear to be endemic.

Type 3 (*Fissispongia* and its relatives) decreases in diversity after the early Permian. However, *Peronidella*-like
forms continue to the end of the Permian in Texas and
elsewhere into the Triassic and beyond.

Type 1 (*Girtyocoelia*), although abundant in the Permian, does not become highly diversified and disappears
from the Texas area in the later Permian. On a worldwide
scale, however, it persists into the Triassic and the lineage
becomes quite diversified in Alpine reefs (Ott, 1967; Dieci
et al., 1970).

Type 5 (*Guadalupia* and its allies) is perhaps the most
characteristic, abundant, and diverse of the pharetronid
groups of the Texas Permian. Curiously, the group (in the
form of *Cystauletes*) disappeared from the area in the later
Pennsylvanian and did not reappear until the early mid-
Permian (Skinner Ranch Formation). At first it was not
very diverse, but quickly developed many new species
(Finks, in prep.), which from their transitional forms appear to be endemic. This also applies to the genus
Cystothalamia Girty, 1909, which appears in the later Permian (Cherry Canyon and Word formations) and shows
clearly its origin from *Guadalupia*. *Cystothalamia* has been
reported elsewhere, in both Permian and Triassic, but at
least some of these occurrences appear to be home-
omorphs.

Type 6, which might be called typical Inozoa, does not
appear in the Texas area until early mid-Permian (Skinner Ranch Formation). It appears at the same time as type
5. type 6 forms do not diversify greatly but are nevertheless numerous and conspicuous, and reach a peak of abundance and diversity in the later Permian. They are obviously related to forms abundant in the Alpine Triassic,
and the group as a whole reaches its peak diversity in the
Mesozoic. Members of this lineage have not been reported
this early in deposits elsewhere, and the possibility remains that they originated in or near the Texas area.

One should add to this group a cloacate form of *Corynella*-

type, that appears in the latest Permian (Capitan Limestone), namely, *"Anthracosycon ficus* var. *capitanense"* of Girty, 1909, which is not an *Anthracosycon* but rather something close to the Triassic *Precorynella* Dieci et al., 1970. Similar forms have been found in the mid-Permian of Sicily (Parona, 1933), and Tunisia (Termier and Termier, 1955), as well as in the Triassic.

The progressive population of the Texas region by increasingly diverse pharetronids during the Pennsylvanian and Permian appears to reflect a combination of immigration of species from outside and local evolution. The immigration in turn must have resulted from contemporaneous evolution elsewhere, for the increase in diversity is a worldwide phenomenon (documented in part in Finks, 1967, 1970, 1971). As for the on-site evolution, it appears to be mainly among types 3, 4, and 5, possibly also type 6. As noted above, types 3, 4, and 6 are structurally related in having a dominant internal trabecular skeleton. Type 5, although structurally similar to type 2, is unique in having stratiform organization and a trabecular exhalant-surface layer. As far as is now known, the Texas region fauna has by far the most diversified representatives of these groups anywhere in the world.

Post-Paleozoic Developments

The uniformly nonspicular and spherulitic calcareous skeleton of the late Paleozoic pharetronids strongly suggests that they are genetically related. It is therefore of interest to see what subsequent developments take place, and what relationship they bear to the Mesozoic pharetronids.

In the Triassic many forms are assignable to the same, or similar, genera, which also have an exclusively spherulitic calcareous skeleton. These include "Sphinctozoa," such as *Enoplocoelia* Steinmann, 1882, a possible type 1 or type 4 descendant based on macrostructure (its spherulite diameters of 20–50 μm also agree with those of the type 1 *Girtyocoelia* but Cuif, 1973, illustrates spherulites of type 4 size, 100–300 μm) and *Thaumastocoelia* Steinmann, 1882, a possible type 1 or other form (it is aporate with cribribullalike structures, but its spherulite diameters of 70–200 μm are like those of type 4). (The spherulite measurements were made by me on topotype specimens.)

The most interesting implications for pharetronid evolution occur among the "Inozoa" of type 6. In the Triassic, there are genera and species of similar gross morphology and similar spherulite size, such as *Hartmanina* Dieci et al., 1974; *Keriocoelia* Cuif, 1974; *Sclerocoelia* Cuif, 1974; *Reticulocoelia* Cuif, 1974; *"Corynella" gracilis* (Muenster) as illustrated by Zittel (1878); and *"Stellispongia" variabilis* (Muenster) Zittel, 1878, observed by me (Figure 5*b*), the last having somewhat smaller spherulites (15–50 μm) as in the flaky forms below. Other species present have spherulites that are not isodiametric but flakelike, in the size-

range 10–60 μm (possibly a different lineage; they are in the size-range of the spherulites of types 1, 2, and 5). These include *Sestrostomella robusta* Zittel, 1878, as observed by me (Figure 6*c*, *d*), and *"Stellispongia" variabilis* (Muenster) as illustrated by Steinmann (1882). Also a Jurassic *"Myrmecidium"* observed by me (Figure 6*e*). In some of the spherulitic Triassic species, Wendt (1974) has found occasional imbedded calcitic monaxons (viz., *Sestrostomella robusta* Zittel, in Wendt, 1974, Figure 6). In the Jurassic and Cretaceous many species have similar flaky bodies surrounding a triradiate or tetraradiate (including "tuning-fork" types), which cores the trabecula. They also may have triradiates embedded paratangentially in an external cortical layer (see Ziegler, 1964). This group includes the type species of such typical Inozoan genera as *Eudea* Lamouroux, 1821; *Stellispongia* d'Orbigny, 1849; *Enaulofungia* de Fromentel, 1860; *Oculospongia* de Fromentel, 1860; *Elasmocoelia* Roemer, 1864; *Blastinia* Zittel, 1878; *Corynella* Zittel, 1878; *Holcospongia* Hinde, 1893; and *Peronidella* Zittel in Hinde (1893) (see Figure 6*f*).

This strongly suggests that most Inozoan pharetronids have a common ancestry among the Permian sponges of type 6, or related lineages, and that spicules typical of the Class Calcarea were gradually added during the Triassic and Jurassic to an originally spherulitic nonspicular skeleton. It is possible that *Murrayona* and *Paramurrayona* are surviving descendants of this group that still retain the flaky spherulites. The flaky spherulites of *Murrayona phanolepis* are in the same size-range (30–60 μm) as those of the Mesozoic pharetronids with flaky spherulites (Vacelet, 1964). On the other hand, *Paramurrayona corticata* has larger flakes (150–400 μm) similar to those of type 6 (Vacelet, 1967). If spherulite size is in fact a conservative character, the two genera would have descended from different lineages within the pharetronid complex. This also implies that the late Paleozoic forms treated here were true Calcarea, in particular Calcinea (see Finks, 1983, for a more extended discussion).

Literature Cited

Cossmann, M. 1909. Rectifications de nomenclature. *Revue critique de Paléozoologie*, 13:67.

Cuif, J. P. 1973. Histologie de quelques Sphinctozoaires (Poriferes) triasiques. *Géobios*, 6:115–125.

———. 1974. Rôle des Sclérosponges dans la faune récifale du Trias des Dolomites (Italie du Nord). *Géobios*, 7:139–153.

Dieci, G., A. Antonacci, and R. Zardini. 1970. Le spugne cassiane (Trias medio-superiore) della regione dolomitica attorno a Cortina d'Ampezzo. *Bolletino della Societá Paleontologica Italiana*, 7:94–155.

Dieci, G., A. Russo, and F. Russo. 1974. Revisione della genere *Leiospongia* d'Orbigny (Sclerospongia triassica). *Bolletino della Societá Paleontologica Italiana*, 13:135–146.

d'Orbigny, A. 1849. Note sur la classe des Amorphozoaires. *Revue et Magazine de Zoologie*, 1:545–550.

Finks, R. M. 1960. Late Paleozoic Sponge Faunas of the Texas Region. The Siliceous Sponges. *Bulletin of the American Museum of Natural History*, 120(1):1–160.

———. 1967. Phylum Porifera Grant, 1836. Pages 333–341 in *The Fossil Record, a Symposium with Documentation*, edited by W. B. Harland. London: Geological Society of London.

———. 1970. The Evolution and Ecologic History of Sponges during Palaeozoic Times. Pages 3–22 in *The Biology of the Porifera*, edited by W. G. Fry, Symposia of the Zoological Society of London, 25. London: Academic Press.

———. 1971. Sponge Zonation in the West Texas Permian Type Section. Pages 285–300 in *Paleozoic Perspectives: A Paleontological Tribute to C. Arthur Cooper*, edited by J. T. Dutro, Jr., Washington, DC: Smithsonian Institution Press.

———. 1983. Pharetronida: Inozoa and Sphinctozoa. Pages 55–69 in *Sponges and Spongiomorphs*. Notes for a short course organized by J. K. Rigby and C. W. Stearn, edited by T. W. Broadhead. Knoxville: University of Tennessee.

Fromentel, M. E. de. 1860. Introduction á l'étude des éponges fossiles. *Mémoires de la Société Linnéenne de Normandie* (series 2), 11:1–50.

Girty, G. H. 1908. On Some New and Old Species of Carboniferous Fossils. *Proceedings of the United States National Museum*, 34:281–303.

———. 1909. The Guadalupian Fauna. *Professional Papers, United States Geological Survey*, 58:1–651.

Hartman, W. D. 1958. A Re-Examination of Bidder's Classification of the Calcarea. *Systematic Zoology*, 7:97–110.

Hinde, G. J. 1893. *Fossil Sponges: Part III, Sponges of the Jurassic Strata*. Pages 189–254. London: Palaeontographical Society, London.

King, R. H. 1933. A Pennsylvanian Sponge Fauna from Wise County, Texas. *Bulletin of the University of Texas*, 3201:75–85.

———. 1938. Pennsylvanian Sponges of North-Central Texas. *Journal of Paleontology*, 12:498–504.

———. 1943. New Carboniferous and Permian Sponges. *Bulletin of the Geological Survey of Kansas*, 47:1–36.

Kirkpatrick, R. 1910. On a Remarkable Pharetronid Sponge

from Christmas Island. *Proceedings of the Royal Society of London*, 83:124–133.

Lamouroux, J. V. F. 1821. *Exposition méthodique des genres de l'ordre des polypiers avec leur description et celle des principales espèces, figurées dans 84 planches*. Paris (private publication). 115 pp.

Ott, E. 1967. Segmentierte Kalkschwämme (Sphinctozoa) aus der alpinen Mitteltrias und ihre Bedeutung als Riffbildner im Wettersteinkalk. *Abhandlungen der Bayerischen Akademie der Wissenschaften, Mathematisch-Wissenschaftliche Klasse* (New series), 131:1–96.

Parona, C. F. 1933. Le spugne della fauna permiana di Palazzo Adriano (Bacino di Sosio) in Sicilia. *Memorie della Societá Geologica Italiana*, 1:1–58.

Roemer, F. A. 1864. Die Spongitarien des norddeutschen Kreidegebirges. *Paläontographica*, 13:1–64.

Steinmann, G. 1882. Pharetronen-Studien. *Neues Jahrbuch für Mineralogie*, 2:139–191.

Termier, H., and G. Termier. 1955. Contribution á l'étude des spongiaires du Djebel Tébaga (extreme sud Tunisien). *Bulletin de la Société Géologique de France* (series 6), 5:613–630.

Vacelet, J. 1964. Etude monographique de l'éponge Calcaire Pharétronide de Méditerrannée *Petrobiona massiliana* Vacelet et Lévi. Les Pharétronides actuelles et fossiles. *Recueil des Travaux de la Station Marine d'Endoume*, 50(34):1–125.

———. 1967. Description d'éponges Pharetronides actuelles des tunnels sous-récifaux de Tuléar (Madagascar). *Recueil des Travaux de la Station Marine d'Endoume*, supplement 6:37–62.

Van de Graaf, W. J. E. 1969. Carboniferous Sphinctozoa from the Cantabrian Mountains, Spain. *Leidse Geologische Mededelingen*, 42:239–257.

Wendt, J. 1974. Der Skelettbau aragonitischer Kalkschwämme aus der alpinen Obertrias. *Neues Jahrbuch für Geologie und Paleontologie, Monatshefte*, 1974:498–511.

Ziegler, B. 1964. Die Cortex der fossilen Pharetronen (Kalkschwämme). *Eclogae geologiae Helveticae*, 57:803–822.

Zittel, K. A. von. 1878. Studien über fossile Spongien, Zweite Abteilung. Monactinellidae, Tetractinellidae und Calcispongiae. *Abhandlungen der Königlich Bayerischen Akademie der Wissenschaften, Mathematisch-Physikalische Klasse*, 2:1–138.

RACHEL A. WOOD
Department of Earth Sciences
University of Cambridge
Downing St., Cambridge, United Kingdom

Position of Mesozoic Stromatoporoids in the Porifera

Abstract

The finding of spicule pseudomorphs within the calcareous skeleton of Mesozoic stromatoporoids confirms poriferan affinity for this group. Additional evidence from functional studies supports this proposal. Studies of spicule type and arrangement allow both a more precise placing within the Porifera and analysis of the relationship between spicule framework and the calcareous skeleton in taxonomic, phylogenetic, and biomineralogical terms.

Comparison of Mesozoic stromatoporoids with Recent sponges shows them to be calcified members of the Class Demospongiae. Closest affinities are to the Orders Haplosclerida, Axinellida and Poecilosclerida, which include the "sclerosponges" *Ceratoporella* spp., *Astrosclera* sp., *Calcifibrospongia* sp. and *Murania* sp. (Upper Triassic–Upper Cretaceous). The varied spicule complements of Mesozoic stromatoporoids indicate that possession of a calcareous skeleton is a convergent feature, and that "stromatoporoids" represent a grade of organization rather than a true taxonomic grouping. As grades of organization, the terms "sclerosponge" and "stromatoporoid" are synonymous. The development of a calcareous skeleton is probably a response to reef building in warm carbonate seas.

> Reasoning by analogy may not be an orthodox method of scientific analysis . . . but many paleontological conclusions have been reached in this way . . . and often cannot be reached in any other. There are a large number of fossils whose only guides to their biological affinities are faint analogies with living forms. Here a paleontologist without a good imagination is lost.
> —G. B. Twitchell

Stromatoporoids are a good example of a fossil group whose place in the natural classification has often been determined by faint analogies. In the absence of any diagnostic features that would finally place them, different workers have been impressed by different analogies and so these fossils have been shunted from one biological group to another.

The genus *Stromatopora* was first described by Goldfuss (1826–1833) who placed it between the Millepores and Madrepores. Rosen (1869) suggested poriferan affinity for stromatoporoids and, using beautifully illustrated examples, concluded that stromatoporoids were calcified horny sponges. A few years later Nicholson (1886–1892) published a monograph on British stromatoporoids, which remains one of the classics in the field. From a comparison of modern calcified hydrozoans, he concluded that stromatoporoids were Hydrozoa. This idea remained almost universally accepted, except by the occasional author who suggested that stromatoporoids were related to corals (Mori, 1982, 1984), Foraminifera (Kirkpatrick, 1912; Hickson, 1934), and Cyanobacteri (Kázmierczak, 1976).

The rediscovery of calcified sponges in the deep forereef of Jamaica renewed interest and controversy in the stromatoporoid problem (Hartman and Goreau, 1966, 1970; Hartman, 1969, 1979). These sponges were placed in a new subclass, the Sclerospongiae. They possess a calcareous skeleton as well as a siliceous spicule one, which bears a system of radiating grooves or canals that represent traces of the exhalant canal system. These stellate structures bear a striking resemblance to stromatoporoid astrorhizae, and the authors proposed that the sclerosponges were "living fossils" and a relict fauna of Paleozoic and Mesozoic forms (Hartman and Goreau, 1970). The authors even showed photographs of proposed spicule "ghosts" within the calcareous skeleton in Mesozoic forms, but these were dismissed as being of diagenetic origin by many stromatoporoid workers.

Morphology

The presence of spicule pseudomorphs in many genera of Mesozoic stromatoporoids confirms poriferan affinity for this group. Functional analysis of the canal system supports this premise, and detailed spicule information allows a more precise placing within the Porifera.

Spicules are preserved as calcite, pyrite, or silica pseudomorphs of siliceous styles (or acanthostyles) and triaxines (or tetraxines), which are variously arranged within the calcareous skeleton. The secondary calcareous skeleton is precipitated upon a primary spicule framework. Three types of spicule organization are identified, which coincide with different ultrastructures of the calcareous skeleton: (1) styles (180 μm long, 8 μm wide), loose fibro-reticulate arrangement, primary irregular and secondary orthogonal fibrous calcareous skeleton; (2) club-shaped styles (135 μm long, 17 μm wide), plumose arrangement, fascicular fibrous calcareous skeleton; (3) styles and triaxines or tetraxines (100 μm long, 10 μm wide; and 120 μm long, 20 μm wide; respectively). Closely packed reticulate arrangement, orthogonal fibrous calcareous skeleton.

FUNCTIONAL MORPHOLOGY

Stromatoporoids possess an intricate canal system, expressed as ramified unwalled spaces within the skeleton that open out to form repeated stellate structures known as astrorhizae on the upper surface (Figure 1a). As the animal grows upward by increments, successive astrorhiza-bearing layers are superimposed. The similarity between the sclerosponge excurrent canal system and the stromatoporoid astrorhizae was first noted by Hartman and Goreau (1966). The excurrent canal system of *Acanthochaetetes wellsi* is shown in Figure 1b. Although sclerosponges are the only forms in which traces of these canals are expressed in the calcareous skeleton, many other sponges, especially thin encrusting demosponges, (e.g., *Spongilla uvirae* and *Potamolepis leubnitziae*) possess the same repeated system of aquiferous units within their soft tissue (Figure 2). The astrorhizae of stromatoporoids and the aquiferous units of many Recent sponges are all based upon the same organization, the rhagon unit, especially that of the derived form where the paragastric cavity has been modified to form excurrent canals (Figure 3a). Stromatoporoids show great variety in construction, organization, and spacing of these units. A stylized stromatoporoid aquiferous unit is shown in Figure 3b.

Figure 4 compares transverse sections of the modern sclerosponge *Calcifibrospongia actinostromarioides* with a section of an Upper Jurassic stromatoporoid, *Burgundia astrotubulata*. The osculum, ostia, and astrorhizae are directly comparable. Similarly, the canal system of *B. wetzeli* (Lower Cretaceous) is formed of stacked rhagon units, where the osculum has been repeatedly truncated by secondarily precipitated tabulae, which sectioned off abandoned parts of the skeleton as the animal grew (Figure 5). This also explains the presence of tabulae across the ostia and within the astrorhizal canals. The position of these secondary tabulae coincides with continuous lamellae, suggesting that this animal possessed a thin veneer of living tissue, perhaps only limited to one interlamellar space.

To conclude, the canal system of stromatoporoids may be considered a poriferan filtration system, based upon the demosponge rhagon unit characteristic of thin encrusting sponges.

SKELETAL MORPHOLOGY

Spicule pseudomorphs of styles (or acanthostyles) have been described from six genera of Mesozoic stromatoporoids (Wood and Reitner, 1986). Table 1 presents details of further spicule findings from three Upper Jurassic genera. Original descriptions and site details are given in the literature cited. The eight genera that contain spicules can be divided into three types, each characterized by different spicule morphologies and arrangements within the cal-

Figure 1. Examples of stromatoporoids and demosponges: *a*, astrohizae of *Shuqraia* sp., H4581A; *b*, traces of excurrent canal system on surface of *Acanthochaetetes wellsi*; *c*, *Dehornella crustans*, H5479, type 2 (Figure 5) spicule arrangement, with club-shaped styles in a plumose orientation, subparallel to microstructural fibers of calcareous skeleton; Makhetesh Haithira, Israel, Kimmeridgian (Upper Jurassic); *d*, *Actinostromarianina lecompti*, H4808a, type 1 (Figure 5) spicule arrangement, styles forming a radial framework, with a primary irregular calcareous skeleton; Alam Abyadh, Arabia, Kimmeridgian; *e*, *Actinostromaria* sp., H5480, type 3 (Figure 5) spicule arrangement, closely packed styles and possibly triaxines, with a primary calcareous skeleton of orthogonal fibrous microstructure, with granular tabulae; Trnovski Gost, Slovenia, N.W. Yugoslavia, Kimmeridgian; *f*, *Actinostromaria* sp., H5480, detail of type 3 spicules; triaxines appear to be placed at junctions of the pillar and pillar lamellae; *g*, *Agelas mauritana*, club-shaped styles and plumose arrangement similar to that of type 1 (Figure 5) stromatoporoid spicules; *h*, *Esperiopsis anomala*, styles trapped in spongin fibers. Specimens *a*, *c–h* held at British Museum (Natural History), photographed in thin section using transmitted light; specimen *b* from collection of J. Vacelet.

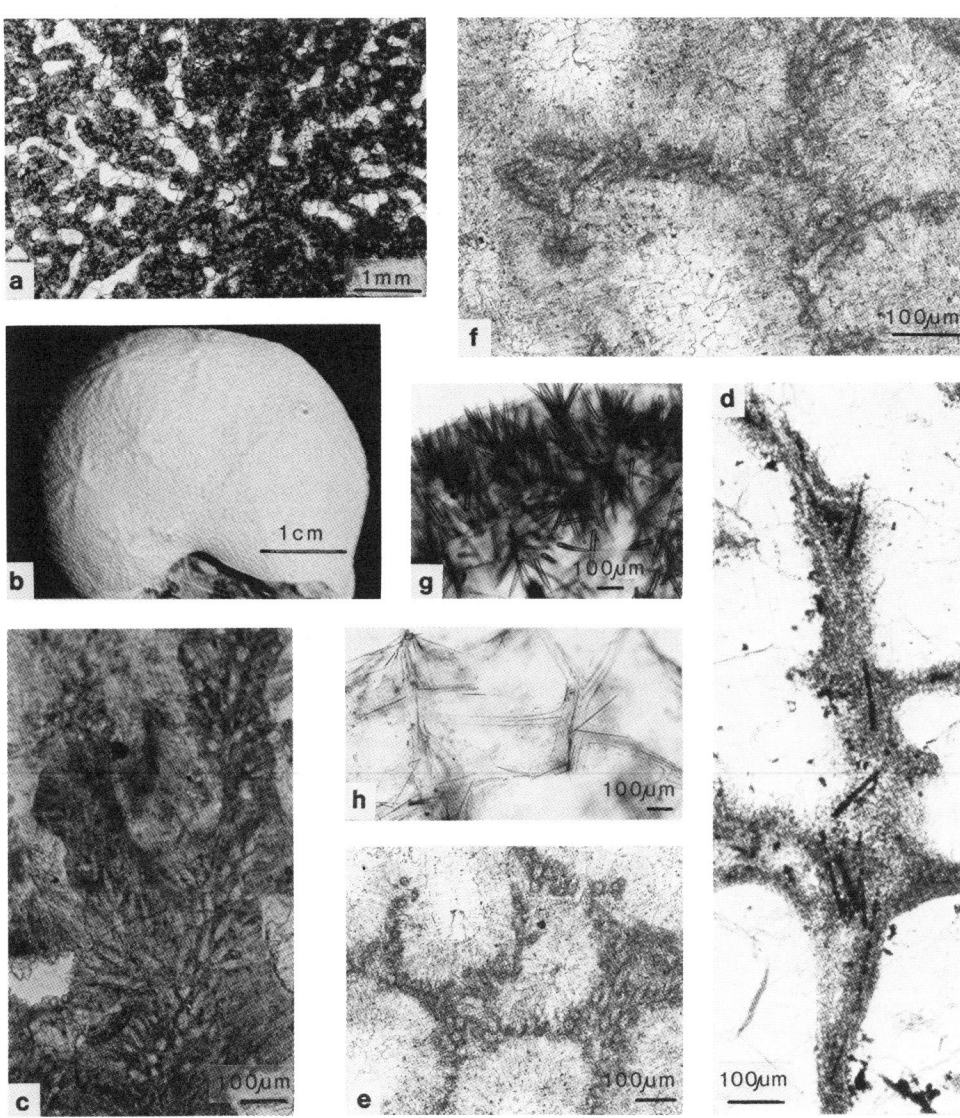

careous skeleton. They are described below and illustrated schematically in Figure 6.

TYPE 1: *Actinostromarianiana lecompti* (Figures 1*d*; 6). A latilaminate form from the Superfamily Actinostromariicae, characterized by orthogonal fibrous microstructure, which is precipitated intermittently as a secondary orthogonal fibrous rim to form the latimaninae. The long, thin styles (maximum, 180 μm long, 10 μm wide) form a radial

arrangement, with an axial zone of densely packed spicules. A primary calcareous skeleton of irregular microstructure forms menisculike structures around the projecting spicules.

TYPE 2: *Dehornella* spp., *Shuqraia* sp., *Steineria* sp., *Promillepora* sp., *Parastromatopora* sp., *Astroporina* spp. (Figures 1*c*; 6). All forms in this type are members of the Superfamily Milleporellicae Hudson. The skeleton is formed by

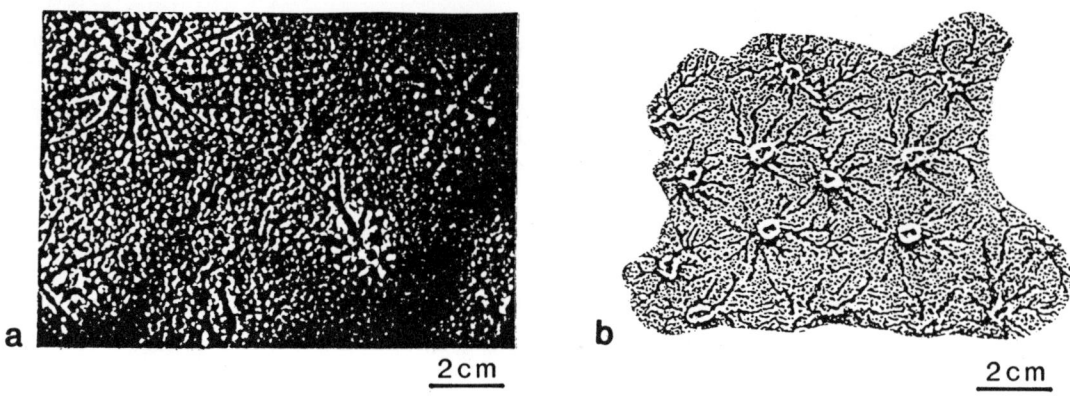

Figure 2. Regular repetition of oscula, bearing stellate excurrent canals (or astrorhizae), in soft-bodied Recent sponges: *a, Spongilla uvirae; b, Potamolepsis leubnitziae.* (After Brien, 1967.)

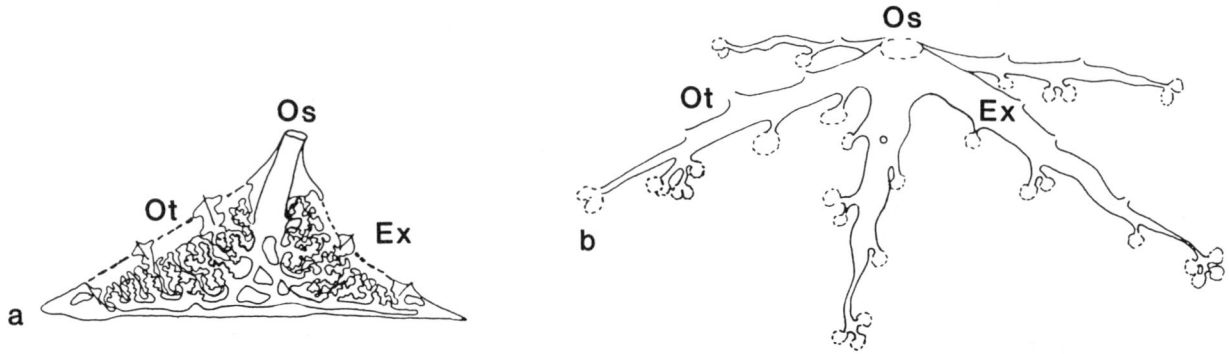

Figure 3. Stylized rhagon aquiferous units: *a,* demosponge unit, where paragastric cavity has become modified to form excurrent canals (after Lévi 1973); *b,* stromatoporoid unit, based upon *Burgundia wetzeli* (Lower Cretaceous). *Ex,* excurrent canals (astrorhizae); *Os,* osculum; *Ot,* osticum.

radiating columns of fascicular fibrous microstructure, connected by tabulae. Club-shaped styles (90–135 μm long, 13.5–17.0 μm wide) radiate upward and outward with tapering tips in a plumose arrangement parallel to the microstructural fibers of the calcareous skeleton, which initiates in tufts from the spicule bases. Spicules are found in the columns only and rarely project into the interskeletal spaces.

TYPE 3: *Actinostromaria* sp. (Figures 1*e, f;* 6). An actinostromarid characterized by a reticulate skeleton of dominant pillars and pillar-lamellae of orthogonal fibrous

Table 1. Specimen and spicule data for three Jurassic stromatoporoids with astrohizae aquiferous system held in the British Museum (Natural History). (O = Oxfordian; L.K. = Lower Kimmeridgian; U.O.–L.K. = Upper Oxfordian–Lower Kimmeridgian)

Species	Age	Locality	Calcareous Skeleton Microstructure	Type	Spicule data Distribution	Mineralogy	Length (μm)	Width(μm)	References
*Astroporina orientalis** H4850a, H4850b	O	Ain Safra, Yanta (Lebanon)	fasicular, fibrous	styles	parallel to microstructural fibers	calcite	120	12 max.	Hudson, 1960
*Actostroma nasri** H4893a	L.K.	Makhtesh Haithira (Israel)	orthogonal, fibrous	styles	parallel to growth axis of skeleton	calcite, pyrite	80 max.	10 max.	Hudson, 1960
Actinostromaria sp. H5480	U.O.–L.K.	Trnovski Gost, Slovenia (Yugoslavia)	orthogonal, fibrous	styles, triaxines ?	parallel to growth axis of skeleton	calcite	100 max. 120 max	10 max. 20 max.	Turnsk, 1966

*Holotype

3d Int. Sponge Conf. 1985

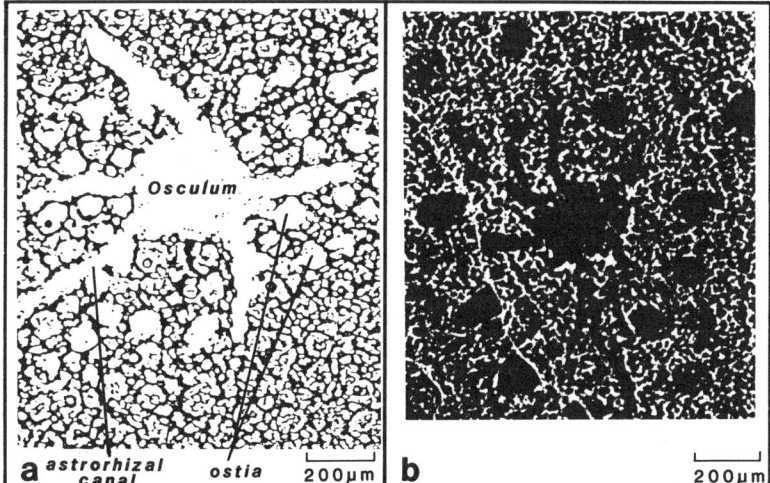

Figure 4. Comparison of transverse sections of aquiferous units of *(a)* Recent sclerosponge, *Calcifibrospongia actinostromarioides*, with *(b)* Upper Jurassic stromatoporoid, *Burgundia astrotubulata*.

microstructure. Within the central zone of the skeletal elements are closely packed styles (100 μm long, 10 μm wide) and triaxine (120 μm long, 20 μm wide) spicules (or possibly tetraxines: these spicules have only been seen in thin section and their three-dimensional nature is uncertain). Triaxine spicules are rare, but appear to be placed at the pillar and pillar-lamellae junctions. The spicules are orientated parallel to the growth axis of the skeleton.

Discussion

The positioning of the spicules in Type 1 *(Actinostromarianina lecompti)* clearly determines the form of the primary irregular calcareous skeleton in a way reminiscent of spongin fibers enmeshing a spicule framework, for example, in *Esperiopsis anomala* (Figure 1h) The primary irregular calcite may have formed by the mineralization of a

network of spongin and may be related to Recent forms where spongin plays a similar role. The radial arrangement of the spicules is similar to that of recent *Reniera* sp., within the Order Haplosclerida, Class Demospongiae. The orthogonal fibrous rim, present periodically and causing latilamination, is a secondary skeletal feature and is possibly related to seasonal environmental fluctuations.

The plumose framework of the Type 2 spicules (members of the Milleporellicae) appear to determine the presence and orientation of the fascicular fibers that form the columns of the calcareous skeleton. The fibers are initiated in tufts at the spicule bases, subparallel to the spicule shafts. The tabulae are secondary infilling tissue and contain no spicules. Type 1 spicules are similar in form, size and distribution, and the calcareous skeleton is similar in microstructure to the "sclerosponge" family Ceratoporellidae, especially to the fossil genus *Murania* sp. There are

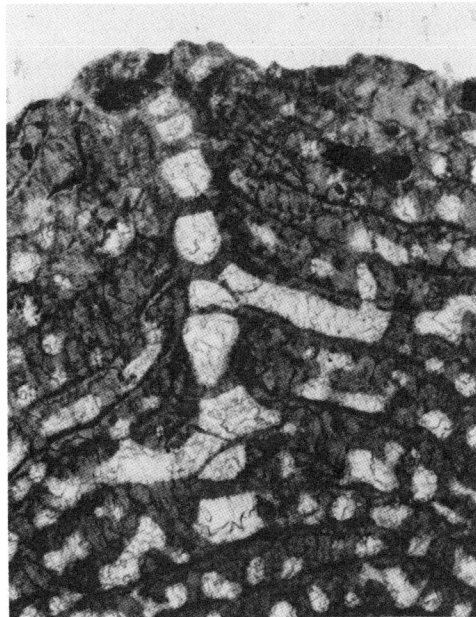

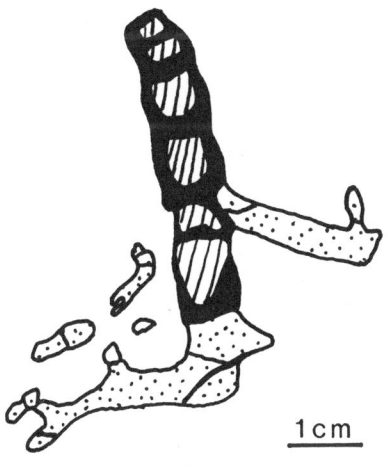

Figure 5. Longitudinal section of Lower Cretaceous stromatoporoid, *Burgundia wetzeli*, showing an arrangement of stacked rhagon aquiferous units. Each unit was sealed off by secondarily precipitated tabulae as the animal grew.

tabulae
osculum
astrorhizal canal

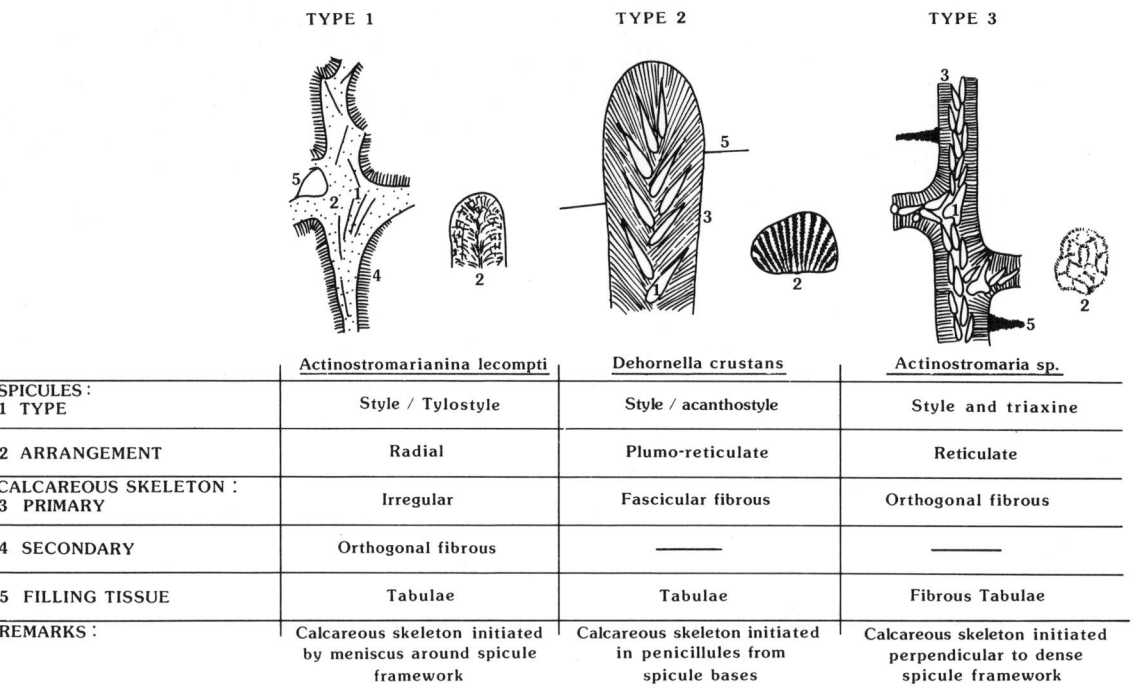

Figure 6. Three skeletal types of Mesozoic stromatoporoid, in terms of spicule type and arrangement, and microstructure of calcareous skeleton. Table below compares homologous structures in the three forms.

also similarities in spicule form and arrangement to the Poecilosclerida, within the Class Demospongiae.

The triaxine (or tetraxine) spicules of *Actinostromaria* sp. (Type 3) show some affinities to the subclass Tetractinomorpha, within the Demospongiae. The three-dimensional morphology of these spicules is unclear; because only calcite pseudomorphs have been found, the possibility of this form being a member of the Calcarea cannot be dismissed.

Mesozoic stromatoporoids appear to produce a secondary calcareous skeleton on a primary spicule framework. The three skeletal types possess different spicule complements, calcareous skeleton microstructures, and differing relationships between the two. The strong correlation between spicule type, arrangement, and calcareous skeleton has important biological consequences. Different modes of biomineralization are required to produce different microstructures, and this suggests profound soft tissue and therefore taxonomic differences.

On the basis of this new spicule information, we can review the taxonomic position of Mesozoic stromatoporoids. Table 2 shows the proposed placing of the spiculate stromatoporoids within the Recent Demosponge taxonomic framework. The placing of sclerosponges is given after Vacelet (1983), and is based upon spicule complement, soft-tissue characteristics, and larval type. Members of the Ceratoporellidae and *Astrosclera* sp. represent calcified members of the Agelasidae, within the Order Axinellida (Figure 1*g*). The Milleporellicae therefore have closest affinities to the Axinellida. *Calcifibrospongia* sp. is a member of the Renieridae, within the Haplosclerida.

The possession of a calcareous skeleton is a convergent

Table 2. Proposed position of Mesozoic "stromatoporoids" and "sclerosponges" (after Vacelet, 1983), in the Recent demosponge classification

Subclass Tectractinomorpha		Subclass Ceractinomorpha	
Poecilosclerida	–*Merlia* sp.	Haplosclerida Renieridae	–*Calicifibrospongia* sp.
Axinellida Agelasidae	–*Astrosclera* sp. –Ceratoporellidae –Milleporellidae (type 2)		
Incertae sedis	–*Actinostromaria* sp. –*Actostroma* sp. (type 3)	Incertae sedis	–*Actinostromarianina* sp. (type 1) –*Vaceletia* sp.

feature in both stromatoporoids and sclerosponges. However, the ultrastructural type of the calcareous skeleton appears to be of taxonomic and phylogenetic significance. Further investigation should make it possible to establish a more precise position and determine to what extent higher taxonomic categories of Mesozoic stromatoporoids are necessary.

The redefinition of Mesozoic stromatoporoids is necessary on two counts: first, stromatoporoids should be considered calcified demosponges, and second, since the word "stromatoporoid" has been made indispensable by widespread use, it should be taken to refer to a "grade of organization" rather than a taxonomic grouping. The proposed redefinition expresses this organizational grade.

> PREVIOUS DEFINITION (TAXONOMIC): Laminar, encrusting, massive, or dendroid layered calcareous organisms. The coenosteum is composed of vertical or radial pillars and horizontal lamellae or tabulae. Astrorhizae may or may not be present. (From Lecompte, 1956.)

> PROPOSED DEFINITION (ORGANIZATIONAL): Calcified demosponges that show a regular repetition of astrorhizal-bearing aquiferous units.

The three species of the genus *Merlia* also support this conclusion. Only one species, *M. normani*, bears a calcareous skeleton, and yet the spicules and soft tissue clearly indicate the congeneric nature of these three forms.

An important consequence of this redefinition is that in organizational terms, the distinction between "sclerosponge" and "stromatoporoid" (in the Mesozoic sense) is purely an arbitrary one.

A similar conclusion was reached by Vacelet (1985) and Reitner (this volume) for the sphinctozooid sponges. Reitner has shown, from spicule complements, that the thalamid skeleton has arisen five separate times in the history of the group.

The development of calcareous skeletons at different grades of organization in different sponge groups may be a response to reef building in warm carbonate seas.

Conclusions

Some members of the Mesozoic "stromatoporoids" possess spicule pseudomorphs and are therefore sponges. The canal and astrorhizal system of stromatoporoids can also be best explained in terms of a poriferan aquiferous filtration system. Originally, siliceous styles (or acanthostyles) and triaxines (or tetraxines) indicate demosponge affinity for the "stromatoporoids." The secondary siliceous skeleton is precipitated on a primary spicule framework. Three different spicule types and arrangements indicate that possession of a calcareous skeleton is a convergent feature. Mesozoic "stromatoporoids" should no longer be used as a taxon, but as an indication of the "grade of

organization" of the calcareous skeleton. As organizational terms, "stromatoporoid" and "sclerosponge" are synonyms. The development of a calcareous skeleton is probably a response to reef building in warm carbonate seas.

Acknowledgments

I wish to thank my supervisors Peter Skelton, Brian Rosen, and Simon Conway Morris for their help and support. Discussion with Jean Vacelet (Marine Station d'Endoume, Marseille) and Joachim Reitner (Institute of Paleontology, Berlin) clarified many points. Thanks to Cedric Shute (British Museum, Natural History), Dick Carlton and the Cartography unit (Open University) for technical support. This work was carried out under an NERC grant, which is gratefully acknowledged.

Literature Cited

Brien, P. 1967. Les éponges; leur naturs metazoaries, leur gastrolation leur état colonial. *Annales Societé Royal Zoologique de Belgique*, 97:197–235.

Goldfuss, G. 1826–1833. *Petrefacta Germaniae* (Düsseldorf), 1:42–82.

Hartman, W. D. 1969. New Genera and Species of Coralline Sponges (Porifera) from Jamaica. *Postilla*, 137:1–39.

———. 1979. A New Sclerosponge from the Bahamas and Its Relationship to Mesozoic Stromatoporoids. Pages 487–475 in *Biologie des Spongiaires*, edited by C. Lévi and N. Boury-Esnault. Colloques internationaux, 291. Paris: Centre National de la Recherche Scientifique.

Hartman, W. D., and T. F. Goreau. 1966. *Ceratoporella*, a Living Sponge with Stromatoporoid Affinities. *American Zoologist*, 6:563–564.

———. 1970. Jamaican Coralline Sponges: Their Morphology, Ecology and Fossil Relatives. Pages 205–243 in *The Biology of the Porifera*, edited by W. G. Fry. Symposia of the Zoological Society of London, 25. London: Academic Press.

Hickson, S. J. 1934. On *Gypsina plana* and on the Systematic Position of the Stromatoporoids. *Quarterly Journal of Microbiological Science*, New Series, 303(76):433–480.

Hudson, R. G. S. 1958. *Actostroma* gen. nov. A Jurassic Stromatoporoid from Makhtesh Hathira, Israel. *Palaeontology*, 1:87–98.

———. 1960. The Tethyan Jurasic stromatoporoids, *Stromatoporina, Dehornella* and *Astroporina*. *Palaeontology*, 2(2):180–199.

Kázmierczak, J. 1976. Cyanophycean Nature of Stromatoporoids. *Nature*, 264:49–51.

Kirkpatrick, R. 1912. On the Stromatoporoid and Eozoon. *Annals and Magazine of Natural History*, 8(X):341–347.

Lecompte, M. 1956. Stromatoporoidea. Pages 108–144 in *Treatise on Invertebrate Paleontology, Part E*, edited by R. C. Moore. Lawrence, Kansas: Society of America and University of Kansas Press.

Lévi, C. 1973. Systematique de la Classe de Demospongia (Demosponges). Pages 577–631 in Traité de Zoologie, 3(1), *Spongiaires*, edited by P. P. Grassé. Paris: Masson et Cie.

Mori, K. 1982. Coelenterate Affinities for Stromatoporoids. *Contributions in Geology,* Stockholm, 37:167–175.

———. 1985. Comparison of Skeletal Structures Among Stromatoporoids, Sclerosponges, and Corals. *Paleontographica America,* 54:354–358.

Nicholson, H. A. 1886–1892. *British Stromatoporoids.* London: Palaeontographical Society. 234 pp.

Rosen, F. B. 1869. Über die Natur der Stromatoporen und die Erhaltung der Hornfaser der Spongien im Fossilen Zustande. *Verhandlungen der Russisch–Kaiserlichen Gesellschaft, St. Petersburgh,* 4(2):1–98.

Turnsek, D. 1966. Upper Jurassic Hydrozoan Fauna from Southern Slovenia. Dissertations, Slovenian Academy of Sciences and Arts, 9:337–428 (in Serbo-Croatian).

Vacelet, J. 1983. Les éponges hypercalcifiées, reliques des organismes constructeurs de récifs du Paláeozoique et du Mésozoique. *Bulletin, Societé de Zoologie de France,* 108:547–557.

———. 1985. Coralline Sponges and the Evolution of Porifera. Pages 1–13 in *The Origins and Relationships of the Lower Invertebrates,* edited by S. Conway Morris et al. London: Systematic Association, Special Volume 28.

Wood, R. A., and J. Reitner. 1986. Poriferan Affinities for Mesozoic Stromatoporoids. *Palaeontology,* 29:469–473.

JOACHIM REITNER
Institute of Paleontology
Freie Universität Berlin
D–1000 Berlin 33, Schwendenerstrasse 8
Federal Republic of Germany

Polyphyletic Origin of the Sphinctozoans

Abstract

The Sphinctozoa are sponges with a thalamid skeleton. The development of this skeletal type occurs six times in the Porifera. There are representatives in the class Calcarea (e.g., *Barroisia*), the class Demospongiae (e.g., *Vaceletia*), the archaeocyathids (e.g., *Dictyosycon*), and in one species of the Hexactinellida, but without a calcareous skeleton *(Casearia articulata)*. The representatives of the "sphinctozoid" demosponges possess a calcareous skeleton of aragonite or high Mg-calcite with trabecular internal structure. Two subclasses of the Demospongiae (Ceractinomorpha, Tetractinomorpha) include species with calcareous segmented bodies. The oldest known species belongs to the Tetractinomorpha. These possess a spicular skeleton integrated within the calcareous skeleton. The scleres are styles and asters. The incorporation of scleres in the calcareous skeleton is a plesiomorphous feature. Further tetractinomorph Sphinctozoa (Murguiathalamida) are known from the Jurassic and mid-Cretaceous. This order possesses megascleres of triaene types within the calcareous skeleton. The first representatives of the ceractinomorph Sphinctozoa *(Stylothalamia)* appear in the Late Triassic. Their morphological patterns are similar to those of the extant genus *Vaceletia*. In most ceractinomorph species spicules are not incorporated in the calcareous skeleton or are lost, which is an apomorphous feature. However, the genus *Vascothalamia* (Late Albian) possesses incorporated fusiform megascleres.

Some Cambrian archaeocyathids (e.g., Archaeosyconiidae) are morphologically similar to the sphinctozoid demosponges. A phylogenetic relation to the sphinctozoid demosponges is possible. The Sphinctozoa are a polyphyletic sponge group for which Sphinctozoida should not be used as a taxonomic term.

Demosponges with calcitic skeletons (stromatoporids, chaetetids, etc.) are commonly found in the Phanerozoic as primary and secondary frame builders in carbonate build-ups. The systematics of these fossil organisms are not fully known. Recent works place these previously uncertain organisms in the phylum Porifera (e.g., Kazmierczak, 1974, 1979; Gray, 1980; Reitner and Engeser, 1983, 1985; Wood and Reitner, 1986).

Stromatoporids and chaetetids are the most common representatives of the calcareous demosponges in carbonate facies. Calcareous demosponges with a thalamid skeleton are very rare in the fossil record, however. A modern representative *(Vaceletia crypta)* of these demosponges can be found in reefs of the Indian and Pacific oceans (Vacelet, 1979a). This modern sphinctozoan has close affinities to some fossil forms (Reitner and Engeser, 1985).

Systematic Position of the Order Sphinctozoa

Steinmann (1882) placed the Sphinctozoa in the class Calcarea, order Pharetronida, with type family Sphaerocoeliidae. Vacelet (1979a), in describing the sphinctozoan *Vaceletia crypta* (see Figure 3), declared the sphinctozoan suborder invalid, because *Vaceletia crypta* is a siliceous sponge. Instead, Vacelet transferred the taxon Sphinctozoa to the class Demospongiae. *Vaceletia crypta* possesses an incubant larva of the parenchymella type characteristic the subclass Ceractinomorpha. Vacelet placed the remaining Sphinctozoa of the class Calcarea into the new order Sphaerocoelida. For example, the genus *Sphaerocoelia* possesses calcitic scleres and should remain in the Calcarea.

Sponges with a Thalamid Calcareous Skeleton

TETRACTINOMORPHA (FIGURES 1, 2).[1] Within the Tetractinomorpha two orders are known to have a thalamid, trabecular calcareous skeleton. The oldest one is Cassianothalamida (Figure 1) from the late Triassic Cassian Beds from northern Italy (Reitner and Engeser, 1985; Reitner, 1987a). The calcareous skeleton of these sponges is constructed of high Mg-calcite that exhibits an irregular microstructure. Beside the trabecular internal structure, the calcareous skeleton has a vesicular tissue found mostly in the ontogenetic older parts of the sponge. The primary spicular skeleton is documented by only few fusiform megascleres (Figure 1*d*) and common aster microscleres (Figure 1*b,c*) embedded in the secondary calcareous skeleton. Sclere arrangement is irregular. The structure of the calcareous skeleton is similar to that of some Mesozoic stromatoporoids, such as *Actinostromaria* (Wood, this volume). Both types differ in the aquiferous systems— Sphinctozoans possess spongocoel (Figure 1*a*), whereas the stromatoporoid aquiferous system is based on the rhagon unit (astrorhizae)—and the segmented skeleton, which is not observed in the stromatoporoids.

---

[1] Note: Most specimens were photographed in thin sections using transmitted light. Specimens are at the Institute of Paleontology, Freie Universität Berlin. Exceptions are noted.

The order Murguiathalamida (Figure 2) is known from the Callovian, Albian, and Cenomanian (Bojko, 1979; Reitner and Engeser, 1985). This order contains tetractinomorph demospongids presumably with an original aragonitic skeleton. The skeletal morphology is variable, as is the structure of the spongocoel wall and the form of the prosopores and apopores. Different types of triaene megascleres are embedded in the calcareous skeleton (Figure 2*b,c*). Microscleres are not present. The sclere arrangement was probably enveloped by aragonite, as in *Calcifibrospongia*. The regular structural arrangement of the scleres (Figure 2*d,e*) is unusual and this cannot be compared with other demospongids. There are similarities to the sclere arrangement of the Hexactinellida, but the triaene scleres indicate demosponge character.

CERACTINOMORPHA (FIGURES 3–6). Ceractinomorph demosponges with a sphinctozoid calcareous skeleton are only found in the order Verticillitida (Reitner and Engeser, 1985). Up to now, sponges have only been found with an aragonite skeleton in which the aragonite crystals are irregularly aligned *(Vaceletia crypta)*. The skeleton itself exhibits a trabecular structure and a vesicular tissue in the ontogenetic older parts, as in the tetractinomorph thalamid demosponges. Scleres are rarely observed. Only the extinct genus *Vascothalamia* (Figure 6) (Reitner and Engeser, 1985) contains fusiform scleres that are embedded in the calcareous skeleton (Figure 6*b,c*). In most cases, these simple monaxonic fusiform spicules are characteristic for the subclass Ceractinomorpha. All other known genera do not exhibit scleres as observed in *Vaceletia crypta* (Figure 3), but it is unclear whether these scleres never existed or whether diagenesis is responsible for this lack. The oldest known representative is *Stylothalamia* found in the Upper Triassic of the Mediterranean realm (Ott, 1967; Senowbary-Daryan, 1978). Apart from recent species, other genera are known from the Jurassic (*Stylothalamia* spp.), from the Cretaceous (*Verticillites* spp., Figure 5; *Menathalamia*, Figure 4; *Vascothalamia*, Figure 6), and from the Tertiary *(Wienbergia, Vaceletia, Marinduqueia)* (Reitner and Engeser, 1985).

CALCAREA (FIGURES 7–9). The taxon Sphinctozoida was originally assigned as a suborder of the order Pharetronida by Steinmann (1882:145). The identifying feature of this group was thought to be the segmented (thalamid) calcareous skeleton (Figures 7,8). All other calcareous sponges were placed into the suborder Inozoa. For Steinmann (1882), the type genus of the Sphinctozoans was *Sphaerocoelia*. This genus possesses calcitic scleres and is certainly a member of the Calcarea. However, it is difficult to assign other Sphinctozoans without affinities to the demosponges on the basis of Steinmann's criteria. No described Paleozoic and Triassic Sphinctozoans have preserved spicules (e.g., *Amblysiphonella*) (Pickett and Jell,

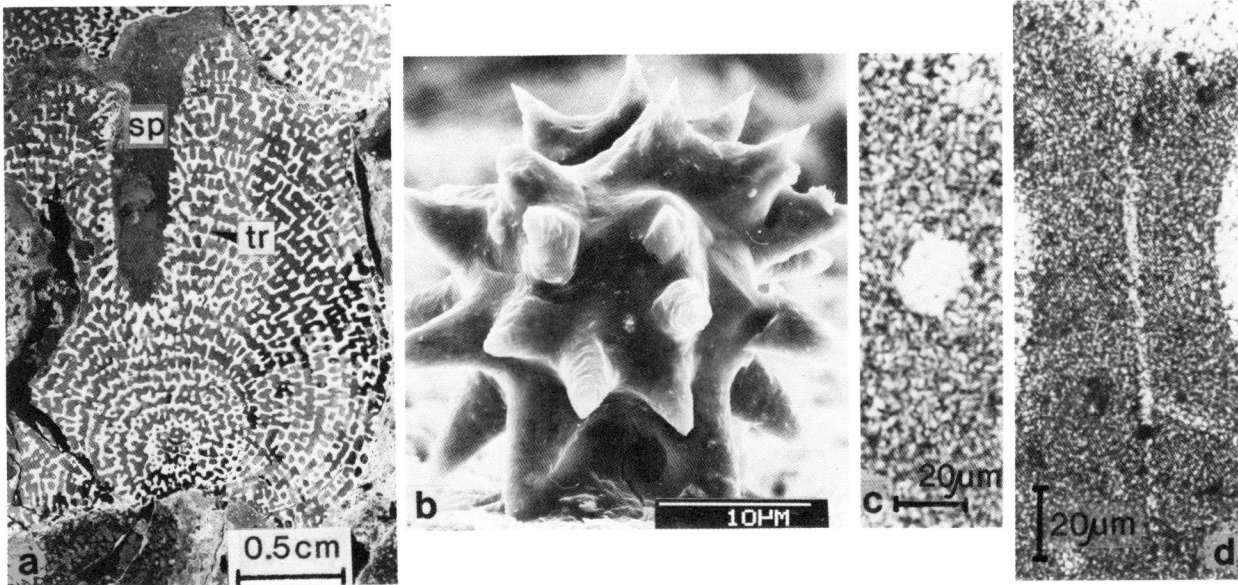

Figure 1. *Cassianothalamia zardini* Reitner (Tetractinomorpha), Upper Triassic of Cassian Beds (northern Italy): *a*, light micrograph, negative print (*sp*, spongocoel; *tr*, trabecula); *b*, scanning electron micrograph of astrose microsclere etched from calcareous skeleton with 1% Titriplex III solution; *c*, astrose microsclere; *d*, monaxon megasclere within calcareous skeleton.

1983; Webby and Rigby, 1985; Seilacher, 1962; Ott, 1967). Therefore it is impossible to definitely assign them to any particular sponge class. Only the Mesozoic Sphinctozoan genera *Barroisia* (Figures 8,9) and *Tremacystia* are certain, because they have calcitic spicules (Reid, 1968)(Figure 9).

ARCHAEOCYATHA (FIGURES 10, 11). The Archaeocyatha have a calcareous skeleton and are limited to the Cambrian. They were established as a phylum consisting of two classes: Regulares and Irregulares (Hill, 1972). These two classes superficially resemble true sponges. This is especially true of the archaeocyathid pore system, which probably possessed a homological function to that of true sponges. The small pores in the outer wall and the larger ones of the inner wall are comparable to the prosopores and the apopores, respectively, in Sphinctozoan sponges. Further similarities to Sphinctozoan sponge morphology appear in the complicated structure of the archaeocyathid inner wall. This structural pattern is comparable to the structure of the spongocoel wall in some thalamid calcareous sponges (e.g., *Verticillites*, *Vascothalamia*) (Reitner and Engeser, 1985) (Figures 10, 11). In addition, all archaeocyathids possess a central cavity similar to the spongocoel of some sponges. Representatives of both classes exhibit in part a segmented skeletal structure (Zhuravleva, 1960). In the class Irregulares, some forms appear similar to demosponges with a calcareous skeleton (Debrenne and Vacelet, 1984; Reitner and Engeser, 1985). The skeletal structure of most of these irregular archaeocyathids is similar to that of the Stromatoporoidea. A few members of the Archaeoyconidae exhibit close morphological af-

finities to the thalamid calcareous demosponges with trabecular skeletal structure.

For example, the thalamid and trabecular calcareous skeleton of the genera *Dictyosycon* (Figure 11), *Tabulacyathus* (Figure 10), *Tabulacyathellus*, *Sphinctocyathus*, and *Hupecyathus* cannot be distinguished from those of sphinctozoid demosponges. In view of these factors, Debrenne and Vacelet (1984) also adopted the sponge model for the archaeocyathids. The archaeocyathids placed in the Regulares class do not resemble the demosponges. Similarities to the hexactinellids certainly exist through the regular skeletal structure, which is comparable to the arrangement of hexactine spicules. Since scleres have never been found in the skeleton of any archaeocyathid, the question of their true affinity to sponges or spongelike forms remains unanswered.

Sponges With a Thalamid Noncalcareous Skeleton (Figure 12).

A hexactinellid sponge, *Casearia articulata*, which possesses a segmented skeleton, is known from the Upper Jurassic of South Germany. However, the thalamid structure is separated in "tabulae" delineating each segment. This feature is not known in the other examples of thalamid skeletons (Müller, 1974).

Diagenetic Problems

The accurate classification of demosponges depends, of course, on the quality of the preserved sclere skeleton, as

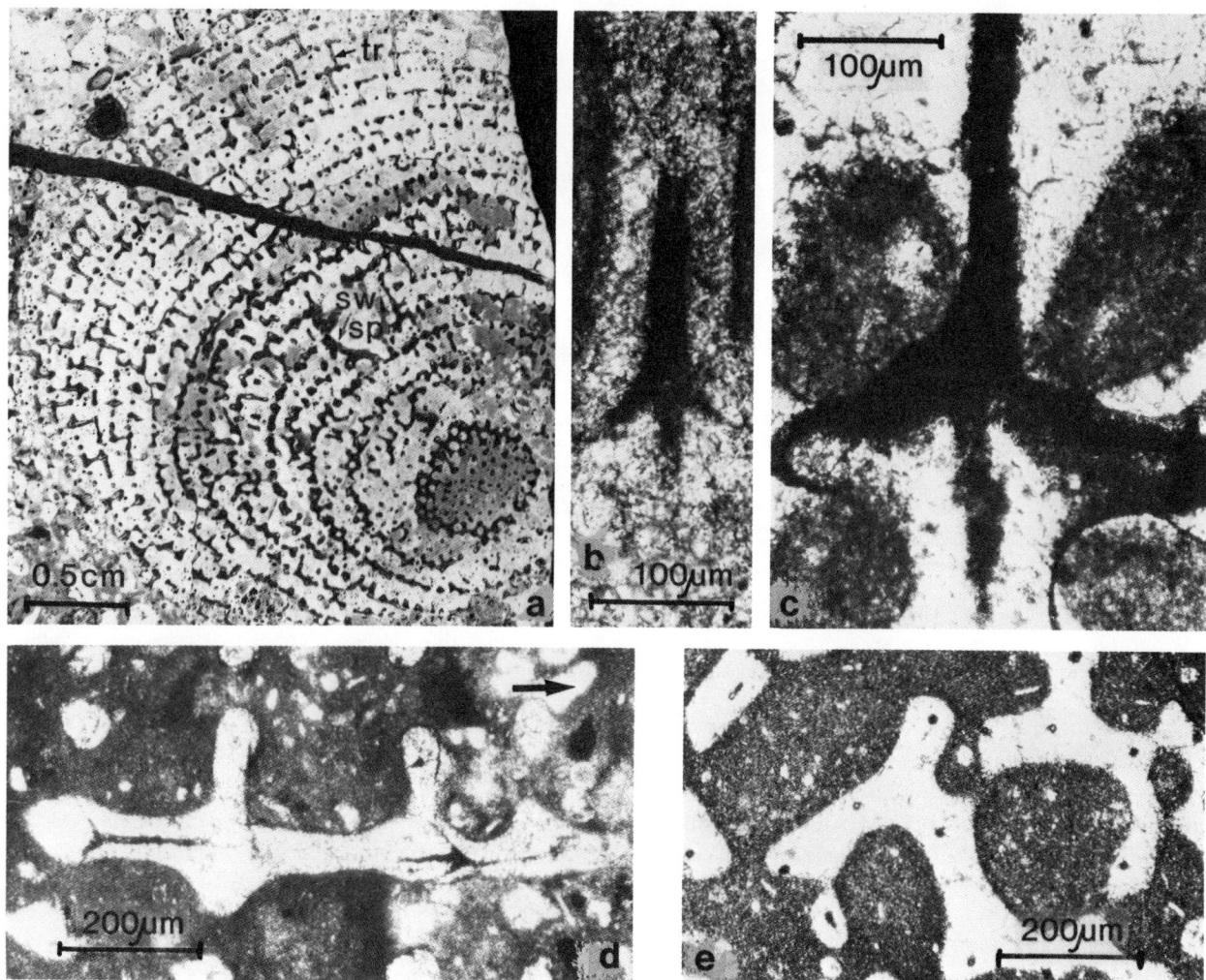

Figure 2. *Boikothalamia*, undescribed species (Murguiathalamida, Tetractinomorpha), Cenomanian hardground of Liencres (northern Spain): *a*, oblique section, negative print (*sp*, spongocoel, *sw*, spongocoel wall, *tr*, trabecula); *b*, simple triaene sclere within a trabecula (pyrite pseudomorph); *c*, anatriaene sclere within a trabecula (pyrite pseudomorph); *d*, vertical section showing arangement of triaene scleres within trabeculae (arrow: direction of surface); *e*, horizontal section showing sclere arangement.

well as the secondary calcareous skeleton. Diagenetic processes generally destroy or alter aragonitic skeletons, whereas calcitic components may be perfectly preserved. This variance in the preservation of sponge features certainly makes it difficult to reconstruct evolutionary patterns. The diagenetic history of each sponge fossil must be taken into account if all sponge characteristics are to be identified correctly. The example below illustrates this problem with respect to the demosponges.

DIAGENESIS OF CALCAREOUS SKELETONS

The calcareous skeletons of the extant demosponge *Vaceletia crypta* is composed of irregularly aligned aragonitic crystals (Vacelet, 1979a; Reitner and Engeser, 1985). Most fossil skeletons of the thalamid ceractinomorphs

(Verticillitida) and a portion of the thalamid tetractinomorphs (Murguiathalamida) are composed of a diagenetic blocky calcite (Figure 2*b,c*) in which the original microstructure has not been preserved (Reitner and Engeser, 1985). The influx of continental vadose waters during diagenesis dissolved the aragonite of the calcareous skeletons. The resultant molds were then cemented with low Mg-calcite under meteoric conditions (Reitner, 1986). Scleractinian corals associated with these demosponges also exhibit this blocky calcite mold cement. Since this same diagenetic process must also have acted upon the original aragonite of these corals, it is probable that the original skeleton of demosponges was also aragonite. In general, the calcitic organisms (e.g., brachiopods) associated with scleractinians and calcareous demosponges do not show any diagenetic alteration.

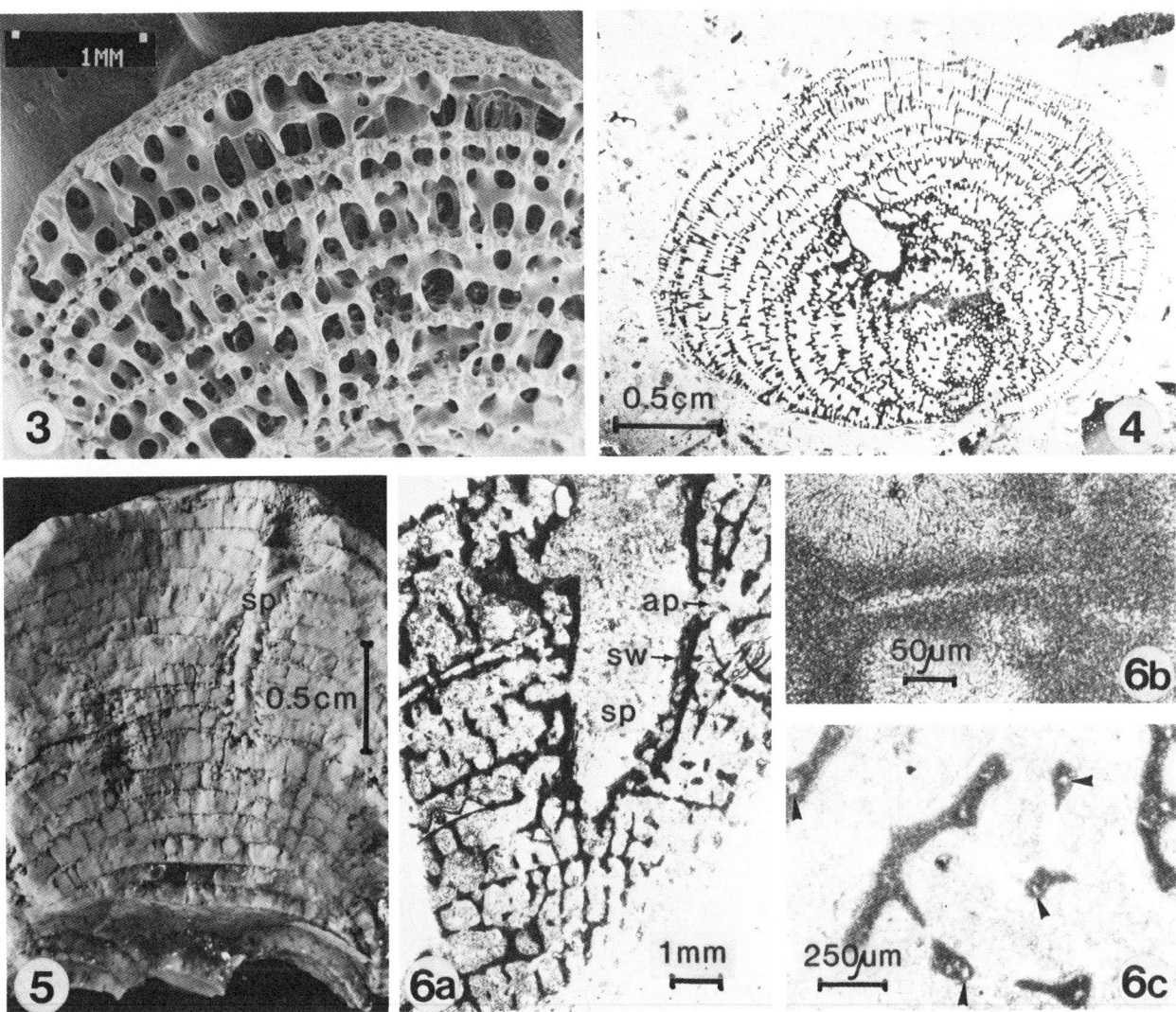

Figures 3–6. Examples of Ceractinomorpha: *3, Vaceletia crypta* (Vacelet), Great Barrier Reef, Australia, scanning electron micrograph; *4, Menathalamia caniegoensis* Reitner and Engeser, Upper Albian (Northern Spain), negative print; *5, Verticillites cretaceous* Defrance, Campanian of Nehou (Northern France) (British Museum Nat. Hist. S 9256); *6, Vascothalamia arayaensis* Reitner and Engeser (holotype), Upper Albian (northern Spain) (*ap*, apopore, *sp*, spongocoel, *sw*, spongocoel wall); *6b*, fusiform sclere within a trabecula; *6c*, horizontal cross section showing sclere arrangement *(arrows)*.

The skeleton of *Vascothalamia*, which is composed of micritic calcite (Figure 6), may also have been aragonite initially. Samples of these sponges from the marls of a paleo forereef (Reitner, 1987b) have apparently undergone a unique diagenetic change. A minor influx of vadose waters caused the neomorphic calcite to crystallize, and the resulting granular mirostructure is comparable to stage 3 in the diagenesis of neomorphic altered aragonite, according to Wendt (1979).

Sponges in the order Cassianothalamida, however, do not have an aragonitic skeleton. The skeleton of these sponges consists of high Mg-calcite with an irregular ultrastructure (Figure 1) (Reitner and Engeser, 1985; Reitner, 1987a). The excellent preservation of this skeleton can be attributed to stable calcite mineralogy and the diagenetic paleoenvironment. These fossil sponges are found in the Upper Triassic Cassian Beds, which contain large blocks (Cipit Limestones) of allochthonous reef blocks. In this position, vadose continental water has little effect, and the calcite and aragonite of the sponge skeletons remained unaltered (Fürsich and Wendt, 1977).

DIAGENESIS OF THE SCLERE SKELETON

Unfortunately, few sclere are preserved in calcareous demosponges. The hydrated opal of the sclere in calcified demosponges is promptly dissolved at a early diagenetic stage, probably as a result of the decay of the organic axial

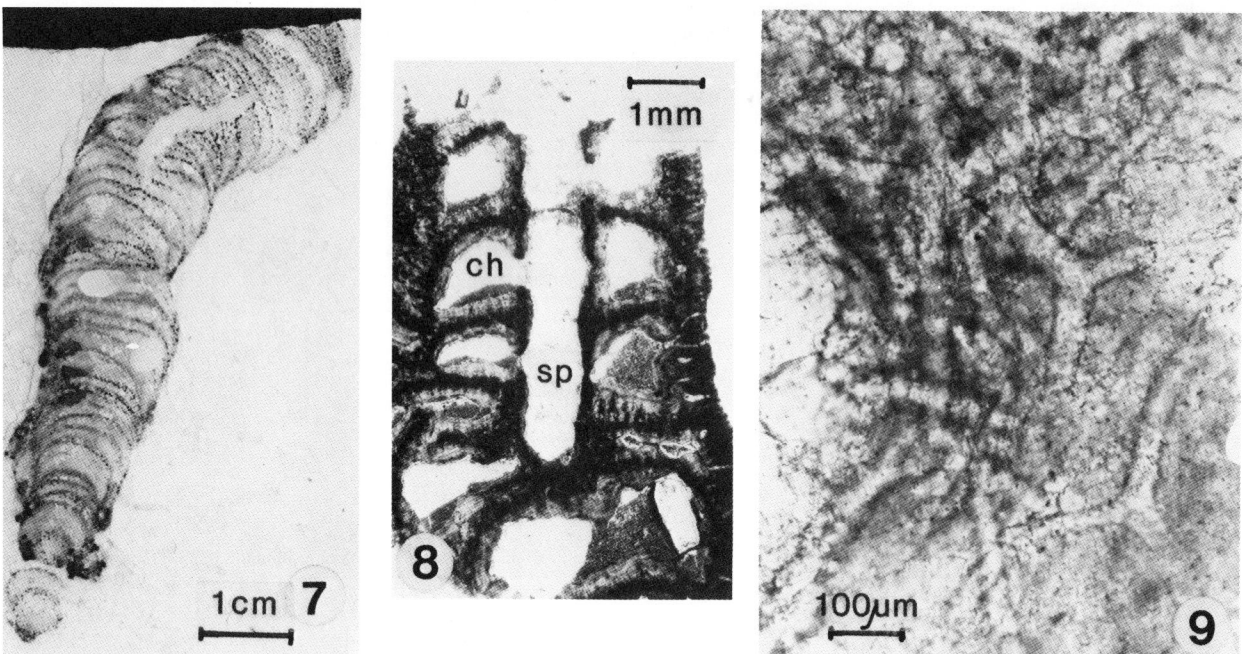

Figures 7–9. Examples of Calcarea: *7, Muellerithalamia extensus* (Lang), Upper Jurassic (Treuchtlinger Marmor, South Germany) (this sponge possesses caltrops and triaene calcitic spicules similar to those in Figure 9); *8, Barroisia* sp., Aptian of Farrington (England) (*ch*, chambers, *sp*, spongocoel); *9, Barroisia gandaraensis* Reitner, Middle Albian of Gandara (northern Spain), showing triaene calcitic spicule arangement.

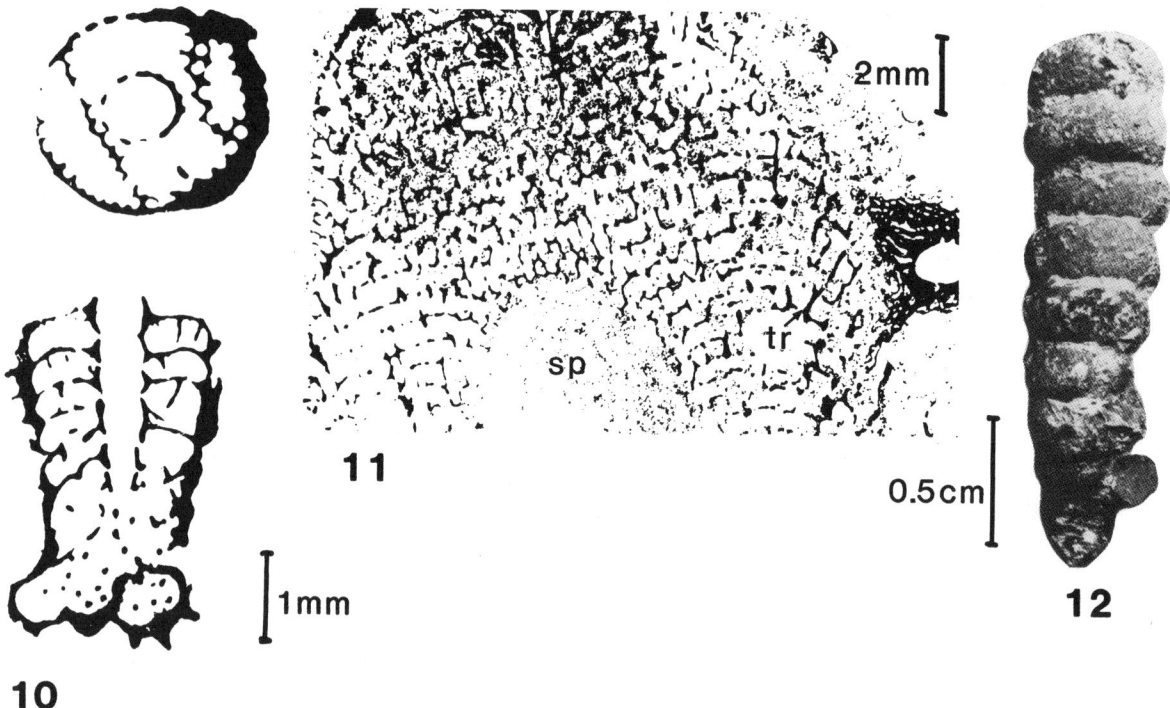

Figures 10–12. Examples of Archaeocyatha and Hexactinellida: *10, Tabulacyathus taylori* Vologdin (Tabulacyathidae, Irregulares), Lower Cambrian (Altai Mountains, Russia) (from Hill, 1972); *11, Dictyosycon gravis* Zhuravleva (Archaeosyconidae, Irregulares), Lower Cambrian (Russia) (from Hill, 1972) (*sp*, spongocoel; *tr*, trabecula); *12, Casearia articulata* Schmidel (Hexactinellida), Upper Jurassic (South Germany) (after Müller, 1974).

filament through bacterial activity (Reitner, 1987c). Hartman (1979) observed this diagenetic process in *Calcifibrispongia*, a stromatoporoid haplosclerid demosponge. In most cases, dissolution produced sclere molds that were then cemented by the epitactical growth of argonite or calcite crystals in the calcareous skeleton. It is nearly impossible to distinguish the original skeleton from the diagenetic "skeleton". Only in special circumstances, as in the Cipit Limestones, were the original sclere cavities cemented with a blocky low Mg-calcite (Figure 1*b-d*) through meteoric diagenesis or promptly filled with a micritic carbonate sediment.

Preserved scleres have also been found in calcareous demosponges associated with hardgrounds (Reitner, 1987b). In this case, scleres have not been replaced with epitactical aragonite or calcite crystals; rather, the molds were cemented by pyrite and thus bacteria are thought to have played a role in sclere dissolution (Figure 2*b-e*). The replacement by pyrite occurred before the aragonite skeleton was dissolved. Following this mineralization, the skeleton was dissolved and cemented by blocky calcite. Thus, the sclere arrangement has been perfectly preserved (Figure 2*d*).

Phylogenetic Aspects of the Sphinctozoid Demosponges

Two of the three subclasses of the Demospongiae have obviously developed a calcareous segmented skeleton independently. The principles of phylogenetic systematics (cladistics) can be used to determine the phylogenetic relationships among these sponges. The characteristic features are identified as either original (plesiomorphic) or derived (apomorphic).

Among the numerous explantions of the general evolution of the demosponges (Hartman et al., 1980) there are three significant theories. One of these (Reid, 1970) is based on the ideas of Dendy (1924). Dendy and Reid argued that megasclere development depends on the diagnostic priority of the component microscleres. In Reid's opinion, the origin of the two fundamental subclasses of the demosponges evolved at the Precambrian-Cambrian boundary out of a sponge group comparable to the Homoscleromorpha. This group, whose extant members include the Plakinidae, is characterized by very small tetractine spicules (calthrops).

The second theory, as advanced by Finks (1970), is that the Calcarea and the Demospongiae evolved from a common ancestor in the Precambrian. Finks postulated that the demospongids derive from a monaxon ancestor, not a tetraxon type, as suggested by Reid (1970). Finks maintained that the genus *Hazelia* is a direct ancestor of the Demospongiae. In addition, he argued that the Homoscleromorpha are derived from a younger sponge group.

These conclusions are questionable, however, in that mid-Cambrian tetraxon scleres can be found with monaxons (van Kempen, this volume). It is more probable that the two main demospongid groups became separated very early in the Cambrian or even in the Precambrian.

The third theory, suggested by van Soest (1984), is that the Homoscleromorpha and the Calcarea have close phylogenetic relationships, as is evident from the similarities between the calthroplike spicules, the amphiblastula larva, and the large choanocyte chambers. In van Soest's opinion, the calthrops in both groups may be a synapomorphy. Therefore he concluded that Ceractinomorpha and Tetractinomorpha are derived from the same ancestor as the Homoscleromorpha and the Calcarea. This theory warrants attention because an amphiblastula stage in *Vaceletia crypta* has been observed during larval development (Vacelet, 1979b). This feature coincides with van Soest's theory. The convergent development of the calcareous skeleton can be demonstrated through cladistics by means of four important features: (1) Segmented skeletons and segmented skeletons with a trabecular internal structure; (2) calcareous skeletons; (3) scleres within the calcareous skeleton, micro- and megascleres within the calcareous skeleton, megascleres within the calcareous skeleton; and (4) sclere reduction. All these features are apomorphous. A detailed cladogram of the phylogenetic relationships of the sphinctozoid Ceractinomorpha can be found in Reitner and Engeser (1985). The cladogram presented here (Figure 13) is based on the theory of van Soest (1984). The phylogenetic dilemma concerning the relationship between the Calcarea and the Homoscleromorpha (see van Soest, 1984) is complicated by the question of the monophyletic origin of the hypercalcified skeleton within the Demospongiae. The demosponges with a calcareous thalamid skeleton exhibit many fundamental differences, especially in their mineralogy, skeletal structure, sclere arrangement, and type of spicules. The development of a calcareous skeleton is apomorphous. The process of biomineralization is simple and probably indirectly influenced by the soft tissue only. The biomineralization is comparable to that in some calcareous algae (Dasycladales), which is initiated by a thin mucus layer of the outer cell membrane (Flajs, 1977). The main pore space is cemented by irregular aragonite crystals. This simple type of biomineralization is observed in the *Vaceletia crypta* in Vacelet's collection.

The phylogenetic relationship between Demospongiae and the class Irregulares (Archaeocyatha) is unclear since scleres have not yet been found in this group. The absence of scleres may indicate that the archaeocyathids do not have generalized scleres or do not have any scleres at all, as is the case in *Vaceletia crypta*. The Irregulares may be a sister group to the Demospongiae (Figure 13). The skeletons of the Archaeosyconidae and Tabulacyathidae exhibit close affinities to the demosponges.

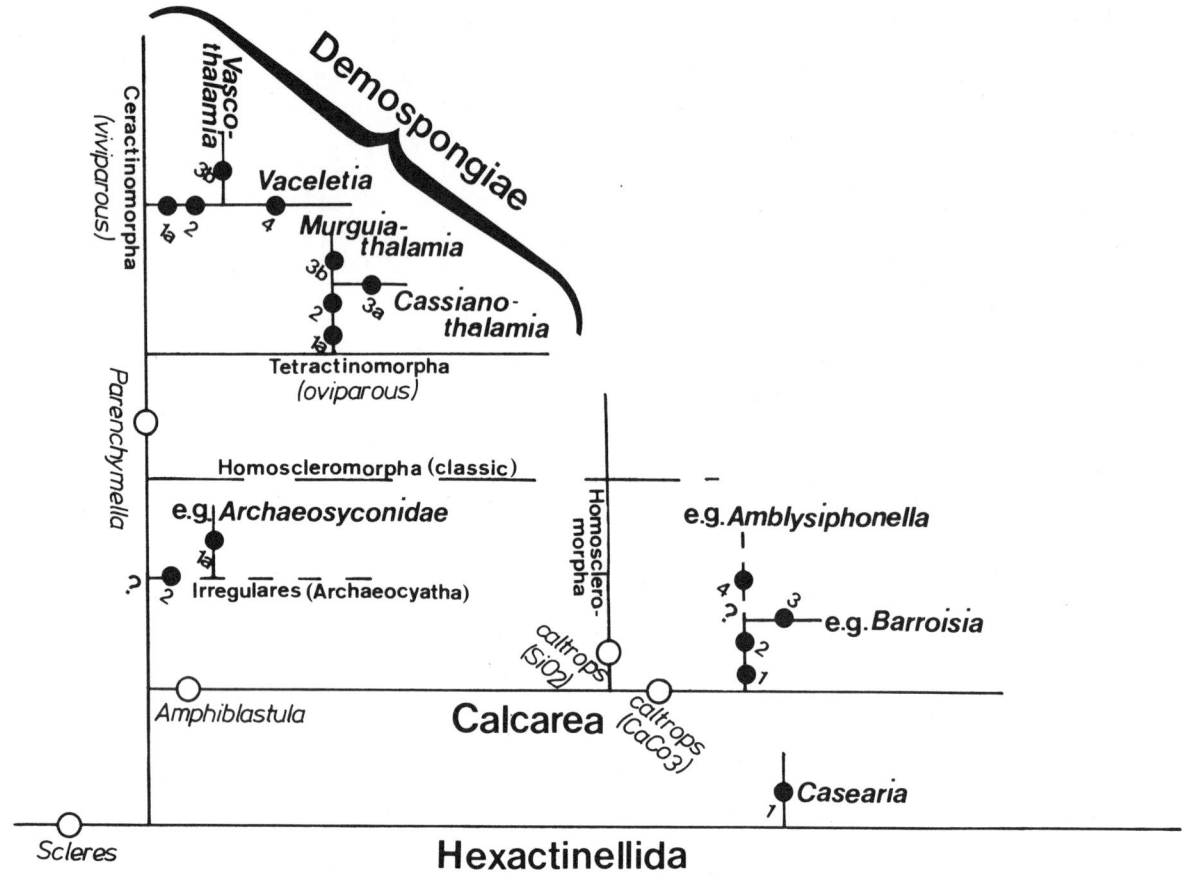

Figure 13. Cladogram based on van Soest's theory demonstrating convergent thalamid calcareous skeletons. (For explanations see text.)

Discussion and Conclusions

In spite of the polyphyletic features of the demosponge sphinctozoans, a few general phylogenetic trends have been established here. The oldest tetractinomorph sphinctozoan *(Cassianothalamia)* exhibits microscleres and some megascleres that are enclosed by a calcareous skeleton. The younger representatives (Murguiathalamida) contain only megascleres built into a regular calcified network. The presence of both sclere types within the calcareous skeleton is a primitive feature, whereas the absence of microscleres is an apomorphous feature.

In the ceractinomorph sphinctozoans, only one form *(Vascothalamia)* possesses monaxonic scleres within the calcareous skeleton. The extant form *(Vaceletia)* and the remaining fossil forms *(Verticillitida)* have no scleres. In this case, scleres no longer need to be present in a calcareous skeleton because sclere skeletons lost their function during later ontogenesis. The reduction of the spicules is an apomorphous feature.

The ceractinomorph sphinctozoans represent the most advanced sponge morphology in the demosponges. This is also true of the nonsphinctozoan ceractinomorphs that

are the most modern forms. The tetractinomorph sphinctozoans exhibit more plesiomorphous features and are relatively more primitive morphologic types.

The archaeocyathid class Irregulares contains forms with a segmented trabecular calcareous skeleton (e.g., Archaeosyconidae). This skeleton type is comparable to the skeleton type observed in the demospongid sphinctozoans. However, the calcareous skeleton is not a reliable criterion for classifying the Archaeosyconidae. The presence or absence of spicules in these sponges must first be confirmed before an organism can be placed with certainty.

The classical sphinctozoans placed in the Calcarea are represented by only a few genera (e.g., *Barroisia, Tremacystia*) that exhibit calcitic scleres. The arrangement of spicules in these forms is comparable to that in fossil and modern inozoans (e.g., *Petrobiona, Murrayona*). Most of the described Paleozoic and Mesozoic sphinctozoans have no spicules (e.g., *Amblysiphonella*). It is impossible to classify these true sponges as members of the Calcarea. Perhaps this feature is comparable to the aspicular skeleton of *Vaceletia crypta*. This problem requires further investigation.

One hexactinellid sponge, *Casearia articulata*, has a

thalamid spicular skeleton. Calcification was not observed here.

Six examples of a thalamid calcareous skeleton were found in the Archaeocyathids, the Calcarea, and Demospongiae; four examples of a thalamid calcareous skeleton with trabecular internal structure were found in the Irregulares (e.g., Archaeosyconidae), the Tetractinomorpha, and the Ceractinomorpha.

The term Sphinctozoida should only be used to describe a special skeletal morphotype, and not a taxon.

Acknowledgments

I wish to thank Beth Gierlowski-Kordesch, Institute of Paleontology, Berlin, and Rachel Wood, Open University, Milton Keynes, England for translation help and for clarifying many critical points. I also thank H. Keupp, University of Bochum, West Germany, for giving me the opportunity to study the specimens from the Upper Jurassic ("Treuchtlinger Marmor") of South Germany.

Literature Cited

Bojko, F. V. 1979. On the Family Verticillitidae Steinmann, 1882, Its Composition and Systematic Position. *Trudy Instituta Geologii Geofiziki Sibirskoe Otdelenie (USSR)*, 481:74–82 (in Russian).

Debrenne, F., and J. Vacelet. 1984. Archaeocyatha: Is the Sponge Model Consistent with Their Structural Organization? *Palaeontographica Americana*, 54:358–369.

Dendy, A. 1924. On an Orthogenetic Series of Growth Forms in Certain Tetraxonid Sponge Spicules. *Proceedings, Royal Society London, Series B*, 97(1925):243–250.

Finks, R. M. 1970. The Evolution and Ecological History of Sponges during Palaeozoic Times. Pages 3–22 in *The Biology of the Porifera*, edited by W. G. Fry. Symposia, Zoological Society of London, 25. London: Academic Press.

Flajs, G. 1977. Die Ultrastrukturen des Kalkalgenskeletts. *Palaeontographica, Abteilung B*, 160:69–128.

Fürsich, F. T., and Wendt, J. 1977. Biostratinomy and Paleoecology of the Cassian Formation (Triassic) of the Southern Alps. *Palaeogeography, Palaeoclimatology, Palaeoecology*, 22:257–323.

Gray, D. I. 1980. Spicule Pseudomorphs in a New Palaeozoic Chaetetid, and Its Sclerosponge Affinities. *Palaeontology*, 23:803–820.

Hartman, W. D. 1979. A New Sclerosponge from the Bahamas and Its Relationship to Mesozoic Stromatoporids. Pages 467–475 in *Biologie des Spongiaires*, edited by C. Lévi and N. Boury-Esnault. Colloques Internationaux, 291. Paris: Centre National de la Recherche Scientifique.

Hartman, W. D., J. Wendt, and F. Wiedenmayer, eds. 1980. *Living and Fossil Sponges*. Notes for a short course. Sedimenta 8. Miami: University of Miami. 274 pp.

Hill, D. 1972. *Archaeocyatha. Treatise of Invertebrate Palaeontology, vol. 1*, edited by C. E. Teichert. Boulder, Colorado: Lawrence. 158 pp.

Kazmierczak, J. 1974. Lower Cretaceous Sclerosponge from the Slovakian Tatra Mountains. *Palaeontology*, 17:341–347.

——. 1979. Sclerosponge Nature of Chaetetids Evidenced by Spiculated *Chaetetopsis favrei* (Deninger, 1906) from the Barremian of Crimea. *Neues Jahrbuch, Geologie und Paläontologie, Monatshefte*, 1979:97–108.

Müller, W. 1974. Beobachtungen an der hexactinelliden Juraspongie *Casearia articulata* (Schmidel). *Stuttgarter Beiträge zur Naturkunde, Serie B*, 12:1–19.

Ott, E. 1967. Segmentierte Kalkschwämme (Sphinctozoa) aus der alpinen Mitteltrias und ihre Bedeutung als Riffbildner im Wetterstein-Kalk. *Bayerische Akademie der Wissenschaften, Mathematisch naturwissenschaftliche Klasse, Abhandlungen Neue Folge*, 131:1–96.

Pickett, J. W., and P. A. Jell. 1983. Middle Cambrian Sphinctozoa (Porifera) from New South Wales. *Memoirs, Association of Australian Paleontologists*, 1:85–92.

Reid, R. E. H. 1968. *Tremacystia, Barroisia*, and the Status of Sphinctozoida (Thalamida) as Porifera. *University Kansas Paleontological Contributions*, 34:1–10.

——. 1970. Tetraxons and Demosponge Phylogeny. Pages 63–89 in *The Biology of the Porifera*, edited by W. G. Fry, Symposia Zoological Society London, 25, London: Academic Press.

Reitner, J. 1986. A Comparative Study of the Diagenesis in Diapir Influenced Reef Atolls and a Fault-Block Reef Platform in the Late Albian of the Vasco-Cantabrian Basin (northern Spain). Pages 186–209 in *Reef Diagenesis*, edited by H. J. Schroeder and B. H. Purser. Berlin: Springer Verlag.

——. 1987a. A New Calcitic Sphinctozoan Sponge Belonging to the Demospongiae from the Cassian Formation (Lower Carnian; Dolomites, Northern Italy) and Its Phylogenetic Relationship. *Geobios*, 20:571–589.

——. 1987b. Mikrofazielle, palökologische und paläogeographische Analyse ausgewählter Vorkommen flachmariner Karbonate im Basko-Kantabrischen Strike Slip Fault-Becken-System(Nordspanien) an der Wende von der Unterkreide zur Oberkreide. Ph.D. Dissertation, University of Tübingen. *Documenta Naturae*, 40:1–239.

——. 1987c. *Buzkadiella erenoensis* n. gen. n. sp. ein Stromatopore mit spikulärem Skelett aus dem Oberapt von Breño (Prov. Guipuzcoa, Nordspanien) und die systematische Stellung der Stromatoporen. *Paläontologische Zeitschrift*, 61:203–222.

Reitner, J., and T. Engeser. 1983. Contributions to the Systematics and Paleoecology of the Family Acanthochaetetidae Fischer, 1970 (Order Tabulospongida, Class Sclerospongiae). *Geobios*, 16:773–779.

——. 1985. Revision der Demospongier mit einem thalamiden. aragonitischen Basalskelett und trabekularer Internstruktur ("Sphinctozoa" Pars). *Berliner geowissenschaftliche Abhandlungen (A)*, 60:151–193.

Seilacher, A. 1962. Die Sphinctozoa, eine Gruppe fossiler Kalkschwämme. *Akademie der Wissenschaften und Literatur Mainz, Abhandlungen der Mathematisch Naturwissenschaftlichen Klasse*, 1961:721–790.

Senowbary-Daryan, B. 1978. Neue Sphinctozoen (segmentierte Kalkschwämme) aus den "oberrhätischen" Riffkalken der nördlichen Kalkalpen (Hintersee/Salzburg). *Senckenbergiana Lethaea*, 59:37–362.

Soest, R., van. 1984. Deficient *Merlia normani* Kirkpatrick, 1908, from the Curaçao Reefs, with a Discussion on the Phylogenetic

Interpretation of Sclerosponges. *Bijdragen tot de Dierkunde*, 54:211–219.

Steinmann, G. 1882. Pharetronen-Studien. *Neues Jahrbuch, Mineralogie, Geologie, Palaeontologie*, 2:139–191.

Vacelet, J. 1979a. Description et affinitès d'un eponge sphinctozoaire actuelle. Pages 483–493 in *Biologie des Spongiaires*, edited by C. Lévi and N. Boury-Esnault. Colloques Internationaux du C.N.R.S., 291. Paris: Centre National de la Recherche Scientifique.

———. 1979b. Quelques stades de la reproduction sexuée d'un eponge sphinctozoaire actuelle. Pages 95–101 in *Biologie des Spongiaires*, edited by C. Lévi and N. Boury-Esnault. Colloques Internationaux, 291. Paris: Centre National de la Recherche Scientifique.

Webby, B. D., and K. Rigby. 1985. Ordovician Sphinctozoan Sponges from Central New South Wales. *Alcheringa*, 9:209–220.

Wendt, J. 1979. Development of Skeletal Formation, Microstructure, and Mineralogy of Rigid Calcareous Sponges from the Late Palaeozoic to Recent. Pages 449–457 in *Biologie des Spongiaires*, edited by C. Lévi and N. Boury-Esnault. Colloques Internationaux du C.N.R.S. 291, Paris: Centre National de la Recherche Scientifique.

Wood, R., and J. Reitner. 1986. Poriferan Affinities of Mesozoic Stromatoporids. *Palaeontology*, 29:469–475.

Zhuravleva, I. T. 1960. Arkheotziaty Sibirskoí platformy. Leningrad: Izdatel (NKIP). 343 pp.

Biochemistry and Chemotaxonomy

BRIAN W. SULLIVAN
D. JOHN FAULKNER*
Scripps Institution of Oceanography (A–012F)
University of California at San Diego
La Jolla, California 92093

Chemical Studies of the Burrowing Sponge *Siphonodictyon coralliphagum*

Abstract

Secondary metabolites together with the toxicity and calcium ion chelating ability of the major metabolites were examined in specimens of *Siphonodictyon coralliphagum* from Belize, the Bahamas, Ponape, and Kwajalein. The most noticeable differences in secondary metabolite composition occurred between specimens burrowing into living coral and those found in dead coral. These differences are thought to arise because *S. coralliphagum* forma *typica* and other forms burrowing in living coral need to maintain a "dead zone" around the oscular chimney of the sponge to prevent aggression by coral polyps. A mechanism by which secondary metabolites may assist in the excavation process is also proposed.

Marine sponges are a prolific source of secondary metabolites, most of which have been studied because of their unique chemical structures or useful pharmacological properties (for a recent review see Faulkner, 1984). Relatively few studies have defined natural roles for sponge metabolites although many functions have been proposed. Bakus and Green (1974) postulated that toxic sponge metabolites evolved in response to fish predation and showed that sponge toxicity was higher in tropical waters, paralleling the increased diversity in fishes. Jackson and Buss (1975) proposed that sponge metabolites inhibit the growth of spatial competitors in reef environments either by killing the competitor or by rendering the substrate around the sponge unsuitable for other organisms. Sponge metabolites may also protect the sponge from surface fouling by inhibiting the growth or development of the larvae of fouling organisms (Bergquist, 1978; Thompson et al., 1985) and might enhance the feeding efficiency of sponges by causing aggregation of bacteria (Bergquist and Bedford, 1978).

*Author to whom correspondence should be addressed.

In the majority of the studies cited above, sponge metabolites have been treated as if they were all similar molecules having similar functions. There is, however, ample evidence that the biological activity of each metabolite is highly dependent on molecular structure, even when the metabolites might appear closely related. It is preferable to study the biological activity of pure compounds rather than crude extracts since sponges probably exude pure compounds or mixtures of specific compounds, as has been demonstrated in the case of *Aplysina fistularis* (Walker et al., 1985). Bioassays performed using crude extracts can be extremely valuable to indicate the presence of biologically active compounds, but it must be recognized that they are of limited value when used in a quantitative manner to compare samples.

In our experience, it has been very difficult to ascribe a specific ecological function to specific sponge metabolites. For example, it is well documented that dorid nudibranchs store selected sponge metabolites that render the nudibranch distasteful to fish (Faulkner and Ghiselin, 1983). Although it is true that the same metabolites should protect the sponge from predation by fishes, is this really the most important function for the metabolites or are they equally useful in preventing fouling (Thompson, 1985)? In order to avoid some of these ambiguities, we have studied the chemistry of burrowing sponges since the effects attributed to chemicals are highly specific.

Sponges of the genus *Siphonodictyon* [now considered to be a junior synonym of *Aka*—Ed.] belong to a small group of sponges that burrow into limestone substrata. Some *Siphonodictyon* sponges have the unique ability to burrow into living coral heads, leaving only the oscular chimneys exposed (Rützler, 1971). The oscular chimney is ringed by a "dead zone" which is devoid of living coral polyps and thus protects the sponge from overgrowth (Figure 1). This is in sharp contrast with *Siphonodictyon* sponges that grow on dead coral and are heavily fouled by coralline and other algae, other sponges, bryozoans, and tunicates. We expected to observe significant differences in the secondary metabolites of the sponges depending on their growth form. Furthermore, we expected to find that the secondary metabolites of species of *Siphonodictyon* would chelate calcium ions if they were involved in the burrowing process.

Methods and Results

Our first investigation of *Siphonodictyon coralliphagum* described the isolation of siphonodictyal-A (see structure *1* in Figure 3) and siphonodictyal-B *(2)* from oscular chimneys collected from the vicinity of Lighthouse and Glover reefs, Belize (Sullivan et al., 1981). The structure initially proposed for siphonodictyal-B was corrected using newer spectroscopic techniques (^{1}H nuclear Overhauser effect difference spectroscopy) that clearly estab-

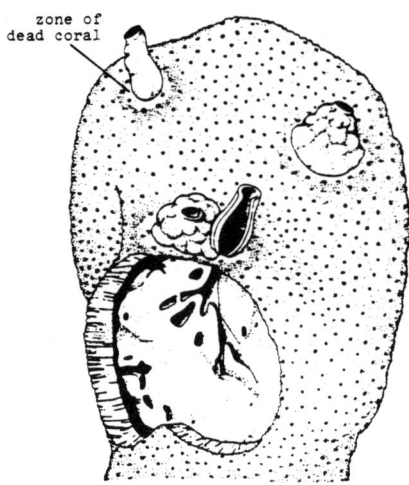

Figure 1. Cutaway view of *Siphonodictyon coralliphagum* growing out of *Stephanocoenia michelinii* (from Rützler, 1971).

lished the substitution pattern about the aromatic ring in *2*. We later reported the isolation of siphonodictidine (Figure 2) from an undescribed species of *Siphonodictyon* from Palau (Sullivan et al., 1983). Siphonodictidine was shown to be toxic to a coral in laboratory experiments. The significant differences in the secondary metabolites of the two species of *Siphonodictyon* provided the impetus for a more detailed study of *Siphonodictyon* sp.

Siphonodictyon coralliphagum from the Caribbean was described as having four distinct morphological forms, forma *typica*, forma *tubulosa*, forma *obruta*, and forma *incrustans*, all belonging to one population (Rützler, 1971). We examined the secondary metabolites of two of the forms: forma *typica*, which excavates in living coral; and forma *tubulosa*, found excavating in dead coral. Samples of each form were collected in Belize and the Bahamas. We also examined specimens of *Siphonodictyon coralliphagum* from Ponape and Kwajalein in the Pacific Ocean.

Eight samples of *Siphonodictyon coralliphagum* forma *typica* and six samples of *S. coralliphagum* forma *tubulosa* were collected at Carrie Bow Cay, Belize, in November 1983. Each sample was extracted separately, and the crude extracts were compared by thin-layer chromatography (tlc) and ^{1}H NMR spectroscopy. (It is important to use more

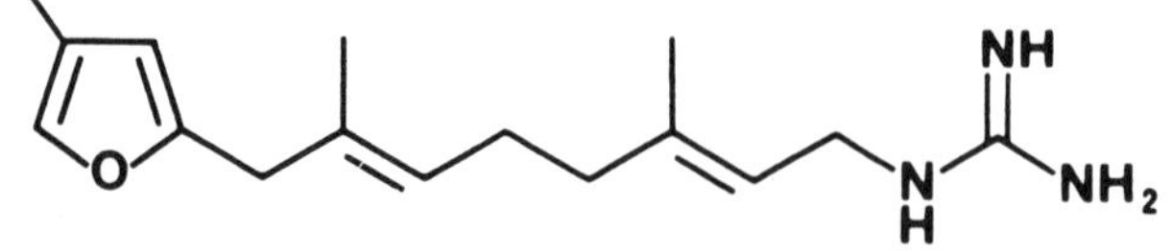

Figure 2. Structure of siphonodictidine from *Siphonodictyon* sp. from Palau.

than one technique for making comparisons since tlc only discriminates between compounds of dissimilar polarity: different compounds often provide identical tlc patterns.) The six samples of f. *tubulosa* were considered indistinguishable and were combined. Of the eight samples of f. *typica* one sample was too small to analyze and the remaining samples were combined into three groups: two (A and B) of three samples each and one (C) unique sample. Subsequent analysis indicated that the three groups differed mainly in the relative proportions of three major metabolites.

Table 1 lists the quantities of each metabolite obtained (Figure 3) from the different samples of *Siphonodictyon coralliphagum*. The collection location and the status of the

Table 1. Isolated yields (% dry weight) of metabolites (Figure 3) from *Siphonodictyon coralliphagum* samples

Sponge form	Collection location	Coral status	Compound 1	2	3	4	5	6	7
unknown	Belize	Live	0.12	0.90					
typica A	Belize	Live		0.11			0.03	0.12	
B		Live		0.21			0.03	0.04	
C		Live		0.20			0.04	0.09	
tubulosa	Belize	Dead						0.09	
typica	Bahamas	Live	3.41				0.25	0.51	
tubulosa	Bahamas	Dead					0.05	0.07	
unknown	Ponape	Live			0.21	0.51	0.10		0.08
unknown	Kwajalein	Dead			0.23		0.05		

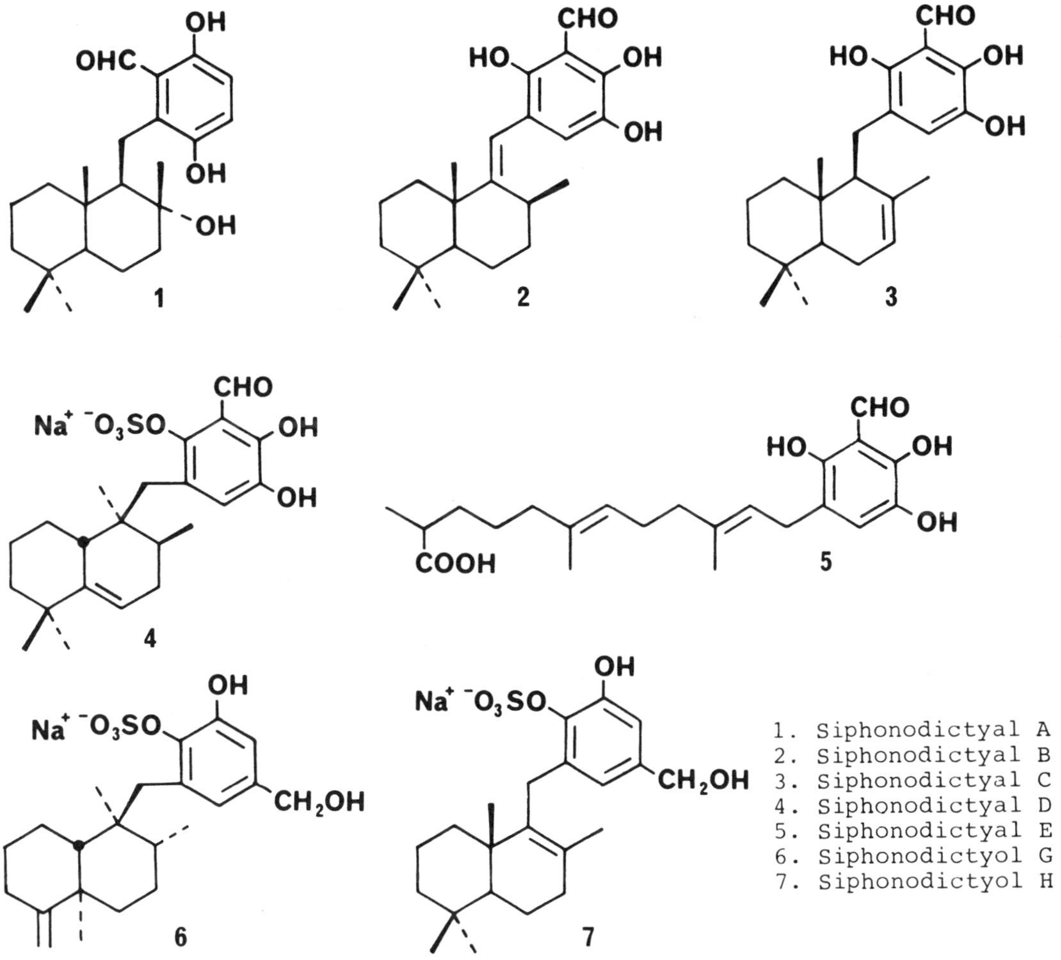

1. Siphonodictyal A
2. Siphonodictyal B
3. Siphonodictyal C
4. Siphonodictyal D
5. Siphonodictyal E
6. Siphonodictyol G
7. Siphonodictyol H

Figure 3. Structures of *Siphonodictyon coralliphagum* metabolites listed in Table 1.

coral substrate (living or dead) are recorded together with the yield of a metabolite expressed as % dry weight of sponge tissue. The structures of the metabolites were elucidated using physicochemical methods, except for that of siphonodictyal D *(4)*, which was determined by X-ray analysis (J. Clardy, pers. comm.). Details of the structural chemistry have been presented elsewhere (Sullivan et al., 1986). The metabolites isolated from *S. coralliphagum* samples comprised between 0.09% and 4.17% dry weight. The higher concentrations of metabolites were always associated with collections that comprised only the oscular chimneys of sponges from living coral heads. Smaller quantities of metabolites are associated with sponge samples living in dead coral. These differences may be due to the oscular chimneys containing less particulate inorganic material or they may represent a true concentration of the metabolites. Siphonodictyal E *(5)* and siphonodictyol G *(6)* were found in both forms of *S. coralliphagum* but siphonodictyal B *(2)* was found only in forma *typica* that grows in living corals. Similarly, siphonodictyal C *(3)* and siphonodictyal E *(5)* were found in both samples of *S. coralliphagum* from the Pacific but siphonodictyal D *(4)* and siphonodictyol H *(7)* were found only in the sample from living corals.

The significance of these results will depend on the toxicity of the metabolites. Leith Webb, James Cook University, has assayed siphonodictyal B *(2)*, siphonodictyal C *(3)*, and siphonodictyol G *(6)* against the coral *Acropora formosa* using the method previously employed for siphonodictidine (Sullivan et al., 1983). She reported that "at high doses, siphonodicytal B *(2)* and siphonodictyol G *(6)* were toxic. At moderate doses, only siphonodictyal B was toxic. Siphonodictyal C *(3)* was not toxic although it gave a response (in a respirometer) that was significantly different from controls. *A. formosa* recovered from the acute shock of siphonodictyal C." The assays suggest that siphonodictyal B is the most toxic of the metabolites, as predicted from the distribution data, but we need more quantitative toxicity data before we can be certain. Rützler (1971) suggested that the "dead zone" around the base of the oscular chimneys was maintained by a flow of mucus down the outside of the oscular chimney and over the surrounding coral tissue. We have shown that the mucus contains the secondary metabolites described above and may therefore serve as a carrier to hold toxic secondary metabolites in contact with the coral tissues.

The burrowing mechanism of *S. coralliphagum* is not completely understood (for a review see Pomponi, 1980). It is generally accepted that the majority of the limestone substrate is removed in the form of particles but a small quantity of calcium carbonate must be dissolved. The dissolution of calcium carbonate into seawater that is saturated or supersaturated with calcium is thermodynamically unfavorable and must involve the expenditure of energy. The siphonodictyals could theoretically serve to

transport calcium ions against the thermodynamic gradient using the hydroquinone-quinone oxidation potential as the energy source. The 2-hydroxy benzaldehyde (salicaldehyde) or 2-hydroxy sulfate groups can chelate calcium ions that would be released by oxidation of the hydroquinone to a quinone (Figure 4). Since we were unable to duplicate the hydroquinone-quinone redox system, we decided to determine the ability of the siphonodictyals to remove calcium ions from an aqueous solution into an organic phase, *n*-butanol. These experiments are by no means quantitative. Equilibration of an aqueous calcium chloride solution with water causes an increase in the concentration of calcium ion in the aqueous phase due to the fact that some water dissolves in the n-butanol. To emphasize the nonquantitative nature of the experiments we have simply reported the readings obtained from the atomic absorption spectrometer, rather than attempting to convert these readings into concentrations (Table 2, Figure 5). These experiments clearly demonstrate that *n*-butanol solutions of siphonodictyals B *(2)*, C *(3)*, D *(4)* and siphonodictyol G *(6)* all removed a significant amount of calcium ions from the aqueous solution. These data support the hypothesis that the secondary metabolites are involved in the transport of calcium ions from the interior of the sponge into seawater. We cannot, however, provide any evidence for or against the involvement of secondary

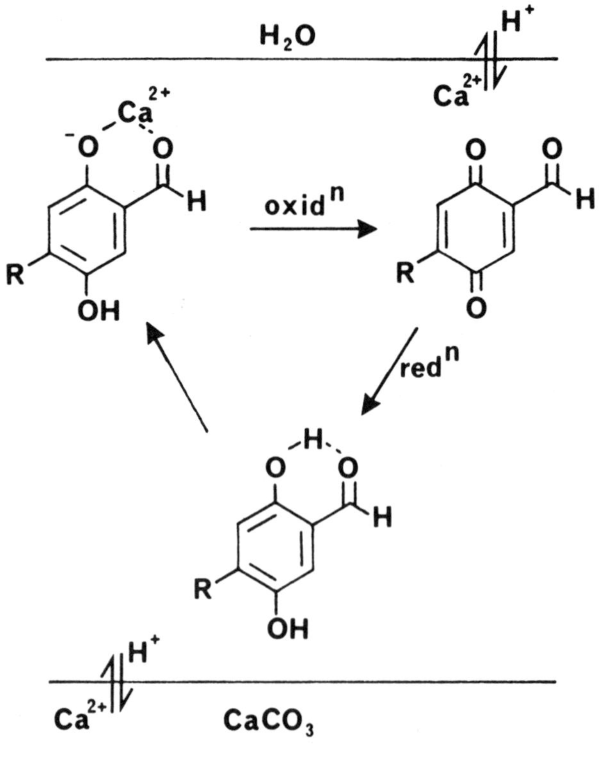

Figure 4. A proposed mechanism for the transport of calcium ions by 2,5-dihydroxybenzaldehydes.

Table 2. Results of calcium chelation experiments

Parameters	Compound				
	2	3	4	6	isozonarol[a]
Concentration of n–BuOH solution (mM)	4.8	5.0	4.6	3.6	11.3
Original Ca^{2+} concentration (ppm)	2	5	5	5	5
Atomic absorption values for Ca^{2+} in aqueous solution:					
(1) after extraction with n–butanol alone	0.103	0.51	0.51	0.51	0.52
(2) after extraction with n–butanol solution of compound	0.027	0.33	0.33	0.36	0.45
Ca^{2+} removal (%)	74	35	35	29	14

[a]Isozonarol is a phenol that can form a calcium salt but lacks the ability to form a chelate

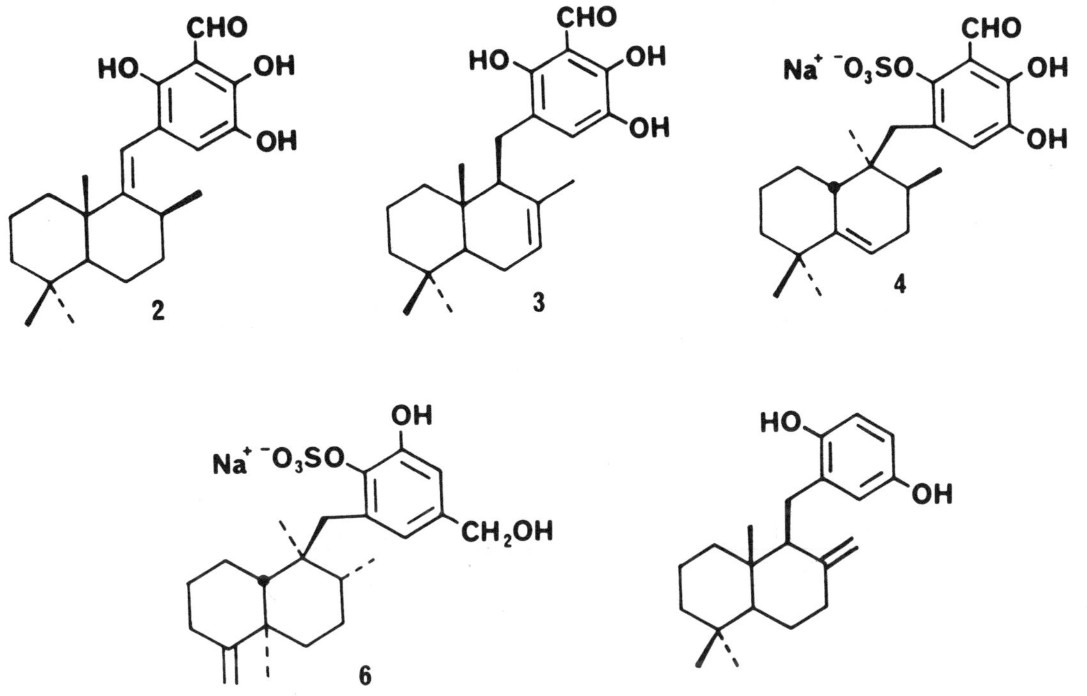

Figure 5. Structure of *Siphonodictyon coralliphagum* metabolites and of isozonarol (control) used in calcium chelation experiments, Table 2.

metabolites in the actual process of dissolution of the calcium carbonate substrate.

Conclusions

This study provides interesting data regarding chemotaxonomy in sponges. The secondary metabolites *1–7* of *Siphonodictyon coralliphagum* are all of similar chemical structure but they clearly differ from siphonodictidine, isolated from an undescribed species of *Siphonodictyon* from Palau. The two *Siphonodictyon* species appear to have evolved different secondary metabolites to perform the essential task of killing coral polyps that threaten to overgrow the exhalant oscular tube of the sponge.

Acknowledgments

This research was supported by the California Sea Grant College Program (NA85AA-D-S9140, R/mp 30) and the National Institutes of Health (AI 11969). This is contribution number 211, Caribbean Coral Reef Ecosystems (CCRE) Program, National Museum of Natural History, Smithsonian Institution, partly supported by the EXXON Corporation.

Literature Cited

Bakus, G. J., and G. Green. 1974. Toxicity in Sponges and Holothurians: A Geographical Pattern. *Science,* 185:951–953.

Bergquist, P. R. 1978. *Sponges.* Berkeley, California: University of California Press. 268 pp.

Bergquist, P. R., and J. J. Bedford. 1978. The Incidence of Anti-bacterial Activity in Marine Demospongiae; Systematic and Geographic Considerations. *Marine Biology,* 46:215–221.

Faulkner, D. J. 1984. Marine Natural Products: Metabolites of Marine Invertebrates. *Natural Product Reports,* 1:551–598.

Faulkner, D. J., and M. T. Ghiselin. 1983. Chemical Defense and Evolutionary Ecology of Dorid Nudibranchs and Some Other Opisthobranch Gastropods. *Marine Ecology Progress Series,* 13:295–301.

Jackson, J. B. C., and L. Buss 1975. Allelopathy and Spatial Competition among Coral Reef Invertebrates. *Proceedings, National Academy of Sciences, U.S.A.,* 72:5160–5163.

Pomponi, S. A. 1980. Cytological Mechanisms of Calcium Carbonate Excavation by Boring Sponges. *International Review of Cytology,* 65:301–319.

Rützler, K. 1971. Bredin-Archbold-Smithsonian Biological Survey of Dominica: Burrowing Sponges, *Siphonodictyon* Berg-quist, from the Caribbean. *Smithsonian Contributions to Zoology,* 77:1–37.

Sullivan, B., P. Djura, D. E. McIntyre, and D. J. Faulkner. 1981. Antimicrobial Constituents of the Sponge *Siphonodictyon coralliphagum. Tetrahedron Letters,* 37:979–982.

Sullivan, B. W., D. J. Faulkner, G. K. Matsumoto, H. Cun-heng, and J. Clardy. 1986. Metabolites of the Burrowing Sponge *Siphonodictyon coralliphagum. Journal of Organic Chemistry,* 51:4568–4573.

Sullivan, B., D. J. Faulkner, and L. Webb. 1983. Siphonodictidine, a Metabolite of the Burrowing Sponge *Siphonodictyon* sp. that Inhibits Coral Growth. *Science,* 221:1175–1176.

Thompson, J. E. 1985. Exudation of Biologically-Active Metabolites in the Sponge *Aplysina fistularis.* I. Biological evidence. *Marine Biology,* 88:23–26.

Thompson, J. E., R. P. Walker, and D. J. Faulkner. 1985. Screening and Bioassays for Biologically-Active Substances from Forty Marine Sponge Species from San Diego, California, USA. *Marine Biology,* 88:11–21.

Walker, R. P., J. E. Thompson, and D. J. Faulkner 1985. Exudation of Biologically-Active Metabolites in the Sponge *Aplysina fistularis.* II. Chemical Evidence. *Marine Biology,* 88:27–32.

MISHELLE P. LAWSON
Department of Zoology
University of Auckland
Private Bag, Auckland, New Zealand
and
Department of Chemistry
Stanford University
Stanford, California 94305

PATRICIA R. BERGQUIST*
Department of Zoology
University of Auckland
Private Bag, Auckland, New Zealand

R. CONRAD CAMBIE
Department of Chemistry
University of Auckland
Private Bag, Auckland, New Zealand

Long Chain Fatty Acids in Sponge Membranes

Abstract

More than half of the fatty acids in a membrane isolate prepared from *Halichondria moorei* (Demospongiae: Ceractinomorpha) were found to be long chain fatty acids (LCFAs) of C_{24}–C_{30} chain lengths. Moreover, most of the LCFAs in the sponge were derived from phospholipids, which themselves are principal components of cell membranes. Therefore, we conclude that in *Halichondria moorei* LCFAs are components of cell membranes.

The cellular origin of long chain fatty acids (LCFAs, C_{24}–C_{30} chain lengths) almost uniformly found in sponge lipid isolates is unknown. The location is of interest because sponges usually contain high proportions of such acids (Bergmann and Swift, 1951; Litchfield et al., 1976; Litchfield and Morales, 1976; Morales and Litchfield, 1976; Dembitskii, 1981a,b; Bergquist et al., 1984; Lawson et al., 1984), but they are seldom found in other organisms (Lehninger, 1975). In sponges, LCFAs can constitute up to 85% of the total fatty acids between C_{14} and C_{30} chain lengths (Bergquist et al., 1984).

Chemical information has pointed to a possible subcellular localization of LCFAs in sponges. The finding that LCFAs are derived from phospholipids (Jefferts et al., 1974; Litchfield and Morales, 1976; Dembitskii et al., 1977; Dembitskii and Neblitsyn, 1980) led Litchfield and Morales (1976) to propose that the LCFAs they had isolated from *Microciona prolifera* were present in the cell membranes because phospholipids themselves are integral components of cellular membranes. However, the idea that sponge cell membranes accommodate such long chain acids was challenged by Lethias and coworkers (Lethias et al., 1983), who argued that (1) the lipid analyses conducted by Litchfield and Morales were based on extracts from unfractionated sponge tissue, the implication being that LCFAs could arise from a nonmembranous source, and (2) in their own studies on seven

---

*Author to whom correspondence should be addressed.

sponge species, Lethias et al. (1983) found no unusual features in the membrane ultrastructure.

To date no attempts have been made to isolate particular fractions of sponge tissue in order to examine the distribution of LCFAs. Therefore, the aim of the present study was to undertake this work and to determine the fatty acid composition of sponge membranes.

Materials and Methods

Halichondria moorei (Class Demospongiae, Subclass Ceractinomorpha) was collected from the low intertidal region at Westmere Reef, Auckland, New Zealand (August 1983). The procedures for subcellular fractionation, thin-section electron microscopy, and fatty acid analysis are given in Lawson et al. (1986). Phospholipids, glycolipids, and neutral lipids were isolated from lipid fractionated and unfractionated tissue by silicic acid column chromatography, as described in Walkup et al. (1981), and the fatty acids of these lipid classes were analyzed following the procedures in the same work.

Results

A microsomal membrane isolate was obtained by centrifugation at 100,000 × g. The composition of this isolate is shown in Figure 1. More than 78% of the microsomal vesicles are free from contaminants. Contamination by ribosomal material (Figure 1b, large arrows) is unavoidable when these procedures are used and is common to microsomal isolates. Ribosomes are nonlipidic, however, and do not affect the fatty acid analyses. Extra-membranous lipid that could influence the results is not a

significant contaminant of the microsomal fraction. Small amounts of such lipid, which is usually highly osmiophilic and stains darkly with osmium, was observed in some sponge cell organelles in unfractionated tissue.

The degree of contamination of sponge membrane microsomes by endosymbiont membranes could not be quantified. Very few bacteria were evident in electron micrographs of unfractionated *Halichondria*. What was considered to be bacteria was observed together with mitochondria and sponge phagosomes and lysosomes in the second pellet, which had been centrifuged at 15,000 × g. Moreover, it is not usual for bacteria and blue-green algae to produce LCFAs (Lehninger, 1975). Sponge cell nuclei were eliminated in the initial centrifugation at 600 × g. Therefore, the microsomal preparation was thought to consist largely of sponge plasma membrane and endoplasmic reticulum.

Fatty acids of C_{14} to C_{28} chain lengths were found in *Halichondria moorei*. Dominate among these were the C_{26} fatty acids, which accounted for more than 40% of the fatty acid total in both the unfractionated tissue and the membrane isolate. The proportions of LCFAs in the membrane isolate (51.1%) were similar to the proportions from unfractionated tissue (51.7%). Of the three lipid classes we examined, the phospholipids contained the highest relative proportions of LCFAs ($\approx$ 74% of the total acids). The sterol ester and glycolipid fractions also contained significant proportions of these acids.

Discussion and Conclusions

This study, which is the first to apply conventional techniques of subcellular fractionation by differential centrifu-

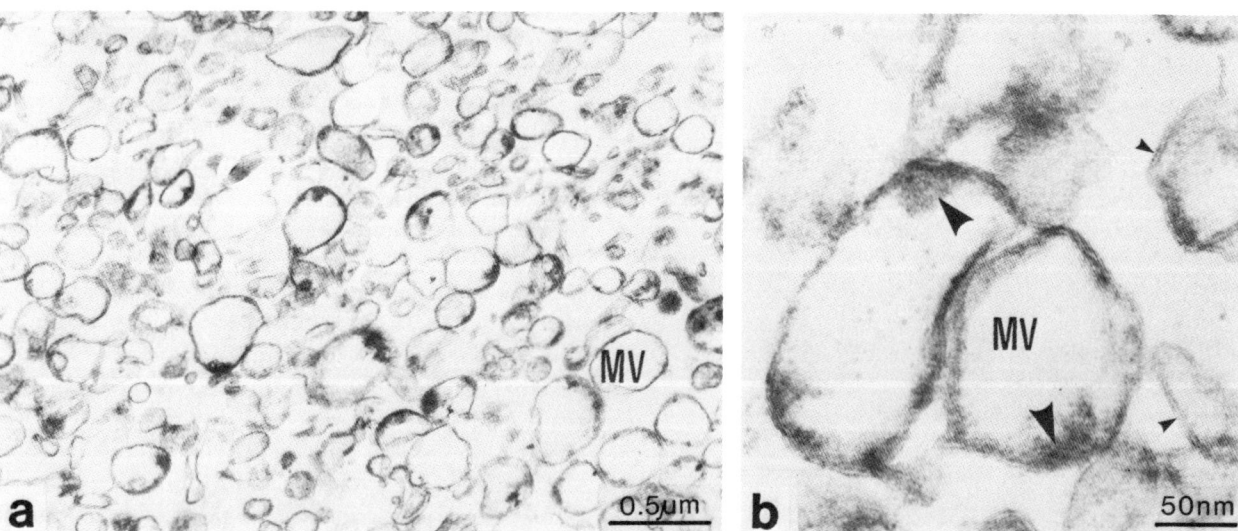

Figure 1. Microsomal membrane preparation from *Halichondria moorei*, stained with uranyl acetate and lead citrate: *a*, magnification, 28,050×; *b*, magnification, 214,500×; large arrows indicate ribosomal material within microsomal vesicles *(MV)*, small arrows point to trilaminate membrane structures.

gation to sponge tissue, has demonstrated that LCFAs are components of cell membranes in *Halichondria moorei* LCFAs constituted approximately half of the total fatty acids in the membranes examined, and most of these acids were derived from phospholipids. These results confirm the hypothesis of Litchfield and Morales (1976) that LCFAs are sponge membrane lipids and do not support the opposing view held by Lethias et al. (1983). Furthermore, present findings indicate that sponge membranes have in some measure a unique chemistry, which differs from the membranes of other organisms that are characterized by the presence of fatty acids with chain lengths up to, and including C_{22} (Lehninger, 1975).

The presence of high proportions of LCFAs in sponge membranes must affect membrane structure and possibly function. In addition to LCFAs, many sponges contain unconventional sterols that show unusual patterns of unsaturation and alkylation in the side chain. Other sponges, notably those belonging to the order Dictyoceratida, contain very low levels of sterol, but high proportions of terpenoids (e.g., pentacyclic sesquiterpenes) that mimic sterol structure and amphipathic properties (Minale, 1976; Bergquist and Wells, 1983). It has been proposed that terpenoids of this type could act as sterol substitutes in the membranes (Bergquist, 1979; Lawson et al., 1986).

Lethias et al. (1983) argued that sponge membranes could not contain LCFAs because they exhibit typical ultrastructure. If one considers chain lengths alone, LCFAs with up to 30 carbon units could produce a membrane bilayer 36% wider than that typical of other organisms. Whether this is the case has yet to be determined empirically, but there are no indications that it is so. However, increased width may not necessarily accompany LCFA occurrence if the acids are bent. Neither the occurrence of LCFAs nor of unconventional sterols in sponge membranes physically precludes them from exhibiting typical ultrastructure and typical responses to fractionation procedures during membrane preparation. The cell membranes of *Halichondria moorei* exhibit classical trilaminar structure (Figure 1*b*, small arrows).

Results similar to those reported for *Halichondria moorei* have subsequently been obtained for other demosponges, *Pseudaxinyssa* sp. and *Reniera* sp. (Lawson et al., 1988b). Approximately half of the acids in the unfractionated tissue of *Reniera* sp. were LCFAs, but almost 90% in the membrane isolate were LCFAs. *Pseudaxinyssa* had a 10% LCFA content in unfractionated tissue, in comparison with approximately 60% in the membrane isolate. Unconventional sterols also occur in *Pseudaxinyssa* (Bergquist et al., 1980) and have been localized to the membrane fraction (Lawson et al., 1988a).

This study should be extended to provide quantitative information on the lipid composition of the cell membranes of additional sponge species. At present the pro-

cedure for the preparation of sponge membrane isolates is being refined with a view to minimizing contamination by endosymbionts and purifying certain subcellular membranes in order to determine individual lipid contents. A major problem exists in identifying membranes of sponge organelles. Electron microscopy alone is inadequate because there is no morphological distinction between many subcellular membranes, for example plasma membrane and smooth endoplasmic reticulum. The ideal method for characterization is enzyme assay; however, membrane-associated enzymes have not yet been characterized for sponges. What remains to be established at present is whether all types of sponge membranes contain LCFAs and whether LCFA is localized at additional sites in sponge tissue.

Acknowledgments

We are grateful to John White for performing the electron microscopy work.

Literature Cited

Bergmann, W., and A. N. Swift. 1951. Contributions to the Study of Marine Products. XXX. Component Acids of Lipids of Sponges. *Journal of Organic Chemistry,* 16:1206–1221.

Bergquist, P. R. 1979. Sponge Chemistry—A Review. Pages 383–392 in *Biologie des Spongiaires,* edited by C. Lévi and N. Boury-Esnault. Colloques Internationaux, 291. Paris: Centre National de la Recherche Scientifique.

Bergquist, P. R., W. Hofheinz, and G. Oesterhelt. 1980. Sterol Composition and Classification of the Demospongiae. *Biochemical Systematics and Ecology,* 8:423–435.

Bergquist, P. R., M. P. Lawson, A. Lavis, and R. C. Cambie. 1984. Fatty Acid Composition and the Classification of the Porifera. *Biochemical Systematics and Ecology,* 12:63–84.

Bergquist, P. R., and R. J. Wells. 1983. Chemotaxonomy of the Porifera: The Development and Current Status of the Field. Pages 1–50 in *Marine Natural Products, Chemical and Biological Perspectives 5,* edited by P. Scheuer. New York: Academic Press.

Dembitskii, B. M. 1981a. Fatty Acid Composition of Fresh-water Sponges of the Class Demospongiae. I. Genus *Lubomirskia. Khimia Prirodnych Soedinenij,* 4:511–513. (Chemistry of Natural Processes, in Russian.)

———. 1981b. Fatty Acid Composition of Fresh-water Sponges of the Class Demospongiae. II. Genera *Swartschewskia* and *Baikalospongia. Khimia Prirodnych Soedinenij,* 4:513–515. (Chemistry of Natural Processes, in Russian.)

Dembitskii, B. M., and B. D. Neblitsyn. 1980. Lipids of Marine Origin II. Comparative Analysis of Phospholipid and Fatty Acid Composition of Marine Sponges from the Japan Sea. *Bioorganitcheskaja Khimia,* 6:1542–1548. (Bioorganic Chemistry, in Russian.)

Dembitskii, B. M., V. I. Svetasher, and V. E. Vaskorsky. 1977. Lipids of Marine Origin I. Unusual Lipid from *Halichondria panicea* Sponge. *Bioorganitcheskaja Khimia,* 3:930–933. (Bioorganic Chemistry, in Russian.)

Jefferts, E., R. W. Morales, and C. Litchfield 1974. Occurrence of *Cis*-5, *Cis*-9-hexacosadienoic and *Cis*-5, *Cis*-9, *Cis*-19-hexacosatrienoic Acids in the Marine Sponge *Microciona prolifera*. *Lipids*, 9:244–247.

Lawson, M. P., P. R. Bergquist, and R. C. Cambie. 1984. Fatty Acid Composition and the Classification of the Porifera. Part 2. *Biochemical Systematics and Ecology*, 12:375–393.

————. 1986. The Cellular Localization of Long Chain Fatty Acids in Sponges. *Tissue and Cell*, 18:19–26.

Lawson, M. P., I. L. Stoilev, J. E. Thomson, and C. Djerassi. 1988a. Cell Membrane Localization of Sterols with Conventional and Unusual Side Chains in Two Demosponges. *Lipids*, 23:750–754.

Lawson, M. P., J. E. Thomson, and C. Djerassi. 1988b. Cell Membrane Localization of Long Chain C_{24}–C_{30} Fatty Acids in Two Marine Demosponges. *Lipids*, 23:741–749.

Lehninger, A. L. 1975. *Biochemistry*. Second Edition. New York: Worth Publishers. 1104 pp.

Lethias, C., R. Garrone, and M. Mazzorana. 1983. Fine Structure of Sponge Cell Membranes: Comparative Study with Freeze-Fracture and Conventional Thin Section Methods. *Tissue and Cell*, 15:523–535.

Litchfield, C., A. J. Greenberg, G. Noto, and R. W. Morales. 1976. Unusually High Levels of C_{24}–C_{30} Fatty Acids in Sponges of the Class Demospongiae. *Lipids*, 11:567–570.

Litchfield, C., and R. W. Morales. 1976. Are Demospongiae Membranes Unique Among Living Organisms? Pages 183–200 in *Aspects of Sponge Biology*, edited by F. W. Harrison and R. K. Cowden. New York: Academic Press.

Minale, L. 1976. Natural Product Chemistry of the Marine Sponges. *Pure and Applied Chemistry*, 48:7–23.

Morales, R. W., and C. Litchfield. 1976. Unusual C_{24}, C_{25}, C_{26} and C_{27} Polyunsaturated Fatty Acids of the Marine Sponge *Microciona prolifera*. *Biochimica et Biophysica Acta*, 431:206–216.

Walkup, R. D., G. C. Jamieson, M. R. Ratcliff, and C. Djerassi. 1981. Phospholipid Studies of Marine Organisms, 2: Phospholipids, Phospholipid-Bound Fatty Acids and Free Sterols of the Sponge *Aplysina fistularis* (Pallas) forma *fulva* (Pallas) (= *Verongia thiona*). Isolation and Structure Elucidation of Unprecedented Branched Fatty Acids. *Lipids*, 16:631–646.

PETER KARUSO
MARK R. HAGADONE
PAUL J. SCHEUER*
Department of Chemistry
University of Hawaii
Honolulu, Hawaii 96822

PATRICIA R. BERGQUIST
Department of Zoology
University of Auckland
Private Bag, Auckland, New Zealand

Chemotaxonomy of the Porifera by Infrared Spectroscopy

Abstract

The supposition that crude organic extracts of small (10 g) samples of sponges may yield infrared (IR) spectra of chemotaxonomic value was tested with 85 identified sponges. When a spectrum of ubiquitous sponge metabolites (fatty acids, sterols) is subtracted from the spectrum of the crude extract, the resulting difference spectrum is then analyzed by computer for prominent absorption peaks and possible functional groups. The computer library is then searched for the 25 closest matches to the unknown spectrum.

Contrary to our original expectations, sponges that contain compounds with distinctive functional groups (e.g., isocyano, isothiocyano) do not give especially good correlations by this method. However, the method appears to have merit for a rapid evaluation of the chemical nature of the metabolites of an unknown sponge and of its closest chemical relatives in the existing library. As with all chemotaxonomic studies, a larger spectral library will greatly enhance the usefulness of this technique.

Infrared radiation occupies the part of the electromagnetic spectrum between the visible range at the high energy end and the microwave range at the low end. Its energy range, between 5×10^{-1} and 4×10^{-3} eV, corresponds to the stretching and bending vibrations of chemical bonds. Physical chemists used the technique for some fifty years after the turn of the century to study fundamental properties of small molecules. One of the major organic problems of World War II was to elucidate the structure of penicillin. It was this research that hinted at the usefulness of infrared spectroscopy (IR) as a structural tool for organic molecules and that resulted in the production of the first commercial spectrometers.

Although molecular vibrations are described by a fundamental relationship between bond strength and vibration frequency (Hooke's law), it was impossible to apply this knowledge to even modest-sized (say, 100 atoms) organic molecules in the 1940s because of the lack of com-

*Author to whom correspondence should be addressed.

puters. As a result, organic chemists developed a valuable empirical method, which was diagnostic for specific functional groups (such as hydroxyl and carbonyl) in the high-energy region (4000–1500 cm^{-1})[1] of the spectrum and which provided a unique fingerprint of virtually all organic molecules in the low-energy portion of the spectrum (1500–600 cm^{-1}), where weak vibrations and rotations of large moieties or of entire molecules are recorded (Table 1). Current computer technology has added a rapid search and comparison capability to IR spectroscopy as a structural tool. One would not expect a method that recognizes functional groups in organic molecules to be a suitable aid in the chemotaxonomy of Porifera. The picture that has been emerging over the past decade of research on sponge metabolites is one of great diversity of carbon architecture coupled with functional group economy (Bergquist and Wells, 1983). Furthermore, the meteoric development of nuclear magnetic resonance spectroscopy (NMR) as a structural tool has all but overshadowed the intrinsic value of IR spectroscopy.

Our involvement with *Halichondria* sp. (Burreson and Scheuer, 1974) and with *Ciocalypta* sp. (Burreson et al., 1975), which elaborate the rare isocyano (-NC) function, has made us sensitive to the diagnostic value of IR spectroscopy. In fact, it was our recognition of the coexistence of isocyano, isothiocyano (-NCS), and isocyanato (-NCO) functions in a crude extract of *Ciocalypta* sp. (Hagadone et al., unpublished) by IR spectroscopy that prompted us to investigate whether infrared spectra of crude sponge extracts might assist in preliminary recognition of at least some sponge taxa.

Materials and Methods

The experimental method is simple—it consists of steeping approximately 10 g of fresh or frozen sponge in ethanol for at least 48 h at room temperature. To the decanted, filtered liquid is added an equal volume of water and of chloroform. The heterogeneous mixture is well shaken in a separatory funnel and allowed to settle. The lower chloroform layer is separated, evaporated, and dried in vacuo. The residue is dissolved in a small volume of chloroform and spread onto a silver chloride plate. Chloroform is evaporated with gentle heating and the IR spectrum is recorded from the resulting film on a Perkin-Elmer 1430 Series spectrometer with 3600 data station.

Analysis of the crude sponge extracts by IR spectrometry presents a number of problems. First, crude ethanolic extracts of sponges contain both organic and inorganic compounds, which give broad and featureless IR spectra (Figure 1a). A partition with chloroform successfully removes the interfering inorganic and highly polar compo-

Table 1. Diagnostic infrared frequencies

Functional group	Wavenumber (cm^{-1})
NC	2175–2130
NCO	2275
NCS	2220–2040
NCCl$_2$	1660–1645
NCHO	1680
CHO	2740, 1725, 1050–1150
CO$_2$H	1725–1695
CONHR	1695–1615
COR	1710
CO$_2$Me	1740, 1015
CO$_2$Et	1740, 1030, 855
AcOR	1740, 1250
γ–lactone	1800–1820
β–lactone	1885–1820
(ring structure)	1780
R–≡–H	3310, 2130, 665–625
R–≡–R	2220

nents and makes sample preparation easier (Figure 1b). Second, all sponges contain large quantities of fatty acids, phospholipids, and varying amounts of sterols, which generally obscure the absorptions of the less abundant secondary metabolites. This problem was overcome by computer subtraction of a reference extract, which was chosen because of its featureless IR and NMR spectra. The resulting spectrum is often of low intensity and must be normalized, and the baseline corrected and smoothed by computer manipulations to obtain an acceptable spectrum (Figure 1c). The region above 2500 cm^{-1} was generally of little value due to the exceedingly variable -OH and -CH absorptions; this region was therefore neglected. Each difference spectrum was reduced to a table of peaks consisting of 10–30 absorption values and corresponding intensities (Table 2). In this study, data for a total of 85

Table 2. Peak table values (in cm^{-1}) computer generated from Figure 1c and possible functional groups.

ν_{max}	Transmission (%)	ν_{max}	Transmission (%)
2179	50.5	1149	30.4
2129	33.4	1123	31.4
1742	67.4	1100	34.0
1670	8.9	1078	36.1
1565	36.8	1031	38.1
1535	28.7	999	46.8
1472	47.5	947	66.0
1449	33.2	895	64.7
1384	18.5	834	70.3
1323	46.9	696	86.4
1273	46.0	654	85.0
1226	45.5		

Possible structural units are aliphatic tertiary amide; multiple bonded nitrogen compound, acetylene or silane; carbonyl compound, possibly ketone, amide or aromatic aldehyde. (These functional group assignments are not always reliable and should be treated with caution)

---

[1] Wavenumber (ν) measured in cm^{-1} is directly proportional to the energy of vibration and inversely proportional to wavelength (λ)

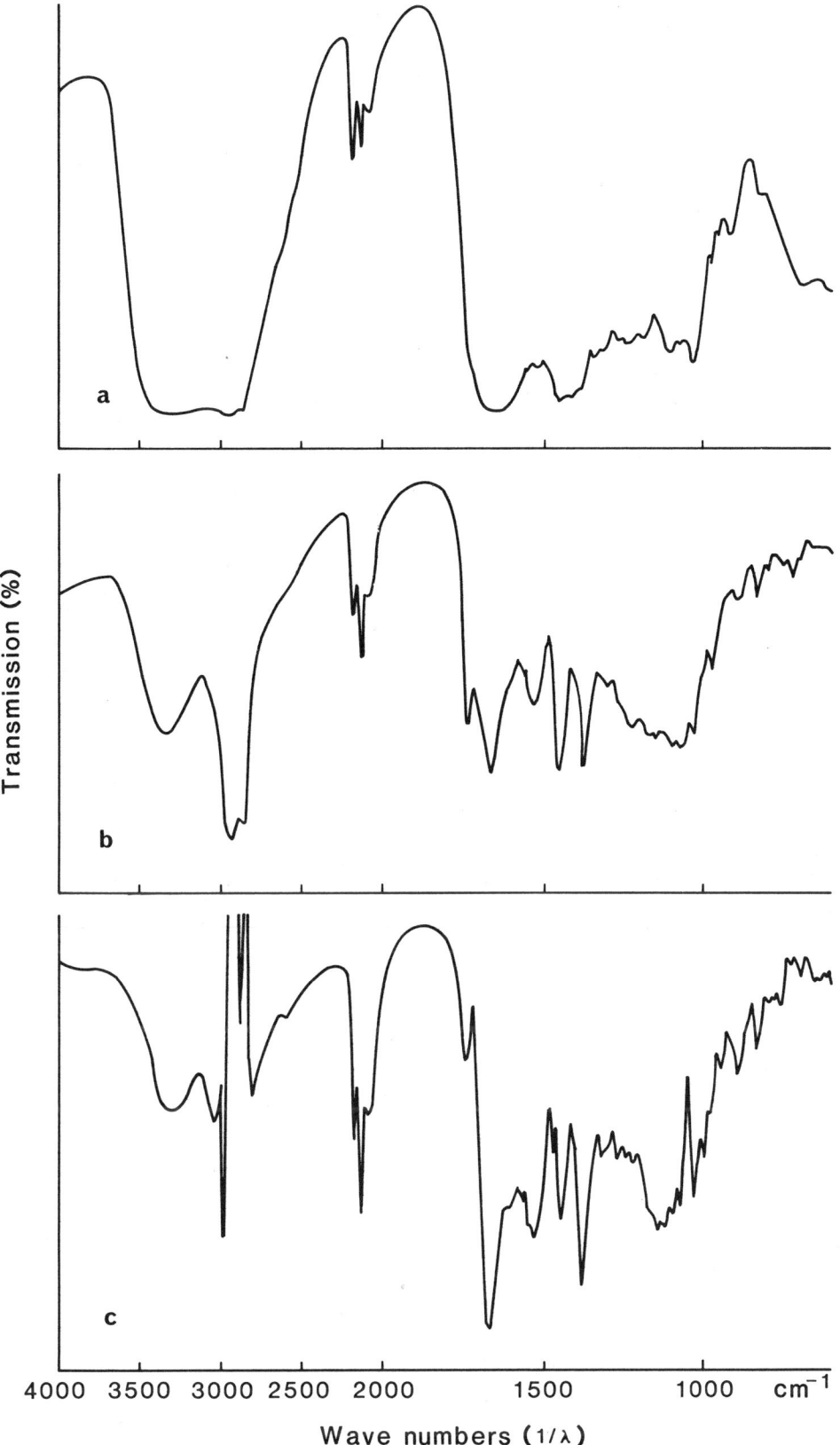

Figure 1. Infrared spectra of a crude extract of *Acanthella* sp. from Guam: *a*, crude ethanolic extract; *b*, chloroform phase of ethanolic extract; *c*, spectrum b minus reference spectrum; note increased intensity of isocyanide (≈ 2100 cm^{-1}) and fingerprint regions (1500–600 cm^{-1}).

sponge samples were used to produce a basic library for computer analysis.

After the peak tables were generated each spectrum was analyzed by computer for possible structural units. The program searches each peak table for characteristic high energy absorptions (Table 1) and stores possible functional group identities with the peak tables (Table 2). This information is invaluable, especially to the nonchemist who desires to evaluate the chemotaxonomic significance of crude IR spectra. For instance, γ-lactones and acetate carbonyl absorptions are characteristlc of many dendroceratid and dictyoceratid sponges, whereas amides are always observed in verongid sponges. The isocyano/isothiocyano/isocyanate absorptions are characteristic of a group of axinellidlike sponges.

Extracts from unknown sponges can now be compared with those in the spectra library. This can be done by two methods, which differ only in the relative weighting given to peak intensities versus position. The method that is least sensitive to peak intensities was found to give the best correlations.

Results and Discussion

Distinctive functional groups, such as isocyanides do not appear to provide particularly good correlation, since the computer interprets the spectrum as a whole and does not afford greater significance to any specific peak. Table 3 shows the computer correlation between the Hawaiian isocyano-containing sponge *Ciocalypta* sp. and the library spectra. Those sponges marked with an asterisk are known to contain isocyanoterpenes. Note that the iso-

Table 3. Closest 20 computer library matches for *Ciocalypta* sp. from Hawaii (sponges marked with asterisk are known to contain isocyanoterpenes)

Correlation[a]	Species	Origin
9–9	*Ciocalypta* sp.	Hawaii
6–3	*Pseudaxinyssa* sp.	Fiji
3–1	*Acanthella* sp. 1	Guam
0–2	Cribrochalina sp. 1	Papua New Guinea
0–1	*Rhaphoxya* sp.	Ponape
0–1	Ircinia sp. 2	Guam
0–1	Dysidea sp. 4	Guam
0–1	Dysidea sp. 4	Guam
0–1	*Densa* sp.	Hawaii
0–1	Phakellia sp.	Papua New Guinea
0–1	Haliclona sp. 1	Papua New Guinea
0–0	Haliclona sp. 4	Hawaii
0–0	Xestospongia sp. 1	Papua New Guinea
0–0	Neofolitispa deanchora	Papua New Guinea
0–0	Cribrochalina sp. 2	Papua New Guinea
0–0	*Phycopsis* sp.	Guam
0–0	Pandaros acanthifolium	Curaçao
0–0	*Acanthella* sp. 2	Guam
0–0	Cinachyra sp.	Papua New Guinea
0–0	Spirastrella sp. 1	Papua New Guinea

[a]The first number denotes peak position correlation and the second number denotes intensity correlation.

Table 4. Closest 15 computer library matches for an unknown *Dysidea* sp.

Correlation[a]	Species	Origin
9–9	*Dysidea* sp. 1	Papua New Guinea
0–6	*Dysidea* sp. 4	Guam
0–6	*Dysidea* sp. 4	Guam
6–4	*Dysidea arenaria*	Hawaii
2–1	*Hippospongia* sp.	Papua New Guinea
0–3	*Dactylospongia* sp.	Hawaii
2–0	*Xestospongia exigua*	Guam
0–2	*Haliclona* sp. 1	Guam
1–1	*Haliclona* sp. 2	Papua New Guinea
2–0	*Pericharax heterorhaphis* (Calcarea)	Papua New Guinea
0–2	*Epipolasis* sp.	Hawaii
0–2	*Hyrtios* (= *Heteronema*) *erecta*	Ponape
0–1	*Clathria* sp. 1	Papua New Guinea
0–2	*Dasychalina* sp.	Papua New Guinea
0–1	*Pseudaxinyssa* sp.	Fiji

[a]The first number denotes peak position correlation and the second number denotes intensity correlation.

cyanide containing *Acanthella* (from Figure 1) is third on the list.

Members of the genus *Dysidea* are well characterized morphologically and can be identified relatively easily, but species determination is another matter. On the basis of chemistry, the genus can be divided into two groups: species containing furanosesquiterpenes and those containing sesquiterpenes linked to phenols (or quinones). These can easily be differentlated by our IR technique. Table 4 shows a correlation between the IR spectrum of an unknown *Dysidea* sp. from Papua New Guinea and the other library spectra. Good correlation between this sponge and several other species of *Dysidea* was observed. *Dysidea arenaria* and *Dysidea* sp. 4 are both known to contain the sesquiterpene phenols avarol and arenarol (Karuso, unpublished). Note that *Dactylospongia* sp. (Luibrand et al., 1979) and *Xestospongia exigua* (Roll et al., 1983) contain, though taxonomically unrelated, related aromatic compounds and hence give a false correlation. *Dysidea fragilis* which contains furanosesquiterpenes (Schulte et al., 1980) gives a good correlation with *Dysidea* sp. 3 (Karuso, unpublished) and *Dysidea herbacea* (Dunlop et al., 1982), both of which also contain furans rather than phenols (Table 5).

Haplosclerid sponges are not noted for their secondary metabolites. It was therefore of interest to see if any correlation among sponges of this order could be obtained. Table 6 shows a correlation for a *Sigmadocia* sp. from Hawaii with our library. Our library contains only three species of *Sigmadocia*, all of which appear in the correlation. It should be noted that the sponges marked with an asterisk are also haplosclerid sponges. The IR and NMR spectra of the extracts gave us no clue to any particular class of organic compound in any of these sponges. This

Table 5. Closest 15 computer library matches for *Dysidea fragilis*

Correlation[a]	Species	Origin
9–9	*Dysidea fragilis*	Hawaii
9–2	*Dysidea* sp. 3	Hawaii
9–2	*Hyattella* sp.	Papua New Guinea
9–0	*Sigmadocia* sp.	Papua New Guinea
9–0	*Sigmadocia* sp. 1	Hawaii
9–0	*Callyspongia* sp.	Hawaii
9–0	*Damiriana hawaiiana*	Hawaii
9–0	*Carteriospongia* sp.	Guam
0–2	*Xestospongia exigua*	Guam
9–0	*Dysidea herbacea*	Papua New Guinea
0–2	*Gelliodes fibulatus*	Papua New Guinea
0–2	*Dasychalina* sp.	Papua New Guinea
9–0	Family Axinellidae	not known
9–0	*Sigmadocia* sp. 2	Hawaii
9–0	*Aplysina lacunosa*	Curaçao

[a]The first number denotes peak position correlation and the second number denotes intensity correlation.

example demonstrates how the computer can "see" relationships that elude the chemist.

A number of problems are inherent in this type of analysis. A few examples will explain some of these. First, the taxonomy of the Porifera is far from well understood. Many orders are polyphyletic (Bergquist, 1978), notably the Lithistida; furthermore, sponge classification is continually under revision. The isocyano-containing terpenes have distinctive IR spectra but have so far been found in nine genera in three orders. The possession of isocyanoterpenes is undoubtedly a good taxonomic marker, but until the systematics has been revised, this very characteristic functional group will be of little chemotaxonomic significance.

Sesquiterpenes linked to phenols also give characteristic IR spectra. However, these compounds are found in taxonomically diverse sponges, such as the Haploscerida (*Xestospongia, Siphonodictyon (= Aka), Adocia*) and the Dic-

Table 6. Closest 15 computer library matches for *Sigmadocia* sp. (sponges marked with asterisk belong to the Haplosclerida)

Correlation[a]	Species	Origin
9–9	*Sigmadocia* sp. 1	Hawaii
9–3	*Sigmadocia* sp.	Papua New Guinea
9–2	*Dysidea* sp. 3	Hawaii
9–2	*Hyattella* sp.	Papua New Guinea
0–2	*Gelliodea fibulatus*	Papua New Guinea
9–0	*Haliclona* sp. 1	Guam
9–0	*Family Niphatidae	Papua New Guinea
9–0	*Carteriospongia* sp.	Guam
9–0	*Hymeniacidon hauraki*	New Zealand
9–0	*Dysidea fragilis*	Hawaii
0–2	*Dasychalina* sp.	Papua New Guinea
9–0	*Spinosella vaginalis*	Curaçao
9–0	*Sigmadocia* sp. 2	Hawaii
9–0	*Gelliodes* sp.	Papua New Guinea
9–0	*Callyspongia* sp.	Papua New Guinea

[a]The first number denotes peak position correlation and the second number denotes intensity correlation.

tyoceratida *(Sarcotragus, Dactylospongia, Fasciospongia, Smenospongia, Dysidea)* (Faulkner, 1984). To make matters worse, not all members of each genus contain these compounds.

Individuals of the same species gathered in different locations or at different times do not always display the same chemistry. This problem is a minor one; it is usually a matter of relative concentration of metabolites rather than totally different chemistry. Extract preparation and peak table reduction must also be consistent to obtain reproducible results.

Conclusions

Notwithstanding these problems, infrared spectroscopy should find application among sponge biologists who perhaps know a little about a particular sponge and want to differentiate among a number of alternatives. Comparison of an IR spectrum with those in our library gives relatively simple access to the gross chemistry of the sponge and hence provides an additional taxonomic character. Sponge chemists also could use this system. A spectrum of an unknown sponge can be compared with library spectra to obtain a list of chemically related sponges, thus giving a clue as to the type of metabolites one may expect from the unknown specimen. It is hoped that this method will eventually be used routinely, as are spicule mounts by biologists and bioassays by chemists. This would help overcome the largest problem we have encountered—an insufficient data base. To this end we invite all sponge biologists and chemists to send us crude extracts or samples of identified sponges to be added to our library.

Acknowledgments

We thank INA Laboratories of Honolulu for the use of the IR instrument and Paul Perkins of Perkin-Elmer for helpful discussions on how to adapt the infrared library software to our purposes. Financial support from the National Science Foundation and the University of Hawaii Sea Grant College program under Institutional Grant NA81AA-D–0070 from NOAA, Office of Sea Grant, U.S. Department of Commerce, is gratefully acknowledged.

Literature Cited

Bergquist, P. R. 1978. *Sponges*. London: Hutchinson. 268 pp.

Bergquist, P. R., and R. J. Wells. 1983. Chemotaxonomy of the Porifera; The Development and Current Status of the Field. Pages 1–46 in *Marine Natural Products: Chemical and Biological Perspectives*, 5, edited by Paul J. Scheuer. London: Academic Press.

Burreson, B. J., C. Christopherson, and P. J. Scheuer. 1975. Co-occurrence of Two Terpenoid Isocyanide-Formamide Pairs in a Marine Sponge. *Tetrahedron*, 31:2015–2018.

Burreson, B. J., and P. J. Scheuer. 1974. Isolation of a Diterpenoid Isonitrile from a Marine Sponge. *Journal, Chemical Society Chemical Communications*, 1035–1036.

Dunlop, R. W., R. Kazlauskas, G. March, P. T. Murphy, and R. J. Wells. 1982. New Furanosesquiterpenes from the Sponge *Dysidea herbacea. Australian Journal of Chemistry*, 35:95–103.

Faulkner, D. J. 1984. Marine Natural Products: Metabolites of Marine Invertebrates. *Journal, Chemical Society, Natural Products Reports*, 1:551–598.

Luibrand, R. T., T. R. Erdman, J. J. Vollmer, and P. J. Scheuer. 1979. Ilimaquinone, a Sesquiterpenoid Quinone from a Marine Sponge. *Tetrahedron*, 35:609–612.

Roll, D. M., P. J. Scheuer, G. K. Matsumoto, and J. Clardy. 1983. Helenaquinone, a Pentacyclic Polyketide from a Marine Sponge. *Journal, American Chemical Society*, 105:6177–6178.

Schulte, G., P. J. Scheuer, and O. J. McConnell 1980. Two Furanosesquiterpenes Marine Metabolites with Antifeedant Properties. *Helvetica Chimica Acta*, 63:2159–2167.

WELTON L. LEE[*]
SARAH W. KLONTZ[**]
Department of Invertebrate Zoology and Geology
California Academy of Sciences
San Francisco, California 94118

Seasonal and Geographic Variability in Sponge Carotenoids

Abstract

In a two-year study of three sponge species in three localities off California the total carotenoid content was found to vary with locality but showed no obvious correlation with season. The species studied were *Plocamia karykina* de Laubenfels, *Ophlitaspongia pennata* (Lambe), and *Haliclona* sp. Distinct species-specific patterns of carotenoids were always present regardless of time or locality. The number of these carotenoids was directly related to the overall synthetic ability of the species. Furthermore, patterns among species that synthesized oxidized carotenoids were made up largely of esterified carotenoids; patterns among species with low synthetic activity were not. An earlier study (Lee and Gilchrist, 1985) that verifies the stability of these patterns suggests that carotenoid esters may have some systematic importance in poecilosclerid sponges. High carotenoid variability in haplosclerid sponges may obviate the use of this biochemical character in these sponges.

Lee and Gilchrist (1985) recently demonstrated the value of using carotenoid patterns as a possible tool in the systematics of the Porifera. The methods they devised allowed them to identify patterns among individual sponge specimens of less than 1 g wet weight and to use the results to compare sponge taxa at all levels. Thus they were able to indicate which of the carotenoids in any given pattern were significant in discriminating between two taxa. The method thus proved to be highly important as a preliminary assessment of relationships between taxa and in delineating specific carotenoids that were important in separating two or more closely related groups. Discriminative analysis can be used to establish general relationships and it provides a means of distinguishing between stable, sponge metabolized carotenoids and the more variable unmodified carotenoids of dietary origin or those derived

[*]Present address: 1235 44th Ave., San Francisco, California 94122.
[**]Present address: 4612 Chestnut St., Bethesda, Maryland 20814.

from sponge symbionts. The evidence obtained to date suggests that carotenoid esters can be used to distinguish between closely related taxa.

Whether morphological or biochemical characters are used in systematics, the nature and extent of variability must be determined before a particular character can be used. Up to now, variability has received little attention in systematic studies based on biochemical characters. In studies of carotenoids, variability has occasionally been determined through the analysis of a few additional samples taken either at a different time or from a second locality.

The sponge taxa investigated by Lee and Gilchrist (1985) showed distinct patterns of carotenoids. What was not clear was whether or not those patterns were stable over time and the same at different localities. The purpose of the present investigation is to establish the nature and extent of carotenoid variability as a function of both time and locality. The study was carried out over a period of about two years at three localities off the coast of California.

Materials and Methods

SPONGES. To keep the number of samples at a manageable level, we focused on three species: *Plocamia karykina*, de Laubenfels, *Ophlitaspongia pennata* (Lambe), and *Haliclona* sp. *A* (Hartman, 1975). The first two species are of the order Poecilosclerida and were investigated in our earlier study (Lee and Gilchrist, 1985), and one of these *(O. pennata)* was the subject of an unpublished study that allowed us to specifically identify its carotenoids. *Haliclona* sp. was included because it provides an example of another order, Haplosclerida.

The localities (Table 1) were chosen for their proximity to the laboratory and because they represent a continuum from a low-energy to a high energy shoreline. The sponges were transported alive on ice to the laboratory, cleaned of all visible surface debris, blotted dry, and processed as indicated in Lee and Gilchrist (1985) or frozen at $-20°C$ until needed. Portions of each specimen were isolated, preserved in 95% ethanol, and assigned California Academy of Sciences voucher numbers (Table 1). Identifications were made by one of the authors (W. L. Lee). Samples showing mixtures of two or more species were discarded.

EXTRACTION OF CAROTENOIDS. We followed the extraction procedures outlined in Lee and Gilchrist (1985), but used only unsaponified fractions in this study as the topic of interest was the relative stability of carotenoid esters, which appeared to be significant in our earlier study. However, small portions of the extracts were saponified in order to verify the presence of carotenoid esters. Saponification was carried out as described in Lee and Gilchrist (1985).

THIN LAYER CHROMATOGRAPHY (TLC). Carotenoids were separated on Analtech silica-gel HL, preabsorbent hard-coat plates (250 μm thick). The solvent system consisted of 20% acetone in petroleum ether. Concentrated carotenoid extracts in petroleum ether were spotted at the origin. Each plate also included a standard "marker spot" consisting of authentic β-carotene. All further procedures were the same as those described by Lee and Gilchrist (1985).

RECORDING OF RESULTS AND ANALYSES. On completion of the TLC runs, photographs were taken and data recorded as described in Lee and Gilchrist (1985). Additional runs of saponified fractions were made to ascertain the position and identity of carotenoid esters. These runs were recorded in a similar manner.

Table 1. Collection sites, California sponge species and California Academy of Science (Invertebrate Zoology) voucher numbers

Locality	Species and voucher numbers	
Dillon Beach (Second Sled Rd., Marin Co.)	*Ophlitaspongia pennata*	(CASIZ 061281−061288)
	Plocamia karykina	(CASIZ 061289−061296)
	Haliclona sp.	(CASIZ 061297−061303)
Great Tide Pool (Pacific Grove, Monterey Co.)	*Ophlitaspongia pennata*	(CASIZ 061304−061311)
	Plocamia karykina	(CASIZ 061312)
	Haliclona sp.	(CASIZ 061313−061319)
Pescadero Point (17 Mile Drive, Carmel, Monterey Co.)	*Ophlitaspongia pennata*	(CASIZ 061320−061325)
	Plocamia karykina	(CASIZ 061326−061331)

Results

Ophlitaspongia pennata

Figure 1 shows the carotenoid patterns at the three study sites. Note that in the Dillon Beach samples (Figure 1*a*) carotenoid bands Y-1 through R-3 occur at the 100% level in all samples. The first of these, Y-1, is β-carotene mixed with a small amount of α-carotene. Next to it is Y-2, which represents an acetylenic pigment, most probably crocoxanthin, allobetaxanthin, or a mixture of the two, both characteristic of the Cryptophyceae (Pennington et al., 1985). The next four pigment bands are carotenoid esters. Apart from these six pigment bands, only an occasional xanthophyll appears in all samples on a single collection date.

Samples from Great Tide Pool and Pescadero Point (Figure 1*b*, *c*) show a similar pattern, except that pigment band Y-2 does not occur in these two localities.

Figure 2 shows the results from all collection sites. The patterns noted above can be seen clearly here; β-carotene and the four carotenoid esters are the only pigments consistently found in all samples. All other pigments vary from a single occurrence at a single site and time to high but variable frequency at all sites and times. There seems to be no seasonal pattern.

Plocamia karykina

Figure 3 shows the results at the two collection sites. Note that the pattern here is similar to that seen in *Ophlitaspongia pennata:* α- and β-carotene (Y-1) and the carotenoid esters (R-1, R-2, R-3) are present in all samples. Here, however, three additional pigments (Y-2, Y-3, Y-8) also occur in all samples. Otherwise the variability is the same as in *Ophlitaspongia.* Figure 4 illustrates the composite results for all collection sites. Here, too, no seasonal pattern is discernible.

Haliclona sp.

The results for each of the two collection sites for this species are presented in Figure 5. The patterns here are quite different from those found in the other two species: considerably fewer total pigments can be discerned in *Haliclona*, and carotenoid esters are not a major component of the pigment complement. Although several pigments do occur in all samples from the Great Tide Pool site, only one of these is found at Dillon Beach (see Figure 6). Note that even the carotenes (Y-1) do not occur in all

samples at all times. No seasonal pattern can be detected and variability is high in these samples.

ANALYSIS OF TOTAL PIGMENT CONCENTRATIONS

Figure 7 presents the total pigment concentration in the three species studied and at all collection sites. No seasonal or geographic pattern can be seen and variability is high. The only pattern of note is that pigment concentration is generally higher in the samples of poeciloscerid sponges than in the representative of Haplosclerida.

Discussion

Despite the small size of our sample, the results clearly show the importance of analyzing biochemical variability and indicate ways in which biochemical investigations can be made more productive and reliable. Perhaps of least value was our investigation of total carotenoid content, since no seasonal or geographic patterns could be discerned and variability was extremely high. These results may well have been influenced by the generally low carotenoid concentrations in sponge tissues and the presence of various numbers of symbionts. The relatively higher pigment concentration in poeciloscerids merely reflects greater synthetic activity in these sponges (Liaaen-Jensen et al., 1982; Lee and Gilchrist, 1985).

Pigment variability was extremely high in all three species. This ranged from a single occurrence of a specific pigment in a species from a given locality, to frequent and almost ubiquitous occurrence. In the latter cases it is possible that a given pigment band was missed if it was present in small concentrations. This suggestion is supported by our observation of distinct differences in pigment concentration even in bands that had 100% occurrence.

One would expect the most variable pigments to be derived from symbionts and unmetabolized pigments from food items, the least variable carotenoids being those metabolized by the sponge. Among carotenoids in this latter group, carotenoid intermediates could also vary greatly. In our samples of *Ophlitaspongia pennata* we found a pigment band that appeared to be related to unmetabolized carotenoids in food. The Dillon Beach samples showed 100% occurrence of carotenoids characteristic of the Cryptophyceae whereas samples from the other two localities showed no sign of these pigments. In the same species, where we know something of the details of pigments present, all of the end products of oxidative metabolism were always present and in the form of carotenoid esters.

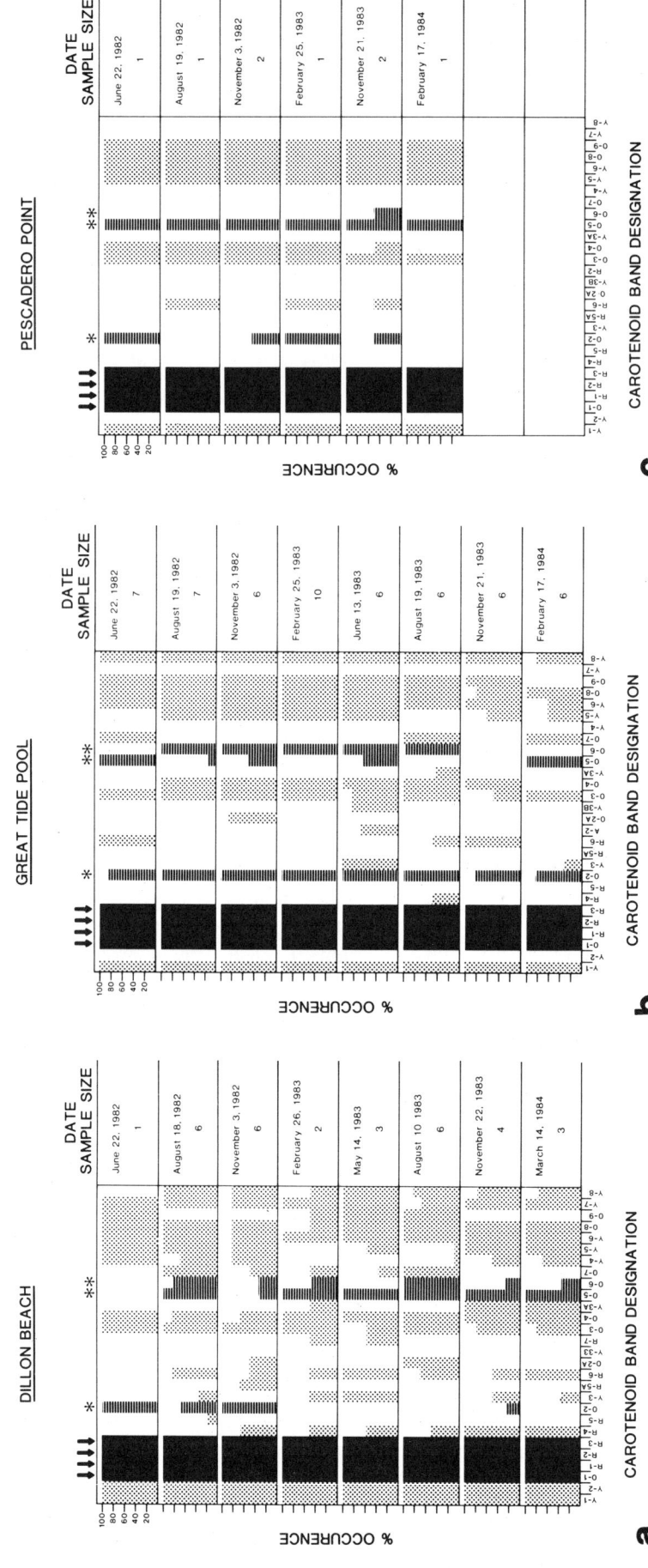

Figure 1. Percentage occurrence of carotenoids of *Ophlitaspongia pennata* at three collecting sites: *a*, Dillon Beach; *b*, Great Tide Pool; *c*, Pescadero Point. Carotenoid esters shown in black and designated by arrows; free carotenoids of these esters shown in horizontal stripes and designated by an asterisk. Carotenoid pigment bands are plotted from least polar on left to most polar on right.

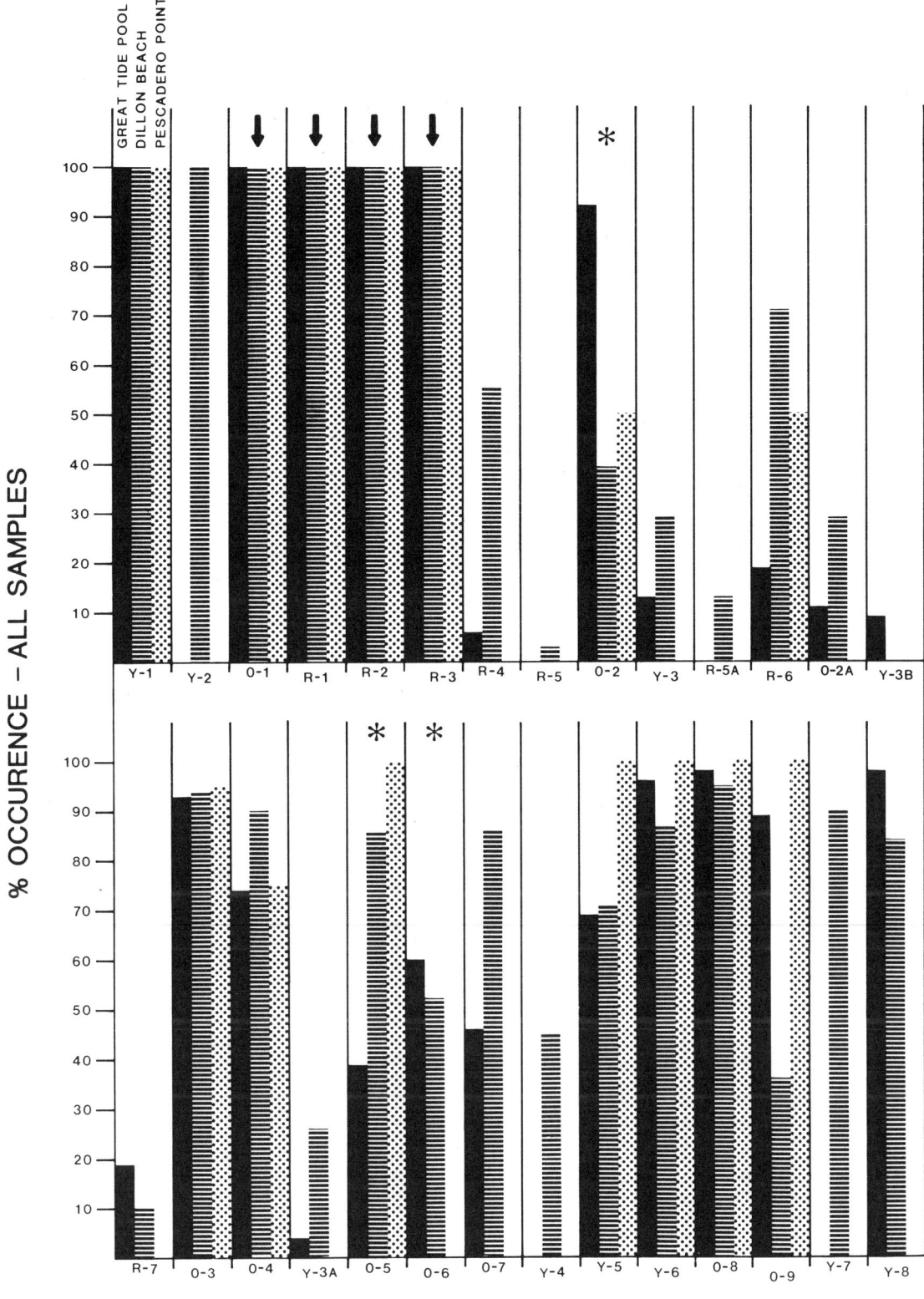

Figure 2. Percentage occurrence of carotenoids of *Ophlitaspongia pennata* at all collecting sites. Samples from Great Tide Pool shown in black, those from Dillon Beach in horizontal stripes, and those from Pescadero Point in dots. Arrows designate carotenoid esters, asterisks designate free carotenoids of these esters. Carotenoid bands plotted from least polar on top left of diagram to most polar on bottom right.

66

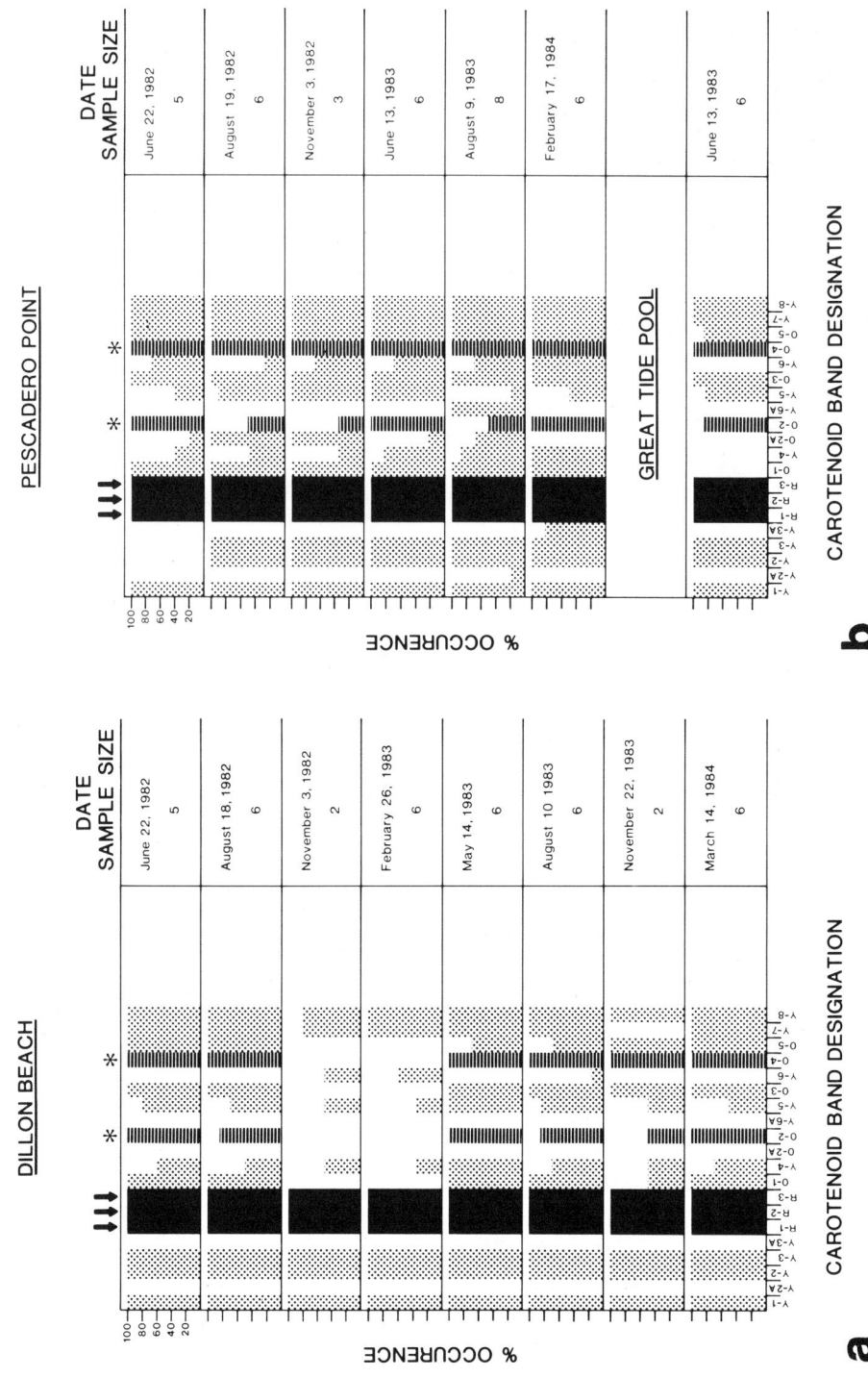

Figure 3. Percentage occurrence of carotenoids of *Plocamia karykina* at three sites: *a*, Dillon Beach; *b*, Pescadera Point and Great Tide Pool. Carotenoid esters shown in black and designated by arrows; free carotenoids of these esters shown in horizontal stripes and designated by an asterisk. Carotenoid pigment bands plotted from least polar on left to most polar on right.

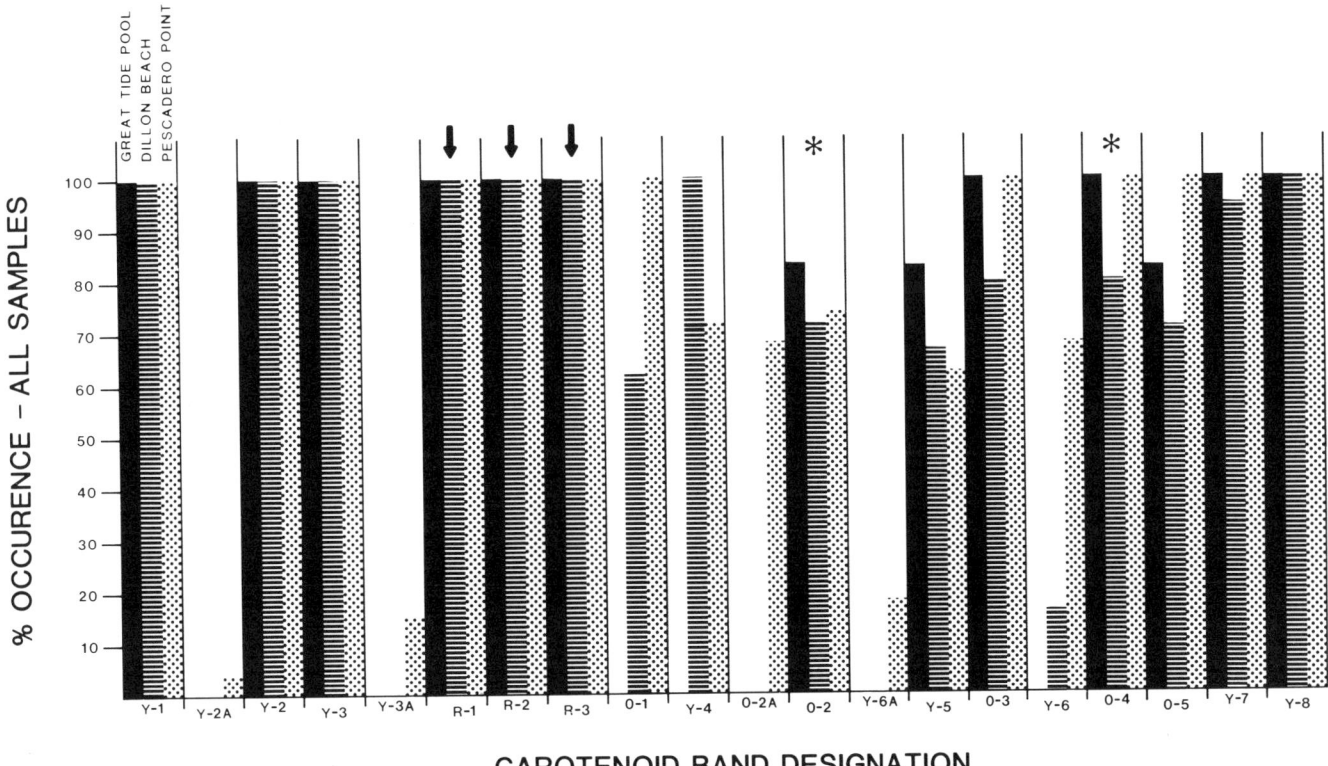

Figure 4. Percentage occurrence of carotenoids of *Plocamia karykina* at all collecting sites. Samples from Great Tide Pool shown in black, those from Dillon Beach in horizontal stripes, and those from Pescadero Point in dots. Arrows designate carotenoid esters, asterisks designate free carotenoids of these esters. Carotenoid pigment bands plotted from least polar on left to most polar on right.

The results of this study confirm an earlier finding (Lee and Gilchrist, 1985) suggesting a parallel between the generation of oxygenated carotenoid end products and the formation of carotenoid esters. In our previous work we had noted that analyses of unsaponified fractions provided more systematically valuable information than that derived from saponified material. This suggested that the carotenoid complement itself might not be as important systematically as the unique pairing of a specific carotenoid with a specific fatty acid. Even if such a relationship proves to be of systematic importance, it will be of little value if variability is high. Our present results point out that pigments that do esterify are the most stable ones. In the poecilosclerid sponges investigated here, virtually all esterified carotenoids were present at all times in all samples. This finding strongly supports our previous suggestion that carotenoid esters in poecilosclerid sponges may well be of systematic value. Further investigation of this possibility is obviously needed.

Whereas our analyses of poecilosclerid sponges suggest that carotenoids might be of use as a systematics tool, the work on Haplosclerida points to the opposite conclusion. Here the few esters present were transitory at best, and only a single pigment band could be consistently found in all the samples. Carotenoids of this group thus appear to have little systematic value.

As noted earlier, however, it is difficult to generalize on the basis of a small sample. In any case, our intent was not to develop systematic criteria for these groups using carotenoids, but rather to investigate carotenoid variability as a means of developing a more useful and meaningful application of biochemical characters as a systematics tool. In that sense, our study has raised some important points. First, it is vital to sample individual sponges. When material is combined (batched) the real relationships and any variability that was present tend to be obscured. Second, biochemical variability must be assessed, preferably before in-depth biochemical analysis is

68

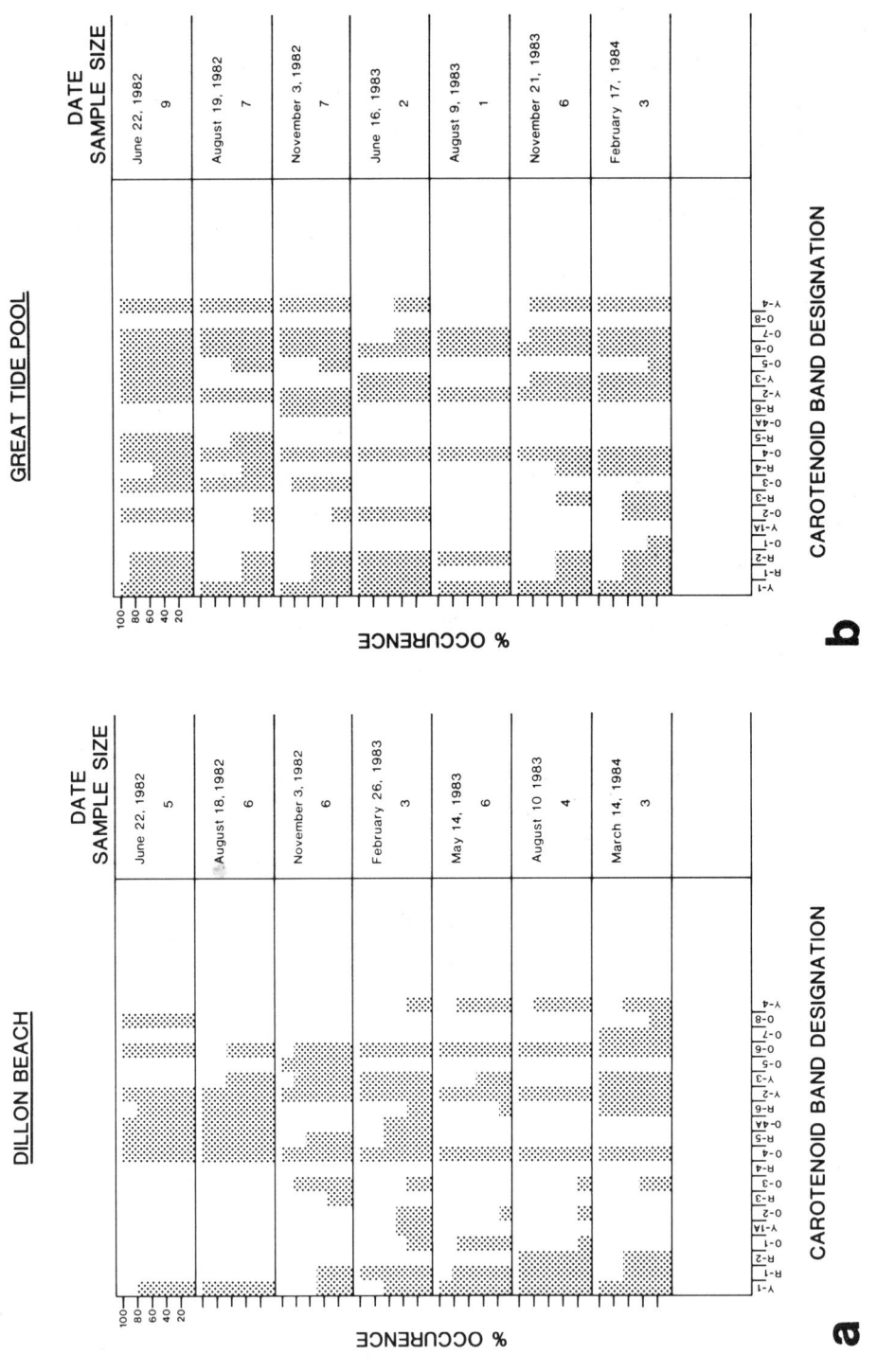

Figure 5. Percentage occurrence of carotenoids of *Haliclona* sp. at two collecting sites: *a*, Dillon Beach; *b*, Great Tide Pool. Carotenoid bands plotted from least polar on left to most polar on right.

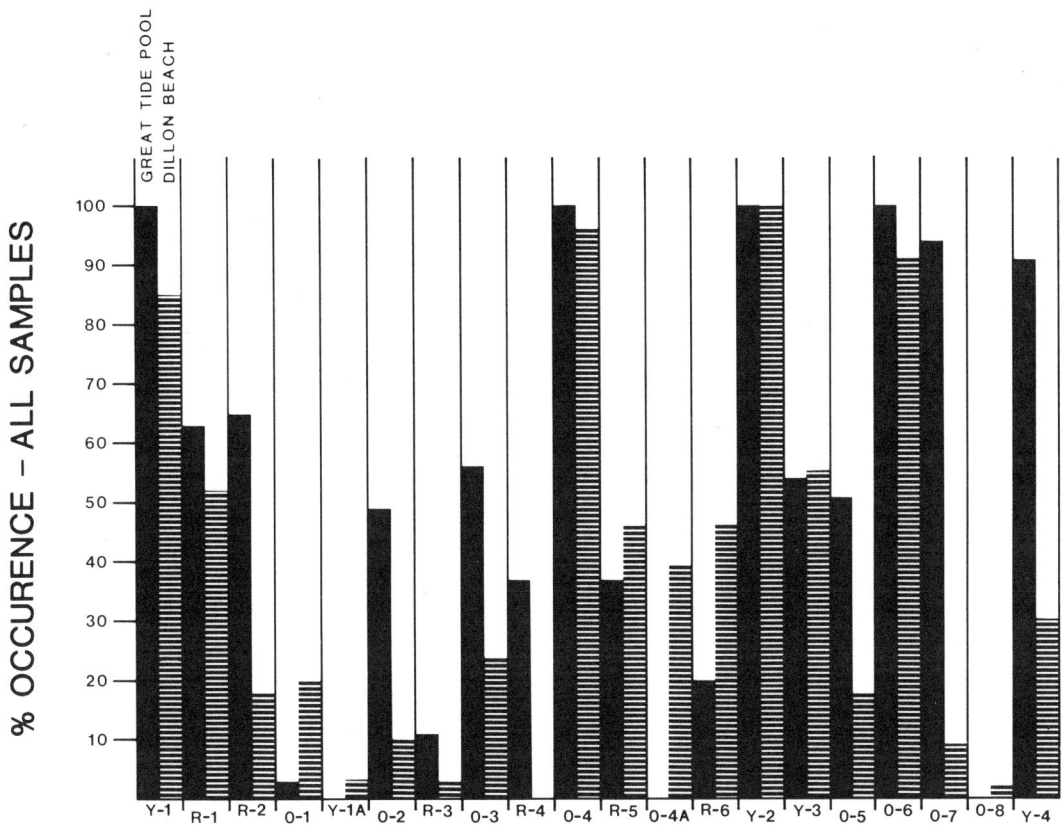

Figure 6. Percentage occurrence of carotenoids of *Haliclona* sp. at all localities. Samples from Great Tide Pool shown in black, those from Dillon Beach in horizontal stripes. Carotenoid bands plotted from least polar on left to most polar on right.

undertaken. Third, if duplicate material is only sampled occasionally in an in-depth study, the results are likely to be erroneous. As Figure 6 shows, considerable variability can occur in different localities. At one locality, four pigment bands occurred in all the samples, whereas at a second locality only about 200 miles away only one pigment could be found in all samples. In several instances, we found that multiple samples of a given species from a particular site exhibited a number of apparently stable pigments. Yet, at other times or at other localities, these same pigments may be absent. This difference may be related to the type of microhabitat in which the specimen was collected. For example, *Haliclona* from the Great Tide Pool is almost always found at the base of moderate-sized boulders in a level, well protected, and very homogenous environment. In contrast, the same species at the Dillon Beach locality can be found in a variety of microhabitats, which include the edges of boulders like those at the Great Tide Pool, crevices in highly exposed areas, and caves, to name a few.

Biochemical variability is generally assumed to be negligible and can be a serious problem in systematics. The problem of potential variability must therefore be addressed before biochemicals can be used with confidence in systematics studies.

Pattern analysis may be used as a preliminary means of establishing the nature and extent of biochemical variability. Such analyses will also give some idea of the relationships between the taxa under study. This approach will enable researchers to avoid expensive, time-consuming and often complicated in-depth studies when they may not be necessary and to undertake more meaningful,

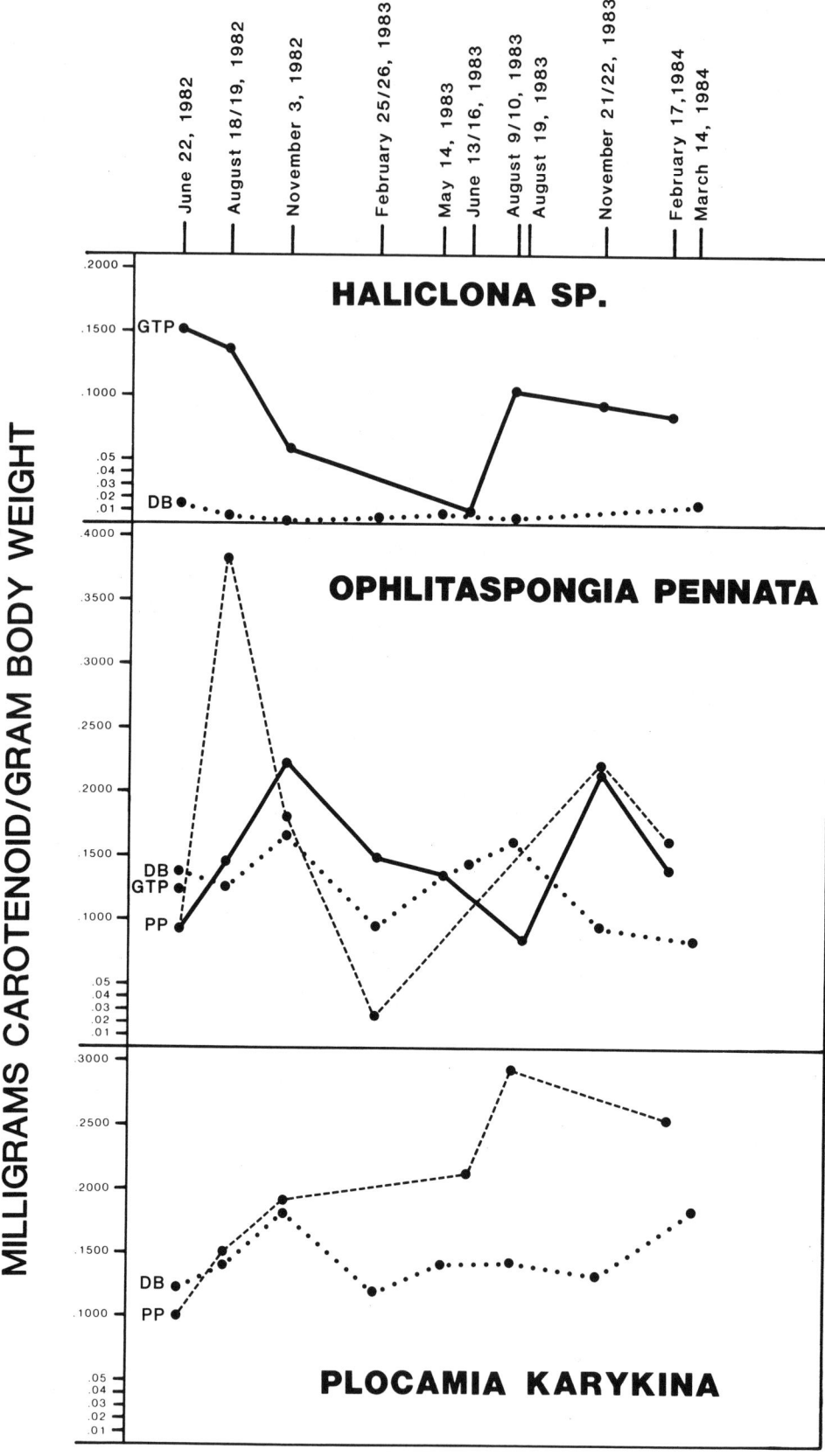

Figure 7. Total carotenoid content of sponge tissue for *Haliclona* sp., *Ophlitaspongia pennata*, and *Plocamia karykina*. Figures are means for all samples at a given collecting site and time.

better planned, and less involved analysis since variability will have been assessed.

Conclusions

The carotenoid complement of the sponge species investigated is characterized by high spatial and temporal variability. The total carotenoid complement per unit body weight is highly variable and shows no seasonality. There appears to be a parallel between the generation of oxygenated carotenoid end products and esterification. Carotenoid esters appear to be highly stable and of some potential systematic value in poecilosclerid sponges. Carotenoids and carotenoid esters are extremely variable in the haplosclerid sponge investigated and appears to have little systematic value in the carotenoid patterns of these sponges. Biochemical variability within a species can be a serious problem and must be addressed before biochemical evidence can be used. Pattern analysis is an effective method of testing for biochemical variability and for making a preliminarily assessment of the distribution of biochemical characters within and between taxa.

Acknowledgments

The authors wish to thank Dustin Chivers, Robert Van Syoc, and Karl Klontz for their help in collecting samples. We also thank Terry Gosliner, Peter Rodda, and Carolyn Sell for critically reading the manuscript, Lynette Cooke for assisting with the illustrations.

Literature Cited

Hartman, W. D. 1975. Phylum Porifera. Pages 32–64 in *Light's manual: Intertidal Invertebrates of the Central California Coast*, edited by R. I. Smith and J. T. Carlton. Berkeley: University of California Press.

Lee, W. L., and B. M. Gilchrist. 1985. Carotenoid Patterns in Twenty-Nine Species of Sponges in the Order Poecilosclerida (Porifera:Demospongiae): A Possible Tool for Chemosystematics. *Marine Biology,* 86:21–35.

Liaaen-Jensen, S., B. Renstrom, T. Ramdahl, M. Hallenstvet, and P. Bergquist. 1982. Carotenoids of Marine Sponges. *Biochemical Systematics and Ecology.* 10(2):167–174.

Pennington, F. C., F. T. Haxo, G. Borch, and S. Liaaen-Jensen. 1985. Carotenoids of Cryptophyceae. *Biochemical Systematics and Ecology,* 13(3):215–219.

PATRICIA R. BERGQUIST
PETER KARUSO
Department of Zoology
University of Auckland
Private Bag, Auckland, New Zealand

R. CONRAD CAMBIE
Department of Chemistry
University of Aukland
Private Bag, Auckland, New Zealand

Taxonomic Relationships within the Dendroceratida: A Biological and Chemotaxonomic Appraisal

Abstract

The diterpenoid chemistry of a number of species of dendroceratid sponges provide further evidence of a close relationship between members of the Dysideidae and Aplysillidae. The distribution and diversity of these metabolites within the dendroceratids is discussed in relation to the histology and morphology of particular organisms. Arguments are put forward in favor of transferring genera between these two groups and restructuring the family level classification of the Dendroceratida.

Previous work has established that certain groups of sponges have characteristic chemical patterns. For example, members of the order Verongida contain brominated metabolites based on tyrosine and species of the order Dictyoceratida contain terpenoids of six different classes. The occurrence of these terpenoid types, alone or in particular combinations, can be used to distinguish different families (Bergquist and Wells, 1983). The distribution of terpenoids is thought to indicate that the order Dendroceratida may be susceptible to chemotaxonomic analysis (Bergquist and Wells, 1983). Dendroceratid sponges have a distinctive external morphology, but their simple construction poses problems in distinguishing families and genera. In addition, the order includes convenience allocations such as the Halisarcidae, which remain in the order mainly because there is no clear alternative location (Bergquist, 1980). We review the chemical literature on the Dendroceratida and introduce new data that permit us to take some preliminary steps toward revising the taxonomy of the order. The genera and families discussed, together with their previous and suggested systematic positions, are given in Table 1.

Results

Darwinella oxeata, a yellow encrusting species from New Zealand waters, contains aplysulphurin (Figure 1*a*) as the major metabolite in addition to three tetrahydro-derivatives. A similar yellow encrusting Australian sponge referred to as *Aplysilla sulfurea*, has yielded aplysulphurin unaccompanied by the tetrahydro-derivatives (Karuso et al., 1984). The correct generic name for the Australian sponge is *Darwinella*, and it differs only slightly from *D. oxeata*. A second, rose-colored species of *Darwinella* from New Zealand has consistently yielded one of the minor tetrahydro-derivatives as the major metabolite (Karuso et al., 1986).

The chemical affinity shown by these three *Darwinella* species came into question when a further roseate encrusting species from Australia was examined. Morphologically, this sponge is very close to the New Zealand species of *Darwinella*, but it has a different terpene profile (Karuso and Taylor, 1986). Twelve diterpenes were isolated from this sponge, ambliofuran (Figure 1*b*) and aplyroseol-1 (Figure 1*c*) being the major metabolites. All other components were based on aplyroseol-1. Two of these metabolites—aplyroseol-1 and aplyroseol-2—have been reported by Schmitz et al. (1985) from the Caribbean sponge *Igernella notabilis*. Our samples of studied *Dendrilla rosea* from New zealand yielded six aplyroseols and four new, but closely related, derivatives, which suggests that the Australian sponge is probably not *Aplysilla* or *Dar-*

winella, but a *Dendrilla*. No specimens were available for histological study, but New Zealand specimens of the two genera are easily mistaken at certain growth stages.

A bright orange encrusting Australian sponge, which at present is assigned to *Darwinella* but possesses spongin spicules an order of magnitude smaller than those found in *Igernella* and in all species of *Darwinella* discussed above, has yielded gracilin-A (Figure 1*d*) as the major metabolite, along with several other derivatives of this compound (Poiner and Taylor, unpublished data). The same compound has been recorded as the major metabolite of *Spongionella gracilis* from the Mediterranean (Mayol et al., 1985).

Chelonaplysilla, with its sandy surface reticulum, is morphologically a well characterized genus. Australian and New Zealand specimens that are deep purple and nearly always thinly encrusting have yielded terpenoids based mainly on two skeletal types, dendrillolide-B (Figure 1*e*) and norrisolide (Figure 1*f*). These compounds had previously been isolated from a Palauan sponge referred to as *Dendrilla* (Sullivan and Faulkner, 1984). Unfortunately, no voucher specimens of the Palauan sponge are available for examination, but they have been described as an erect, deep purple sponge with a dendritic skeleton. These features suggest that it could be *Chelonaplysilla*, not *Dendrilla*. Further evidence to support this view comes from Hawaii, where a deep purple, almost black, *Chelonaplysilla* sp. grows as a small erect sponge. The Hawaiian sponge also contains the same compounds as the Palauan, Australian, and New Zealand specimens (Karuso, unpublished data).

Discussion

All the terpenoids referred to above are related chemically and can be derived from the hypothetical diterpenoid ether spongiane (Figure 1*g*). However, three distinct subgroups with differing biogenesis can be recognized.

The aplyroseol compounds isolated from *Dendrilla rosea*, *Igernella notabilis*, and the rose-colored *Dendrilla* from Australia are simply oxidative elaborations of spongiane. They are characterized by the presence of a D-ring lactone and the unusual bridging lactol between C17 and C15. Most show oxygenation at C7 and (or) C6.

The aplysulphurin diterpenes found in *Darwinella oxeata*, *Darwinella* sp., and *Spongionella gracilis* result from ring B cleavage of spongiane with concomitant methyl migration from C8 to C7. Loss of one carbon, probably by decarboxylation, would yield the C19 gracilin-A skeleton. Further oxygenation and dehydrogenation of the skeletons would eventually lead to the observed metabolites (Figure 2).

The *Chelonaplysilla* metabolites, dendrillolide-B, and norrisolide, can be produced from spongiane by two alternative rearrangements of ring B, both of which result in cleavage of ring C. This process is initiated by abstracting

Table 1. Present and suggested arrangement of genera mentioned in this work

Present	Suggested
Order Dendroceratida	Order Dendroceratida
Family Aplysillidae	Family Aplysillidae
Genus *Aplysilla*	Genus *Aplysilla*
Dendrilla	*Dendrilla*
Darwinella	*Darwinella*
Chelonaplysilla	New Genus
Pleraplysilla	*Chelonaplysilla*
	Igernella
Family Dictyodendrillidae	Family Dictyodendrillidae
Genus *Dictyodendrilla*	Genus *Dictyodendrilla*
Igernella	
Family Halisarcidae[a]	Family Halisarcidae[a]
Genus *Halisarca*	Genus *Halisarca*
Order Dictyoceratida	Order Dictyoceratida
Family Dysideidae	Family Dysideidae
Genus *Dysidea*	Genus *Dysidea*
Spongionella	*Pleraplysilla*
	Family New Family
	Genus *Spongionella*

[a]Convenience allocation only

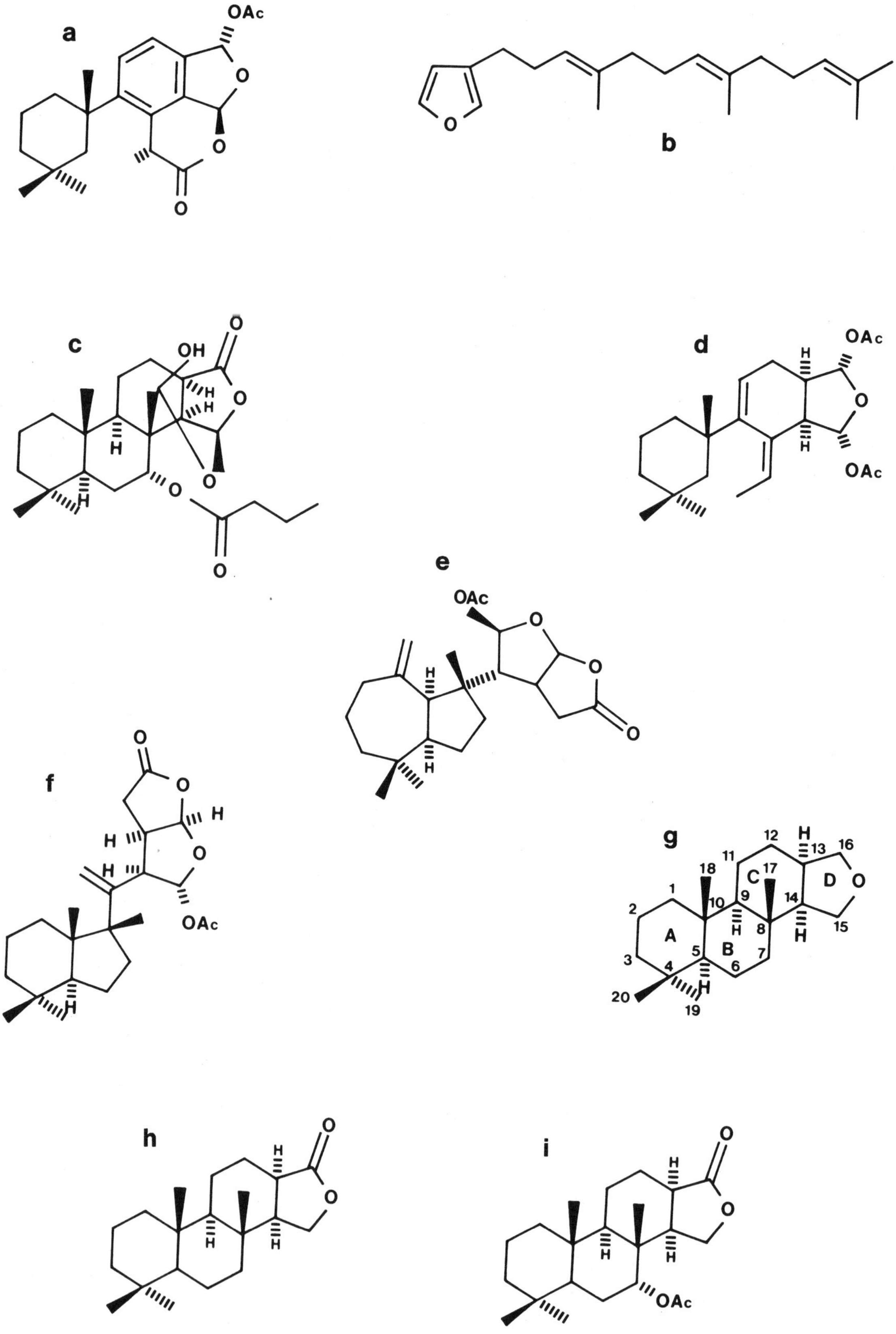

74

P.R. BERGQUIST, P. KARUSO, AND R.C. CAMBIE

Figure 1. Structure of sponge metabolites: *a*, aplysulphurin; *b*, ambliofuran; *c*, aplyroseol-1; *d*, gracilin-A; *e*, dendrillolide-B; *f*, norrisolide; *g*, spongiane; *h*, *i*, spongiane-related metabolites from *Chelonaplysilla violacea*.

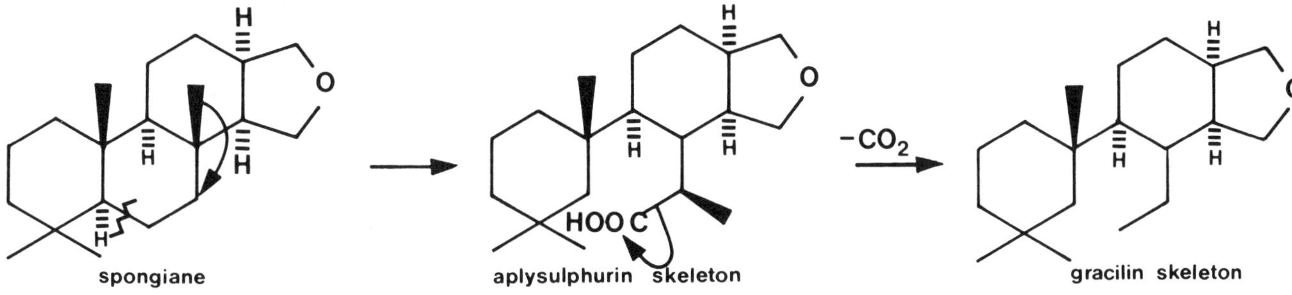

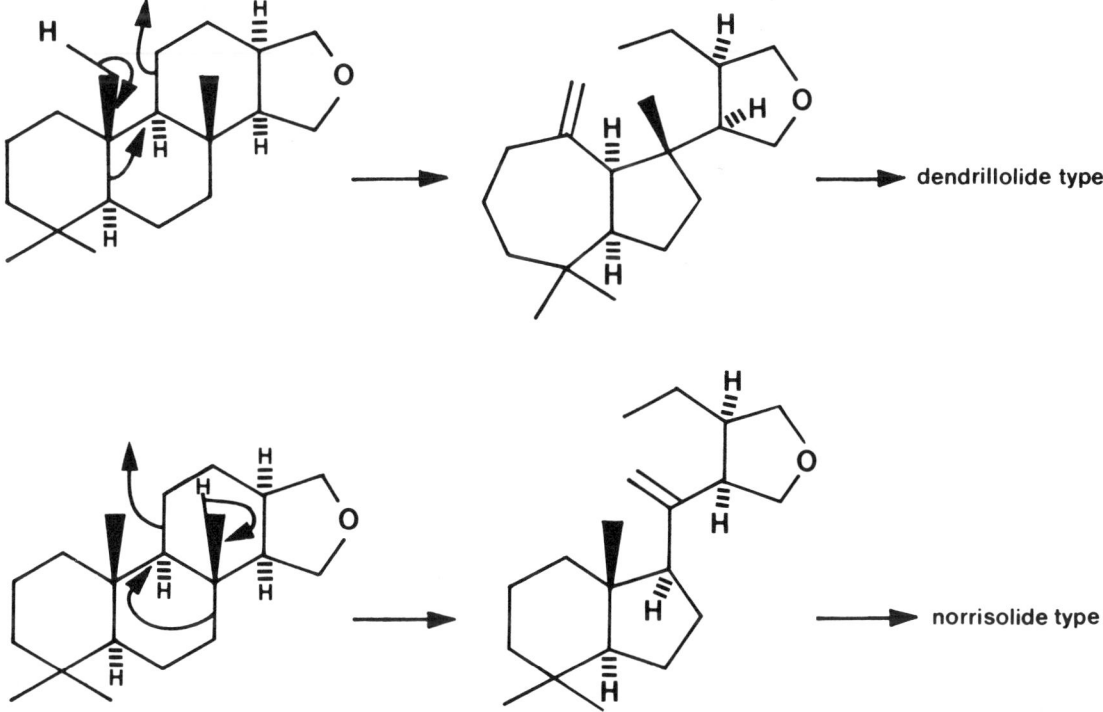

Figure 2. Biosynthetic relationships between spongiane, aplysulphurins, and gracilins.

a hydrogen atom from C10 and C8-methyl groups. Migration of adjacent bonds yields the two skeletal types observed in *Chelonaplysilla* (Figure 3). That this is the biogenetic sequence is substantiated by the isolation of unrearranged spongiane-type diterpenoids in very minor quantities. For instance, two spongiane metabolites (Figure 1*h, i*) are constituents of both Australian and New Zealand collections of *Chelonaplysilla violacea*. The latter compound is also a minor constituent of *Denrilla rosea* and the rose *Dendrilla* from Australia.

Spongiane diterpenoids have also been reported in two species of the Spongiidae (Dictyoceratida). Dictyoceratid sponges are sufficiently distinct morphologically from the Dendroceratida to exclude any possibility of a relationship, and the compounds do not occur widely within the former group. There are also subtle differences in the chemistry of the diterpenoids from the two sources. Iso-

gathalactone (Figure 4*a*) and aplysillin (Figure 4*b*) from *Spongia officinalis* (Cimino et al., 1974, 1982; Gonzales et al., 1982, 1984) possess a C11(12) double bond, functionalized in aplysillin by the addition of acetic acid. This double bond is always present in the *Spongia* metabolites but never in the dendroceratid diterpenoids. An as yet unnamed genus of Spongiidae from the Great Barrier Reef yielded metabolites with the general structure of spongiatriol (Figure 4*c*) always having a furan as ring D and being heavily oxygenated on ring A (Kazlauskas et al., 1979; Capelle et al., 1980).

The only additional diterpenoids isolated from sponges lacking a mineral skeleton are from *Dysidea amblia* (Walker and Faulkner, 1981). Among other diterpenoids, this sponge contained dehydroambliol (Figure 4*d*), which has also been found in *Dendrilla* sp. from Palau (probably *Chelonaplysilla*), and ambliofuran, which was found in the

Figure 3. Derivation of dendrillodides and norrisolides from spongiane.

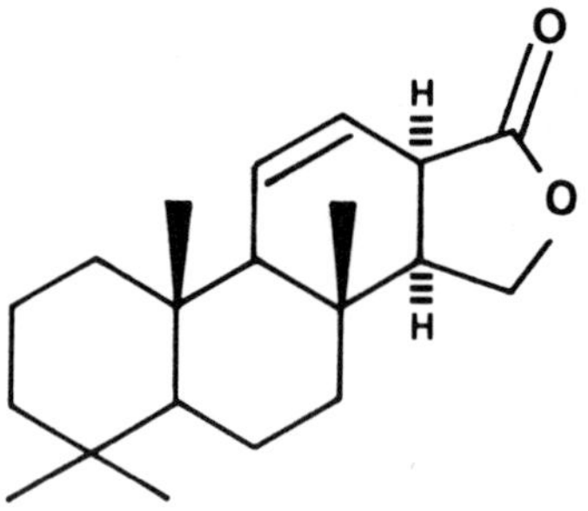

Figure 4. Diterpenoids from Dictyoceratida: *a*, isogathalactone; *b*, aplysillin; *c*, spongiatriol; *d*, dehydroambliol.

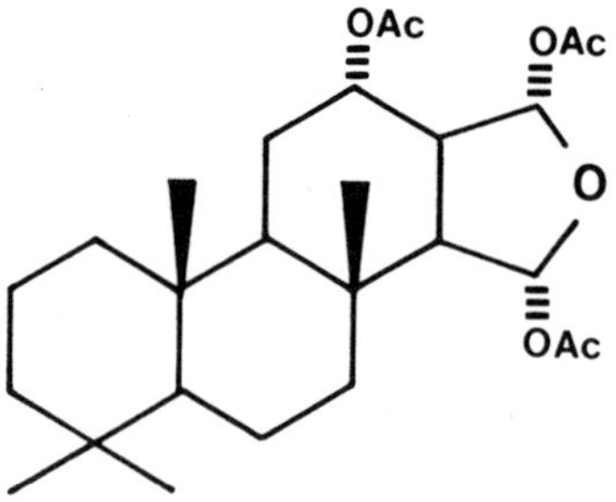

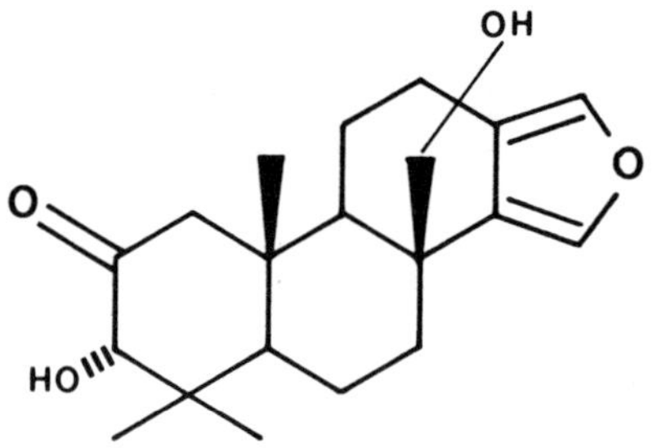

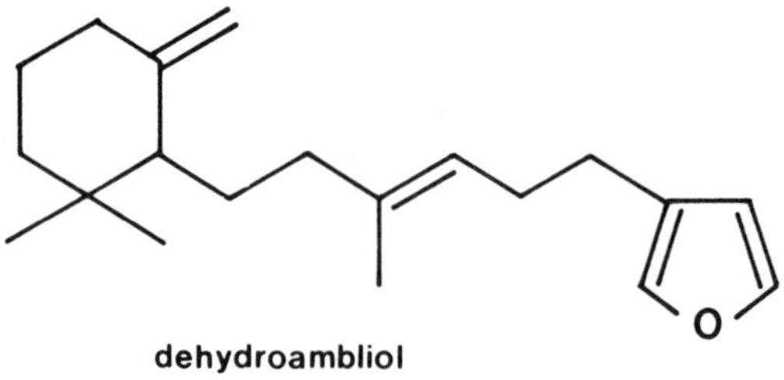

same Palauan sponge as well as in *Dendrilla rosea* from New Zealand and two *Darwinella* species, one from Australia and one from New Zealand. The chemistry of *Dysidea amblia* appears to have affinities with that of members of the Aplysillidae (Dendroceratida) and Dysideidae (Dictyoceratida). Interestingly, aspects of histology, such as choanocyte chamber structure, surface morphology with strong conule development, and fiber development where in both cases the fiber constitutes a minor component of the tissue volume reinforce this suggestion of affinity between the two groups. Unfortunately, again there is ambiguity in the chemical literature. A second collection of *Dysidea amblia* yielded typical *Dysidea* sesquiterpenoids in conjunction with the diterpenoids (Walker et al., 1984). At this stage we cannot rule out the possibility that this sponge has been misidentified or contaminated with a colorless epizoic aplysillid. Therefore, an effort should be made to recollect *Dysidea amblia* and subject it to careful morphological scrutiny.

The diterpenes discussed above all have the same basic structure however, the sponges from which these metabolites have been derived are not taxonomically homogeneous as the literature would suggest (Table 1). We believe that in certain cases the generic identification reported in

the chemical literature is in error. In addition, some aspects of the present classification of the dendroceratids need to be reconsidered. A reexamination of the morphology and histology of several dendroceratid sponges has been undertaken in conjunction with our chemical study, but this work is incomplete at present.

Among the long-standing problems of generic relationships that need to be addressed, the status and affinity of the genus *Spongionella* is of particular concern. At present, *Spongionella* is classified within the Dictyoceratida (Dysideidae), but its affinities with that group have often been questioned (Bergquist, 1980). From the numerous chemical studies of the Dysideidae it appears that furanosequiterpenoids or sesquiterpene quinones are present in this group throughout the world (Bergquist and Wells, 1983). The gracilin-type diterpenoids that occur in *Spongionella gracilis* are very distinct from these compounds. Unfortunately, although we can reappraise the structure and histology of *S. gracilis*, the type species of *Spongionella*—*S. pulchella*—remains elusive as it is known with certainty only from the dry type specimen that was a drift specimen from the coast of Ireland. Recent finds from the Mediterranean (Vacelet, 1959) do not seem to be *S. pulchella* and may be better assigned to *S. gracilis*. To determine the correct use of the generic name *Spongionella*, we should work only with skeletal characters. However, such an undertaking is beyond the scope of this chapter.

What the chemical and morphological data compiled to date do indicate is that a group of very closely related genera exists that is almost equivalent to the present Aplysillidae. This group consists of *Dendrilla*, *Darwinella*, *Igernella*, and *Chelonaplysilla*. We cannot comment on *Aplysilla* since there are no taxonomically verifiable records of chemical characteristics for this genus, and the possibility—indeed probability—of confusion with *Darwinella* in field collections must be acknowledged. These genera share basic diterpenoid characteristics, but each is in some measure distinctive. It will be interesting to sample further species to confirm the pattern and to aid in evaluating the significance of the biosynthetic differences between the various terpenoid types.

The family Dictyodendrillidae, members of which have a distinctly aplysillid histology and fiber structure in conjunction with a regular, reticulated skeleton, remains distinct. *Dictyodendrilla cavernosa* is the only species of this family to have been examined chemically, and the mixture of C30 aldehydes that it yields is as yet inseparable. The genus *Igernella* was placed in the Dictyodendrillidae (Bergquist, 1980) because it has a reticulate rather than rigidly dendritic skeleton. However, there is no real similarity between the fine, regular skeletal reticulum of *Dictyodendrilla* and the coarse, sparse, and irregular fiber pattern of *Igernella* (Topsent, 1905). The same superficial fibro-reticulation, in which certain areas appear more like fenestrated spongin plates than branching fibers, occurs in *Darwinella oxeata* and *Darwinella* sp. from New Zealand. This structural affinity, in conjunction with the large fibrous spicules present in the two genera, as well as the chemical parallels between the two genera, argues for transferring *Igernella* from the Dictyodendrillidae to the Aplysillidae. The family diagnoses should therefore be amended, as we expect to do shortly.

In our collection, the orange *Darwinella* from Australia is unusual. It has a sparse but typical aplysillid dendritic skeleton augmented by very small fiber spicules. The major metabolite is gracilin-A which is also the major component of *Spongionella gracilis*. It can be argued convincingly that the genus *Darwinella*, as presently construed, contains two generic entities with differing fiber and fibrous spicule morphology. Neither group shares any skeletal similarity with *Spongionella gracilis*. Moreover, histology, choanocyte chamber structure, spherulous cell characteristics and mesohyl organization indicate that *Darwinella* species are dramatically distinct from *S. gracilis*. The latter has the sparsely cellular, diffuse, weakly collagenous mesohyl characteristic of the Dysideidae. However, this should be confirmed by electron microscopy.

We must conclude in this case that the coincident diterpenoid chemistry once again reveals a close relationship

between the Dysideidae and Aplysillidae as presently construed. As Dendy (1905) has observed of the Dysideiidae (Spongeliidae) "This family may be retained as a matter of convenience, but it is, as I have already indicated, logically impossible to separate it sharply from the Aplysillidae, for the genus Megalopastas, on the one hand, and Schulze's *Spongelia spinifera*, on the other, are strictly intermediate between the two groups."

In this comment, Dendy was referring to the presence of a reticulate skeleton in an otherwise aplysillid sponge (his concept of *Megalopastas*) and to the detritus-cored fibers in *Pleraplysilla* (= *Spongelia*) *spinifera*. Other points of synonymy have been made regarding these examples (Bergquist, 1980; Vacelet, 1959), but the suggestion at the familial level is still valid. The concept of a dendritic skeleton as an ordinal discriminator is unsustainable, and the occurrence of large eurypylous choanocyte chambers in dysideiid and dendroceratid sponges may indeed reflect a common ancestry. Clearly, this is a complex issue in which many points of histology, ultrastructure, and skeletal morphology must to be taken into account before it can be resolved.

Another important point to note concerns the genus *Pleraplysilla*. There is every possibility that *Pleraplysilla minchini*, the type species of *Pleraplysilla* (Topsent, 1905) may not be conspecific with *P. spinifera*. The latter, in Mediterranean specimens, has typical *Dysidea* chemistry, but no chemistry is known for Atlantic coast specimens that would equate to *P. minchini*. This genus deserves careful reexamination, both chemically and structurally. *Pleraplysilla spinifera* could be a dysideiid with vestigial skeleton, while *P. minchini* could be a true aplysillid. At present, too little is known of their morphology and chemistry to comment further on this possibility.

The literature is now accumulating rapidly as chemists and biologists turn their attention to dendroceratid sponges. More information is needed on different species, especially with respect to histology and chemistry. As this investigation has shown many times over, it is imperative to maintain adequate reference material.

Literature Cited

Bergquist, P. R. 1980. A Revision of the Supraspecific Classification of the Orders Dictyoceratida, Dendroceratida, and Verongida (Class Demospongiae). *New Zealand Journal of Zoology*, 7:443–503.

Bergquist, P. R., and R. J. Wells. 1983. Chemotaxonomy of the Porifera: The Development and Current Status of the Field. Pages 1–50 in *Marine Natural Products, Chemical and Biological Perspectives*, edited by P. J. Scheuer. New York: Academic Press.

Capelle, N., J. C. Braekman, D. Daloze, and B. Tursch. 1980. Studies of Marine Invertebrates XLIV. Three New Spongian Diterpenes from *Spongia officinalis*. *Bulletin, Societés Chimiques Belges*, 89:399–404.

Cimino, G., S. De Stefano, D. De Rosa, and L. Minale. 1974. Isogathalactone, a Diterpene of a New Structural Type from the Sponge *Spongia officinalis*. *Tetrahedron*, 30:645–649.

Cimino, G., R. Morrone, and G. Sodano. 1982. New Antimicrobial Diterpenes from the Sponge *Spongia officinalis*. *Tetrahedron Letters*, 23:4139–4142.

Dendy, A. H. 1905. Report on the Sponges Collected by Professor Herdman at Ceylon in 1902. Pages 59–246 in *Report to the Government of Ceylon on the Pearl Oyster Fisheries of the Gulf of Manaar*, 3. London: Royal Society.

Gonzales, A. G., V. Darias, and E. Estevez. 1982. Contribution to the Biological Study of *Spongia officinalis*. *Farmaco*, 37:179–183.

Gonzales, A. G., D. M. Estrada, J. D. Martin, V. S. Martin, C. Perez, and R. Perez. 1984. New Diterpenes from *Spongia officinalis*. *Tetrahedron*, 40:4109–4113.

Karuso, P., P. R. Bergquist, J. S. Buckleton, R. C. Cambie, G. R. Clark, and C. E. F. Rickard. 1986. Constituents of Morphologically Similar Sponges. *Australian Journal of Chemistry*, 39:1643–1653.

Karuso, P., B. W. Skelton, W. C. Taylor, and A. H. White. 1984. The Constituents of Marine Sponges I. The Isolation from *Aplysilla sulphurea* (Dendroceratida) of Aplysulphurin and the Determination of Its Crystal Structure. *Australian Journal of Chemistry*, 37:1081–1093.

Karuso, P., and W. C. Taylor. 1986. The Constituents of Marine Sponges II. The Isolation from *Aplysilla rosea* Barrois (Dendroceratida) of Aplyroseol–1 and Related Diterpenes (Aplyroseol–2 to 6). *Australian Journal of Chemistry*, 39:1629–1641.

Kazlauskas, R., P. T. Murphy, R. J. Wells, K. Noak, W. E. Oberhansli, and P. A. Scholholzer. 1979. A New Series of Diterpenes from Australian *Spongia* Species. *Australian Journal of Chemistry*, 32:867–880.

Mayol, L., V. Piccialli and D. Sica. 1985. Gracilin-A, a Unique Non-Diterpene Metabolite from the Marine Sponge *Spongionella gracilis*. *Tetrahedron Letters*, 26:1357–1360.

Schmitz, F. J., J. S. Chang, M. B. Hossain, and D. van der Helm. 1985. Marine Natural Products: Spongiane Derivatives from the Sponge *Igernella notabilis*. *Journal of Organic Chemistry*, 50:2862–2865.

Sullivan, B., and D. J. Faulkner. 1984. Metabolites of the Marine Sponge *Dendrilla* sp. *Journal of Organic Chemistry*, 49:3204–3206.

Topsent, E. 1905. Etude sur les Dendroceratida. *Archives de Zoologie Expérimentale et Générale*, 4(3):clxxi–cxcii.

Vacelet, J. 1959. Répartition générale des Eponges et systématique des Eponges cornées de la région de Marseille et de quelques stations Méditerranéenes. *Recueil des Travaux de la Station Marine d'Endoume*, 26(16):39–101.

Walker, R. P., and D. J. Faulkner. 1981. Diterpenes from the Sponge *Dysidea amblia*. *Journal of Organic Chemistry*, 46:1098–1102.

Walker, R. P., R. M. Rosser, D. J. Faulkner, L. S. Bass, C. H. He, and J. Clardy. 1984. Two New Metabolites of the Sponge *Dysidea amblia* and Revision of the Structure of Ambliol B. *Journal of Organic Chemistry*, 49:5160–5163.

Immunology and Chemical Defense

GRADIMIR N. MISEVIC
MAX M. BURGER
Department of Biochemistry
Biocenter, University of Basel
Klingelbergstrasse 70
CH–4056 Basel, Switzerlan
and
Marine Biological Laboratory
Woods Hole, Massachusetts 02543

Multiple Low-Affinity Carbohydrates as the Basis of Cell-Cell Recognition in *Microciona prolifera*

Abstract

Species-specific cell aggregation in the marine sponge *Microciona prolifera* is mediated by a proteoglycanlike aggregation factor (MAF) of $M_r = 2 \times 10^7$ via a cell-binding and a self-association domain. After extensive trypsin digestion, over 60% of the MAF mass was converted into a glycopeptide fragment of $M_r = 10,000$ showing cell-binding activity. However, the apparent affinity was 13,000 times lower than that of native MAF. The reconstitution of binding affinity in the same order of magnitude as native MAF was achieved by cross-linking the glycopeptide into the polymers of MAF size. We therefore propose that MAF cell association is based on highly polyvalent interactions of low affinity glycopeptide sites.

A monoclonal antibody directed against MAF (Block 1) was found to selectively inhibit activity of the self-association domain only. Since the Block 1 antibody showed specific binding to a fraction of protein-free oligosaccharides that are present in 1,100 copies in MAF, highly repeating carbohydrate structures could be considered the active sites of the self-association domain. The monomeric glycans did not display measurable self-binding activity. However, after they were cross-linked into polymers of MAF size the self-association could be reconstituted. This indicates that the highly polyvalent, low-affinity carbohydrate–carbohydrate interactions are mediating self-association of MAF. Functional activity of the cell binding and the self-association domain of *Microciona prolifera* aggregation factor is based on the multiple low-affinity interactions of two probably different carbohydrate structures of MAF. This differs from the single and higher affinity interactions of most adhesion molecules.

Specificity of cellular recognition is the basis of many fundamental biological processes such as fertilization, morphogenesis, and immune defense. Some of the first experimental evidence demonstrate the existence of such a cell-cell recognition phenomenon was obtained by species-specific sorting of reaggregating sponge cells (Wilson, 1907; Galtsoff, 1929). Sponges are still very attractive for study of cell-cell recognition at the molecular level because of the simplicity of the assay system employing aggregation of either live or fixed dissociated cells without detectable damage of surface components.

In the last two decades it has been discovered that at least three components are necessary for species-specific reaggregation of dissociated cells from the marine sponge *Microciona prolifera* (Poecilosclerida): (1) a large peripheral membrane proteoglycanlike molecule of $M_r = 2 \times 10^7$, called the MAF-*Microciona* aggregation factor (Humphreys, 1963; Henkart et al., 1973); (2) a cell surface receptor for MAF, known as the base plate (Weinbaum and Burger, 1973); and (3) 10 mM Ca^{2+} ions.

Chemical analyses have shown that 60% of the MAF weight is composed of carbohydrates and 40% of protein (Henkart et al., 1973; Misevic et al., 1982). High negative charge is due to the presence of glucuronic acid ammounting to 10% of MAF mass and to some sulfate and glycine. On electron micrographs the large MAF proteoglycanlike complex appears as a sunburst with a circle of 800 Å and 16 radiating arms each 1,100 Å long (Humphreys et al., 1977).

Functional tests revealed that MAF mediates species-specific reaggregation of *Microciona prolifera* cells via two functional domains, a cell-binding domain and a MAF self-interaction domain (Jumblatt et al., 1980). The cell binding domain has a stringent species-specificity and does not require physiological 10 mM Ca^{2+}. Specificity of the self-interaction domain has only been suggested and more experimental evidence is necessary. In contrast to cell binding of MAF, self-association is active only in presence of 10 Mm Ca^{2+} (Jumblatt et al., 1980).

Methods and Results
ISOLATION OF A CELL-BINDING FRAGMENT FROM MAF

In an attempt to isolate the cell-binding domain, dissociation procedures, which irreversibly inactivate self-association of MAF, were combined with trypsin treatment (Figure 1). Two subunits, purified after dissociation with urea and EDTA (UE 1 and UE 2), and four tryptic fragments (T-124, T-70, T-27, and T-10), all showed species-specific binding to the homotypic cells, thus indicating the presence of cell-binding activity (Table 1). Specificity of binding has been confirmed by competitive inhibition of the intact ^{125}I-MAF binding to cells by subunits and fragments (Misevic et al., 1982; Misevic and Burger, 1982; Burger and Misevic, 1986).

Since each of the isolated subunits and fragments showed activity of the cell-binding domain, their binding affinity to homotypic cells was determined. Interestingly, the association constants (K_a) decreased proportionally with decreasing molecular weight of the fragments,

Table 1.   Binding specificity (in % bound) of ^{125}I MAF and MAF fragments to cells of three sponge species

MAF fragments	M. prolifera	C. celata	M. fusca
MAF	33	2.2	2.1
U–1500	59	2.7	2.3
U–250	54	0.6	1.6
T–124	31	8.1	7.9
T–70	29	7.3	6.8
T–27	31	7.8	6.7
T–10	45	12.0	11.1

Binding of iodinated MAG and MAG fragments of *Microciona prolifera*, *Cliona celata*, and *Mycale fusca* (107/ml) was carried out under standard assay conditions for 20 min at room temperature (Misevic et al., 1982; Jumblatt et al., 1980). After incubation, cells were layered on and centrifuged through 0.1% bovine serum albumine with 10% sucrose in $Ca^{2+}Mg^{2+}$–free seawater. Supernatants were aspirated and the pellet counted in a Packard gamma spectrometer.

whereas a reverse relationship was found for B_{max} = maximal number of fragments bound per cell (Figure 2). A possible explanation could be that smaller subunits or fragments have fewer cell-binding sites and therefore lower binding affinity to cells; thus the cell-binding domain of MAF would be highly polyvalent. The notion that the cell-binding domain of MAF may be composed of multiple low-affinity cell-binding sites contrasts with the single and higher affinity interactions thought to characterize most adhesion molecules.

In order to test the hypothesis that the MAF cell-binding domain is highly polyvalent, the maximal amount of 10,000-dalton fragment (T-10) recovered after extensive and repeated trypsin digestion of MAF was quantified. Since T-10 is the smallest isolated fragment with the lowest binding affinity to cells, it may well be the cell-binding unit. Upon the first trypsin treatment, MAF fragments were separated by gel filtration on a Sephacryl S-200 column and the content of protein, neutral hexoses, and uronic acid was determined (Figure 3). Retrypsinization of the fractions larger than T-10 gave an almost identical profile of fragments. A third digestion of void volume fractions, however, resulted in no further fragmentation. After the first trypsin treatment, 31% of the total MAF mass (protein, neutral hexoses, and uronic acid) was converted into the T-10 fragment and an additional 32% was converted after the second digestion. Since 63% of the 2×10^7-dalton MAF is composed of the T-10 fragment ($M_r = 10,000$), the native MAF has 1,300 repeats of T-10 sites. We therefore propose that the major part, but not the entire MAF molecule, is built up of T-10 fragments.

BIOCHEMICAL PROPERTIES OF THE CELL-BINDING GLYCOPEPTIDE FRAGMENT

Ultracentrifugational analysis in the presence or absence of 6M guanidine hydrochloride, gel filtration, and SDS-gel electrophoresis suggests that the T-10 glycopeptide is

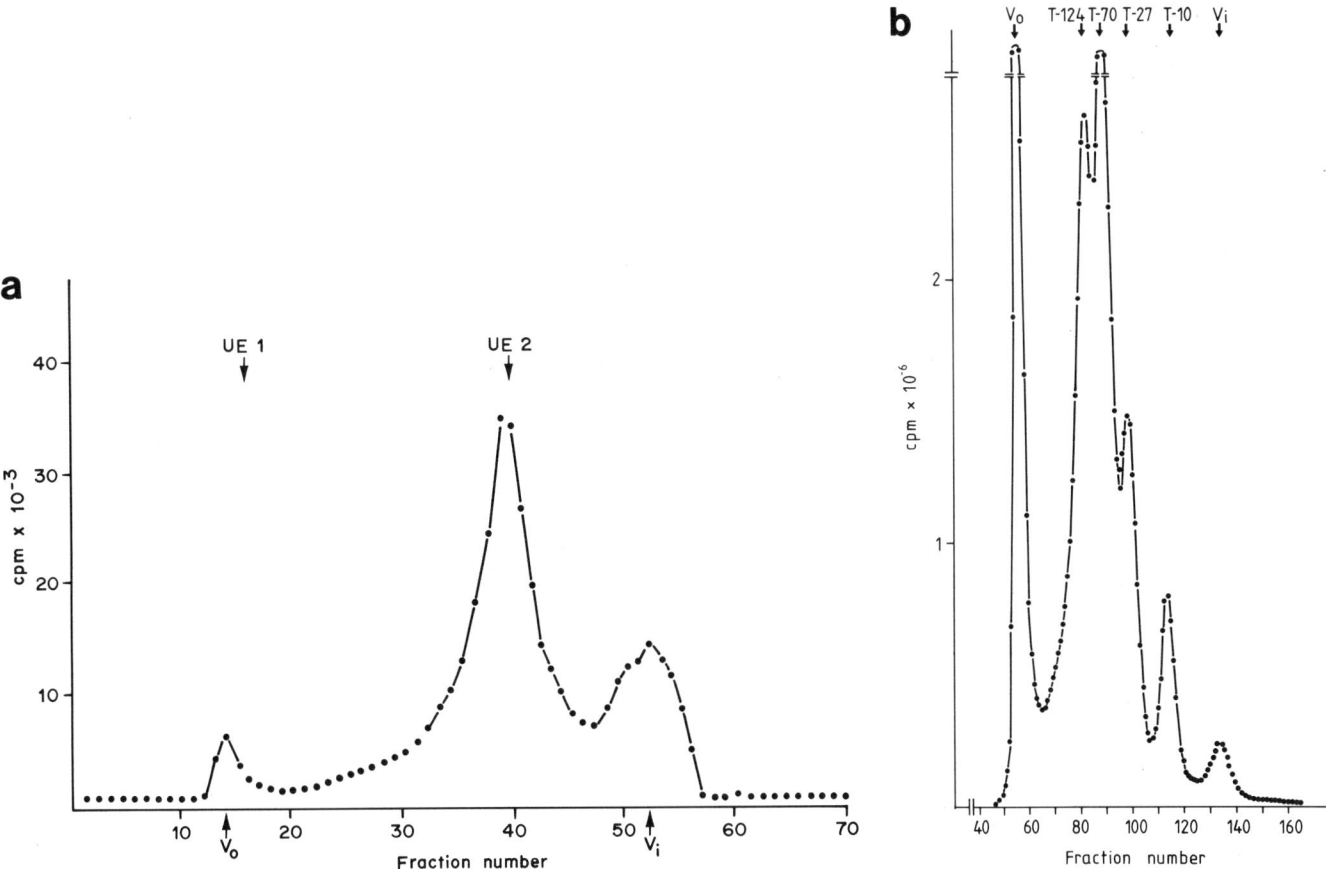

Figure 1. Gel filtration after different treatments of ^{125}I-MAF: a, 5 μg of ^{125}I-MAF dissociated with urea and EDTA (UE1, UE2) were applied to Sepharose 4B column (75 × 0.8 cm) and eluted with Ca^{2+} Mg^{2+}-free seawater; b, 0.5 mg of ^{125}I-MAF were treated with trypsin for a period of 8 h and MAF fragments separated by gel filtration on a Sephacryl S-300 column (90 × 1.6 cm). (1-ml fractions were collected and counted in a Packard gamma counter.)

homogeneous in size. Only minor charge micro-heterogeneity could be observed by ion exchange chromatography (Misevic and Burger, 1986).

The smallest T-10 cell binding fragment of MAF had 60% carbohydrate and 40% protein by weight and is therefore a glycopeptide. Composition of amino acids and sugar was found to be similar to that of native MAF (Tables 2, 3). However, the lower fucose content of T-10 is due to the presence of the large trypsin-resistant domains in MAF containing 60% fucose by weight. These large fucose-rich molecules are found to account for the rest of 30–40% of the total MAF mass. Carbohydrate analysis of the purified T-10 glycopeptide has revealed the unique feature that fucose and uronic acid together with mannose are linked to the peptide portion of the molecule. Preliminary isolation of the pure protein-free oligosaccharide

Table 2. Amino acid composition of MAF and of the T–10 glycopeptide fragment

	Amino acids (%)															
	Asx	Thr	Ser	Glu	Pro	Gly	Ala	Val	Met	Ile	Leu	Tyr	Phe	His	Lys	Arg
MAF	16	10	6	11	8	7	7	7	2	5	7	3	4	3	1	3
T–10	17	9	6	10	5	8	7	7	4	7	7	3	3	1	1	5

For analysis, dried samples taken up in 6 M HCl were sealed under vacuum in small test tubes. Hydrolysis was carried out at 110° C for 24 h. Analysis was performed on a Durrum D–500.

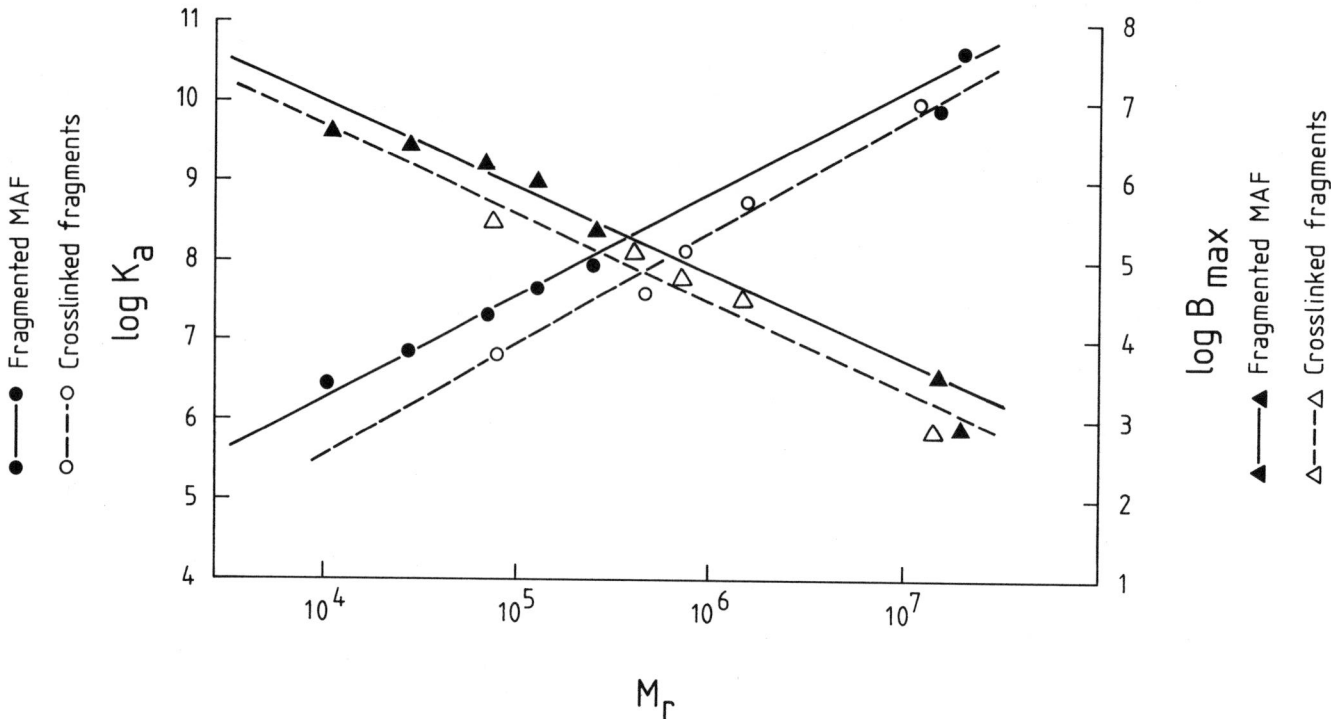

Figure 2. Comparison of binding data for MAF fragments and for reconstituted polymers of T-10 fragment showing dependence of K_a and B_{max} on molecular weights (M_r) of MAF fragments and T-10 reconstituted polymers.

chains indicate that these monosaccharides are components of the same glycan, with N-glycosidic linkage to asparagine.

RECONSTITUTION OF THE HIGH BINDING AFFINITY BY CROSS-LINKING OF THE GLYCOPEPTIDE FRAGMENT

In order to provide the final evidence for the polyvalency of the MAF cell-binding domain, reconstitution of the high affinity equal to that of the native MAF was approached by cross-linking the low affinity glycopeptide fragment. Two different methods were employed for polymerization of T-10: (1) the bifunctional reagent diepoxybutane was used in combination with the polyfunctional cross-linking reagent glutaraldehyde; and (2) mild periodate oxidation of T-10, which introduces aldehyde groups into the carbohydrate portion of the molecule, was followed by addition of glutaraldehyde. It is important to note that none of the conventional bifunctional cross-linking reagents specific for amino or amino and carboxyl groups were able to cross-link T-10 to a high degree, although they were very efficient with lysozyme, which was used as a control. This was probably due to the small amount of the basic amino acids present in T-10. Novel cross-linking procedures were therefore needed to obtain polymers of MAF size. Polymers of T-10 cross-linked by diepoxybutane and glutaraldehyde (EG polymers) as well as periodate oxidized T-10 cross-linked with glutaraldahyde (PG polymers) were separated by gel filtration on a Bio-Gel A-1.5 m (Figure 4a). The void volume fractions were then further analyzed on a Bio-Gel A-15 m, and polymers of MAF size could be isolated in the excluded volume fractions (Figure 4b).

Each of the polymers isolated showed species-specific binding to homotypic cells, indicating that cell binding activity was not destroyed by cross-linking (Misevic and Burger, 1982, 1986). The affinity of binding was measured for each of the EG polymers and for the PG polymers of $M_r > 1.5 \times 10^7$. The apparent association constants of the intermediate EG (1–5) polymers with $M_r = 1 \times 10^5$, 6×10^5, 9×10^5, 2×10^6, and above 1.5×10^7 increased proportionally to their size (Figure 2). The largest poly-

Table 3. Carbohydrate composition of MAF and of the T–10 glycopeptide fragment

	Carbohydrates (%)				
	Fuc	Man	Gal	GlcNAc	Uronic acid
MAF	26.0	7.7	23.3	18.5	25.5
T–10	14.1	10.9	25.4	23.1	26.5

For analysis the dry sample was hydrolyzed with 0.5 M HCl in methanol at 85° C for 16 h. The carbohydrate composition was determined on a Perkin Elmer 900 gas chromatograph.

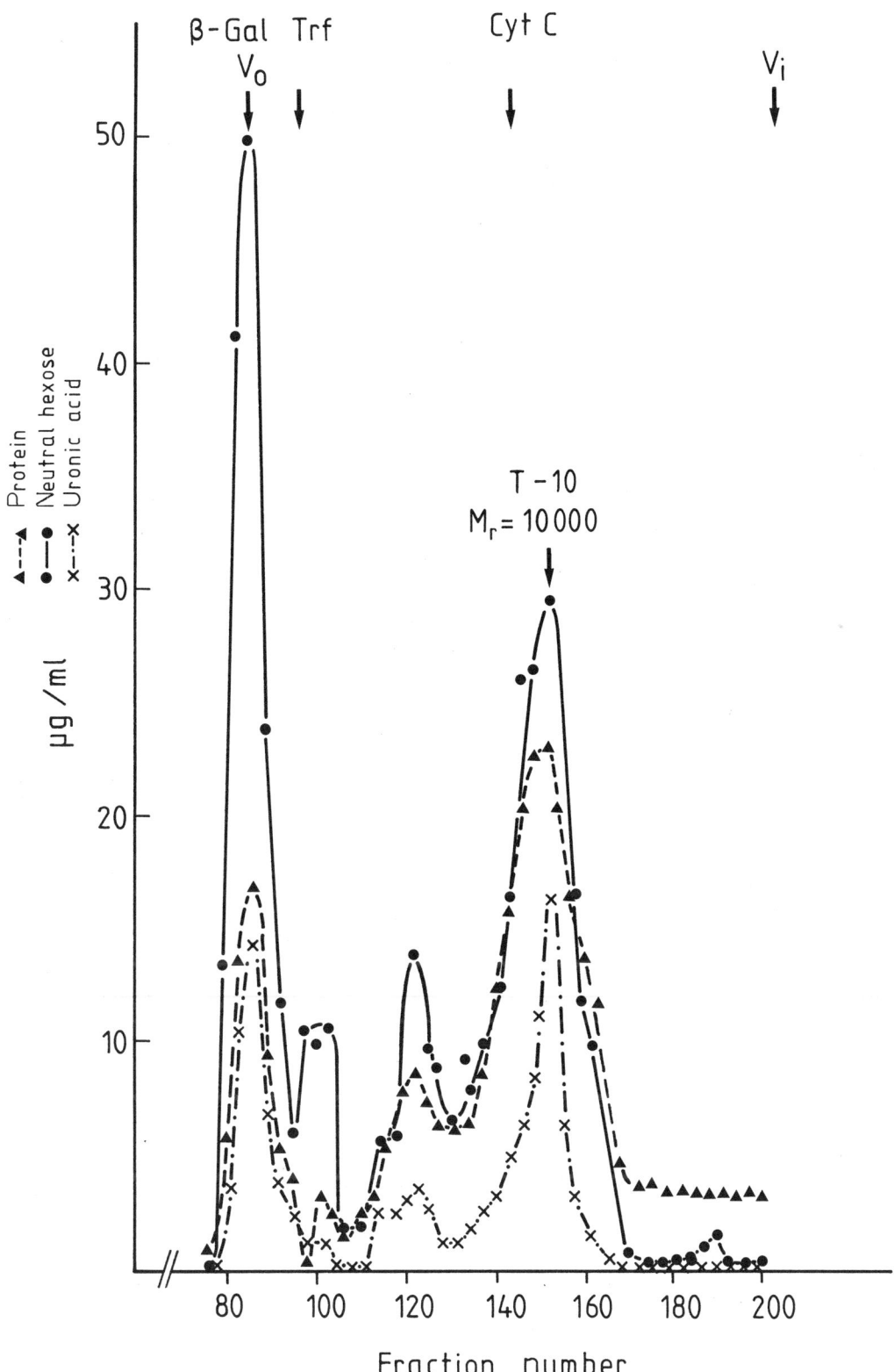

Figure 3. Gel filtration of trypsin-digested MAF on Sephacryl S-200; after first trypsin treatment of MAF, 2.5 mg of digested material were applied to an S-200 column (115 × 1.5 cm) and eluted with 0.25 M NH_4HCO_3; neutral hexoses, uronic acid, and protein content of every third fraction (1 ml each) was determined, and total amount of recovered material in M_r = 10,000 fragment was calculated in terms of percentage of total mass of digested MAF; apparent molecular weight was determined from migration of molecular weight standards, cytochrome c (Cyt C) M_r = 12,384, transferrin (Trf) M_r = 90,000 (both obtained from Sigma), and β-galactosidase (β-Gal) M_r = 959,000 (from Worthington).

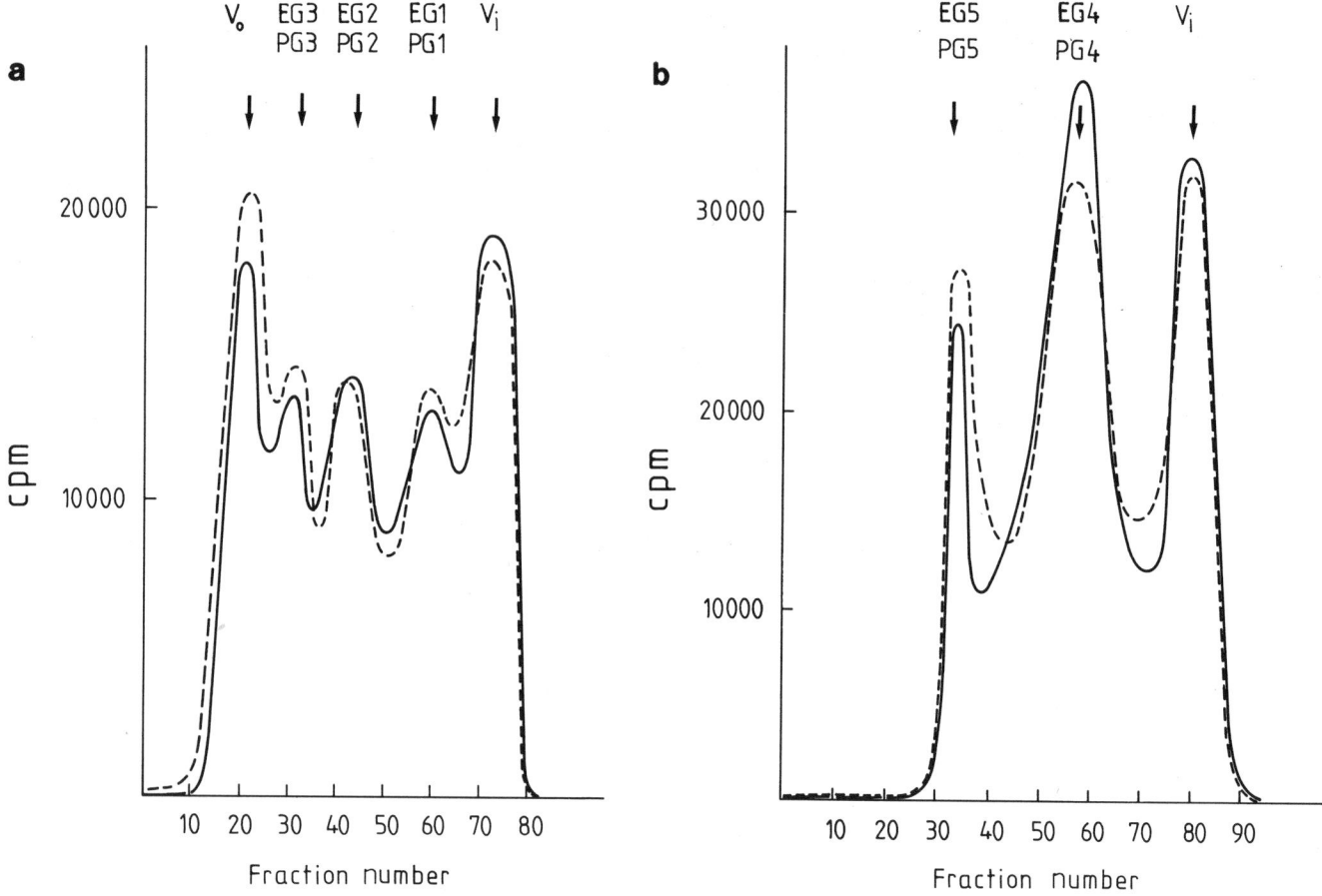

Figure 4. Results of Gel filtration of polymerized T-10 glycopeptide fragment: *a*, 200 μg (2.5 × 10⁴ cpm/μg) of diepoxybutane and glutaraldehyde T-10 polymers (EG polymers, *solid line*) and 210 μg (2.5 × 10⁴ cpm/μg) of periodate oxidized and glutaraldehyde cross-linked T-10 (PG polymers, *dashed line*) were applied to a Bio-Gel A-1.5 m column (1 × 46 cm) and radioactivity of 50 μl from each fraction was measured; *b*, 150 μg (2.5 × 10⁴ cpm/μg) of void volume fractions from Bio-Gel A–1.5 m column were separated using a Bio-Gel A–15 m column (1 × 46 cm) and radioactivity of 100 μl of each fraction was measured (same symbols as in *a*). (The same molecular weight standards were used as in Figure 3.)

mers EG5 and PG5 with molecular weight above 1.5 × 10⁷, which is in the range of the native MAF molecule, showed binding affinity in the same order of magnitude as MAF. Since 80% of the reconstituted polymer mass is composed of T-10, the polymer contains 1,200 repeats of this fragment, which is also almost equivalent to the number of potential binding sites in MAF. We have therefore succeeded in reconstituting the high affinity binding of MAF by cross-linking low affinity T-10 glycopeptide into the size of the native MAF.

Since the T-10 glycopeptide has several amino acids, the question arose whether only a carbohydrate portion is important for cell binding. Oligosaccharides were isolated from MAF containing only one asparagine per mole of the glycan. Cross-linking them into the size of native MAF showed that high binding affinity could also be reconstituted from pure oligosaccharides.

Proteolytic conversion of MAF into the T-10 glycopeptide as well as reconstitution of the original binding af-

finity by cross-linking this glycopeptide or pure oligosaccharides into MAF showed that high binding affinity and specificity of MAF is due to high polyvalency of the cell-binding site, which consists of carbohydrate. This novel type of interaction enabling cell recognition and adhesion via multiple low-affinity binding of carbohydrates may also be used in higher organisms by proteoglycan or proteoglycanlike molecules and especially in cases of cell-matrix interactions occurring during embryonal path recognition.

IDENTIFICATION OF THE SELF-INTERACTION DOMAIN OF MAF

Monoclonal antibodies were produced against the purified aggregation factor in order to find an antibody that would selectively block self-association of MAF. It was necessary to use this approach to identify a self-association domain because dissociation procedures irreversibly

3d Int. Sponge Conf. 1985

inactivate the MAF self-interaction domain. The system used to test self-association of MAF was based on the aggregation of MAF-coated agarose beads. This was the most suitable assay because apparent precipitation of MAF by antibodies or even their Fab fragments could be avoided. Aggregation of MAF beads occur only in the presence of 10 mM Ca^{2+}, as in the case of MAF in solution. After initial screening of 21 clones reacting with MAF, Fab fragments of the Block 1 clone were found to inhibit aggregation of MAF beads in the presence of 10 mM Ca^{2+} in a concentration-dependent manner (Figure 5). Among several negative clones that showed no inhibition even when a molar ratio to MAF of 5000:1 was used, C-16 clone was chosen as a control since it was found to have similar binding properties to MAF as Block 1. Specificity of the blocking effect of the Block 1 antibody toward self-association of MAF is suggested, since neither of these two clones showed inhibitory activity to cell binding of MAF.

The binding assay of the Block 1 and C-16 antibody and their Fab fragments to MAF attached to ELISA dishes showed that both of these antibodies have very similar affinity toward MAF (Table 4). Interestingly, the number of binding sites was over 1,000 for each of the Fab fragments per MAF molecule, thus indicating the extreme repetitivity of both epitopes. The block 1 antibody could completely inhibit self-association of MAF only when almost all of the 1,000 epitope sites were occupied. Conversely, the C-16 antibody showed no inhibitory activity, although it had the same affinity and even more binding sites. This result suggests that the blocking effect is proba-

Table 4. Binding characteristics of anti-MAF monoclonal antibodies to MAF

[125]I-Antibody		No. of binding sites (moles/mole)	Ka (M)	Binding ratio	
				Block 1	C-16
Block 1	IgG3	350	1.0×10^8	0.9	0.0[a]
C-16	IgG2b	370	8.4×10^7	0.0	1.0
Block 1	Fab	1,100	2.0×10^6	0.9	0.0
C-16	Fab	2,000	1.0×10^6	0.0	0.9

[a] 0.0 = no competition of binding even with 500-fold excess.
Number of binding sites for monoclonal antibodies and their Fab fragments as well as their association constants (K_a) calculated from binding data. In competition experiments subsaturating concentration of antibodies were used: binding of 1 μg ^{125}I-antibodies ($1.3-1.5 \times 10^4$ cpm/μg) or 10 μg of their ^{125}I-Fab fragments ($2-3 \times 10^4$ cpm/μg) to 40 ng MAF was measured in presence of an equal amount of homologous or heterologous antibodies or their Fab fragments. Values expressed as ratio of amount of iodized antibody or Fab fragment bound to MAF at subsaturating concentration to amount of nonlabeled antibody or Fab fragment causing 50% binding inhibition of iodinated ligands.

bly not due to steric hindrance, but rather to direct interaction of the Block 1 Fab fragments with the highly polyvalent self-association site of MAF.

The binding of both monoclonal antibodies to multiple subunits of MAF could have been due to the carbohydrate nature of the highly polyvalent antigenic site (Figure 6). In order to test this hypothesis, MAF was digested with pronase, and oligosaccharides were isolated by gel filtration and ion exchange chromatography (Finne and Krusius, 1982). After N-acetylation of glycans with ^{3}H acetic anhydride (Wheat, 1966), immunobinding assay was performed (Chard, 1980). The Block 1 antibody

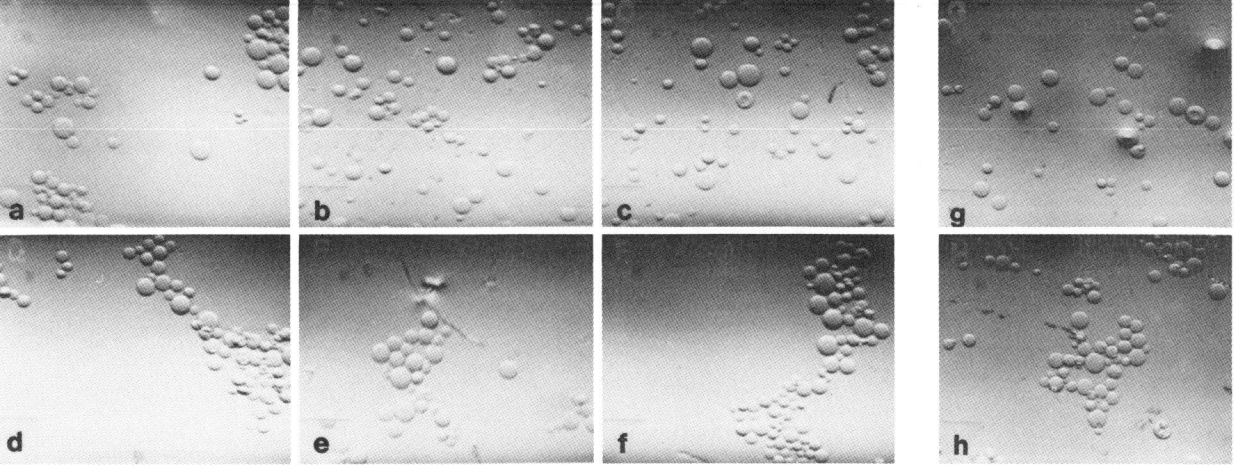

Figure 5. Inhibition of MAF-Sepharose bead aggregation by Fab fragments from monoclonal antibodies; suspensions of 20 μl of MAF-Sepharose beads (5×10^4 beads/ml, 0.36 ng MAF per bead) were incubated with Fab fragments from monoclonal antibodies in Ca^{2+} Mg^{2+}-free seawater for 2 h at room temperature; upon raising Ca^{2+} concentration to 10 mM (a-f,h), beads were gently mixed in a moist chamber for 10 min at room temperature and inhibition of aggregation was examined in a Zeiss Axiomat microscope: a, 1 μg of Fab from Block 1; b, 10 μg of Fab from Block 1; c, 20 μg of Fab from Block 1; d, 1 μg of Fab from C-16; e, 10 μg of Fab from C-16; f, 20 μg of Fab from C-16; g, control, no Fab added, Ca^{2+} 2 mM; h, control, no Fab added, Ca^{2+} 10 mM.

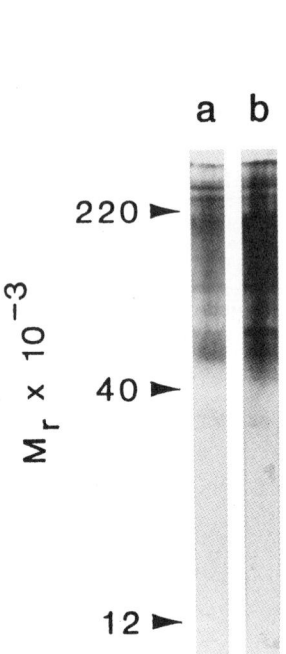

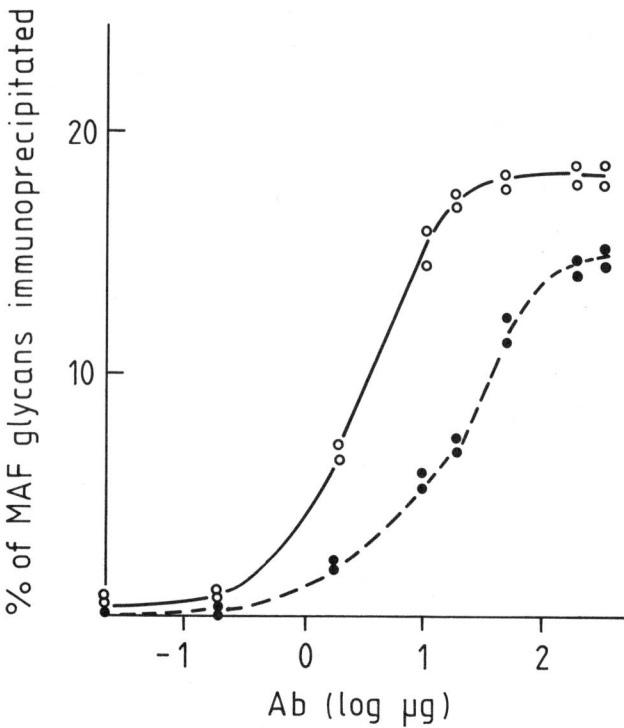

Figure 6. Polyacrylamide gel electrophoresis of MAF and with monoclonal antibodies; 10 μg of purified MAF were subjected to SDS gel electrophoresis on 7–20% gradient polyacrylamide gel which was blotted on nitrocellulose sheet; strips were cut and incubated for 30 min in 1% BSA in phosphate buffered saline; 5 μg of Block 1 antibody in 1% BSA in antibody in the same buffer saline were added in lane a, and 5 μg of C-16 antibody in the same buffer were added in lane b; after 2 h incubation both strips were washed and rabbit antimouse antibodies, coupled with peroxidase, were added; after 2 h, both strips were again washed and bound antibody was stained using chloronaphthol as a substrate for peroxidase.

Figure 7. Immunobinding of MAF glycopeptides: 200 ng of N-acetylated MAF glycopeptides (5.6×10^4 cpm/μg) were incubated in Ca^{2+} Mg^{2+}-free seawater containing 1% of BSA with different amounts of monoclonal antibodies for 12 h at 4°C and amount bound to antibodies was determined by polyethylene glycol precipitation method (*solid line*, Block 1; *dashed line*, C-16).

bound at saturation maximally 19% of the oligosaccharides and C-16 15% (Figure 7). Both antibodies, added as a mixture, precipitated 36% from the total material indicating that the epitopes are probably localized on the different oligosaccharide chains. Since there was also no binding competition of MAF between Block 1 and C-16 antibodies, they indeed recognized two different carbohydrate epitopes (Table 4). The carbohydrate nature of the antigenic sites for both antibodies was also confirmed by the ability of the oligosaccharides to specifically block binding of the iodinated antibodies to MAF attached to the ELISA dish. Inhibition of 50% binding of 0.5 μg ^{125}I-Block 1 or ^{125}I-C-16 was achieved by 0.1 μg of MAF or 1–1.5 μg of MAF oligosaccharides. The antigenic sites of MAF glycans are not of the simple saccharide or known glycosaminoglycan structure, since none of the monosaccharides (L-fucose, D-galactose, D-mannose, N-acetyl-D-glucosamine, N-acetyle-D-galactosamine, D-glucuronic acid, D-galacturonic acid) at 0.5 M concentration or glycosaminglycans (chondroitin sulfate A, chondroitin sulfate C, dermatan sulfate, heparin, heparan sulfate, and hyaluronic acid) at a concentration of 10 mg/ml were able to inihibit binding of either antibody to MAF.

RECONSTITUTION OF SELF-ASSOCIATION ACTIVITY BY CROSS-LINKING MAF OLIGOSACCHARIDES

The epitope of the Block 1 antibody was localized in the fraction of the total oligosaccharide that is highly repetitive in the MAF molecule. The self-association activity of

this MAF monomeric glycans was tested in a self-aggregation assay. However, no measurable self-association or binding of the ^{3}H-glycans to MAF could be detected in the presence or absence of 10 mM Ca^{2+} (Table 5). Also, glycans were unable to inhibit self-aggregation of ^{125}I-MAF. Since the negative result could be due to the very low, nonmeasurable affinity interactions of the monomeric oligosaccharides, reconstitution of the activity was tested by cross-linking them into the polymers of MAF size. In this case, the prepolymerized glutaraldehyde was used for cross-linking the MAF glycans. The polymers of $M_r > 1.5 \times 10^7$ were isolated in the void volume fractions on a Bio-Gel A-15 m column and tested for self-association activity. The glycans cross-linked to the size of the native MAF indeed showed self-aggregation as well as binding to MAF only in presence of 10 mM Ca^{2+}, indicating that polyvalent interactions of carbohydrates are essential for the reconstitution of the activity (Table 5).

Discussion and Conclusions

Immunobinding data on the Block 1 antibody to MAF glycans, together with the reconstitution of the self-

Table 5. Self association and MAF binding activity of cross–linked MAF glycopeptides (expressed as % coaggregation)

Substances tested	Without MAF		With MAF	
	2mM Ca^{2+}	10mM Ca^{2+}	2mM Ca^{2+}	10mM Ca^{2+}
^{125}I–MAF	5	60	6	67
^{3}H–glycopeptides	1	1	1	1
^{3}H–cross–linked glycopeptides	5	24	12	50

^{125}I–MAF (1 μg), ^{3}H–glycopeptides (50 μg), or ^{3}H–cross–linked glycopeptides (1 μg) were incubated in the presence of 2 or 10 mM Ca^{2+} with or without 10 μg of unlabeled MAF. Percentage of molecules coaggregated was determined after centrifugation at 10,000 × g.

association activity by cross-linking these monomeric oligosaccharides into the polymers of MAF size, showed that self-association of the *Microciona* aggregation factor is based on the novel principle of multiple low-affinity interactions (Figure 8). This differs, as in the case of the cell-binding domain, from the commonly found single higher-affinity interactions of adhesion molecules, like contact sites A in slime molds (Gerisch, 1976; Beug et al., 1970; Müller and Gerisch, 1978), neural adhesion molecule (Edelman, 1983; Grumet et al., 1983), liver adhesion molecule (Gallin et al., 1983), glial adhesion molecule (Grumet and Edelman, 1984), desmosomes (Garrod and Nicol, 1981; Franke et al., 1983), and surface lectins and transferases (Harrison and Chesterton, 1980; Roseman, 1970; Blackburn and Schnaar, 1983; Pless et al., 1983). Since the pure oligosaccharides containing at most one asparagine per glycan chain showed self-association activity in the reconstituted polymeric form, carbohydrate-carbohydrate interaction must be the basis for these highly polyvalent self-association interactions (Figure 8). Such carbohydrate-carbohydrate interactions have thus far not been found in self-association or binding of adhesion or recognition molecules to their receptors; rather, these are protein-protein or protein-carbohydrate nature.

The finding that multiple low-affinity carbohydrate interactions are important in cell recognition and adhesion of the marine sponge *Microciona prolifera* raises the question whether the same mechanism is operational in other multicellular organisms. Such interactions are most likely to occur during cell-substratum recognition. Since proteoglycans have been found in extracellular matrix, and since they have multiple repeats of glycosaminoglycan chains, we consider them good candidates for mediating specific interactions between cell and matrix via multiple low affinity binding sites. The advantage of polyvalent molecular interactions of proteoglycans lies first in the possibility that minor chemical modifications of the active sites may result in change their overall specificity, and second may result may result in the reversibility of adhesion enabling cell migration over a specific pathway. Although such a role of proteoglycans is still hypothetical, an approach similar to the one described here may be used to study the involvement of multiple low-affinity interactions of proteoglycan carbohydrates in cell recognition processes of higher Metazoans.

Acknowledgments

This work was supported by the Swiss National Foundation for Scientific Research, Grant No. 3.269.82, the Ministry of the City and Canton of Basel, and a European Molecular Biology Organization short term fellowship. We are grateful to Jukka Finne for critical discussion, Kjell Tullberg for help with English grammar, and Verena Schlup for technical assistance and helpful comments.

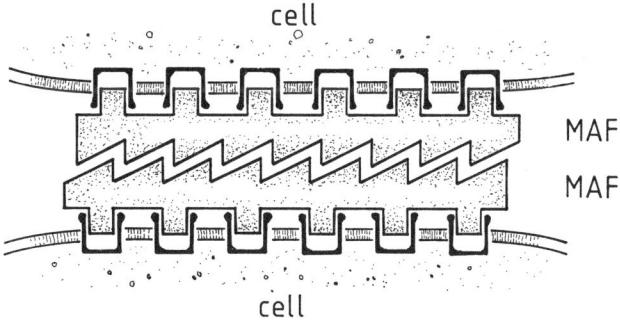

Figure 8. Model of species-specific aggregation in the marine sponge *Microciona prolifera*. Species-specific recognition and adhesion of dissociated cells are mediated by an aggregation factor via its cell-binding and self-association domain. Both of these functional properties of MAF are based on a multiple low-affinity interaction of MAF carbohydrates (*square junctions*, carbohydrates representing cell-binding site; *zig-zag junctions*, carbohydrates representing self-association site).

Literature Cited

Beug, H., Gerisch, G., Kempff, S., Riedel, V., and G. Cremer. 1970. Specific Inhibition of Cell Contact Formation in *Dictyostelium* by Univalent Antibodies. *Experimental Cell Research*, 63, 147–158.

Blackburn, C. C., and R. L. Schnaar. 1983. Carbohydrate-specific Cell Adhesion is Mediated by Immobolized Glycolipids. *The Journal of Biological Chemistry*, 258, 1180–1188.

Burger, M. M., and G. N. Misevic. 1986. Cell Encounter: Molecular and Biological Aspects. Pages 3–26 in *Cellular and Molecular Control of Direct Cell Interactions in Developing Systems* edited by H.-J. Marthy. Nato-ASI, New York and London: Plenum.

Chard, T. 1980. Ammonium Sulfate and Polyethylene Glycol as Reagents to Separate Antigen from Antigen-Antibody Complexes. *Methods in Enzymology*, 70:280–291.

Edelman, G. M. 1983. Cell Adhesion Molecules. *Science*, 219:450–457.

Finne, J., and T. Krusius. 1982. Preparation and Fractionation of Glycopeptides. *Methods in Enzymology*, 83:269–277.

Franke, W. W., H. Mueller, S. Mittnacht, H. P. Kapprell and J. L. Jorcsno. 1983. Significance of Two Desmosome Plaque-Associated Polypeptides of Molecular Weights 75,000 and 83,000. *The EMBO Journal*, 2:2211–2215.

Gallin, W. J., G. M. Edelman, and B. A. Cummingham. 1983. Characterization of L-CAM, a Magor Cell Adhesion Molecule from Embryonic Liver Cells. *Proceedings, National Academy of Science, U.S.A.* 80:1038–1042.

Galtsoff, P. S. 1929. Heteroagglutination of Dissociated Sponge Cells. *Biological Bulletin*, 57:250–260.

Garrod, D. R., and A. Nicol. 1981. Cell Behaviour and Molecular Mechanisme of Cell-Cell Adhesion. *Cambridge Philosophical Society, Biological Reviews*, 56:199–242.

Gerisch, G. 1976. Receptors Mediating Cell Aggregation in *Dictyostelium discoideum*. Pages 67–72 in *Surface Membrane Receptors* edited by R. A. Bradshaw, W. A. Frazier, R. C. Merrel, D. I. Gottlieb, and R. A. Hogue-Angeletti. New York: Plenum.

Grumet, M., and G. M. Edelman. 1984. Heterotypic Binding Between Neuronal Membrane Vesicles and Glial Cells is Mediated by a Specific Adhesion Molecule. *The Journal of Cell Biology*, 98:1746–1756.

Grumet, M., U. Rutishauser, and G. M. Edelman. 1983. Neuron–Glia Adhesion is Inhibited by Antibodies to Neural Determinants *Science*, 222:60–62.

Harrison, F. L., and C. J. Chesterton. 1980. Erythroid Developmental Agglutinin is a Protein Lectin Mediating Specific Cell Adhesion between Differentiating Rabbit Erythroblasts. *Nature*, 286:502–504.

Henkart, P., S. Humphreys, and T. Humphreys. 1973. Characterization of Sponge Aggregation Factor. A Unique Proteoglycan Complex. *Biochemistry*, 12:3045–3050.

Humphreys, S., T. Humphreys, and J. Sano. 1977. Organization and Polysaccharides of Sponge Aggregation Factor. *Journal of Supramolecular Structure*, 7:339–351.

Humphreys, T. 1963. Chemical Dissolution and in Vitro Reconstruction of Sponge Cell Adhesion. *Developmental Biology*, 8:27–47.

Jumblatt, J. E., V. Schlup, and M. M. Burger. 1980. Cell-Cell Recognition: Specific Binding of *Microciona* Sponge Aggregation Factor to Homotypic Cells and the Role of Calcium Ions. *Biochemistry*, 19:1038–1042.

Misevic, G. N., and M. M. Burger. 1982. The Molecular Basis of Species Specific Cell-Cell Recognition in Marine Sponges, and a Study on Organogenesis During Metamorphosis. Pages 193–209 in *Embryonic Development, Part B*, edited by M. M. Burger, and R. Weber. New York: Alan R. Liss.

———. 1986. Reconstitution of High Cell Binding Affinity of a Marine Sponge Aggregation Factor by Cross-linking of Small Low Affinity Fragments in to Large Polyvalent Polymer. *The Journal of Biological Chemistry*, 261:2853–2859.

Misevic, G. N., J. E. Jumblatt, and M. M. Burger. 1982. The Molecular Basis of Species Specific Cell-Cell Recognition in Marine Sponges, and a Study on Organogenesis During Metamorphis. *The Journal of Biological Chemistry*, 257:6931–6936.

Müller, K., and G. Gerisch. 1978. A Specific Glycoprotein as the Target Site of Adhesion Blocking Fab in Aggregating *Dictyostelium* Cells. *Nature*, 274:445–449.

Pless, D. D., Y. Chuan Lee, S. Roseman, and R. L. Schnaar. 1983. Specific Cell Adhesion to Immobilized Glycoproteins Demonstrated Using New Reagents for Protein and Glycoprotein Immobilization. *The Journal of Biological Chemistry*, 258:2340–2349.

Roseman, S. 1970. The Synthesis of Complex Carbohydrates by Multiglycosyltransferase Systems and Their Potential Function in Intercellular Adhesion. *Chemistry and Physics of Lipids*, 5:270–297.

Weinbaum, G., and M. M. Burger. 1973. Two component System for Surface Guided Reassociation of Animal Cells. *Nature*, 244:510–512.

Wheat, R. W. 1966. Analysis of Hexosamines in Bacterial Polysaccharides by Chromatographic Procedures. *Methods in Enzymology*, 8:60–78.

Wilson, H. V. 1907. On Some Phenomena of Coalescence and Regeneration in Sponges. *Journal of Experimental Zoology*, 5:245–258.

IAN S. JOHNSTON
Department of Biology
Northwestern College
Orange City, Iowa, 51041

Behavior of Reaggregating Cell Suspensions and the Phenomenon of Allograft Rejection in *Callyspongia diffusa* and *Toxadocia violacea*

Abstract

A fluorescent vital stain was used to monitor the progress of reaggregation in both xenogeneic (mixed interspecific) and allogeneic (mixed intraspecific) mixtures of dissociated cells from the tissues of the tropical marine demosponges, *Callyspongia diffusa* and *Toxadocia violacea*. Xenogeneic mixtures gave rise to species-specific aggregates whereas allogeneic mixtures gave rise to mixed, or chimeric, initial aggregates. Under culture conditions in which isogeneic, or homogeneic, aggregates further developed into new functional sponge individuals, the principal allogeneic aggregates all began to die between 24 and 48 hours after reaggregation was initiated. Allogeneic aggregation therefore leads rapidly to mutually directed cytotoxic reactions that are presumed to correlate with the immunological phenomenon of allograft rejection in whole sponges.

The tropical marine demosponge *Callyspongia diffusa* (Ridley) is capable of responding in different ways to tissue contact with isogeneic, allogeneic, or xenogeneic sponge tissue (Hildemann et al., 1979; Bigger et al., 1983). Isogeneic contact (i.e., between two pieces of tissue taken from the same sponge colony) results in compatible sustained tissue fusion; allogeneic contact (i.e., between tissue taken from two genetically different colonies of the same sponge species) results in mutually directed cytotoxic graft rejection (Johnston and Hildemann, 1983). Xenogeneic contact (i.e., between tissue taken from different species) results in different kinds of tissue reactions, depending on the species combination, but all are characterized by the absence of sustained tissue fusion (see Bigger et al., 1983). These phenomena are presumed to demonstrate a degree of immunological competence that has features in common with transplantation immunity in higher organisms (Hildemann et al., 1980).

There has been some speculation that graft rejection in sponges, particularly allograft rejection, may well be correlated with events of recognition and adhesion previously studied in reaggregating sponge cells (Evans and Curtis, 1979). This suggestion stems from H. V. Wilson's (1907) classical studies on coalescence and regeneration. Wilson showed that whole sponges can reform from suspensions of their cells, and if cells from two species are intermingled, the regenerated individuals each consist of cells from only one species type. Much is now known about the chemical nature and specific ion requirements of coupling molecules, called aggregation factors, that function in the reaggregation of dissociated sponge cells (see review by Müller, 1982). Might these same cell surface molecules function in the recognition phase of allograft rejection?

One purpose of the present study was to investigate whether Wilson's species specificity of reaggregation could be extended to allospecific reaggregation from allogeneic mixtures of dissociated sponge cells. Second, the fate of allogeneic mixtures of dissociated cells was followed to determine when, or if, a correlate of the cytotoxic phase of allograft rejection would become apparent.

Materials and Methods

Colonies of *Callyspongia diffusa* (Ridley) and *Toxadocia violacea* de Laubenfels were collected from patch reefs in Kaneohe Bay, Oahu, Hawaii, and held in a running-seawater system at the Hawaii Institute of Marine Biology. The allogeneic nature of a pair of C. diffusa colonies was established by grafting together a sample branch from each and requiring that allograft rejection occur within 8 days (see Jokiel et al., 1982).

Sponge tissue was mechanically dissociated by squeezing it through a pouch made up of four layers of 60-μm-mesh plankton netting; 0.5 cm³ of sponge was squeezed into 10 ml of filtered seawater (0.4-μm-pore membrane filter). The suspensions were then immediately pipetted into multiwell plates (16 mm wells, Costar type 3524; 0.5 ml cell suspension/well) or into 35 mm tissue culture Petri plates (Falcon type 3001; 2 ml/plate). The suspensions were incubated at 24°C (ambient seawater temperature during this time ranged between 25° and 26°C), and once the sponge cells had sedimented, the seawater medium was gently replaced every 4 to 6 hours. In certain cases *Callyspongia diffusa* cells were vitally stained with fluorescein isothiocyanate (FITC) prior to mechanical dissociation; 0.5 cm³ of sponge tissue was incubated for 30 minutes in 50 ml of 60 mg% FITC in filtered seawater,

followed by a 60-minute rinse in running seawater to wash out any unincorporated FITC.

Cell suspensions were transferred into their incubation vessels so that the final volume and density of cells was always equal. Replicated combinations of the following cell suspensions and mixed-cell suspensions were set up and monitored for up to 14 days: isogeneic cells (cells derived from a single sponge individual to serve as a control for reaggregation), allogeneic cell mixtures (intraspecific mixtures), and xenogeneic cell mixtures (interspecific mixtures). Replicates of each of these three combinations were also set up using FITC-labeled cells from one particular *Callyspongia diffusa* individual. The presence or absence of FITC-labeled cells in individual cellular aggregates was monitored by UV-fluorescence microscopy.

Results

The mechanical dissociation technique used in this experiment produced a suspension of predominately single cells with occasional tissue fragments which in the case of *Callyspongia diffusa* could be as large as whole choanocyte chambers (40 μm diameter). When viability of cells in one such suspension was tested with propidium iodide staining, 16 out of 650 cells examined proved to be damaged or dying; this amount of cell damage was considered experimentally acceptable.

Isogeneic control suspensions from both sponge species rapidly reaggregated during the first 24 hours into increasingly larger spherical cell masses (up to 0.5 mm in diameter). By 48 hours most of the larger aggregates had begun to flatten down onto the plastic substratum and metamorphose into functional sponge individuals complete with aquiferous canals and choanocyte chambers. These sponges showed further coalescence over the next 12 days as their dynamic growth processes brought them into contact with each other. These sponges flourished during the 14-day culture period, although some of the smallest aggregates never did metamorphose.

For the first 4 to 6 hours allogeneic mixtures of FITC-labeled and unlabeled *Callyspongia diffusa* cells aggregated at the same rate as the isogeneic controls. These initial aggregates all contained a mixture of labeled and unlabeled cells (Figure 1). Xenogeneic mixtures of FITC-labelled C. diffusa cells and unlabeled *Toxadocia violacea* cells formed smaller initial aggregates during the first 6 to 12 hours, but each aggregate contained only labeled cells or alternatively only unlabeled cells; that is, each aggregate was made up of cells from only one sponge species (Figure 2). Vital staining of sponge cells with FITC was not interferred with by autofluorescence of unlabeled cells and could be successfully followed for about 48 hours, although during that time the fluorescent intensity of individual cells and whole aggregates decayed appreciably.

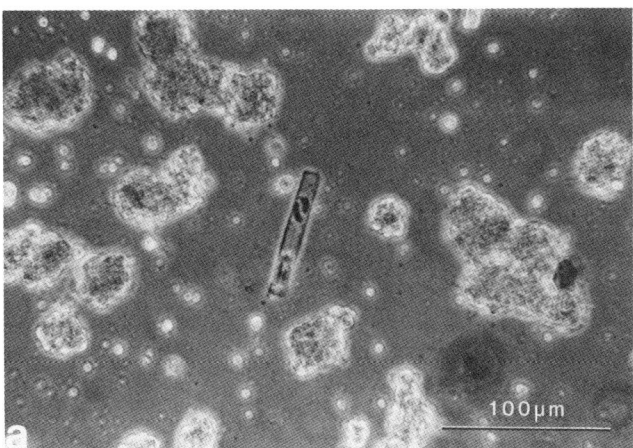 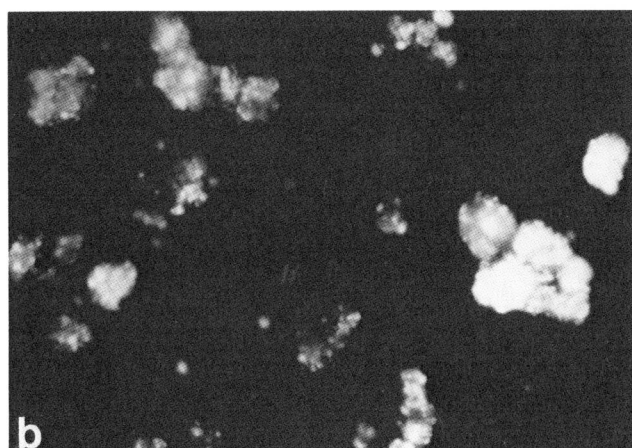

Figure 1. Four-hour aggregates formed from a mixture of FITC-labeled cells from one *Callyspongia diffusa* individual and unlabeled cells from an allogeneic individual; phase-contrast *(a)* and UV-fluorescence *(b)* images of the same field of view. Note that all aggregates are partly labeled, although some are more intensely fluorescent than others; that is, both cell types were not always equally represented in a single aggregate.

After 4 to 6 hours, allogeneic *Callyspongia diffusa* aggregates failed to grow in size at the same rate as isogeneic controls, seldom exceeding 0.2 mm by 24 hours. After 24 hours the aggregates began to take on a fuzzy appearance, and by 48 hours propidium iodide staining of squash mounts of these fuzzy aggregates indicated that more than half of the cells were either damaged or dead. The aggregates showed further decay over the next 6 to 8 days until they were reduced to masses of debris colonized by bacteria, protozoa, and in some cases filamentous algae. At 8 to 14 days very small (40 μm) living aggregates were noted in peripheral areas of the Petri plates away from the debris of the initial large aggregates.

After 24 hours the aggregates resulting from the xenogeneic cell mixtures progressed in the same way as

the isogeneic controls, except that they were consistently smaller than the controls and a smaller proportion of them metamorphosed into functional sponges. The coexisting *Callyspongia diffusa* and *Toxadocia violacea* individuals could be distinguished by their different pigments (*C. diffusa* a deep blue-purple, *T. violacea* a dark reddish violet). Some of these coexisting sponges survived the 14-day culture period, but many of these particular aggregates and metamorphosed sponges died after 6 to 14 days.

Discussion

When dissociated cells from a *Callyspongia diffusa* individual are mixed with those from a *Toxadocia violacea* individual, the resultant aggregates are species specific. From the

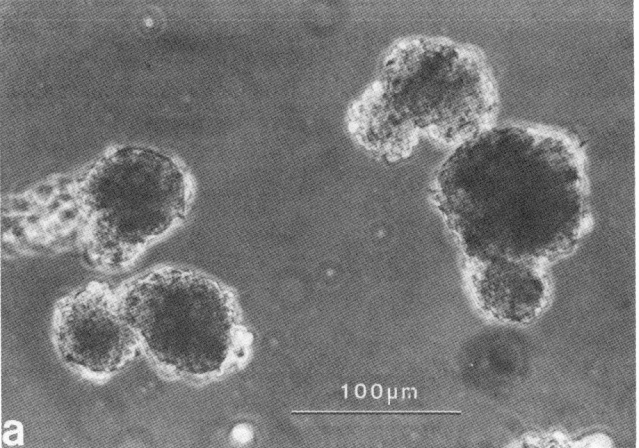 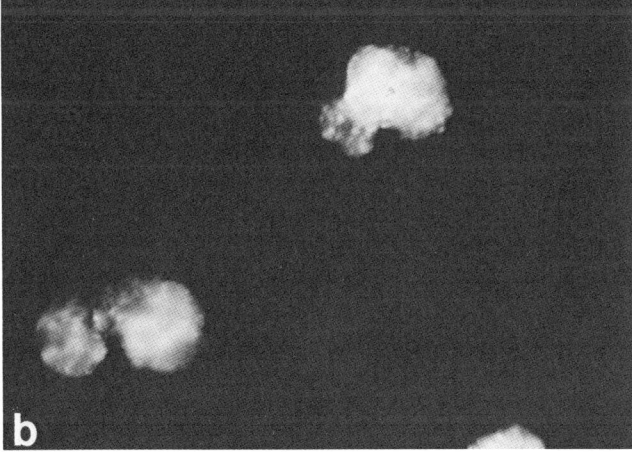

Figure 2. Four-hour aggregates formed from a mixture of FITC-labeled cells from one *Callyspongia diffusa* individual and unlabeled cells from a *Toxadocia violacea* individual; phase-contrast *(a)* and UV-fluorescence *(b)* images of the same field of view. Note aggregates in which none of the cells are FITC-labeled. Fluorescent aggregates are presumed to consist of monospecific, isogeneic, *C. diffusa* cells, and nonfluorescing aggregates are made up of *T. violacea* cells.

3d Int. Sponge Conf. 1985

initial stage of adhesion of one cell to another in the nucleation of an aggregate, there is discrimination between "self" and "not-self" (not-self in this case is represented by a cell from another sponge species). This observation of the fate of interspecific mixtures merely repeats that of Wilson (1907). However, in the context of this present study it also provided an important control indicating the successful use of FITC as a vital stain for cells originating from one individual. Over a period of 48 hours unlabeled *T. violacea* cells did not pick up any FITC from labeled *C. diffusa* cells, even though they were incubated together; it is presumed that the same was true for mixtures of labeled and unlabeled *C. diffusa* cells. Previous attempts to label sponge cells involved in reaggregation have used radioactive compounds, namely, H^3-leucine (McClay, 1971) or H^3-thymidine (De Sutter and Van de Vyver, 1979), but in both cases there was considerable transfer of the label to initially unlabeled cells.

Intraspecific (allogeneic) mixtures of *Callyspongia diffusa* cells give rise to chimeric aggregates containing cells from more than one original sponge individual. Therefore, in the initial reaggregation of dissociated allogeneic cells species-specific cellular adhesion apparently overrides the potential recognition of adjacent cells as "allo-not-self." Since this study was carried out, Zea and Humphreys (1985) have developed a similar vital staining technique (using the fluorescent dye Hoechst 33324) and have shown that allogeneic mixtures of *Microciona prolifera* cells also do not discriminate against each other during initial reaggregation.

Previous observations of whole grafts of *Callyspongia diffusa* showed some initial differences in tissue surface morphology at the interface of isografts versus allografts (Johnston and Hildemann, 1983). This was sometimes apparent as early as 6 hours after grafting and certainly by 24 hours; that is, allo-not-self recognition leads to distinct morphological correlates in the first few hours after grafting. It is not clear whether allo-not-self recognition is instantaneous when allogeneic cells come into contact, or whether this recognition develops at some time in the next few hours, subsequently leading to specific responses on the part of individual cells or the whole sponge. The fact that the rate of coalescence of allogeneic aggregates in the present study slowed down after 6 hours (relative to isogeneic controls) may indicate that allorecognition already occurred during the first 6 hours. Since reaggregation, presumably involving aggregation factors, does not show allospecificity, the same aggregation factors are unlikely to be involved in allorecognition.

In *Callyspongia diffusa*, whole sponges grafted together in allogeneic combinations show a series of tissue responses during the first 3 days (e.g., a "bridging" response) followed later by a full-blown cytotoxic rejection response in which tissues die back from the graft interface (Johnston and Hildemann, 1983). The death of chimeric alloaggre-

gates (at 24 to 48 hours) in the present study may well correlate with cytotoxic rejection phenomena in whole sponge allografts. Cell death is certainly apparent much earlier in the case of alloaggregates (versus allografts), but the intimate intermixing of cells in an alloaggregate may well intensify or acccelerate the cytotoxic rejection process.

A wide variety of other sponge species have been subjected to reaggregation assays, particularly in interspecific combinations, and both species-specific and species-nonspecific reaggregation have been noted (e.g., McClennan, 1970). However, a failure to consider that individuals of the same species might be either allogeneic or clonally identical (isogeneic) to one another confuses these results. Furthermore, the isolation or characterization of "strains" of a particular sponge (e.g., in *Ephydatia fluviatilis* by Van de Vyver, 1970) raises the question of whether interstrain mixtures might represent truly intraspecific or even interspecific combinations. Interstrain mixtures of *E. fluviatilis* cells give rise initially to mixed aggregates, but after several days the cells in the aggregates segregate back out until each aggregate contains cells from only one strain (Van de Vyver, 1975). I would interpret this as being similar to the situation involving allogeneic *Callyspongia diffusa* cells, except that after the formation of allogeneically mixed aggregates the subsequent allorejection phenomenon in *E. fluviatilis* consists of detachment of allogeneic cell partners from each other rather than the generation of mutually directed cytotoxic reactions. Overt cytotoxic allograft rejection reactions are apparently not found in all sponge species, but they are certainly not restricted to *C. diffusa* and *Toxadocia violacea* (e.g., *Axinella polypoides*, Buscema and Van de Vyver, 1984).

Conclusions

Cells derived from mechanically dissociated *Callyspongia diffusa* tissues do not discriminate among allogeneic partners when they first adhere to each other during reaggregation; they do, however, discriminate xenogeneic partners and fail to adhere to such cells. Aggregates made up of allogeneically mixed *C. diffusa* cells fail to metamorphose into functional sponge individuals (in contrast to isogeneic aggregates) and instead are killed by cytotoxic histoincompatibility reactions occurring within each aggregate.

Acknowledgments

This study was supported by NIH grant AI 15705 to William Hildemann (deceased) and by faculty travel and research grants from Northwestern College. Field work was aided by Paul Jokiel and manuscript preparation by Charles Bigger.

Literature Cited

Bigger, C. H., P. L. Jokiel, and W. H. Hildemann. 1983. Cytotoxic Transplantation Immunity in the Sponge *Toxadocia violacea. Transplantation*, 35:239–243.

Buscema, M., and G. Van de Vyver. 1984. Cellular Aspects of Alloimmune Reactions in Sponges of the Genus *Axinella* I. *Axinella polypoides. Journal of Experimental Zoology*, 229:7–17.

De Sutter, D., and G. Van de Vyver. 1979. Cell Recognition Properties of Isolated Sponge Cell Fractions. Pages 217–224 in *Biologie des Spongiaires*, edited by C. Lévi and N. Boury-Esnault. Colloques Internationauxdu C.N.R.S., 291. Paris: Centre National de la Recherche Scientifique.

Evans, C. W., and A. S. G. Curtis. 1979. Graft Rejection in Sponges: Its Relation to Cell Aggregation Studies. Pages 211–215 in *Biologie des Spongiaires*, edited by C. Lévi and N. Boury-Esnault. Colloques Internationaux du C.N.R.S., 291. Paris: Centre National de la Recherche Scientifique.

Hildemann, W. H., C. H. Bigger, I. S. Johnston, and P. L. Jokiel. 1980. Characteristics of Transplantation Immunity in the Sponge *Callyspongia diffusa. Transplantation*, 30:362–367.

Hildemann, W. H., I. S. Johnston, and P. L. Jokiel. 1979. Immunocompetence in the Lowest Metazoan Phylum: Transplantation Immunity in Sponges. *Science*, 204:420–422.

Johnston, I. S., and W. H. Hildemann. 1983. Morphological Correlates of Intraspecific Grafting Reactions in the Marine Demosponge *Callyspongia diffusa. Marine Biology*, 74:25–33.

Jokiel, P. L., W. H. Hildemann, and C. H. Bigger. 1982. Frequency of Intercolony Graft Acceptance or Rejection as a Measure of Population Structure in the Sponge *Callyspongia diffusa. Marine Biology*, 71:135–139.

MacLennan, A. P. 1970. Polysaccharides from Sponges and their Possible Significance in Cellular Aggregation. Pages 299–324 in *The Biology of the Porifera*, edited by W. G. Fry. Symposium, 25. London: Zoological Society of London, 25:205–243.

McClay, D. R. 1971. An Autoradiographic Analysis of the Species Specificity during Sponge Cell Reaggregation. *Biological Bulletin*, 141:119–130.

Müller, W. E. G. 1982. Cell Membranes in Sponges. *International Review of Cytology*, 77:129–181.

Van de Vyver, G. 1970. La Non Confluence Intraspecifique Chez les Spongiaires et la Notion d'Individu. *Annales d'Embryologie et de Morphogenese* 3:251–262.

———. 1975. Phenomena of Cellular Recognition in Sponges. *Current Topics in Developmental Biology*, 10:123–140.

Wilson, H. V. 1907. On Some Phenomena of Coalescence and Regeneration in Sponges. *Journal of Experimental Zoology*, 5:245–258.

Zea, S., and T. Humphreys. 1985. Self-Recognition in the Sponge *Microciona prolifera* (Ellis and Solander) Examined by Histocompatibility and Cell Reaggregation Experiments. *Biological Bulletin*, 169:538.

GYSELE VAN DE VYVER
MARCO BUSCEMA
Laboratoire de Biologie Animale et Cellulaire
Université Libre de Bruxelles
50 av. F. D. Roosevelt
1050 Bruxelles, Belgium

Diversity of Immune Reactions in the Sponge *Axinella polypoides*

Abstract

Autografts of *Axinella polypoides* are compatible, but allografts and xenografts are always incompatible. The present work investigates the cellular aspects of the different rejection processes against non-self. Allogeneic incompatibility always results in antagonistic rejection. This process is characterized by extensive archeocyte invasion, phagocytosis, and individual cell lysis. Xenogeneic histocompatibility reactions range from weak to strong, depending on the species combination. Grafted with sponges of the same genus *(A. damicornis* or *A. verrucosa)*, *A. polypoides* displays chronic rejection, which consists of a progressive anatomical isolation of the parabionts by secretion of a collagen barrier between the partners. Acute rejection, involving deep tissue necrosis in the zone of contact, is observed when *A. polypoides* is grafted with *Crambe crambe* or with *Hemimycale columella*. When put into contact with inert materials used as controls, *A. polypoides* reacts differently according to the nature of the substrate. These results and earlier results with other species lead to the conclusion that sponges, and particularly *A. polypoides,* not only discriminate "self" from "non-self", but also discriminate between different "non-self". Such discriminative capabilities were unsuspected or doubted in sponges until now. They are quite remarkable considering that these organisms are the lowest metazoans.

The mechanisms that allow cells to recognize "self" from "non-self" are found throughout the animal kingdom, from Protozoa and lower invertebrates to vertebrates. These mechanisms, which have evolved into sophisticated internal defense systems in higher vertebrates, integrate components that at the outset served functions not exclusively related to defense.

Phagocytosis, for instance, is used by animals of all groups including Protozoa for a variety of purposes, such as nutrition, waste disposal, or defense. This situation is particularly obvious in sponges, which, as filter organisms, phagocytose bacteria and organic particles to feed themselves (Rasmont, 1961; Reiswig, 1975; Mank and Kilian, 1979; Vacelet, 1979). Moreover, in sponges phagocytosis occurs in most rejection processes, as shown in allorecognition and xenorecognition experiments (De Sutter and Van de Vyver, 1979; Buscema and Van de Vyver, 1984a,b; 1985).

Comparative immunology stressed the diversity of defense mechanisms and immune reactions throughout the animal kingdom, but until recently diversity of reactions within a same zoological group was thought to be restricted to vertebrates. However, by analyzing graft rejection mechanisms between marine sponges, down to the ultrastructural level, we pointed out the variety of immune processes that can be developed by sponges, depending on the species, the strains, or the individuals brought into contact.

These results prompted us to undertake an exhaustive study of the responses to foreignness of one sponge species used as a reference species: *Axinella polypoides*.

Material and Methods

SPONGES. Five species were used. Three of them belong to the same genus: *Axinella polypoides*, *A. damicornis* and *A. verrucosa*. The two others, *Crambe crambe* and *Hemimycale columella*, are taxonomically remote. They were found within scuba diving range of the laboratory where all the experiments were conducted (Station de recherches sous-marines et océanographiques, Calvi, Corsica).

Axinella polypoides (Axinellida, Axinellidae) is an orange ramose sponge, 30 to 100 cm high, with numerous interconnected digitations. It lives on rocky substrates below 30 m deep. *A. verrucosa* is also an orange ramose sponge, 5 to 10 cm high. It lives on rocky substrates below 25 m deep. *A. damicornis* looks quite similar to *A. verrucosa* but its branches are less elongated and more palmate. It is very common on rocky substrates below 5 m deep. *Crambe crambe* (Poecilosclerida, Crambidae) is a red, flat encrusting species 0.2 to 1 cm thick that can grow to a diameter of 50 cm. It is very common from 1 to 20 m on overhangs and in clefts. *Hemimycale columella* (Halichondriida, Hymeniacidonidae) is a dark pink encrusting sponge. Its size varies from 3 to 20 cm in diameter, with an average thickness of 0.5 cm. It can be found below 5 m on overhangs and in clefts.

GRAFTING TECHNIQUE. Grafts were made by parabiosis according to the procedure used by Buscema and Van de Vyver (1984a,b). They were examined daily. Since encrusting sponges cannot be collected and kept healthy in an aquarium, the experiments using *Crambe crambe* and *Hemimycale columella* were carried out in situ at depths of 3 to 9 m.

TREATMENT FOR MICROSCOPY. Specimens were prepared for ultrastructural studies after 1, 3, 6, 10, and 15 days, and in a few cases after 2 months. The zone of contact was

dissected and fixed for 2 h using 3% glutaraldehyde and postfixed for 1 h in 1% osmium tetroxide (Buscema and Van de Vyver, 1984a,b). For the experiments with encrusting sponges, transfer in fixative was made in situ because the tissues are extremely fragile

Results

COMPLETE FUSION. Complete fusion occurs in all the autografts. It involves an important reorganization of the superficial tissues during the days following grafting. The pinacoderms of the two pieces brought into contact disappear within a few hours and fusion of the ectosomes is effective after one day (Figure 1). After 10 days, choanocyte chambers begin to form in the median region. The complete continuity of all parabiont tissues is achieved after 2 weeks.

ANTAGONISTIC REJECTION. Extensive archeocyte invasion, cell lysis, and phagocytosis are typical of this process, which characterizes the *Axinella polypoides* allograft rejection. Paradoxically, the rejection process begins with the strong adherence of the parabionts. Macroscopically, a dense and sharp median band demarcates the two sponges. Microscopically, this band corresponds to a dense accumulation of cells, mainly archeocytes, collencytes, and spherulous cells that have migrated from deep tissues into the zone of contact (Figure 2).

Ultrastructural observations clearly show that cell migration is followed by phagocytosis. It is achieved by the accumulated archeocytes and directed against either intact healthy cells of all types or previously lysed cells. Cell accumulation is maximal after three days of contact. Later, the intensity of phagocytosis is so important that it leads to the progressive decrease of the cell density. Consequently, the adhesion between the parabionts is weakened and they separate from each other. The reaction is limited to the former zone of contact, leaving a scar of denuded skeleton network.

NONFUSION. Nonfusion, or almost complete indifference, is observed when *Axinella polypoides* is grafted against *A. verrucosa*. Indeed, at any time after grafting, the parabionts fall apart if the thread that binds them together is removed. Microscope observations show that the rejection process is restricted to the disappearance of the pinacoderms in the zone of contact. In this species combination, the reaction is reciprocal since *A. verrucosa* reacts in the same way as *A. polypoides*.

CHRONIC REJECTION. Chronic rejection consists of a progressive anatomical violation of the parabionts by secretion of a collagen barrier between them. *Axinella polypoides* exhibits an almost complete indifference when grafted against *A. damicornis*, since its reaction is restricted to the

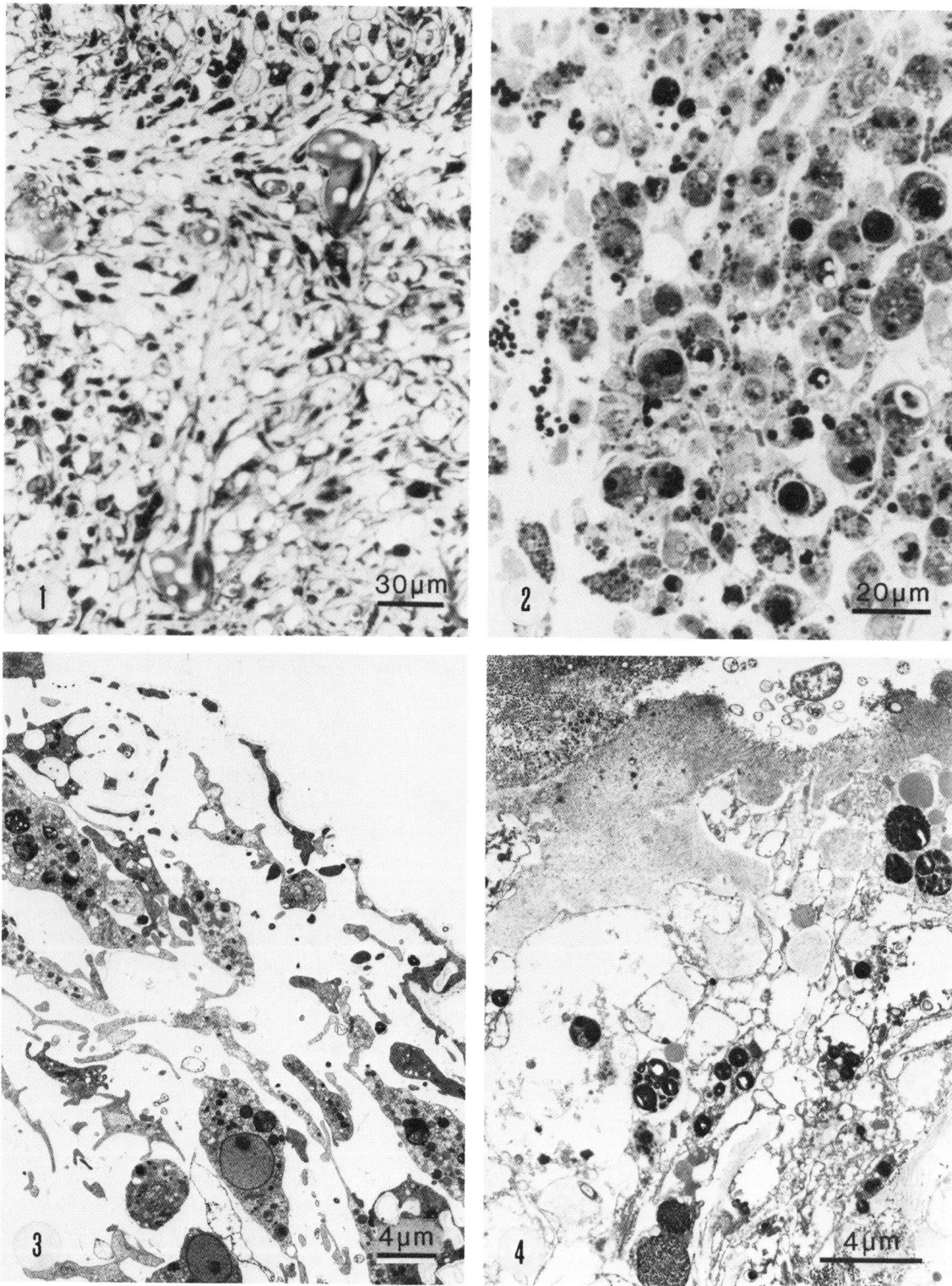

Figures 1–4. Histological results of grafting in *Axinella polypoides*: *1*, one-day autograft; complete continuity of ectosomes is already achieved; zone of fusion is approximately horizontal in middle of picture; *2*, one-day allograft; dense cells, mainly archeocytes, contain numerous large secondary lysosomes and are very heterogeneous; *3*, six-day xenograft from *A. damicornis*; pinacoderm of *A. polypoides* has become rough and discontinuous; underlying cells are much more irregular and numerous than in a normal ectosome; *4*, six-day xenograft from *A. damicornis* pinacocytes are no longer distinct at surface of *A. damicornis*, which is covered by a dense and heterogenous deposit of collagen fibers.

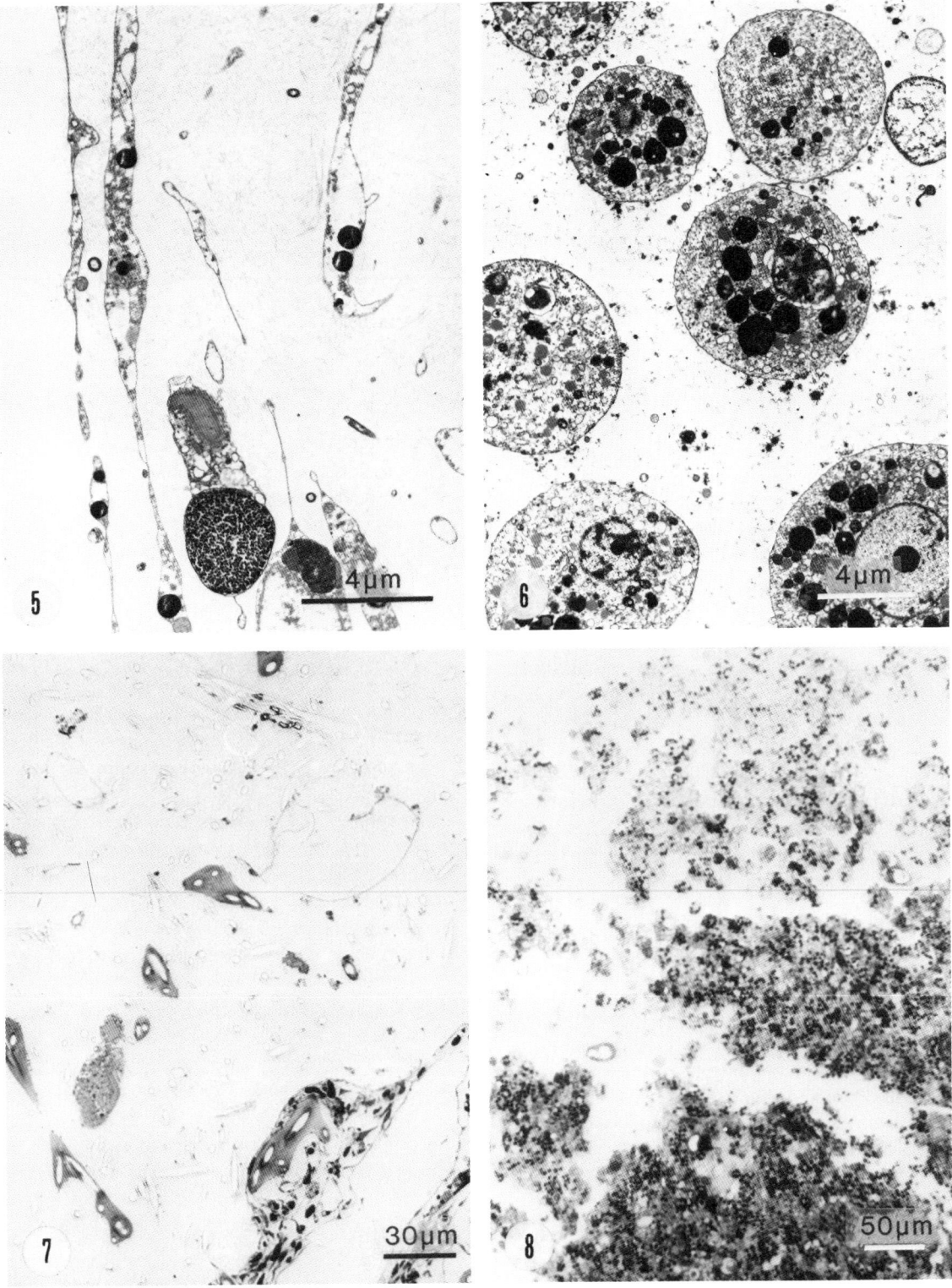

Figures 5–8. Histological results of grafting in *Axinella polypoides: 5*, normal surface of *A. damicornis*, showing extremely flat pinacocytic processes (left); cells and collagen network of mesohyl appear much less dense than in Figure 4; *6*, xenograft with *Crambe crambe*; after one day, all cells of ectosome in *A. polypoides* begin to round up and undergo very fast necrosis; *7*, xenograft with *Crambe crambe*; after 10 days a pinacoderm is reconstituted under layer of spicules remaining from total tissue necrosis; choanocyte chambers lie immediately under this new pinacoderm; *8*, surface of *Axinella polypoides* after 10 days of contact with granite; a dense layer of spherulous cells has formed; all usual ectosomal structures have disappeared.

disappearance of the pinacoderm (Figure 3). However, note that, in this combination, the reaction—or more precisely the absence of reaction—is not reciprocal. Indeed, after a few days of contact, *A. damicornis* reacts by secreting large amounts of collagen between the partners (Figures 4, 5).

ACUTE REJECTION. Acute rejection is rapid, strong and consists of a reciprocal reaction. It involves tissue necrosis of great depth (Figure 6). This type of rejection has been observed when *Axinella polypoides* is grafted against *Crambe crambe* and *Hemimycale columella*. In both combinations, complete disorganization of the superficial tissues begins within the first day of contact. A gradient of progressive reactions appears afterwards, running from the surface of the parabionts toward the inner tissues. A few days after grafting, *A. polypoides* appears to be reduced to its spicule framework upon more than 500 μm. After 6 to 10 days, the necrotic process stops and a new pinacoderm is reconstituted under the stripped off skeleton (Figure 7).

The strong and generalized necrosis reaction makes "acute rejection" quite different from "antagonistic rejection" observed in *Axinella polypoides* allografts. Indeed, both processes involve cell lysis and necrosis, but in antagonistic rejection, target cells are individually lysed by killer cells, and necrosis is never generalized to large portions of tissue.

CONTROLS WITH INERT MATERIALS. We tested the reactions of *Axinella polypoides* when put into contact with four materials: PVC foam, glass, cork, and granite.

When put into contact with PVC foam, *Axinella polypoides* reacts very weakly. After 10 days, the pinacoderm disappears, but the ectosome looks unchanged. The reaction against glass is stronger but temporary. After 6 days, the pinacoderm is destroyed and the ectosome is invaded by numerous cells containing many newly formed phagosomes. However, after 10 days of contact, the ectosome and the choanosome are completely regenerated. Only the pinacoderm is not yet renewed.

Contact with cork induces severe and long-lasting modifications. About 4 days after the initial contact, the totality of the ectosome and a large part of the choanosome become completely disorganized. Both tissues are invaded by numerous collencytes, archeocytes, and spherulous cells.

Contact with granite induces the most severe reactions. After 6 days, the ectosome and the choanosome are completely disorganized. Both tissues are invaded by large phagocytosing archeocytes and many spherulous cells. No recovering was observed (Figure 8). It is important to notice that the four reactions described above are different and can easily be distinguished from one another.

Discussion

The synthetic analysis of the *Axinella polypoides* responses to foreignness emphasizes the diversity of processes pointed out recently in allogeneic and xenogeneic combinations (Buscema and Van de Vyver, 1983; Van de Vyver and Barbieux, 1983; Buscema and Van de Vyver, 1984a,b,c). Indeed, *A. polypoides* sponges appeared to be capable of 5 different reactions which range from almost complete indifference to severe damages. Complete fusion is observed in all parabiosis autografts, and antagonistic rejection in all allografts, whereas three types of responses are observed for parabiosis xenografts: nonfusion, chronic rejection, and acute rejection. Moreover, when put into contact with inert materials, *A. polypoides* reacts differently, according to its nature. The rejection processes involved may be similar for both species paired, but not necessarily. For instance, in the association between *A. polypoides* and *A. damicornis*, only the latter is responsible for building of the collagen barrier.

Such diversity of responses reveals discrimination and specificity to a high degree. Both aspects are clear manifestations of immunity in sponges. However, this specificity is different or differently expressed from that suggested by the accelerated second set graft rejection described in *Callyspongia diffusa* (Hildemann et al., 1979). The type of specificity pointed out in *Axinella polypoides* takes into account the ability of sponges to discriminate not only "self" from "non-self", but also between different non-self, either alive or inert.

In sponges, there are no blood vascular or organ systems. This particular situation, characteristic of diploblastic organisms, may probably explain why, despite the great diversity of reactions, the basic pattern of response to foreignness always involves an initial and direct cellular contact between the partners. These observations give some evidence that specific cell-mediated immunity arose very early in evolution, perhaps with sponges.

In higher invertebrates and in vertebrates, histocompatibility is assued by immunocompetent cells. In sponges, all the techniques used to study cell recognition—noncoalescence, cell aggregation, and grafts—suggest that all the cell types are involved in immune processes. Nevertheless, depending on the response type, the key role may be played by specific cell types. For example, in antagonistic rejection, cell lysis depends on cytotoxic interactions between archeocytes since these cells appear to achieve both cytotoxicity or killing activity and phagocytosis. In chronic rejection, which involves an important deposit of collagen material, the main role is played by collencytes.

Conclusions

The discriminative capacities of *Axinella polypoides* make clear that cell immunity in sponges is based on the same fundamental reactions as those found in more evolved animals: isolation of the foreign individuals, phagocytosis, and cytotoxicity. However, the most striking fact learned about sponge immunity using *A. polypoides* as a reference species, is that the type of rejection depends not only on the species concerned, but also and essentially on the species association. Such capacity is astonishing, particularly if we keep in mind that sponges are the most primitive Metazoa and that some phyla higher in the evolution–such as nemerteans or arthropods–do not even show allogeneic recognition.

Literature Cited

Buscema, M., and G. Van de Vyver. 1983. Variability of Allograft Rejection Processes in *Axinella verrucosa*. *Developmental and Comparative Immunology,* 7:401–405.

———. 1984a. Cellular Aspects of Alloimmune Reactions in Sponges of the Genus *Axinella*. I. *Axinella polypoides*. *Journal of Experimental Zoology,* 229:7–17.

———. 1984b. Cellular Aspects of Alloimmune Reactions in Sponges of the Genus *Axinella*. II. *Axinella verrucosa* and *Axinella damicornis*. *Journal of Experimental Zoology,* 229:19–32.

———. 1984c. Allogeneic Recognition in Sponges: Development, Structure, and Nature of the Nonmerging Front in *Ephydatia fluviatilis*. *Journal of Morphology,* 181:291–303.

———. 1985. Cytotoxic Rejection of Xenografts between Marine Sponges. *Journal of Experimental Zoology,* 235:297–308.

De Sutter, D., and G. Van de Vyver. 1979. Cell Recognition Properties of Isolated Sponge Cell Fraction. Pages 217–224 in *Biologie des Spongiaires,* edited by C. Lévi and N. Boury-Esnault. Colloques Internationaux du C.N.R.S. 291. Paris: Centre National de la Recherche Scientifique.

Hildeman, W. S., I. S. Johnston, and P. L. Jokiel. 1979. Immunocompetence in the Lowest Metazoan Phylum. Transplantation Immunity in Sponges. *Science,* 204:420–422.

Mank, A., and E. F. Kilian. 1979. The Ingestion and Digestion of Food of the Freshwater Sponge *Spongilla lacustris*. Pages 353–360 in *Biologie des Spongiaires* edited be C. Lévi and N. Boury-Esnault. Colloques Internationaux du C.N.R.S. 291. Paris: Centre National de la Recherche Scientifique.

Rasmont, R. 1961. Une technique de culture des éponges d'eau douce en milieu contrôlé. *Annales, Societé Royale de Zoologie de Belgique,* 91:147–146.

Reiswig, H. M. 1975. Bacteria as Food for Temperate-Water Marine Sponges. *Canadian Journal of Zoology,* 53:582–589.

Vacelet, J. 1979. La place des Spongiaires dans les systèmes trophiques marins. Pages 259–270 in *Biologie des Spongiaires,* edited by C. Lévi and N. Boury-Esnault. Colloques Internationaux du C.N.R.S. 291. Paris: Centre National de la Recherche Scientifique.

Van de Vyver, G., and B. Barbieux. 1983. Cellular Aspects of Allograft Rejection in Marine Sponges of the Genus *Polymastia*. *Journal of Experimental Zoology,* 227:1–7.

GERALD J. BAKUS
BRUCE SCHULTE
STEVEN JHU
MINTURN WRIGHT
Department of Biological Sciences
University of Southern California
Los Angeles, California 90089–0371

GERARDO GREEN*
PATRICIA GOMEZ
Instituto de Ciencias del Mar y Limnologia
Universidad Nacional Autonoma de Mexico
A.P. 70–305, Mexico 20, DF, Mexico

Antibiosis and Antifouling in Marine Sponges: Laboratory versus Field Studies

Abstract

The results of two methods of extracting sponges for tests on antimicrobial activity were not significantly different, nor were the results of antimicrobial activity between replicate pairs of species. Five of 19 species of sponges (26%) showed antimicrobial activity in the laboratory. This low value is attributed to possible seasonal differences in metabolite production. None of these five species inhibited larval settlement of fouling organisms in the field. Evidence that two other sponge species may inhibit larval settlement suggests that some antifouling compounds may not be antibiotic. Field studies indicate that positional effects of the sun, adjacent panels, and water movement may be especially important variables. The question is raised whether the results of antimicrobial activity by medically important bacteria and fungi can be extapolated to the marine environment, whether laboratory antibiosis in marine bacteria can be extrapolated to field conditions, and whether tropical sponge metabolites can be compared with temperate marine bacteria to test ecological and evolutionary hypothesis.

Although numerous laboratory studies have been conducted on antibiosis in marine organisms, the problems that can arise in this area of research have received little critical attention. The purpose of this discussion is to identify these problems and to compare the results of laboratory tests with those of experiments conducted in the field.

Methods and Materials

Marine sponges were collected by scuba diving at depths of 3 to 13 m at the University of Mexico Puerto Morelos marine station located about 30 km south of Cancun

---

*Deceased.

(21°N 86°W) (Green et al., this volume, Figure 1). The distinctive physiographic features of the study site are its lagoon (approximately 600 m wide and up to 6 m deep), the reef crest or boulder zone, and the forereef which gradually slopes out to sea (Jordan, 1979; Jordan et al., 1981). Sponges were taken by hand between January 17 and 26, 1985, during which time the surface water temperature was about 25°C. The specimens were frozen and shipped to the University of Southern California (USC) where laboratory bioassays were conducted between February 28 and March 3, 1985. Field experiments with plywood panels were conducted at the USC Fish Harbor Laboratory in Los Angeles Harbor between April 27 and June 5, 1985.

Replicate antimicrobial assays were performed on six types of human pathogens (consisting of five species of bacteria and one yeast) plus two types of marine bacteria. Nutrient agar plates of Tryptic Soy Agar (TSA) were prepared for the human pathogens, and nutrient seawater agar was used for the marine bacteria. Six-mm-diameter filter paper discs were placed on the inoculated agar surface. Sponge extracts were made by two methods: (1) 0.5 g sponge was homogenized by mortar and pestle in 10 ml of methanol and centrifuged, and 0.1 ml of the supernatant was added five times to each of eight discs per sponge, and (2) a 25-g piece of sponge was placed in 60 ml methanol and left for 24 h. The supernatant (0.1 ml) was spotted five times on eight discs per sponge. Plates were incubated for 24 h at 37°C (human pathogens) and 30°C (marine bacteria). Zones of inhibition of bacterial growth around the disc were measured (mm) and compared with the following controls: (1) blank disc, (2) penicillin 10 units, (3) tetracycline (30 μg), and (4) methanol.

The first extraction procedure described above was followed for the field experiments, except that the quantity of sponge was increased (from 0.5 g to 25 g) because a large volume of extract was required for each fouling panel. Crude extract from the sponge was mixed with powdered abietic acid and dichloromethane in the ratio of 5 g abietic acid to 10 ml dichloromethane to 4 ml sponge crude extract. This mixture was poured into four grooves (approximately 14 ml each) in each of six 8-in. × 8-in. panels of 3/8-in. plywood during the period of April 15–25, 1985. The panels were assembled into 21 frames (15 sponges, 6 controls) of six panels apiece and placed along two sides of a floating dock in Fish Harbor, Los Angeles Harbor, on April 27, 1985. The frames were deployed at a depth of 1 m and were 2 m apart (Figure 1). They were roped together to ensure that the sun struck each panel at right angles to the frame edge, since differences in sunlight from one side of a panel to another can have significant consequences on algal and larval settlement (Bakus and Kawaguchi, 1984; Bakus, in press). Unfortunately, there was insufficient time to determine if algal cover was a possible confounding factor in this series of experiments. The frames were

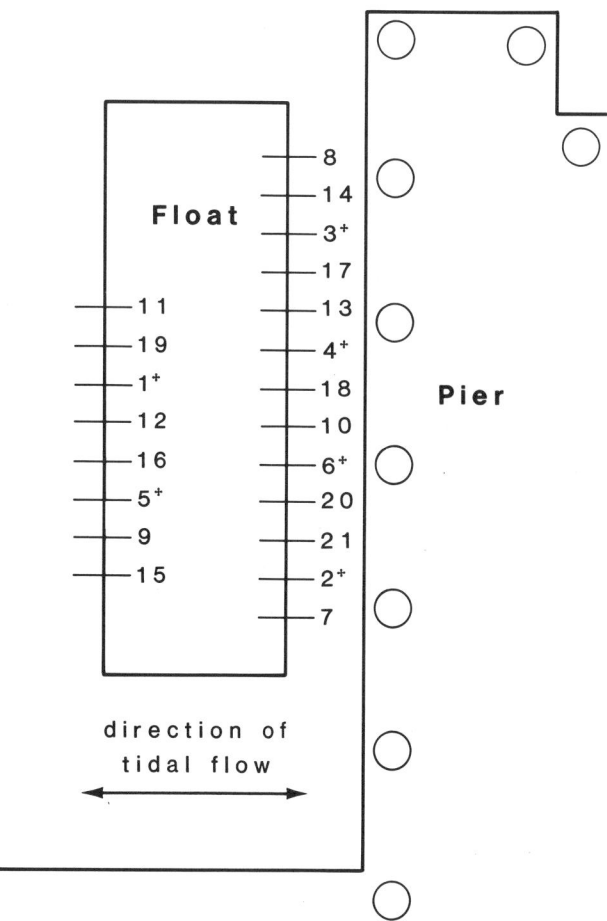

Figure 1. Deployment of frames in Los Angeles Harbor (+ = control).

checked for tangles weekly, removed on June 5, 1985, and fixed in 70% ethanol.

A grid measuring 18 cm × 18 cm was constructed. A panel was submerged in an enamel tray and the grid centered above it. Four seasonally common and abundant harbor species were selected for antifouling studies: *Bugula neritina* (Bryozoa), *Botryllus? tuberatus* (Ascidiacea), *Janua (Dexiospira) brasiliensis* (formerly *Spirorbis* sp.) (Polychaeta), and *Hydroides pacificus* (Polychaeta). Initially, individual fouling organisms were counted with the aid of a dissecting microscope: it took 6 to 8 h to record species abundance on one side of a panel. Because of the time that would be required to count organisms on 252 panel sides, we employed the point-center quarter method to determine density which allowed us to process a panel side in 20–30 minutes (Unesco, 1984). A correction factor was used for missing point-center quarter data (Warde and Pertanka, 1981).

Methods of extracting crude extracts and replicate pairs were compared using a chi-square contingency test (Bernstein, 1966; Sokal and Rohlf, 1981). The Statistical Analysis System (SAS) was used to analyze the field data.

First, a Wilcoxon rank sum test was performed by species to compare the densities on the two sides of each frame. Second, a General Linear Models (GLM) procedure was performed by species on the ranked density data. Finally, a Duncan's multiple-range test was run to determine which frames differed significantly from each other.

Results

The point-center quarter method consistently gave lower density values than did actual count (Table 1). This trend is less evident for larger colonial animals (e.g., *Bugula*); however, the actual density was two to eight times as great as that estimated from the point-center quarter method for small, abundant orgainsms (i.e., *Janua*, *Hydroides*). Since this effect is evident for all species and comparisons are made within species, the results should be acceptable although variability might be reduced.

There was no significant difference in the results (chi-square, $p < 0.05$) between the two methods of obtaining crude extracts for tests on antimicrobial activity (Table 2). Antimicrobial activity did not differ significantly (chi-square, $p < 0.01$) between replicate pairs of species. The laboratory antimicrobial assays yielded only 5 of 19 identified sponge species (26%) with bioactivity (Tables 2, 3).

Distributions of the four fouling species were not normal ($p < 0.01$), as expected. The standard deviation of densities for *Janua* and *Hydroides* were very high. The Wilcoxon rank sum test shows that, in general, the two sides of each frame did not differ significantly with respect to densities of settled larvae for each species. Although densities for certain species did differ significantly on some frames, ($p < 0.05$), there was no pattern to these differences and so the two sides of each frame were treated as a unit, in keeping with the original experimental design.

The General Linear Models (GLM) procedure performed on the ranked density data indicated a significant

Table 2. Antimicrobial screening using two extraction methods. Values are inhibition zone diameters equal to or greater than 1 mm beyond the disc (*S.a.* = *Staphlococcus aureus*, Gram positive; *B.s.* = *Bacillus subtilis*, Gram positive; 0 = no antibiosis)

Sponge species	Extraction Method[a]	Microorganisms	
		S.a.	B.s.
Ircinia felix	1	0	0
	2	1.0	3.5
Ircinia felix	1	0	0
	2	3.5	4.0
Ircinia campana	1	0	0
	2	1.0	1.0
Ircinia strobilina	1	1.5	0
	2	1.0	1.0

[a]Method 1: Sponge homogenated and crude extract applied to discs (see text); method 2: sponge left in methanol overnight then supernatant applied to discs (see text).
Note: The following sponges showed no antibiosis with either extraction method: *Xestospongia tierneyi*, *Iotrochota birotulata*, *Amphimedon rubens*, *Spinosella vaginalis*, and *Spheciospongia vesparia*; the following organisms remained unaffected by the experiment: *Escherichia coli*, Gram negative; *Enterobacter aerogenes*, Gram negative; *Pseudomonas aeruginosa*, Gram negative; *Candida albicans*, yeast; *Vibrio anguillarum*, Gram negative marine bacterium; *Vibrio harveyi* (B–392), Gram negative marine bacterium.

difference between the frames ($p < 0.01$). A Duncan's multiple-range test was initiated to determine where these differences existed. This test showed that, for the four species examined, the control frames (1–6) were not significantly different except for frame 1 for *Hydroides* (Table 4). Each experimental frame was compared to the control frames. Table 5 shows the statistically significant differences in the density of marine fouling organisms. If these values are compared with the location of the frames (Figure 1), most of the differences can be accounted for by the location of experimental and control panels on opposite sides of the floating dock or by the considerable distances between them (> 10 m). Interestingly, the frames at the

Table 1. Comparison of densities of fouling organisms by actual count and by the point–center quarter method

Panel No.	Species	A, Density by count (No. indiv./100 cm²)	B, Density by point–center quarter method (No. indiv./100 cm²)	A/B
5[a]	*Bugula*	7.9	6.0	1.3
	Janua	180.0	41.6	4.3
	Hydroides	179.0	25.2	7.1
10[b]	*Bugula*	5.5	3.8	1.4
	Janua	182.0	75.6	2.4
	Hydroides	344.0	43.5	7.9

[a]Actual count from two panels or four sides
[b]Actual count from six panels or 12 sides

Table 3. Antimicrobial effects of marine sponges from Puerto Morelos, Mexico. Values are inhibition zone diameters equal to or greater than 1 mm beyond the disc (S.a. = *Staphlococcus aureus*, Gram positive; B.s. = *Bacillus subtilis*, Gram positive; E.c. = *Escherichia coli*, Gram negative; E.a. = *Enterobacter aerogenes*, Gram negative; C.a. = *Candida albicans*, yeast; V.a. = *Vibrio anguillarum*, Gram negative marine bacterium; 0 = no antibiosis)

Sponge species	Microorganisms					
	S.a.	B.s.	E.c.	E.a.	C.a.	V.a.
Ptilocaulis spiculifer	4.0	5.0	3.5	2.0	8.53	0
Ircinia campana	1.0	1.0	0	0	0	0
Ircinia sp., spec. 1	1.0	0	0	0	0	0
Thalysias juniperina, spec.1	0	1.5	0	0	0	0

Note: 208 bioassays; methodology 1 (see text). The following sponges showed no antibiosis: *Ulosa ruetzleri*, *Dysidea etheria* (2 specimens), *Amphimedon compressa*, *Amphimedon rubens* (2 specimens), *Pseudaxinella lunaecharta* (2 specimens), *Holopsamma helwigi* (2 specimens), *Niphates amorpha*, *Niphates* sp., *Smenospongia aurea*, *Siphonodictyon coralliophagum* (2 specimens), *Callyspongia fallax*, *Ircinia felix*, *Ircinia* sp. spec. 2, *Thalysias juniperina* spec. 2, and three unidentified sponge species; The following microorganisms remained unaffected by the experiment: *Pseudomonas aeruginosa*, Gram negative; *Vibrio harveyi* (B–392), Gram negative marine bacterium.

ends of the right side of the dock (i.e., frames 7 and 8) had, on average, the highest densities of the four species; and in general the frames on the right-hand side had higher densities than those on the left.

None of the sponges with antimicrobial metabolites inhibited the settling of the marine invertebrate larvae examined. Frame pair 5 and 19 differed in densities of *Botryllus* and *Hydroides*, while frame pair 5 and 15 differed in densities of *Bugula*. The sponge *Pseudaxinella lunaecharta* (frame 19) apparently inhibited the settling of the polychaete *Hydroides* and the ascidian *Botryllus*, while the sponge *Dysidea etheria* (frame 15) apparently inhibited the settling of the bryozoan *Bugula*. Frame pairs 6 and 7, and 6 and 17, also differed in densities of *Hydroides*, but in this

case *Ircinia campana* (frame 7) and *Smenospongia aurea* (frame 17) showed much higher densities of the polychaete than did the control frame (Table 5). On the basis of the arrangement of frames (Figure 1), we conclude that frame 15 (experimental) and possibly 19 (experimental) can be accepted as significantly reducing animal densities. It should be noted, however, that frame 5 (exposed) lacked dichloromethane as a solvent. Frame 4, an enclosed control frame, was recovered with a fish in it. How this happened and what effect it might have had on the results are unknown.

Discussion and Conclusions

We found no significant difference in the results of antimicrobial activity between replicate pairs of species. However, one specimen of *Thalysias juniperina* showed slight antibiosis (1.5 mm) against *Bacillus subtilis*, yet a replicate specimen did not. This is not unusual since Amade et al. (1982) found that one specimen of the sponge *Hyrtios eubamma* inhibited *Staphlococcus aureus*, whereas two other specimens did not because they lacked the antimicrobial compound puupehenone. Thompson (1985) noted that two antimicrobial metabolites from the sponge *Aplysina fistularis* occurred in intertidal sponges but not in specimens at a depth of 5 to 15 m. Perhaps the occurrence of antibacterial compounds in *Hyrtios eubamma* may also be a function of depth.

In the laboratory bioassays, only 5 of 19 identified species of sponges (26%) showed antimicrobial activity. This low incidence of antimicrobial activity is attributed to possible seasonal differences in antimicrobial metabolite production, antimicrobial activity was low in January or winter (see Green et al., this volume).

The assumption was made that antimicrobial metabo-

Table 4. Fouling panels (identified by frame numbers) showing statistically significant differences in the density of marine fouling organisms as compared to control panels (Duncan's multiple-range test on a GLM procedure of the ranked data, 12 panel sides per frame; data from the point-center quarter method)

Control frame no.	Control treatment[a]	Fouling organisms			
		Bugula	*Botryllus*	*Janua*	*Hydroides*
1	CFO,AA,DM	–	7,20	8,13	2,3,7,8,10 13,14,17
2	CFO,AA,DM	–	19	15,19	1,19
3	CFO,AA,DM	15,19	7	11,15,19	1,11,15,19
4	CFC,AA,DM	–	19	–	19
5	CFO,AA	15	19	–	19
6	CC,AA,DM	–	11,15,19	–	7,17

[a]CFO = control frame open (exposed); CFC = control frame closed (caged); CC = cage control; AA = abietic acid; DM = dichloromethane
Note: Controls 2, 3, 4, 5, and 6 were alike; control 1 was different (P < 0.01).

Table 5.   Density of fouling organisms on sponge extract treated test panels, Los Angeles Harbor, determined by the point–center quarter sampling method (12 panel sides per sponge species or control)

Frame no.	Treatment	Fouling organisms (no. indiv. or colonies / 100 cm^2)			
		Bugula	Botryllus	Janua	Hydroides
	Controls[a]:				
1	CFO,AA,DM	4.3	3.1	61.6	11.3
2	CFO,AA,DM	4.6	5.7	168.2	86.4
3	CFO,AA,DM	11.7	4.3	171.6	81.9
4	CFC,AA,DM	5.2	7.9	133.3	58.3
5	CFO,AA	9.6	5.3	84.8	52.9
6	CC,AA,DM	4.2	8.7	67.9	39.3
Mean and SD (n = 72)		6.6±7.3	5.8±5.7	114.6±116.9	55.0±62.8
	Extracted sponges:				
7	Ircinia campana	6.8	10.5	258.9	187.4
8	Ircinia strobilina	9.8	9.0	232.8	92.8
9	Ircinia felix	3.7	3.4	100.5	37.1
10	Spheciospongia vesparia	5.0	10.0	146.4	105.1
11	Amphimedon rubens	6.4	2.1	75.3	27.9
12	Ircinia strobilina	7.7	4.1	125.2	72.5
13	Ircinia felix	26.0	4.2	231.8	94.7
14	Ptilocaulis spiculifera	9.5	5.2	237.2	237.2
15	Dysidea etheria	2.9	2.1	42.7	19.6
16	Holopsamma helwigi	3.8	3.8	63.7	36.6
17	Smenospongia aurea	16.8	4.5	121.3	242.9
18	Amphimedon rubens	6.8	5.5	126.5	82.2
19	Pseudaxinella lunaecharta	2.9	3.4	41.7	15.7
20	Niphates sp.	4.5	11.8	145.4	37.7
21	Holopsamma helwigi	8.6	7.0	124.0	52.0
Mean and SD (n = 180)		7.5±16.1	5.8±6.5	138.2±175.7	89.4±183.2

[a]CFO = control frame open (exposed); CFC = control frame closed (caged); CC = cage control (sides open); AA = abietic acid; DM = dichloromethane.

lites disclosed by laboratory experiments should interfere with the initial stages of marine succession (i.e., formation of the initial bacterial slime) because the initial stages involve the permanent adhesion of bacteria to a surface newly exposed to seawater (Bakus et al., 1986). The sponges *Pseudaxinella lunaecharta* (frame 19) and *Dysidea etheria* (frame 15) lacked antimicrobial activity. Since *Pseudaxinella lunaecharta* appeared to inhibit the settling of the polychaete *Hydroides* and the ascidian *Botryllus*, while *D. etheria* apparently inhibited the settling of the bryozoan *Bugula*, we concluded that some antifouling compounds may not be antibacterial. They may directly affect marine invertebrate larvae either by preventing them from permanently settling or by killing them soon after settling. Both Thompson et al. (1985) and McCaffrey and Endean (1985) have found that sponges possessing antimicrobial properties are generally free of fouling organisms.

As is evident from the results, positional effects appear to have played a major role in larval settlement among the frames. The frames on the right-hand side of the dock (Figure 1) are subjected first to tidal inflow, and these panels consistently have a higher density of settled larvae. Almost all the density differences between the frames occur across the dock and not because of the extract treatment. This pattern might be considered an upstream phenomenon; that is, the frames initially bathed by the tidal flow have a richer pool of settling larvae than do the frames downstream. Furthermore, the frames on the right side of the dock with no bordering frames on one end (frames 7 and 8) consistently had very high species densities; however, the two sides of each of these frames did not differ significantly.

The assumption that bacteria are a requisite for marine succession appears to be valid, depending on which species are settling and the inorganic and organic nature of the substratum. Many taxa can settle on unfilmed surfaces, but they prefer to settle on filmed ones (Crisp, 1984; Little, 1984; Mitchell and Kirchman, 1984). A microbial film is not required for the settling of abalone *(Haliotis)* or the bryozoan *Bugula neritina* (Morse, 1984; Woollacott, 1984). Barnacles may attach before filamentous bacteria and diatoms do (Little, 1984).

The sponge *Ptilocaulis spiculifer*, well known for its strong antibiosis (Wright, 1984), inhibited the marine bacterium *Vibrio anguillarum* but not *Vibrio harveyi* (Table 3). Generally 66% to 98% of marine bacteria isolated have been asporogenous Gram-negative rods (Sugahara et al., 1984). *Vibrio, Pseudomonas,* and *Acromonas* appear to be among the most abundant genera (Bergquist and Bedford, 1978; Sugahara et al., 1984). Burkholder and Ruetzler (1969) and McCaffrey and Endean (1985) found that Gram-positive bacteria were inhibited by sponge extracts more than Gram-negative bacteria; Bergquist and Bedford (1978) found that Gram-negative bacteria were inhibited considerably more than Gram-positive bacteria by sponge extracts and that activity against marine bacteria is more frequent than that against nonmarine species; and Amade et al. (1982) discovered that an equal number of Gram-positive and Gram-negative bacteria were sensitive to sponge extracts. McCaffrey and Endean (1985) found that Gram-positive bacteria were particularly sensitive to sponge extracts.

Obviously, the results of antimicrobial activity by medically important bacteria and fungi cannot be extrapolated to the marine environment to test ecologically and evolutionary hypotheses—a point aptly made by Bergquist and Bedford (1978). Moreover, even the extrapolation of information from laboratory tests with marine bacteria to bacteria in the sea is questionable, in view of the results with laboratory antibiosis in *Ptilocaulis spiculifer*. Furthermore, results from antifouling tests conducted in a warm temperate region with natural compounds from a tropical region are invalid ecologically and evolutionarily because tropical metabolites evolved in response to natural selective pressures in the tropics. Such experiments are done out of necessity or for convenience, a potential benefit being the accidental discovery of a strong antifouling compound that may be of commercial importance. Our extraction methods were not designed to simulate the action of a sponge in its natural environment, but rather to maximize the quantity and types of compounds attainable from a given piece of sponge. Given the variety of contradictory features of marine antibiosis in sponges and differences in the results between laboratory and field experiments, it is clear that considerable research on marine antibiosis in the field is needed to clarify what appears to be a bewildering array of antimicrobial activities.

Acknowledgments

We deeply appreciate the help of Eric Jordan, director of the Puerto Morelos marine facility. We are also grateful to Eduardo Martin, Michael Temkin, Gretchen Lambert, and Donald Reish for helping us with identifications. Mohammad Yazdandoust counted fouling organisms. Sue Lieberman and Michael Temkin assisted in data analysis. John Coll and Edward Ruby kindly reviewed the manuscript. This work was supported in part by grant no. N00014-84-K-0375 (to GJB), Office of Naval Research, U.S. Navy, and University of Southern California grant no. BRSG SO7 RR07012-16 (to GJB), awarded by the Biomedical Research Support Grant Program, Division of Research Resources, National Institutes of Health.

Literature Cited

Amade, Ph., D. Pesando, and L. Chevolot. 1982. Antimicrobial Activities of Marine Sponges from French Polynesia and Brittany. *Marine Biology,* 70:223–228.

Bakus, G. J. In press. Practical and Theoretical Problems in the Use of Fouling Panels. *Proceedings, International Conference on Marine Biodeterioration,* edited by M.-F. Thompson, S. Nagabhushanam, and R. Nagabhushanam. January 16–20, 1986. Goa, India. 20 pp.

Bakus, G. J., and M. Kawaguchi. 1984. Toxins from Marine Organisms: Studies on Antifouling. Pages 43–46 in *Toxins, Drugs, and Pollutants in Marine Animals,* edited by L. Bolis et al. Berlin: Springer.

Bakus, G. J., N. M. Targett, and B. Schulte. 1986. Chemical Ecology of Marine Organisms: An Overview. *Journal of Chemical Ecology,* 12:951–987.

Bergquist, P. R., and J. J. Bedford. 1978. The Incidence of Antibacterial Activity in Marine Demospongiae: Systematic and Geographic Considerations. *Marine Biology,* 46:215–221.

Bernstein, A. L. 1966. *A Handbook of Statistics Solutions for the Behavioral Sciences.* New York: Holt, Rinehart, and Winston. 145 pp.

Burkholder, P. R., and K. Ruetzler. 1969. Antimicrobial Activity of Some Marine Sponges. *Nature,* 222:938–984.

Crisp, D. J. 1984. Overview of Research in Marine Invertebrate Larvae, 1940–1980. Pages 103–126 in *Marine Biodeterioration: An Interdisciplinary Study,* edited by J. D. Costlow and R. C. Tipper. Annapolis, MD: Naval Institute Press.

Jordan, E. 1979. An Analysis of Gorgonian Community in a Reef Calcareous Platform on the Caribbean Coast of Mexico. *Anales, Centro Ciencas del Mar y Limnologica (Mexico),* 6:87–96.

Jordan, E., M. Marino, O. Moreno, and E. Martin. 1981. Community Structure of Coral Reefs in the Mexican Caribbean. Pages 303–308 in *The Reef and Man,* edited by E. D. Gomez et al. Proceedings of the Fourth International Coral Reef Symposium, Manila. Qezon City, Philippines: Marine Science Center.

Little, B. J. 1984. Succession in Microfouling. Pages 63–67 in *Marine Biodeterioration: An Interdisciplinary Study,* edited by J. D. Costlow and R. C. Tipper. Annapolis, MD: Naval Institute Press.

McCaffrey, E. J., and R. Endean. 1985. Antimicrobial Activity of Tropical and Subtropical Sponges. *Marine Biology,* 89:1–8.

Mitchell, R., and D. Kirchman. 1984. The Microbial Ecology of Marine Surfaces. Pages 49–56 in *Marine Biodeterioration: An Interdisciplinary Study,* edited by J. D. Costlow and R. C. Tipper. Annapolis, MD: Naval Institute Press.

Morse. D. E. 1984. Biochemical Control of Larval Recruitment and Marine Fouling. Pages 134–140 in *Marine Biodeterioration: An Interdisciplinary Study,* edited by J. D. Costlow and R. C. Tipper. Annapolis, MD: Naval Institute Press.

Sokal, R. R., and F. J. Rohlf. 1981. *Biometry.* San Francisco: W. H. Freeman and Co. 859 pp.

Sugahara, I., L. C. Lim, and K. K. Hooi. 1984. Heterotrophic Bacterial Population in Tropical Coastal Waters. *Bulletin, Japanese Socety of Scientific Fisheries,* 50:1385–1394.

Thompson, J. E. 1985. Exudation of Biologically-Active Metabolites in a Sponge *(Aplysina fistularis).* I. Biological Evidence. *Marine Biology,* 88:23–26.

Thompson, J. E., R. P. Walker, and D. J. Faulkner. 1985. Screening and Bioassays for Biologically-active Substances from For-

ty Marine Sponge Species from San Diego, California, U.S.A. *Marine Biology,* 88:11–21.

Unesco. 1984. *Comparing Coral Reef Survey Methods.* Unesco Reports, Marine Science 21. Paris: Unesco. 170 pp.

Warde, W., and J. W. Pertanka. 1981. A Correction Factor Table for Missing Point-Center Quarter Data. *Ecology,* 62:491–494.

Woollacott, R. M. 1984. Environmental Factors in Bryozoan Settlement. Pages 149–154 in *Marine Biodeterioration: An Interdisciplinary Study,* edited by J. D. Costlow and R. C. Tipper. Annapolis, MD: Naval Institute Press.

Wright, J. L. C. 1984. Biologically Active Marine Metabolites: Some Recent Examples. *Proceedings, Nova Scotian Institute of Science,* 34:133–161.

GERARDO GREEN[*]
PATRICIA GOMEZ
Instituto de Ciencias del Mar y Limnologia
Universidad Nacional Autonoma de Mexico
A.P. 70–305, Mexico 04510, D.F. Mexico

GERALD J. BAKUS
Department of Biological Sciences
University of Southern California
Los Angeles, California 90089–0371

Antimicrobial and Ichthyotoxic Properties of Marine Sponges from Mexican Waters

Abstract

Marine sponges collected from shallow Mexican waters of the Pacific, the Gulf of Mexico, and the Caribbean were tested for antibiosis and ichthyotoxicity. Of the 9 species found near Mazatlán, 6 have antimicrobial and 2 have ichthyotoxic properties; of the 5 species found near Zihuatenejo, 4 have 7 antimicrobial and 1 has ichthyotoxic properties; of the 19 species found near Veracruz, 7 have antimicrobial and 7 have ichthyotoxic properties; and of the 22 species from the area of Puerto Morelos, 11 have antimicrobial and 10 have ichthyotoxic properties. Seasonal variation exists in the level of activity of antimicrobial properties, and antibiosis increases during the warmest part of the year. Several hypotheses are proposed to explain these results.

Various research groups have been testing marine sponges for antimicrobial substances ever since they were discovered by R. F. Nigrelli and colleagues in the late 1950s (Burkholder, 1973; Green, 1977; Bergquist and Bedford, 1978; Amade et al., 1982; Thompson et al., 1985). These efforts have gained widespread attention because commercial antibiotics are gradually becoming less effective in controlling infectious diseases, owing to their indiscriminate use and because microorganisms are capable of developing a resistance to antibiotics through natural selective processes. Researchers have also been looking for other active substances such as antiviral, cytotoxic, and antitumoral compounds (Sigel et al., 1970; Stempien, 1970; Rinehart et al., 1981). They have all used basically the same chemical, microbiological, physiological, and cellular techniques, but have collected organisms at different times of the year and have employed different methods of preserving and storing the sponges for study. We present the results of our research conducted during the past

---

[*]Deceased; please direct correspondence to P. Gomez.

10 years and compare our findings with those of previous studies.

Materials and Methods

Sponges were collected by SCUBA diving at depths of 1–20 m in four localities along the coast of Mexico (Figure 1): the Bay of Mazatlán, Sinaloa, Pacific Ocean, during May 1981; the Bay and adjacent waters of Zihuatenejo, Guerrero, Pacific Ocean, during March 1982; the waters around the Port of Veracruz, Veracruz, Gulf of Mexico, during May 1979; and Puerto Morelos, Quintana Roo, Caribbean, during July 1976 and April 1979. The sponges were frozen immediately in dry ice, transported to our laboratory in Mexico City, and transferred to deep freezers. Voucher specimens were kept for identifications, fixed in formalin and preserved in alcohol, in our collections at the Instituto de Ciencias del Mar y Limnologia, Universidad Nacional Autonoma de Mexico. Identifications

were based on Wiedenmayer (1977) and Van Soest (1978, 1980, and 1984).

An extract was prepared for each sponge in the following manner: A volume of 25 ml of each sponge, cut into small pieces, was homogenized with 50 ml of methanol in a blender for 10 minutes and then centrifuged for 15 minutes at 3000 rpm. For the ichthyotoxic tests, 25 ml of the supernatant from the methanolic extract were evaporated and resuspended in 200 ml of aquarium water. A test fish (*Lebistes* sp., guppy) was placed in the solution and its behavior timed and recorded. The control fish was placed in 200 ml of the same aquarium water along with the residue from 25 ml of evaporated methanol. For the antimicrobial tests, one ml of the supernatant was applied to a sterile filter paper disk (5 mm diameter) and allowed to dry at room temperature. Control disks were prepared by adding methanol only. The disks were placed on plates seeded with the following microorganisms: *Staphlococcus aureus*, *Streptococcus pneumonie*, *Streptococcus pyogenes*, *Shigella*

Figure 1. Map of Mexico with localities of sponges collected for experimental study marked.

3d Int. Sponge Conf. 1985

dysenteriae, Cryptococcus neoformans, Sporotrix schenckii, and *Candida albicans.* The plates were incubated at 37°C and the results were recorded at 24h and 48h. The ichthyotoxic and antimicrobial tests were both replicated at the same time.

Results

Results of the antimicrobial and ichthyotoxic tests are presented in Tables 1–4. Of the 9 species collected near Mazatlán, 6 (67%) have antimicrobial and 2 (22%) have ichthyotoxic properties; of 5 species found near Zihuatenejo, 4 (80%) have antimicrobial and 1 (20%) has ichthyotoxic properties; of the 19 species from the area of Veracruz, 7 (36%) have antimicrobial and 7 (36%) have ichthyotoxic properties; and of 22 species from Puerto Morelos, 11 (50%) have antimicrobial and 10 (45%) have ichthyotoxic properties.

Table 1. Ichthyotoxic and antimicrobial activities of marine sponges from Puerto Morelos, collected during July 1976 (*) and April 1979 (+) (T.t. = Toxicity test; S.a. = *Staphylococcus aureus*; S.t. = *Salmonella typhi*; S.d. = *Shigella dysenteriae*)

Sponge species	Test			
	T.t.	S.a.	S.t.	S.d.
Epipolasidae (Epipolasida)				
Species no. 1	0	0	0	0
Plakinidae (Homoscleromorpha)				
Plakortis zygomorpha	+ +	+ +	+	0
Spirastrellidae (Hadromerida)				
Anthosigmella varians	0	0	0	0
Spheciospongia vesparia	+	+,**	0	***
Agelasidae (Axinellida)				
Agelas clathrodes	+ +	***	0	0
Axinellidae (Axinellida)				
Pseudaxinella lunaecharta	+ +	0	0	0
Esperiopsidae (Poecilosclerida)				
Iotrochata birotulata	0	0	0	0
Mycalidae (Poecilosclerida)				
Mycale angulosa	+ + +	+	0	0
Psammasidae (Poecilosclerida)				
Holopsamma helwigi	0	0	0	0
Species no. 1 (Haplosclerida)	0	0	0	0
Niphatidae (Haplosclerida)				
Amphimedon compressa	+ + +	+,**	**	0
Niphates digitalis	0	0	0	0
Niphates amorpha	0	+	0	0
Siphonodictyon coralliphagum	+ + +	+ + +	0	0
Callyspongiidae (Haplosclerida)				
Callyspongia vaginalis	0	0	0	0
Callyspongia fallax	+ +	+	0	0
Thorectidae (Dictyoceratida)				
Ircinia felix	+ +	0	0	0
Ircinia strobilina	0	*	**	0
Ircinia campana	0	0	0	0
Dysideidae (Dictyoceratida)				
Dysidea etheria	+ +	0	0	0
Aplysinidae (Verongida)				
Aplysina cauliformis	0	+ + +	+ + +	+ + +
Aplysinellidae (Verongida)				
Pseudoceratina crassa	0	+ + +	0	0

Note: For the toxicity test: + + + = death of fish within 15 min.; + + = death of fish within 16 to 90 min.; + = death of fish from 91 to 360 min. For the antimicrobial test, radius of the halo of inhibition from the edge of the disk: + + +,*** = more then 10 mm; + +,** = from 5 to 10 mm; +,* = less than 5 mm; 0 = no activity.

Table 2. Antimicrobial and ichthyotoxic activities of marine sponges from Veracruz, collected during May 1979 (T.t. = Toxicity test; S.a. = *Staphylococcus aureus*; S.t. = *Salmonella typhi*; S.d. = *Shigella dysenteriae*)

Sponge species	Test			
	T.t.	S.a.	S.t.	S.d.
Tethyidae (Hadromerida)				
Tethya actinia	0	0	0	0
Agelasidae (Axinellida)				
Agelas dispar	0	+ +	+	0
Species no. 1 (Poecilosclerida)	+ +	0	0	0
Esperiopsidae (Poecilosclerida)				
Iotrochata birotulata	0	0	0	0
Desmapsamma anchorata	0	0	0	0
Mycalidae (Poecilosclerida)				
Mycale laevis	0	0	0	0
Psammasidae (Poecilosclerida)				
Holopsamma helwigi	0	0	0	0
Niphatidae (Haplosclerida)				
Amphimedon compressa	+ + +	+ +	+	0
Amphimedon viridis	+ + +	0	+ +	0
Niphates erecta	0	0	0	0
Callyspongiidae (Haplosclerida)				
Callyspongia vaginalis	0	0	0	0
Callyspongia armigera	+	+	0	0
Petrosiidae (Haplosclerida)				
Xestospongia subtriangularis	0	0	0	0
Halisarcidae (Dendroceratida)				
Halisarca purpurea	0	0	0	0
Thorectidae (Dictyoceratida)				
Ircinia felix	+ + +	0	0	0
Ircinia strobilina	0	0	0	0
Aplysinidae (Verongida)				
Aplysina fistularis insularis	+	+ +	+ +	+ + +
Verongia rigida	0	+ + +	+ + +	+ +
Aplysinellidae (Verongida)				
Pseudoceratina crassa	+ +	+ +	+ +	0

For the toxicity test: + + + = death of fish within 15 min.; + + = death of fish within 16 to 90 min.; + = death of fish from 91 to 360 min. For the antimicrobial test, radius of the halo of inhibition from the edge of the disk: + + + = more then 10 mm; + + = from 5 to 10 mm; + = less than 5 mm; 0 = no activity.

Discussion

ICHTHYOTOXICITY

Our results of the ichthyotoxic tests differ somewhat—both with regard to the sponges found to be ichthyotoxic and their level of toxicity—from those reported for species from the Caribbean and adjacent waters by Stempien et al. (1970), Burkholder (1973), Green (1977), and Bakus and Thun (1979). We attribute these differences to problems in the identification of sponges and perhaps to the fact that sponges were collected during different times of the year. If some of the sponge metabolites are playing multiple roles in the field (i.e., in antipredation and antifouling—see McCaffrey and Endean, 1985; Thompson, 1985), then natural selection for these metabolites may also result in variable ichthyotoxicity, as explained later in the discussion.

ANTIMICROBIAL ACTIVITY

Antimicrobial activity was evident in 36–80% of the sponges studied, but the highest value is probably spurious because it is based on the smallest sample size (i.e.,

Table 3. Antimicrobial and ichthyotoxic activities of marine sponges from Mazatlán, collected during May 1981 (T.t. = Toxicity test; *S.a.* = *Staphylococcus aureus*; *S.p.* = *Streptococcus pneumoniae; S.py.* = *Streptococcis pyogenes*; *S.t.* = *Salmonella typhi*; *S.d.* = *Shigella dysentariae*; *C.n.* = *Cryptococcus neoformans*; *S.s.* = *Sporothrix schenckii*; *C.a.* = *Candida albicans*)

Sponge species	Test								
	T.t.	S.a.	S.p.	S..py.	S.t.	S.d.	C.n.	S.s.	C.a.
Mycalidae (Poecilosclerida)									
Zygomycale parishii	+	0	0	0	0	0	0	0	0
Mycale microsigmatosa	0	0	0	0	0	0	0	0	0
Myxillidae (Poecilosclerida)									
Damiriana hawaiiana	0	0	0	0	0	0	0	0	0
Species no. 1 (Haplosclerida)	0	0	0	+ +	0	0	0	0	0
Species no. 2 (Haplosclerida)	0	0	+ +	+ +	0	0	0	0	0
Haliclonidae (Haplosclerida)									
Sigmadocia caerula	0	0	+	+ +	0	0	0	0	0
Callyspongiidae (Haplosclerida)									
Callyspongia sp.	0	0	+ +	+ +	0	0	0	0	0
Spongiidae (Dictyoceratida)									
Hyatella intestinalis	0	0	0	+ +	0	0	0	0	0
Aplysinidae (Verongida)									
Aplysina sp.	+ + +	+ + +	+ +	0	+ +	+ +	0	0	0

Note: For the toxicity test: + + + = death of fish within 15 min., + + = death of fish from 16 to 90 min., + = death of fish from 91 to 360 min. For the antimicrobial test: radius of the halo of inhibition from the edge of the disk: + + + = more than 10 mm, + + = from 5 to 10 mm, + = less than 5 mm. 0 = no activity.

five species of sponges tested from Zihuatenejo). These results are not remarkable by themselves. However, if the data for Puerto Morelos are compared at different times of the year, particularly during the winter season (Bakus et al., this volume), a strong seasonal trend is suggested. Three of four species of sponges collected in July 1976 with antimicrobial activity lacked antimicrobial activity in January 1985, all six species of active sponges collected in April 1979 lacked antibiosis in January 1985, and three of four active species of sponges collected in July 1976 lacked antimicrobial activity in April 1979. Whether or not such a pattern exists in other tropical areas or even temperate regions is unknown.

Considerable variation exists in the microbial activity of sponges reported in the literature (Table 5). Although some of this activity is the probable result of individual variation, different extraction techniques (Thompson et al., 1985), and the fact that the samples are from different depths (Thompson, 1985), part of the variation may be due to seasonal differences in metabolite production. Whether symbiotic cyanophytes contribute to antimicrobial production in sponges is unknown (McCaffrey and Endean, 1985).

There are several possible reasons for the increase in antimicrobial activity during the warmest season of the year. Bacteria on coral reefs derive from the sediment

Table 4. Antimicrobial and ichthyotoxic activities of marine sponges from Zihuatanejo collected during March 1982 (T.t. = Toxicity test; *S.a.* = *Staphylococcus aureus*; *S.p.* = *Streptococcus pneumoniae; S.py.* = *Streptococcis pyogenes*; *S.t.* = *Salmonella typhi*; *S.d.* = Shigella dysentariae; *C.n.* = *Cryptococcus neoformans*; *S.s.* = *Sporothrix schenckii*; *C.a.* = *Candida albicans*)

Sponge species	Test								
	T.t.	S.a.	S.p.	S.py.	S.t.	S.d.	C.n.	S.s.	C.a.
Mycalidae (Poecilosclerida)									
Zygomycale parishii	0	0	0	0	0	0	0	0	0
Haliclonidae (Haplosclerida)									
Haliclona sp.	+ +	+ +	+ + +	+ + +	+ +	+	0	0	+ + +
Callyspongiidae (Haplosclerida									
Callyspongia sp.	0	+	+ +	+ +	+	+ +	0	0	0
Aplysinidae (Verongida)									
Aplysina fulva	0	+ +	0	+ +	+ +	+	0	0	0
Aplysina sp.	0	+ +	+	+	+ +	+ +	0	0	0

Note: For the toxicity test: + + + = death of fish within 15 min., + + = death of fish from 16 to 90 min., + = death of fish from 91 to 360 min. For the antimicrobial test: radius of the halo of inhibition from the edge of the disk: + + + = more than 10 mm, + + = from 5 to 10 mm, + = less than 5 mm. 0 = no activity.

Table 5. Literature survey of antibiosis in marine sponges

Locality	Number of Sponge Species	Antibiosis (%)	Reference
Mediterranean	31	58	Burkholder & Ruetzler, 1969
Puerto Rico	20	20–75	Burkholder & Ruetzler, 1969
Caribbean	777[a]	60	Burkholder & Ruetzler, 1969
Great Barrier Reef, Australia	464[a]	62	Burkholder & Ruetzler, 1969
Queensland, Australia	24	79	McCaffrey & Endean, 1985
New Zealand	30	87	Bergquist & Bedford, 1978
Polynesia	15	47	Amade et al., 1982
Brittany	7	29	Amade et al., 1982
Southern California	9	55	Green, 1977
Southern California	40	70	Thompson et al., 1985
Northern California	12	92	Zaro, 1982
Mazatlán, Mexico	9	67	Green et al., this study
Zihuatenejo, Mexico	5	80	Green et al., this study
Veracruz, Mexico	12	67	Green, 1977
Veracruz, Mexico	19	36	Green et al., this study
Puerto Morelos, Mexico	22	50	Green et al., this study
Puerto Morelos, Mexico	19	26	Bakus et al., this volume

[a]Number of samples, chiefly sponges.

(10^5–10^9 bacteria/g dry sediment), the water column (10^4–10^6 bacteria/ml seawater), fish fecal pellets (10^{10} bacteria/g dry feces), and coral mucus (bacteria = 2.3% of organic carbon and 6.06% of organic nitrogen), among other sources (e.g., decomposing organisms) (Bernard, 1981; Jen and Bell, 1982; Moriarty, 1982; Chartock, 1983; Gottfried and Roman, 1983; Hoppe et al., 1983; Geesey et al., 1984; Sugahara et al., 1984). Bacterial populations increase more rapidly at higher temperatures in the tropics (Jen and Bell, 1982). Decomposition is much more rapid on coral reefs than in other marine habitats (Louis H. DiSalvo, written comm.). The increase in the standing crops of bacteria in seawater and sediments on the Great Barrier Reef of Australia off Townesville, Australia, during the warmest season (29°C) is about an order of magnitude greater than during the coldest season (23°C) (Clive Wilkinson, pers. comm.). With increased light intensities, primary production may go up and more photosynthates may be released by benthic algae and cyanophytes, so that a richer energy source may be available for marine bacteria during the warmer season (Rudolfo Iturriaga, pers. comm.). Thus the encounter rate between sponges and bacteria is greater during the warmest months. Although bacteria are consumed in large numbers by tropical sponges (Reiswig, 1971), broken or damaged sponges may be exposed to larger numbers or more varieties of pathogenic bacteria or fungi during the warm season. The increase in microbial metabolites at this time would be an adaptive response to this increase in pathogens.

An alternative hypothesis is that the bacteria are competing among themselves for dominance and produce more antimicrobial metabolites during the warm season. However, Louis H. DiSalvo (written comm.) suggests that marine bacteria compete in the sense of having a specific enzyme for an available substrate or a selection for high growth rate rather than producing more antimicrobial metabolites. A third alternative is that the production of metabolites is simply related to a Q10 effect, whereby higher temperatures increase metabolic rates and antimicrobial metabolite production in sponges. Still another alternative is that cyanophytes and algae symbiotic in sponges may be producing or enhancing the production of antimicrobial metabolites, especially during the warmer season. Metabolites may also have other functional roles in addition to antibiosis (i.e., antipredation, antifouling) and thus if predation or fouling increase during the warmer months, there will be a concomitant increase in antibiosis (McCaffrey and Endean, 1985; Thompson, 1985). Considerable field research will be needed to determine whether these hypotheses can explain seasonal differences in antimicrobial metabolite production in sponges.

Acknowledgments

We deeply appreciate the help of Eric Jordan, director of the Puerto Morelos marine facility, Francisco Flores, Fernando Gonzalez, Virginia Ordaz, Francisco Cruz, Armando Garduno, and Miguel Angel Bolanos. We are grateful to Louis H. DiSalvo (Universidad del Norte, Coquimbo, Chile), a pioneer in bacterial research on coral reefs, for answering our query concerning tropical marine microbiology. This work was supported in part by grant

114

N00014–84–K–0375 (to GJB), Office of Naval Research, U.S. Navy, and grant BRSG SO7 RR07012–16 (to GJB), awarded by the Biomedical Research Support Grant Program, Division of Research Resources, National Institutes of Health.

Literature Cited

Amade, Ph., D. Pesando, and L. Chevolot. 1982. Antimicrobial Activities of Marine Sponges French Polynesia and Brittany. *Marine Biology*, 70:223–228.

Bakus, G. J., and M. A. Thun. 1979. Bioassays on the Toxicity of Caribbean Sponges. Pages 417–422 in *Biologie des Spongiaires*, edited by C. Lévi and N. Boury-Esnault. Colloques Internationaux du C.N.R.S. 291. Paris: Centre National de la Recherche Scientifique.

Bergquist, P. R., and J. J. Bedford. 1978. The Incidence of Antibacterial Activity in Marine Demospongiae: Systematic and Geographic Considerations. *Marine Biology*, 46:215–221.

Bernard, P. 1981. Bacterial Counts in the Lagoon and Great Marine Cul-de-sac and Its Coastal Zone, Guadeloupe, French Antilles. *Journal of Experimental Marine Biology and Ecology*, 50:197–212.

Burkholder, P. R. 1973. Ecology of Marine Antibiotics and Coral Reefs. Pages 117–182 in *Biology and Geology of Coral Reefs*, II, Biology 1, edited by O. A. Jones and R. Endean. New York: Academic Press.

Burkholder, P. R., and K. Ruetzler. 1969. Antimicrobial Activity of Some Marine Sponges. *Nature*, 222:938–984.

Chartock, M. A. 1983. The Role of *Acanthurus guttatus* (Bloch and Schneider, 1801) in Cycling Algal Production to Detritus. *Biotropica*, 15:117–121.

Geesey, G. G., G. V. Alexander, A. C. Miller, and R. N. Bray. 1984. Fish Fecal Pellets are a Source of Minerals for Inshore Reef Communities. *Marine Ecology Progress Series*, 15:19–26.

Gottfried, M., and M. R. Roman. 1983. Ingestion and Incorporation of Coral Mucus Detritus by Reef Zooplankton. *Marine Biology*, 72:211–218.

Green, G. 1977. Antibiosis in Marine Sponges. *FAO Fisheries Report*, 200:199–205.

Hoppe, H-G., K. Gocke, D. Zamorano, and R. Zimmerman. 1983. Degradation of Macromolecular Organic Compounds in a Tropical Lagoon, Cienega-Grande, Colombia, and Its Ecological Significance. *Internationale Revue der Gesamten Hydrobiologie*, 68:811–824.

Jen, W-C., and R. G. Bell. 1982. Influence of Temperature and Time of Incubation on the Estimation of Bacterial Numbers in Tropical Surface Waters. *Water Research*, 16:601–604.

McCaffrey, E. J., and R. Endean. 1985. Antimicrobial Activity of Tropical and Subtropical Sponges. *Marine Biology*, 89:1–8.

Moriarty, O. J. W. 1982. Feeding of *Holothuria atra* and *Stichopus chloronotus* on Bacteria Organic Carbon and Organic Nitrogen in Sediments of the Great Barrier Reef, Australia. *Australian Journal of Marine and Freshwater Research*, 33:255–264.

Reiswig, H. 1971. Particle Feeding in Natural Populations of Three Marine Demospongiae. *Biological Bulletin*, 141:568–591.

Rinehart, K. L., Jr., et al. [26 authors]. 1981. Marine Natural Products as Sources of Antiviral, Anti-microbial, and Actineoplastic Agents. *Pure and Applied Chemistry*, 53:795–817.

Sigel, M. M., L. L. Wellham, W. Lichter, L. E. Dudeck, J. L. Gargus, and A. H. Lucas. 1970. Anticellular and Antitumor Activity of Extracts from Tropical Marine Invertebrates. Pages 281–294 in *Food-drugs from the Sea, Proceedings 1969*, edited by H. W. Youngken. Washington, D.C.: Marine Technology Society.

Soest, R. W. M. van. 1978. Marine Sponges from Curaçao and other Caribbean Localities. Part I. Keratosa. *Studies of the Fauna of Curaçao and other Caribbean Islands*, 56 (179):1–94.

———. 1980. Marine sponges from Curaçao and other Caribbean Localities. Part II. Haplosclerida. *Studies of the Fauna of Curaçao and other Caribbean Islands*, 62 (191):1–173.

———. 1984. Marine Sponges from Curaçao and other Caribbean Localities. Part III. Poecilosclerida. *Studies of the Fauna of Curaçao and other Caribbean Islands*, 66 (199):1–167.

Stempien, M. F., G. D. Ruggieri, R. F. Nigrelli, and J. T. Cecil. 1970. Physiologically Active Substances from Extracts of Marine Sponges. Pages 295–305 in *Food-drugs from the Sea, Proceedings 1969*, edited by H. W. Youngken. Washington, D. C.: Marine Technology Society.

Sugahara, I., L. C. Lim, and K. K. Hooi. 1984. Heterotrophic Bacterial Population in Tropical Coastal Waters. *Bulletin, Japanese Society of Scientific Fisheries*, 50:1385–1394.

Thompson, J. E. 1985. Exudation of Biologically-Active Metabolites in a Sponge *(Aplysina fistularis)*. I. Biological Evidence. *Marine Biology*, 88:23–26.

Thompson, J. E., R. P. Walker, and D. J. Faulkner. 1985. Screening and Bioassays for Biologically-active Substances from Forty Marine Sponge Species from San Diego, California, U.S.A. *Marine Biology*, 88:11–21.

Wiedenmayer, F. 1977. *Shallow Water Sponges of the Western Bahamas*. Basel und Stuttgart: Birkhäuser Verlag. 287 pp.

Zaro, B. A. 1982. Marine Sponges: A Source of Novel Antibiotics. *Proceedings, Western Pharmacological Society*, 25:11–13.

JOELLE HUYSECOM
GYSELE VAN DE VYVER
Laboratoire de Biologie Animale et Cellulaire
Université Libre de Bruxelles
50 av. F. D. Roosevelt
1050 Bruxelles, Belgium

JEAN-CLAUDE BRAEKMAN
DESIRE DALOZE
Laboratoire de Chimie Bioorganique
Université Libre de Bruxelles
50 av. F. D. Roosevelt
1050 Bruxelles, Belgium

Chemical Defense in Sponges from North Brittany

Abstract

The present work investigates the chemical defense of sponge species collected along the North Brittany coast. Organic extracts of 27 species were tested for ichthyotoxicity as well as for toxicity to dissociated cells of *Ephydatia fluviatilis* (spongotoxicity). Of the tested species, 48% were found to be active in both tests. In order to check the effectiveness of sponge toxins in discouraging predation, food-choice experiments were conducted using the dorid nudibranch *Archidoris pseudoargus*. The preliminary results obtained with 5 sponge species are presented.

Many sponge extracts are reported to show bactericidal or cytotoxic activity (Jackson and Buss, 1975; Green, 1977; Bergquist and Bedford, 1978; Bakus, 1981; Amade et al., 1982; Thompson et al., 1985). An increasing number of the secondary metabolites responsible for these biological activities have now been isolated and identified. Furthermore, a few recent studies have concluded that the ecological role of sponge secondary metabolites may be to discourage predation or to protect sponges against spatial competitors and fouling organisms (Groweiss et al., 1983; Sullivan et al., 1983; Thompson, 1985; Braekman and Daloze, 1986).

Sponge predators belong to a great variety of zoological groups (Sará and Vacelet, 1973). The most important predators in shallow marine tropical ecosystems are fishes, whereas in temperate or cold-water ecosystems invertebrates are predominantly in this role (Green, 1977). Although toxicity to fishes has been well documented, considerably less is known about the effect of sponge toxins on invertebrate predators and grazers (Walker et al., 1980; Thompson, 1985; Thompson et al., 1985).

Most dorid nudibranchs are known to be important sponge eaters (Todd, 1981). These sponge predators range from strict specialists (monophagous species) such as *Peltodoris atromaculata*, which feeds on *Petrosia ficiformis*, to polyphagous predators such as *Archidoris pseudoargus*.

This last species is reported to feed on at least 15 sponge species (Just and Tendal, 1983) .

To study the potential chemical defense of temperate water sponges against invertebrate predators, we screened for the organic extracts of 27 sponge species for toxicity and designed food-choice experiments were using the dorid nudibranch *Archidoris pseudoargus*. Our objective was to determine whether the food preferences of this sponge predator coincide with the absence of toxic metabolites in sponges.

Material and Methods

SPONGES. Twenty-seven sponge species were collected in the intertidal zone or by dredging in the Roscoff area (48°43′ N), of North Brittany, France. Specimens were cut into small pieces and preserved in methanol and then tested for toxicity. Sub-samples were dipped into 70% ethanol and utilized for species identification.

EXTRACTION PROCEDURE. The methanolic extracts of the fresh sponges were filtered and the resulting filtrates were concentrated by evaporation under reduced pressure. The residual aqueous solutions were then extracted successively with dichloromethane (extract A) and dichloromethane/ethanol 3:2 (extract B). Evaporation to dryness of the remaining aqueous solution yielded extract C containing mostly inorganic salts.

ICHTHYOTOXICITY TEST. *Lebistes reticulatus* (Poeciliidae) has been used successfully in previous toxicity studies (Alberrici et al., 1982). In this study each sponge extract was tested on two fishes at a concentration of 50 mg/l of tapwater. Toxic activity was defined by the death of the fishes, either after 12 hours (++) or after 24 hours (+).

SPONGOTOXICITY TEST. This test evaluates the effect of sponge extracts on the viability of dissociated sponge cells, particularly on their capacity to aggregate and to reconstitute functional sponges (Huysecom et al., in press). *Ephydatia fluviatilis* of strain δ (Van de Vyver, 1970, 1975) grown from gemmules in petri dishes containing mineral medium (Rasmont, 1961) was used. After 7 days of incubation, sponges were mechanically dissociated by pipetting. After the spicules and the empty gemmule shells were eliminated, the dissociated cells were centrifuged (10 minutes at 1000 RPM) and resuspended in mineral medium at a final concentration of 10^6 cells/ml. For each test, 0.1 mg of the extract to be tested was added to 1 ml of cell suspension placed on a microtest plate. The extracts were considered as highly toxic (+++) if they prevented any aggregation and killed the cell suspension within two hours; moderately toxic (++) if they allowed cell aggregation but the aggregates degenerated before the time of settlement; and weakly toxic (+) if aggregates

formed and remained healthy but were unable to settle. Nontoxic extracts did not disturb normal cell aggregation or settlement.

FOOD CHOICE TEST. Our food-choice test using the polyphageous dorid nudibranch *Archidoris pseudoargus*, a common species along the North Brittany coasts, was adapted from the procedure used by Bloom (1976) and Rowel-Rahier (1983). The food preference of *A. pseudoargus* was tested in 5 of the 27 sponge species screened for toxicity. Before the experiments, the mollusks were starved for 7 days to ensure that all spicules from previous feedings had been voided from the dorid's digestive tract. For each experiment, 2 sponge species consisting of pieces of approximately 1 cm^3 were made available to a specimen of *A. pseudoargus* for a period of 5 hours. The mollusks were tested individually with all binary combinations of the 5 sponge species, each combination being repeated 6 times. For each trial, a relative score was allotted to each of the two sponges tested. These scores correspond to the following categories: much more eaten, slightly more eaten, equally eaten, slightly less eaten, much less eaten. These categories were assigned arbitrary numerical scores of +2, +1, 0, −1, and −2 respectively. The relative amount of each sponge eaten by one nudibranch was estimated by the proportion of characteristic sponge spicules of each species in the mollusk feces. For each combination of two species, a mean score was obtained by averaging the scores of the six similar experiments. For each sponge species, the scores obtained in the different combinations were summed and the total used to determine the rank in palatability.

Results

SCREENING FOR TOXICITY. Before the food-choice experiments, organic extracts of the 27 sponge species were screened for toxic activity by two bioassays, one using fishes, the other sponge cell suspensions (Table 1). Only results obtained on extracts A and B were reported since extract C proved to be consistently non toxic. The 27 species investigated belong to 2 orders of calcarean sponges and 7 orders of demosponges. All are common along the North Brittany coast. Of the 27 species, 13 (48%) were found to be toxic to *Lebistes reticulatus*. The same percentage was toxic to *Ephydatia fluviatilis* cell suspensions. Ten species appeared to be toxic both to *L. reticulatus* and *E. fluviatilis*, three species were found to be lethal only for the fish, and three others were toxic only to the *E. fluviatilis* cells. The remaining 11 species were completely innocuous in the two tests.

FOOD CHOICE EXPERIMENTS. The results reported here pertain to 5 of the 27 sponge species tested for toxicity. These species were collected in the same biotopes as the mollusk.

Table 1. Screening results for toxic sponge metabolites in solution (extracts A and B) against fish (*Lebistes*) and sponge (*Ephydatia*) + + + = highly toxic; + + = moderately toxic; + = weakly toxic; and – = nontoxic (see methods).

Sponge Species	L. reticulatus		E. fluviatilis	
	A	B	A	B
Calcinea				
Clathrina coriacea	+	–	–	–
Calcaronea				
Sycon ciliatum	–	–	–	+ + +
Grantia compressa	+	–	+ + +	–
Leuconia johnstoni	+	–	+ + +	–
Astrophorida				
Pachymatisma johnstoni	–	+ +	+ + +	+ + +
Hadromerida				
Tethya aurantium	–	–	–	–
Cliona celata	–	+ +	–	+ + +
Polymastia mamillaris	+	–	+ + +	+ + +
Polymastia robusta	–	–	+ +	+ +
Ficulina ficus	–	–	–	–
Axinellida				
Axinella dissimilis	–	–	–	–
Phakellia ventilabrum	+	+	+ + +	+ + +
Raspallia hispida	+	–	–	–
Raspallia ramosa	–	–	–	–
Halichondrida				
Halichondria panicea	–	–	–	–
Halichondria bowerbanki	–	–	–	–
Hymeniacidon sanguinea	–	–	–	–
Hemimycale columella	–	–	–	–
Poecilosclerida				
Myxilla rosacea	–	+	+	+ +
Anchinoe fictitius	–	+	+ + +	+ + +
Pronax plumosum	–	–	–	–
Ophlitaspongia seriata	+ +	–	–	–
Haplosclerida				
Haliclona elegans	–	–	–	–
Haliclona indistincta	–	–	–	–
Haliclona viscosa	+	–	+ +	+ + +
Adocia simulans	–	–	+ +	–
Dictyoceratida				
Dysidea fragilis	+	–	+ + +	+ +

Two species *(Grantia compressa* and *Pachymatisma johnstoni)* are toxic, and the remaining three *(Ficulina ficus, Halichondria panicea* and *Pronax plumosum)* are nontoxic. Figure 1 shows the rank in palatability of the 5 species eaten by *Archidorus pseudoargus.* Two nontoxic sponges, *Halichondria panicea* and *Ficulina ficus,* were favored by the dorids. In contrast, the 3 other species (the 2 highly toxic sponges *Grantia compressa* and *Pachymatisma johnstoni* and the nontoxic one *Pronax plumosum)* had a negative score.

Discussion and Conclusions

Despite the great taxonomic difference in the materials tested, the results obtained in both spongotoxicity and ichthyotoxicity tests were remarkably consistent across 21 of the 27 species studied. These findings suggest that toxic secondary metabolites of sponges have a wide spectrum of bioactivities that provide a multidirectional defense.

For the species tested, no relationship could be established between toxicity and general systematic position of the sponges. Moreover, the presence of toxic metabolites

does not appear to be related to the absence of alternative defense mechanism: for example, *Pachymastisma johnstoni,* one of the most toxic species collected, is characterized also by cortical spherasters that provide a superficial armor.

One of the ecological advantages attributed to the production of toxins in sponges is the chemical defense against predation. Other defense mechanisms may be used by sponges, such as cryptic habitat, structure and density of the mineral skeleton, or production of repulsive but nontoxic factors.

The effect of toxins on invertebrate predators or on mobile animals that damage sponge tissue when feeding on fouling organisms associated with sponges has recently been documented by Walker et al. (1980), Thompson

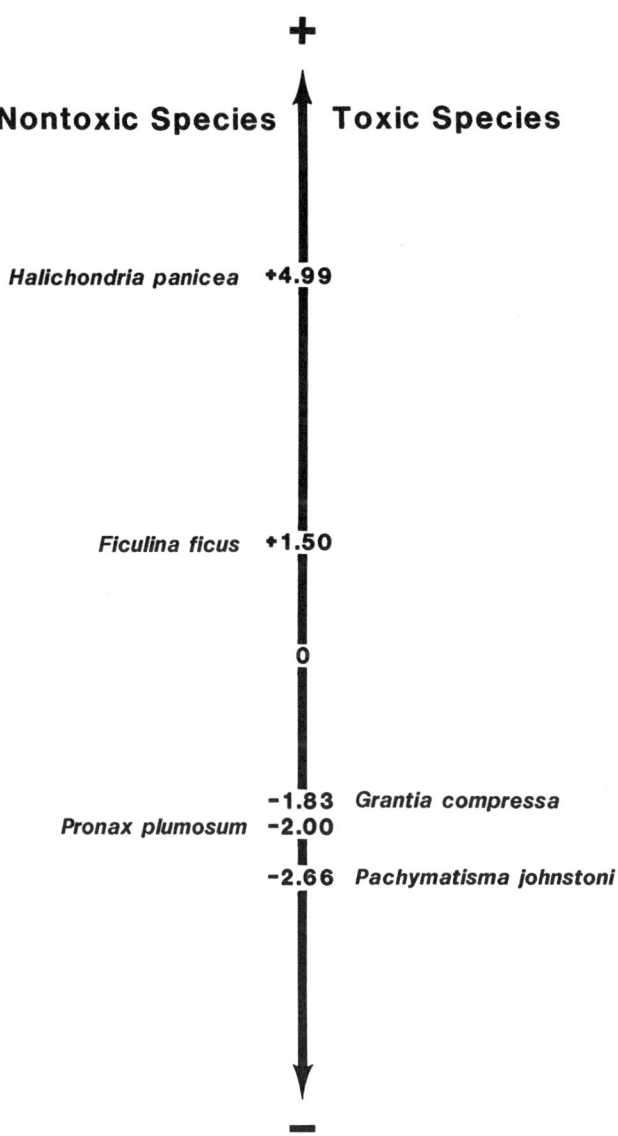

Figure 1. Rank of palatability of 5 sponge species presented to *Archidoris pseudoargus.*

(1985), and Thompson et al. (1985). While testing the toxicity of natural exuded allelochemicals of the sponge *Aplysina fistularis*, Thompson (1985) found behavior modifications in a wide range of invertebrate organisms. However, no toxicity was observed against dorid nudibranchs.

In order to check the effectiveness of sponge toxins in discouraging predation by nudibranchs, 5 of the 27 species we tested for toxicity were used to examine the food preference of a polyphageous dorid. Our laboratory studies focused on the food choices of *Archidorus pseudargus* when placed in the presence of 2 toxic sponge-species and 3 nontoxic one. In natural conditions, diets are the result of complex interactions between predator abilities and preferences, prey availability, and morphological or chemical characteristics. In laboratory feeding trials, however, the number of parameter involved are restricted. Consequently they tend to produce more positive responses because the predator is tested outside the context of its normal food-searching behavior. Moreover, the nudibranch, if limited to a choice between two unpalatable sponges, may be compelled to eat something it would normally reject.

Our results were obtained from only 5 sponge species and therefore do not provide a definitive answer to the question of whether the food preferences of *Archidoris pseudoargus* coincide with the absence of toxic metabolites in sponges. Despite the small size of the sample tested, however, the low palatability of two toxic species *(Grantia compressa* and *Pachymatisma johnstoni)* suggests that sponge toxins may be effective in discouraging predation by *Archidoris*.

Acknowledgments

This study was supported by grants from the Fonds pour la recherche fondamentale collective (FRFC no. 2451585) and the Ministère de l'Emploi et du Travail (CST 20564). The authors are indebted to G. Lassere, L. Cabioch, and G. Deroux of the Roscoff Marine Station (Brittany, France) for provision of laboratory facilities and services. We also thank A. Remacle and M. Mertens for technical assistance.

Literature Cited

Albericci, M., J. C. Braekman, D. Daloze, and B. Tursch. 1982. The Chemistry of Three Norsesterterpene Peroxides from the Sponge *Sigmosceptrella laevis*. *Tetrahedron*, 38(13):1881–1890.

Amade, Ph., D. Pesando, and L. Chevolot. 1982. Antimicrobial Activities of Marine Sponges from French Polynesia and Brittany. *Marine Biology*, 70:223–228.

Bakus, G. J. 1981. Chemical Defense Mechanisms on the Great Barrier Reef, Australia. *Science*, 211:497–498.

Bergquist, P. R., and J. J. Bedford. 1978. The Incidence of Antibacterial Activity in Marine Demospongiae, Systematic and Geographic Considerations. *Marine Biology*, 46:215–221.

Bloom, S. A. 1976. Morphological Correlations between Dorid Nudibranch Predators and Sponge Prey. *The Veliger*, 18(3):289–301.

Braekman, J. C., and D. Daloze. 1986. Chemical Defense in Sponges. *Pure and Applied Chemistry*, 58:357–364.

Green, G. 1977. Ecology of Toxicity in Marine Sponges. *Marine Biology*, 40:207–215.

Groweiss, A., U. Shmueli, and Y. Kashman. 1983. Marine Toxins of *Latrunculia magnifica*. *Journal of Organic Chemistry*, 48:3512–3516.

Huysecom, J., G. Van de Vyver, J. C. Braekman, and D. Daloze. In press. Screening and Bioassays for Toxic Substances in Sponges from North Brittany. *Vie et Milieu*.

Jackson, J. B. C., and L. W. Buss. 1975. Allelopathy and Spatial Competition among Coral Reef Invertebrates. *Proceedings, National Academy of Science, USA*, 72:5160–5163.

Just, H. J., and O. S. Tendal. 1983. On the Sponge–Diet of *Archidoris pseudoargus* (Rapp, 1827). *The Veliger*, 25:403–404.

Rasmont, R. 1961. Une technique de culture des éponges d'eau douce en mi]ieu controlé. *Annales, Societé Royale Zoologique de Belgique*, 91:147–156.

Rowel-Rahier, M. 1983. Relationships between *Salix* Species and Their Herbivores: The Example of *Phratora vitellinae* (Coleoptera: Chrysomelidae). Ph.D. Dissertation, University of Basel. 172 pp.

Sarǵ, M., and J. Vacelet. 1973. Ecologie des Démosponges. Pages 462–576 in *Traité de Zoologie. III. Spongiaires*, edited by P.-P. Grassé. Paris: Masson.

Sullivan, B., D. J. Faulkner, and L. Webb. 1983. Siphonodictidine, A Metabolite of the Burrowing Sponge *Siphonodictyon* sp. That Inhibits Coral Growth. *Science*, 221:1175–1176.

Thompson, J. E. 1985. Exudation of Biologically-Active Metabolites in the Sponge *Aplysina fistularis*. I. Biological Evidence. *Marine Biology*, 88:23–26.

Thompson, J. E., R. P. Walker, and D. J. Faulkner. 1985. Screening and Bioassays for Biologically Active Substances from Forty Marine Sponge Species from San Diego, California, USA. *Marine Biology*, 88:11–22.

Todd, C. D. 1981. The Ecology of Nudibranch Molluscs. *Oceanography and Marine Biology, Annual Review*, 19:141–234.

Van de Vyver, G. 1970. La non-confluence Intraspécifique chez les spongiaires et la notion d'individu. *Annales d'Embryologie et de Morphogenése*, 3:251–262.

———. 1975. Phenomena of cellular recognition in sponges. *Current Topics of Developmental Biology*, 10:123–140.

Walker, R. P., J. E. Thompson, and D. J. Faulkner. 1980. Sesterterpenes from *Spongia idia*. *Journal of Organic Chemistry*, 45:4976–4979.

Cell Structure
and Mobility

ROBERT GARRONE
CLAIRE LETHIAS
Histologie Expérimentale (UA CNRS 244)
ICBMC, Bât. 403, Université Claude Bernard
69622 Villeurbanne Cedex, France

Freeze-Fracture Study of Sponge Cells

Abstract

Information on the molecular architecture of sponge cell membranes and on the biological processes closely related to the change in membrane organization was obtained using freeze-fracture techniques. Conventional methods show that intramembrane protein complexes are randomly distributed (most frequently) or form clusters or ordered groupings. Some of them have been related to known structures (ciliary necklaces, desmosome plates, choanocyte collar coats). Others (lines and highly ordered plates, rosettes) are not clearly understood. High protein densities can be seen in *Chondrosia* spherulous cell membranes and in the silicalemma of freshwater sponge sclerocytes. Membrane pits are observed in the whole plasma membrane of *Verongia* spherulous cells and in the plasma membrane depressions of *Ephydatia* sclerocytes.

Protein organization did not appear to be related to tight or gap junctions. The use of a cholesterol-binding agent, filipin, prior to freeze-fracturing made it possible to study the localization and distribution of membrane sterols in *Ephydatia*, which were found to be abundant and homogeneously distributed, except in membrane areas containing organized protein arrangements.

Freeze-fracture replication is now a well-established method of studying the molecular architecture of cells, especially of membranes. The path fracture of frozen cells runs preferentially along the hydrophobic planes of membranes and reveals two new faces, which contain intramembrane particles that are considered possible membrane protein macromolecules. Important biological events are closely related to modifications of the organization, density, and distribution of these intramembrane particles (Branton, 1971).

Although the transmission electron microscope (TEM) has been widely used to study sponges, very few researchers have employed freeze-fracturing (Bergquist and Green, 1977; Garrone et al., 1980; Lethias et al., 1983). Nevertheless, some characteristics of freeze-fractured sponge membranes merit closer attention. Garrone et al. (1980) and Lethias et al. (1983), for example, have shown the classical size and distribution of the intramembrane particles on the membrane faces, the presence of a flagellar necklace (also confirmed by Bardele, 1983), and the apparent lack of conventional tight or gap junctions. They have also pointed out irregular patches of particles, which probably correspond to desmosomes, and ordered arrangements (rows, lines, rosettes, rhombic arrays, and plates), the significance of which is not fully understood.

This study reviews earlier work and expands upon it by reporting on the detection and localization of membrane sterols in freshwater sponges. These compounds can now be studied using freeze-fracture methods in conjunction with the probe filipin (an antibiotic extracted from *Streptomyces filipinensis*). Filipin binds to beta-hydroxy-cholesterols and induces a deformation that is easily visible on freeze-fracture replicas (see the reviews by Severs and Robenek, 1983; Miller, 1984).

Materials and Methods

Several individuals of the following species were studied: *Oscarella lobularis* Schmidt, *Chondrosia reniformis* Nardo, *Chondrilla nucula* Schmidt, *Tethya lyncurium* Lamarck, *Hemimycale columella* Bowerbank, *Axinella damicornis* Esper, *Axinella polypoides* Schmidt, *Haliclona elegans* Bowerbank, *Ephydatia muelleri* Lieberkühn, and *Verongia cavernicola* Vacelet. All of these marine sponges were collected in the Mediterranean Sea, near Marseille or Banyuls (France). The freshwater species *(E. muelleri)* was collected near Lyon or were obtained in the laboratory from gemmules.

The technique of freeze-fracturing has been described in details elsewhere (Lethias et al., 1983). Sponges were fixed 1 hour in glutaraldehyde in a sodium cacodylate-hydrochloric acid buffer, pH 7.4, at 4°C. The concentrations of the aldehyde and of the buffer were adjusted to a final osmotic pressure of 1100 mOsM (marine sponges) or 80 mOsM (freshwater sponge). Specimens were then washed in the same buffer and incubated for 1 hour in 25% buffered glycerol.

For the detection of sterols, fixed specimens were incubated for 1 hour at room temperature in a 50 $\mu g \cdot ml^{-1}$ solution of filipin (Upjohn, Kalamazoo, USA) in a sodium cacodylate-hydrochloric acid buffer; they were then washed in buffer and incubated in glycerol as described above.

All the samples were mounted on copper holders, frozen in nitrogen slush (−210°C), and then fractured in a Reichert-Jung CF 190 unit. The exposed faces were immediately shadowed at 45° with carbon-platinum and backed with carbon. The replicas were cleaned with sodium hypochlorite and mounted on 600-mesh grids before TEM examination with Hitachi 12A, Philips 300, or Jeol 1200 EX at the Centre de Microscopie Electronique Appliquée à la Biologie et à la Géologie, Université Claude Bernard.

Results

The cells of all the species studied were uniformly fractured in a classical way (Figures 1, 2). The nucleus, the nuclear membrane and its pores, and the Golgi apparatus are easily recognizable (Figures 1, 3). Most of the cell membranes, especially the Golgi membranes, show a high density of intramembrane particles (7 to 8 nm in diameter). Between the cells, collagen fibrils and symbiotic bacteria are also unequivocally identifiable (Figures 2, 4).

In general, the intramembrane particles are isolated or occur in very small groups (Figure 5). They have an average density of 500 particles per μm^2 on the protoplasmic (P) face and 160 particles per μm^2 on the exoplasmic (E) face. Areas of high intramembrane particle density are often observed on the pinacocyte membranes (Figure 6); they probably correspond to desmosomes. Besides this unordered grouping, lines of particles are sometimes observed on the plasma membrane of internal cells (Figure 7). These lines can be loosely (Figure 7) or densely organized, with a chevron pattern, to form plates (Figure 8).

The choanocytes were studied in detail because they are easiest to identify on the freeze-fracture replicas of sponge cells. In the species studied, the base of the flagellum possesses a necklace (Figures 9, 10) composed of three rows of particles. In some cases, a fourth row is also visible, although it is less defined. The microvilli, which make up the collar, always show two parallel lines of particles of about 500 nm long and spaced about 25–70 nm apart, usually 60 nm (Figures 11, 12). These lines are located on the internal side of the microvilli, facing the flagellum. The membrane domain situated between the lines contains fewer particles than the other areas of the microvilli membrane (Figures 11, 12). In *Ephydatia muelleri*, in all the individuals studied so far (either collected in the field or hatched in the laboratory), the choanocyte membrane domain located at the bottom of the collar shows a crown of rosettes, each of which is formed by the grouping of six particles (Figure 10). Because of the frequent coalescence of particles, the rosettes appear to be formed by four particles (Figure 10, inset).

The distribution of the rosettes is not homogeneous; they often form more or less regular radiating stripes. Rosettes have also been found on choanocytes of the marine sponge *Oscarella lobularis* (Figure 13). This is an exception, however, because in general, however, choanocytes in marine sponges are without rosettes.

Some unusual features have also been observed on the membranes of special cells and siliceous sclerocytes. For example, the spherulous cells of *Chondrosia* (Figure 14) have unusual intramembrane particle density in their plasma membrane, notably, in their spherule membranes (Figure 14, inset). In contrast, the density of intramembrane particles in the spherulous cells of *Verongia* (Figure 15) is normal but the plasma membrane (P-face) is pitted with small depressions. In spite of the complex structure of the cystancyte inclusion in *Ephydatia* (Figure 16), no special features are revealed after freeze-fracturing (Figure 17). Lastly, the plasma membrane of the siliceous macrosclerocytes of *Ephydatia* shows pitted depressions (Figure 18), while their silicalemma (Figure 19) contain the highest density of intramembrane particles recorded in sponges (6000 per μm^2 on the P-face).

Following filipin incubation and freeze-fracturing, the membranes of *Ephydatia* cells show the small protrusions (approximately 25 nm in diameter) characteristic of this cholesterol-binding agent. However, the nuclear envelope and some intracellular membranes are poorly affected (Figure 20). The distribution of filipin deformations is homogeneous for most cells of *Ephydatia*. They cover the entire surface of pinacocytes (Figure 21) and of internal cells. One exception is represented by the choanocytes. The membrane domains located at the base of the collar microvilli (Figure 22) or at their distal end (Figure 23) are less sensitive to filipin. Similarly, a slight-to-negative response occurs near the rosettes (Figure 24).

Discussion

As emphasized in previous reports (Garrone et al., 1980; Lethias et al., 1983), freeze-fracture studies of sponge cells yield both expected and surprising results. In this study, the membranes of sponge cells were fractured in a fashion similar to all other cells and their particle contents and densities are not unusual. They reacted to filipin in the same manner as other cells (Severs and Robenek, 1983): the response was positive for the plasma membrane, endocytotic vesicles; slight-to-negative for most intracellular membranes, especially for the nuclear envelope; and moderate in areas containing intramembrane particle grouping (lines or rosettes). Whether or not such findings can be interpreted as a direct reflection of sterol distribution in membranes remains to be demonstrated. It would be interesting to follow up on these observations on marine sponges and attempt to correlate them with biochemical measurements of sterol content.

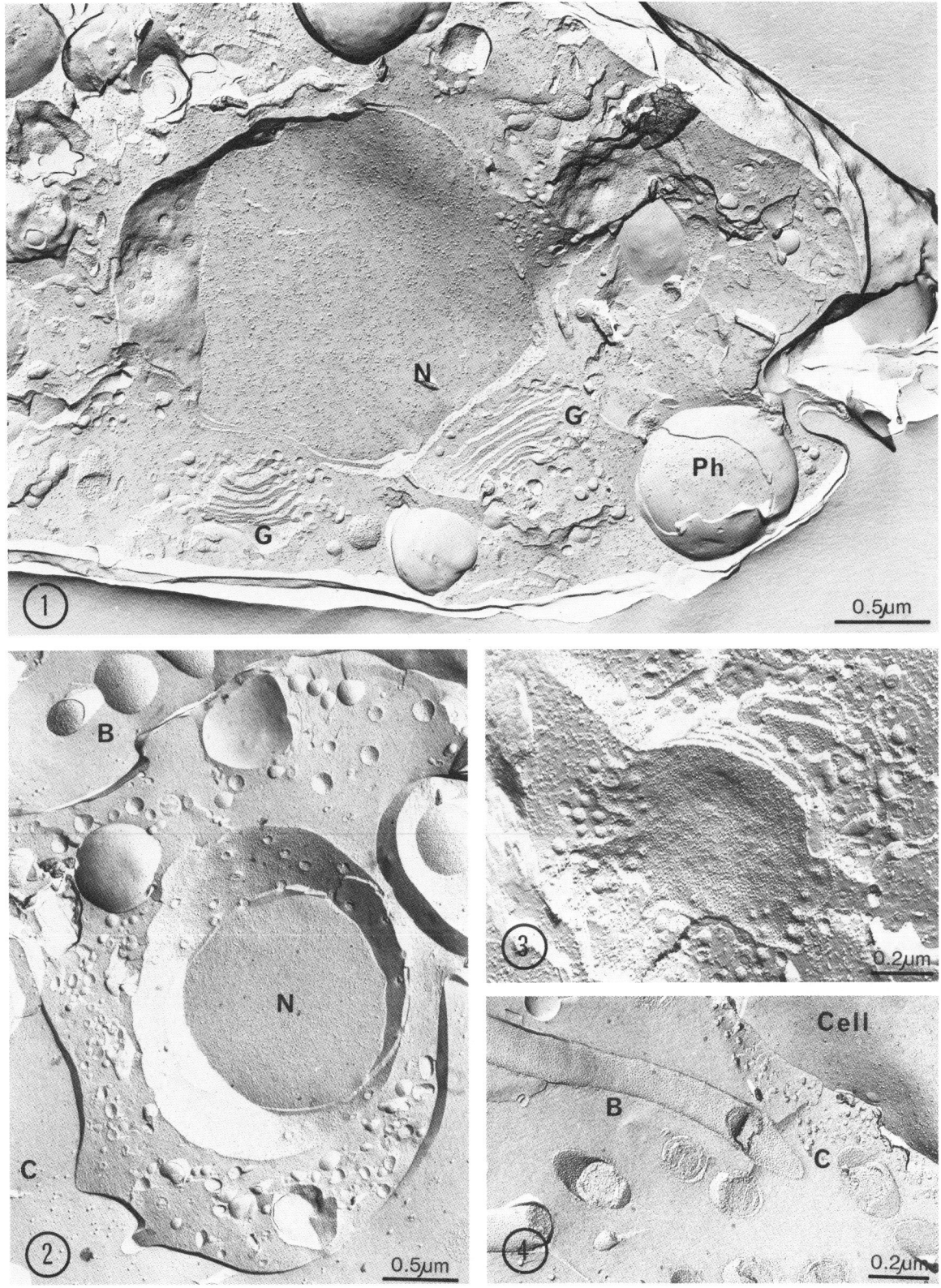

Figures 1–4. Freeze-fracture electron micrograph of sponge cells: *1, Ephydatia muelleri* archeocyte; *2, Verongia cavernicola* archeocyte, collagen fibrils in cross section; *3, Ephydatia muelleri*, Golgi saccule; *4, Hemimycale columella*, symbiotic bacteria and cross section of collagen fibrils surrounding a cell. *B*, symbiotic bacteria; *C*, cross sections of collagen fibrils; *G*, golgi saccules; *N*, nucleus; *Ph*, phagosomes.

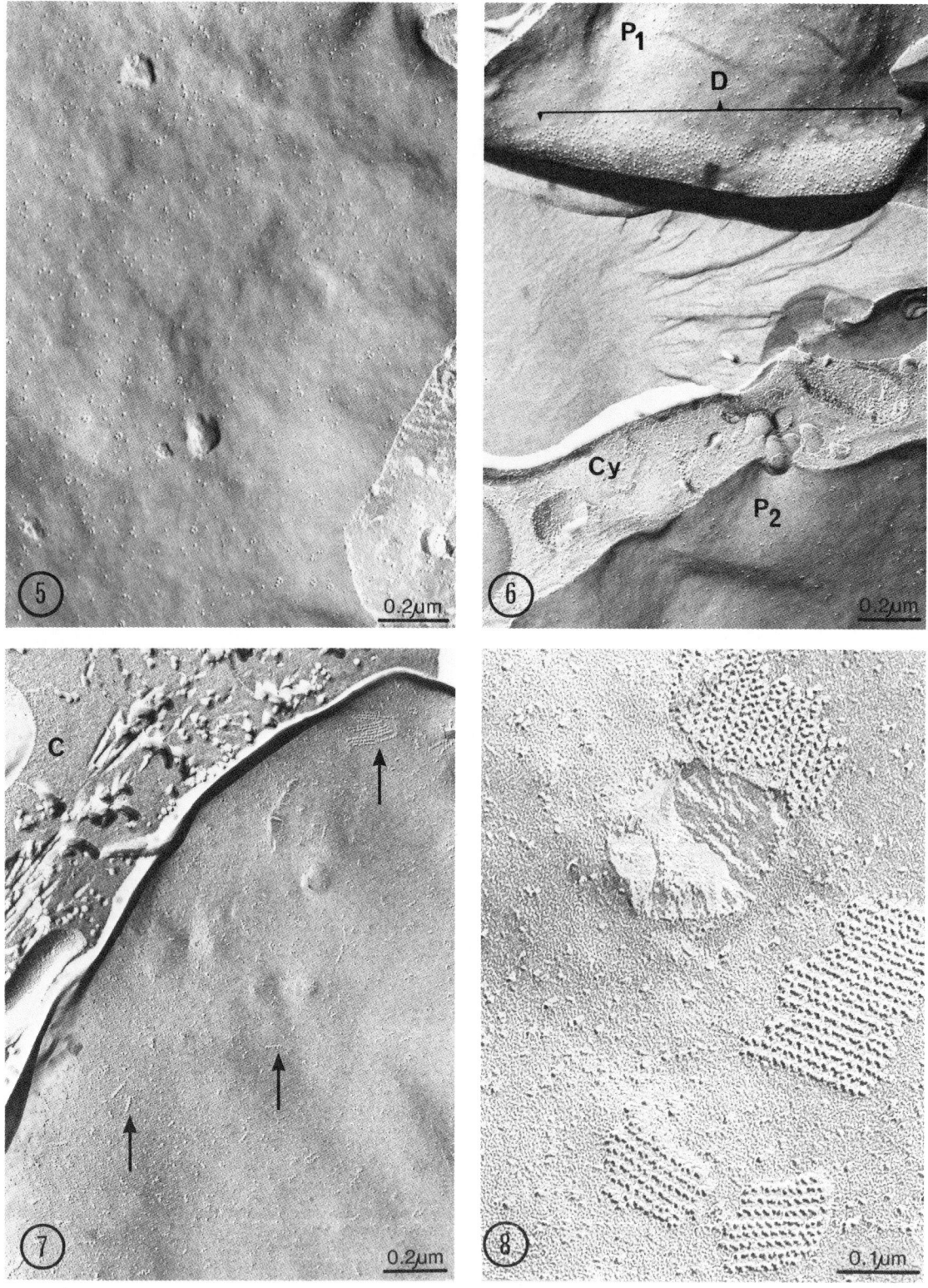

Figures 5–8. Freeze-fracture electron micrographs of sponge cells: *5, Ephydatia muelleri*, intramembrane particles on a P-face membrane of an exopinacocyte; *6, Ephydatia muelleri*, grouping of particles *(D)* corresponding to a desmosome on the P-face of a pinacocyte; *7, Chondrilla nucula* mesohyl cell; note lines of particles *(arrows)* on P-face membrane; *8, Chondrilla nucula* mesohyl cell; lines of particles form plates exhibiting a chevron pattern. *C*, collagen fibrils; *Cy*, cytoplasm of pinacocyte; *D*, grouping of particles corresponding to a desmosome; P_1 and P_2, plasma membrane (P-face) of two pinacocytes.

3d Int. Sponge Conf. 1985

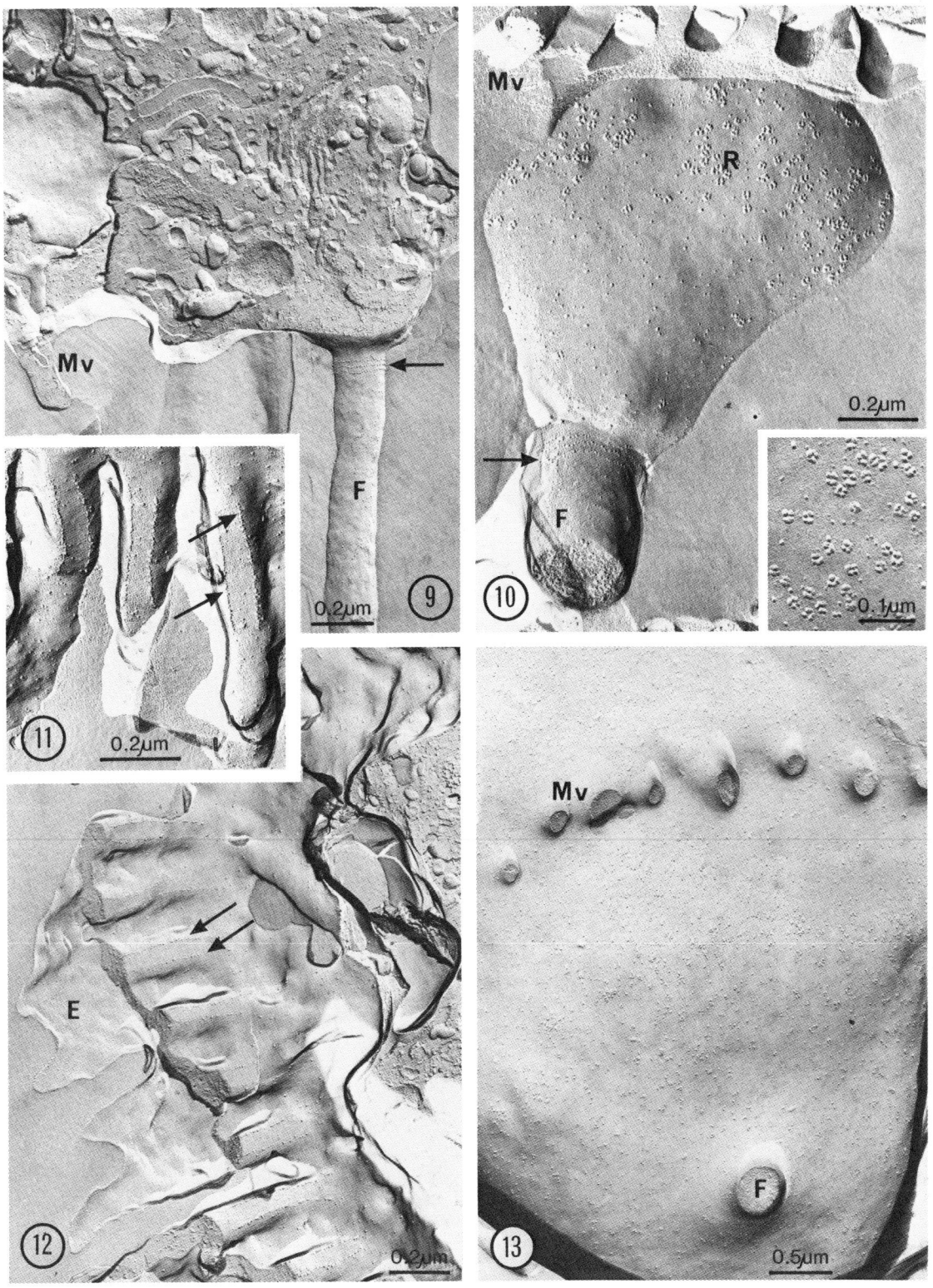

Figures 9–13. Freeze-fracture electron micrographs of sponge cells: *9, Ephydatia muelleri* choanocyte (note flagellar necklace, *arrow*); *10, Ephydatia muelleri* choanocyte; arrow indicates flagellar necklace, inset shows a rosette; *11, Ephydatia muelleri*, parallel lines *(arrows)* on microvilli of a choanocyte collar; *12, Ephydatia muelleri*, parallel lines *(arrows)* on microvilli at base of a choanocyte collar; *13, Oscarella lobularis* choanocyte. *E*, E-face of membrane; *F*, flagellum (cross section in *13*); *Mv*, microvilli; *R*, rosettes.

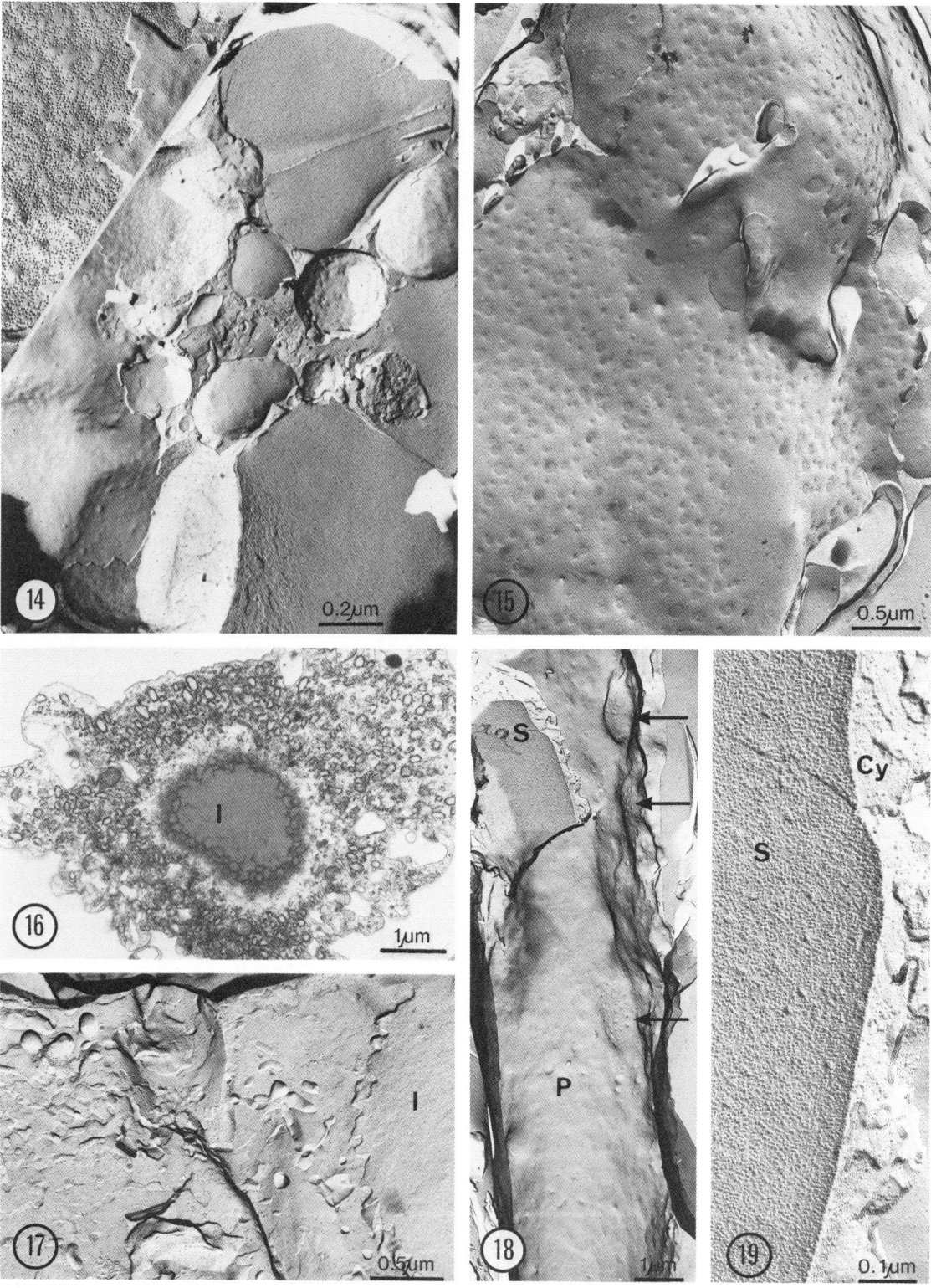

Figures 14–19. Freeze-fracture and thin section electron micrographs of sponge cells: *14, Chondrosia reniformis* spherulous cell; inset shows membrane of a spherule; *15, Verongia cavernicola* spherulous cell; note numerous pits on plasma membrane P-face; *16, Ephydatia muelleri* cystancyte (thin section); *17, Ephydatia muelleri* cystancyte (freeze-fractured); *18, Ephydatia muelleri* sclerocyte; note pitted depressions *(arrows); 19, Ephydatia muelleri* sclerocyte silicalemma showing high intramembrane particle density. *Cy,* cytoplasm; *I,* large inclusion; *P,* plasma membrane; *S,* silicalemma.

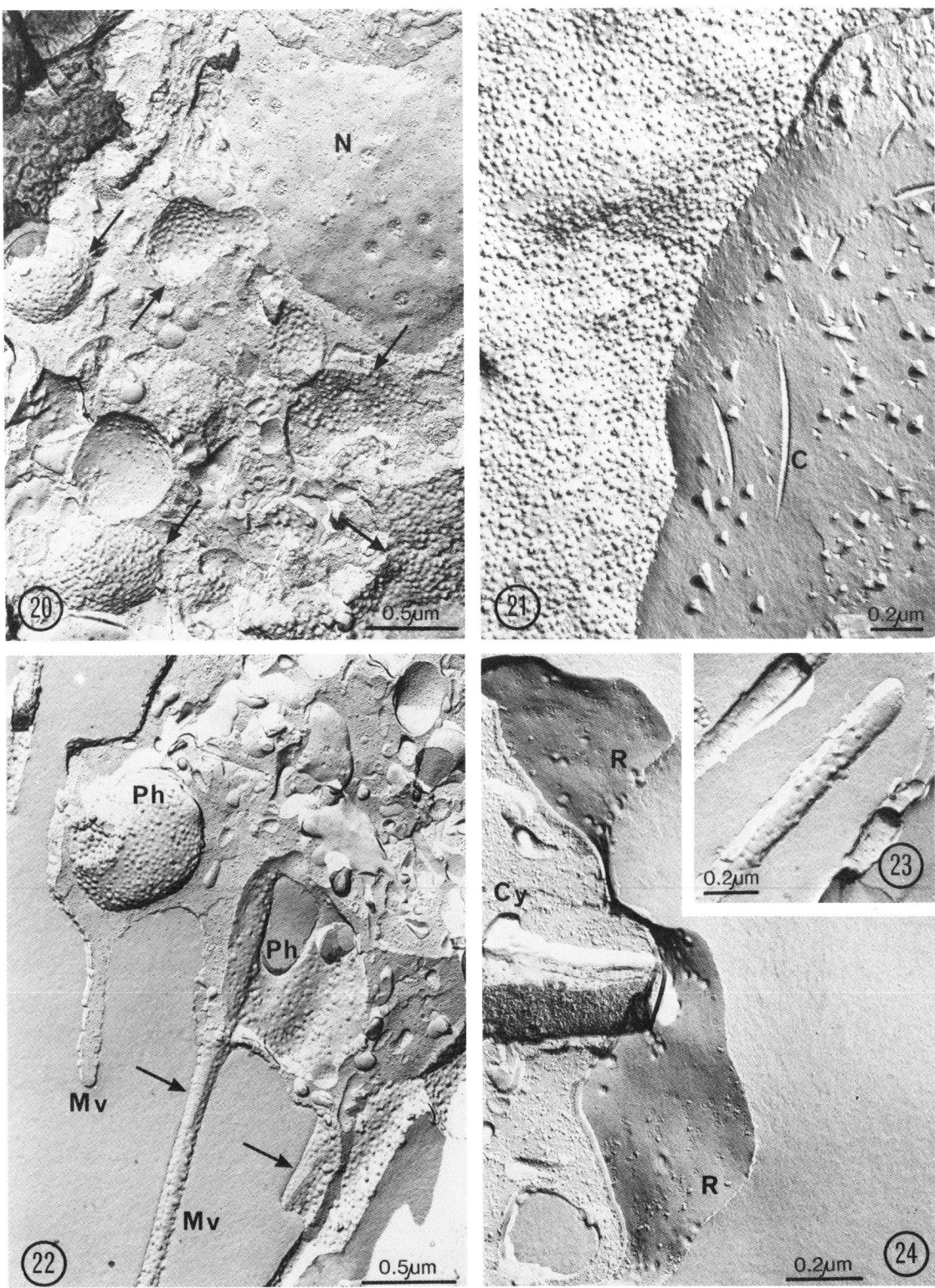

Figures 20–24. Freeze-fracture electron micrographs of sponge cells: *20, Ephydatia muelleri* archeocyte, filipin incubation; phagosome membranes *(arrows)* show characteristic deformations induced by sterol-binding agent; *21, Ephydatia muelleri* pinacocyte, characteristic effect of filipin in this plasma membrane; *22, Ephydatia muelleri* choanocyte, reduced effect of filipin at base of collar microvilli *(arrows); 23, Ephydatia muelleri* choanocyte, note reduced effect of filipin at distal end of a microvillus; *24, Ephydatia muelleri* choanocyte, reduced effect of filipin in rosette area. *C,* collagen fibrils; *Cy,* cytoplasm; *Mv,* microvilli; *N,* nucleus; *Ph,* phagosomes; *R,* rosettes print on E-face membrane.

Some intramembrane particle arrangements can be correlated to known structures or functions reasonably well. These include the lines on the choanocyte microvilli, which probably represent the intramembrane part of surface glycoproteins, and the high density of particles on the silicalemma of sclerocytes, which should correspond to the Si-pump function of this membrane (Simpson et al., 1983). For other particle groupings, however, the interpretation is less clear. This is the case for the choanocyte rosettes and for the lines and plates on mesohyl cells. The former structures might be related to the genesis of the glycoproteic microfibrils of the cell coat in a manner comparable to the synthesis of cellulose microfibrils (Staehelin and Gibbing, 1982). Note that the microfibrillar architecture of the cellulose cell wall in the plant is membrane-mediated through strings of particles and particle rosettes and is oriented by membrane-associated microtubules. The latter might be adhesive devices between cells and external proteins (Lethias et al., 1983; Garrone, 1984).

In addition to drawing attention to the high density of intramembrane particles in the spherulous cells of *Chondrosia* (Lethias et al., 1983), we have demonstrated an intriguing structure for the spherulous cells in *Verongia:* the pitted membranes. These structures resemble the pitted depressions found in the plasma membrane of sclerocytes. A similar function in cell permeability is possible.

One important question remains unanswered, however: Do tight or gap junctions exist in sponges? Although the classical structures unequivocally associated with these junctions have not been found, unexpected particle arrangements or highly labile junctions may well exist on restricted membrane areas.

Acknowledgments

We thank A. Callé for her help in freeze-fracturing sponges incubated with filipin. We are grateful to R. Got for the gift of filipin, and to A. Bosch for photographic plates. We also thank C. Péchoux for typing this manuscript.

Literature Cited

Bardele, C. F. 1983. Comparative Freeze-Fracture Study of the Ciliary Membrane of Protists and Invertebrates in Relation to Phylogeny. *Journal of Submicroscopic Cytology,* 15:263–267.

Bergquist, P. R., and C. R. Green. 1977. An Ultrastructural Study of Settlement and Metamorphosis in Sponge Larvae. *Cahiers de Biologie Marine,* 18:289–302.

Branton, D. 1971. Freeze-Etching Studies of Membrane Structure. *Philosophical Transactions, Royal Society, London B,* 261:133–138.

Garrone, R. 1984. Formation and Involvement of Extracellular Matrix in the Development of Sponges, a Primitive Multicellular System. Pages 461–477 in *The Role of Extracellular Matrix in Development,* edited by R. L. Trelstad. New York: Alan R. Liss.

Garrone, R., C. Lethias, and J. Escaig. 1980. Freeze-fracture Study of Sponge Cell Membranes and Extracellular Matrix. Preliminary Results. *Biologie Cellulaire,* 38:71–74.

Lethias, C., R. Garrone, and M. Mazzorana. 1983. Fine Structure of Sponge Cell Membranes: Comparative Study with Freeze-fracture and Conventional Thin Section Methods. *Tissue and Cell,* 15:523–535.

Miller, R. G. 1984. The Use and Abuse of Filipin to Localize Cholesterol in Membranes. *Cell Biology, International Reports,* 8:519–535.

Severs, N. J., and H. Robenek. 1983. Detection of Microdomains in Biomembranes. An Appraisal of Recent Development in Freeze-fracture Cytochemistry. *Biochimica et Biophysica Acta,* 737:373–408.

Simpson, T. L., R. Garrone, and M. Mazzorana. 1983. Interaction of Germanium (Ge) with Biosilification in the Freshwater Sponge *Ephydatia mülleri:* Evidence of Localized Membrane Domains in the Silicalemma. *Journal of Ultrastructure Research,* 85:159–174.

Staehelin, L. A., and T. H. Gibbing. 1982. Membrane-mediated Control of Cell Wall Microfibrillar Order. Pages 133–147 in *Developmental Order: Its Origin and Regulation,* edited by S. Subtelny and P. B. Green. New York: Alan R. Liss.

BRUNO BURLANDO
ELDA GAINO
Istituto di Zoologia
Università di Genova
via Balbi 5, I–16126 Genova, Italy

Cytoskeleton and Motility of *Clathrina cerebrum* Dissociated Cells

Abstract

Cultured cells obtained from dissociation of the calcareous sponge *Clathrina cerebrum* were studied by fluorescence microscopy, scanning electron microscopy, and transmission electron microscopy. After dissociation and settling, cells are rounded and show cytoplasmic processes such as lamellipodia and scleropodia. Scleropodia are peculiar thin extensions characterized by a prominent actin core; they function as attachment sites and promote cellular spreading. Spread out cells assume a fibroblastlike shape and show no motility. The cytoskeleton, the major component of which is actin, is organized into either a reticular meshwork of filaments or a ramified system of bundles. The reticular arrangement is associated with phenomena of cell motility, such as lamellipodium extension. The formation of parallel arrays of actin filaments is related to the protrusion of scleropodia. The cytoskeleton of fully spread out cells is composed mainly of filament bundles. Patterns of cell behavior may be correlated with cytoplasmic processes involved in cell locomotion and reaggregation.

Advances in experimental methods have been responsible for much of the recent progress in research on the cytoskeleton. The use of fluorescent probes, for example has improved our understanding of the cellular distribution of cytoskeletal components (Lazarides and Weber, 1974; Goldman et al., 1975; Weber et al., 1975; Lazarides, 1976). Furthermore, electron microscopic analyses of whole-mounted cells extracted with cytoskeleton-preserving buffers have made it possible to examine cytoskeletal structures directly (Small and Langanger, 1981; Schliwa and Van Blerkom, 1981; Schliwa, 1982; Letourneau and Ressles, 1983).

Our knowledge of the cytoskeleton of sponge cells is still very poor, however (see the review by Simpson, 1984). In early studies in this field, prominent microfilament bundles were detected by electron microscopy in cells named myocytes (Bagby, 1966). More recently, the presence of actin was shown in basopinacocytes of the freshwater sponge *Ephydatia* (Pavans de Ceccatty, 1981). Treatment of dissociated sponge cells with cytochalasin B has suggested that microfilaments are involved in both cell migration and reaggregation (Gaino et al., 1979). In addition, prominent actin structures have been identified in cells of a calcareous sponge by fluorescence microscopy (Burlando et al., 1984), and the organization of the cytoskeleton has been revealed by whole-mount electron microscopy (Gaino et al., 1985). This study concentrated on light and electron microscopic data in order to evaluate the nature and function of cytoskeletal components in dissociated cells of the calcareous sponge *Clathrina cerebrum*.

Materials and Methods

Specimens of the calcareous sponge *Clathrina cerebrum* were collected along the coast of the Promontory of Portofino (Ligurian Sea), at 3–5 m depth.

CELL CULTURE. Cells were dissociated by squeezing small fragments of the sponge through a fine mesh cloth, and were cultured in filtered natural seawater in 50 mm Falcon plastic dishes (Beckton Dickinson).

SCANNING ELECTRON MICROSCOPY. Dissociated cells were plated on glass coverslips and fixed 2 h after dissociation with 2.5% glutaraldehyde in artificial seawater (pH 7.5) for 0.5 h at 4°C. Cells were subsequently washed in artificial seawater, dehydrated with ethanol, and critical-point dried from CO_2. Specimens were covered with gold using a SEM Coating Unit E 5000 (Polaron Equipment Ltd.) and observed in a Philips 500 SEM.

FLUORESCENCE MICROSCOPY. Dissociated cells were plated on glass coverslips and fixed 1.5 h after dissociation in 3.7% formaldehyde in artificial seawater (pH 7.6) for 10 min at room temperature. Coverslips were then washed in artificial seawater and rinsed in phosphate-buffered saline (PBS) (pH 7.6). Thereafter, cells were incubated with fluorescein-labeled phalloidin (F-PHD) (a gift of P. C. Marchisio, Universitá di Torino) for 45 min at 37°C, rinsed in PBS, and washed in distilled water. Coverslips were mounted in PBS (pH 7.8) containing 30% Mowiol (Hoechst) and 2% n-propyl-gallate (Sigma Chemical Co.) on microscope slides. Cells were observed with a Leitz microscope equipped with epifluorescent illumination using a Planapo 63× oil-immersion objective (Carl Zeiss).

WHOLE-MOUNT MICROSCOPY. Dissociated cells were plated on 200-mesh nickel electron microscope grids supporting a polylysine-treated, carbon-coated formvar film. Two hours after dissociation, cells were briefly washed in artificial seawater, rinsed in PHEM buffer (60 mM PIPES, 25 mM HEPES, 10 mM EGTA, 2 mM MgCl, pH 6.9), and extracted with PHEM containing 0.1% Triton X-100 for 2 min at room temperature. Some grids were treated with 1 mg/ml heavy meromyosin (HMM) in glycerol for 5 min and were carefully washed in PHEM buffer. Both HMM-treated and untreated specimens were then fixed with 1% glutaraldehyde in PHEM containing 0.5% tannic acid for 25 min, washed with buffer, post-fixed with 0.5% OsO_4 for 30 min, and dehydrated in ethanol. Cells were critical point dried from CO_2, covered with a film of carbon, and observed in a Zeiss EM 9 electron microscope.

Results

When the cells of *Clathrina cerebrum* are dissociated and settle on a surface, they tend to coalesce and form aggregation masses. After dissociation and plating, cells are generally rounded and show particular microextensions named scleropodia (Burlando et al., 1984). Upon attachment, they either maintain a rounded shape and show motility or begin spreading out. Motile cells produce lamellipodia, which, together with scleropodia, are involved in cell migration and cell-to-cell contact formation.

Scleropodia are thin elongated microextensions by which cells firmly adhere to the substratum (Figure 1a). Staining with F-PHD and fluorescent microscopic observations reveal the presence of a prominent actin core within these protuberances (Figure 1b). In extracted, whole-mounted cells observed by electron microscopy, scleropodia (200–500 nm in diameter) appear to consist of a thick bundle of filaments that diverge at the base and are in continuity with bundles of the cell body (Figure 1c).

Lamellipodia are typical of motile cells and protrude from anywhere around the cell perimeter (Figure 2a); their motility gives rise to ruffles and occasionally causes the entire lamina to fold (Figure 2b). Protruding lamellipodia show a diffuse actin-dependent fluorescence and sometimes engulf the actin core of scleropodia (Figure 2c). They are occasionally crossed by ridges representing scleropodial remnants (Figure 2d). The cytoskeleton of the lamina consists mainly of a peripherally thickened meshwork of 10–15 nm filaments with condensation centers (Figure 3a).

Cell spreading involves the progressive engulfment of scleropodia by the flattening cell body. The microextensions induce the cell to assume a stellate shape by transforming their core into a system of divergent fibers (Figure 3b). These fibers are branched and intersect the meshwork of filaments. In fully spread-out cells, the filament meshwork almost disappears and the cytoskeleton is arranged into a network of ramified bundles; the large marginal fibers give the cell a fibroblastlike shape (Figure 3c).

After treatment with HMM, the reticular and the fi-

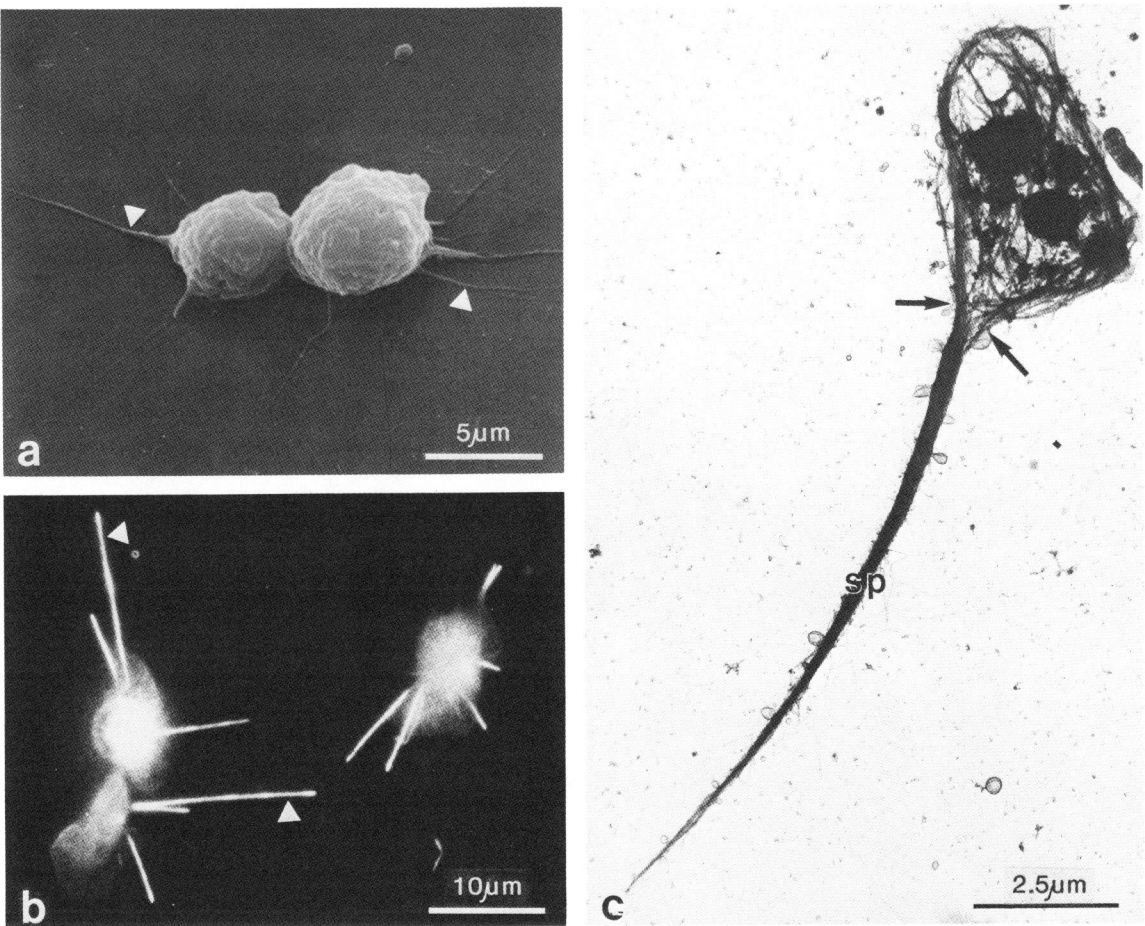

Figure 1. *Clathrina cerebrum*: *a*, scanning electron micrograph showing two dissociated cells with scleropodia *(arrowheads)*; *b*, dissociated cells stained with fluorescein-labeled phalloidin (F-PHD) and observed by epifluorescence microscopy; scleropodia *(arrowheads)* clearly shown by their higly fluorescent actin core; *c*, transmission electron micrograph of an extracted, whole-mounted cell with a long scleropodium *(sp)*; scleropodium shows a filamentous core connected with bundles *(arrows)* running within cell body.

brous components of the cytoskeleton become decorated (Figure 3*d*).

Discussion

The in vitro cytoskeletal patterns of dissociated cells of *Clathrina cerebrum* are comparable to those of cultured vertebrate cells (Stossel, 1984). In these sponge cells, prominent actin fluorescence and extensive HMM-dependent decoration suggest that actin is a major component of the cytoskelof ton.

Whole-mount microscopy shows that the cytoskeleton is essentially arranged in either a meshwork of filaments or a system of bundles. The meshwork is particularly developed in rounded cells and might be strictly related to motility. It shows specialized regions such as marginal thickening and condensation centers. The latter are similar to the actin foci and probably represent attachment sites (Ryder et al., 1984; Trotter, 1981). The system of bundles running within the cell body is connected with the cores of scleropodia. These structures originate from the arrangement of actin into parallel filaments. The intersection of the reticular meshwork and the system of bundles at the base of scleropodia demonstrated by cellular spreading suggests their possible interaction. Cell spreading involves the arrangement of the cytoskeleton into an extensive network of ramified fibers that can reach the size of fibroblast stress fibers. During this process, actin takes on a more structural function, in contrast to its otherwise dynamic role.

The in vitro behavior of *Clathrina cerebrum* cells is influenced by their cytoplasmic protuberances. Scleropodia contribute to cell-to-substratum attachment and probably also play an exploratory role comparable to that of fibroblast filopodia (Albrecht-Buehler, 1976). They could also be involved in cell aggregation by transducing infor-

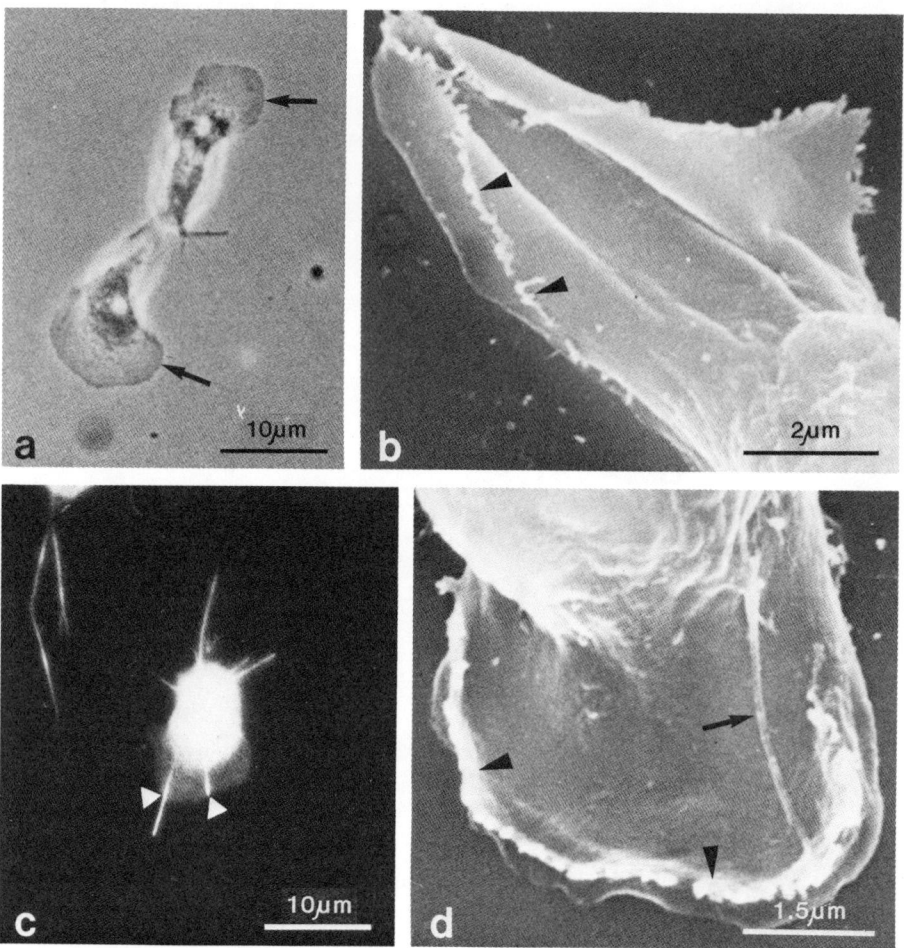

Figure 2. *Clathrina cerebrum: a,* phase-contrast micrograph of two dissociated cells producing lamellipodia *(arrows); b,* scanning electron micrograph of a lamellipodium characterized by prominent motility; arrowheads point to ruffles; *c,* F-PHD-stained cell showing a lamellipodium engulfing fluorescent scleropodial cores *(arrowheads); d,* scanning electron micrograph of a lamellipodium including a scleropodial remnant *(arrow);* arrowheads point to ruffles.

mation among cells, as do stereocilia of inner ear cells (Flock and Cheung, 1977). During cell locomotion, protrusion and withdrawal of scleropodia could depend on a polymerization-depolymerization mechanism involving their filamentous core. Actin-myosin interaction probably causes lamellipodium motility by triggering contraction of the reticular meshwork. The dynamics of cytoplasmic extensions is probably related to phenomena of nonrandom migration pointed out in these cells (Gaino et al., 1986).

The highly structured cytoskeleton of *Clathrina cerebrum* cells is unusual and could be related to the loosely assembled cellular organization of this asconoid sponge. In fact, the stability of the sponge epithelia could depend, at least in part, on the cytoskeletal structures of their cells.

Conclusions

Dissociated cells of the calcareous sponge *Clathrina cerebrum* are rounded and produce rodlike scleropodia and lamellipodia. They can occasionally undergo spreading, during which they show no motility and assume a polygonal shape.

The cytoskeleton consists mainly of actin structures arranged in either a reticular meshwork of 10–15 nm filaments or a ramified system of bundles.

Scleropodia have a core of actin filaments. Lamellipodia are supported by a peripherally thickened meshwork of microfilaments.

The dynamics of scleropodia and lamellipodia is involved in cell locomotion and reaggregation.

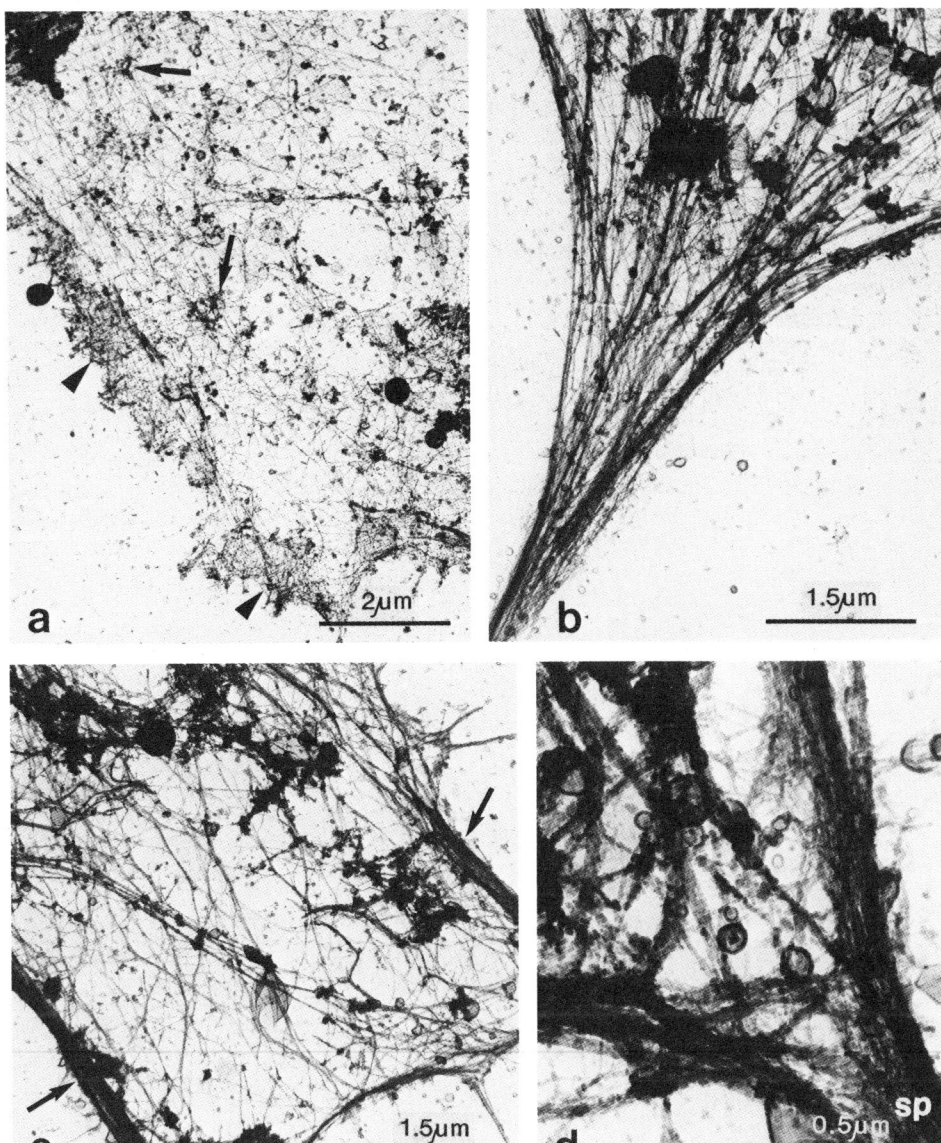

Figure 3. *Clathrina cerebrum*: *a*, transmission electron micrograph of an extracted lamellipodium showing a peripherally thickened *(arrowheads)* meshwork of filaments converging on condensation centers *(arrows)*; *b*, part of an extracted spreading cell showing base of a scleropodium transforming into a system of divergent fibers; *c*, transmission electron micrograph of an extracted spread-out cell showing a network of ramified bundles and large peripheral fibers *(arrows)*; *d*, part of an extracted, HMM-labeled cell showing extensive decoration of cytoskeleton; *sp*, scleropodium.

Acknowledgments

We thank Maria Agnese Sabatini from the Istituto di Anatomia Comparata, Università di Modena, for her help with the scanning electron microscope.

Literature Cited

Albrecht-Buehler, G. 1976. The Function of Filopodia in Spreading 3T3 Mouse Fibroblasts. Pages 247–264 in *Cell Motility*, edited by R. Goldman et al. Cold Spring Harbor, New York: Cold Spring Harbor Laboratory.

Bagby, R. M. 1966. The Fine Structure of Myocytes in the Sponges *Microciona prolifera* (Ellis and Solander) and *Tedania ignis* (Duchassaing and Michelotti). *Journal of Morphology*, 118:167–182.

Burlando, B., E. Gaino, and P. C. Marchisio. 1984. Actin and Tubulin in Dissociated Sponge Cells. Evidence for Peculiar Actin-Containing Microextensions. *European Journal of Cell Biology*, 35:317–321.

Flock, A., and H. Cheung. 1977. Actin Filaments in Sensory Hairs of the Inner Ear Receptor Cells. *Journal of Cell Biology*, 75:339–343.

Gaino, E., B. Burlando, and M. Sarà. 1979. Le espansioni cellulari in dissociati di *Clathrina clathrus* (Schmidt) (Porifera, Calcispongiae) ed effetti della citocalasina B e della colchicina. *Bollettino dei Musei e degli Istituti Biologici, Università di Genova*, 47:35–54.

Gaino, E., B. Burlando, M. A. Sabatini, and P. Buffa. 1985. Cytoskeleton and Morphology of Dissociated Sponge Cells. A Whole-Mount and Scanning Electron Microscopic Study. *European Journal of Cell*, 39:328–332.

Gaino, E., L. Zunino, B. Burlando, and M. Sarà. 1986. The Locomotion of Dissociated Sponge Cells: A Cell-by-Cell, Time-Lapse Film Analysis. *Cell Motility*, 5:463–474.

Goldman, R. D., E. Lazarides, R. Pollack, and K. Weber. 1975. The Distribution of Actin in Non-Muscle Cells. *Experimental Cell Research*, 90:333–344.

Lazarides, E. 1976. Actin, α-Actinin and Tropomyosin Interaction in the Structural Organization of Actin Filaments in Non-Muscle Cells. *Journal of Cell Biology*, 68:202–219.

Lazarides, E., and K. Weber. 1974. Actin Antibody: The Specific Visualization of Actin Filaments in Non-Muscle Cells. *Proceeding, National Academy of Science, USA*, 71:2268–2272.

Letourneau, P. C., and A. H. Ressler. 1983. Differences in the Organization of Actin in the Growth Cones Compared with the Neurites of Cultured Neurons from Chick Embryos. *Journal of Cell Biology*, 97:963–973.

Pavans de Ceccatty, M. 1981. Demonstration of Actin Filaments in Sponge Cells. *Cell Biology, International Reports*, 5:945–952.

Ryder, M. I., R. N. Weinreb, and R. Niederman. 1984. The Organization of Actin Filaments in Human Polymorphonuclear Leukocytes. *Anatomical Record*, 209:7–20.

Schliwa, M. 1982. Action of Cytochalasin D on Cytoskeletal Networks. *Journal of Cell Biology*, 92:79–91.

Schliwa, M., and J. Van Blerkom. 1981. Structural Interaction of Cytoskeletal Components. *Journal of Cell Biology*, 90:222–235.

Simpson, T. L. 1984. *The Cell Biology of Sponges*. New York: Springer. 662 pp.

Small, J. V., and G. Langanger. 1981. Organization of Actin in the Leading Edge of Cultured Cells: Influence of Osmium Tetroxide and Dehydration on the Ultrastructure of Actin meshwork. *Journal of Cell Biology*, 91:695–705.

Stossel, T. P. 1984. Contribution of Actin to the Structure of the Cytoplasmic Matrix. *Journal of Cell Biology*, 99:15s–21s.

Trotter, J. A. 1981. The Organization of Actin in Spreading Macrophages. *Experimental Cell Research*, 132:235–248.

Weber, K., R. Pollak, and T. Bibring. 1975. Antibody Against Tubulin: The Specific Visualization of Cytoplasmic Microtubules in Tissue Culture Cells. *Proceedings, National Academy Science, USA*, 72:459–463.

L. COURTNEY SMITH*
WILLIAM H. HILDEMANN**
Department of Microbiology and Immunology
School of Medicine
University of California
Los Angeles, California 90024

Cellular Morphology of *Callyspongia diffusa*

Abstract

Callyspongia diffusa, an Indo-Pacific sponge, shows a cell-mediated cytotoxic rejection response to artificially grafted allogeneic tissues. Basic to the understanding of this cellular interaction is a characterization of the cells normally found within this species. Histological analysis of the tissues of this sponge reveals pinacocytes, choanocytes, archeocytes, spherulous cells, acid mucopolysaccharide-positive cells, acidophilic granular cells, globoferous cells, sclerocytes, and germ cells.

Many different cell types from various sponge species have been morphologically characterized. Some distinct cell types, such as pinacocytes, choanocytes and archeocytes are found in all sponges. However, other cell types, including many of the mesohyl cells, are not found in all sponge species. Consequently, conclusions drawn from cytological analyses of one sponge cannot be assumed to apply to all other species.

Recently, *Callyspongia diffusa*, an Indo-Pacific sponge, has been used to investigate allograft rejection responses (Hildemann et al., 1979, 1980; Bigger et al., 1981, 1982; Johnston et al., 1981; Johnston and Hildeman, 1983; Smith and Hildemann, 1986 a,b). Because this response is a cell mediated cytotoxic reaction to non-self tissues, an understanding of the normal component of cell types within this species becomes important. In the present study, the morphological identifications of the cell types normally found within *C. diffusa* are presented. They include pinacocytes, choanocytes, archeocytes, spherulous cells, acid mucopolysaccharide positive (AMP+) cells, acidophilic granular cells, globoferous cells, sclerocytes, and germ cells.

Materials and Methods

Callyspongia diffusa is a brilliant purple sponge found on reef crests in Kaneohe Bay, Oahu, Hawaii. Several large

*Present address, Division of Biology, California Institute of Technology, Pasadena, California 91125.
**Deceased; 1927–1983.

specimens were collected from the fringing reefs around Coconut Island and from a patch reef near the Kaneohe Yacht Club. Sponges were maintained in unfiltered running seawater at the Hawaii Institute of Marine Biology (University of Hawaii) and used within 4 days of collection. Sponges were fixed overnight in Bouins fixative, followed by an extensive rinsing in filtered seawater and then in tap water. The specimens were routinely dehydrated and embedded in paraplast (Monoject Scientific, Dehand, FL). The spicular skeleton of the sponge made sectioning at room temperature impossible. Consequently, to harden the paraplast, the rough trimmed blocks were frozen (−25°C) and cut at 6–8 μm in a cryostat. Test sections were stained with 1% toluidine blue before deparaffinizing to assess tissue orientation. Selected sections were routinely stained with hematoxylin and eosin (H & E), periodic acid Schiff (PAS), alcian blue with fast red as a counter stain, and Masson's trichrome stain. The sections were photographed on Plus-X film in a Zeiss or Olympus photomicroscope.

Results

TISSUE ORGANIZATION. The tissues of *Callyspongia diffusa* are supported by the anastomosing, spongin-encased, spicular skeleton. The body of the animal is divided into two general regions: the choanosome and the ectosome (Figure 1). The choanosome encompasses the inner regions and contains the choanocyte chambers or water pumping structures, the smaller aquiferous canals, and the mesohyl. The ectosome is located in the superficial regions and includes the exopinacoderm or outer layer of cells, and the subdermal spaces located just below the exopinacoderm. The ectosome normally contains few choanocyte chambers, but in *C. diffusa* there are many areas in which the choanosome extends between the subdermal spaces to reach the outer covering.

The aquiferous canal system is an interconnection of incurrent and excurrent canals, at the junction of which are found the choanocyte chambers (Figure 1). Normal physiological functions of this system include food pro-

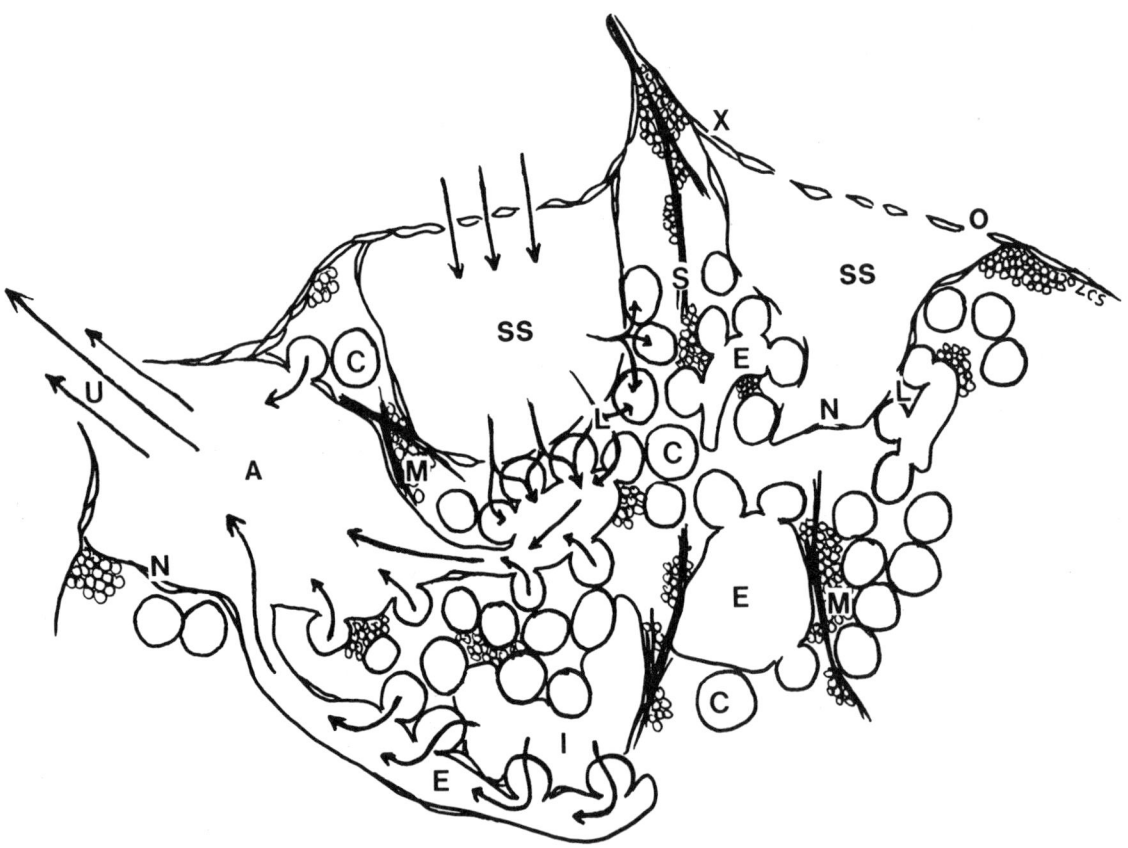

Figure 1. Pathway of water flow through *Callyspongia diffusa* begins as water *(arrows)* enters through incurrent pores or ostia, often grouped together in sieve-like complexes in exopinacoderm. From ostia, which lead into subdermal spaces and then to incurrent canals, water passes into lacunar spaces and then between cell bodies of spherically arranged choanocytes. After passing through chamber, water enters excurrent canal. Several excurrent canals join to form atrium located below excurrent pore or oscula. *A*, atrium; *C*, choanocyte; *E*, excurrent canal; *I*, incurrent canal; *L*, lacunar space; *N*, endopinacocyte; *O*, ostia; *S*, skeleton; *SS*, subdermal space; *U*, osculum; *X*, exopinacoderm.

curement, elimination, gas exchange, and gamete release (Simpson, 1984). An excellent account of the pathway of water flow in *Callyspongia diffusa* has been presented by Johnston and Hildemann (1982), and is briefly reviewed in Figure 1.

Sponges lack defined organ systems. The only anatomically discrete tissue is the epithelium, which is composed of two cell types: pinacocytes and choanocytes. These cells cover the sponge surface, line the aquiferous canal system, and are in constant, direct contact with the ambient water. The remaining cells, which are located in the mesohyl, are bounded on all sides by the epithelium and constitute the true internal regions of the sponge body. The cell types characterized in *C. diffusa* are presented in Tables 1–3.

PINACOCYTES. The pinacocytes cover the external and internal surfaces of the sponge. They are flattened, fusiform, slightly basophilic cells with small, centrally located nucleolate nuclei (Figure 2), and according to Johnston and Hildemann (1982) are irregular in shape and about 20 μm in diameter. The exopinacocytes line the outer covering of the sponge while the morphologically identical endopinacocytes line the aquiferous canal system. The external surface layer, the exopinacoderm, is composed of a double layer of cells: the outer exopinacocytes and the inner endopinacocytes that line the subdermal spaces and atria. These cells may line up edge to edge, or they may overlap. The exopinacoderm located over the subdermal spaces has many pores, or ostia, grouped together in sievelike structures that allow water to enter the aquiferous system. Each ostium appears to be an intracellular structure of a single porocyte (Johnston and Hildemann, 1982).

CHOANOCYTES. The spherical choanocyte chambers are located at the points where the incurrent and excurrent canals intersect. Choanocytes are rounded and slightly basophilic cells with condensed, basally located nuclei (Figure 3). Each choanocyte has a single flagellum oriented toward the center of the chamber which beats to create a flow of water through the aquiferous system. Around each flagellum, there is a ring or collar of microvilli that acts as the fine filter for collecting food particles brought in by the water current. (The collar and flagella on the choanocytes are not easily visualized by methods employed here; see Johnston and Hildemann, 1982). It is surprising that these cells, which function to capture food by filtration and phagocytosis and normally exhibit many phagosomes (Willenz, 1980), contain no cytoplasmic inclusions. Lack of inclusions may be a result of recent improvements in water clarity in Kaneohe Bay, which have reduced particulate food supplies.

3d Int. Sponge Conf. 1985

MESOHYL. The mesohyl is a loose association of cells bounded on all sides by pinacocytes or choanocytes. The several cell types found in this region include archeocytes, several types of spherulous cells, acid mucopolysaccharide positive (AMP+) cells, acidophilic granular cells, globoferous cells, sclerocytes, and germ cells.

ARCHEOCYTES. Archeocytes are easily identified by their large nucleolate nuclei (Figure 4). These cells are very heterogeneous; the cytoplasmic basophilia ranges from almost clear to dark blue and the cytoplasmic inclusions, which are also basophilic, can be numerous or completely absent. The archeocytes, which comprise 84.4% of the total cells in the mesohyl, are generally considered to be amoeboid, phagocytic cells.

SPHERULOUS CELLS. The spherulous cells have small, condensed, anucleolate nuclei that are located in the center of the cell. The cytoplasm is filled with large spherical inclusions (the basis for the cell name). There are three types of spherulous cells in *Callyspongia diffusa*, which are based on differences in spherule-staining properties (Table 1). By H & E, they are acidophilic, neutrophilic, and mixed. The acidophilic spherulous cells (Figure 5), which consist of 7.4% of the total mesohyl, are found scattered throughout the sponge and in clusters (containing 8.9 ± 5.6 cells) located just below the exopinacoderm where the mesohyl reaches the surface (Figure 6). This distribution has been noted in other species (Bretting and Königsmann, 1980; Bretting et al., 1983). The neutrophilic spherulous cells, which compose 7.8% of the mesohyl, are also scattered throughout the sponge, but they are never found clustered in groups, nor do they appear to release their spherules (Figure 7). The rare, mixed spherulous cells are usually found on the endopinacoderm (Figure 8).

OTHER MESOHYL CELLS. Several other rare mesohyl cell types include AMP+ cells, acidophilic granular cells, sclerocytes, and globoferous cells. The large AMP+ cells are best characterized by intense cytoplasmic staining with alcian blue, indicating the presence of acid mucopolysaccharides in the cytoplasm (Figure 9). The nuclei are small and in some cells appear stellate. These cells are located randomly within the mesohyl and are also found slightly flattened against the endopinacoderm inside aquiferous canals. Acidophilic granular cells have irregular shapes and small but prominent nucleolate nuclei, and are full of small acidophilic granules (Figure 10). They are usually located on the endopinacoderm, although they have been noted in the mesohyl. Globoferous cells, found rarely in the mesohyl, are identified by a large clear eccentric area and large basophilic spherical vacuoles that obscure the nuclei (Figure 11). Spicule secreting sclerocytes (Simpson and Vacaro, 1974), have only been identified in *Callyspongia diffusa* in preparations of dissociated cells by

the presence of intracellular microscopic spicules (not illustrated). This cell type has not been noted in histological
preparations, perhaps because the growing tips of the
sponge branches were not used in the experimental procedures.

GERM CELLS. In addition to the somatic cells, *Callyspongia
diffusa* contains sperm, eggs, and embryos (Table 2). In the
male sponge, the sperm are found in cysts located
throughout the mesohyl, but always near an aquiferous
canal (Figure 12). Within each cyst, cell division appears

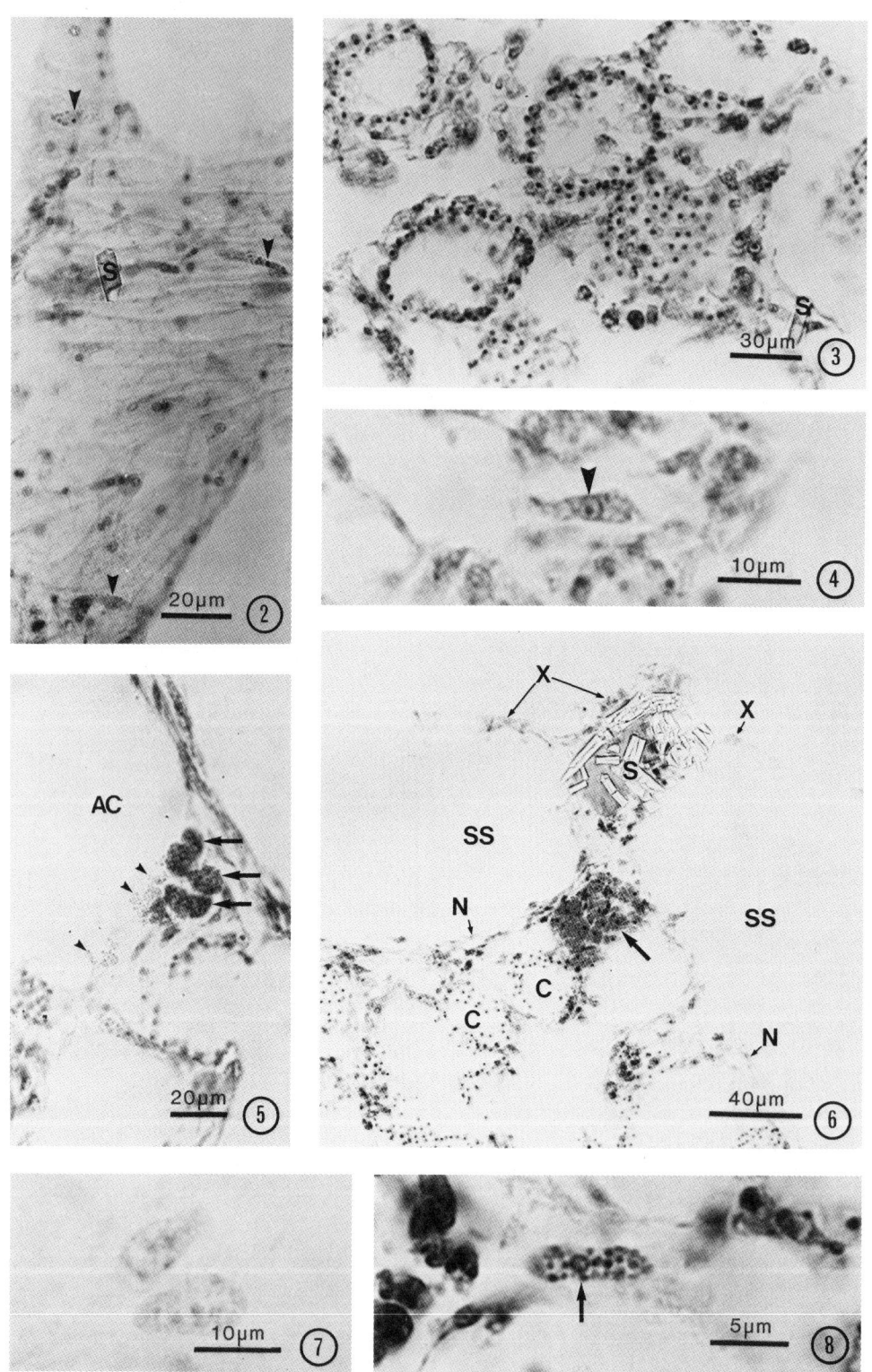

Figures 2–8. Histology of
Callyspongia diffusa: *2*, pinacocytes in an "en face" view (arrowheads show mixed spherulous cells), H & E stain; *3*,
choanocyte chambers located
within choanosome; flagellae
and microvillar collars not visible; H & E stain; *4*, an archeocyte shows prominent nucleolate nucleus *(arrowhead)*, H & E
stain; *5*, acidophilic spherulous
cells *(arrows)*; spherules *(arrowheads)* can be seen in aquiferous canal, H & E stain; *6*, a
group of acidophilic spherulous
cells *(arrow)* in ectosome near
exopinacoderm; H & E stain; *7*,
neutrophilic spherulous cells, H
& E stain; *8*, a mixed spherulous
cell *(arrow)* on endopinacoderm,
trichrome stain. *AC*, aquiferous
canal; *C*, choanocyte chamber;
N, endopinacoderm; *S*, spicule
fragments; *SS*, subdermal
space; *X*, exopinacoderm.

Table 1. Cell types in Callyspongia (+ = PAS probably positive, + + = PAS distinctly positive, ND = not determined, ? = not observed)

| Cell type | Histological staining properties | | | | Nuclear morphology | Cytoplasmic inclusions | Anatomical location | Dimensions (μm,±SD) | |
	H & E	Trichrome	Alcian blue/fast red	PAS				Nucleus	Cell (lxw)
Pinacocytes	pale blue	pale pink	moderately blue nucleolate, central	+	small, lines canals	none	covers surface,	2.02 (± 1.33)	20 (diam.)[a]
Choanocyte	pale blue to clear	pale pink	blue green	+ +	condensed, basal	none	chambers in the choanosome	ND	2.93 × 2.93
Archeocyte	pale blue to dark blue	pink to red	moderately blue central	+	large, nucleolate,	few to many	mesohyl	3.29	15.50 × 4.35 (± 0.66)
Acidophilic spherulous cell	red spherules	red spherules	lavender blue	+	condensed, anucleolate, central	spherules	mesohyl	ND	21.43 × 5.95 (± 7.12)(± 1.84)
Neutrophilic spherulous cell	clear spherules	blue spherules	lavender blue	+	condensed, anucleolate, central	spherules	mesohyl	ND	14.74 × 5.40 (± 3.49)(± 0.88)
Mixed spherulous cell[b]	red and clear spherules	red and blue spherules	?	+	condensed, anucleolate, central	spherules	mesohyl	ND	ND
AMP + cell	pink	blue	dark blue	+	condensed, anucleolate, central	large vacuoles	mesohyl, endopinacoderm	ND	14.32 × 10.50
Acidophilic granular cell[b]	red granules	?	lavender blue	+	moderate, nucleolate, central	many small inclusions	mesohyl, endopinacoderm	ND	5.69 × 4.79 (± 0.50)(± 0.02)
Globoferous cell[b]	?	blue	?	?	obscured by inclusions	many, large, with unstained area	mesohyl	ND	ND

[a]From Johnston and Hildemann, 1982.
[b]Very few cells observed and measured.

to be synchronous while different stages of sperm development are present in different cysts. The mean size of the sperm cyst tends to increase as the sperm inside mature (Table 3). Fertilization occurs within the female sponge after sperm are released from the male and brought into the female with the incurrent water flow (Figure 13). Eggs and embryos are located in the central core of female sponges within the mesohyl (Figure 14). Eggs differ greatly in size, have cytoplasmic inclusions and large nuclei with prominent nucleoli (Figure 13). Parenchymatous embryos develop within females in the same regions where the eggs are located. They are eventually released as ciliated, free swimming larvae (Figure 14) representing the dispersal stage of the sponge life cycle.

Table 2. Germ cells in *Callyspongia* (+ + = PAS distinctly positive, – = PAS negative, ND = not determined, extremely variable depending on the developmental stage)

| Cell type | Histological staining properties | | | | Nuclear morphology | Cytoplasmic inclusions | Anatomical location | Diameter (μm,±SD) | |
	H & E	Trichrome	Alcian blue/fast red	PAS				Nucleus	Cell (l × w)
Sperm	not measurable	not measurable	not measurable	–	very small	none	within cysts in the mesohyl near a canal	See Table 3	
Egg	Cytoplasm: pink Inclusions: pink/blue Nucleus: pink/purple	Cytoplasm: red/purple Inclusions: pink Nucleus: red	Cytoplasm: blue Inclusions: pink/red Nucleus: pink/red	+ +	very large, prominent nucleolus	many inclusions, various sizes	central part of the mesohyl	5.4 (±1.99) (nucleolus = 2.37 [±0.86])	35.4 × 22.2 (±18.7)(±12.04)
Embryo	same as egg	same as egg	same as egg	+ +	mostly nucleolate	yolk	same as egg	ND	ND

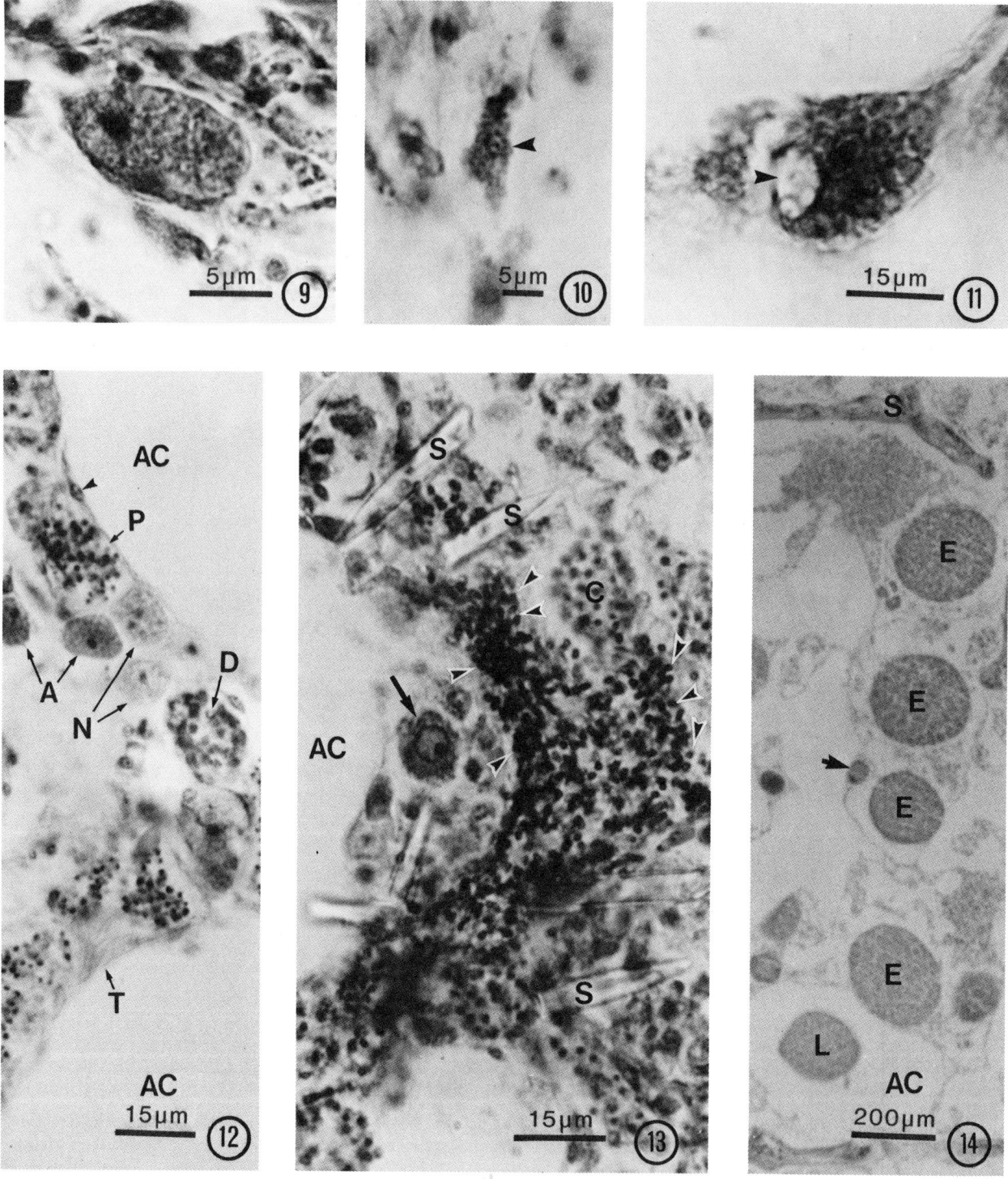

Figures 9–14. Histology of *Callyspongia diffusa: 9,* acid mucopolysaccharide positive (AMP+) cell in mesohyl, trichrome stain; *10,* acidophilic granular cell *(arrowhead)* on endopinacoderm, H & E stain; *11,* globoferous cell with acentric clear area *(arrowhead),* trichrome stain; *12,* sperm cysts in various stages of development (arrowhead indicates endopinacocyte nucleus), H & E stain; *13,* sperm *(arrowheads)* in choanosome of female sponge near egg *(arrow),* H & E stain; *14,* egg *(arrow),* embryos, and larva in choanosome of a female sponge; larva located within aquiferous canal; alcian blue with fast red stain. *A,* acidophilic spherulous cells; *AC,* aquiferous canals; *C,* choanocyte chamber; *D,* dividing sperm; *E,* embryo; *L,* larva; *N,* neutrophilic spherulous cells; *P,* sperm without tails; *S* skeleton, spicule fragments; *T,* sperm tails.

Table 3.   Sperm cysts in *Callyspongia*

Sperm Maturity	Cyst length, μm (± SD)	Cyst width, μm (± SD)
Dividing sperm	18.84 (± 2.16)	13.00 (± 3.66)
Sperm without tails	22.88 (± 3.18)	19.12 (± 2.02)
Sperm with tails	25.84 (± 7.76)	17.34 (± 5.20)

Discussion

The general organization of the typical sponge body consists of choanocyte chambers located at the interfaces of incurrent and excurrent streams of water that pass through a more or less well developed canal system. The unique structure of the choanocyte, with a single flagellum surrounded by a circle of microvilli, or collar, is remarkably constant from one species to another and is a hallmark of the poriferan phylum (Simpson, 1984). Yet, there are some basic anatomical variations within the leuconoid Demospongia. Orientations of the choanocyte chambers differ, implying different strategies of food capture for different species. For example, the choanocytes in *Callyspongia diffusa* and *Reniera* sp. (Langenbruch, 1983) are connected by short basal protrusions (Johnston and Hildemann, 1982) and form spherical chambers that "hang" in the lacunar (incurrent) spaces (Simpson, 1984). Here the choanocytes are the primary cells involved in particle capture and phagocytosis before passing the nutrients to the mesohyl cells for digestion. On the other hand, in a related species *(Ephydatia fluviatilis)* the flow of water passes from the incurrent canal into the mesohyl, where particles are phagocytosed directly by the mesohyl cells before entering the choanocyte chamber (Weissenfels, 1976). Thus, the long evolutionary history of the poriferans has allowed ample time for many successful structural "experiments" based on a flagellated-cell-powered filter feeding organization (Bergquist, 1978).

This long evolutionary history is also reflected in the great diversity of sponge cell morphology and function. For example, all pinacocytes function as epithelial cells in sponges, but in different species they can have quite different morphologies. In *Microciona prolifera*, the exopinacocyte morphology shows an external glycocalyx and a "T" configuration where the nucleus is located in the club shaped cell body that extends into the dermal mesohyl. The endopinacocytes in this species are uncoated and fusiform (Bagby, 1970). In *Callyspongia diffusa*, however, all pinacocytes are uncoated, fusiform, and identical to each other. Conversely, archeocytes in different sponge species

have a similar morphology, but they may be quite diverse in function. Variability in the sizes and number of cytoplasmic inclusions in the archeocytes of *C. diffusa* seems to indicate only different stages in the digestion of phagocytosed food particles, but other possibilities for archeocyte function can be inferred from this study. *Callyspongia diffusa* secretes silicious spicules and encases them in spongin, and in some situations, such as graft rejection and inflammation (Smith and Hildemann, 1986a) will secrete small amounts of sponge collagen. Such activities suggest the presence of sclerocytes, spongocytes, lophocytes or collencytes, but these cell types have not been identified in this species by their defined morphological characteristics (see Simpson, 1984). The possibility remains that the archeocytes in *C. diffusa* perform all of these functions. This has been suggested previously from studies on *M. prolifera*, where sclerocytes appear morphologically identical to archeocytes (Simpson, 1978).

In general, the activities of archeocytes in *Callyspongia diffusa* may include: (a) secretion of the skeleton, (b) phagocytosis of food particles and injury induced cell debris (Smith and Hildemann, 1986a), and (c) regeneration of damaged or lost cells (Smith and Hildemann, 1986a). In cytotoxic reactions to allogeneic contact, only a subpopulation of archeocytes seems to be directly involved at the graft interface (Smith and Hildemann, 1986b). This suggests that the archeocytes in *C. diffusa* are not a homogeneous population of cells capable of performing all types of functions, but that they may be a conglomerate of functionally differentiated subpopulations that includes a pluripotential stem cell.

The remaining mesohyl cells are an assemblage of diverse, poorly characterized, and confusing cell types (Simpson, 1984). The variety of interpretations of previous morphological investigations of sponge cells has added to the confusion already imposed by existing sponge diversity. In many instances, analyses have not progressed beyond simple cellular identifications. Such problems can be illustrated with several of the mesohyl cells in *Callyspongia diffusa*. There is a disagreement on the identification of acidophilic spherulous cells in another *Callyspongia* species. Pomponi (1976) calls this cell type a granular cell. The mixed spherulous cells may be a distinct cell type, but they may also be transitional forms between the acidophilic and neutrophilic spherulous cells, or remnants of acidophilic spherulous cells after spherule release. The identification of AMP+ cells is not certain. In some aspects, these cells are morphologically similar to AMP+ cells identified by Simpson (1963, 1968), and in others, they are similar to spongocytes characterized by Garrone and Pottu (1973). The acidophilic granular cells are also confusing. They seem most similar to eosinophilic amoebocytes in calcareous sponges (Simpson, 1984), and they also resemble some descriptions of gray cells (Boury-Esnault and Doumenc, 1979) although eosinophilia is not typical of most gray cells (Boury-Esnault, 1977). Final identification of these cell types will have to await more detailed analysis of cells from many sponge species.

Functional considerations of the less common mesohyl cells are not numerous. However, interest in the spherulous cells has yielded some interesting speculations. A nonspecific defense role for acidophilic spherulous cells in *Callyspongia diffusa* is suggested from the location of groups of these cells close to the surface, near incurrent subdermal spaces (Figure 6). In addition, their mass release of spherules into the canal system (Figure 5) may have a protective function. These speculations are supported by an antibiotic activity of spherulous cell contents from other marine sponges (Thompson et al., 1983). Alternatively, spherulous cells may also be involved in spongin secretion. Neutrophilic spherulous cells in *C. diffusa* stain blue with Masson's trichrome stain, as do the skeletal spongin and small patches of mesohyl collagen. These cells may have spongin secreting activity, even though they do not appear to be associated with the spongin casing around the skeleton nor with the collagen deposits as has been observed for spongocytes and spherulous cells in other species (Garrone and Pottu, 1973; Bretting et al., 1983). It is noteworthy that these two activities, nonspecific defense and spongin secretion, are found in morphologically similar spherulous cells in *Axinella polypoides* (Bretting et al., 1983), whereas, in *C. diffusa* these activities may be localized in different types of spherulous cells.

Conclusions

Future studies of sponge cell biology may show that a variety of morphologically different cells in different species perform similar functions and, conversely, that morphologically similar cells can perform very different functions even in the same species. In light of the age of this phylum and the various successful adaptations by sponges to challenges presented by different environments, final and definitive characterizations of sponge cell types may eventually have to be based primarily on the functions that the cells perform with secondary importance given to cellular morphology and anatomical locations.

Acknowledgments

This work was supported by National Institutes of Health grants AI 15075 and AI 19470. The authors are greatly indebted to C. H. Bigger for his assistance in all aspects of this study. Thanks are also extended to D. Doyle for his photographic expertise. V. L. Scofield, J. E. Neigel, and L. Muscatine, offered helpful suggestions and criticisms during preparation of the manuscript.

Literature Cited

Bagby, R. M. 1970. The Fine Structure of Pinacocytes in the Marine Sponge *Microciona prolifera* (Ellis and Solander). *Zeitschrift für Zellforschung und Mikroskopische Anatomie*, 105:579–594.

Bergquist, P. 1978. *Sponges*. Berkeley, California: University of California Press. 268 pp.

Bigger, C. H., W. H. Hildemann, P. L. Jokiel, and I. S. Johnston. 1981. Afferent Sensitization and Efferent Cytotoxicity in Allogeneic Tissue Responses of the Marine Sponge *Callyspongia diffusa*. *Transplantation*, 31:461–464.

Bigger, C. H., P. L. Jokiel, W. H. Hildemann, and I. S. Johnston. 1982. Characterization of Alloimmune Memory in a Sponge. *Journal of Immunology*, 129:1570–1572.

Boury-Esnault, N. 1977. A Cell Type in Sponges Involved in the Metabolism of Glycogen. The Gray Cell. *Cell and Tissue Research*, 175:523–539.

Boury-Esnault, N., and D. A. Doumenc. 1979. Glycogen Storage and Transfer in Primitive Invertebrates: Demospongea and Actiniaria. Pages 181–192 in *Biologie des Spongiaires*, edited by C. Lévi and N. Boury-Esnault. Colloques Internationaux du C.N.R.S. 291. Paris: Centre National de la Recherche Scientifique.

Bretting, H., G. Jacobs, C. Donadey, and J. Vacelet. 1983. Immunohistochemical Studies on the Distribution and the Function of the D-galactose Specific Lectins in the Sponge *Axinella polypoides* (Schmidt). *Cell and Tissue Research*, 229:551–571.

Bretting, H., and K. Königsmann. 1980. Investigations on the Lectin-Producing Cells in the Sponge *Axinella polypoides*. *Cell and Tissue Research*, 201:487–497.

Garrone, R., and J. Pottu. 1973. Collagen Biosynthesis in Sponges. Elaboration of Spongin by Spongocytes. *Journal of Submicroscopic Cytology*, 5:199–218.

Hildemann, W. H., C. H. Bigger, I. S. Johnston, and P. L. Jokiel. 1980. Characterization of Transplantation Immunity in the Sponge *Callyspongia diffusa*. *Transplantation*, 30:362–367.

Hildemann, W. H., I. S. Johnston, and P. L. Jokiel. 1979. Immunocompetence in the Lowest Metazoan Phylum: Transplantation Immunity in Sponges. *Science*, 204:420–422.

Johnston, I. S., and W. H. Hildemann. 1982. Cellular Organization in the Marine Demosponge *Callyspongia diffusa*. *Marine Biology*, 67:1–7.

———. 1983. Morphological Correlates of Intraspecific Grafting Reactions in the Marine Demosponge *Callyspongia diffusa*. *Marine Biology*, 74:25–33.

Johnston, I. S., P. L. Jokiel, C. H. Bigger, and W. H. Hildemann. 1981. The Influence of Temperature on the Kinetics of Allograft Reactions in a Tropical Sponge and a Reef Coral. *Biological Bulletin*, 160:280–291.

Langenbruch, P. F. 1983. Body Structure of Marine Sponges. I. Arrangement of the Flagellated Chambers in the Canal System of *Reniera* sp. *Marine Biology*, 75:319–325.

Pomponi, S. A. 1976. A Cytological Study of the Haliclonidae and the Callyspongiidae (Porifera, Demospongia, Haplosclerida). Pages 215–235 in *Aspects of Sponge Biology*, edited by F. W. Harrison and R. R. Cowden. New York: Academic Press.

Simpson, T. L. 1963. The Biology of the Marine Sponge *Microciona prolifera* (Ellis and Solander). I. A Study of Cellular Function and Differentiation. *Journal of Experimental Zoology*, 154:135–147.

———. 1968. The Structure and Function of Sponge Cells. *Bulletin, Peabody Museum of Natural History*, 25:1–141.

———. 1978. The Biology of the Marine Sponge *Microciona prolifera* (Ellis and Solander). III. Spicule Secretion and the Effect of Temperature on Spicule Size. *Journal of Experimental Marine Biology and Ecology*, 35:31–42.

———. 1984. *The Cell Biology of Sponges*. New York: Springer. 662 pp.

Simpson, T. L., and C. A. Vacaro. 1974. An Ultrastructural Study of Silica Deposition in the Freshwater Sponge *Spongilla lacustris*. *Journal of Ultrastructure Research*, 47:296–309.

Smith, L. C., and W. H. Hildemann. 1986a. Allograft Rejection, Autograft Fusion and Inflammatory Responses to Injury in *Callyspongia diffusa* (Porifera; Demospongia). *Proceedings, Royal Society, London B*, 266:445–464.

———. 1986b. Allogeneic Cell Interactions during Graft Rejection in *Callyspongia diffusa* (Porifera; Demospongia); a Study with Monoclonal Antibodies. *Proceedings, Royal Society, London B*, 266:465–477.

Thompson, J. E., K. D. Barrow, and D. J. Kaulkner. 1983. Localization of Two Brominated Metabolites, Aerothionin and Homoaerothinon, in Spherulous Cells of a Marine Demosponge, *Aplysina fistularis (Verongia thiona)*. *Acta Zoologica*, 64:199–210.

Weissenfels, N. 1976. Bau und Funktion des Süßwasserschwamms *Ephydatia fluviatilis* L. (Porifera). III. Nahrungsaufnahme, Verdauung und Defäkation. *Zoomorphology*, 85:73–88.

Willenz, P. 1980. Kinetic and Morphological Aspects of Particle Ingestion by the Freshwater Sponge *Ephydatia fluviatilis* L. Pages 163–178 in *Nutrition in the Lower Metazoa*, edited by D. C. Smith and Y. Tiffon. Oxford: Pergamon Press.

JEAN VACELET
Centre d'Océanologie de Marseille
Station Marine d'Endoume
13007 Marseille, France

Storage Cells of Calcified Relict Sponges

Abstract

The solid calcareous skeleton of three living calcified sponges contains accumulations of cells with numerous storage granules that display ultrastructural features similar to those of the thesocytes of marine demosponge gemmules. In two demosponges—the living chaetetids *Merlia normani* and *Acanthochaetetes wellsi*—these cells resemble the thesocytes of the Suberitidae. They are enclosed in the calcareous skeleton, at the bottom of pseudocalices, where they form moniliform cylinders. The cell masses communicate with the choanosome only through a small hole in the calcareous tabulae or are isolated by a collagenous partition. In *Petrobiona massiliana*, which is a calcisponge, the cells occur in small canals of the massive skeleton and possess numerous reserve inclusions.

These cells do not build the calcareous skeleton. Rather, they are thought to be thesocytelike cells. Their evolution from normal cells seems to be related to the fact that they are isolated from well-irrigated endosomal tissue, as is the case in gemmule formation. These masses of storage cells are not true gemmules, however, as they are devoid of an organic envelope, and cannot be interpreted as free reproductive bodies. However, they seem to be dormant bodies involved in regenerative processes or in wintering. Such structures in living representatives of fossil groups may explain the discontinuous growth that is characteristic of a number of calcareous fossil reef-builders and may provide some clues to their past ecology.

Calcified "coralline" sponges are living fossils that offer an exciting opportunity to study the biological characters of organisms that were important reef builders in Paleozoic and Mesozoic seas (stromatoporoids, chaetetids, sphinctozoids, pharetronids). One of the first tasks in such an endeavor is to determine the significance of the masses of cells living in the basal cavities of the calcareous skeleton in 3 of the 15 species presently known in Recent seas.

These masses have been described in light microscopy in a calcisponge, *Petrobiona massiliana* Vacelet and Lévi (Vacelet, 1964) and in two demosponges, *Merlia normani* Kirkpatrick (Kirkpatrick, 1911) and *Acanthochaetetes wellsi* Hartman and Goreau (Vacelet, 1981). Two interpretations have emerged from these earlier studies. Kirkpatrick (1911) speculated that the cell masses are groups of calcocytes that secrete the solid calcareous skeleton, whereas I suggested that they are storage cells. A preliminary electron microscope study of some poorly preserved crypt cells of *M. normani* (Vacelet, 1980) has indicated that they are remarkably similar to gemmular archeocytes of the marine demosponge *Suberites domuncula* (Carrière et al., 1974) and thus confirmed that they are storage cells.

A more complete study of the ultrastructure of the cellular masses of these sponges has been undertaken in order to elucidate their relationship with the calcareous skeleton. If these cells are actually calcocytes, it would be interesting, from both a biological and a paleontological point of view, to know how they build such calcareous skeletons, which are highly unusual in modern sponges but were extremely important in ancient reefs. If they are storage cells, it would also be useful to know if they did exist in fossil forms and to compare them with gemmules of some marine and freshwater demosponges in order to explain their functions in both modern and ancient calcified sponges.

Material and Methods

The sponges examined in this study were collected by SCUBA diving in marine caves: *Merlia normani* was found in the south of Italy (Golfo di Taranto), *Acanthochaetetes wellsi* on the Australian Great Barrier Reef, and *Petrobiona massiliana* in the Mediterranean Sea near Marseille. Specimens were fixed immediately after collection in 2.5% glutaraldehyde in seawater or in a buffer composed of 0.4 M cacodylate and seawater, 1:1 (20 h). They were then decalcified in EDTA or in RDO (Du Page Kinetic Lab.), 10% in seawater, during 1 to 3 days. After complete demineralization, the tissue was postfixed in 2% osmium tetroxide in seawater and embedded in araldite. Thin sections were cut with a diamond knife and stained with uranyl acetate and lead citrate, and then were examined with a Phillips EM 300 electron microscope.

For *Petrobiona massiliana*, DNA was stained on grids by the Feulgen-thiosemicarbazide-silver proteinate method (Lewis and Knight, 1977) on tissue fixed in glutaraldehyde and demineralized in RDO. Specimens of this sponge were kept in aquaria in running seawater for two months (March and April). One specimens was fixed each week in order to follow the fate of the storage cells that migrate in these conditions.

Some mineralized specimens were embedded in araldite and prepared as polished thin sections.

Results

Merlia normani (FIGURES 1A, 2)

Masses of cells are lodged within the basal crypts of the calcareous skeleton. Their morphology and general histology have been accurately described by Kirkpatrick (1911). They form several layers joined by an isthmus of tissue passing through small holes in the center of horizontal partitions, or tabulae, of the skeletal pseudocalices (Figure 1, 2*a*). The crypt tissue is orange and contains a single cell type, collagen fascicles, and spicules.

In transmission electron microscopy (TEM), the crypt cells closely resemble gemmular archeocytes described in the marine sponge *Suberites domuncula* by Carrière et al. (1974). They are large cells (Figure 2*d*), 10 to 20 μm in diameter, with a nucleus 2.5 to 3 μm in diameter, which rarely contains a small nucleolus. The cytoplasm is crowded with complex, heterogeneous storage granules, measuring up to 3 μm. These platelets (Figure 2*b, c*) contain homogeneous lipidic inclusions, stacked membranes, inclusions with a periodic striation, and clear areas containing glycogen particles. The fact that the remains of a nucleus have been observed in one platelet suggests that they may be formed by phagocytosis, as in *S. domuncula* (Connes, 1975). Their outline is irregular, and they are surrounded by multiple membranes. Most of their inclusions are also found free in the cytoplasm. Mitochondria and Golgi complex were not observed.

The dense collagen fascicles are composed of smooth fibrils belonging to the smooth type described by Garrone (1978). They fill all the spaces between the crypt cells.

The top layers of the crypt tissue contain more collagen and spicules than the ones below. Their cells are less dense and are relatively small and elongated. The cells of the deeper layers are large and more closely packed. They contain a greater number of inclusions and glycogen particles; their nucleus is anucleolated. They are often poorly preserved, owing to the difficulty of reaching them with fixative fluids.

Intermediate stages between crypt cells and choanosome cells have been observed in the deeper parts of the choanosome (Figure 2*e*), especially between the deeper choanocyte chambers and the first tabula. Here, some choanocytes contain complex granules, which were not observed in choanocytes from more superficial tissue. Complex granules are also present in archeocytes, which are smaller than crypt cells, however, and possess a large nucleolated nucleus.

Crypt tissue is absent in *Merlia deficiens* Vacelet, which is devoid of a calcareous skeleton. However, cells similar to the intermediate cells are found in basal parts of the sponge, especially in areas where the sponge tissue occurs inside small cavities of the substratum; it seems that these cavities are not closed off enough to ensure a complete evolution of archeocytes into crypt cells.

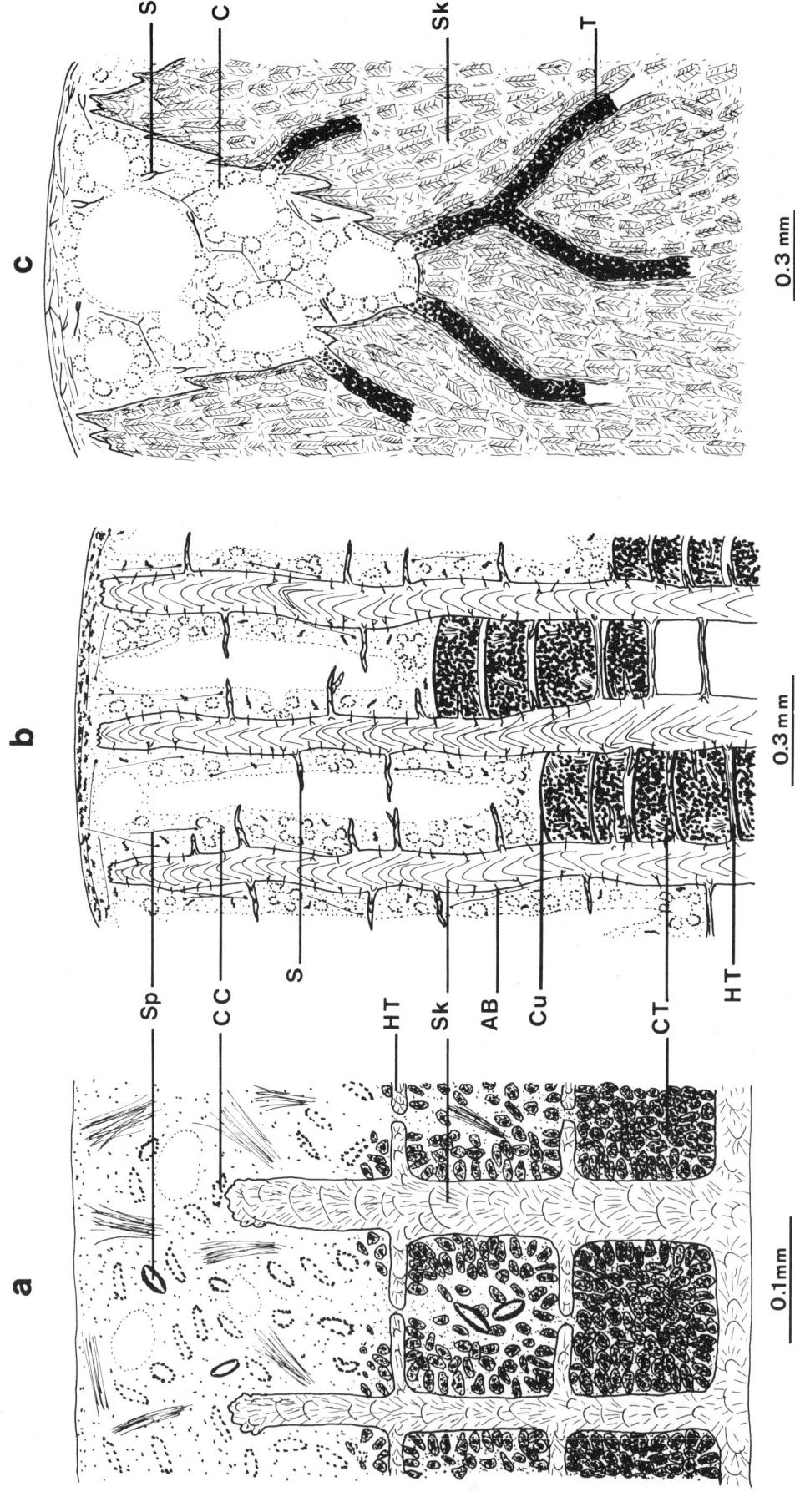

Figure 1. Vertical sections of skeleton and living tissue of three calcified sponges possessing masses of storage cells: *a*, *Merlia normani*; *b*, *Acanthochaetetes wellsi*; *c*, *Petrobiona massiliana. AB*, anchoring collagen bundles; *CC*, choanocyte chamber; *CT*, crypt tissue; *Cu*, cuticle; *HT*, horizontal tabulae; *S*, spine; *Sk*, calcareous skeleton; *Sp*, spicules; *T*, tract of storage cells.

3d Int. Sponge Conf. 1985

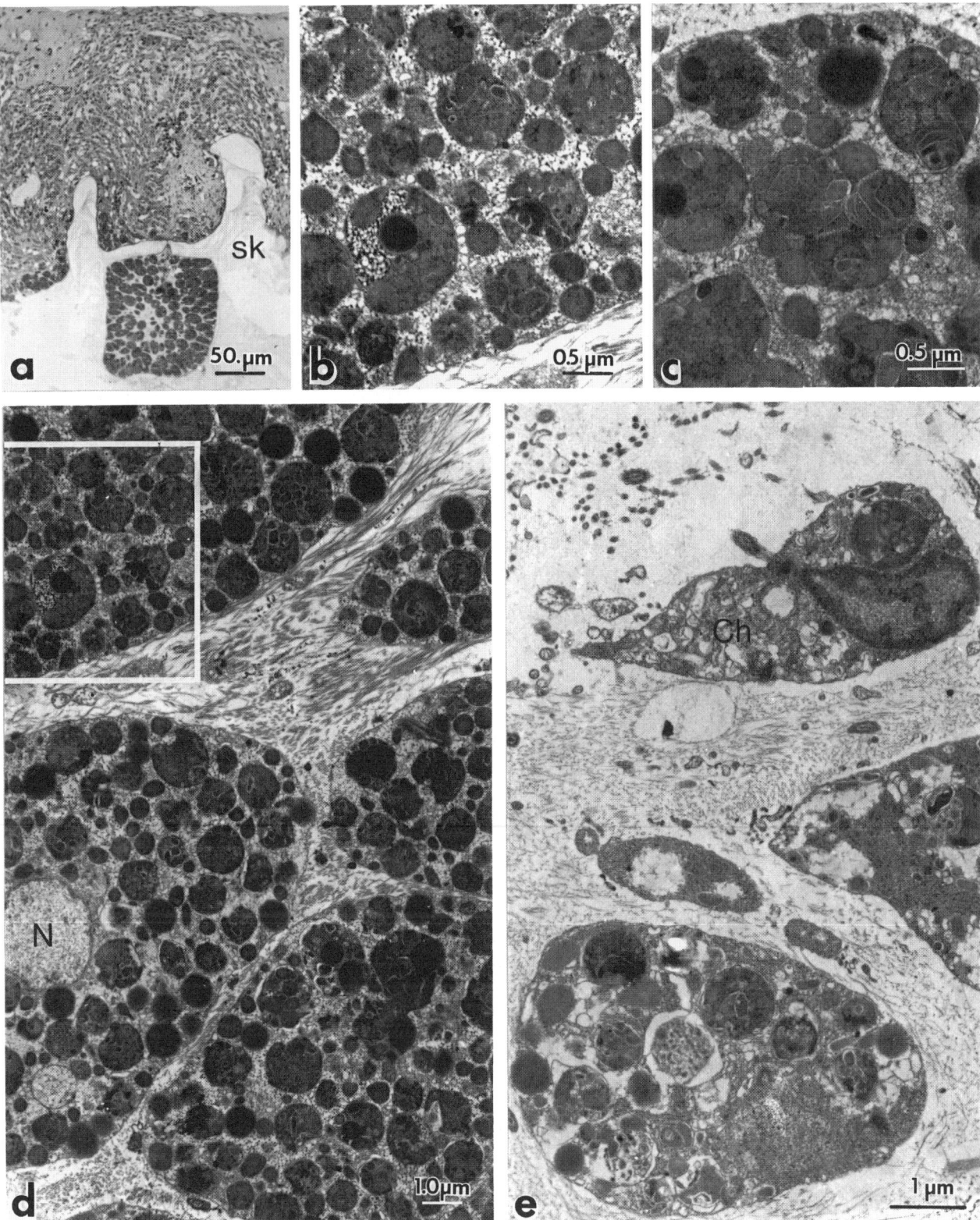

Figure 2. *Merlia normani: a,* semithin, vertical section of decalcified specimen; *b,* enlarged area of *d* (note presence of glycogen particles inside storage granule); *c,* storage granules of crypt cell; *d,* crypt cells in top layer; *e,* choanocyte and archeocytes accumulating storage granules in basal part of choanosome. *Ch,* choanocyte; *N,* nucleus; *Sk,* skeleton.

Acanthochaetetes wellsi (FIGURES 1B, 3)

Masses of cells were not observed by Hartman and Goreau (1975) when they described this survivor of the chaetetid corals, and they seem to be absent in some specimens (Vacelet, 1981). When they do occur, they are found in the basal part of only some pseudocalices. A preliminary description of their collagen components can be found in Vacelet and Garrone (1985).

As in *Merlia normani*, the cell masses occur in layers (Figures 1b, 3a) 1 to 1.3 mm thick and are composed of thesocytelike cells. However, they differ in several respects. The layers may be more numerous (up to 20), and the top ones are separated only by a thin cuticle (Figure 3a, c, e), 0.1 to 0.2 μm thick. Furthermore, there is no visible link between two adjacent layers; therefore, they are weakly attached to each other and, after decalcification, the moniliform cylinder they form is not very coherent, as opposed to what is observed in *M. normani*. Deeper in the pseudocalicle, successive layers are separated by horizontal calcareous tabulae. The first of these is incomplete (Figure 3a), whereas the deeper ones are complete and close off the empty cavities of the basal skeleton. Still deeper, these empty closed cavities are filled in by a calcareous matrix. Some of the deeper layers of cell masses appear to be inside completely closed cavities; however, this is difficult to confirm in the available specimens.

The cell masses are white in decalcified specimens (reflected light) and comprise crypt cells, collagen bundles, and spicules. They are anchored in the skeleton by dense bundles of collagen fibrils running perpendicular to the walls of the calicles (Vacelet and Garrone, 1985). These bundles enter the skeleton through canals 1 μm in diameter and up to 65 μm long. These anchoring bundles seldom occur in the innermost basal masses.

Two types of crypt cells are present: (1) rare spherulous cells (Figure 3d) containing dense ovoid, homogeneous inclusions, 2 μm in diameter, which also occur in the choanosome, and (2) gemmular archeocytelike cells (Figure 3e). These closely resemble those of *Merlia normani*. They are up to 15 μm in diameter, and their clear nucleus (2 to 2.5 μm) is usually anucleolated. The cytoplasm contains no apparent mitochondria or Golgi complex, but is crowded by complex platelets, 1 to 2 μm in diameter, which are similar to those described in *M. normani*. There are no obvious differences between the cells of the successive layers. Intermediate stages between crypt cells and choanosome cells were not observed.

The collagen bundles are very dense, and their fibrils show an unusually sharp banding pattern with a periodicity of 63 nm. Cells are separated from each other either by these bundles or by empty spaces.

Spicules are represented by microscleres (spirasters). The tylostyle megascleres of the other tissue have never been found. The masses are sometimes penetrated by spines of the calcareous skeleton.

Petrobiona massiliana (FIGURES 1C, 4)

In this species, cells accumulate within thin canals of the massive calcareous skeleton and form long tracts (Figure 1c). According to previous light microscope studies (Vacelet 1962, 1963, 1964), the tracts are formed by "thesocytes" containing storage products and cytoplasmic DNA granules. The tracts also contain rare eosinophilic amoebocytes (Gallissian and Vacelet, 1985) and microdiactines. In winter or in adverse environmental conditions, as in aquaria, tract cells can be found in the choanosome. This has been interpreted as a migration from the tracts toward the other tissue, where presumably "thesocytes" provide choanocytes with storage products or differentiate into other cell types. However, it may also be that tissue regression is occurring and that choanocytes are evolving into storage cells that will later accumulate in tracts. In the present study, specimens kept several weeks in an aquarium were found to have fewer tracts, or even no tracts at all. This indicates that tract cells actually migrate toward the choanosome and that this phenomenon does not correspond to the filling of the canals by newly formed tract cells.

Tract cells are surrounded by loose intertwined collagen fibrils that form capsules. Because the cells are small, an empty space occurs between them at the beginning of the tract (Figure 4c). They are large and closely packed together in the deep parts (Figure 4a); however, the cell limits are always visible in TEM, and the cells are always mononucleate. What was thought to be a syncytial stage in the deep parts of the tracts as seen in light microscopy (Vacelet, 1964) is actually due to the extremely thin space between two adjacent cells or to poor preservation of the deeper cells of these long, narrow canals.

Tract cells have a nucleus (3 μm in diameter) that generally contains a small nucleolus. The outline of the nuclear envelope is irregular. The chromatic material is adjacent to the nuclear envelope and also forms central irregular masses. The cytoplasm is filled with several types of inclusions (Figure 4a, b): (1) Some are round, homogeneous lipidic spherules. (2) others are heterogeneous platelets (0.5 to 2.5 μm in diameter) surrounded by a membrane with an irregular outline. The material in most of these is separated from the membrane by a clear space. This material is highly diversified and quite different from that of the platelets found in *Merlia normani* and in *Acanthochaetetes wellsi*; it is often cloudy, with a long, narrow lacuna or dark, polygonal areas; other platelets contain multivesicular or fibrillar elements. These platelets appear to be very sensitive to preservation conditions and their membrane is often disrupted. (3) Some inclusions

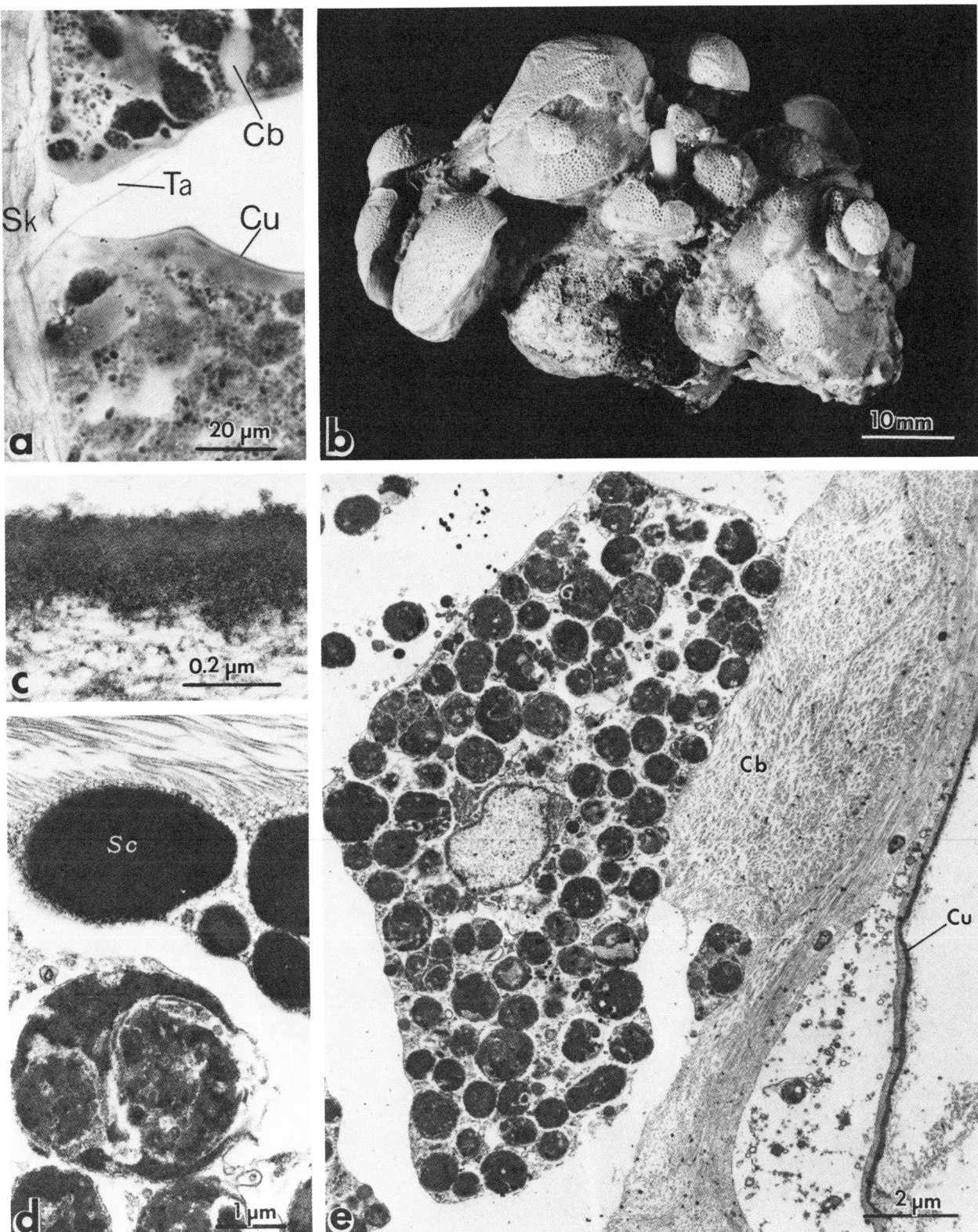

Figure 3. *Acanthochaetetes wellsi: a*, Two successive layers of cell masses, polished section; *b*, group of specimens growing on dead skeletons; *c*, close-up view of cuticle; *d*, spherulous cell and platelets of thesocytelike cell; *e*, thesocytelike cell in top layer (note thick collagen bundles and cuticle at upper part of cell mass). *Cb*, collagen bundle; *Cu*, cuticle; *Sk*, skeleton; *Sc*, spherulous cell; *Ta*, incomplete tabula.

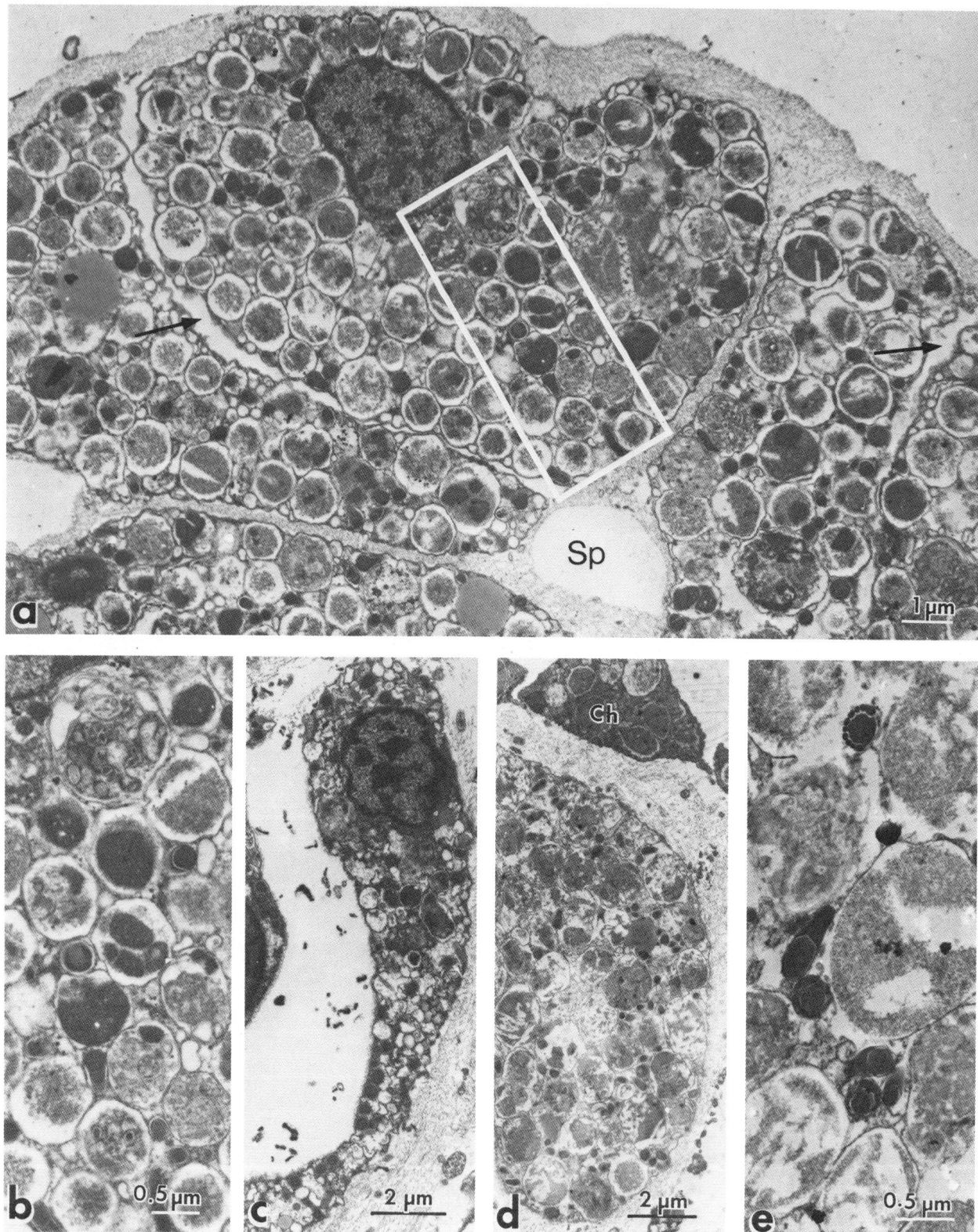

Figure 4. *Petrobiona massiliana: a,* tract cells; note empty space *(arrows)* between two cells within a collagen capsule and trace of a microdiactine spicule; *b,* enlarged area of *a; c,* two tract cells inside a collagen capsule at beginning of a tract; *d,* tract cell migrating into choanosome; *e,* Feulgen-positive, dark granules in a tract cell. *Ch,* choanocyte; *Sp,* spicule.

consist of abundant clear vesicles 0.2 μm in diameter. (4) Still others are small dark granules, round (0.2 to 0.4 μm) or ovoid (0.70 by 0.15 μm). Many are composed of a round central granule surrounded by a row of small granules, or by a round granule enclosed in an ovoid one (Figure 4e). Their contrast is enhanced by the Feulgen reaction. Some irregular granules are also stained by this reaction inside the large platelets. They likely correspond to the cytoplasmic Feulgen-positive granules previously described (Vacelet, 1963), the significance of which is unknown.

In the deep parts of the tracts, the cells contain more lipidic inclusions. The top cells of the tracts (Figure 4c), which are smaller, contain only a few smaller inclusions. The storage cells observed in the choanosome of specimens kept for several weeks in an aquarium are similar to those found in the tracts (Figure 4d).

Discussion

The striking similarities of the crypt cells of *Merlia normani* and *Acanthochaetetes wellsi* and the thesocytes of the gemmules of *Suberites domuncula* indicate that they probably have a similar function and are storage cells that are able to regenerate a complete sponge. The same may be true of the tract cells of *Petrobiona massiliana*. However, in this representative of the calcisponges, the inclusions are somewhat different from those found in demosponges and they cannot be compared to other calcisponges, as gemmules are unknown in this class. No ultrastructural observation confirms Kirkpatrick's hypothesis that in *M. normani* crypt cells are calcocytes which build up the flakes of the calcareous skeleton by becoming completely calcified. Moreover, it is unlikely that these cells secrete the skeleton as the rigid skeleton cannot be built up by cells that lie inside basal cavities rather than in the areas of growth. The skeletal mode of growth remains unknown in these three sponges.

Although the cells of these masses are very similar to thesocytes of marine demosponges, they cannot be considered true gemmules as they are devoid of a complete organic envelope and, from their structure, they clearly cannot be free reproductive bodies. However, in *Acanthochaetetes wellsi*, the cuticle of the upper part of each successive mass resembles the spongin coat of true gemmules and has to be digested in order to free the cells. Another difference of minor importance is that the cells here are surrounded by collagen fibrils, which are absent inside true gemmules.

Nevertheless, there could be some functional similarities between gemmules and these masses, which are more organized than reduction bodies. It appears likely that they are dormant structures involved in the regeneration of the superficial and choanosomal living tissue after

partial degenerescence. In *Petrobiona massiliana*, tract cells can migrate into the choanosome in adverse environmental conditions, and this could be interpreted as an internal germination. This has not been observed in the two other species, which have not been studied as closely. However, it is clear that growth can be discontinuous in these three sponges. Young individuals often grow on dead skeletons (Figure 3b) and discontinuity lines are common in the skeleton. Such discontinuous growth could be explained by a regeneration from crypt or tract cells after the death of the superficial living tissue in temporary adverse environmental situations. Discontinuity lines are well known in fossil chaetetids and stromatoporoids and could be due to the presence of dormant bodies. The supposition that such dormant bodies exist has been based on direct observations of several fossils related to sponges or to favositids (Bodergat, 1975; Stel, 1978; Termier and Termier, 1979). In the past, such regenerative abilities may have been of great importance in shallow water reef-building organisms. Their survivors in Recent seas, which use cryptic habitats as refugia, might rely on dormant bodies as an important means of survival since the trophic resources in such a habitat are generally inadequate and unreliable (Harmelin et al., 1985), and the organism could thereby withstand periods of trophic shortage.

Interestingly, these sponge cell masses are somewhat similar to the "vesicles" composed of undifferentiated large cells and surrounded by a chitinous lining with an apical pore that have been described in a calcified hydroid, *Janaria mirabilis*, by Cairns and Barnard (1984). They were interpreted either as storage bodies or as a symbiont. A detailed comparison based on cytological studies of these vesicles would be of great interest.

Only a few species of marine sponges form gemmules (Simpson and Fell, 1984), whereas among calcified forms, 3 of the 15 known Recent species possess these presumably dormant structures. This remarkably high proportion could be related to their past and present ecology, as mentioned above. However, it may also be directly related to the presence of a solid skeleton. In these three sponges, the skeleton provides an opportunity for portions of living tissue that can trigger the transformation of choanocytes or archeocytes into thesocytelike cells to become isolated. The piling up of tissue components in areas where water is no longer circulating and where an inducer may accumulate is thought to be an important process in true gemmule formation (Rasmont, 1963; Simpson and Fell, 1974). These conditions of tissue condensation are not found in other calcified sponges in which the skeleton is reticulate (*Astrosclera, Murrayona*) or devoid of horizontal partitions (ceratoporellids, *Vaceletia*).

The resistance of these structures to adverse environmental factors (high temperatures, drying, exposure to fresh water), and the factors and processes of their "germination" are unknown. An experimental study would be of a great interest, as it would throw light on the ecology of ancient reef builders such as chaetetids.

Acknowledgments

Thanks go to C. Wilkinson for fixing the Great Barrier Reef specimens and to C. Bézac and M. R. Causi for their technical help.

Literature Cited

Bodergat, A. M. 1975. *Ptychochaetes (Varioparietes) resurgens* nov. sp. (Cnidaria, Chaetetida) du Burdigalien du bassin rhodanien (Miocène, France). *Geobios*, 8:291–301.

Cairns, S. D., and J. L. Barnard. 1984. Redescription of *Janaria mirabilis*, a Calcified Hydroid from the Eastern Pacific. *Southern California Academy of Science Bulletin*, 83:1–11.

Carrière, D., R. Connes, and J. Paris. 1974. Ultrastructure et nature chimique de la coque et du vitellus gemmulaire chez l'Eponge marine: *Suberites domuncula* (Olivi) Nardo. *Comptes Rendus de l'Académie des Sciences, Paris*, 278:1577–1580.

Connes, R. 1975. Mode de formation de certains systèmes membranaires au niveau des plaquettes vitellines de thésocytes d'une Démosponge marine: *Suberites domuncula* (Olivi) Nardo. *Comptes Rendus de l'Académie des Sciences, Paris*, 281:1851–1854.

Gallissian, M. F., and J. Vacelet. 1985. Ultrastructure des amoebocytes éosinophiles d'Eponges calcaires Calcaronées. *Comptes Rendus de l'Académie des Sciences, Paris*, 300(4):151–156.

Harmelin, J. G., J. Vacelet, and P. Vasseur. 1985. Les grottes sous-marines obscures: un milieu extrême et un remarquable biotope refuge. *Téthys*, 11(3–4):214–229.

Hartman, W. D., and T. F. Goreau. 1975. A Pacific Tabulate Sponge, Living Representative of a New Order of Sclerosponges. *Postilla*, 167:1–21.

Kirkpatrick, R. 1911. On *Merlia normani*, a Sponge with a Siliceous and Calcareous Skeleton. *Quarterly Journal of Microscopical Science*, 56:657–702.

Lewis, P. R., and D. P. Knight. 1977. *Staining Methods for Sectioned Materials, In Practical Methods in Electron Microscopy* 5 (1), edited by A. M. Glauert. Amsterdam: North-Holland. 311 pp.

Rasmont, R. 1963. Le rôle de la taille et de la nutrition dans le déterminisme de la gemmulation chez les Spongillides. *Developmental Biology*, 8:243–271.

Simpson, T. L., and P. E. Fell. 1974. Dormancy among the Porifera: Gemmule Formation and Germination in Freshwater and Marine Sponges. *Transactions of the American Microscopical Society*, 93(3):544–577.

Stel, J. H. 1978. Growth and Reproduction in the Tabulate *Favosites forbesi* from the Silurian of Gotland. *Geologiska Föreningens i Stockholm Förhandlingar*, 100:181–188.

Termier, H., and G. Termier. 1979. Conditions d'ensilage des produits de survie chez les Tabuliatomorphes et les Ischyrosponges. *Comptes Rendus de l'Académie des Sciences, Paris*, 288:383–386.

Vacelet, J. 1962. Existence de formations de réserve chez une Eponge Calcaire Pharétronide. *Comptes Rendus de l'Académie des Sciences, Paris*, 254:2425–2426.

———. 1963. Acide désoxyribonucléique dans le cytoplasme de cellules á réserves d'une Eponge Calcaire Pharétronide. *Archives d'Anatomie microscopique et de Morphologie expérimentale*, 52(4):591–600.

———. 1964. Etude monographique de l'Eponge Calcaire Pharétronide de Méditerranée, *Petrobiona massiliana* Vacelet and Lévi. Les Pharétronides actuelles et fossiles. *Recueil de Travaux de la Station marine d'Endoume*, 34(50):1–125.

———. 1980. Squelette calcaire facultatif et corps de régénération dans le genre *Merlia*, Eponges apparentées aux Chaetétides fossiles. *Comptes Rendus de l'Académie des Science, Paris*, 290:227–230.

———. 1981. Eponges hypercalcifiées ("Pharétronides," "Sclérosponges") des cavités des récifs coralliens de Nouvelle Calédonie. *Bulletin du Muséum national d'Histoire naturelle*, 3A:313–351.

Vacelet, J., and R. Garrone. 1985. Two Distinct Populations of Collagen Fibrils in a "Sclerosponge" (Porifera). Pages 183–189 in *Biology of Invertebrate and Lower Vertebrate Collagens*, edited by A. Bairati and R. Garrone. NATO series A, 93. New York: Plenum.

LOUIS DE VOS
Laboratoire de Biologie Animale et Cellulaire
Université Libre de Bruxelles
50 av. F. D. Roosevelt
1050 Bruxelles, Belgium

NICOLE BOURY-ESNAULT
JEAN VACELET
Centre d'Océanologie de Marseille
Station Marine d'Endoume
13007 Marseille, France

The Apopylar Cell of Sponges

Abstract

A new cell type, the apopylar cell, with morphological features of both choanocytes and pinacocytes was revealed in an SEM study of Homoscleromorpha. These intermediate cells form a rim at the junction between the choanocyte chambers and the pinacoderm. They were also found in species of three orders of Ceractinomorpha—Dendroceratida, Dictyoceratida, and Haplosclerida—but not in Tectractinomorpha. The most typical aspect of the apopylar cells was observed in *Oscarella lobularis* in which the microvillar fringe was morphologically very similar to the choanocyte collar. In its most reduced form (in *Cacospongia scalaris* and *Ephydatia fluviatilis*) the fringe appeared as a thin and narrow velum without any microvillar expansion. Cytological evidence suggest that apopylar cells are actually modified choanocytes. Physiologically, the role of these cells may be to help regulate water flow through the sponge.

Our comparative study of the choanosome of sponges (Boury-Esnault et al., this volume) led us to pay particular attention to the contact zone between choanocyte chambers and canals (De Vos et al., 1984). In a previous work on Homoscleromorpha (Boury-Esnault et al., 1984), we described a particular cell type linking choanocytes and apopinacocytes at the junction between choanocyte chambers and exhalant canals. We called this an apopylar cell. It displays morphological characteristics intermediate between those of choanocytes and pinacocytes, possesses a flagellum and an unfolded collar of microvilli.

Earlier, Weissenfels (1980) mentioned the presence of "cone cells" at the apopyle of choanocyte chambers in *Ephydatia fluviatilis*. Although a precise ultrastructural description was not recorded, these cone cells appear homologous to the apopylar cells of Homoscleromorpha. More recently, Langenbruch et al. (1985) observed the presence of a flagellum on the cone cells of *Petrosia ficiformis*.

We investigated the extent of the apopylar cell in the phylum by examining the two other subclasses of Demospongea–Tetractinomorpha and Ceractinomorpha.

Material and Methods

Sponge specimens were collected by scuba diving in the Marseille area. The sponges were fixed in situ with a mixture of OsO_4 and $HgCl_2$ in a ratio of 6:1 (Johnston and Hildemann, 1982). Specimens were dehydrated in increasing concentrations of alcohol. Before critical point drying, sponge fragments were cooled in liquid nitrogen and fractured with a precooled razor blade. After thawing in ethanol, the fragments were dried using CO_2 as the transient fluid. The fragments were mounted on aluminum stubs with the fractured face exposed. Specimens were sputter coated with gold and examined in an ISI-DS 130 scanning electron microscope (SEM).

Results

The species that we examined belong to five orders of Tetractinomorpha (including Agelasidae) and six orders of Ceractinomorpha, including two species of uncertain affinity (Table 1). Apopylar cells could not be found in any of the species of Tetractinomorpha, but were observed in the Ceractinomorpha in all species of Dictyoceratida and Dendroceratida, and also in the freshwater sponge *Ephydatia fluviatilis* (Haplosclerida). Considerable morphological variations have been observed in the species studied.

Table 1. List of species examined (* = apopylar cell present)

Subclass	Order	Species
Tetractinomorpha	Astrophorida	*Geodia conchilega*
		Thenea sp.
	Spirophorida	*Cinachyra* sp.
	Hadromerida	*Suberites domuncula*
		Terpios fugax
		Spirastrella cunctatrix
		Tethya aurantium
		Cliona viridis
		Spheciospongia vesparia
		Acanthochaetetes wellsi
	Axinellida	*Axinella polypoides*
		Acanthella acuta
		Pseudaxinella lunaecharta
		Higginsia tethyoides
? (subclass uncertain)	Agelasida	*Agelas oroides*
		Ceratoporella nicholsoni
Ceractinomorpha	Poecilosclerida	*Anchinoe fictitius*
		Hamigera hamigera
		Grayella sp.
	Halichondriida	*Hemimycale columella*
	Haplosclerida	*Ephydatia fluviatilis*
	Verongiida	*Aplysina aerophoba*
	Dictyoceratida	*Spongia nitens*
		Spongia officinalis
		Spongia virgultosa
		Phyllospongia sp.
		Cacospongia scalaris
		Ircinia oros
		Ircinia fasciculata
	Dendroceratida	*Dysidea tupha*
		Dysidea pallescens
		Chelonaplysilla noevus
		Pleraplysilla spinifera
	incertae sedis	*Halisarca dujardini*
		Thymosia guernei

In Dysideidae, where the choanocyte chambers are rather large (Boury-Esnault et al., this volume) the apopyle is limited by more than 10–12 apopylar cells (Figure 1). The flagellum of these cells is the same length as that of the apopinacocytes. The cells are triangular in cross section and have free lateral lamellipodia (Figure 1b). One is in direct contact with choanocytes, the second, opposite, is attached to the apopinacocytes, and the third protrudes freely in the direction of the choanocyte chamber lumen. This free lamellipodium tapers into a double row of thin microvilli. The top area of the apopylar cells is 4–6 times smaller than that of the apopinacocytes.

In Aplysillidae, the chambers are large and in general the apopylar cells resemble those in Dysideidae: they have a flagellum, are triangular in cross section, and one edge is in contact with choanocytes, one is in contact with apopinacocytes, and one is free. The cell bodies are particularly narrow and stretch around the apopyle. The average dimensions are 30 μm in length and 3 μm in width, and only 5–6 cells surround the apopyle. The lamellipodium of the free border of the cell is a narrow fringe of slender microvilli (Figure 1c).

Halisarcidae have tubular branched choanocyte chambers, which are the largest known chambers in Demospongea. The apopyle is limited by a rim of 5–8 apopylar cells having a very short flagellum, whereas the apopinacocytes lack a flagellum (Figure 1d). They have the characteristic triangular shape. Their free inner border is reduced to an irregular lamellipodium with a few short filopodia (Figure 1e). They extend around the apopyle and are about 15 μm long and 5 μm wide.

In the Dictyoceratida families Spongiidae and Thorectidae, the chambers are relatively small and the apopinacocytes are always flagellated.

The apopylar cells of Spongiidae display the characteristic triangular shape, but they are thin. Their thickest part does not exeed the diameter of their nucleus, and 3–4 cells surround the apopyle. The free border is a small reduced velum interrupted over small distances (Figure 2a). The side of the cells facing the choanocyte is covered by a discrete layer of cell coating (Figure 2b). The flagellum is present but is shorter than that of the apopinacocytes.

The same type of apopylar cells has been observed in Thorectidae. They show a narrow, stretched velum on which lean the collars of the most proximal choanocytes (Figure 2c, d). Two to three apopylar cells surround the apopyle.

The chambers of the freshwater sponge *Ephydatia fluviatilis* (Haplosclerida, Spongillidae), are spherical, and the apopyle is surrounded by 3–4 apopylar cells. The lamellipodium of the free edge is turned inward and rests on the collars of the most proximal choanocytes (Figure 2e). Their flagellum hangs into the lumen of the choanocyte chamber, in contrast to that of the other spe-

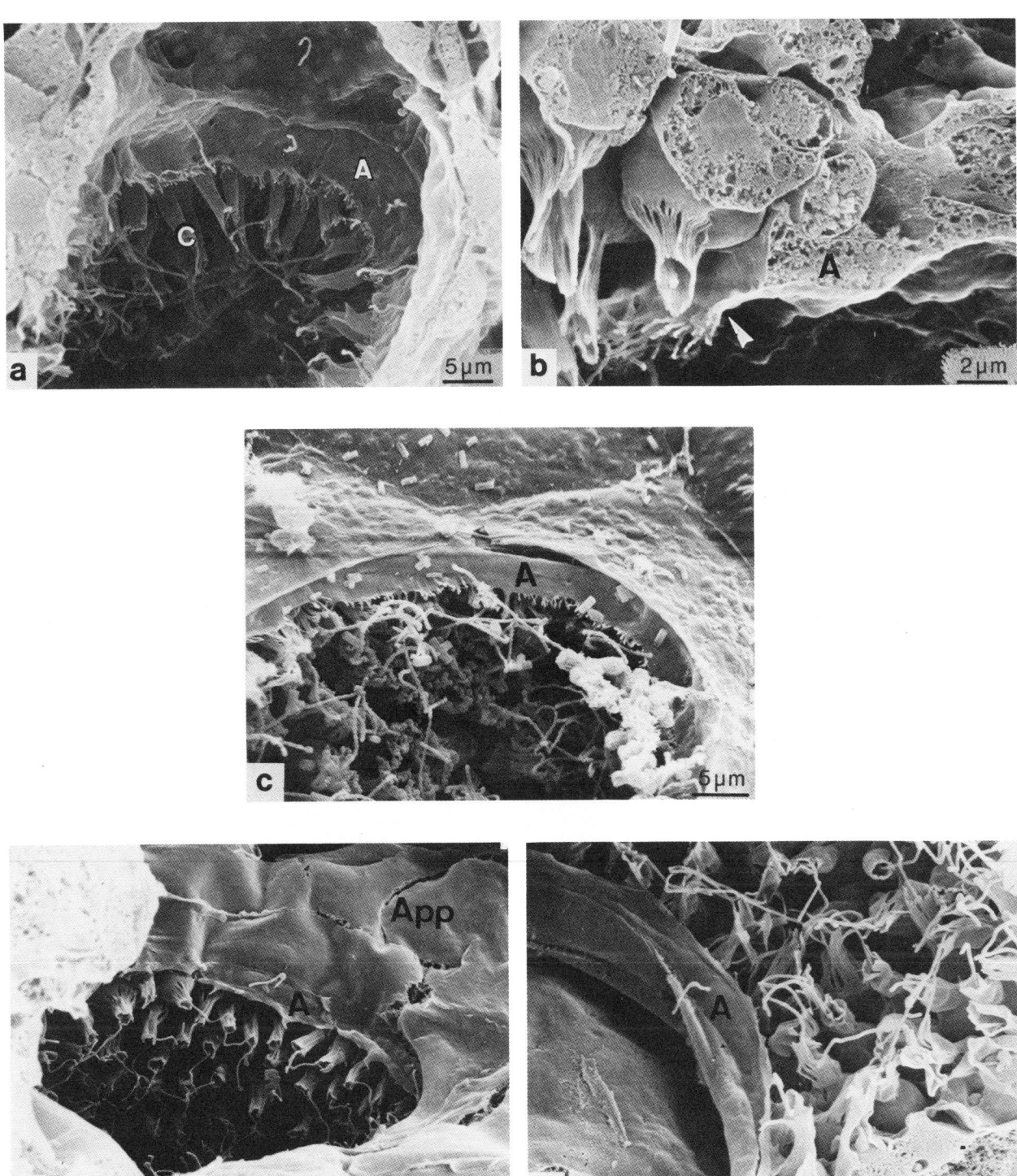

Figure 1. Structure of apopylar cells in Ceractinomorpha (SEM): *a, Dysidea pallescens* (Dysideidae), apopyle seen from exhalant canal showing rim of apopylar cells; *b, D. pallescens*, cryofracture through apopyle showing apopylar cell and choanocytes (arrow points to lamellipodia with microvilli of apopylar cell); *c, Chelonaplysilla noevus* (Aplysillidae), partial view of apopylar opening (note flat rim of apopylar cells); *d, Halisarca dujardini* (Halisarcidae), partial view of an apopyle (note difference in size between apopylar cells and apopinacocytes); *e, H. dujardini*, side view showing thin and narrow lamellipodia of apopylar cell. *A*, apopylar cell; *App*, apopinacocyte; *C*, choanocyte.

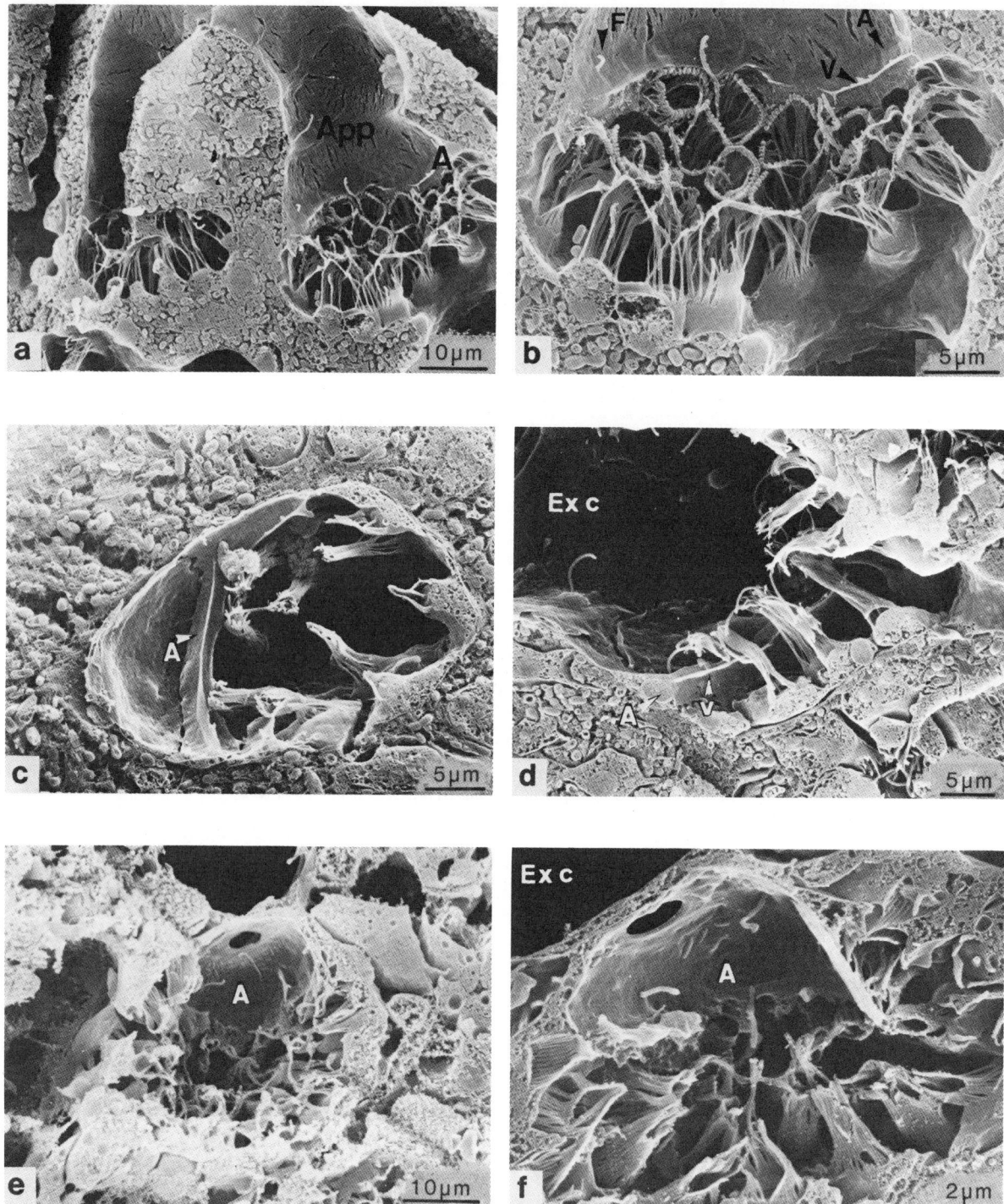

Figure 2. Structure of apopylar cells in Ceractinomorpha (SEM): *a, Spongia nitens* (Spongiidae), two choanocyte chambers (note difference in lengths between flagella of apopylar cell and of apopinacocyte in aphodus); *b*, detail of *a* at higher magnification shows features of apopylar cell, such as very reduced velum, triangular shape in cross section, and flagellum; *c, Cacospongia scalaris* (Thorectidae), view of an oblique section through apopyle—one apopylar cell is visible; *d, C. scalaris*, cryofracture through choanocyte chamber showing shape of an apopylar cell and its diaphragmatic velum *(arrow); e, Ephydatia fluviatilis* (Spongillidae), general view of a choanocyte chamber—a large area of inner surface around apopyle covered by apopylar cells; *f*, detail of *e* at higher magnification shows bacteria on surface of apopylar cell (notice irregular free edge of apopylar cell.) *A*, apopylar cell; *App*, apopinacocyte; *C*, choanocyte; *Exc*, exhalant canal; *F*, flagellum; *V*, velum.

cies. Bacteria were attached to the surface of the apopylar cells in all the specimens examined (Figure 2*f*).

Discussion

Our recent observations together with our previous work on Homoscleromorpha (Boury-Esnault et al., 1984) can be summed up as follows:

> Apopylar cells occur at the apopyle of choanocyte chambers, where they form the boundary between apopinacocytes and choanocytes.
> They possess a central flagellum.
> They are triangular in cross section.
> Their free edges face inward toward the choanocyte chambers and taper into cytoplasmic extensions. The morphology of these extensions varies from an unfolded collar of microvilli *(Oscarella, Corticium)* to a thin velum *(Halisarca, Ephydatia).*

Three types of apopyle organization can be discerned.

The first pattern is marked by the absence of apopylar cells, and there is a direct contact between choanocytes and apopinacocytes. This situation was found in *Geodia*, for example (Figure 3*a*). In the second pattern, apopylar cells form the junction between choanocytes and apopinacocytes, and their flagellum is directed toward the exhalant canal (Figure 3*b, c*). This arrangement was observed in Homoscleromorpha, Dysideidae, Aplysillidae, and Halisarcidae, the third pattern is found in *Ephydatia fluviatilis.* The apopylar cells are completely deflected inward so that they are invisible from the outside and make the apopylar opening appear cone shaped from the inside (Figure 3*d*). The same pattern has been reported in *Petrosia ficiformis* (Langenbruch et al., 1985).

The origin of this cell type is still poorly understood. Sollas (1888) suggested that choanocytes are transformed into endopinacocytes, as was later observed by Diaz (1974). In Homoscleromorpha (Boury-Esnault et al., 1984) the apopylar cells of *Oscarella* and *Corticium* have a fringe with the same number of microvilli as the collar of the choanocytes. Moreover, the apopylar cells of *Corticium* contain the same kind of cytoplasmic inclusions as the choanocytes. In *Halisarca*, apopylar cells are flagellated,

whereas apopinacocytes are not; apparently this is also the case in *Petrosia* (Langenbruch et al., 1985). All these data demonstrate that apopylar cells have more characters in common with choanocytes than with apopinacocytes and strongly suggest that they are likely to be derived from choanocytes.

The same hypothesis has been proposed for *Ephydatia*. (Weissenfels, 1980, 1981) and *Reniera* (Langenbruch, 1983) but without much cytological evidence, except for specific localization. Indeed, the localization of the so-called cone cells in *Ephydatia* corresponds to what we observed in apopylar cells. However, we did not find pore cells in the endopinacoderm, as described by these authors. The pinacocytes in direct contact with apopylar cells are normal apopinacocytes.

Apopylar cells have been described in Homoscleromorpha, and in the Ceractinomorpha orders Dictyoceratida, Dendroceratida, Halisarcidae, Haplosclerida, and Petrosiida. Thus far they have not been found in the other orders of Ceractinomorpha (Halichondriida, Poecilosclerida and Verongiida), but too few species have been examined to determine whether they exist there as well. It must be emphasized that these cells, which have long been overlooked, are extremely difficult to recognize in TEM or in semithin sections, and that good photographs of properly preserved material are needed to identify them in SEM.

In Tetractinomorpha, no special cells were observed at the apopyle of the studied species (orders Hadromerida, Tetractinellida, and Axinellida). However, in the order Hadromerida (including the "sclerosponge" *Acanthochaetetes wellsi*), we consistently found "central cells" (Connes et al., 1971; Reiswig and Brown, 1977) that perhaps play a similar function.

The localization of these cells suggests that they play a role in controlling the water current. As it passes through the sponge canal system, the water current must cross a number of valves and regulation systems: the ostia, the prosopyle, the apopylar rim or the central cell, and the osculum. Ostia, pore cells, and osculum are reported to have contractile properties, and actin filaments have been demonstrated in pinacocytes (Pavans de Ceccatty, 1981). TEM observations are now being made in an effort to

Figure 3. Patterns of apopyle organization in sponges: *a, Geodia; b,* and *c,* Homoscleromorpha, Dysideidae, Aplysillidae, and Halisarcidae (*c* shows pattern when an aphodus is present); *d, Ephydatia fluviatilis.* *A,* apopylar cell; *C,* choanocyte; *P,* pinacocyte.

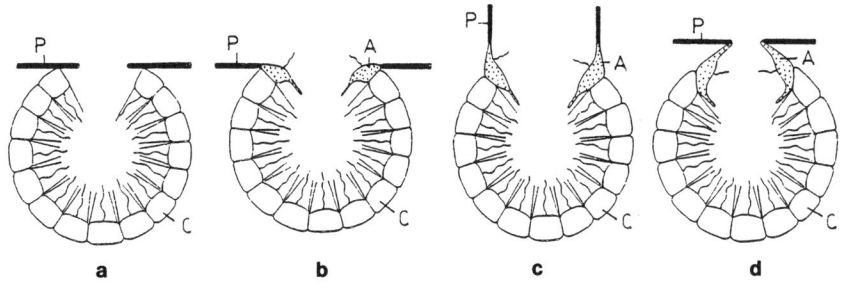

determine whether contractile microfilaments exists in the apopylar cells. The eventual morphological and functional homologies between apopylar cells and "central cells" obviously require further investigations.

Acknowledgments

This work was supported by FRFC grant 2.4525.80 and funds from NATO grant 85–0134. We are grateful for the technical assistance of Chantal Bézac, Marie-Rose Causi, Emile De Cock, and François Lambert.

Literature Cited

Boury-Esnault N., L. De Vos, C. Donadey, and J. Vacelet. 1984. Comparative Study of the Choanosome of Porifera. I. The Homoscleromorpha. *Journal of Morphology*, 180:3–17.

Connes, R., J. P. Diaz, and J. Paris. 1971. Choanocyte et cellule centrale chez la Démosponge *Suberites massa* Nardo. *Comptes Rendus de l'Académie des Sciences, Paris*, 273(18):1590–1593.

De Vos, L., N. Boury-Esnault, J. Vacelet, and C. Donadey. 1984. La cellule apopylaire chez les Spongiaires. Colloque, Société française Microscopie électronique. *Biology of Cell*, 51:42a.

Diaz, J. P. 1974. De l'origine de certains endopinacocytes à partir de choanocytes chez la Démosponge *Suberites massa* Nardo.

Bulletin, Société Zoologique de France, 99(4):687–693.

Johnston, I. S., and W. H. Hildemann. 1982. Cellular Organization in the Marine Demosponge *Callyspongia diffusa*. *Marine Biology*, 67:1–7.

Langenbruch, P.-F. 1983. Body Structure of Marine Sponges. 1. Arrangement of the Flagellated Chambers in the Canal System of *Reniera* sp. *Marine Biology*, 75, 319–325.

Langenbruch P.-F., T. L. Simpson, and L. Scalera-Liaci. 1985. Body Structure of Marine Sponges. 3. The Structure of Choanocyte Chambers in *Petrosia ficiformis* (Porifera, Demospongiae). *Zoomorphology*, 105:383–387.

Pavans de Ceccatty, M. 1981. Demonstration of Actin Filaments in Sponge Cells. *Cell Biological International Reports*, 5(10):945–952.

Reiswig H. M., and M. J. Brown. 1977. The Central Cells of Sponges. *Zoomorphology*, 88:81–94.

Sollas, W. J. 1888. Report on the Tetractinellida Collected By H.M.S. Challenger During the Years 1873–1876. *Report, Voyage of H.M.S Chalenger, Zoology*, 25:1–458.

Weissenfels, N. 1980. Bau und Funktion des Süsswasserschwamms *Ephydatia fluviatilis* L. (Porifera). VII. Die Porocyten. *Zoomorphology*, 95:27–40.

———. 1981. Bau und Funktion des Süsswasserschwamms *Ephydatia fluviatilis* L. (Porifera). VIII. Die Entstehung und Entwicklung der Kragengeisselkammern und ihre Verbindung mit dem ausführenden Kanalsystem. *Zoomorphology*, 98:35–45.

Developmental
Biology

HEATHER R. KAYE
Redpath Museum
McGill University
859 Sherbrooke St. W.
Montreal, Quebec H3A 2K6, Canada

Reproduction in West Indian Commercial Sponges: Oogenesis, Larval Development, and Behavior

Abstract

Oogenesis in four commercial West Indian sponge species—*Hippospongia lachne, Spongia barbara, S. cheiris,* and *S. graminea*—was studied using light and electron microscopy. Eggs and embryos develop asynchronously in localized endosomal nurseries of these viviparous and gonochoristic species. Meiotic division produces four secondary oocytes, which are fertilized before undergoing major growth by phagocytosis and the transfer of nutrients through cytoplasmic bridges. Symbiotic bacteria are dispersed between blastomeres of all young and maturing embryos and internal cells of cytodifferentiated parenchymella larvae. During cleavage, umbilici form between embryos and nurse cell layers, and function in the transfer of symbiotic bacteria from the maternal parent. Free-swimming larvae display directional swimming with constant rotation and negative phototaxis. There is no evidence of substrate selection or orientation by larvae.

The genera *Spongia* L. and *Hippospongia* Schulze, (Dictyoceratida) include all of the commercially important bath sponges of the West Indian fishery. Epidemic diseases devastated entire populations of these species four and five decades ago but in the interim there has been a resurgence of these populations. Despite these interesting circumstances, few investigators since have looked at the biology and ecology of these once extremely valuable economic commodities. Most of the literature on West Indian commercial sponge species deals with gross morphology and fisheries. Thus, little is known of their reproduction, embryology and larval development, metamorphosis, and other aspects of their biology.

There are relatively few studies on reproductive processes in the dictyoceratids (Tuzet and Pavans de Ceccatty, 1958; Liaci et al., 1971), and much of what is known about processes has been inferred from gross structural observations in the course of general studies on reproduction. Although electron microscopy has now made it possible to study in detail the specific events during reproductive processes, only one such study has been carried out on the dictyoceratids (Gaino et al., 1984).

Recent electron microscope studies have revealed the uniform presence of immense symbiotic bacterial populations in four commercial West Indian sponge species. Bacteria isolated from these sponges are specific sponge symbionts and differ from ambient seawater bacteria (Kaye et al., 1985). Intercellular bacterial symbionts have also been observed in the embryos and brooded larvae of these species.

The present study was undertaken as a first step in analyzing specific reproductive processes in the same four commercial West Indian sponge species and includes information on sexual differentiation, oogenesis, embryonic and larval development, transfer of bacterial symbionts, and larval behavior. Other investigations in progress are focusing on larval settlement and metamorphosis, reproductive cycles, and spermatogenesis in these four species. Light and electron microscope studies, together with laboratory experiments, have been employed to help elucidate the reproductive processes of these economically important members of the phylum Porifera.

Material and Methods

The Four West Indian commercial sponge species under investigation are *Hippospongia lachne* de Laubenfels, *Spongia barbara* Duchassaing and Michelotti, *Spongia cheiris* de Laubenfels and Storr, and *Spongia graminea* Hyatt. These sponges are rounded, black to dark gray, and common members of the fauna of Biscayne Bay, Florida (average depth 2.5 m). The collection site (25°38′ N; 80°12′ W) is characterized by a coarse sandy bottom, much of which is covered by eel and turtle grasses, small scattered corals, gorgonians, and other sponge species. A total 104 specimens were collected and analyzed histologically over a three-year period during and between reproductive cycles.

Endosomal tissue samples were removed in situ from the lower two-thirds of the individuals. A small (20 mm³) sample from the central region of the sponge was then fixed in situ by placing the tissue in a syringe, replacing the ambient seawater, with 2.5% glutaraldehyde in seawater. Back in the laboratory, samples were placed in vials of fresh fixative for 16 h, and then rinsed in three changes of fresh seawater for 10 min each. A smaller subsample (ca. 5 mm³) from each specimen was post-fixed for 1 h in 1% osmium tetroxide in seawater, dehydrated in an alcohol series, cleared in propylene oxide, and embedded in Spurr epoxy resin. The remaining fixed tissue from each specimen was dehydrated in an alcohol series, cleared in xylene, and embedded in paraffin.

Epoxy blocks were sectioned on a Porter-Blum ultramicrotome MT-2B at 1 μm (glass knife) and 0.1 μm (diamond knife). Semithin sections were mounted on glass slides and stained for 1 min at 60°C with a mixture (equal parts by volume) of methylene blue (1% in 1% Na-borate solution) and azure II (1% in distilled water). Ultrathin sections were floated onto formvar and carbon-coated single-slot copper grids and stained for 20 min in saturated aqueous uranyl acetate and for 15 min in lead citrate. Sections were photographed with a Phillips 410 transmission electron microscope. Paraffin blocks were sectioned at 10 μm and stained with hematoxylin-eosin. Sections were viewed and measurements of reproductive elements recorded under a compound microscope with an ocular micrometer accurate to 0.5 μm.

During the reproductive period, large pieces of tissue were collected in plastic bags and returned to the laboratory, where they were placed in running seawater aquaria. Released larvae were observed and photographed under a WILD M5A dissecting microscope.

Fifty free-swimming larvae of each species were separately placed in a black plexiglass box (32.5 × 17.5 × 10 cm) filled with ambient seawater. Responses to a unidirection light source (tungsten lamp with irradiance of 0.04 μein·cm⁻²·sec⁻¹ at larval swimming depth in a darkened room) were observed and recorded under 3 separate trials. To eliminate behavioral modifications due to habituation and temperature, the location of the light source was varied between trials and the light applied for only 30–60 seconds at a time.

Substrate selection was tested by offering clean and "aged" glass and asbestos cement (transite) slides (25 × 75 × 1 mm) for settlement surfaces. Slides were "aged" by suspending them in running-seawater troughs for 2 days to be coated by a film of bacteria and algae.

Results

The four sponge species investigated are viviparous, brooding their embryos and larvae within maternal tissues. Samples contained either male, female, or no reproductive elements. Male and female elements were never observed in the same specimen. Oocytes and embryos develop asynchronously during the reproductive cycle, as do spermiocysts. These species may therefore be gonochoric, although successive hermaphroditism cannot be ruled out in that the same individuals were not sampled consecutively.

Gamete production and embryo development are localized in patches or "nurseries" of endosomal tissue. These nurseries contain 20 or more oogeneic elements (Figure 1a) and are located in the lower two-thirds and toward the central part of the sponges, where much of the tissue may be involved at any time. In the immediate vicinity of the nurseries the mesohyl is disrupted with a

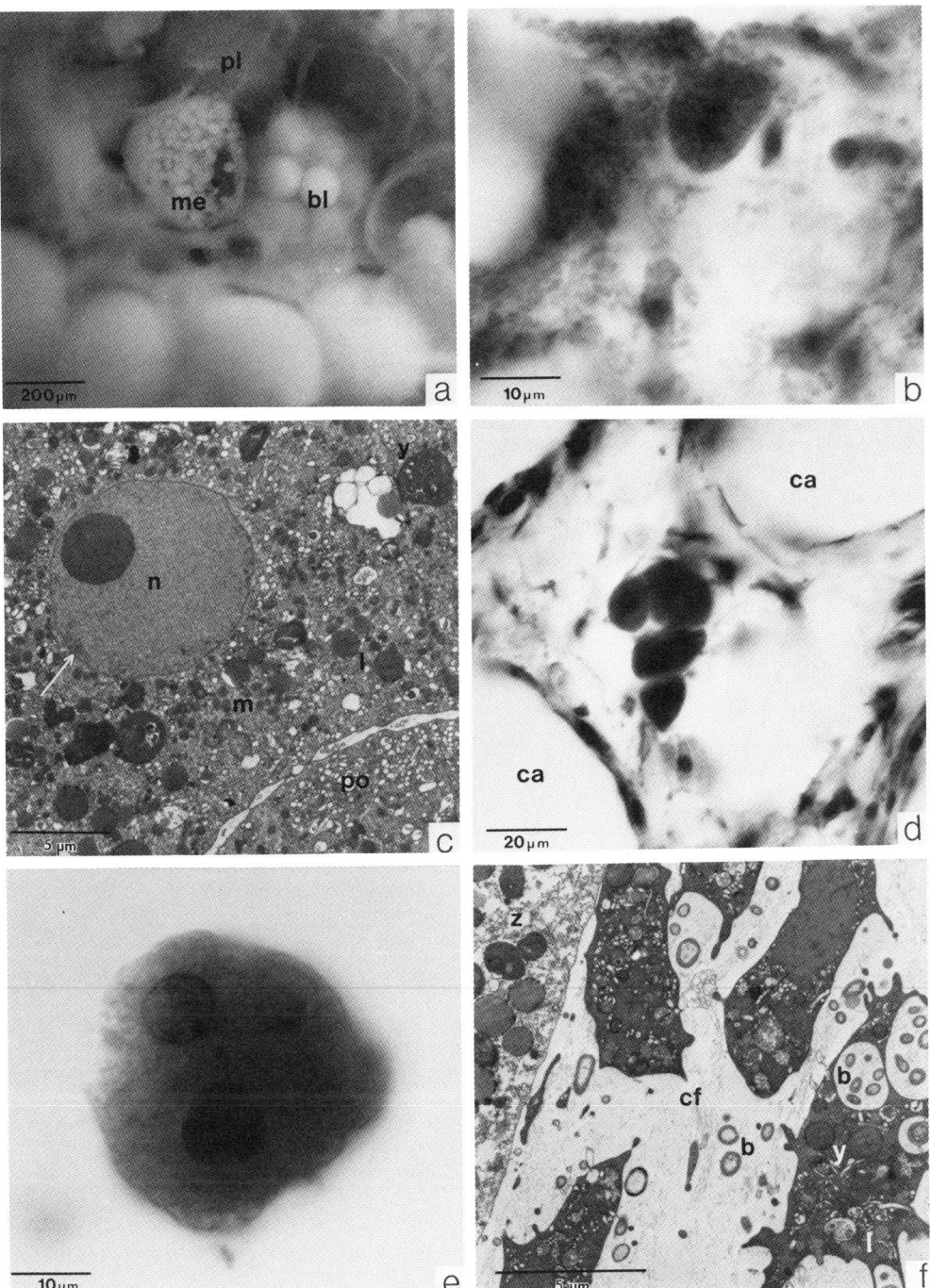

Figure 1. Photomicrographs of sponge development: *a*, nursery in endosomal tissue of *Hippospongia lachne*; *b*, light micrograph showing separation of chromatids in late anaphase stage of mitosis in endosomal tissue of *H. lachne* (note triangular shape of cell and its vesicular cytoplasm); *c*, electron micrograph of primary oocyte in *Spongia graminea*; part of another primary oocyte *(po)* visible in bottom right-hand corner (arrow indicates dense granular material around periphery of nuclear membrane); *d*, light micrograph showing separation of four ova in late telophase II of meiosis in an endosomal nursery of *H. lachne*, (*ca*, cavities remaining after release of differentiated parenchymella larvae); *e*, light micrograph of male and female pronuclei in secondary oocyte (ovum) of *Spongia barbara*; *f*, electron micrograph of nurse cells phagocytosing bacteria and collagen fibrils from mesohyl of *S. graminea*. *b*, bacteria; *bl*, blastomeres of young embryos; *cf* collagen fibrils; *l*, lipid; *m*, cluster of mitochondria; *me*, mature embryo; *n*, nucleolated nucleus; *pl*, parenchymella larva; *y*, yolk; *z*, zygote.

decrease in cell numbers, choanocyte chambers, and symbiotic bacteria. Endosomal tissue outside nursery areas and the ectosome remain undisturbed.

The sequence of events in oogenesis and development are inferred from linking together static images clearly similar or very nearly identical to one another in series. For convenience, the processes are described as active. Analyses of specimens undergoing oogenesis suggest that oogonia differentiate directly from archaeocytes within the mesohyl. These cells are initially triangular, 10–15 μm in diameter, and contain a distinctly enlarged nucleus (4–6 μm) with a single prominent nucleolus (ca. 2 μm) and vesicular cytoplasm. They undergo mitosis (Figure 1b) to produce primary oocytes which become ovoid or spherical and grow to 20–25 μm. These primary oocytes maintain the distinct nucleolated (3 μm) nucleus (7–9 μm) of the oogonia, but the cytoplasm becomes dense. Often two discrete nucleoli were observed, as well as dense granular material around the periphery of the nuclear membrane. Archaeocytes begin to move in from the surrounding mesohyl to form a limiting layer of nurse cells around the oocyte.

The cytoplasm of these primary oocytes already contains yolk granules, glycogen, and lipid (Figure 1c). However, there is no evidence of phagocytosis and transfer of material from cells in the mesohyl. In the final phase (telophase II) of the maturational divisions in these oocytes, four closely associated secondary oocytes (ova, Figure 1d) are consistently produced. Polar body formation and fertilization were not observed, but male and female pronuclei were seen in secondary oocytes; thus, fertilization may occur shortly after the second meiotic division (Figure 1e).

The zygote grows to about 300 μm in diameter before cleavage occurs. During this time there is subsequent apposition of nurse cell layers (2–3) to the initial limiting layer and these cells are seen phagocytosing bacteria, collagen fibrils, and cells from the mesohyl (Figure 1f). The nurse cells are about 10 μm in length, have a dense nucleus (ca. 4 μm), and contain cytoplasmic inclusions of yolk, lipid, phagosomes, and glycogen. These cells make intimate contact, by means of microvilli, with the surface of the zygote and often form cytoplasmic bridges in which the membranes of both lose their integrity (Figure 2a). Large yolk spheres, phagosomes, glycogen and lipid accumulate within the growing zygote. In the latter stages, as the zygote increases in size, nurse cells are phagocytosed directly, and the remaining cells become progressively elongated and flattened.

At this stage a species difference was observed. Zygotes and nurse cells in *Spongia cheiris* contain extremely electron-dense cytoplasmic inclusions which are concentrated around the membrane of the zygote and are about the same size as the yolk granules, 1–2 μm, (Figure 2b). They are also dispersed in the blastomeres of the embryo and in the internal cells of the differentiating larvae. These inclusions are very similar in appearance to, but larger than, the pigment granules observed in the flagellated epithelial layer of differentiating larvae in all four species.

Cleavage is total and equal, and results in solid, translucent-white stereoblastulae. As cleavage proceeds, a single layer of flattened nurse cells becomes apposed to the embryonic membrane. Each of these cells has a large nucleus (many are binucleated) and dense cytoplasm. Collagenouslike connections (umbilici) were observed between this single layer of nurse cells and the other nurse cell layers, which were separated by a cavity to accommodate the increase in volume of the cleaving embryo (Figure 2c). Symbiotic bacteria were observed dispersed between blastomeres and inside the embryonic membrane of all of the young and maturing embryos. No other cells were observed in the embryos (Figure 2d). During cleavage, the large yolk granules are broken down and become less distinct.

At the end of cleavage, the blastomeres differentiate, resulting in the development of a cytodifferentiated parenchymella larva possessing two distinct regions: a central cellular mass, and a peripheral region consisting of several cellular layers. The central cellular mass consists of collagen fibrils, symbiotic bacteria, and archaeocytes in a lose arrangement. The archaeocytes (ca. 10 μm) contain a diffuse nucleus, numerous phagosomes, small yolk and lipid granules, and, in the larvae of *Spongia cheiris*, the extremely electron-dense inclusions noted earlier. The peripheral region consists of small uniflagellate cells (ca. 4 μm) forming a tightly packed and pigmented columnar epithelial layer and underlying layers of large amoeboid cells (ca. 5 μm), which contain densely staining nuclei, phagosomes, and small yolk and lipid granules (Figure 2e). Umbilici were not observed in differentiated larvae.

Parenchymella larvae of all four species have an average size of 420 μm by 350 μm when released (Figure 2f). The free-swimming larvae are ovoid with dark gray pigmentation, and their posterior regions are encircled by a black pigmented ring of cells bearing long cilia (ca. 80 μm). Shorter cilia (ca. 16 μm) are dispersed over the entire surface of the larvae. Released larvae, propelled by the long posterior cilia, displayed directional swimming with constant lateral rotation for a period of 24–48 h, and were negatively phototactic. When light was applied, the larvae would immediately swim out of the path of the light and into the darkened regions of the box. This same response was recorded for all larvae on all three trials. No geotactic behavior was observed as larvae would swim at all depths in the box. Prior to settlement and metamorphosis, the larvae entered a creeping phase that was interrupted by sporadic episodes of swimming. This phase lasted for 2–8 hours, during which time the larvae demonstrated no particular taxis to light. Settlement of the larvae occurred 26–56 h after release.

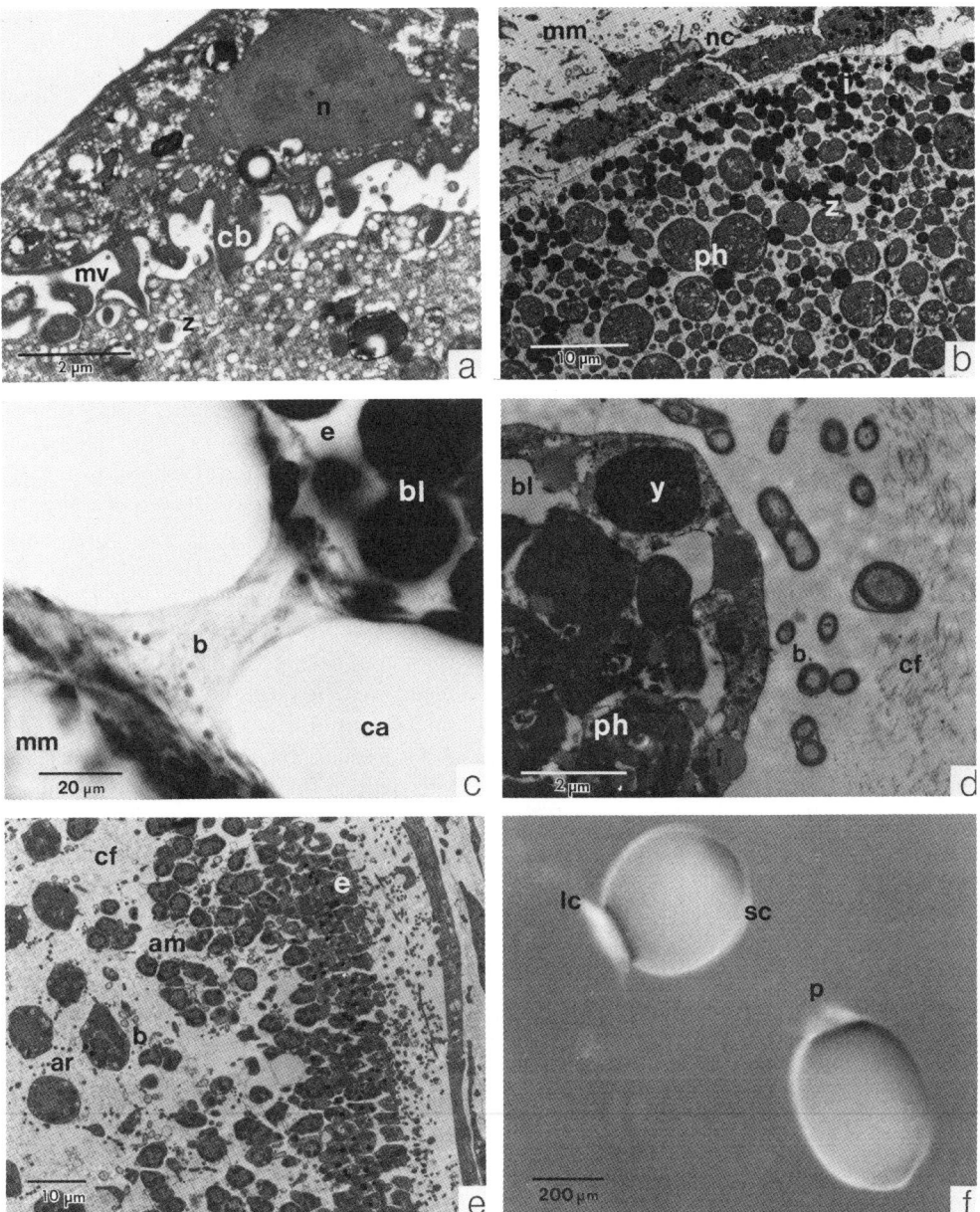

Figure 2. Photomicrographs of sponge development: *a*, electron micrograph of a nurse cell feeding a zygote in *Spongia barbara* by formation of microvilli and cytoplasmic bridges, which transfer material to growing zygote; *b*, electron micrograph of a zygote in endosomal tissue of *S. cheiris;* note presence of extremely electron-dense inclusions in surrounding nurse cells and concentrated around membrane of zygote; *c*, light micrograph of an umbilicus connecting an embryo of *Hippospongia lachne* to parental tissue; bacteria are extracellularly transferred from maternal mesohyl to cleaving embryo; *d*, electron micrograph of symbiotic bacteria from maternal parent of *H. lachne* outside a blastomere in an embryo; blastomere contains lipid, phagosomes, and yolk; *e*, electron micrograph of a cytodifferentiated parenchymella larva of *S. barbara* showing peripheral region consisting of small uniflagellated cells of pigmented epithelial layer, and underlying layers of larger amoeboid cells undergoing mitosis; central region contains symbiotic bacteria and collagen fibrils dispersed between archaeocytes containing small lipid and yolk granules; note flattened layer of nurse cells (on right-hand side) around larva; *f*, live parenchymella larvae of *H. lachne;* posterior ends encircled by black pigmented ring of cells bearing long cilia which steer the swimming larvae, and short cilia cover remainder of larval surface. *a*, ameoboid cell; *ar*, archaeocyte; *b*, bacteria; *bl*, blastomere; *ca*, cavity in mesohyl accomodating growth of cleaving embryo; *cb*, cytoplasmic bridge; *cf*, collagen fibrils; *e*, embryo; *i*, inclusions; *l*, lipid; *lc*, long cilia; *mm*, maternal mesohyl; *mv*, microvilli; *n*, nucleus of nurse cell; *nc*, nurse cell; *p*, posterior end; *ph*, phagosome; *sc*, short cilia; *y*, yolk; *z*, zygote.

There was no evidence of substrate selection, by type or orientation, by the larvae of these four species. There were equal numbers of larvae on the upper and under surfaces of clean or "aged" glass and clean or "aged" transite (four-factor ANOVA, including interaction effects, always $p >$ 0.15). A few larvae metamorphosed on the surface film of the water, but did not survive beyond their initial flattening. Larvae appeared to exhibit gregarious behavior as they tended to settle near each other.

Discussion

Recent and extensive literature reviews on reproduction in sponges have demonstrated that sexual differentiation among demosponges varies greatly (Fell, 1983; Reiswig, 1983; Simpson, 1984), but two consistent patterns emerge from this variability. Demosponges that broadcast eggs or embryos (i.e., that are oviparous) have separate sexes (are gonochoristic). Only two exceptions to this pattern have been reported, *Tetilla* sp. (Liaci et al., 1976) and *Verongia* (= *Aplysina*) *aerophoba* (Liaci et al., 1971), both belonging to genera in which other species are oviparous and gonochoristic (Egami and Ishii, 1956; Watanabe, 1978; Reiswig, 1973). All tetractinomorphs whose mode of sexual reproduction is known (with the exception of *Tetilla* sp.) exemplify this form of sexuality. The second pattern that emerges is a preponderance of demosponges that brood embryos and larvae (i.e., are viviparous) to be hermaphroditic. Within the ceractinomorphs, specifically among the keratose sponges, the pattern of viviparity with gonochorism is considered an exception.

The present study demonstrates that *Hippospongia lachne*, *Spongia barbara*, *S. cheiris*, and *S. graminea* are viviparous, and all available evidence suggests that these species are gonochoric. However, increasing evidence of sex reversal and of a high degree of interspecific variability in sexual differentiation among sponges (Fell, 1970; Diaz, 1973; Van de Vyver and Willenz, 1975; Elvin, 1976; Gilbert and Simpson, 1976; Fell and Jacob, 1979) suggests that intrinsic interactions between genetic or environmental factors may determine sexual differences.

Sponges lack localized discrete reproductive organs, and in most cases large areas of mesohyl are involved in reproduction. However, as in other viviparous species that produce large-yolked eggs (Lévi, 1951), oocytes and embryos of the four species studied here develop asynchronously in clusters within the basal or central zone of endosomal tissue. Similar asynchronous development of oogeneic elements grouped in small clusters has been reported in other sponges (Lévi, 1951; 1956; Bergquist and Sinclair, 1968).

The general pattern of oogenesis and larval development in *Hippospongia lachne*, *Spongia barbara*, *S. cheiris*, and

S. graminea is similar to that reported for other demosponges (Levi, 1956; Tuzet and Pavans de Ceccatty, 1958; Simpson, 1968). In the present study, observations of gross and fine structures have provided a supporting and detailed account of these processes.

Transformation or differentiation of choanocytes and choanocyte chambers into oogonia was not observed. Archaeocytes with prominent nucleolated nuclei and mitotically active cells (intermediate in size and appearance between archaeocytes and oocytes) were observed in the vicinity of nurseries. This study and light microscope studies of oogenesis in other viviparous demosponges (Leveaux, 1941; Lévi, 1956; Simpson, 1968) suggest that archaeocytes are the anlagen of oogonia. Ultrastructural studies of oogenesis in an oviparous demosponge, *Suberites massa* (Diaz et al., 1973, 1975; Diaz, 1979), and a viviparous demosponge, *Halisarca dujardini* (Korotkova and Aisenstadt, 1976), have shown that oogonia are derived from choanocytes. It should be noted that all conclusions regarding the origin of female germ cells are based upon interpretation of fixed tissues. Conclusive experimental work has not yet been possible but is desirable. Present knowledge of oogonial genealogies indicates that species differences occur in this process as in other biological and ecological phenomena in these lower metazoans.

The dense granular material observed around the nuclear membrane of developing oocytes has also been reported in a number of other demosponges (Tuzet, 1947; Liaci and Sciscioli, 1967; Diaz et al., 1975; Gallissian and Vacelet, 1976; Diaz, 1979). The nature of the material is not known, but it has been suggested that its presence may serve as a useful mark of developing oocytes (Simpson, 1984). The presence of two distinct nucleoli in the nuclei of developing oocytes has also been reported in an oviparous demosponge (Gaino et al., 1987). If further studies reveal this to be a consistent feature, perhaps it, too, can serve as a mark of developing oocytes.

Two phases of growth in oocytes were first described in spongillids by Leveaux (1941) and have since been reported in a number of viviparous sponges (Lévi, 1956; Tuzet and Pavans de Ceccatty, 1958; Aisenstadt and Korotkova, 1976). In the present study, the first phase, "le petit accroissement," does not involve the transfer of material from cells in the mesohyl. The small yolk granules, glycogen, and lipid are probably synthesized by the oocytes themselves, utilizing dissolved substances acquired through pinocytosis or diffusion. The second phase of growth, "le grand accroissement," initially involves the transfer of materials from cells through cytoplasmic bridges, and later direct phagocytosis of nurse cells by zygotes.

Phagocytic nutrition of oocytes or zygotes has also been reported in other viviparous demosponges (Tuzet and Pavans de Ceccatty, 1958; Fell 1969; Aisenstadt and

Korotkova, 1976). Alternatively, a nonphagocytic growth process has been reported in oocytes of oviparous demosponges (Gallissian and Vacelet, 1976; Lévi and Lévi, 1976; Diaz, 1979). These nutritional differences may reflect the apportioning of resources to reduce energy loss in gametic wastage. For example, during oogenesis sponges may allocate energy to the phagocytic nurturing of zygotes, or energy may be directed toward the production of large numbers of small unfed oocytes. The direction of energy in gametic nutrition would then depend on the strategy (i.e., viviparous vs. oviparous) adopted by an individual to ensure reproductive success. Crucial information on the growth and nutrition of oocytes and zygotes will only be gained through autoradiographic and biochemical studies.

Although all phases of meiosis were not observed, the persistence of tetrads of dividing oocytes in the first phase of growth suggests that meiosis occurs without the formation of polar bodies and extensive layers of nurse cells. Although fertilization was not observed, the presence of presumed male and female pronuclei in oocytes having the size and appearance of those that had completed meiosis suggests that after initial growth and meiosis, fertilization occurs and then zygotes undergo "le grand accroissement." Other studies of the relationship between fertilization and meiosis have offered different interpretations of the order of this sequence of events (Duboscq and Tuzet, 1944; Tuzet, 1947; Tuzet and Pavans de Ceccatty, 1958; Tuzet and Paris, 1964). However, none of these studies has documented meiosis clearly. Gallissian (1980) has presented the only ultrastructural evidence of sperm cell transfer (via a carrier cell), and to date no ultrastructural studies have documented meiosis. The great problem in trying to understand these two very important events in the sexual process is that dynamic events, occurring in a brief moment of time, must be interpreted from static images.

In the present study, symbiotic bacteria were observed dispersed between blastomeres in all of the young and maturing embryos and within the central mass of cells in larvae. Oocytes and zygotes, however, were devoid of these elements. The presence of symbiotic bacteria in the embryos, together with the establishment of umbilici at this stage of development, suggests that the umbilici function as the pathway by which maternal intercellular symbionts are extracellularly transferred to progeny. The bacteria, which have the same appearance as those of the parent, showed no signs of digestion, and many were undergoing division. No other somatic cells are incorporated into oocytes, zygotes, or embryos. Only two other studies have clearly demonstrated the transfer of bacteria symbionts from one generation to the next (Gallissian and Vacelet, 1976; Lévi and Lévi, 1976). Transfer in these oviparous sponges took place directly by oocyte ingestion of bacteria from parental mesohyl, or indirectly by embryo incorporation of somatic cells containing bacteria. The present study is the first to record the transfer of bacteria between generations in viviparous sponges and reveals a unique mode of transmitting symbionts to progeny.

Various behavioral patterns and preferences for settlement substrate type and orientation have been reported for sponge larvae (Bergquist and Sinclair, 1968; Bergquist et al., 1970; Fell, 1976). In contrast, the larvae in this study displayed no unique responses with respect to substrate. The relationship between larval responses and survival is not clearly evident since settlement success depends on many environmental variables. However, specific larval responses probably represent mechanisms leading to habitat specialization and niche partitioning of adult populations.

Conclusions

Hippospongia lachne, Spongia barbara, S. cheiris, and *S. graminea* are viviparous and probably gonochoristic. Gametes and and embryos develop asynchronously within localized nurseries of endosomal tissue. Oogonia differentiate directly from archaeocytes and undergo a single oogonial division to produce primary oocytes. These oocytes autosynthesize yolk and lipid by utilizing soluble substances and then undergo meiosis, producing four secondary oocytes.

Zygotes undergo major growth through transfer of nutrients via cytoplasmic bridges and phagocytosis of nurse cells. Archaeocytes move from the adult mesohyl and form extensive layers of nurse cells around growing zygotes. Cleavage is total and equal. Umbilici are formed between the embryo and nurse cell layers and these function as pathways for the transfer of symbiotic bacteria from the maternal parent. At the end of cleavage, blastomeres differentiate, and the pigmented parenchymella larva that results has a flagellated epithelium and a mass of internal cells and symbiotic bacteria.

Free-swimming larvae possess a ring of cells bearing long flagella. This ring aids in directional swimming with constant lateral rotation. The larvae demonstrate negative phototaxis until shortly before they enter a creeping phase, during which time they show no particular taxis to light. Settlement of larvae occurs 26–56 hours after release. There is no evidence for substrate selection or orientation.

Acknowledgments

I thank M. Neuwirth for assistance with electron micrograph interpretations, and H. M. Reiswig for advise on the manuscript. The Rosenstiel School of Marine and Atmospheric Sciences, University of Miami, provided laboratory facilities. This study was supported by a grant from the Royal Bank/McGill International/Caribbean Development Fund to H. M. Reiswig and the author and an NSERC (Canada) operating grant to H. M. Reiswig.

Literature Cited

Aisenstadt, T. B., and G. P. Korotkova. 1976. A Study of Oogenesis in the Marine Sponge *Halisarca dujardini*. II. Phagocytic Activity of the Oocytes and Vitellogenesis. *Tsitologiya*, 18:818–823 (in Russian).

Bergquist, P. R., and M. Sinclair. 1968. The Morphology and Behaviour of Larvae of Some Intertidal Sponges. *New Zealand Journal of Marine and Freshwater Resources*, 2:426–473.

Bergquist, P. R., M. E. Sinclair, and J. J. Hogg. 1970. Adaptation to Intertidal Existence: Reproductive Cycles and Larval Behavior in Demspongiae. Pages 247–271 in *The Biology of the Porifera*, edited by W. G. Fry. London: Academic Press.

Diaz, J-P. 1973. Cycle sexuel de deux démosponges de l'étang de Thau: *Suberites massa* Nardo et *Hymeniacidon caruncula* Bowerbank. *Bulletin, Societé Zoologique de France*, 98:145–146.

———. 1979. Variations, différenciations et fonctions des categories cellulaire de la démosponge d'eau saumaîtres, *Suberites massa* Nardo, au cours du cycle biologique annuel et dans biologique annuel et dans les conditions experimentales. Thése, Université de la Science et Technologie, Languedoc. 332 pp.

Diaz, J-P., R. Connes, and J. Paris. 1973. Origine de la lignée germinal chez une démosponge de l'étang de Thau: *Suberites massa* Nardo. *Compte Rendus, Academie de Sciences (Paris)*, 227:661–664.

———. 1975. Etude ultrastructurale de l'ovogénèse d'une demosponge: *Suberites massa* Nardo. *Biologie Cellulaire*, 24:105–116.

Duboscq, O., and O. Tuzet. 1944. L'ovogénèse, la fécondation et les premiers stades du développement de *Sycon elegans* Bowerbank. *Archives de Zoologie Expérimentale et Générale*, 83:445–459.

Egami, N., and S. Ishii. 1956. Differentiation of Sex Cells in United Heterosexual Halves of the Sponge, *Tethya serica*. *Annotations in Zoology (Japan)*, 29:199–201.

Elvin, D. W. 1976. Seasonal Growth and Reproduction of an Intertidal Sponge, *Haliclona permollis* (Bowerbank). *Biological Bulletin*, 151:108–125.

———. 1969. The Involvement of Nurse Cells in Oogenesis and Embryonic Development in the Marine Sponge, *Haliclona ecbasis*. *Journal of Morpholology*, 127:133–150.

———. 1970. The Natural History of *Haliclona ecbasis* de Laubenfels, a Siliceous Sponge of California. *Pacific Science*, 24:381–386.

———. 1976. The Reprodution of *Haliclona loosanoffi* and Its Apparent Relationship to Water Temperature. *Biological Bulletin*, 150:200–210.

———. 1983. I. Porifera. Pages 1–29 in *Reproductive Biology of Invertebrates, Vol. 1: Oogenesis, Oviposition, and Oosorption*, edited by K. G. and R. G. Adiyodi. Chichester: John Wiley and Sons.

Fell, P. E., and W. F. Jacob. 1979. Reproduction and Development of *Halichondria* sp. in the Mystic Estuary, Connecticut. *Biological Bulletin*, 156:62–75.

Gaino, E., B. Burlando, P. Buffa, and M. Sará. 1987. Ultrastructural Study of the Mature Egg of *Tethya citrina* Sará and Melone (Porifera, Demospongiae). *Gamete Research*, 16:259–265.

Gaino, E., B. Burlando, L. Zunino, M. Pansini, and P. Buffa. 1984. Origin of Male Gametes from Choanocytes in *Spongia officinalis* (Porifera, Demospongiae). *International Journal of Invertebrate Reproduction*, 7:83–93.

Gallissian, M–F. 1980. Etude ultrastructurale de la fécondation chez *Grantia compressa*. *International Journal of Invertebrate Reproduction*, 2:321–329.

Gallissian, M–F., and J. Vacelet. 1976. Ultrastructure de quelques stades de l'ovogénèse de spongiaires du genre *Verongia* (Dictyoceratida). *Annales des Sciences Naturelles, Zoologie (Paris)*, 18:381–404.

Gilbert, J. J., and T. L. Simpson. 1976. Sex Reversal in a Freshwater Sponge. *Journal of Experimental Zoology*, 195:145–151.

Kaye, H. R., R. A. MacLeod, and H. M. Reiswig. 1985. Nutritional Characteristics and Antibiotic Sensitivities of Bacteria Isolated from Four Commerical West Indian Sponge Species. Pages 153–158 in *Proceedings, Fifth International Coral Reef Congress, Tahiti*, 5.

Korotkova, G. P., and T. B. Aisenstadt. 1976. A Study of the Oogenesis of the Marine Sponge, *Halisarca dujardini*. I. The Origin of the Oogonia and Early Stages of Oocyte Development. *Tsitologiya*, 18:549–555 (in Russian).

Leveaux, M. 1941. Contribution á l'étude histologique de l'ovogénèse et de la spermatogenese des Spongillidae. *Annales, Societé Royale Zoologique de Belgique*, 72:251–269.

Lévi, C. 1951. Existence d'un stade grégaire transitoire au cours de l'ovogénèse des spongiaires, *Halisarca dujardini* (Johnston) et *Oscarella lobularis* (O.S.). *Compte Rendus Academie de Science, (Paris)*, 233:826–828.

———. 1956. Etude des *Halisarca* de Roscoff. Embryologie et systématique des Démosponges. *Archives de Zoologie Expérimentale et Générale*, 93:1–181.

Lévi, C., and P. Lévi. 1976. Embryogénèse de *Chondrosia reniformis* (Nardo), Démosponge ovipare, et transmission des bactéries symbiotes. *Annales de Sciences Naturelles, Zoologie (Paris)*, 18:367–380.

Liaci, L., and M. Sciscioli. 1967. Osservazioni sulla maturazione sessuale di un Tetractinellidae: *Stelleta grubii* O.S. (Porifera). *Archivio Zoologico*, 52:169–177.

Liaci, L. Scalera-, M. Sciscioli, A. Matarrese, and C. Giove. 1971. Osservazioni sui cicli sessuali di alcune keratosa (Porifera) e loro interesse megli studi filogenetici. *Atti, Societa Peloritana Scienze Fisiche, Matematiche e Naturali*, 17:33–52.

Liaci, L. Scalera-, M. Sciscioli, and G. Piscetelli. 1976. Raffronto tra il comportamento sessuale di alacune Ceractinomorpha. *Rivista di Biologia*, 66:135–153.

Reiswig, H. M. 1973. Population Dynamics of Three Jamaican Demospongiae. *Bulletin of Marine Science*, 223:191–226.

———. 1983. I. Porifera. Pages 1–21 in *Reproductive Biology of Invertebrates, 2: Spermatogenesis and Sperm Function*, edited by K. G. and R. G. Adiyodi. Chichester: John Wiley and Sons.

REPRODUCTION IN WEST INDIAN COMMERCIAL SPONGES

Simpson, T. L. 1968. The Biology of the Marine Sponge *Microciona prolifera* (Ellis and Sollander). II. Temperature-Related Annual Changes in Functional and Reproductive Elements with a Description of Larval Metamorphosis. *Journal of Experimental Marine Biology and Ecology*, 2:252–277.

———. 1984. *The Cell Biology of Sponges*. New York: Springer. 662 pp.

Tuzet, O. 1947. L'ovogénèse et la fécondation de l'éponge calcaire *Leucosolenia (Clathrina) coriacea* Montagu et de l'éponge siliceuse *Reniera elegans* Bowerbank. *Archives de Zoologie Expérimental et Générale*, 85:127–148.

Tuzet, O., and J. Paris. 1964. La spermatogénèse, l'ovogénèse, la fécondation et les premiers stades du développement chez *Octavella galangaui*. *Vie et Milieu*, 15:309–327.

Tuzet, O., and M. Pavan de Ceccatty. 1958. La spermatogénèse, l'ovogénèse, la fécondation et les premiers stades du développement d'*Hippospongia communis* Lamarck (= *H. equina* O.S.). *Bulletin Biologique*, 92:331–348.

Van de Vyver, G., and P. Willenz. 1975. Experimental Study of the Life-cycle of the Freshwater Sponge, *Ephydatia fluviatilis* in Its Natural Surroundings. *Wilhelm Roux's Archives*, 177:41–52.

Watanabe, Y. 1978. The Development of Two Species of *Tetilla* (Demosponge). *Natural Science Reports, Ochanomizu University*, 29:71–106.

KEIKO TANAKA-ICHIHARA*
Shimoda Kita High School
Shimoda, Shizuoka 415, Japan

YOKO WATANABE
Department of Biology
Ochanomizu University
Otsuka, Tokyo 112, Japan
and
Tateyama Marine Laboratory
Ochanomizu University
Tateyama, Chiba 294–03, Japan

Gametogenic Cycle in *Halichondria okadai*

Abstract

The formation of gametes in *Halichondria okadai* from the Nabeta Bay (Shimoda, Japan) was examined by light microscopy from 1981 to 1985. Sexual reproduction in *H. okadai* takes place once a year in summer. Male reproductive cells were found in the specimens collected from June to September. Female reproductive cells were observed in the specimens collected from April to October. Parenchymella larvae were released from July to early September. In June and July, almost all the specimens possessed either male or female cells. Individual sponges were never found to have both sexual elements at the same time. Although *H. okadai* may not exhibit contemporaneous hermaphroditism, gonochorism or successive hermaphroditism does occur.

During the reproductive period, the number of choanocyte chambers dropped sharply. At the end of this period, choanocyte chambers were gradually regenerated and normal aquiferous systems were reconstituted. This shows that choanocytes may be involved in the formation of gametes.

As is well known, sponges do not possess true gonads or specific reproductive ducts. In most species, gametogenesis occurs at a specific time of year. A major part of the sponge is involved in reproduction (Fell, 1974). In some cases, the tissues regress as sexual reproduction proceeds (Diaz, 1973; Chen, 1976). With respect to the origin and the timing of the differentiation of gametes, sexual reproduction should be considered through the gametogenic cycle rather than the reproductive period. Several researchers have investigated the gametogenic cycle of sponges (Simpson, 1968; Simpson and Gilbert, 1973; Diaz, 1973; Fell and Jacob, 1979). The life cycle of *Halichondria* sp. in the Mystic Estuary, Connecticut, has been studied in detail (Fell and Jacob, 1979). Reproduction in this sponge is very complicated. In one year, a population may consist predominantly of distinct males and females, whereas in another year, most of the population may be contemporaneous hermaphrodites. Reproductive ele-

*Present address: Department of Biomedical Polymer Science, Institute for Comprehensive Medical Science, Fujita-Gakuen Health University School of Medecine, Toyoake, Aichi 470–11, Japan.

ments are found in the postdormant specimens from May to July, while in the postlarval specimens the reproductive period begins in July and continues at least into October. The reproductive periods of postdormant and postlarval specimens are always separated from each other.

In Japan, little work has been done on reproduction in *Halichondria* except for postlarval development in *H. okadai* (Watanabe, 1976). This is one of the common species in the shallow intertidal zone of central and west Japan, where it is frequently found on rocks. It reproduces in the summer, when individuals are often exposed to the sun at low tide. It is not dormant in winter. Its gametogenic cycle may differ somewhat from that of the Connecticut species mentioned above.

We investigated sexual reproduction in *Halichondria okadai* over a five-year period using histological procedures to analyze our samples. The present study of the gametogenic cycle in *Halichondria* represents a first step toward elucidating gametogenesis in this species.

Materials and Methods

This study was carried out from 1981 to 1985 in Nabeta Bay, Shimoda (34°40′ N, 138°57′ E). Samples of sponges were collected once or twice a month and made into histological sections according to the following procedures. All the sections were observed under a light microscope.

For histological studies, some specimens were fixed with Bouin's fixative, dehydrated through a butanol series, and embedded in Tissue Prep (Fisher Scientific Company, U.S.A.). Series of 6 μm sections were made and stained with Mayer's haematoxyline, eosin, and alcian blue 8GX at pH 2.6. Some specimens were fixed in 2.5% glutaraldehyde in 0.2M cacodylate buffer, pH 7.2, and postfixed in 1% osmium tetroxide in 0.2M cacodylate buffer. After treatment by 2.5% hydrofluoric acid to dissolve spicules specimens were dehydrated through an ethanol series and embedded in low-viscosity epoxy resin after Kushida (1980). Sections 2 μm thick were made and stained with toluidine blue.

Results

Seasonal changes in structure of the tissues were observed in the histological sections. The tissue of nonreproductive specimens shows a normal aquiferous system, with many choanocyte chambers and some archeocytes (Figure 1a, b). Specimens from the next period have small oocytes scattered in the tissue among the choanocyte chambers (Figure 1c). As the oocytes grow larger, they accumulate yolk granules. They are fertilized in situ and then start to develop. At the climax of reproduction, female specimens contain various stages of oocytes, embryos, and larvae throughout the mesohyl (Figure 1d) the number of choanocyte chambers is less than that in the nonreproductive period. In the males, spermatic cysts begin to fill the sponge tissue, which takes on the appearance of testis (Figure 1e). Each spermatic cyst shows a different stage in spermatogenesis. The number of choanocyte chambers drops sharply. After the sperm or larvae are released, the choanocyte chambers regenerate, and the normal aquiferous systems are formed.

Table 1 shows the reproductive elements that occurred each month over the five years of the study. Specimens containing only female reproductive elements (such as oocytes, embryos, and larvae) and those containing only male reproductive elements (such as spermatogenic cells and sperm) are identified as female and male, respectively. Females are observed from April to October. Males are observed from June to early September. Contemporaneous hermaphroditic specimens containing both female and male elements were never observed. In consequence, reproductive specimens possess either female or male elements. In *Halichondria okadai*, the sexes are separate, perhaps as a result of gonochorism or successive hermaphroditism.

Shimoda has a mild climate. The water temperature in Nabeta Bay is lowest in February and highest in August. The monthly mean temperature exceeds 20°C between June and October. Figure 2 shows the percentage of reproductive specimens and the monthly mean water temperature for the period under investigation. Female elements begin to appear in April, always before the male elements. Gametes form when the water temperature rises. During June and July, almost all the specimens contain reproductive cells. Embryos begin to appear in June and larvae in July. In fact, parenchymella larvae were released from the female specimens collected and kept in the aquarium in July, August, and early September. After late August, degenerated oocytes are found. Embryos are never seen in the sections of specimens collected after late September, but larvae are seen in the sections of October.

Table 1. Occurrence of reproductive elements in *Halichondria okadai* between 1981 and 1985 (c.herm. = contemporaneous hermaphrodite)

	Jan	Feb	Mar	Apr	May	Jun	Jul	Aug	Sep	Oct	Nov	Dec
No. specimens	14	14	13	19	21	48	45	27	16	22	18	12
No. females	0	0	0	2	3	29	25	11	7	2	0	0
No. males	0	0	0	0	0	17	19	4	2	0	0	0
No. c.herm.	0	0	0	0	0	0	0	0	0	0	0	0

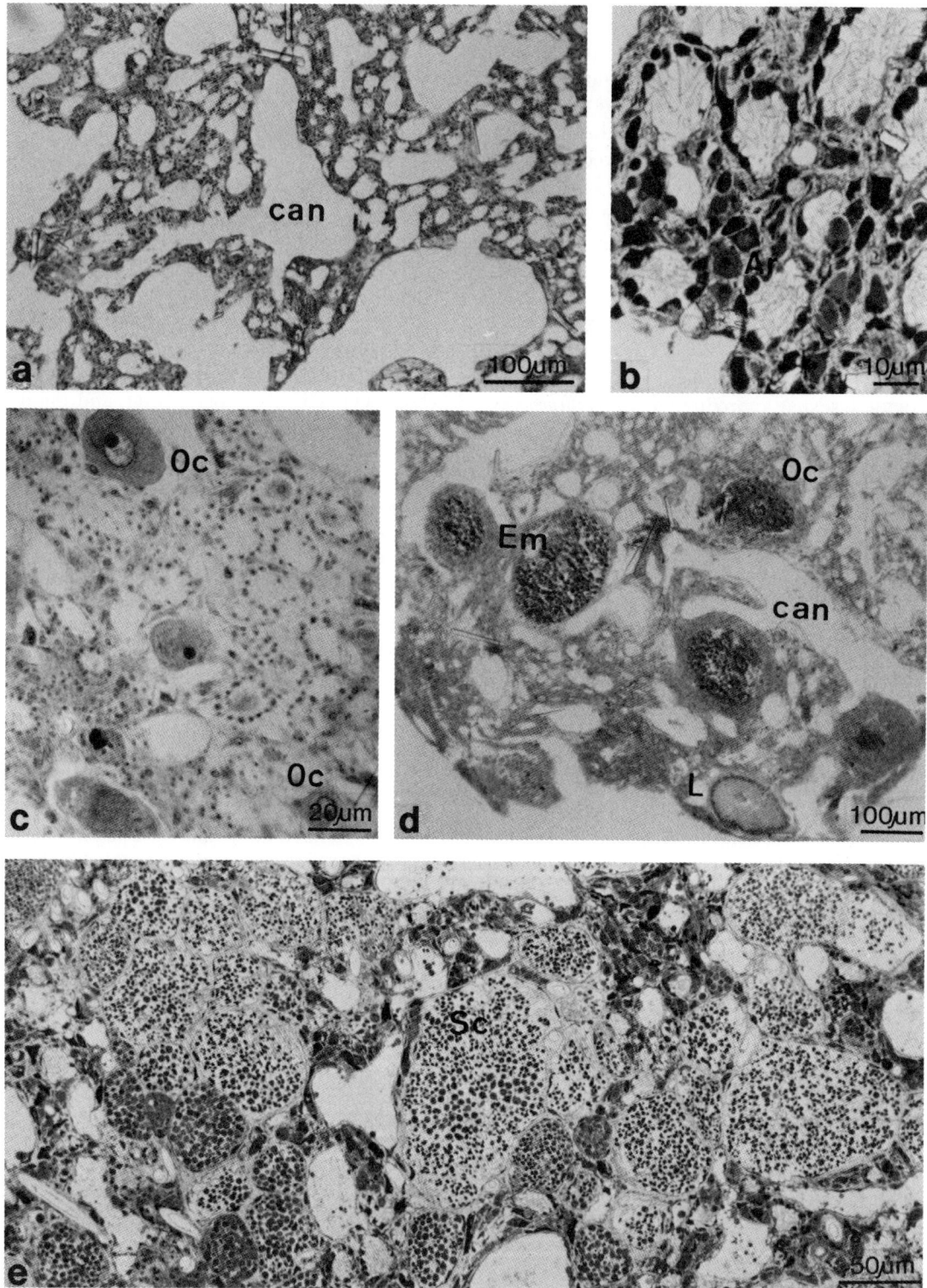

Figure 1. Seasonal changes of tissue in *Halichondria okadai*: *a*, nonreproductive specimen collected in January, with normal aquiferous system; *b*, specimen collected in January with mesohyl and choanocyte chambers; *c*, reproductive specimen collected in June, with small oocytes among choanocyte chambers; *d*, specimen collected in August, with oocytes, embryos, and a larva; *e*, specimen collected in late June, with tissue filled with spermatic cysts (note sharp drop in number of choanocyte chambers). *Ar*, archeocyte; *can*, canal; *Em*, embryo; *L*, larva; *Sc*, spermatic cyst.

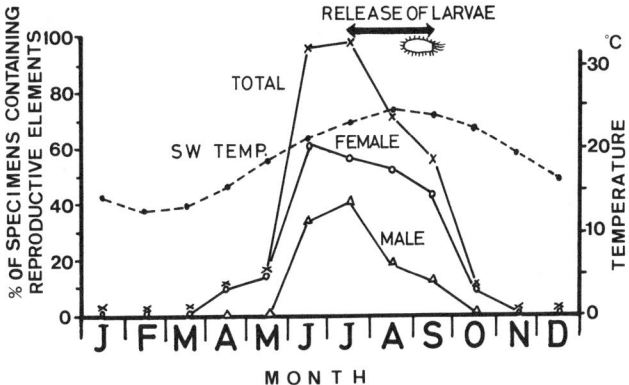

Figure 2. Occurrence of reproductive elements in *Halichondria okadai (solid lines)* and average seawater temperatures *(dashed line)* for each month in Nabeta Bay, Shimoda, from 1981 to 1985. Temperature readings are taken from record kept in Shimoda Marine Research Center, University of Tsukuba, Japan.

Male specimens collected in September contain a small number of spermatic cysts with sperm. None of the reproductive elements can be recognized from November to May. These observations are summarized in Figure 3.

Discussion

It has been confirmed that *Halichondria okadai* reproduces once a year in summer. In our research, no specimens exhibited female and male reproductive elements at the same time. Sponges are reported to have various patterns of gametogenesis (Fell, 1974; Sará, 1974; Fell, 1983), but *H. okadai* is unlikely to be a contemporaneous hermaphrodite. As shown in Figure 2, the period of oogenesis almost coincides with that of spermatogenesis. Thus, the sex of an individual sponge may not change within one reproductive period, as it is the case in gonochorism or successive hermaphroditism. *Spongilla lacustris* appears to be gonochoristic (Simpson and Gilbert, 1973), but some specimens appear to undergo sex reversal from one year to the next (Gilbert and Simpson, 1976). We do not yet have

sufficient data to distinguish between gonochorism and successive hermaphroditism. As Fell (1983) noted, it is necessary to study a sponge population over an extended period of time and to use a variety of approaches before sexuality can be confirmed.

Halichondria is thought to have two species of contemporaneous hermaphroditic sponges, *H. panicea* (Fell, 1974) and *Halichondria* sp. (Fell and Jacob, 1979), although the latter seems to exhibit labile sexuality. However, there are conflicting reports concerning sexuality in some species, such as *Spongilla lacustris* (Leveaux, 1941; Simpson and Gilbert, 1973), and *Hymeniacidon sanguinea (H. caruncula)* (Sará, 1961; Diaz, 1973). Simpson (1984) states that sex in sponges is highly labile and therefore may be physiologically rather than genetically determined. It may be natural to find various patterns of sexuality in *Halichondria*.

It has been shown that gametes are formed during the period of rising water temperature. In many species, gametogenesis occurs during the warmer months (Simpson, 1984). Water temperature in Nabeta Bay begins to rise in March and reaches a peak in August. Our observations that oocytes and spermatic cysts occur in April and June, respectively, support the view that gametogenesis is correlated with rising water temperature (Simpson, 1968; Fry, 1971; Elvin, 1976). However, temperature may not be the only factor to affect gametogenesis, because spermatogenesis decreases considerably in August. It seems that gametogenesis is also related to endogenous factors and to environmental factors other than temperature. As already mentioned, oocytes in the specimens collected after late August appear to be degenerated, and embryos are not present in the specimens collected after late September. This suggests that fertilization may end in August. This timing seems to coincide with the end of spermatogenesis. It is likely that unfertilized oocytes may break down after August. Therefore, the occurrence of fertilization may be controlled by the period of spermatogenesis in *H. okadai*. These findings support the view of Reiswig (1983) that a combination of internal factors and external factors governs reproduction in sponges inhabiting stressful shallow intertidal habitats.

We were unable to distinguish the postlarval specimens from the overwintered specimens of *Halichondria okadai*. Further investigations are required to determine whether the postlarval specimens have reproductive elements or not.

Tissue regression was observed during the reproductive phase: the number of choanocyte chambers appeared to decrease at the beginning of reproduction and regenerate at the end of it. In the female specimens, this phenomenon was observed in the choanocyte chambers around the oocytes. In the male specimens, choanocyte chambers were considerably reduced and the tissues of the entire sponge were filled with spermatic cysts. This suggests that

MONTH

J F M A M J J A S O N D

Oogenesis	
Spermatogenesis	
Embryos and Larvae	
Release of Larvae	
Tissue Regression	
Postreproductive Regeneration	
Functional tissue	

Figure 3. Gametogenic cycle in *Halichondria okadai* in Nabeta Bay.

3d Int. Sponge Conf. 1985

the formation of gametes is closely related to the decrease of choanocyte chambers, which may be caused by several factors: choanocytes may be phagocytized by other cells for nutritional purposes or they may be differentiated into the other types of cells necessary in reproduction, or they may play both roles. Although we have not yet obtained enough direct evidence, we suspect that choanocytes may play an important role in gamete formation.

Conclusions

Sexual reproduction in *Halichondria okadai* takes place once a year in summer in Nabeta Bay, Shimoda, and may reflect gonochorism or successive hermaphroditism. The number of choanocyte chambers decreases during reproduction, and increases at the end of the period. Choanocytes probably play an important role in reproduction.

Acknowledgments

We wish to express our thanks to Hiroshi Watanabe, Professor of the Shimoda Marine Research Center, the University of Tsukuba, for his critical comments on the manuscript and for his encouragement. We also thank the staff of the center, who kindly facilitated our use of the laboratory. Contribution no. 485 of the Shimoda Marine Research Center. This work was supported in part by a Grant-in-Aid for Scientific Research form the Ministry of Education, Science and Culture, Japan.

Literature Cited

Chen, W. 1976. Reproduction and Speciation in *Halisarca*. Pages 113–139 in *Aspects of Sponge Biology*, edited by F. W. Harrison and R. R. Cowden. New York: Academic Press.

Diaz, J. 1973. Cycle sexuel de deux démosponges de l'étang de Thau: *Suberites massa* Nardo et *Hymeniacidon caruncula* Bowerbank. *Bulletin de la Société Zoologique de France*, 98:145–156.

Elvin, D. W. 1976. Seasonal Growth and Reproduction of an Intertidal Sponge, *Haliclona permolis* (Bowerbank). *Biological Bulletin*, 151:108–125.

Fell, P. E. 1974. Porifera. Pages 51–132 in *Reproduction of Marine Invertebrates I*, edited by A. C. Giese and J. S. Pearse. New York: Academic Press.

———. 1983. Porifera. Pages 1–29 in *Reproductive Biology of Invertebrates I: Oogenesis, Oviposition, and Oosorption*, edited by K. G. Adiyodi and R. G. Adiyodi. Chichester: John Wiley and Sons.

Fell, P. E., and W. F. Jacob. 1979. Reproduction and Development of *Halichondria* sp. in the Mystic Estuary, Connecticut. *Biological Bulletin*, 156:62–75.

Fry, W. G. 1971. The Biology of Larvae of *Ophlitaspongia seriata* from Two North Wales Populations. Pages 155–178 in *Fourth European Marine Biological Symposium*, edited by D. J. Crisp. Cambridge: Cambridge University Press.

Gilbert, J. J., and T. L. Simpson. 1976. Sex Reversal in a Freshwater Sponge. *Journal of Experimental Zoology*, 195:145–151.

Kushida, H. 1980. An Improved Embedding Method Using ERL4206 and Queto1653. *Journal of Electron Microscopy*, 29:193–194.

Leveaux, M. 1941. Contribution à l'étude histologique de l'ovogénèse et de la spermatogénèse des Spongillidae. *Annales, Société Royale Zoologique de Belgique*, 72:251–269.

Reiswig, H. M. 1983. Porifera. Pages 1–21 in *Reproductive Biology of Invertebrates 2: Spermatogenesis and Function*, edited by K. G. Adiyodi and R. G. Adiyodi. Chichester: John Wiley and Sons.

Sarà, M. 1961. Ricerche sul gonocorismo ed ermafroditismo nei Poriferi. *Bollettino di Zoologia*, 28:47–59.

———. 1974. Sexuality in the Porifera. *Bollettino di Zoologia*, 41:327–348.

Simpson, T. L. 1968. The Biology of the Marine Sponge *Microciona prolifera* (Ellis and Sollander). II. Temperature-Related Annual Changes in Functional and Reproductive Elements with a Description of Larval Metamorphosis. *Journal of Experimental Marine Biology and Ecology*, 2:252–277.

———. 1984. *The Cell Biology of Sponges*. New York: Springer. 662 pp.

Simpson, T. L., and J. J. Gilbert. 1973. Gemmulation, Gemmule Hatching and Sexual Reproduction in Fresh-water Sponges. I. The Life Cycle of *Spongilla lacustris* and *Tubella pennsylvanica*. *Transactions of the American Microscopical Society*, 92:422–433.

Watanabe, Y. 1976. Postlarval Development of the Demosponge *Halichondria okadai*. *Zoological Magazine*, 85:340.

MARIE-FRANCE GALLISSIAN*
JEAN VACELET
Centre d'Océanologie de Marseille
Station Marine d'Endoume
13007 Marseille, France

Fertilization and Nutrition of the Oocyte in the Calcified Sponge *Petrobiona massiliana*

Abstract

Transmission electron microscope observations of some stages of the oocyte development of *Petrobiona massiliana* have revealed some added information on the unusual phenomena witnessed in the reproduction of this relict calcified sponge in earlier light microscope studies. The spermatozoon is conventionally transferred by a modified choanocyte, the carrier cell, to the oocyte at an early stage of its development. The material from numerous nurse cells, corresponding to the modified choanocytes of a whole chamber, are then transferred to the oocyte through the intermediary of the carrier cell, which remains partly enclosed in the oocyte cytoplasm during its major growth. This carrier cell becomes hypertrophied, while its nucleus undergoes profound changes and accumulates dense granules. This transmission of material from nurse cells to the oocyte via the carrier cell apparently involves a large, dense organite, which seems to correspond to the considerably enlarged spermatozoon. At the end of the oocyte growth, the carrier cell degenerates, while the dense organite becomes fibrillar.

These phenomena, especially the trophic role of the structures involved in fertilization and the modifications of the spermatozoon, appear to be unique in the animal kingdom.

The original carrier-cell system of the fertilization of sponges of the Class Calcarea has been well described in light microscopy studies (see Fell, 1974, and Simpson, 1984, for reviews). A recent electron microscope study of *Grantia compressa* by Gallissian (1980) confirmed that the inclusion transferred to the oocyte by the carrier cell is actually a modified, encysted spermatozoon ("spermiocyst").

Unusual phenomena occurring during the development and the fertilization of the oocyte in *Petrobiona massiliana* Vacelet and Lévi, a Mediterranean survivor of the

---

*Present address: Université de Provence, Centre St. Charles Hydrobiologie, 1 Place Victor Hugo, F-13331 Marseille Cedex 3, France.

so-called pharetronid sponges, (Calcaronea) have been described and tentatively interpreted in a light microscope study (Vacelet, 1964) as follows. The oocyte receives the carrier cell and the spermatozoon at a very early stage of its development. The carrier cell remains partly engulfed by the oocyte throughout its growth and is surrounded by nurse cells, which are modified choanocytes from a whole chamber. This cell greatly enlarges as its nucleus accumulates dense granules and becomes a large chromatic mass, but then degenerates when the oocyte reaches full growth. The nutrients from the nurse cells apparently enter the oocyte through the carrier cell, which is attached to a large organite inside the oocyte cytoplasm. This organite, which has the shape of a truncated sphere, is remarkably hard and could not be cut in paraffin sections; it was tentatively identified as an enlarged spermiocyst, and the spermatozoon was thought to be inside a capsule, which remained very dense until oocyte development was complete. (For further details and cytochemical results see Vacelet, 1964.)

In our opinion, these phenomena, which seem to imply that the fertilization structures—that is, both the carrier cell and the spermiocyst—also act as feeding structures of the oocyte, warranted an electron microscope study. However, the scarcity of developing oocytes and technical difficulties in this highly calcified sponge makes it difficult to undertake a detailed study.

Material and Methods

Petrobiona massiliana Vacelet and Lévi was collected by SCUBA diving in marine caves near Marseille during the peak months of reproduction (May to October) over a period of several years.

Considerable difficulties arose in this study. Like in other representatives of the Class Calcarea, this sponge is not easily preserved to the extent required for ultrastructural studies. Special, long decalcification procedures must be employed and the tissue of the living sponge cannot be observed owing to extensive calcification. Moreover, oocytes and embryos are always scarce in the tissue; only one or two stages are usually present in the same specimen. Several of the intermediate stages also seem to be very short. In addition, the sponge lives in dark caves that are not easy to reach and is difficult to keep alive in an aquarium. These conditions have, up to now, made it impossible to undertake a cytochemical study in electron microscopy and to observe the early stages of fertilization.

Our specimens were fixed immediately after collection in 2.5% glutaraldehyde in seawater or in a buffer composed of 0.4 M sodium cacodylate and seawater 1/1 (20 h). They were then decalcified in ethylendiamine tetracetic acid (EDTA) or in RDO (Du Page Kinetic Laboratory,

Downers Grove, Illinois), 10% in sea water, for 3 days. After complete demineralization, the tissue was postfixed in 2% osmium tetroxide in sea water and embedded in Araldite. Thin sections were cut using a diamond knife, stained with uranyl acetate and lead citrate, and then examined with a Phillips EM 300 electron microscope. Semithin sections were stained in toluidin blue.

Results

We did not observe the early stages of the fertilization, which were rarely seen in previous studies under a light microscope and are presumably very short.

In the first stage observed here (Figure 1), the oocyte measures 20 to 40 μm in maximum diameter. The carrier cell lies in a concavity of its cytoplasm and is surrounded by nurse cells (Figure 1a), which are larger than choanocytes and have larger inclusions and a well-developed Golgi complex. The carrier cell consists of a rounded cell body and an extension that is engulfed in the oocyte cytoplasm. This cell increases in size at the same time as the oocyte. Its body, 15 to 25 μm in diameter, contains a nucleus, 4 μm in diameter, that at the beginning of this stage has a typical appearance and has a large nucleolus (1.25 μm). At the end of this stage, this nucleus measures up to 11 μm, and its nucleolus 3 μm. Dense masses then begin to appear in the cytoplasm toward the outer layer of the nuclear envelope (Figure 1c). The cytoplasm of the cell body contains numerous diverse inclusions and is denser than the oocyte and choanocyte cytoplasm. The extension of the carrier cell inside the oocyte contains a bundle of microfilaments that is parallel to its axis and ends in the "spermiocyst" (Figure 1a,d). This so-called spermiocyst is a dense structure, 7 to 10 μm in diameter, attached to the extremity of the extension of the carrier cell close to the oocyte nucleus. In sections, this organite appears to be homogeneous. It always has the shape of a truncated sphere, with two lateral expansions and a central conical one. The membranes between the carrier cell and the oocyte cytoplasm seem to disappear around the spermiocyst, which is surrounded by a finely granular layer, about 0.4 μm thick; this layer may be a densification of the oocyte cytoplasm (Figure 1b).

In a later stage, when the oocyte is about 65 to 75 μm in diameter (Figure 2a), the carrier cell measures up to 25 μm and its nucleus is highly modified (Figure 2a,c,d). A central homogeneous sphere is surrounded by smaller ones embedded in a dense matrix. The nucleoplasm is restricted to a narrow area between the dense central mass and the nuclear envelope; it contains several finely granular masses that resemble nucleoli. Irregular masses of a similar granular material are located outside the nuclear envelope, where they constitute a complete crown. Ex-

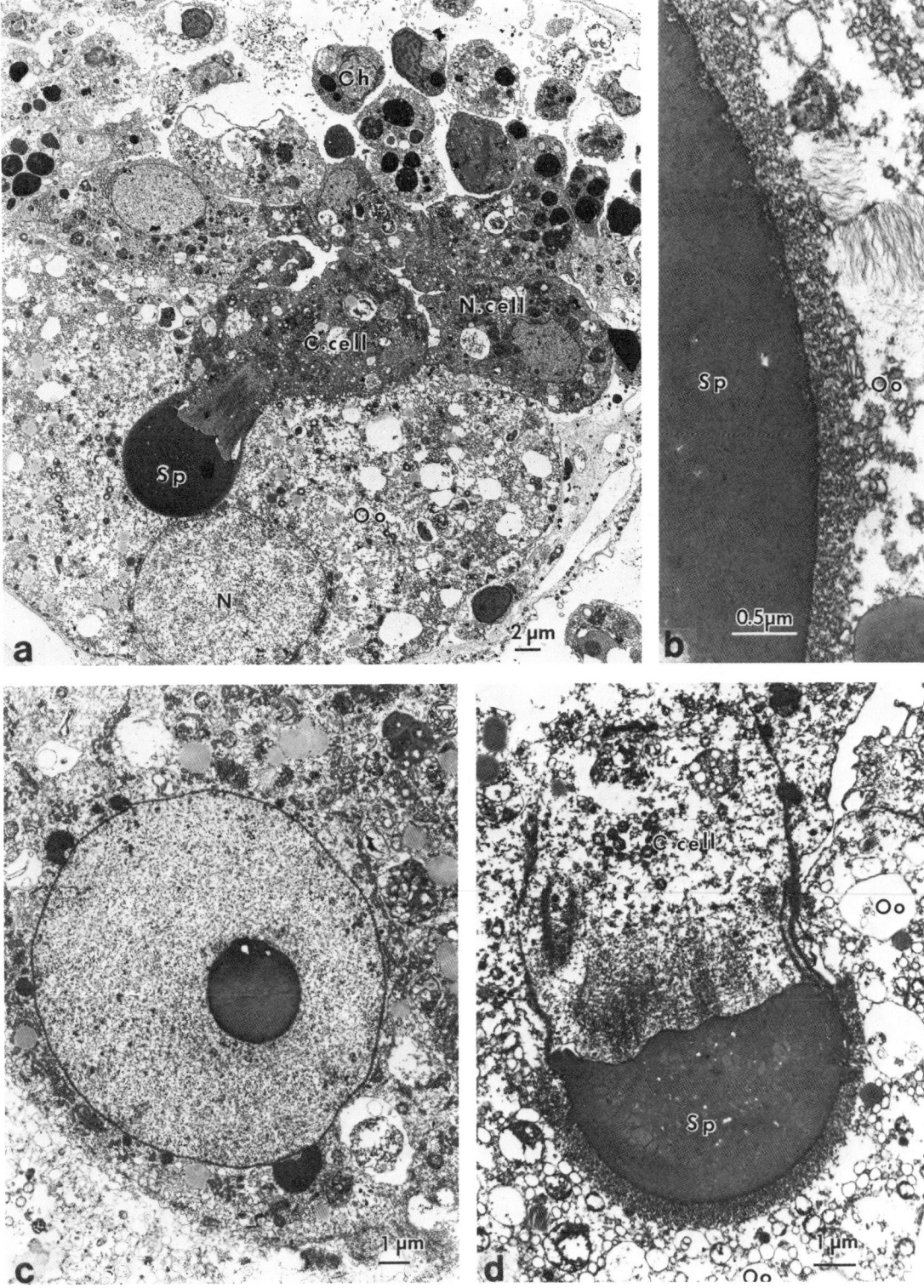

Figure 1. *Petrobiona massiliana*, fertilization complex in oocyte, 40–45 μm in diameter: *a*, oocyte and its carrier cell surrounded by nurse cells and choanocytes; spermiocyst is inside oocyte cytoplasm, close to its nucleus; *b*, spermiocyst and its surrounding layer inside oocyte; *c*, nucleus and surrounding cytoplasm of a carrier cell; *d*, a spermiocyst and extension of carrier cell inside an oocyte (note fibrillar content of carrier cell). *C. cell*, carrier cell; *Ch*, choanocyte; *N*, oocyte nucleus; *N. cell*, nurse cell; *Oo*, oocyte; *Sp*, spermiocyst.

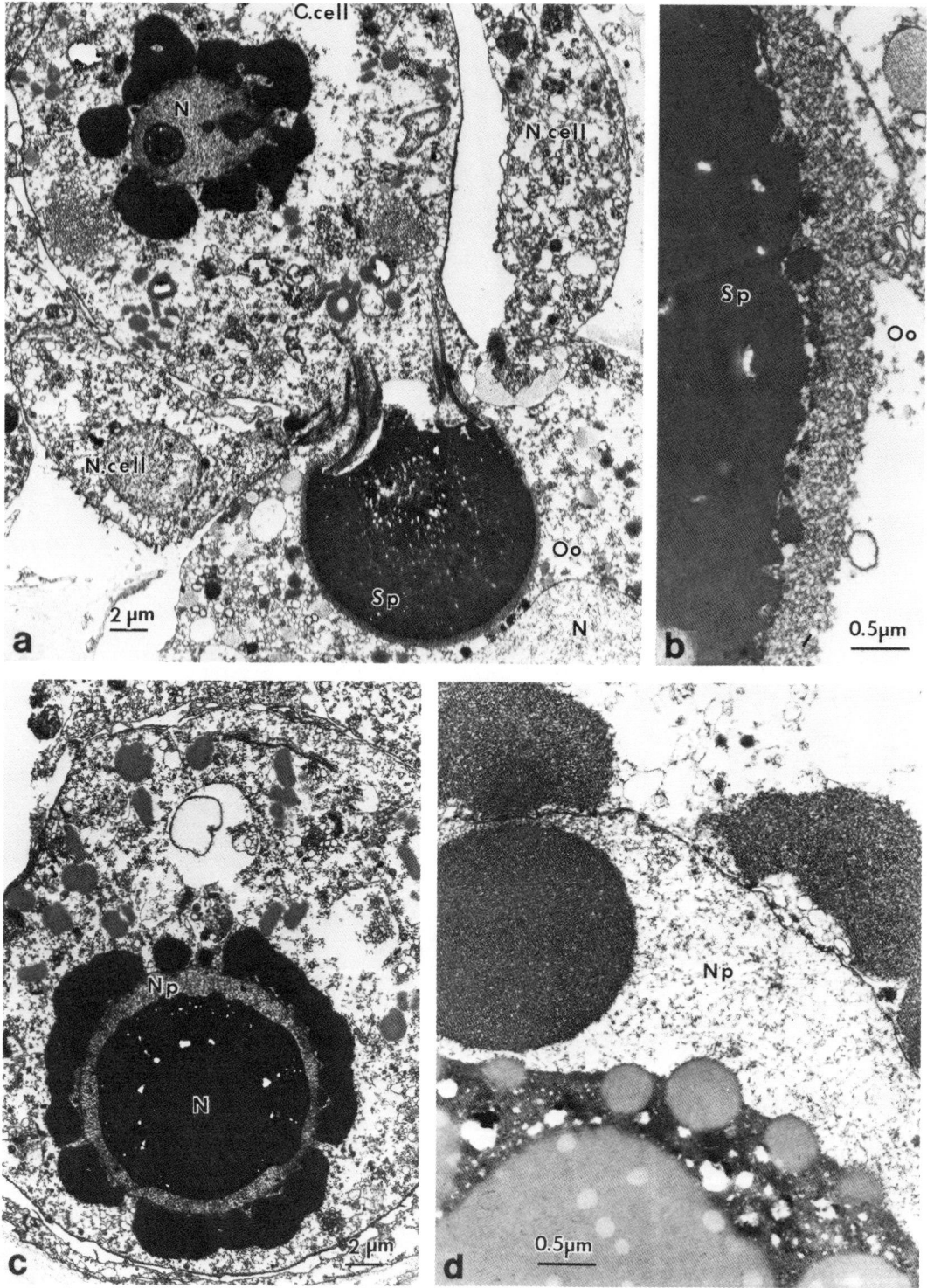

Figure 2. *Petrobiona massiliana*, fertilization complex in oocyte, 65–75 μm in diameter: *a*, oocyte and its carrier cell, the nucleus of which undergoes accumulation of granules; spermiocyst has its peculiar shape; *b*, spermiocyst beginning to decondense, and its surrounding layer in oocyte cytoplasm; *c*, carrier cell and its nucleus at a later stage than in *a*; *d*, carrier cell nucleus, with its diverse granules in nucleoplasm and outside nuclear envelope. *C. cell*, carrier cell; *N*, nucleus; *N. cell*, nurse cell; *Np*, nucleoplasm; *Oo*, oocyte; *Sp*, spermiocyst.

changes seem to occur between inner and outer masses, the latter being more irregular in shape and attached to the nuclear envelope by some points only. The carrier cell is still surrounded by nurse cells, some of which are also in direct contact with the oocyte (Figure 2a). Its cytoplasm still contains diverse inclusions and vesicles. Now the spermiocyst (Figure 2a,b) is enlarged up to 16 to 18 μm, and has the same shape as in the earlier stage. However, its homogeneous matrix seems to decondense. Scattered empty spaces and superficial indentations appear, with some pieces of dense material being separated.

In the next stage (Figure 3), the oocyte is about 80 μm in diameter and contains numerous, striated vitelline inclusions (Gallissian, 1981). Nurse cells are rare, and the carrier cell undergoes degeneration. Its cytoplasm has disappeared. However, a dense structure, most probably the remains of the modified nucleus of the carrier cell, is still attached to the oocyte (Figure 3a,c,f). This structure consists of an aggregation of spheres, with a dense central area surrounded by a finely granular zone. The spermiocyst is now heterogeneous. It is still surrounded by a granular layer. The dense, homogeneous matrix forms a thin layer with an irregular internal outline, which surrounds a fibrillar central area (Figure 3b,d,e,f). Vitelline inclusions may occur in this central area. In some cases, the central area contains a dispersed, granular material surrounded by a membrane, which closely resembles the oocyte nucleoplasm (Figure 3f).

Discussion

Our electron microscope observations confirm earlier descriptions of the development and fertilization of this sponge and permit better observations of some structures. It is now evident that the granules that progressively fill the nucleus of the carrier cell appear both inside and outside the nuclear envelope. Another important point is that the so-called spermiocyst is not a hollow capsule containing the spermatozoon, as supposed, but rather a homogeneous dense organite that decondenses at the end of oocyte growth. However, the possibility cannot be ruled out that a small vesicle contining a tiny spermatozoan has been overlooked, as we did not examine serial sections of this dense organite.

However, ultrastructural observations do not lead to a complete understanding of the phenomenon. The interpretation that is tentatively proposed here has to be confirmed by cytochemical data at the ultrastructural level and by observations of the early stages of fertilization. It must be emphasized that spermatogenesis and ultrastructure of the free spermatozoon are unknown in the class Calcarea. Fertilization processes have never been observed in living material from other calcareous sponges,

owing to technical obstacles that are even more insurmountable here.

Nonetheless, we speculate that the spermatozoon may be the so-called spermiocyst itself. The sperm DNA might be dispersed among a dense protein matrix. When decondensation occurs, the fibrillar or granular material that then appears in the central area and that is morphologically similar to the oocyte nucleoplasm could be the male pronucleus. This interpretation needs to be confirmed by cytochemical DNA localization. A previous cytochemical study on paraffin sections showed that the "spermiocyst", which was too hard to be cut by this technique, presented negative metachromasy (staining green with toluidin blue) and was Feulgen negative. However, this result is not conclusive, as the sperm DNA would be very diluted within this large organite and most probably undetectable by the Feulgen reaction.

This interpretation may explain the observations in both light and electron microscopy. Thus, the overall process, although considerably more complex, would not be fundamentally different from that known in other calcareous sponges: namely, a modified encysted spermatozoon is brought to the oocyte by a carrier cell. However, aspects of the process remain puzzling. The spermatozoon, which is only slightly modified in *Grantia compressa* where its main components are still recognizable (Gallissian, 1980), would be strongly modified here and would form this homogeneous, very hard organite, 18 μm in diameter. In addition, the structures involved in fertilization may also be involved in oocyte nutrition and may play a transfer role between nurse cells and the oocyte, as an entire choanocyte chamber is transferred to the oocyte through both the carrier cell and the spermatozoon itself. Such modifications of a spermatozoon and such participation of fertilization structures in oocyte feeding both appear to be highly unusual in the animal kingdom (Baccetti, 1985). However, that is not altogether the case in other calcareous sponges: in *Sycon elegans* (Duboscq and Tuzet, 1944), for example, the carrier cell is surrounded by four nurse cells, and a situation resembling that in *Petrobiona massiliana* has been observed in *Achramorpha nivalis* (Vacelet, 1964). Therefore, the seemingly unusual phenomena observed here could well reflect the complex development of processes that are merely suggested in some other representatives of the calcaronean Calcarea. Note that this extreme complexity occurs in a living fossil, which is expected to be more primitive, and not in the structurally very simple recent Calcarea.

Alternatively, the present observations may indicate that the "spermiocyst" is only a structure involved in the nutrient transfer from nurse cells to the oocyte, and that the sperm cell itself is somewhere within the carrier cell—where inclusions are numerous (but all very different from a conventional sperm cell)—or is the carrier cell itself. It

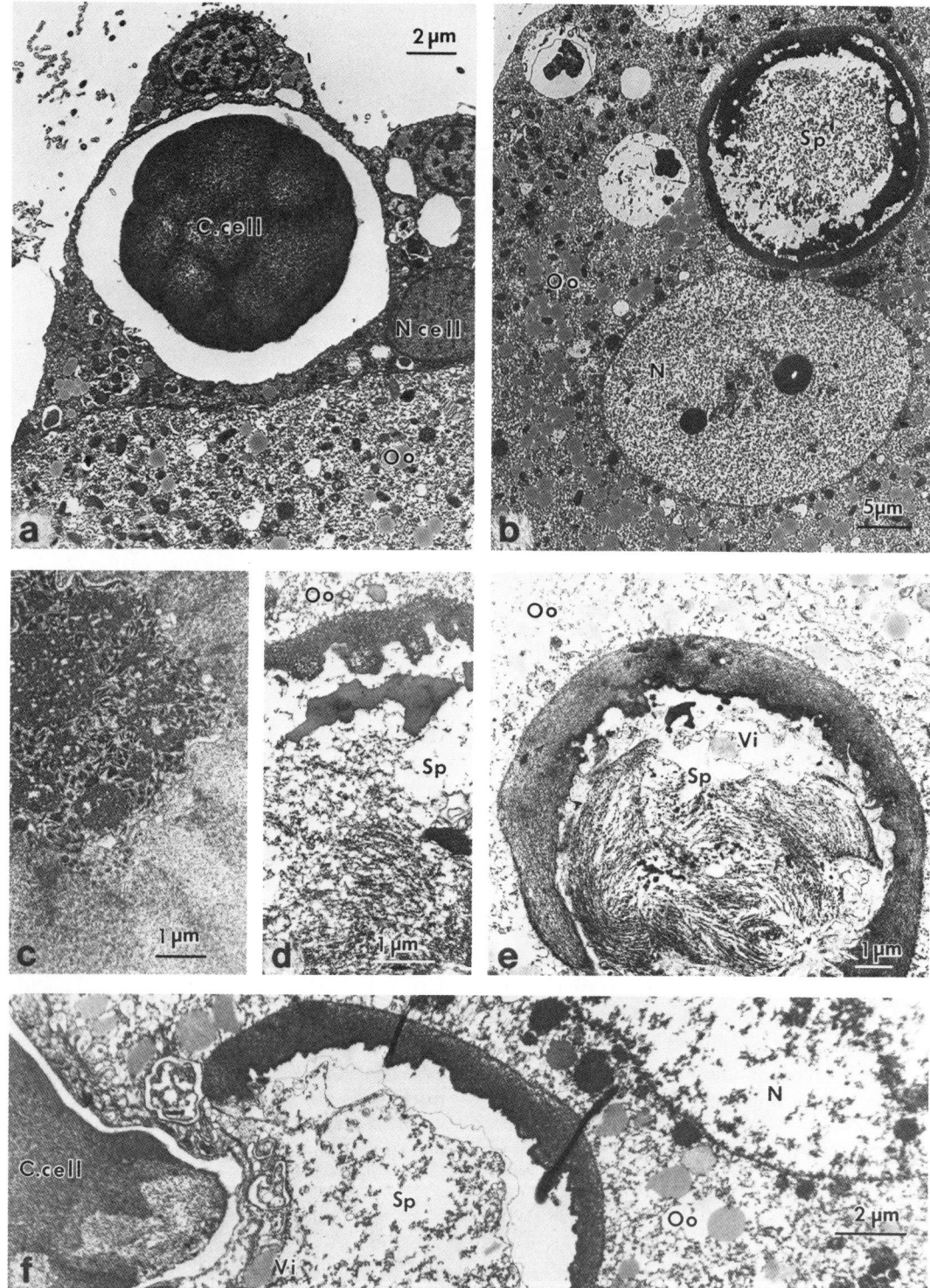

Figure 3. *Petrobiona massiliana*, fertilization complex in a mature oocyte, 80 μm in diameter: *a*, oocyte and remains of modified nucleus of carrier cell; *b*, oocyte and decondensed spermiocyst; *c*, remains of nucleus of carrier cell; *d*, a decondensed spermiocyst; *e*, a decondensed spermiocyst (note vitelline inclusion in central area); *f*, oocyte, spermiocyst, and remains of carrier cell. *C. cell*, carrier cell; *N*, nucleus; *N. cell*, nurse cell; *Oo*, oocyte; *Sp*, spermiocyst; *Vi*, vitelline inclusion.

may also be that we are dealing here only with oocyte feeding and not with fertilization, which would occur later by some unknown process. These alternative interpretations are just as speculative as the first one, and are less in agreement with what is known about fertilization in other calcareous sponges.

Acknowledgments

We are grateful to C. Bézac and M. R. Causi for their technical help. We also thank J. G. Harmelin, who collected many of the specimens examined in this study.

Literature Cited

Baccetti, B. 1985. Evolution of the Sperm Cell. Pages 3–58 in *Biology of Fertilization, 2: Biology of the Sperm*, edited by C. B. Metz and A. Monroy. New York: Academic Press.

Duboscq, O., and O. Tuzet. 1944. L'ovogenèse, la fécondation et les premiers stades du développement du *Sycon elegans* Bow. *Archives de Zoologie expérimentale et générale*, 83:445–459.

Fell, P. E. 1974. Porifera. Page 51–132 in *Reproduction of Marine Invertebrates, I*, edited by A. C. Geise and J. S. Pearse. New York: Academic Press.

Gallissian, M. F. 1980. Etude ultrastructurale de la fécondation chez *Grantia compressa*. *International Journal of Invertebrate Reproduction*, 2:321–329.

———. 1981. Etude ultrastructurale de l'ovogenèse chez quelques Eponges calcaires (Porifera, Calcarea). *Archives de Zoologie expérimentale et générale*, 122:329–340.

Simpson, T. L. 1984. *The Cell Biology of Sponges*. New York: Springer. 662 pp.

Vacelet, J. 1964. Etude monographique de l'éponge calcaire pharétronide de Méditerranée, *Petrobiona massiliana* Vacelet et Lévi. Les Pharétronides actuelles et fossiles. *Recueil de Travaux de la Station marine d'Endoume*, 34(50):1–125.

GRADIMIR N. MISEVIC
VERENA SCHLUP
MAX M. BURGER
Department of Biochemistry
Biocenter, University of Basel
Klingelbergstrasse 70
CH–4056 Basel, Switzerland
and
Marine Biological Laboratory
Woods Hole, Massachusetts 02543

Larval Metamorphosis of *Microciona prolifera:* Evidence against the Reversal of Layers

Abstract

The developmental fate of the flagellated epithelial cells of the larvae from the marine sponge *Microciona prolifera* was investigated using two approaches: (1) epithelial cells of the free-swimming larvae were selectively iodinated and their life cycle during metamorphosis was followed by light and electron microscopical autoradiography; and (2) the larval metamorphosis was observed by time-lapse cinematography in combination with scanning electron microscopy. The results of both approaches showed that upon attachment of the larvae to the substrate, the flagellated epithelium is first invaginated and subsequently phagocytosed by archeocytes. This led to the conclusion that the flagellated epithelial cells of the free-swimming larval form do not give rise to the choanocytes that build up the flagellated chamber system of the adult sponge. Thus, metamorphosis of the parenchymella larvae of *M. prolifera* is not associated with inversion of embryonic layers, as previously suggested. This finding supports the view that sponges should indeed be classified as Metazoa.

Cell differentiation and consequently their spatial organization into tissues and organs are the major events occurring during embryogensis. Our understanding of the mechanisms of these very complex and multistep processes has been dependent upon identification of differentiation pathways of embryonal stem cells. However, study of cell fate mapping is associated with analyses of many different cell types constantly interacting with each other. Such developmental studies using sponges involve fewer cell types and a simpler cellular organization of adult tissues and therefore provide a good experimental model system that may be more likely to result in greater knowledge of the fundamental principles of cell differentiation.

The classification of sponges as descendants of the first and most primitive multicellular organisms has in the past few decades been a matter of discussion for evolutionary biologists. The criteria used for systematics were as

182

usual based on cytological and biochemical characteristics of the adult organism, as well as on the sequences of morphological changes and differentiation processes during embryonal development. Differences in interpretation in developmental studies have given rise to two contradictory opinions about the classification of sponges. Delage (1892, 1898), Codreanu (1970), and Hadzi (1963) suggested that sponges are not Metazoans and should be placed into a separate subkingdom (Enantiozoa). On the other hand, Bergquist (1978), Meewis (1938), Brien (1972), and Tuzet (1973) argued that sponges should be classified as proper Metazoa. The crucial point of disagreement between the two opposing views is whether the inversion of embryonic cell layers and occurs during larval metamorphosis. Delage (1892) proposed that metamorphosis of the parenchymella-type larva is associated with the inversion of embryonic layers and therefore (Delage, 1898) that sponges had to be separated from Metazoa. The layer inversion theory is based on the findings that highly differentiated flagellated epithelial cells of free swimming parenchymella-type larvae migrate inward after settlement and during metamorphosis become choanocytes, which finally form "flagellated" (choanocyte) chambers (Delage, 1892; Codreanu, 1970; Hadzi, 1963). In contrast, the second theory suggests that the flagellated larval epithelium consists of terminally differentiated cells that do not give rise to choanocytes (Bergquist, 1978; Meewis, 1938; Brien, 1972; Tuzet, 1973).

The study of development and metamorphosis of sponge larvae not only clarifies the processes of cell differentiation, but also helps to improve the natural classification of sponges and our understanding of evolution of these early multicellular organisms.

Material and Methods

SPONGES. Live specimens of *Microciona prolifera* (Poecilosclerida) were collected by the Supply Department of the Marine Biological Laboratory in the Woods Hole, Massachusetts, area during the months of June and July. To stimulate release of larvae, sponges were placed in a glass beaker containing one liter of seawater for 5–15 min. Then sponges were transferred back to the aquarium with running sea water and the larvae were collected individually by glass pipette and placed into 20 ml filtered seawater.

IODINATION OF FREE-SWIMMING LARVAE. About 100 freshly released larvae were collected in 0.5 ml filtered seawater and placed in a glass vial coated with Iodo-gen (Pierce), which was used as a catalyst for the iodination reaction (Markwell and Fox, 1978; Misevic and Burger, 1982). The live larvae were labeled in the presence of 1 mCi Na^{125}I for 5 min at room temperature. Unreacted ^{125}I was washed away with four changes of 20 ml filtered seawater at 4°C. After washing, the larvae were transferred to chambers

made of two teflon rings holding a 0.2 μm-pore size Nucleopore filter at the bottom (1.5 cm inner diameter).

FIXATION AND EMBEDDING OF LARVAE. Larvae were fixed in 3% glutaraldehyde in cacodylate buffer (0.1 M Na-cacodylate, pH 7.2, 1.75% NaCl, and 10 mM CaCl$_2$) for 1 h at room temperature. After three washes with cacodylate buffer, larvae were postfixed in 1% OsO$_4$ in cacodylate buffer for 1 h at room temperature. Following dehydration, larvae were embedded in Epon 812 and sectioned.

AUTORADIOGRAPHY. Sections of larvae embedded in Epon 812 were coated with Ilford L4 autoradiographic emulsion and exposed for 10–20 days. Semithin sections (1–4 μm) were developed in Kodak D-19, thin sections in Kodak Microdol X. For scanning electron microscopy, larvae were fixed as outlined above, dried after dehydration by the critical-point method, and coated with platinum.

TIME-LAPSE CINEMATOGRAPHY. Freshly collected larvae (10–15) were placed into a microinjection chamber, as described by Kiehart (1981). Time lapse cinematography of freshly attached larvae was performed using a video-intensifying camera.

Results

FATE OF FLAGELLATED LARVAL EPITHELIUM DURING METAMORPHOSIS

In order to test the hypothesis of inversion of embryonic layers during larval metamorphosis, flagellated epithelial cells of the free-swimming larvae of the marine sponge *Microciona prolifera* were selectively iodinated and their course of development followed by light and electron microscope autoradiography.

Autoradiographs of tissue sections of iodinated free-swimming larvae of *Microciona prolifera* at the light and electron microscope level revealed that over 98% of the label is localized in the epithelial cells (Figures 1a-c). The background level inside the larvae corresponded to only 1–2% of the total radioactivity, indicating that only flagellated epithelium was labeled.

Iodinated larvae were cultured in Nucleopore filter chambers. The chambers, each containing 1–5 larvae in 0.5 ml filtered seawater, were kept in tissue culture plates also filled with filtered seawater. Cultures were grown at 20°C, which is equal to the temperature of the sea during June and July. Most larvae were still free swimming 12 h after release and tests showed that radioactivity remained localized within the epithelium, indicating that no metabolic transfer of the label to the other cells occurred. Between 12 h and 24 h, the larvae started to attach and

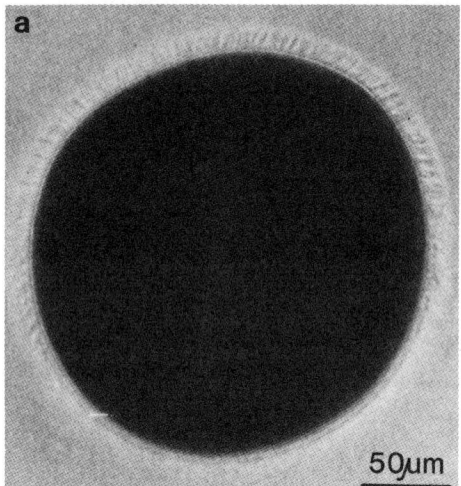

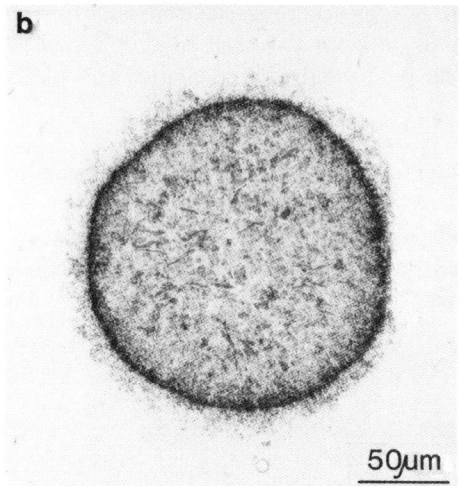
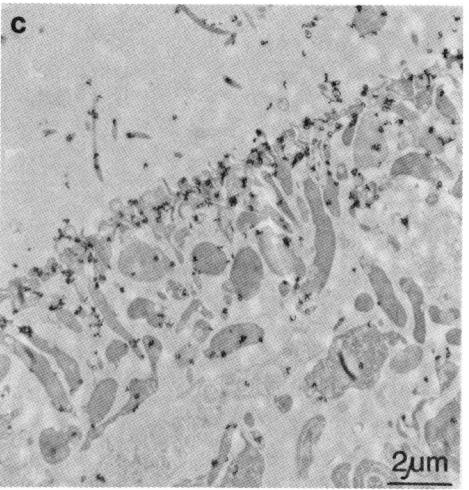

Figure 1. *Microciona prolifera*, free-swimming larva: *a*, light micrograph; *b*, light microscope autoradiograph of a semithin section of surface-iodinated flagellated epithelial cells; *c*, electron microscope autoradiograph of thin section of larva in *b*.

spread. In order to follow the fate of the flagellated epithelial cells precisely, it was crucial to examine the larvae exactly during these transition phases, between attachment and early spreading (Figure 2a).

Autoradiography at this stage revealed that most of the labeled flagellated epithelium cells are located in the central core mass, just above the attachment site (Figures 2b,c). It is interesting to note that initial adherence of larvae is accomplished by a portion of flagellated epithelium (Figure 2c).

During this early period of spreading, as well as in later stages, there was no difference in the time of development between labeled and nonlabeled larvae, indicates that iodizing has no effect on the process of metamorphosis.

Most larvae entered the spreading phase 25 h after release (Figure 3a). At this time, the label associated with invaginated epithelium also started to appear in archeocytes surrounding these epithelial cells (Figure 2b). As the spreading of larvae progressed, label accumulated in archeocytes, while the flagellated epithelium disappeared. After 48 h all larvae had spread on the substrate, and autoradiographic grains could only be seen inside the

archeocytes (Figure 3b). Electronmicrographs revealed that the radioactive label was now located entirely in archeocytes phagosomes (Figure 3c). This finding makes it unlikely that simple metabolic transfer of the label from intact flagellated cells to archeocytes took place. Phagosomes containing the label were small at an early phase of spreading, they must subsequently fuse into larger vesicles. It can be concluded that the invaginated epithelium of free-swimming larvae is eventually phagocytosed by archeocytes.

Three to four days after release the larvae had metamorphosed into a fully developed sponge (Figure 4a). At this stage, the archeocytes had excreted part of the label into the extracellular space within the sponge and part into the external medium (Figure 4b). The newly formed choanocytes building up the choanocyte chambers did not show the presence of any label exceeding the background level at either the light or electron microscopical level (Figure 4b, c). Thus the flagellated epithelium does not appear to give rise to choanocytes but represents a terminally differentiated cell line that is phagocytosed by archeocytes upon larval settlement.

3d Int. Sponge Conf. 1985

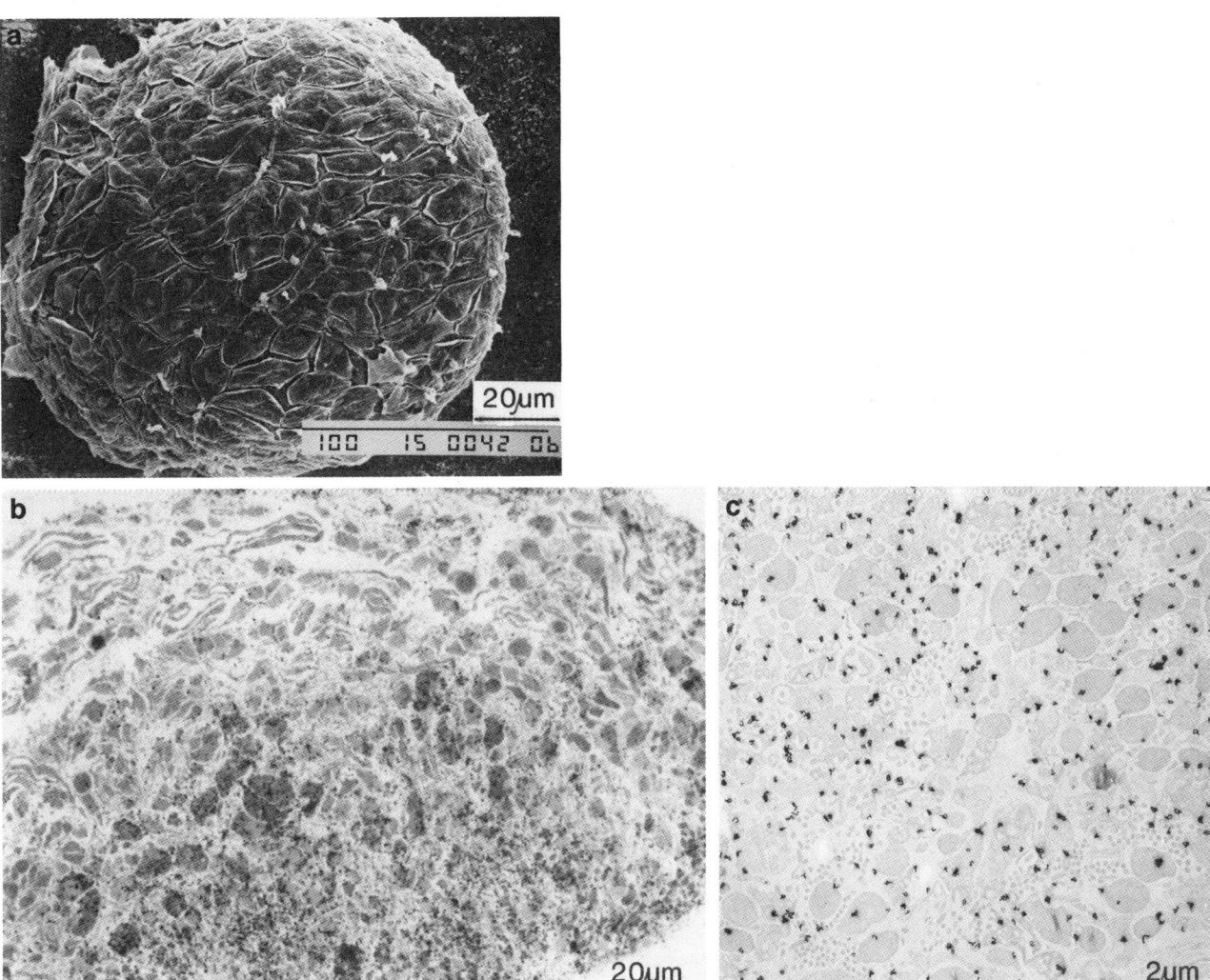

Figure 2. *Microciona prolifera*, attaching larva, 16 h after release: *a*, scanning electron micrograph of larva attaching to Nucleopore filter substrate; *b*, light microscope autoradiograph of semithin section; *c*, electron microscope autoradiograph of thin section of larva in *b*.

TIME-LAPSE CINEMATOGRAPHY AND SCANNING ELECTRON MICROSCOPY OF LARVAL ATTACHMENT

Time-lapse cinematography of early events of larval attachment and spreading was also used to follow the fate of the flagellated epithelium. Invagination of the flagellated cells was seen immediately after attachment to the substrate, but no shedding of cells or flagella was observed. The process of inward translocation of the epithelium was associated with regular pulsing contractions with periods of 2–5 min. However, the mechanism of this invagination still remains unknown. Either the flagellated epithelium of the free swimming larva is pushed into the interior, as the ectoderm of the new sponge is forming, or it is simply pulled into the larval body by contraction of the central cell mass.

In addition, scanning electron microscopy was em-

ployed to determine whether some of the flagellated epithelial cells were shed. Since the larvae were attached to Nucleopore filters (0.2 μm pore diameter), the exchange of fixatives, washing buffers, and dehydration solutions was performed by filtration. Therefore, cells or flagella released by the larvae would have remained on the filter and been detected by scanning electron micrography. The fact that none were found supports the results of autoradiography and time-lapse cinematography, which suggest that the flagellated epithelium is invaginated and then phagocytized.

Discussion

The controversy surrounding the classification of sponges as Metazoa stems from disagreements about the occurrence of inversion of embryonal layers during meta-

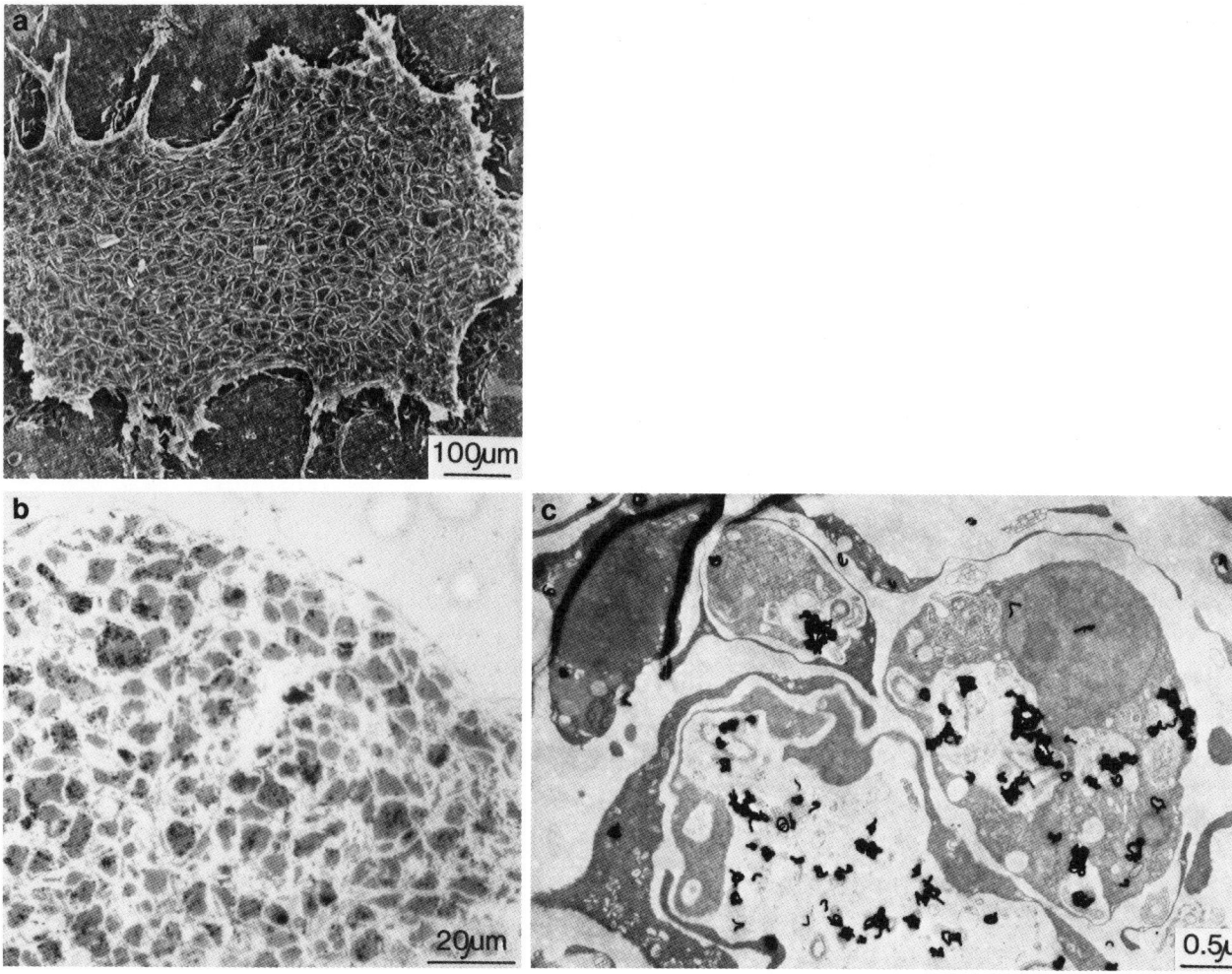

Figure 3. *Microciona prolifera*, spread larva, 48 h after release: *a*, scanning electron micrograph; *b*, light microscope autoradiograph of semithin section; *c*, electron microscope autoradiograph of thin section of larva in *b*.

morphosis of parenchymella-type larvae. Histological investigations and time-lapse cinematography of the fate of the flagellated epithelial cells of *Microciona prolifera* larvae reveal that these cells do not give rise to choanocytes, which later form the chamber system. The inversion of germ layers may therefore not occur in sponges. Early investigators, observing invagination of the flagellated epithelial cells into the interior cell mass during settlement, concluded that these cells transformed into choanocytes. However, experimental evidence of the fate of the internalized cells was missing in all of these studies. We have demonstrated here that the flagellated epithelium is phagocytized upon invagination and does not give rise to choanocytes. A similar observation was reported by Bergquist (1978:242), who also pointed out that invagination may not precede phagocytosis in some species.

Experiments showing full morphogenesis of the larvae from which the core cell mass was surgically removed do not exclude the presence of archeocytes and other cell types as contaminants of the flagellated epithelia (Borojevic, 1966). Therefore, this method does not serve to prove that choanocytes are formed from flagellated epithelia. We have recently shown that the core cell mass separated from the flagellated epithelium will give rise to fully developed sponges in the same time as attached larvae. However, as in experiments where the epithelium has been isolated, the development of manipulated larvae may not reflect real metamorphosis, but rather a process comparable to the reconstitution of adult sponges from dissociated cells.

We have demonstrated that an inversion of the embryonal layers does not occur during metamorphosis of *Microciona prolifera* larvae. Our experiments with this sponge together with histological evidence indicate that sponges have the same basic developmental pattern as other multicellular animals. Thus sponges are true Metazoans.

3d Int. Sponge Conf. 1985

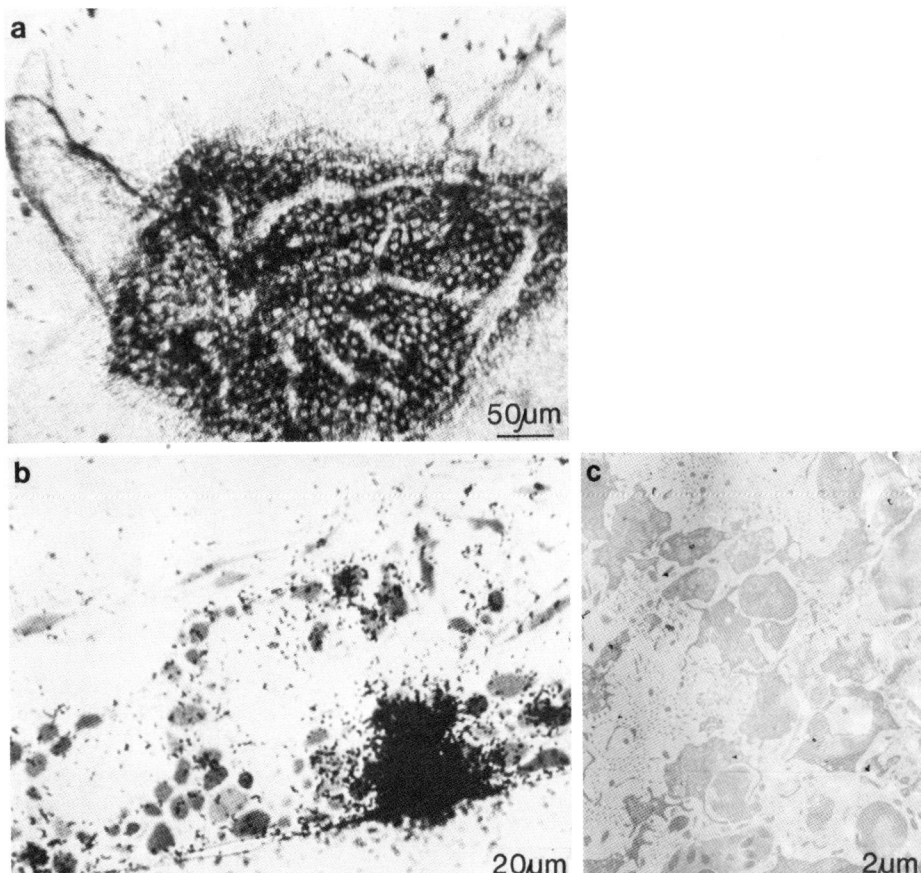

Figure 4. *Microciona prolifera*, fully developed young sponge: *a*, light micrograph of live sponglet; *b*, light microscope autoradiograph of semithin section prepared from specimen in *a*; *c*, electron microscope autoradiograph of thin section prepared from specimen in *a*.

Acknowledgments

This work was supported by the Swiss National Foundation for Scientific Research, Grant No. 3.269.82 and the Ministry of the City and Canton of Basel. We are grateful to Melvin and Evelin Spiegel for performing the time-lapse cinematography, to Theo Schäfer for expert help with electron microscope autoradiography, and to Kjell Tullberg for improving the English grammar.

Literature Cited

Bergquist, P. R. 1978. *Sponges*. London: Hutchinson. 268 pp.

Borojevic, R. 1966. Etude expérimentale de la différenciation des cellules de l'éponge au cours de son dévelopment. *Developmental Biology*, 14:130–153.

Brien, P. 1972. Les feuillets embryonnaires des éponges. *Bulletin, Academie royale de Belgique, Series 5*, 58:715–732.

Codreanu, R. 1970. Grands problèmes controverses de l'évolution phylogénétique des metazoires. *Année Biologique*, 9:671–709.

Delage, Y. 1892. Embriogénie des éponges. Développment postlarvaire des éponges siliceuses et fibreuses marines et d'eau douce. *Archives de Zoologie Expérimentale et Général*, 10:345–598.

——. 1898. Sur la place des Spongiaires dans la classification. *Comptes Rendus, Académie des Sciences, Paris*, 126:545–548.

Hadzi, J. 1963. *The Evolution of the Metazoa*. Oxford, London, New York, Paris: Pergamon Press. 499 pp.

Kiehart, D. P. 1981. Studies on the In Vivo Sensitivity of Spindle Microtubules to Calcium Ions and Evidence for a Vesicular Calcium Sequestering System. *Cell Biology*, 88:604–617.

Markwell, M.-A. K., and F. Fox. 1978. Surface-specific Iodination of Membrane Proteins of Viruses and Eucariotic Cells using 1,3,4,6,-tetrachloro–3α,6α-diphenylglycoluril. *Biochemistry*, 17:4807–4817.

Meewis, H. 1938. Contribution à l'étude de l'embriogenèse des Myxospongidae *Halisarca lobularis*. *Archives de Biologie*, 50:3–66.

Misevic, G. N., and M. M. Burger. 1982. The Molecular Basis of Species-specific Cell-Cell Recognition in Marine Sponges, and a Study on Organogenesis during Metamorphosis. Pages 193–209 in *Embryonic Development, B: Cellular Aspects* edited by M. M. Burger and R. Weber. New York: Alan R. Liss.

Tuzet, O. 1973. Introduction et place des spongiaires dans la classification. Pages 1–26 in *Traité de Zoologie, 3, Spongiaires*, edited by P.-P. Grassé. Paris: Masson.

HAZIME MIZOGUCHI
Division of Biology
Junior College of Rissho University
Kumagaya City, Saitama 360–01, Japan

YOKO WATANABE
Department of Biology
Ochanomizu University
Otsuka, Tokyo 112, Japan
and
Tateyama Marine Laboratory
Ochanomizu University
Tateyama, Chiba 294–03, Japan

Collagen Synthesis in *Ephydatia fluviatilis* during Its Development

Abstract

Collagen synthesis was examined in the freshwater sponge *Ephydatia fluviatilis* during its development. Inhibitors of collagen synthesis were employed to determine the role of collagen during morphogenesis of developing sponges. After gemmule germination of [^{14}C]proline-loaded sponges, the radioactivity of [^{14}C]hydroxyproline residues found in the hot trichloroacetic acid-extracted protein fraction (collagen fraction) was reduced by azetidine 2-carboxylic acid, an analogue of proline, or α,α'-dipyridyl, an inhibitor of prolylhydroxylase. The rate of [^{14}C]proline incorporation into the collagen fraction and the production of [^{14}C]hydroxyproline residues increased up to the stage of hatching and did not decrease significantly thereafter. Azetidine 2-carboxylic acid inhibited the fixation of the sponge to the substratum as well as the formation of choanocyte chambers. In addition, α,α'-dipyridyl blocked hatching and fixation of the sponges to the substratum. Collagen, which is produced in freshwater sponges after gemmule germination, probably contributes to the attachment of the sponges to their substratum and to the following morphogenesis, such as the development of the aquiferous system.

It is well known that collagenous proteins are present in sponges. For instance, basopinacoderm contains collagen fibrilis. It is also known that skeletons are composed of siliceous spicules and collagenous spongin. Furthermore, collagen fibrils have been found in the intercellular space of the mesh structure of sponges (see the review in Garrone, 1984). These observations suggest that sponge morphogenesis during development is supported by collagen synthesis.

In the present study, we measured the change in the rate of hydroxylation of proline residues in collagen in the freshwater sponge *Ephydatia fluviatilis* during its development. In addition, we examined the effects of the inhibitors of collagen synthesis, such as azetidine 2-carboxylic acid and α,α'-dipyridyl (Hurch and Chvapil, 1965; see the reviews in Garrone, 1978; Olsen, 1981) on the development of sponges in order to obtain information on the role of collagen in morphogenesis.

Materials and Methods

Gemmule-bearing specimens of *Ephydatia fluviatilis* were collected from the Yokotone-gawa River in Ibaragi prefecture, Japan. The gemmules, attached to the parent sponge tissue, were stored in the dark at 4°C until the beginning of the experiments.

Gemmules were isolated from sponge tissue and cleaned with 1% hydrogen peroxide and were washed 10 times with M-medium (Rasmont, 1961). Batches of 300 gemmules were incubated at 25°C in sterilized glass test tubes.

At the time indicated in Figure 1, [^{14}C]proline (final concentration: 0.2 μCi/ml, specific activity: 285 mCi/mmole) (Radiochemical Centre, Amersham, England) was added and sponges were incubated at 25°C for 3 h. The sponges exposed to [^{14}C]proline, were washed twice with ice-chilled M-medium and suspended in 5% ice-cold trichloroacetic acid (TCA). The suspended material was homogenized in a glass homogenizer with a motor-driven Teflon pestle in an ice bath. After centrifugation at 3,000 rpm for 10 min, the pellet obtained was washed twice with ice-cold 5% TCA, twice with 95% ethanol saturated with CH$_3$COONa, twice with ethanol-ether (3:1), and sus-

pended again in 5% TCA. This suspension was heated in a water bath at 95°C for 75 min (Fitch et al., 1955) and centrifuged at 3,000 rpm for 10 min. An aliquot of the supernatant fluid was dialyzed against deionized water and the dialyzed aliquot, which contained gelatin, was analyzed for radioactivity. The rate of [^{14}C]proline incorporation was indicated as dpm of [^{14}C]proline incorporated into protein the hot TCA-soluble fraction per mg of sponge protein per 3 hr. Subsequently, the [^{14}C]hydroxyproline residue in the dialyzed supernatant was separated by the method described by Peterkofsky and Prockop (1962). The radioactivity was estimated with the aid of a liquid scintillation counter (Aloka, LSC-700, Aloka Co., Tokyo). Protein was estimated by the method of Lowry et al. (1951).

Azetidine 2-carboxylic acid (Sigma Chemical Co., USA) or α,α'-dipyridyl (Kanto Chemical Co., Tokyo) was added to the cultures just after gemmule incubation. Then sponges were successively cultured until control sponges reached full development (124 h after incubation at 25°C). The sponges in these cultures were examined under a light microscope and photographed.

Results

Figure 1 shows the change in the rate of [^{14}C]proline incorporation into the collagen fraction and the peptidyl proline hydroxylation in freshwater sponges during development. The rate of peptidyl proline hydroxylation is expressed as the ratio of [^{14}C]hydroxyproline residue to total radioactivity in hot TCA-extracted proteins per mg of sponge protein per 3 h. The rate of [^{14}C]proline incorporation into the collagen fraction and the peptidyl proline hydroxylation increased until hatching and did not decrease significantly thereafter.

Figure 2 shows the effects of azetidine 2-carboxylic acid, an analogue of proline, on peptidyl proline hydroxylation in freshwater sponges. Treatment of the sponges with azetidine 2-carboxylic acid was started just after gemmule incubation. The rates of peptidyl proline hydroxylation were estimated at 124 h after incubation at 25°C. Figure 3 shows examples of sponges thus treated. At concentrations of azetidine 2-carboxylic acid below 0.2 mM, the rate of peptidyl proline hydroxylation did not decrease (Figure 2) and morphogenesis in the azetidine 2-carboxylic acid-treated sponges (Figure 3a) was almost the same as in the sponges kept in M-medium alone (Figure 3e). At concentrations above 0.3 mM, the hydroxylation rate of peptidyl proline in collagen decreased gradually (Figure 2) and the spread of basopinacocytes and (or) the formation of an aquiferous system were inhibited (Figure 3b-d). At a concentration of 0.5 mM, some gemmules were attached to each other and some were attached to the substratum but there was no development of basal pinacoderm (Figure 3c). At a concentration of 1.0 mM of azetidine 2-car-

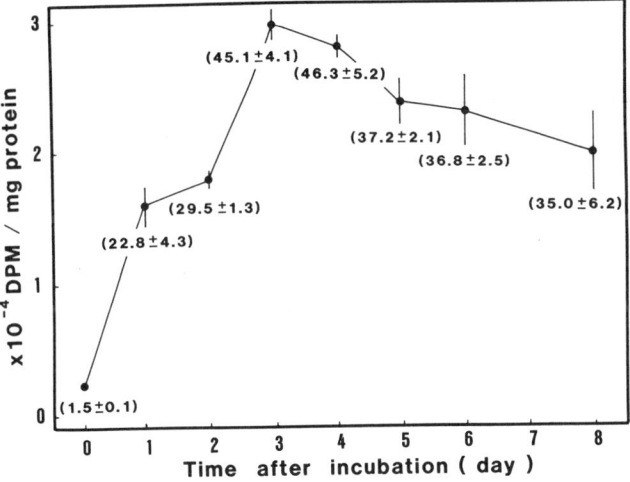

Figure 1. Changes in rate of [^{14}C]proline incorporation and hydroxylation of proline residues in collagen in *Ephydatia fluviatilis* during development. Sponges were exposed to [^{14}C]proline (0.2 μCi/ml) for 3 h at 25°C. Hot TCA-extracted protein was then analyzed for total [^{14}C]radioactivity. Radioactivity of [^{14}C]hydroxyproline residues was also measured. Solid circles show radioactivity of [^{14}C]proline incorporation into hot TCA-extracted protein fraction. Values in parentheses indicate amount of peptidyl proline hydroxylation. Vertical bars indicate standard error of mean.

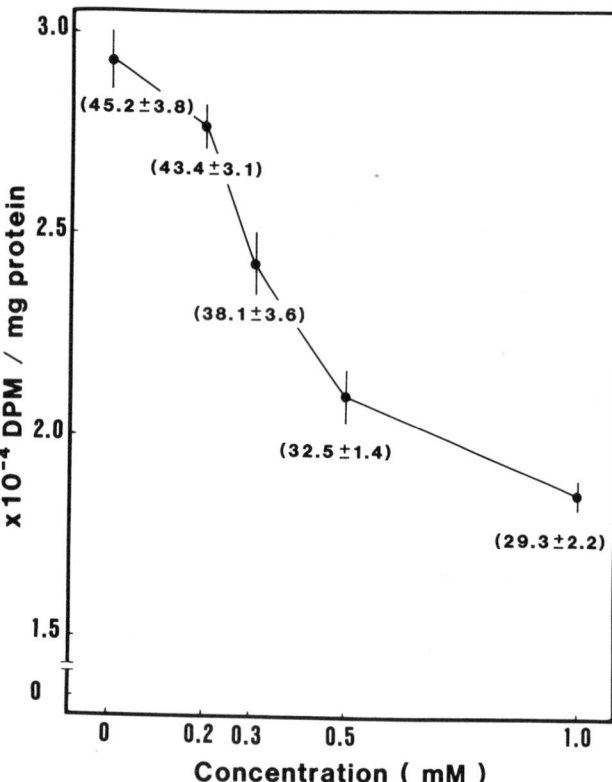

Figure 2. Effects of azetidine 2-carboxylic acid on hydroxylation of [^{14}C]proline residue in collagen in *Ephydatia fluviatilis*. Treatment of sponges with azetidine 2-carboxylic acid was started just after incubation; 124 h after incubation sponges were exposed to [^{14}C]proline for 3 h at 25°C. Acid-extracted proteins were analyzed for [^{14}C]proline and [^{14}C]hydroxyproline residues. Solid circles show radioactivity of [^{14}C]proline incorporation into collagen fraction. Value in parentheses indicates amount of peptidyl proline hydroxylation. Vertical bars indicate standard error of mean.

boxylic acid, the rate of peptidyl proline hydroxylation in sponges was markedly lower than observed at concentrations below 0.3 mM (Figure 2) and the gemmules did not attach to the substratum (Figure 3d). Sponges were also treated with azetidine 2-carboxylic acid for 50 h at 25°C from the time of hatching. At a concentration of 1.0 mM, the sponges did not develop further. At a concentration of 0.5 mM, gemmules were attached to each other and some gemmules were attached to the substrate and a basopinacoderm developed. As the concentration of azetidine 2-carboxlyic acid was increased, the rate of collagen synthesis decreased.

As is well known, α,α'-dipyridyl is an inhibitor of prolylhydroxylase. Its effects on collagen synthesis in *Ephydatia fluviatilis* are shown in Table 1. Treatment with this reagent at 25°C continued from the time of gemmule incubation until 124 h after the incubation. Then, collagen synthesis was measured by the method described

above. It was found that as concentration of α,α'-dipyridyl increased, collagen synthesis decreased. When α,α'-dipyridyl was at a concentration 1.0 mM, gemmules were hatched but did not attach to the substratum. At a concentration of 0.5 mM, gemmules were attached to the substratum, but the spread of pinacoderm was poor. At a concentration of 0.25 mM, gemmules were attached to each other and some gemmules were attached to the substratum. They were inhibited from further development.

Discussion

Collagen biosynthesis in the freshwater sponge *Ephydatia muelleri* has been examined by the electron microscope autoradiograpic method (Garrone and Pottu, 1973). This work has suggested that spongocyte syntheizes spongin, which is a form of collagen and combines the inorganic spicules. Furthermore, a study of the rate of protein synthesis in *Ephydatia fluviatilis* (Efremova, 1970) suggests that the rate of collagen metabolism is slow (see Garrone, 1978). However, little is yet known about change in the rate of collagen synthesis in freshwater sponge during development or the role of collagen during the initial spreading of the basopinacoderm and the following morphogenesis. In the present study, the rate of hydroxylation of proline residues in sponge collagen increased after gemmule germination up to the time of hatching and did not decrease significantly thereafter. A marked increase in the rate of collagen synthesis during gemmule germination (the period between the initiation of gemmule incubation and hatching) suggests that the collagen synthesized during this period contributes to the spreading of basopinacoderm, because the inhibitors of collagen synthesis, such as azetidine 2-carboxylic acid and α,α'-dipyridyl (Hurch and Chvapil, 1965; see the reviews in Garrone, 1978; Olsen, 1981) also inhibited the spreading basopinacoderm. Collagen synthesis after hatching did not decrease significantly, although the inhibitors of collagen synthesis

Table 1. Effect of α,α'-dipyridyl on the hydroxylation of peptidyl proline in *Ephydatia fluviatilis*.

Addition of α,α'-dipyridyl	Total [^{14}C]radio–activity x 0.01 dpm per mg protein	[^{14}C]hydroxyproline/ [^{14}C]–proline x 100 dpm per mg protein
None	236.4 ± 17.5	36.9 ± 4.8
1.0 mM	170.3 ± 31.4	24.7 ± 3.2
0.5 mM	181.4 ± 16.6	27.1 ± 3.8
0.25mM	223.8 ± 15.9	35.8 ± 4.4

Control sponge kept in M–medium were used 124 h after gemmule incubation at 25° C. α,α'–dipyridyl was added to sponges at initiation of gemmule incubation. Sponges were exposed to [^{14}C]proline for 3 h at 25° C. Acid–extracted proteins were analyzed for total [^{14}C]radioactivity and [^{14}C]hyroxyproline radioactivity.

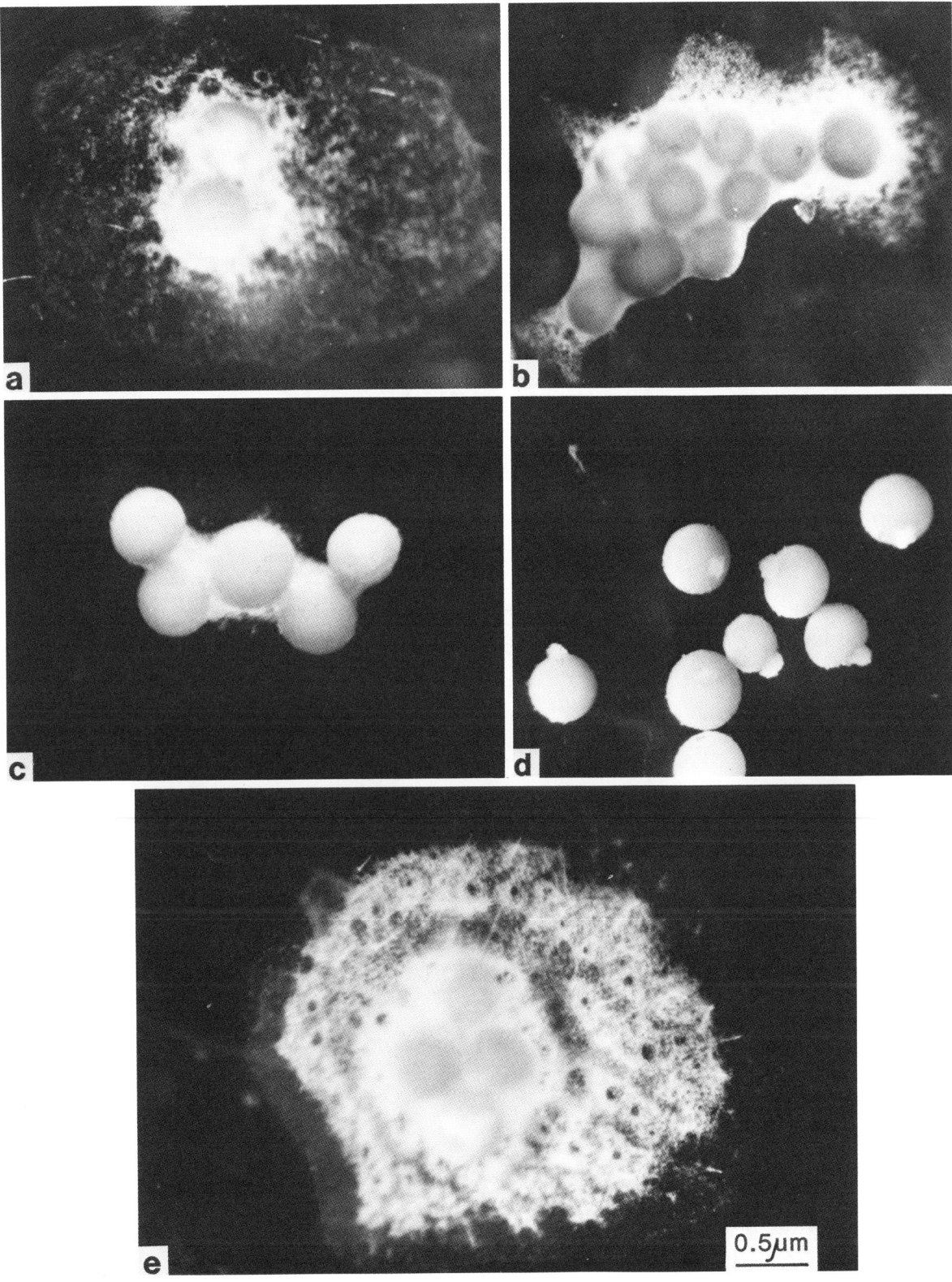

Figure 3. Photographs of *Ephydatia fluviatilis* kept in M-medium containing azetidine 2-carboxylic acid from shortly after gemmule incubation until control sponges were fully developed (124 h after incubation at 25°C). Photographs show sponges treated with azetidine 2-carboxylic acid at different concentrations: *a*, 0.2 mM; *b*, 0.3 mM; *c*, 0.5 mM; *d*, 1.0 mM; and *e*, control sponges.

inhibited certain aspects of morphogenesis, such as the formation of aquiferous systems and spicules. It has been reported that the rate of collagen metabolism in freshwater sponges is slow (see Garrone, 1978). Hence, it seems that the markedly increased collagen synthesis during germination influenced morphogenesis after hatching. We suggest that collagen synthesis after hatching contributes to cell movement in the mesohyl and subsequent morphogenesis of the aquiferous system. This hypothesis requires further investigation.

Conclusions

The results of this study indicate that collagen produced in freshwater sponges after gemmule germination plays a role in the attachment of sponges to the substratum and in their subsequent morphogenesis, such as the formation of the aquiferous system.

Acknowledgment

We are grateful to Professor Ikuo Yasumasu, Waseda University, for his encouragement and cooperation.

Literature Cited

Efremova, S. 1970. Prolification Activity and Synthesis of Protein in the Cell of Freshwater Sponges during Development after Dissociation. Pages 399–412 in *The Biology of Porifera*, edited by W. G. Fry. London: Academic Press.

Fitch, S. M., M. L. R. Hankness, and R. D. Hankness. 1955. Extraction of Collagen from Tissue. *Nature*, 176:163.

Garrone, R. 1978. *Phylogenesis of Connective Tissue. Morphological Aspects and Biosynthesis of Sponge Intercellular Matrix*. Basel: S. Karger. 250 pp.

———. 1984. Formation and Involvement of Extracellular Matrix in the Development of Sponges. A Primitive Multicellular System. Pages 461–477 in *The Role of Extracellular Matrix in Development*, edited by R. L. Treland. New York: Alan R. Liss.

Garrone, R., and J. Pottu. 1973. Collagen Biosynthesis in Sponges: Elaboration of Spongin by Spongocytes. *Journal of Submicroscopical Cytology*, 5:199–218.

Hurch, J., and M. Chvapil. 1965. Influence of Chelating Agents on the Biosynthesis of Collagen. *Biochimica et Biophysica Acta*, 97:361–363.

Lowry, O. H., N. J. Rosebrough, A. L. Farr, and R. J. Randall. 1951. Protein Measurement with the Folin Phenol Reagent. *Biochemical Journal*, 153:259–264.

Oslen, B. R. 1981. Collagen Biosynthesis. Pages 139–178 in *Cell Biology of Extracellular Matrix*, edited by E. D. Hay. New York and London: Plenum Press.

Peterkofsky, B., and D. J. Prockop. 1962. A Method for the Simultaneous Measurement of the Radioactivity of Proline-^{14}C and Hydroxyproline-^{14}C in Biological Materials. *Analytical Biochemistry*, 4:400–406.

Rasmont, R. 1961. Une technique de culture des éponges d'eau en milieu centrale. *Annales, Société Royale de Zoologie Belgique*, 91:147–156.

YOKO WATANABE
Department of Biology
Ochanomizu University
Otsuka, Tokyo 112, Japan
and
Tateyama Marine Laboratory
Ochanomizu University
Tateyama, Chiba 294–03, Japan

YOSHIKI MASUDA
Department of Biology
Kawasaki Medical School
Kurashiki, Okayama 701–01, Japan

Structure of Fiber Bundles in the Egg of *Tetilla japonica* and Their Possible Function in Development

Abstract

Fiber bundles that envelop the unfertilized egg of *Tetilla japonica* may act as a support for cell migration and the formation of cell layers during the earliest stage of development of this sponge. The eggs of *T. japonica* are released from the osculum and fertilization takes place outside the body. They develop directly into adult sponges without a larval period. The fibers are drawn into the perivitelline space and are progressively absorbed by the embryo during its development. An examination of the structure of the fiber bundles and their nature yields evidence of their possible function in development.

Although there are many oviparous sponge species, little is known about their reproduction, including egg structure and development. In several species, spawned eggs have been observed as envelopes containing somatic cells or as collagenous capsules (Warburton, 1961; Lévi and Lévi, 1976; Reiswig, 1976; Gallissian and Vacelet, 1976). The fate and function of these envelopes during the course of development have yet to be clarified.

Spawned mature eggs of two species of *Tetilla, T. serica* and *T. japonica*, (Spirophorida) are surrounded with fiber bundles that are as long as the radius of the egg (Watanabe, 1960, 1978). In *T. serica*, these fibers display 17 nm striated banding (Endo et al., 1967). The eggs of *Verongia cavernicola* are surrounded with collagenous capsules, but their function is still unknown (Gallissian and Vacelet, 1976). In our research, it became clear that the fiber bundles around the egg of *T. japonica* also have a striated banding pattern like that of *T. serica*. These fibers show a structure similar to mesohyl collagen. They denature and become invisible upon acid treatment or heating. This strongly indicates a collagenous nature.

Sponges are the most primitive multicellular animals

with collagen. Sponge bodies are known to contain various form of collagen: mesohyl collagen, skeletal spongin, and basal spongin (Garrone, 1978, 1982, 1984, 1985), and one species of Homosclerophorida is reported to have a basal laminalike structure (Donadey, 1979; Boury-Esnault et al., 1984; Garrone, 1984). Collagen fibers are thought to have various functions in the development of sponges. The secretion of basal spongin and subsequent fixation to the substratum is necessary for normal development of larvae, asexual gemmules and reaggregating cell mass (Borojevic, 1971). This strong fixation by the developing young sponge is said to generates the physical tensions needed for morphogenesis (Garrone, 1984).

The fiber bundles of *Tetilla japonica* are completely enveloped by the young sponge during the process of development (Watanabe, 1978). In this species, the eggs adhere to the substratum soon after fertilization and develop into adult sponges without going through a larval stage. If the fibers surrounding the egg are collagen, they may well function as a support in cell movement and pinacocyte differentiation during the first stage of the development of this sponge.

Materials and Methods

Tetilla japonica was collected at Tateyama Bay near the Tateyama Marine Laboratory, Ochanomizu University. The sponges were cultured under running sea water in laboratory tanks. To obtain unfertilized eggs, we isolated and cultured the female sponges in separate tanks from male sponges, and eventually they spawned spontaneously. The eggs are easily fertilized by adding to the egg suspension a small amount of seawater taken from tanks used in culturing male sponges.

For scanning electron microscopy, the eggs were fixed with 1% osmium tetroxide in seawater. After dehydration in a graded series of ethanol, the specimens were transferred to a Hitachi HCP-1 critical point drying apparatus, coated with gold palladium in an Eiko IB-3 ion coater, and observed with a Hitachi HHS-2R scanning electron microscope. Samples for transmission electron microscopy were fixed with 1% osmium tetroxide or 2.5% glutaraldehyde, post-fixed with 1% osmium tetroxide, and embedded in Epon. Thin sections were stained with uranyl acetate and lead citrate and were examined with a Hitachi HS-9 electron microscope.

Results

LIGHT MICROSCOPE OBSERVATIONS OF LIVING MATERIALS. The unfertilized egg of *Tetilla japonica* radiated fiber bundles around itself that were as long as its radius (Figure 1a). After fertilization these fiber bundles were drawn into the perivitelline space by the rotating motion of the egg as soon as the fertilization membrane was formed. Within 10 minutes after fertilization, the fibers were completely taken into the perivitelline space, and were still visible (Figure 1b). Eleven hours after fertilization, outer cells of the embryo began to project pseudopodia along the fibers and migrate into the perivitelline space toward the fertilization membrane (Figure 1c). Upon reaching the fertilization membrane, the cells flattened and differentiated into pinacocytes (Figure 1d), and formed pinacoderm just inside the fertilization membrane.

ELECTRON MICROSCOPE OBSERVATIONS. With the scanning electron microscope, the fibers could be seen radiating in bundles from the egg surface at their base, but at their distal ends they were separated into distinct fibrils (Figure 2). Transmission electron microscopy showed that at the base of fiber bundles there were trapezoidal projections of egg cytoplasm where fibers radiated in bundles. Micellar substance was observed on the egg surface (Figure 3). Under high magnification, fibrils were measured to be 13–17 nm in diameter, and striated banding patterns were observed. This banding was approximately 41 nm (Figure 3b). Intermediate banding was faintly visible at about 20 nm (Figure 3b, insert). Figure 4a shows the newly formed fertilization membrane one minute after fertilization. By then, fiber bundles had passed through the fertilization membrane. The micellar substance that had been found on the surface of an unfertilized egg was observed on the outer surface of the fertilization membrane (Figure 4a,b). Six minutes after fertilization, the fiber bundles were almost completely taken into the perivitelline space, though some fibrous structure still remained (Figure 4b).

OTHER OBSERVATIONS. When the unfertilized eggs were immersed in 0.2% perchloric acid or put in hot seawater at 90°C for one minute, fiber bundles became invisible. Simultaneously, eggs equidistant from each other came closer together (Figure 5). The fibrils were not affected when exposed to collagenase.

Discussion

It has been reported that the spawned eggs of several oviparous sponges are surrounded by an envelope containing somatic cells. However, little is known about the function of these envelopes during embryonic development. In *Cliona celata* (Warburton, 1961), *Hemectyon ferox* (Reiswig, 1976), and *Chondrosia reniformis* (Lévi and Lévi, 1976), the eggs are released and carry somatic cells around their surface. In *Verongia cavernicola*, the eggs are surrounded by spherulous cells enclosed within a collagenous capsule (Gallissian and Vacelet, 1976). Various functions have been reported for the cells and envelopes around the eggs. The somatic cells

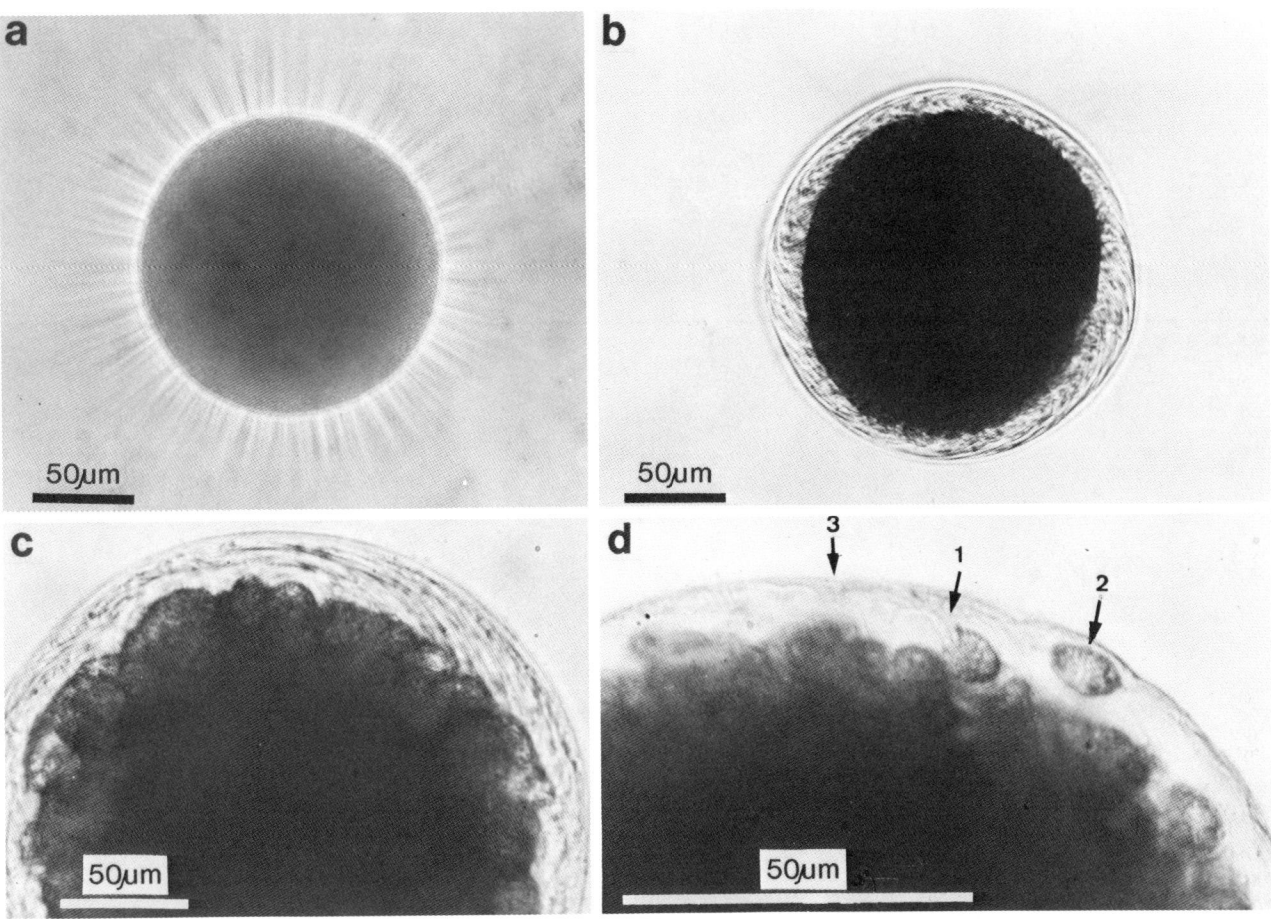

Figure 1. The egg of *Tetilla japonica* and the process of pinacoderm formation observed with light microscopy: *a*, unfertilized egg (phase contrast); *b*, 10 min after fertilization, fibers still visible in perivitelline space; *c*, 11 h after fertilization, outer cells began to migrate into perivitelline space; *d*, 36 h after fertilization, cells project pseudopodia *(arrow 1)* and migrate into perivitelline space *(arrow 2)*, and flatten out just under fertilization membrane and differentiate into pinacocytes *(arrow 3)*.

of *C. celata, H. ferox,* and *C. reniformis* are brought into the interior of the larvae in the course of development, where they serve a nutritive function. However, the function of the envelope in *V. cavernicola* is unknown, although some believe that it plays a protective role.

In many viviparous sponges, eggs are enclosed within the follicular epithelium, nurse cells, and follicle cells, and they receive a nutritive supply through these cells from maternal sponges during embryonic development (Fell, 1983; Simpson, 1984). In *Haliclona loosanoffi,* the embryo grows to a diameter about five times as large as that of the egg during development (Hartman, 1958). It is assumed that the envelopes around the eggs of oviparous sponges supply some materials necessary for development, such as the nutriments viviparous sponges receive from maternal sponges during embryonic development. Since the envelopes of *Tetilla japonica* eggs were completely taken into the embryonic sponges during development (Watanabe,

1978), it may be that they supply some substance necessary for development.

Envelopes of *Tetilla japonica* consist of fiber bundles. The ultrastructure of these fibers display striated banding similar to the distinctive pattern of collagen fibrils (Figure 3*b*). The pattern is also similar to that of mesohyl collagen fibrils. These fibrils were not affected by a collagenase, but they denatured and became invisible upon heating (Figure 5*b*) or being treated with acid. This indicates that collagen denatured into gellatine and the fibrous structure became invisible. We could not obtain enough quantities of eggs to quantify collagen chemically but we asked Dr. K. Yoshizato to analyze amino acids from a very small sample. He detected proline and hydroxyproline from unfertilized eggs, which strongly suggests that the fibrils constituting the fiber bundles are collagen.

Sponges are the most primitive animal to contain collagen, and they are also the first true multicellular animal.

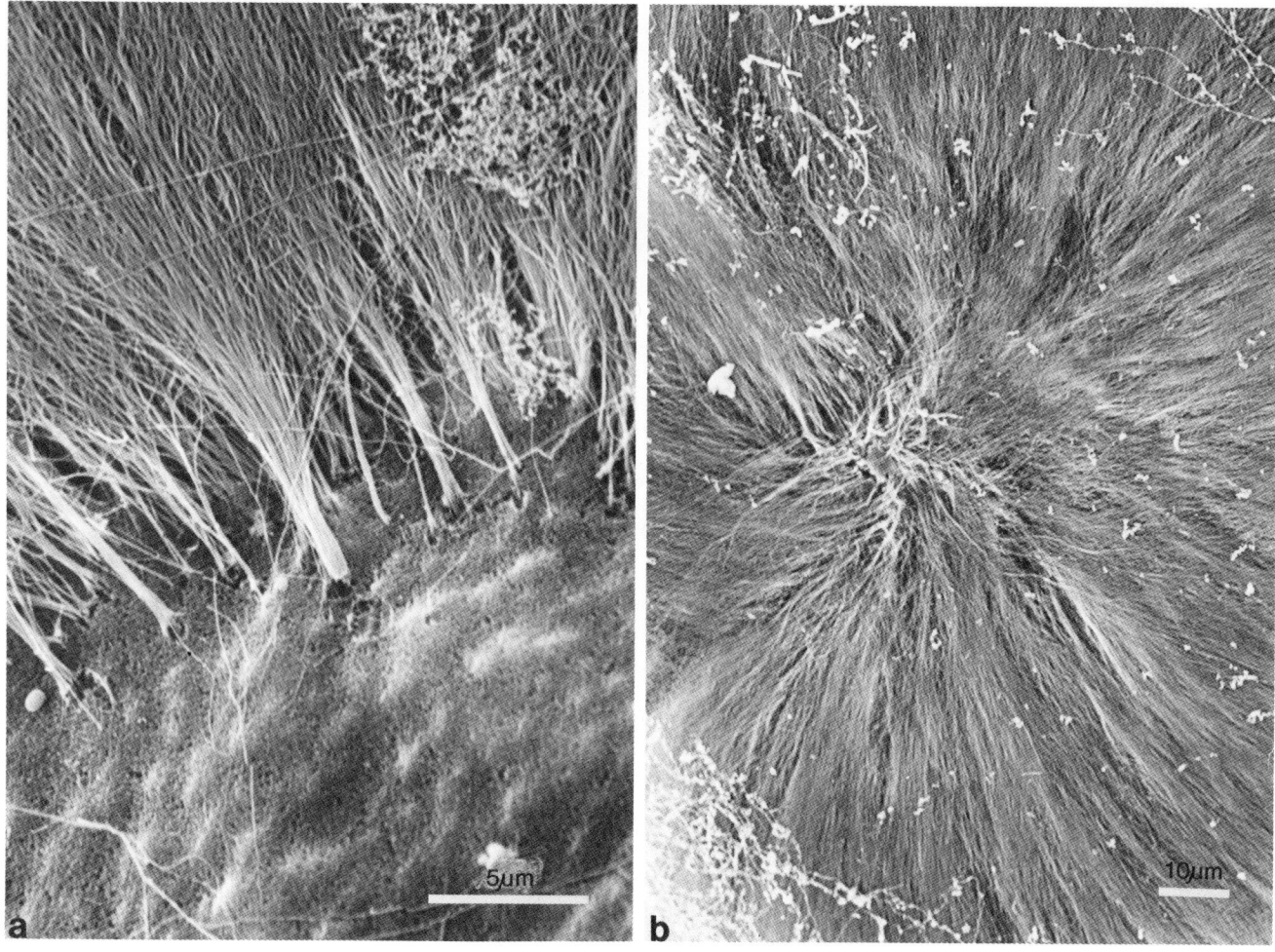

Figure 2. Fibers of unfertilized egg of *Tetilla japonica* seen with a scanning electron microscope: *a*, base of fibers radiating from egg surface in bundles; *b*, distal ends of fibers separate into distinct fibrils.

Garrone (1984) has pointed out: "The shift from unicellular, independent life to a multicellular, integrated organism has probably been accompanied by the invention of collagen, some one billion years ago." Cells migrate from the interior of fixed larvae or gemmules, differentiate into pinacocytes and adhere, as they spread on the substratum, by the secretion of collagen. Sponges are known to have a large amount of collagen in the mesohyl. Sponges also need collagen arranged as spongin fibers to connect spicules and construct a skeletal framework. One species is even known to have a basal laminalike structure (Garrone, 1978, 1984; Donadey, 1979; Boury-Esnault et al., 1984). Collagen fibrils probably help support cell movement in the mesohyl and on the substratum and also function as the base of the cell layer giving rise to the multicellular system.

During pinacoderm formation in *Tetilla japonica*, the fibrils in the perivitelline space may serve as support for cell migration and cell layer formation (Figure 1*d*). The eggs of *T. japonica* adhere to the substratum soon after fertilization and develop directly into adult sponges without a larval period. During ontogeny the embryos of *T. japonica* neither have the opportunity to receive any substance from the maternal sponge, nor have the opportunity to synthesize necessary substances during their earliest stage of development. Collagen probably acts as an extracellular matrix that is the basis of cell layer formation needed to maintain a multicellular system.

Conclusions

The fiber bundles of the unfertilized egg of *Tetilla japonica* display a characteristic ultrastructure similar to that of

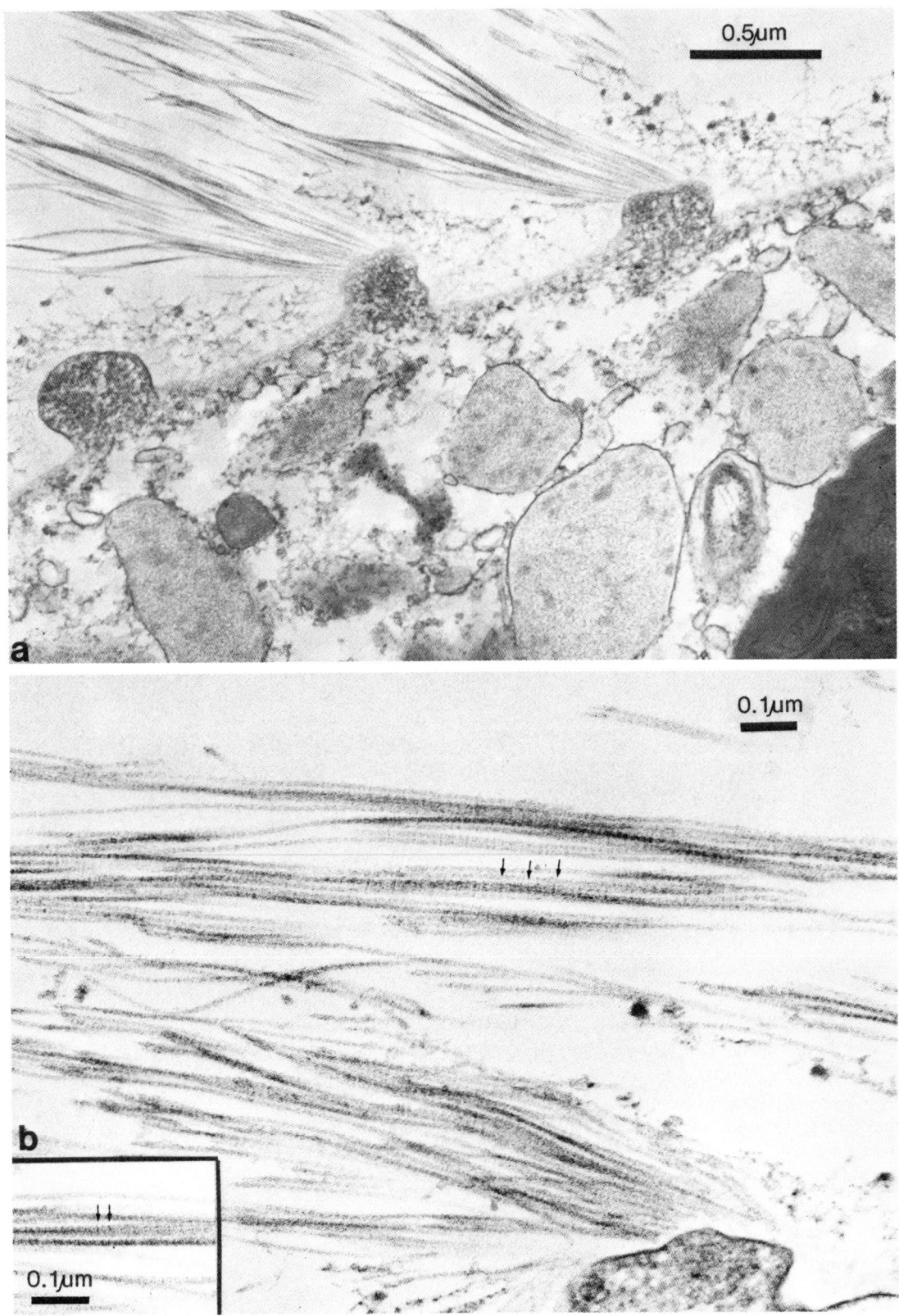

Figure 3. Transmission electron micrographs of an unfertilized *Tetilla japonica* egg: *a*, fibers radiate from egg surface in bundles, trapezoidal projections are seen at base of bundles, micellar substance is seen on egg surface; *b*, fibers under high magnification showing 41 nm striated banding (arrows) and (insert) a fibril with intermediate banding at about 20 nm (arrows).

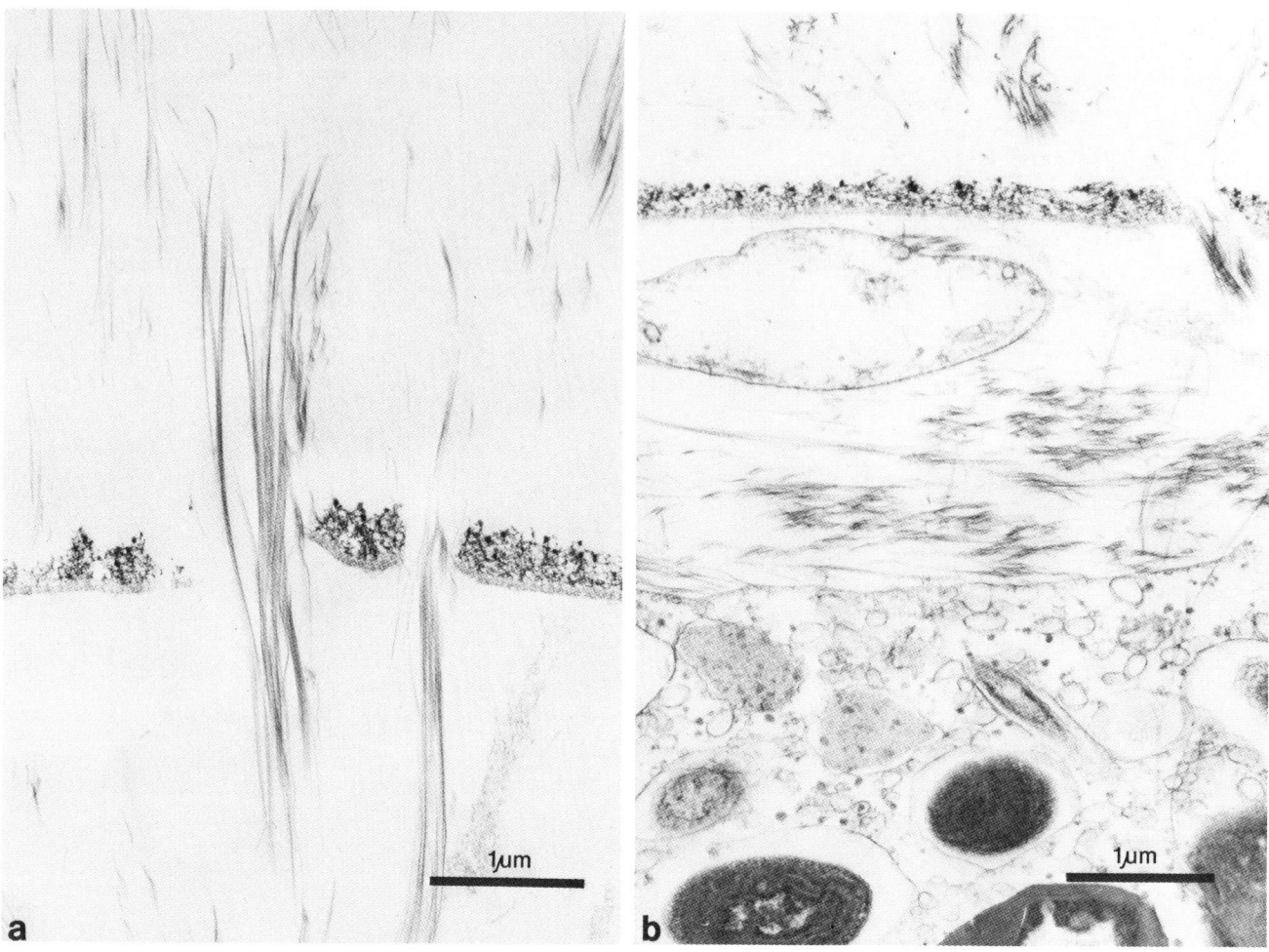

Figure 4. Transmission electron micrographs of a fertilized *Tetilla japonica* egg: *a*, fertilization membrane in cross section 1 min after fertilization, fibers passing through fertilization membrane; *b*, 6 min after fertilization, fibers have been taken into perivitelline space.

collagen fibers. These fibrils are also taken to be collageneic in nature in view of the denaturation of the fibers upon heating or acid treatment. The fiber bundles are taken into the embryo during development. The cells constituting exopinacoderm migrate along the fibers projecting pseudopodia. *T. japonica* skips the larval stage and develops directly into an adult sponge. It is assumed that fiber bundles around the unfertilized egg function to supply collagen, which is necessary for the migration and differentiation of cells related to morphogenesis in the first step of the development of nonlarval oviparous sponges.

Acknowledgments

We wish to thank N. Usui, associate professor of Teikyo University, for the scanning electron microscope photographs. We are grateful to K. Yoshizato, associate professor of Tokyo Metropolitan University, for performing the amino acid analysis.

Literature Cited

Borojevic, R. 1971. Le comportment des cellules d'éponge lors de processus morphogénétique. *Annales de Biologie*, 10:533–545.

Boury-Esnault, N., L. De Vos, C. Donadey, and J. Vacelet. 1984. Comparative Study of the Choanosome of Porifera. I. The Homoscleromorpha. *Journal of Morphology*, 180:3–17.

Donadey, C. 1979. Contribution à l'étude cytologique de de deux Démosponges Homosclerophorides: *Oscarella lobularis* (Schmidt) et *Plakina triopha* Schulze. Pages 165–172 in *Biologie des Spongiaires*, edited by C. Lévi and N. Boury-Esnault. Colloques Internationaux, 291. Paris: Centre National de la Recherche Sciéntifique.

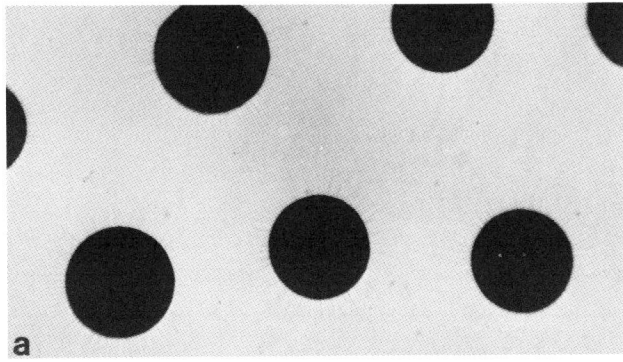

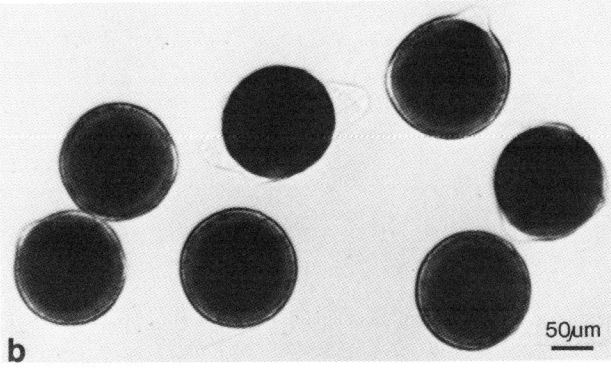

Figure 5. Unfertilized eggs of *Tetilla japonica*: *a*, untreated, note even spacing; *b*, after 1 min at 90°C, fibers become invisible, and eggs come closer together.

Endo, Y., Y. Watanabe, and S. Hiramoto. 1967. Fertilization and Development of *Tetilla serica*, a Tetraxonian Sponge. *Japanese Journal of Experimental Morphology,* 21:40–60.

Fell, P. E. 1983. Porifera. Pages 1–29 in *Reproductive Biology of Invertebrates, 1: Oogenesis, Oviposition, and Oosorption,* edited K. G. Adiyodi and R. G. Adiyodi. New York: John Wiley and Sons.

Gallissian, M. F., and Vacelet, J. 1976. Ultrastructure de quelques stades de l'ovogénèse de spongiaires du genre *Verongia* (Dictyoceratida). *Annales des Sciences Naturelles, Zoologie et Biologie Animale,* 18:381–404.

Garrone, R. 1978. *Phylogenesis of Connective Tissue. Morphological Aspects and Biosynthesis of Sponge Intercellular Matrix.* Basel: S. Karger. 250 pp.

———. 1982. Le collagène des invertebrés: matrice intercellulaire, squelette ou revêtement? *Bulletin de la Société Zoologique de France,* 107:387–399.

———. 1984. Formation and Involvement of Extracellular Matrix in the Development of Sponges, a Primitive Multicellular System. Pages 461–477 in *The Role of Extracellular Matrix in Development,* edited by R. L. Trestad. New York: Alan R. Liss.

———. 1985. The Collagen of the Porifera. Pages 157–177 in *Biology of Invertebrate Collagens,* edited by A. Bairati and R. Garrone. NATO ASI Series A, Life Sciences 93. New York: Plenum Press.

Hartman, W. D. 1958. Natural History of the Marine Sponges of Southern New England. *Bulletin, Peabody Museum of Natural History,* 12:1–155.

Lévi, C., and Lévi, P. 1976. Embryogénèse de *Chondrosia reniformis* (Nardo). Démosponge ovipare, et transmission des bactéries symbiotes. *Annales des Sciences Naturelles, Zoologie et Biologie Animale,* 18:367–380.

Reiswig, H. M. 1976. Natural Gamete Release and Oviparity in Caribbean Demospongiae. Pages 99–112 in *Aspects of Sponge Biology,* edited by P. W. Harrison and R. R. Cowden. New York: Academic Press.

Simpson, T. L. 1984. *The Cell Biology of Sponges.* New York: Springer. 662 pp.

Warburton, F. E. 1961. Inclusion of Parental Somatic Cells in Sponge Larva. *Nature,* 191:1517.

Watanabe, Y. 1960. Outline of Morphological Observations on the Development of *Tethya serica* Lebwohl. *Bulletin of Marine Biological Station of Asamushi Tōhoku University,* 10:145–148.

———. 1978. The Development of Two Species of *Tetilla* (Demospongiae). *Natural Science Report, Ochanomizu University,* 29:71–106.

SANDRINO HOLVOET
GYSELE VAN DE VYVER
Laboratoire de Biologie Animale et Cellulaire
Université Libre de Bruxelles
50 av. F. D. Roosevelt
1050 Bruxelles, Belgium

Skeletogenesis in *Ephydatia fluviatilis* Grown in the Presence of Puromycin and Hydroxyurea

Abstract

Unfed sponges grown from gemmules differentiate sclerocytes only during a short period after hatching. If the silicate concentration of the culture medium is increased, fewer spicules are produced. The biological structures required for skeletogenesis regulation were investigated in sponges cultivated in mineral medium containing either hydroxyurea (Hu) or puromycine (Pu). Hu-treated sponges, despite the absence of an aquiferous system, developed a normal tridimensional network of spicules embedded in a matrix of spongin. In contrast, Pu-treated sponges lacked a functional skeleton. Spicules remained isolated because no perispicular spongin sheath was secreted to join them up into a network. Quantitative analysis showed that Hu-treated sponges and Pu-treated sponges produced 75% and 55%, respectively, of the number of spicules produced by the controls. However, the silicate-dependent inhibition remained effective in both types of sponges tested. Our experiments demonstrate that the aquiferous system and the perispicular spongin are not required in sclerocyte differentiation and that they do not interfere with the silicate-dependent regulation.

Detailed information is now available on spongillid morphogenesis as a result of the histological work by Brien (1932), and subsequent work by Harrison and Cowden (1975), Pottu-Boumendil and Pavans de Ceccaty (1976), and Höhr (1977). In contrast, little is known about the systems that regulate cell differentiation (Rasmont, 1975; Garrone, 1984).

Quantitative analysis of skeletogenesis in the sponge *Ephydatia fluviatilis* appears to be a suitable approach to the study of such mechanisms. Indeed, the spicules in demosponges are individually and intracellularly secreted by single sclerocytes. After extrusion of their spicule, the sclerocytes degenerate, whereas the spicules are carried by spongocytes and are integrated into a tridimensional

200

network through the secretion of perispicular spongin (Garrone, 1969; Garrone and Pottu, 1973; Weissenfels and Landschoff, 1977). This means that the number of spicules secreted at any time of the sponge development corresponds to the number of sclerocytes that have become differentiated from archeocytes since the beginning of sponge morphogenesis. As a result, it is possible to follow the quantitative evolution of sclerocyte differentiation during the development of *E. fluviatilis* merely by counting spicules. This type of quantitative approach has been used to demonstrate that the differentiation of archeocytes into sclerocytes is controlled by a system in which silicates act as inhibitory factors (Pé, 1973).

In order to identify the biological structures that might be involved in sclerocyte regulation, we tried to determine which structures can be supressed in sponge without affecting regulation of sclerocyte differentiation. Antimetabolites appeared to offer the means of doing so. Indeed, hydroxyurea is known to specifically inhibit the differentiation of the aquiferous system of *Ephydatia fluviatilis* (Rozenfeld and Rasmont, 1976; Garrone and Rozenfeld, 1981; Tanaka and Watanabe, 1984), and puromycine has been reported to disturb the integration of spicules into a tridimensional network (Rozenfeld, 1980).

Material and Methods

SPONGES. *Ephydatia fluviatilis* (Spongillidae) of strains α and δ (Van de Vyver, 1970; 1975) were used. For the quantitative experiments, batches of 10 gemmules were incubated in 10 ml polystyrene Petri dishes. The gemmules were carefully separated from each other in order to obtain individual sponges grown from single gemmules (1-gml sponges). The sponges were cultivated at 20°C and were not fed during the experiments. For all the experiments, day 0 of the time scale was determined by the hatching time of the control gemmules.

CHEMICALS. Hydroxyurea (Hu) (Calbiochem) or puromycin (Pu) (Sigma) were dissolved in M medium (Rasmont, 1961) at final concentrations of 100 μg/ml (Hu medium) or 4 μg/ml (Pu medium). In experiments designed to test simultaneously the effects of silicate concentration and antimetabolites, the concentration of sodium silicate in the M medium (15 mg/l silicon) was either reduced to 1.5 mg/l or increased to 60 mg/l.

SPICULE COUNTING. The technique designed by Pé (1973) was used. That is, the culture medium in petri dishes containing sponges whose spicules were to be counted was replaced by a few drops of concentrated nitric acid. This treatment destroys the sponges except for the spicules, which adhere to the bottom of the petri dishes. After acid evaporation, the spicules were counted using a projection

microscope. Ten sponges were used for each experimental point, with 95% consistent results. The mean size of the spicules was measured 12 days after hatching (95% consistent results); each sample consisted of 125 spicules.

CELL DISSOCIATION. Sponges were scraped from petri dishes by means of a sharpened spatula and then dissociated by gentle pipetting in calcium- and magnesium-free medium (CMF medium). The gemmule shells were removed by brief sedimentation, and the spicules were eliminated by filtration through nylon gauze (25 μm mesh).

CELL FRACTIONATION. Archeocytes were isolated by applying a rapid method for cell separation—adapted from the procedures of De Sutter and Buscema (1977) and De Sutter and Van de Vyver (1977)—using a discontinuous density gradient composed of four layers (12%, 10%, 6%, and 4%) of Ficoll (a high molecular weight polysaccharide) in CMF/HU medium. After centrifugation for 7 minutes at 3,000 × g, pure archeocytes were recovered from the 10%–12% interface by perforating the bottom of the tube. The purity and homogeneity of the archeocyte fraction were checked by phase-contrast microscopy of living cells.

AGGREGATION. Archeocyte suspensions were briefly washed in CMF/HU medium and then resuspended in M/HU medium at a average concentration of 10^6 cells/ml. Petri dishes containing 10 ml of cell suspension were placed on a gyratory shaker (radius, 2.5 cm) and rotated at 60 rpm.

Results

DEVELOPMENT OF SPONGES. Control gemmules in M medium hatched after 72 hours of incubation at 20°C (day 0). The first step of morphogenesis consisted in differentiation of pinacoderms delimiting the sponges. This was followed by the appearance of choanocyte chambers and aquiferous canals; the osculum was formed around day 4 (Figure 1). Young intracellular spicules were already present on day 1. Two days later, the first extracellular spicules were embedded in a sheath of perispicular spongin and organized themelves to form a tridimensional network.

Gemmules raised in the Hu medium hatched, but with a 24 hour delay. Morphogenesis of Hu-treated sponges was similar to that described for the controls except for the aquiferous system which never differentiated (Figure 2).

In contrast, although there was no delay in hatching, sponge development in Pu-treated gemmules was seriously disrupted. Very few choanocytes were differentiated and the spicules remained isolated in the mesohyl instead of being integrated into a perispicular spongin matrix (Figure 3). Moreover, from day 6 isolated spicules could be observed on the bottom of the petri dish all around the sponges.

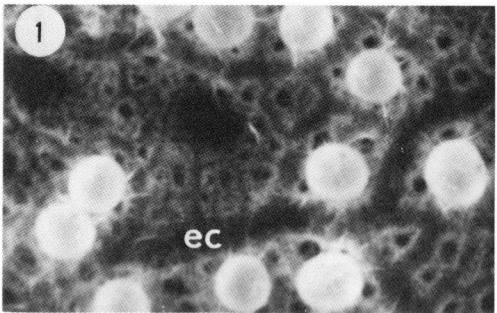
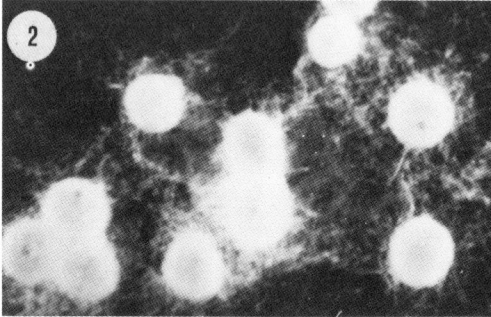
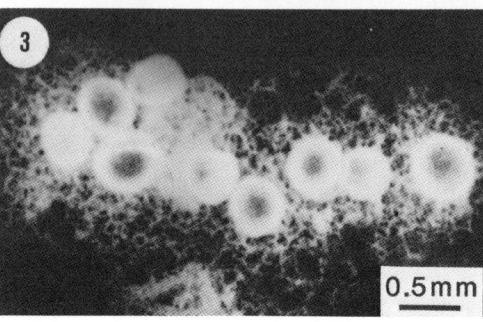

Figures 1–3. Photomicrographs of cultivated *Ephydatia fluviatilis: 1,* view of a fully developed sponge (control) after 7 days of incubation; *ec,* excurrent canals; *2,* 7-day-old sponge incubated in 100 μg/ml hydroxyurea; aquiferous system is not differentiated; *3,* 7-day-old sponge incubated in 4 μg/ml puromycin; excurrent canals are lacking.

KINETICS OF SPICULE FORMATION. Figure 4 compares the number of spicules produced by 1-gml control sponges and by 1-gml Hu treated sponges in the course of time. If both curves show the same general pattern, it is obvious that the number of spicules secreted by the Hu-treated sponges never overstepped three quarters of the corresponding number in the controls. At the steady state, reached 4 days after hatching in both situations, the number of spicules was 195 ± 24 for the controls and only 131 ± 27 for the Hu-treated sponges.

Figure 5 compares the number of spicules produced by 1-gml control sponges and by 1-gml Pu-treated sponges. In the latter, the maximum number (113 ± 22 spicules) was reached only 5 days after hatching. A nonsignificant

decrease in the number of spicules was observed between day 5 and day 11.

EFFECT OF SODIUM SILICATE CONCENTRATION ON SPICULE PRODUCTION. Table 1 shows that variation in silicate concentration has similar effects on Pu- and Hu-treated sponges. A fourfold increase in silicate concentration in the Hu or Pu medium reduced the number of spicules secreted. In contrast, a tenfold decrease in silicate concentration produced a dramatic increase in the number of spicules.

SKELETOGEN POTENTIAL OF CELLS ISOLATED FROM SPONGES TREATED WITH HU AND PU. In order to determine whether the arrest in spicule production observed in the quantitative experiments was caused by some internal mechanism or depletion of cell skeletogen potential, we tested cell suspensions prepared from either control, Hu-, or Pu-treated sponges that had completed their skeletogenesis for their morphogenetic potential. Two types of cell suspensions were used; complete cell suspensions and pure archeocyte suspensions. Since no choanocytes were differentiated in Hu-treated sponges, the corresponding complete cell suspensions did not contain cells of this type.

Table 2 summarizes the evolution of cell suspensions in mediums of different aggregation (i.e., M, Hu, and Pu mediums) over the period of one week. Both the control and treated sponges that developed from cells suspended in M medium constituted aggregates able to settle. Settlement was always followed by differentiation of a new skeleton and organization of a functional aquiferous system. Similarly, complete cell suspensions or archeocyte suspensions put in Hu medium reconstituted sponges having a spicule skeleton. However, cells from Hu-treated sponges did not develop an aquiferous system.

In the presence of Pu, complete cell suspensions from control or Pu-treated sponges were able to aggregate, but no further evolution was observed. Reversal of Pu inhibition was tested with 7-day-old treated aggregates. Aggregates transferred in M medium settled and their cells reconstituted a skeleton and a functional aquiferous system.

Table 1. Mean number of *Ephydatia fluviatilis* spicules (± SD) secreted by 1–gml Hu– and Pu–treated sponges as a function of the medium silicate concentration ($p = 0.05$)

Treatment	SiO$_2$ Concentration (mg/ml)		
	1.5	15	60
Hu–treated sponges	211±16 (100%)	155±19 (74%)	133±14 (63%)
Pu–treated sponges	76±9 (100%)	57±7 (75%)	44±5 (58%)

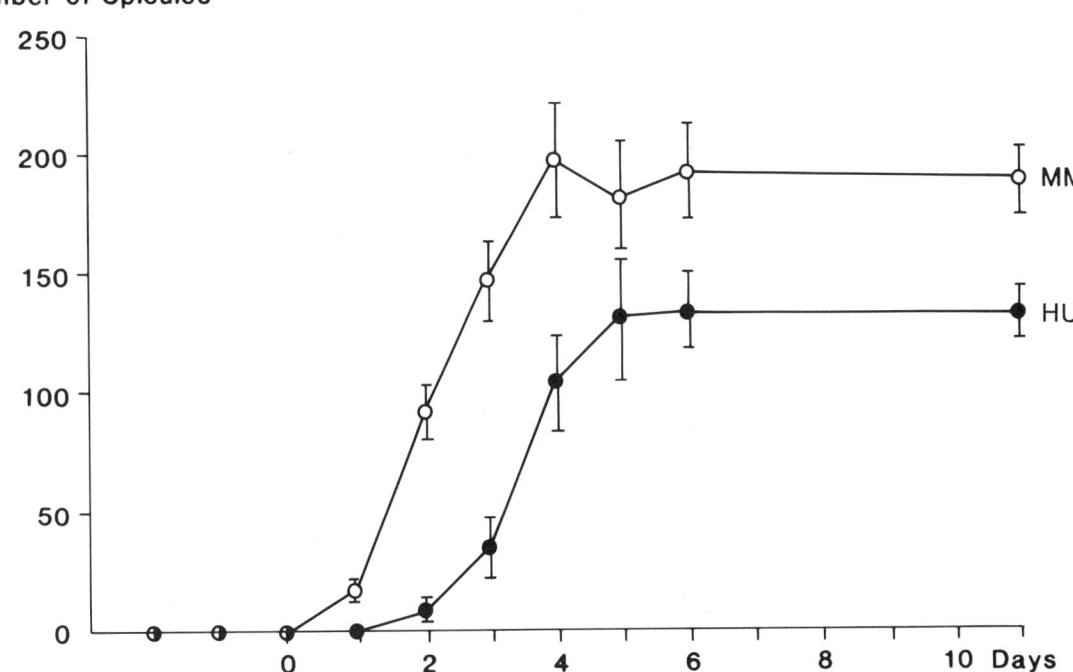

Figure 4. Number of spicules produced during development of one gemmule into a sponge in mineral medium *(open circle, curve MM)* and in 100 μg/ml hydroxyurea *(solid dots, curve HU); (p = 0.05)*.

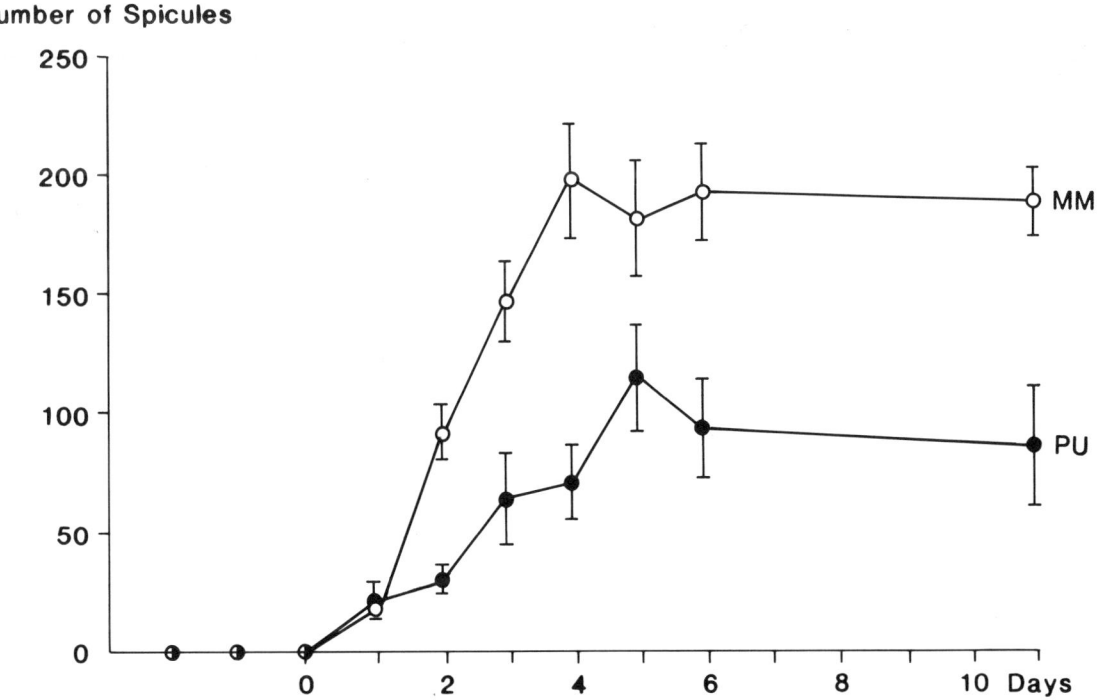

Figure 5. Number of spicules produced during development of one gemmule into a sponge in mineral medium *(open circle, curve MM)* and in 4 μg/ml puromycin *(solid dots, curve PU); (p = 0.05)*.

Table 2. Influence of culture medium on development of *Ephydatia fluviatilis* dissociated cells (cc susp. = complete cell suspension; archeoc. = archeocyte cell suspension; + = activity or development; – = no activity or development; 0 = not tested)

Aggregation medium	Cell origin	Cell types	Aggregation	Settlement	Spicules	Aquiferous system
M	Control	cc susp.	+	+	+	+
		archeoc.	+	+	+	+
	Hu–treated sponges	cc susp.	+	+	+	+
		archeoc.	+	+	+	+
	Pu–treated sponges	cc susp.	+	+	+	+
		archeoc.	0	0	0	0
Hu Medium	Control	cc susp.	+	+	+	+
		archeoc.	0	0	0	0
	Hu–treated sponges	cc susp.	+	+	+	–
		archeoc.	+	+	+	–
Pu Medium	Control	cc susp.	+	–	–	–
		archeoc.	0	0	0	0
	Pu–treated sponges	cc susp.	+	–	–	–
		archeoc.	0	0	0	0

Discussion

During the past 10 years, a number of investigators have shown that antimetabolites provide clues to the morphogenesis of freshwater sponges (Pottu-Boumendil and Pavans de Ceccatty, 1976; Rozenfeld and Rasmont, 1976; Garrone, 1978; Rozenfeld, 1980; Mazzorana and Garrone, 1983; Mazzorana et al., 1984). In the present work, Hu- and Pu-treated sponges were used as models in studying the role of the aquiferous system and the perispicular spongin in the regulation of skeletogenesis.

Our findings can be summarized as follows: The skeleton of puromycine-treated sponges appeared to undergo incomplete development. Indeed, if spicules were secreted, they were not integrated into a tridimensional network by a sheath of perispicular spongin. Like Rozenfeld (1980), we found that some of these isolated spicules were rejected by the sponges. Our results could be interpreted either as an inhibition of spongocyte differentiation or as a disturbance of the cooperation between the skeleton—building cells. An alternative hypothesis involving direct inhibition of the perispicular spongin secretion appears unlikely since the production of collagen materials associated with the exo- and basopinacoderms is not inhibited in Pu-treated sponges (Van de Vyver et al., 1986).

The sclerocyte differentiation from archeocytes stops when a critical number of spicules has been secreted in Hu- and Pu-treated sponges as well as in control sponges. However, Hu- and Pu-treated sponges, respectively, produced 75% and 55% of the number of spicules produced by the controls.

Pé (1973) has demonstrated that the number of spicules produced in functional sponges is inversely related to the silicate concentration of the medium. We obtained similar results with Hu- and Pu-treated sponges. Despite the morphological differences between the experimental sponges–the absence of an aquiferous system in Hu-treated sponges and of a perispicular spongin sheath in Pu-treated sponges–they reacted in the same way to variations in the medium silicate concentration. These results clearly demonstrate that neither the aquiferous system nor the perispicular spongin matrix are required for silicate inhibition.

Once isolated from Hu-treated sponges having achieved their skeletogenesis, archeocytes again start differentiating into sclerocytes. Pé (1973) has reported that although the renewal of the culture medium maybe inefficient in restoring spicule production in functional sponges, feeding of sponges can induce the recovery of their skeletogenesis. Fractionation of archeocytes has made it possible to extend the analysis of skeletogenesis regulation. Indeed, during the morphogenesis of sponges from isolated archeocytes, skeletogenesis started up again without external feeding. Moreover, since phagocytic activity is very low in pure archeocyte aggregates (Van de Vyver and Buscema, 1977), the internal food contribution due to the engulfing of cell debris can be considered negligible. Consequently, we can assume that the termination of sclerocyte differentiation is not due to exhaustion of the stem cell potential or of the stem cell food reserves, but to internal inhibition. The persistence of numerous vitelline platelets in the cytoplasm of the Hu–treated archeocytes (Rozenfeld and Rasmont, 1976) corroborates this assertion.

Cells dissociated in presence of Pu are able to form

aggregates, but settlement of these aggregates is reversibly inhibited by the protein synthesis inhibitor. This suggests that the morphogenetic movements involved in settlement are greatly dependent on protein synthesis. Unfortunately, the settlement inhibition induced by Pu made it impossible to use dissociation-neomorphogenesis experiments to analyze the cell skeletogen potential of Pu-treated sponges having achieved skeletogenesis.

Conclusions

Our experiments on hydroxyurea-treated sponges demonstrate that the regulation system of sclerocyte differentiation remains intact despite the absence of the aquiferous system. Furthermore, the study of puromycine-treated sponges shows that at least one part of skeletogenesis regulation—that is, silicate dependent inhibition—is effective in sponges devoid of a perispicular spongin matrix.

Thus, the morphogenetic inhibition produced by antimetabolites analyzed through the quantitative study of spicule production constitutes a useful tool for investigating the mechanisms involved in the regulation of sponge skeletogenesis.

Acknowledgments

This study was supported in part by a grant from the Institut pour l'Encouragement de la Recherche Scientifique dans l'Industrie et l'Agriculture (IRSIA 840364). The authors are indebted to N. Cardon, E. De Cock, and F. Lambert for their help in preparing the illustrations.

Literature Cited

Brien, P. 1932. Contribution à l'étude de la régénération naturelle chez les Spongillidae—*Spongilla lacustris* (L)—*Ephydatia fluviatilis* (L). *Archives de Zoologie Expérimentale et Générale*, 74: 461–506.

De Sutter, D., and M. Buscema. 1977. Isolation of a Highly Pure Archeocyte Fraction from the Fresh-water Sponge *Ephydatia fluviatilis*. *Wilhelm Roux's Archives of Developmental Biology*, 183: 149.

De Sutter, D., and G. Van de Vyver. 1977. Aggregation Properties of Different Cell Types of the Freshwater Sponge *Ephydatia fluviatilis*. *Wilhelm Roux Archives of Developmental Biology*, 181:151–161.

Garrone, R. 1969. Collagène, spongine et squelette minéral chez l'éponge *Haliclona rosea* (O.S.) (Demosponges; Haploscleride). *Journal de Microscopie*, 8:581–598.

———. 1978. *Phylogenesis of Connective Tissue. Morphological Aspects and Biosynthesis of Sponge Intercellular Matrix.* Basel: S. Karger. 250 pp.

———. 1984. Formation and Involvement of Extracellular Matrix in the Development of Sponges, a Primitive Multicellular System. Pages 461–477 in *The Role of Extracellular Matrix in Development*, edited by R. L. Trelstad. New York: Alan R. Liss.

Garrone, R., and J. Pottu. 1973. Collagen Biosynthesis in Sponge Elaboration of Spongin by Spongocytes. *Journal of Submicroscopical Cytology*, 5:199–218.

Garrone, R., and F. Rozenfeld. 1981. Electron Microscope Study of Cell Differentiation a and Collagen Synthesis in Hydroxyurea-Treated Freshwater Sponges. *Journal of Submicroscopical Cytology*, 13:127.

Harrison, F. W., and R. R. Cowden. 1975. Cytochemical Observations of Gemmule Development in *Eunapius fragilis* (Leidy). Porifera Spongillidae. *Differentiation*, 4:99–109.

Höhr, D. 1977. Differenzierungsvorgänge in der keimenden Gemmula von *Ephydatia fluviatilis*. *Wilhem Roux's Archive of Developmental Biology*, 182:329.

Mazzorana, M., and R. Garrone. 1983. Reversible Cell Scattering in Developing Sponges Induced by Penicillamine. *Journal of Submicroscopical Cytology*, 15:695–703.

Mazzorana, M., R. Garrone, N. Martel, and H. Yamasaki. 1984. Specific Binding and Biological Effects of Tumor Promoting Phorbol Esters on Sponges. *Biology of the Cell*, 52:27–34.

Pé, J. 1973. Etude quantitative de la regulation du squelette chez une éponge d'eau douce. *Archives de Biologie*, 84:147–173.

Pottu-Boumendil, J., and H. Pavans de Ceccaty. 1976. Mouvements cellulaires et morphogénèse de l'Eponge *Ephydatia mulleri* (Lieb). *Bulletin Societé Zoologique de France*, 101:31–38.

Rasmont, R. 1961. Une technique de culture des Eponges d'eau douce en milieu controls. *Annales, Societé Royale de Zoologique de Belgique*, 91:147–156. 1975. Freshwater Sponges as a Material for the Study of Cell Differentiation. Page 141 in *Current Topics in Developmental Biology*, edited by A. A. Moscona and A. Monroy. New York: Academic Press.

Rozenfeld, F. 1980. Effects of Puromycin on the Differentiation of the Freshwater Sponge, *Ephydatia fluviatilis*. *Differentiation*, 17:193–198.

Rozenfeld, F., and R. Rasmont. 1976. Hydroxyurea, An Inhibitor of the Differentiation of Choanocytes in Freshwater Sponges and a Possible Agent for the Isolation of Embryonic cells. *Differentiation*, 7:53.

Tanaka K., and Y. Watanabe. 1984. Choanocyte Differentiation and Morphogenesis of Choanocyte Chambers in the Freshwater Sponge *Ephydatia fluviatilis*, after Reversal of Developmental Arrest Caused by Hydroxyurea. *Zoological Science*, 1:561.

Van de Vyver, G. 1970. La non confluence intraspécifique chez les spongiaires et la notion d'individu. *Annales de Embryologie et de Morphologènese*, 3:251–262.

———. 1975. Phenomena of Cellular Recognition in Sponges. Pages 123–140 in *Current Topics in Developmental Biology*, edited by A. A. Moscona and A. Monroy. New York: Academic Press.

Van de Vyver, G., and M. Buscema. 1977. Phagocytic Phenomena in Different Types of Freshwater Sponge Aggregates. Pages 3–7 in *Developmental Immunobiology*, edited by J. B. Solomon and J. D. Horton. Amsterdam: Elsevier/North-Holland Biomedical Press.

Van de Vyver, G., S. Holvoet, and J. Huysecom. 1986. Inhibition of the Allorejection in the Freshwater Sponge *Ephydatia fluviatilis*. *Developmental and Comparative Immunology*, 10:429–435.

Weissenfels, V. N., and H. Landschoff. 1977. Bau und Funktion des Süsswasserschwamms *Ephydatia fluviatilis* (L) (Porifera) 4. Die Entwicklung der monaxialen SiO₂-Nadeln in Sandwich-Kulturen. *Zoologische Jahrbücher, Anatomie*, 98:355.

SANDRINO HOLVOET
GYSELE VAN DE VYVER
Laboratoire de Biologie Animale et Cellulaire
Université Libre de Bruxelles
50 av. F. D. Roosevelt
1050 Bruxelles, Belgium

Effects of 2,2'-Bipyridin on Skeletogenesis of *Ephydatia fluviatilis*

Abstract

Freshwater sponges, *Ephydatia fluviatilis*, were grown in a mineral medium containing 2,2'-bipyridin (Bip), a Fe^{++} chelator known to inhibit collagen secretion, at final concentrations of 5 μg/ml, 4 μg/ml, and 3 μg/ml. At 5 μg/ml, Bip appeared to be toxic since it prevented morphogenesis of the sponges. At 4 μg/ml, Bip selectively affected skeletogenesis. Interference with silicification of the spicules gave rise to small bulbous spicules, and spicule production was reduced by 50%. At 3 μg/ml, Bip affected only the morphology of the spicules, which appeared to be bulbous. But, when the silicate concentration of the medium was lowered tenfold, the number of spicules was drastically reduced. The effect of an Fe^{++} chelator (Bip) on the skeletogenesis of *E. fluviatilis* led us to conclude that enzymes with Fe^{++} cofactors are directly involved in the biosilicification of the axial filament. As Bip acts on the number of secreted spicules also, the regulation system of sclerocyte differentiation could be Fe^{++}-dependent too.

The skeleton of *Ephydatia fluviatilis* consists of a tridimensional network of siliceous spicules joined together by a sheath of spongin fibers (Garrone and Pottu, 1973). The experiments of Jørgensen (1944) and Pé (1973), have shown that sclerocyte differentiation is regulated by a system in which silicates act as an inhibitory factor. Although it is well established that the extracellular matrix of freshwater sponges plays some role in cell differentiation (Garrone, 1984), little is known about the role of the perispicular organic matrix in the regulation of skeletogenesis.

One of the first studies to tackle this problem has been concerned with sclerocyte differentiation in freshwater sponges devoid of perispicular spongin (Holvoet and Van de Vyver, this volume). Development of such sponges can be induced by adding puromycin (4 μg/ml) in the culture medium (Rozenfeld, 1980). Analysis of the number of spicules secreted by sponges treated with puromycin has shown that the absence of perispicular spongin does not affect the regulation process depending on the silicate concentration of the culture medium (Holvoet and Van de Vyver, this volume).

In sponges, the neomorphogenesis of either complete cell suspensions (among others, Wilson, 1907; Brien, 1937; Lévi, 1960; Borojevic and Lévi, 1964; Van de Vyver, 1971) or purified cell types (Huxley, 1921; Borojevic, 1966; De Sutter and Van de Vyver, 1977) appears to be one of the best ways to study cell differentiation. Unfortunately, puromycin prevents the neomorphogenesis of dissociated cells. As a result, we investigated the regulation of sclerocyte differentiation using other inhibitors that would have a more direct effect on the secretion of spongin. In the present study, we tested 2,2'-bipyridin, an Fe^{++} and molybdemum chelator, known for its inhibitory effect on the secretion of collagen (Juva et al., 1966).

Material and Methods

SPONGES. *Ephydatia fluviatilis* (Spongillidae) of strains α and β (Van de Vyver, 1970, 1975) were used. For the quantitative experiments, we incubated batches of 10 gemmules in 10 ml polystyrene Petri dishes. The gemmules were carefully separated from each other in order to obtain individual sponges grown from single gemmules (1-gml sponges). The sponges were cultivated at 20°C and were not fed during the experiments. For all the experiments, day 0 on the time scale corresponds to the hatching time of the control gemmules.

In order to compare experimental sponges having very different shapes, we defined a standard corresponding to the number of cells contained in a 1-gml control 9-day-old sponge. This standard is referred to here as the "biomass" of the sponge.

The cellular concentration of dissociated sponge suspensions was measured with a haemocytometer.

CHEMICALS. Gemmules were incubated in M medium (Rasmont, 1961) to which 5, 4, or 3 μg/ml 2,2'-bipyridin (Bip) had been added. In the experiments designed to test the effects of silicate concentration and 2,2'-bipyridin simultaneously, the sodium silicate concentration of the M medium (15 mg/l Silicon) was reduced to 1.5 mg/l (SiP medium).

SPICULE COUNTING. Spicules were counted by the method introduced by Pé (1973). The culture medium in Petri dishes containing sponges whose spicules were to be counted was replaced by a few drops of concentrated nitric acid. This treatment destroyed all but the spicules, which adhered to the bottom of the Petri dishes. After acid evaporation, the spicules were counted using a projection microscope. Ten sponges were used for each experimental point, with 95% consistent results. Each sample consisted

of 125 spicules, and the length of the spicules was measured 12 days after hatching (95% consistent results).

Results

CONTROL. The control sponges hatched after 3 days of incubation at 20°C in M medium. Subsequently, choanocytes chambers and aquiferous canals became differentiated, and the osculum appeared around day 4 (Figure 1a). Skeletogenesis began on day 1 with the differentiation of sclerocytes. At that time, the sclerocytes dispersed in the mesohyl were recognizable by their intracellular spicule. These young spicules were already silicified and resistant to nitric acid treatment. By day 3, the first extracellular spicules could be seen embedded in a sheet of perispicular spongin (Figure 1d,g). The kinetics of spicule production could be divided into two distinct periods. During the first period, which lasted three days, the number of spicules increased to about 190 spicules by day 4 (Figure 2). This was followed by a steady state, which was maintained until the end of the experiment at day 11. At that time, no more sclerocytes containing an intracellular spicule could be observed. The biomass of the control sponges was of 10^5 cells per 1-gml sponge.

Bip, 5 μg/ml. Gemmules placed in M medium to which 5 μg/ml Bip had been added hatched one day later than the controls. Archaeocytes issued from these gemmules remained undifferentiated and degenerated during the following days.

Bip, 4 μg/ml. At this concentration, the treated sponges and the controls developed at about the same pace (Figure 1c). Both were functional around day 4. The biomass of the treated sponges was estimated to be 80% of that of the controls.

In contrast, the skeletogenesis was qualitatively and quantitatively disturbed. Cells containing an axial filament were observed in the hours following hatching, but silicification of the filaments was slowed down and silicified spicules resistant to nitric acid treatment could not be counted until two days later. Mature spicules were characterized by bulbous structures, and their mean length (115 ± 7 μm) was about 40% lower than that of the controls. However, the spicules were embedded in a perispicular spongin sheath and integrated into a tridimensional network (Figure 1f,i).

The number of spicules produced by the treated sponges was 50% of the number produced by the control sponges cultivated in M medium. However, the period of spicule production lasted 4 days in both cases. The inhibitory effect of Bip became evident during the second day. The speed of spicule production, in comparison with that of the controls, dropped to 48%, 72%, and 32% on the second, third, and fourth days, respectively.

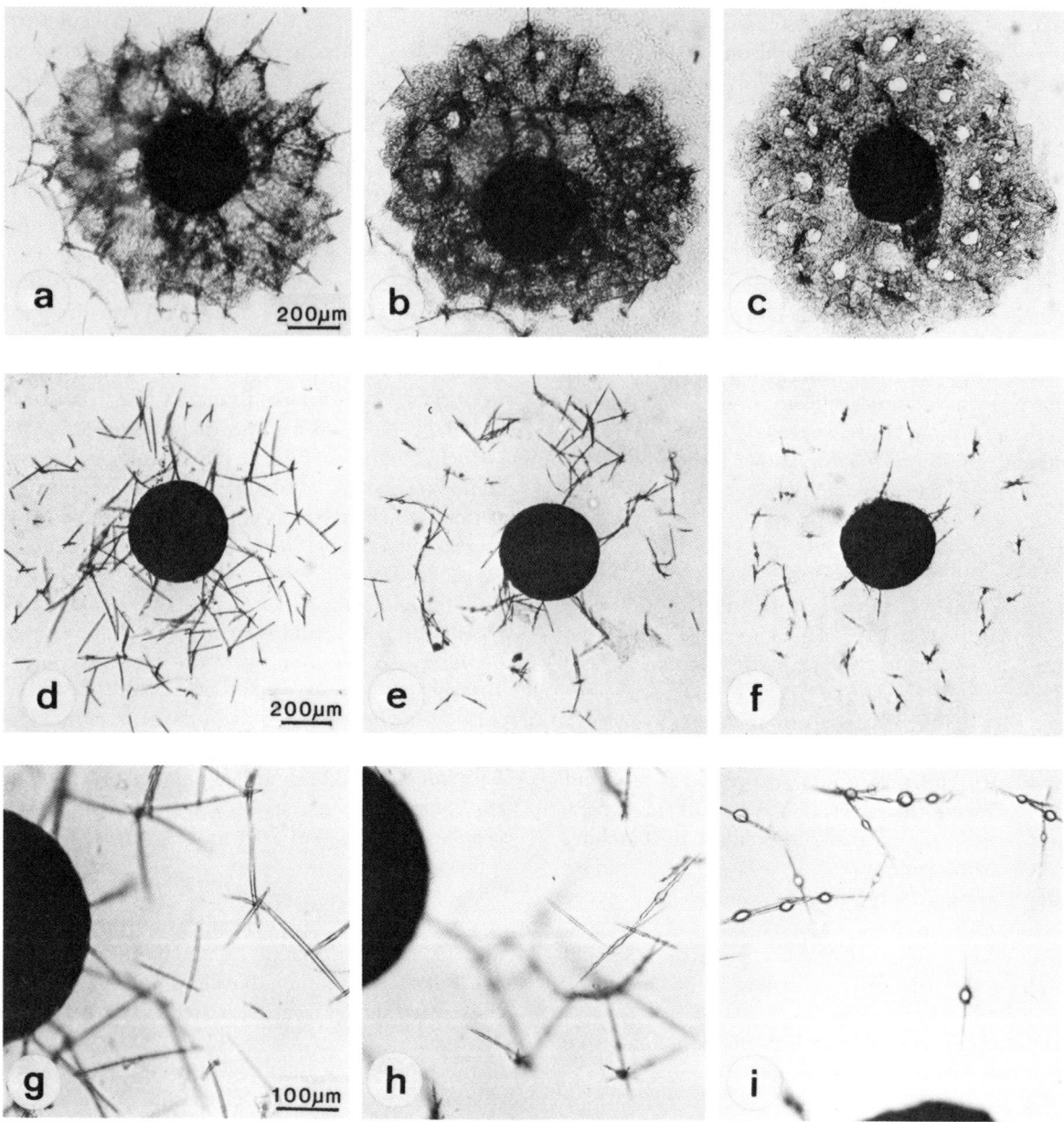

Figure 1. Cultivated *Ephydatia fluviatilis* and details of skeleton: *a, d, g*, M medium; *b, e, h*, 3 µg/ml Bip; *c, f, i*, 4µg/ml Bip.

In order to determine whether the existence of a sclerocyte population would eventually be blocked at an immature level, we examined the mesohyl of 11-day old treated sponges in vivo using light microscopy. No cells presenting either an axial filament or an intracellular spicule could be detected.

Reversibility of inhibition was tested in treated sponges that had stopped producing spicules (day 6). Five days after their transfer in M medium (day 11), the original number of spicules (89 ± 14) had increased by about 100. At that time, the total number of spicules produced (223 ± 30) was not significantly different from that of the controls (186 ± 14).

Bip, 3 µg/ml. The kinetics of spicule production and the total number of spicules secreted by 1-gml sponges were not significantly different from those observed in the controls (Figure 2). The length of the spicules (173 ± 5 µm) was hardly affected, but their general shape was aberrant and appeared bulbous (Figure 1*b,e,h*).

In order to determine whether action of 2,2'-bipyridin is related to the silicate concentration of the medium, we

Number of Spicules

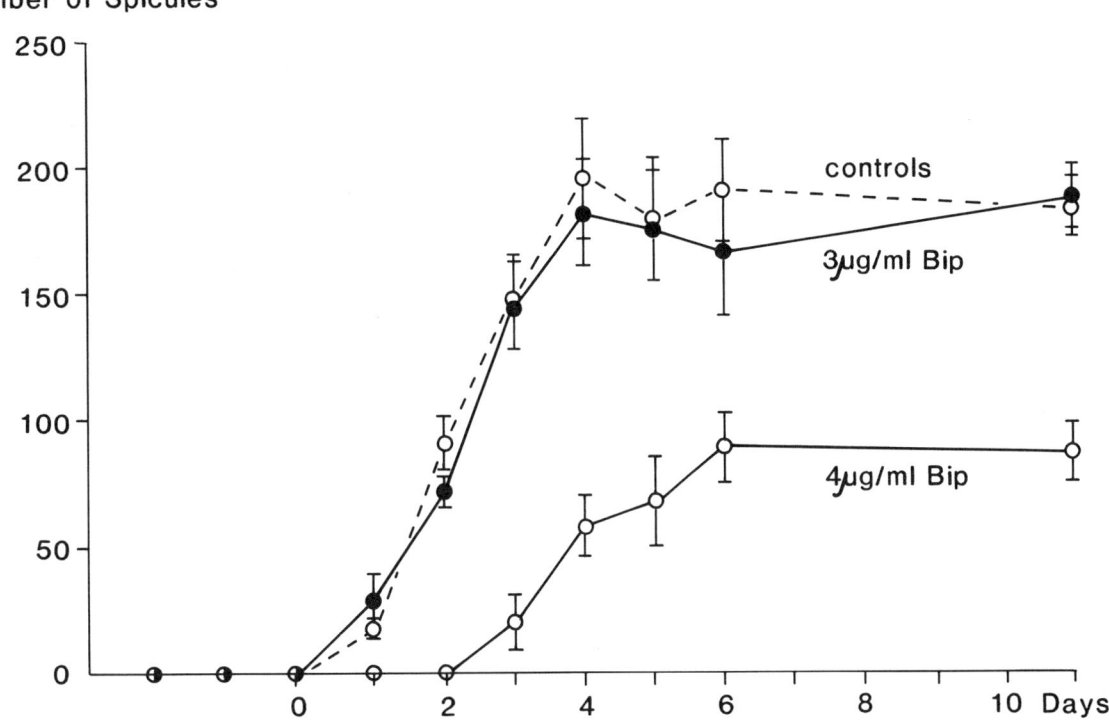

Figure 2. Kinetics of spicule production by 1-gml *Ephydatia fluviatilis* cultivated in M medium or in M medium with 3 μg/ml Bip or 4 μg/ml Bip.

cultivated sponges in a silicate-poor medium containing 3 μg/ml Bip (Table 1). Compared with sponges cultivated in M medium, sponges grown in SiP medium produced spicules. In the presence of Bip, however, the number of secreted spicules was lower in the SiP medium than in M medium. When these results were considered alongside the Bip/Si ratio, the inhibition rate was less than 10% in M medium (Bip/Si ratio of 0.0008), but reached 40–50% in the SiP medium (Bip/Si ratio of 0.008).

Discussion and Conclusions

In earlier work, (Garrone, 1978), the Fe^{++} chelator activity of 2,2'-bipyridin has been used successfully to block

Table 1. Mean number of spicules (± SD) secreted by 1 gml sponges as a function of the medium silicate concentration (p = 95)

Treatment	SiO₂ Conc. (mg/ml)	
	1.5	15
control	282 ± 48 (100%)	223 ± 38 (100%)
3 μg/ml Bip	158 ± 22 (56%)	208 ± 23 (93%)

the secretion of collagen by lophocytes in marine sponges. We tried to use Bip for the same purpose in this study. Unfortunately, despite the concentrations tested, we were unable to inhibit the perispicular spongin secretion. However, Bip turned out—somewhat unexpectedly—to be an excellent tool for studying the skeletogenesis of freshwater sponges.

At 4 μg/ml, Bip does not affect the general morphogenesis of the sponges and has little influence on the proliferative capacities of gemmular Archaeocytes. In contrast, the skeletogenesis is greatly disrupted. The number of spicules appears reversibly reduced to 50% and their morphology looks aberrant.

Interestingly, very close concentrations of Bip induced quite different biological effects. At 5 μg/ml, Bip appeared to be highly toxic and inhibited the morphogenesis of the sponges. At 4 μg/ml, it affected the number, the size, and the shape of the spicules, whereas at 3 μg/ml, it only affected the morphology of the spicules.

Spicules of *Ephydatia fluviatilis* grown in the presence of Bip look like those of various freshwater sponges cultivated in the presence of germanium (Elvin, 1972; Simpson et al., 1979, 1983), an element known to be a competitor of silicon in different biosilicification processes (Azam and Volcani, 1981). However, a strong effect on spicule morphology is already obvious when the Bip/Si ratio reaches 0.0012, whereas to obtain comparable effects using germanium, Simpson et al. (1979, 1983) had to use a ratio of at least 0.01.

Our experiments showed that the inhibitory effect of Bip on the spicule production is inversely related to silicate concentration of the medium. This relationship led us to hypothesize that the interference between Bip and skeletogenesis could occur at the level of either silicate receptors or transport systems.

The specific effect of an Fe^{++} chelator (Bip) on the skeletogenesis of *Ephydatia fluviatilis* led us to suggest that enzymes with Fe^{++} cofactors are specifically involved in the biosilicification of the axial filament. As Bip also acts on the number of secreted spicules, the regulation system of sclerocyte differentiation could be Fe^{++} dependent as well.

Acknowledgments

This study was supported in part by a grant from the Institut pour l'Encouragement de la Recherche Scientifique dans l'Industrie et l'Agriculture (IRSIA 840364). The authors are indebted to N. Cardon, E. De Cock, and F. Lambert for assistance in preparing the illustrations.

Literature Cited

Azam, F., and B. E. Volcani. 1981. Germanium-Silicon Interactions in Biological Systems. Pages 43–67 in *Silicon and Siliceous Structures in Biological Systems*, edited by T. L. Simpson and B. E. Volcani. New York: Springer.

Borojevic, R. 1966. Etude experimentale de la différenciation des cellules de l'éponge au cours de son développement. *Developmental Biology*, 14:130–153.

Borojevic, R., and C. Lévi. 1964. Etude au microscope électronique des cellules de l'éponge *Ophlitaspongia seriata* (Grant) au cours de la réorganisation après dissociation. *Zeitschrift für Zellforschung*, 64:708–725.

Brien, P. 1937. La réorganisation de l'éponge après dissociation par filtration et phenomenes d'involution chez *Ephydatia fluviatilis*. *Archives de Biologie*, 48:185–268.

De Sutter, D., and G. Van de Vyver. 1977. Aggregation Properties of Different Cell Types of the Freshwater Sponge *Ephydatia fluviatilis*. *Wilhelm Roux Archive for Developmental Biology*, 181:151–161.

Elvin, D. W. 1972. Effect of Germanium upon Development of Siliceous Spicules of Some Freshwater Sponges. *Experimental Cell Research*, 72:551–553.

Garrone, R. 1978. *Phylogenesis of Connective Tissue. Morphological Aspects and Biosynthesis of Sponge Intercellular Matrix*. Basel: S. Karger. 250 pp.

———. 1984. Formation and Involvement of Extracellular Matrix in the Development of Sponges, a Primitive Multicellular System. Pages 461–477 in *The Role of Extracellular Matrix in Development*, edited by R. L. Trelstad. New York: Alan R. Liss.

Garrone, R., and J. Pottu. 1973. Collagen Biosynthesis in Sponge Elaboration of Spongin by Spongocytes. *Journal of Submicroscopical Cytology*, 5:199–218.

Huxley, J. S. 1921. Further Studies on Reconstitution Bodies and Free Tissues Culture in *Sycon*. *Quarterly Journal of Microscopical Sciences*, 65:293–322.

Jørgensen, C. 1944. On the Spicule Formation of *Spongilla lacustris* (L) 1. The Dependence of the Spicule Formation on the Content of Dissolved and Solid Silicilic Acid of the Milieu. *Det Kongelige Danske Videnskabernes Selskab Biologiske Meddelelser*, 19:1–21.

Juva, K., D. J. Prockop, G. W. Cooper, and J. W. Lash. 1966. Hydroxylation of Proline and the Intracellular Accumulation of a Polypeptide Precursor of Collagen. *Science*, 152:92–94.

Lévi, C. 1960. Reconstitution de l'éponge *Ophlitaspongia seriata* (Grant) à partir de suspensions cellulaires. *Cahiers Biologie Marine*, 1:353–358.

Pé, J. 1973. Etude quantitative de la regulation du squelette chez une éponge d'eau douce. *Archives de Biologie*, 84:147–173.

Rozenfeld, F. 1980. Effects of Puromycin on the Differentiation of the Freshwater Sponge: *Ephydatia fluviatilis*. *Differentiation*, 17:193–198.

Rasmont, R. 1961. Une technique de culture des éponges d'eau douce en milieu controle. *Annales, Societé Royale de Zoologie de Belgique*, 91:147–156.

Simpson. T. L., R. Garrone, and M. Mazzorana. 1983. Interactions of Germanium (Ge) with Biosilicification in the Freshwater Sponge *Ephydatia mulleri*: Evidence of a Localized Membrane Domains in the Silicalemma. *Journal of Ultrastructure Research*, 85:159–174.

Simpson, T. L., L. M. Refolo, and M. Kaby. 1979. Effects of Germanium on the Morphologq of Silica Deposition in a Freshwater Sponge. *Journal of Morphology*, 159:343–354.

Van de Vyver, G. 1970. La non confluence intraspécifique chez les spongiaires et la notion d'individu. *Annales d'Embryologie et de Morphogènese*, 3:251–262.

———. 1971. Mise en evidence d'un facteur d'aggregation chez l'éponge d'eau douce *Ephydatia fluviatilis*. *Annales d'Embryology et de Morphogènese*, 4:373–381.

———. 1975. Phenomena of Cellular Recognition in Sponges. Pages 123–140 in *Current Topics in Developmental Biology*, edited by A. A. Moscona and A. Monroy. New York: Academic Press.

Wilson, H. V. 1907. On Some Phenomena of Coalescence and Regeneration in Sponges. *Journal of Experimental Zoology*, 5:245–258.

W. CLIFFORD JONES
School of Animal Biology
University College of North Wales
Bangor, Gwynedd, LL57 2UW, United Kingdom

J. J. ROY PEARSON
South Devon College of Arts and Technology
Newton Road, Torquay
Devon, TQ2 5BY, United Kingdom

Effects of Metabolic Inhibitors and Nutrients on Spicule Formation in Juveniles of *Sycon ciliatum* and *Grantia compressa*

Abstract

Larval eclosion, settlement, and metamorphosis can take place in seawater containing 10^{-4}M diamox, but spicule production is completely inhibited. Subsequent transfer to experimental solutions makes it possible to determine the effect of these on spicule formation. The addition of L-aspartate (0.1 mM) or an equimolar mixture of L-aspartate, L-glutamate, and D-glucose (0.1 mM and 2 mM) to artificial seawater improved spicule formation. In normal seawater, procaine (20 mM), sodium azide (1 mM), 2,4-dinitrophenol (1 mM), caffeine (10 mM), rotenone (5 mM), KCN (1 mM), and EDTA (10 mM) inhibited spicule development. These results supported the hypothesis that a Ca^{++}-pump is involved and that energy in the form of ATP is required. The fact that ouabain (1.37 mM) was only partly inhibitory suggests that Na^+-Ca^{++} exchange is not directly involved. Significant differences from normal seawater were also obtained after the addition of 6-mercaptopurine (0.3 mM), but not 5-fluorodeoxyuridine (0.1 mM), actinomycin-D (0.008 mM), or puromycin (0.01 mM). Thus cell division and extensive cell differentiation just before spicule secretion do not appear to be essential. In addition, 2,2′-dipyridyl (1 mM) and β-aminopropionitrile fumarate (1.5 mM) were inhibitory, but not tunicamycin (0.5 μg/ml). Hydroxy-L-proline (1.14 mM) stimulated production. These results suggest that the formation of collagenlike materials, but not glycoproteins, facilitates spicule formation.

Juvenile sponges of the calcareous sponge *Sycon ciliatum* have been used to study the rate of production and growth of spicules under experimental conditions (Jones, 1971, 1979; Ledger, 1976; Jones and Ledger, 1986). The juveniles were obtained by allowing larvae to settle and metamorphose on coverslips, which were then immersed in the experimental media after decalcification in bicarbonate-deficient artificial seawater. The method was somewhat unsatisfactory in that complete decalcification could not be guaranteed in all of the juveniles. It was harmful to prolong the immersion in the acidic seawater beyond the stage when all calcite had been dissolved; furthermore, it was unusual for all the juveniles on a coverslip to be of the same age and to have the same spicule complement. Judging the optimal time to end decalcification was therefore difficult, and corroded spicule remnants, which could hinder the recognition and measurement of newly forming spicules, particularly at the early stages of growth, could still be present in some juveniles. However, the discovery that 10^{-3} M diamox, the specific inhibitor of the enzyme carbonic anhydrase (E.C. 4.2.1.1), completely inhibits spicule formation in *Sycon* juveniles (Ledger, 1976; Jones and Ledger, 1986) has afforded a better method of ensuring that the juveniles are completely free of spicules at the start of an experiment. The new method and other improvements in technique are presented here, together with the results of further investigations into the mechanism of spicule secretion.

Materials and Methods

The collapse of the floating landing stage at Menai Bridge, Anglesey, North Wales, in 1978 removed a particularly convenient site for collecting ecloding specimens of *Sycon ciliatum* and *Grantia compressa*. Thus specimens had to be obtained by diving and shore collecting and then had to be maintained in an aquarium tank ($60 \times 45 \times 25$ cm) until larval eclosion took place. The sponges were laid on a horizontal piece of muslin suspended just above the bottom of the tank, so that they would have all-round access to the seawater, half of which was changed daily. Nevertheless, the larvae from aquarium sponges seemed less satisfactory than those from freshly collected sponges and difficulties arose in getting enough larvae to settle on the coverslips. To ensure good settlement one must have concentrated suspensions of well-nourished larvae in fresh seawater.

When the sponges started to release larvae, they were immersed in a solution of 10^{-4} M diamox (acetazolamide; Sigma, U.S.A.) in natural, filtered seawater in 100 ml crystallizing dishes. Preliminary experiments had established that diamox at this concentration inhibits spicule formation without apparently harming eclosion, settlement, or metamorphosis.

The sponges were cut into slices 3 mm thick to facilitate the rapid release of larvae, and larval suspensions were gleaned at intervals of one or two hours, the medium being replaced by fresh diamox-seawater. The suspensions were mixed together and, if too dense, were diluted appropriately with fresh diamox-seawater. The total suspension was then distributed between a number of Petri dishes containing coverslips. We used plastic 10×10 cm Petri dishes, the bottoms of which were lined with 25 specially cut 19.5 mm sided coverslips, set edge to edge. About 8 weeks before use the surface of the coverslips had been smeared by rubbing them with the distal ends of *Fucus serratus* fronds, and the dishes had been filled with seawater and stored in order to encourage the formation of a bacterial film on the coverslips. These were again lightly smeared with algal fronds immediately before use. Both sponges are known to settle on *Fucus serratus* (Boaden et al., 1976). We added 50 ml of the larval suspension to each dish and left the dishes in complete darkness in a constant temperature cabinet for at least four days at 16°C. Juveniles thus obtained were used in experiments as soon as possible after the four days because fewer spicules are produced when the interval is lengthened (Jones, 1979), probably owing to depletion of the nutrient reserves carried over from the larval stage.

After each coverslip was briefly drained against a piece of filter paper and rinsed twice in filtered natural seawater, it was immersed in the experimental medium, which was changed after one hour to ensure removal of the diamox. The medium was again replaced after 11 hours. For the experiments involving metabolic inhibitors, the media were made up using filtered natural seawater. Where appropriate, serial dilution was employed. Sonication over ice facilitated dissolution of some of the reagents. Artificial seawater (ASW; Jones, 1979) was used in some of the experiments involving nutrients. At the end of the experiment the coverslips were fixed by immersion in 90% methanol for 2 min, rinsed in distilled water buffered to pH 7.2, stained in ammoniacal picrocarmine for about 30 min (Friedlander's stain: 150 ml distilled water, 3 g carmine, 3 ml strong ammonia, and 300 ml saturated picric acid solution), dehydrated, cleared, and finally mounted on microscope slides using D.P.X. (BDH Chemicals, Poole, England). The spicules in a random sample of juveniles were counted and the longest monaxon and longest unpaired ray of the triradiate spicules were accurately measured.

Results
EFFECT OF ADDING ORGANIC NUTRIENTS TO THE MEDIUM

Initially spicule production in artificial seawater was not satisfactory, possibly because the larvae derived from aquarium-maintained sponges were not as well nourished as those ecloded by sponges collected directly from the sea. Consequently we decided to test the effect of adding organic nutrients to the seawater.

Table 1 shows the results when L-aspartic acid (or its sodium salt), L-glutamic acid (or its sodium salt), D-glucose, or a mixture of all three, was present at the stated concentrations in the artificial seawater. Three ex-

Table 1. Effect in spicules of adding nutrients to the culture medium (u.r. = unpaired ray; – = inapplicable)

Media	n	Monaxons			Triradiates		
		Maximum number /juv.	Mean number /juv. (SD)	Maximum length (μm)	Maximum number /juv.	Mean number /juv. (SD)	Maximum length of u.r. (μm)
Sycon ciliatum, July 1982							
Start control	20	0	0	–	0	0	–
Artificial seawater	15	15	3.73 (5.08)	56.7	0	0	–
ASW + 3 nutrients	10	43	25.9 (9.99)	77.0	4	1.6 (1.17)	27.0
ASW + aspartic acid	20	70	42.85 (16.60)	112.2	13	6.5 (3.35)	57.5
ASW + glutamic acid	20[a]	7	1.35 (1.93)	95.8	2	0.2 (0.52)	27.4
ASW + glucose	20	28	2.55 (6.33)	78.0	0	0	–
Natural seawater	10	105	75.6 (28.43)	113.5	16	11.4 (3.66)	59.4
Sycon ciliatum, July 1983							
Start control	10	0	0	–	0	0	–
Artificial seawater	10	9	0.9 (2.85)	28.4	3	0.5 (1.08)	4.1
ASW + 3 nutrients	10	71	40.2 (25.24)	89.2	11	4.1 (3.60)	36.5
Natural seawater	10	64	24.9 (24.97)	90.5	5	1.63 (1.78)	3.8
Grantia compressa, July 1983							
Start control	30	0	0	–	0	0	–
Artificial seawater	20	0	0	–	0	0	–
ASW + 3 nutrients	13	25	11.23 (9.28)	72.5	21	11.15 (4.79)	30.1
Natural seawater	20	18	7.55 (5.08)	82.1	23	8.95 (5.93)	46.5

[a]Some aberrant spicules

Experimental protocol: *Sycon ciliatum* from aquarium, July 1982; diamox–seawater used; settlement time 4–5 days; nutrients were L–aspartic acid, L–glutamic acid and D–glucose, each at a concentration of 0.1 mM; temperature of experiment = 16.8° C; pH at end was 7.82–7.91. 20 ml medium used per petri dish; juveniles fixed after 45 h in media. *Sycon ciliatum* from aquarium July 1983; diamox–seawater used; settlement time was 4 days; nutrients were monosodium L–aspartate, sodium L–glutamate and D–glucose, each at a concentration of 2 mM; temperature of experiment = 15.8° C; 20 ml medium used per petri dish; juveniles fixed after 43 h in media; pH at end was 7.89–7.91. *Grantia compressa* from aquarium July 1983, diamox–seawater used; settlement time 4–5 days; nutrients as in 1983 experiment above; 25 ml medium per petri dish; temperature of experiment = 16.3° C; juveniles fixed after 72 h in experimental media; pH at end was 7.39–7.70

periments were carried out, two on juveniles of *Sycon* and the third on juveniles of *Grantia*. Spicule production improved substantially when the mixture or aspartate alone was used, but not when glucose or glutamate alone was added. The glutamate result seems anomalous and might have been due to small-sample error. During the tests with *Sycon* in July 1982, the spicule production in artificial seawater enriched by the 0.1 mM mixture of nutrients was not as good as in natural seawater, but in tests on *Sycon* and *Grantia* in July 1983 that used the nutrients at 2 mM concentration, production was marginally better.

The addition of organic nutrients to the artificial seawa-ter, and at times to the natural seawater, helped improve spicule production considerably. Consequently the media used in later experiments were usually enriched in a similar way.

Later the effect of adding hydroxy-L-proline to natural seawater was studied, because the use of metabolic inhibitors had suggested that the formation of collagenlike materials facilitated spicule production. The result is shown in Table 2. More spicules were produced, but it is not known whether the hydroxyproline was acting as a general nutrient or whether it was being used specifically for the formation of collagen.

Table 2.   Effect of hydroxy–L–proline on spicule development (u.r. = unpaired ray)

Media	n	Monaxons			Triradiates		
		Maximum number /juv.	Mean number /juv. (SD)	Maximum length (μm)	Maximum number /juv.	Mean number /juv. (SD)	Maximum length of u.r. (μm)
Start control	60	2	0.03 (0.26)	34.2	1	0.02 (0.13)	1.4
Natural seawater	47	21	4.83 (5.90)	106.7	2	0.09 (0.35)	38.3
Hydroxy–L–proline + NSW	60	30	15.48 (8.32)	97.1	5	0.67 (1.16)	47.9

Experimental protocol: *Sycon ciliatum* from aquarium, July 1983; diamox–seawater was used; settlement time 4.5 and 6 days; the concentration of the hydroxy–L–proline was 1.14 mM in 10 ml of medium; the pH values were 7.99–8.01 initially, 7.93–7.94 at end; temperature of experiment = 16.2° C; juveniles fixed after 64 h in experimental medium

EFFECT OF METABOLIC INHIBITORS ON SPICULE GROWTH AND PRODUCTION

Table 3 gives the results of two experiments carried out in July 1982 using *Sycon* juveniles. The fact that the mitochondrial inhibitors, potassium cyanide, sodium azide and rotenone, and the uncoupler 2,4-dinitrophenol all inhibited spicule formation suggests that adenosine triphosphate (ATP) is required for spicule secretion. However, because the viability of the cells themselves is doubtful in these conditions, it would be unwise to conclude that ATP participates directly in spicule formation. Even after only 23 hours in the 5 mM rotenone-seawater, some of the juveniles were disintegrating.

Ethylenediaminetetraacetic acid (EDTA) likewise is inhibitory and, in view of its ability to chelate divalent cations, calcium should probably be supplied to the sponge in ionic form. Procaine also inhibited spicule formation, possibly because it can bind to acidic phospholipids in cell membranes (like those associated with the Ca^{++}-pump of the vertebrate sarcoplasmic reticulum). Since ouabain, which blocks Na^+/K^+-ATPase, was only slightly inhibitory, the uptake of Ca^{++} does not appear to be mediated by exchange with Na^+. However, the difference in mean number of spicules produced between the ouabain- and control-seawater was significant at the 0.1% level, and this together with the fact that many of the spicules were abnormal, suggests that the oubain-sensitive Na^+/K^+-ATPase is necessary for normal spicule formation.

Caffeine, which causes a transient release of stored Ca^{++} ions from the sarcoplasmic reticulum of striated muscle fibers (Weber and Herz, 1968; Ebashi et al., 1969), was strongly inhibitory at 10 mM concentration (Table 3).

Table 4 shows the results of adding some more metabolic inhibitors that might yield clues to the mechanisms of spicule secretion. In comparison with natural seawater, 5-fluorodeoxyuridine (which blocks thymidylate synthetase and thus DNA synthesis), actinomycin-D (which blocks transcription by inhibiting RNA synthesis, and thus indirectly blocks protein synthesis) and puromycin (which inhibits the incorporation of amino acid into protein) did not cause any significant changes in spicule production. Thus, cell division and differentiation immediately prior to spicule formation do not appear to be essential, if we assume that the chemicals gain access to the scleroblasts. The juveniles were immersed in the experimental media for 44 hours, during which time spicule production and hence sclerocyte activation would have been continuing in the natural seawater. Possibly the differentiation of the sclerocytes had already occurred during the four day settlement period, before the experiment was started. However, 6-mercaptopurine (which can block nucleic acid synthesis and also the formation of the co-enzyme nicotinamide adenine dinucleotide) did appear to have an inhibitory action, perhaps by reducing ATP production.

Tunicamycin (which inhibits N-linked glycoprotein synthesis) also has no significant effect on spicule formation, so that the concomitant formation of glycoproteins is probably not required. However, 2,2′-dipyridyl (which inhibits prolyl and lysyl hydroxylase) and β-aminopropionitrile fumarate (which inhibits lysyl oxidase, which is required for the cross-linking of collagen fibrils) were definitely inhibitory. This suggests that the formation of collagenlike protein is essential for spicule production.

The experiment was repeated after the three organic nutrients—monosodium-L-aspartate, sodium-L-glutamate, and D-glucose—had been added to the media at 2 mM concentration. The results are given in Table 5. Spicule production in general was improved by the presence of the nutrient mixture and there was substantial agreement with the results in Table 4. Some juvenile disintegration was evident after the treatment with β-aminopropionitrile fumarate. After the treatment with 2,2′-dipyridyl, β-Aminopropionitrile fumarate, and 6-mercaptopurine, most of the spicules were abnormal.

Table 3. Effects of inhibitors on spicule development in *Sycon* juveniles (u.r. = unpaired ray; — = inapplicable).

Media	n	Monaxons Maximum number /juv.	Mean number /juv. (SD)	Maximum length (µm)	Triradiates Maximum number /juv.	Mean number /juv. (SD)	Maximum length of u.r. (µm)
				Experiment I			
Start control	51	0	0	—	0	0	—
Natural seawater	30	86	53.03 (20.49)	121.7	18	9.63 (4.45)	65.7
Procaine, 20 mM	60[a]	1	0.02 (0.13)	45.1	0	0	—
Procaine, 100 mM	60[b]	1	0.02 (0.13)	35.6	0	0	—
Sodium azide, 1 mM	60[c]	8	1.70 (1.99)	41.0	0	0	—
Sodium azide, 1 mM	60[c]	8	1.47 (2.11)	47.9	0	0	—
				Experiment II			
Start control	20	1	0.05 (0.22)	46.5	0	0	—
Natural seawater	36	43	20.81 (10.19)	99.9	10	4.50 (2.75)	58.8
Dinitrophenol, 1 mM	33	1	0.06 (0.24)	20.5	0	0	—
Dinitrophenol, 5 mM	0	0	0	—	0	0	—
Caffeine, 10 mM	32	1	0.03 (0.18)	23.3	0	0	—
Rotenone, 1 mM	40	14	2.15 (2.82)	71.1	1	0.15 (0.36)	4.1
Rotenone, 5 mM	34[a]	0	0	—	0	0	—
Ouabain, 4 mM	37[d]	35	12.0 (9.98)	94.4	8	2.11 (1.91)	30.1
KCN, 1 mM	14[a]	0	0	—	0	0	—
EDTA, 10 mM	40	1	0.02 (0.16)	19.1	1	0.02 (0.16)	57.5

[a] A few juveniles disintegrating
[b] Many juveniles disintegrating
[c] All spicules abnormal
[d] Many spicules abnormal

Experimental protocol: experiment I — *Sycon ciliatum* from aquarium, July 1982; diamox–seawater used; settlement time 6.25 days; inhibitors made up in fresh filtered natural seawater; pH values were 7.95–8.00 initially, 7.78–7.97 at end; 20 ml of medium per petri dish, changed at 2 and 16 h; the temperature of the experiment = 16.8° C; the juveniles were fixed after 38 h in the media; experiment II — same procedures as I except settlement time, 5 days; pH start 7.88–8.08, pH end 7.70–7.85; fixation after 23 h.

EFFECT OF 10⁻⁴M DIAMOX ON SPICULE PRODUCTION

The data on the "start controls" in Tables 1–5 reveal that spicule production is almost completely inhibited by the diamox. Despite being in the diamox-seawater for five days or more, very few, if any, spicules were produced. In fact, only nine short dubious monaxons and two dubious triradiates were found in the 381 juveniles examined on the start-control coverslips. Furthermore, an additional 545 juveniles used as start controls for some other experiments not reported here exhibited no spicules whatsoever. However, the diamox did not affect the ability of the juveniles to produce normal spicules when returned to natural seawater.

VARIATIONS IN RESULTS OBTAINED WITH NATURAL SEAWATER

A comparison of the controls used for experiments carried out at different times reveals some inconsistencies in spicule production and the maximal lengths of rays formed. Such inconsistencies have been noted before (Jones and Ledger, 1986) and are probably caused by differences in the age of the juveniles, differences in nutrient content of the larvae, unequal sample sizes and slight differences in pH, temperature, or freshness of the seawater used. Valid quantitative comparisons can only be made between results within each particular experiment.

Table 4.   Effect of inhibitors on spicule development in *Sycon* juveniles (u.r. = unpaired ray; — = inapplicable).

Media	n	Monaxons Maximum number /juv.	Mean number /juv. (SD)	Maximum length (μm)	Triradiates Maximum number /juv.	Mean number /juv. (SD)	Maximum length of u.r. (μm)
Start control	130	2	0.05 (0.29)	43.8	1	0.01 (0.08)	1.4
Natural seawater	107	46	5.13 (7.67)	112.2	2	0.06 (0.27)	38.3
6-Mercaptopurine, 0.3 mM	120	1	0.03 (0.18)	49.2	0	0	—
5–Fluorodeoxyuridine, 0.1 mM	110	33	5.10 (6.60)	132.7	1	0.06 (0.24)	21.9
Actinomycin–D, 7.97 μM	120[a]	30	3.15 (5.20)	112.2	2	0.05 (0.25)	31.5
Puromycin dichloride, 9.73 μM	120[a]	21	2.5 (4.03)	125.9	1	0.03 (0.18)	31.5
Tunicamycin, 0.5 μg/ml	122	66	3.95 (8.01)	112.2	4	0.05 (0.38)	32.8
2,2'–Dipyridyl, 1.0 mM(3.48)	109[c]	18	2.75	72.5	0	0	—
β–Aminopropionitrile fumarate, 1.5 mM	120[c]	0	0	—	0	0	—

[a]Many spicules abnormal
[b]Juveniles unhealthy, stain weakly; many spicules abnormal
[c]All juveniles disintegrating
Experimental protocol: *Sycon ciliatum* from aquarium, July 1983; combined results for two overlapping replicate experiments; inhibitor solutions made up separately for each; diamox–seawater used; settlement time 4.5 and 6 days (1st experiment), 5.5 and 7.25 days (2nd experiment); coverslips taken at random for both experiments; solutions replaced after 1 h and 25 h (1st) and 1.5 h and 25 h (2nd experiment); 20 ml medium per petri dish; temperature of experiment = 16.2° C; the pH was 7.85–8.08 initially, 7.93–8.01 at end (1st), 7.90–8.01 initially, 8.01–8.08 at end (2nd); juveniles fixed at 64 h in both experiments.

Discussion

EFFECT OF ADDING NUTRIENTS

The results in Table 1 afford clear evidence that spicule production is stimulated by the addition of aspartic acid to the artificial seawater. Production can also be improved when natural seawater is enriched by the addition of hydroxyproline or a mixture of aspartate, glutamate, and glucose (Tables 2 and 5). Such enrichment was not necessary in previous experiments utilizing *Sycon* juveniles (Jones, 1971, 1979; Jones and Ledger, 1986), in which the larvae were obtained from sponges freshly collected from the sea. Presumably the maintenance of the adult sponges for several days in an aquarium before larval eclosion had a deleterious effect. Although half of the aquarium seawater was changed daily, the supply of nutrients could not have been as great as for sponges living in their natural habitat. Possibly the larvae would have acquired smaller nutritional reserves from their parent sponges, and they in turn would have given rise to settled juveniles with a diminished capacity for spicule production. An earlier study (Jones, 1979) has shown that the rate of spicule production declines the longer the juveniles are left in the settle-

ment Petri dishes before experimental use. During this time one would have expected the juveniles to subsist largely on their reserves. It is unlikely that the diamox treatment was the cause of the difference between the later and earlier experiments; the inhibitory action of diamox on carbonic anhydrase is reversible (the dissociation constant of the drug-carbonic anhydrase complex is about 10^{-8}M in nearly all secretory tissues; Maren, 1977) and, even if this were not so, it appears that the amount of carbonic anhydrase in *Sycon* is small (Jones and Ledger, 1986) so that replacement of the enzyme would probably not have overtaxed the nutritional reserves of the juveniles.

The confirmation that calcareous sponges can take up dissolved organic substances is of interest, particularly when one considers that bacteria are sparsely distributed, if at all present, in the mesohyl of calcareous sponges (Jones, 1966; Ledger, 1976). That siliceous sponges can take up dissolved organic substances from the ambient medium has been known for some time, but whether the uptake is directly mediated by the sponge cells themselves or by symbiotic bacteria is not clear (Wilkinson and Garrone, 1980).

Table 5. Effect of inhibitors plus nutrients on spicule development in *Sycon* juveniles (u.r. = unpaired ray; — = inapplicable).

Media		Monaxons			Triradiates		
	n	Maximum number /juv.	Mean number /juv. (SD)	Maximum length (μm)	Maximum number /juv.	Mean number /juv. (SD)	Maximum length of u.r. (μm)
Start control	60	0	0	—	0	0	—
Natural seawater	58	66	25.62 (15.22)	104.0	13	4.26 (3.17)	46.5
6–Mercaptopurine, 0.3 mM	60*	24	7.95 (6.91)	98.5	8	2.42 (2.02)	50.6
5–Fluorodeoxyuridine, 0.1 mM	60	66	30.25 (17.89)	112.2	11	4.75 (2.94)	49.2
Actinomycin–D, 0.008 mM	60	61	27.43 (15.05)	112.2	12	4.92 (3.14)	50.6
Puromycin dichloride, 0.0097 mM	60	68	27.35 (16.45)	125.9	12	5.42 (2.78)	50.6
Tunicamycin, 0.5 μg/ml	60	64	26.13 (15.72)	116.3	13	4.55 (2.96)	52.0
2,2'–Dipyridyl, 1 mM	60[a]	7	1.08 (1.78)	41.0	0	0	—
β–Aminopropionitrile fumarate, 1.5 mM	60[b]	4	0.07 (0.52)	68.4	1	0.02 (0.13)	9.6

[a]Most spicules abnormal
[b]Some juveniles disintegrating, most spicules abnormal
Experimental protocol: *Sycon ciliatum* from aquarium, July 1983; diamox–seawater used; settlement time 6 days; nutrients (monosodium L–aspartate, sodium L–glutamate and D–glucose, each at a final concentration of 2 mM) were added to all media; 20 ml medium per Petri dish, changed at 1.5 and 25 h; the pH values were 8.02–8.09 initially, 7.95–8.07 at end; experimental temperature = 16.1° C; the juveniles were fixed after 44 h in the experimental media.

EFFECT OF METABOLIC INHIBITORS

Some clues to the mechanism of spicule formation can be gleaned from the experiments involving inhibitory substances. However, caution must be exercised in interpreting the results. Lack of inhibition could be due to the inability of the substance to penetrate the sponge cells, although this is unlikely in the case of inhibitors known to affect the cells of other animals. Also, it may be difficult to distinguish between a specific inhibition of the calcite-forming mechanism and toxic effects on other processes in the organism.

The inhibitory action of EDTA, which chelates Ca^{++} ions, indicates that the calcium for spicule secretion is normally taken up in ionic form. One would expect a membrane Ca^{++} pump to be involved in spicule production, because even though the concentrations of Ca^{++} and CO_3^{--} ions in the surface waters of the sea are greater than the solubility product of calcite (Harvey, 1955), the spicules are formed intercellularly in a space that is apparently sealed off from the mesohyl-matrix by the presence of septate junctions between the enveloping sclerocytes (Ledger, 1975; Ledger and Jones, 1977). The free Ca^{++} concentration in the cytoplasm of cells generally is regarded as being low (10^{-8} M or less; Martonosi, 1984), so that ions would have to be secreted into the space against a concentration gradient, if we assume that cytoplasmic

Ca^{++} ions in the sclerocytes are the direct source of the Ca^{++} used. Likewise one would expect to find Ca^{++} pumps in the cell membranes of the pinacocytes and choanocytes bounding the mesohyl, to enable Ca^{++} ions to be secreted into the matrix enveloping the sclerocytes, if we assume that these do not possess a store of Ca^{++} for spicule secretion (it is possible that such a store could be acquired during the time when the sclerocyte precursor cells formed part of the pinacoderm). That Ca^{++} pumps are involved is suggested by the need for ATP, which was shown by the effects of the mitochondrial inhibitors, particularly when there was little evidence of cellular damage (for example, when 1 mM rotenone or 1 mM sodium azide was employed). However, sodium azide is also a partial inhibitor of carbonic anhydrase at a concentration of 2 mM (Meldrum and Roughton, 1933).

The inhibitory action of 20 mM procaine certainly is suggestive of a Ca^{++} pump, because this binds to acidic phospholipids in cell membranes, thereby displacing Ca^{++} (Papahadjopoulos, 1972). Such phospholipids are generally associated with Ca^{++} pumps in the plasma membranes of cells and of the sarcoplasmic reticulum of muscle fibers (Martonosi, 1984). However, Coffe et al. (1985) have recently demonstrated that procaine has a dual effect on the organization of the cytoskeletal system in sea urchin eggs; at low concentrations (0.2–2.5 mM) it stimulates the formation of cytoplasmic fibrillar components, whereas at higher concentrations (10 mM) it inhibits the polymerization of tubulin, causing the disassembly of microtubules. It has been suggested that cytoplasmic microtubules participate in the mechanism of uniplanar growth of spicule rays (Jones, 1985), so that the inhibitory action of 20 mM procaine on spicule formation may have been indirect, by preventing microtubule formation. Thus, although it is probable that a Ca^{++} pump participates in the spicule-forming process, the evidence for this from the effect of the inhibitors is merely supportive, and not fully conclusive.

Another possible pathway for Ca^{++} uptake from the ambient medium would be by means of a Na^+-Ca^{++} exchange system, not coupled to ATP-hydrolysis, as in excitable cell membranes (Martonosi, 1984). However, the fact that ouabain was only slightly inhibitory indicates that this is not so. Kingsley (1984) likewise concluded that a ouabain-sensitive Na^+/K^+-ATPase is probably not involved in the uptake of Ca^{++} by the epithelium of the gorgonian *Leptogorgia*.

Caffeine was tried, because it causes a transient release of stored Ca^{++} ions from the sarcoplasmic reticulum of striated muscle fibers (Weber and Herz, 1968; Ebashi et al., 1969). It also increases the cAMP inside cells by inhibiting phosphodiesterase (Perry, 1974). It inhibited spicule formation in the sponge juveniles at 10 mM concentration. It is known that cAMP inhibits the germination of freshwater sponge gemmules (Simpson and Rodan, 1976; Harrison et al., 1979) and is localized on the membranes of vitelline platelets and on the spicule-forming silicalemma (Harrison et al., 1981). It is thus likely to be involved in the mobilization of nutrient reserves and in transport mechanisms. Details of its precise effect on the sponge cells must await the results of further research.

Two sclerocytes are required for the formation of a monaxon or ray of a triradiate, but it is still not known whether the pair of cells is derived by the division of a single scleroblast or by random association (Jones, 1970). The experiment designed to investigate the effect of mitotic inhibitors, namely, those blocking DNA synthesis, gave somewhat conflicting results in that 5-fluorodeoxyuridine had no significant effect on spicule formation, whereas 6-mercaptopurine was strongly inhibitory. The former inhibits thymidylate synthetase, and so blocks DNA synthesis specifically, while metabolites of the latter can have a variety of effects, including the slowing down of synthesis of adenine and guanine monophosphates and thus of all nucleic acids. These metabolites can also inhibit the formation of certain coenzymes, particularly, nicotinamide adenine dinucleotide, and so could interfere with the supply of ATP. Thus although the inhibitory action of 6-mercaptopurine can be explained by its multiple effects on nucleotide metabolism, the absence of inhibition by 5-fluorodeoxyuridine would suggest that mitotic division is not an essential prerequisite of spicule formation.

It might be argued that some scleroblasts in the diamox-treated juveniles could have divided already and thus could have been available to start spicule formation in the 5-fluorodeoxyuridine medium. However, the juveniles left for 64 hours in this medium, and the spicule numbers produced in comparison with those of the seawater control suggest that sclerocyte-pair formation would have continued throughout this period. Note that neither 5-fluorodeoxyuridine nor 6-mercaptopurine has any effect on spicule production by embryonic micromeres of the sea urchin *Strongylocentrotus purpuratus* cultured in vitro in the presence of 4% horse serum (Mintz et al., 1981).

The slightly inhibitory effect on spicule production of actinomycin (which blocks transcription by inhibiting RNA synthesis) and puromycin (which inhibits the incorporation of amino acid into protein) suggests that sclerocyte differentiation is necessary, but not extensive, just before normal spicule formation. The fact that many of the spicules produced in the presence of these two antibiotics were aberrant (Table 4) supports the view that the concomitant manufacture of some protein is necessary, either directly or indirectly, for normal spicule formation. By contrast, Rozenfeld (1980) has found that puromycin (7 μg/ml) prevents the differentiation of archaeocytes into sclerocytes and other cells in hatching gemmules of the siliceous freshwater sponge *Ephydatia fluviatilis*, yet does not inhibit the formation of normal megascleres within archaeocytes. In the case of the embryonic cells of the sea urchin, the fact that actinomycin-D blocks spicule formation indicates that transcription is required (Mintz et al., 1981).

We also found that 2,2′-dipyridyl was part inhibitory and caused the production of aberrant monaxons, while β-aminopropionitrile was completely so. The former inhibits, besides succinic dehydrogenase, the enzymes prolyl and lysyl hydroxylase, which are required for the manufacture of hydroxyproline and hydroxylysine, whereas the latter inhibits lysyl oxidase, which is required for the cross-linking of collagen fibrils. The inhibitory effects thus suggest that the formation of collagenlike materials is necessary for normal spicule production. This conclusion is supported by the stimulatory action of hydroxyproline, although direct proof of the incorporation of this hydroxyproline into collagen is lacking. Possibly it could be obtained by the use of radioactively labeled hydroxyproline. Tunicamycin, on the other hand, appears to have had little effect on spicule formation. Because it inhibits N-linked glycoprotein synthesis, one can conclude that the formation of glucosamine-derived proteins is not required for spicule production. The spicules are known to be enveloped by a thin elastic membrane throughout their life, but there is no convincing evidence supporting the presence of organic matter within the spicule calcite (Jones, 1967). Possibly the sheath is composed of collagenlike materials; certainly collagen fibrils are laid down around the sheath when the spicule rays are fully formed (Ledger and Jones, 1977). Spicule formation by cultured embryonic sea urchin micromeres differs in that it is blocked by tunicamycin (0.5 μg/ml), which also inhibits the incorporation of ³H-glucosamine into macromolecules. The evidence here suggests that primary mesenchyme cells synthesize one or more hydroxyproline-containing proteins during skeletal formation, but probably not collagen; while 2,2'-dipyridyl and β-aminopropionitrile fumarate block spicule formation, the hydroxyproline-containing polypeptides are resistant to collagenase and characteristic subunits of vertebrate and invertebrate collagen are not released by pepsin-digestion (Mintz et al., 1981).

More recently, Blankenship and Benson (1984) showed that spicule production by sea urchin micromeres is almost completely inhibited by 2,2'-dipyridyl when the cells are cultured on plastic. However, the inhibition can be overcome by using collagen instead of plastic as the culture substratum. Their results thus indicate that collagen is required for normal spicule formation in the echinoderm larva, not because it is a constituent of the spicule calcite, but because it must be present in the extracellular matrix. A similar conclusion would also be acceptable for the secretion of sponge spicules.

Conclusions

Although 10⁻⁴ M diamox in seawater does not interfere with larval eclosion or settlement, it prevents spicule production. The addition of 0.1 mM L-aspartate, or a mixture of L-aspartate, L-glutamate, and D-glucose (equimolar proportions, 0.1–2 mM), or 1.14 mM hydroxyproline, to the seawater enhances spicule production in juvenile calcareous sponges. The fact that 10 mM EDTA inhibits calcite secretion suggests that calcium is taken up in ionic form. Mitochondrial inhibitors (1 mM cyanide, 5 mM rotenone, 1 mM sodium azide, and 1 mM 2,4-dinitrophenol) inhibit spicule formation, as do 20 mM procaine and 10 mM caffeine; 1.37 mM ouabain is partly inhibitory. It is probable that Ca⁺⁺ pumps are involved in calcite secretion. Whereas 0.1 mM 5-fluorodeoxyuridine had no significant effect on spicule formation, 0.3 mM 6-mercaptopurine was strongly inhibitory. Cell division is not essential just before spicule formation. Neither 8 μM actinomycin-D nor 10 μM puromycin affected significantly the numbers of spicules produced, but many of the spicules produced in their presence were abnormal. Apparently, extensive differentiation of the sclerocytes prior to spicule secretion is not required, but some production of protein is necessary, either directly or indirectly, if normal spicule formation is to occur. While 1 mM 2,2'-dipyridyl was partly inhibitory, 1.5 mM β-aminopropionitrile fumarate was almost completely so; thus the concomitant formation of collagenlike proteins appears to be necessary for spicule formation. The fact that tunicamycin (0.5 μg/ml) did not inhibit spicule production indicates that glycosamine-derived proteins were not required.

Literature Cited

Blankenship, J., and S. Benson. 1984. Collagen Metabolism and Spicule Formation in Sea Urchin (Strongylocentrotus purpuratus) Micromeres. Experimental Cell Research, 152:98–104.

Boaden, P. J. S., R. J. O'Connor, and R. Seed. 1976. The Fauna of a Fucus serratus L. Community: Ecological Isolation in Sponges and Tunicates. Journal of Experimental Marine Biology and Ecology, 21:249–267.

Coffe, G., G. Foucault, M. N. Raymond, and J. Pudles. 1985. Dual Effect of Procaine in Sea Urchin Eggs. Inducer and Inhibitor of Microtubule Assembly. Experimental Cell Research, 156:175–181.

Ebashi, S., M. Endo, and I. Ohtsuki. 1969. Control of Muscle Contraction. Quarterly Review of Biophysics, 2:351–384.

Harrison, F. W., E. M. Rosenberg, D. Davis, and T. L. Simpson. 1981. Correlation of Cyclic AMP and Cyclic GMP Immunofluorescence with Cytochemical Patterns during Dormancy Release and Development from Gemmules in Spongilla lacustris L. (Porifera:Spongillidae), Journal of Morphology, 167:53–63.

Harrison, F. W., T. L. Simpson, and E. Rosenberg. 1979. Immunofluorescent Realization of Cyclic AMP and Cyclic GMP during Dormancy Release and Development from Gemmules of Spongilla lacustris L. (Porifera: Spongillidae). Pages 47–51 in Biologie des Spongiaires, edited by C. Lévi and N. Boury-Esnault. Colloques Internationaux du C.N.R.S. 291. Paris: Centre National de la Recherche Scièntifique.

Harvey, H. W. 1955. The Chemistry and Fertility of Sea Waters. Cambridge: Cambridge University Press. 224 pp.

Jones, W. C. 1966. The Structure of the Porocytes in the Calcareous Sponge Leucosolenia complicata (Montagu). Journal of the Royal Microscopy Society, 85:53–62.

———. 1967. Sheath and Axial Filament of Calcareous Sponge Spicules. Nature, 214:365–368.

———. 1970. The Composition, Development, Form and Orientation of Calcareous Sponge Spicules. Pages 91–123 in The Biology of the Porifera, edited by W. G. Fry. Symposia of the Zoological Society of London. London: Academic Press.

———. 1971. Spicule Formation and Corrosion in Recently Metamorphosed Sycon ciliatum (O. Fabricius). Pages 301–320 in Fourth European Marine Biology Symposium, edited by D. J. Crisp. Cambridge: Cambridge University Press.

———. 1979. Spicule Growth and Production in Juvenile Sycon ciliatum. Pages 67–77 in Biologie des Spongiaires, edited by C. Lévi and N. Boury-Esnault. Colloques Internationaux du C.N.R.S. 291. Paris: Centre National de la Recherche Scièntifique.

———. 1985. Spicule Form in Calcareous Sponges (Porifera: Calcarea). The Principle of Uniplanar Curvature. Journal of Zoology, 204:571–584.

Jones, W. C., and P. W. Ledger. 1986. The Effect of Diamox and Various Concentrations of Calcium on Spicule Secretion in the Calcareous Sponge Sycon cilatum. Comparative Biochemical Physiology, 84A:149–158.

Kingsley, R. J. 1984. Spicule Formation in the Invertebrates with Special Reference to the Gorgonian *Leptogorgia virgulata*. *American Zoologist*, 24:883–891.

Ledger, P. W. 1975. Septate Junctions in the Calcareous Sponge Sycon *ciliatum*. *Tissue and Cell*, 7:13–18.

———. 1976. Aspects of the Secretion and Structure of Calcareous Sponge Spicules. Ph.D. Thesis, University of Wales. 125 pp.

Ledger, P. W., and W. C. Jones. 1977. Spicule Formation in the Calcareous Sponge *Sycon ciliatum*. *Cell and Tissue Research*, 181:553–567.

Maren, T. H. 1977. Use of Inhibitors in Physiological Studies of Carbonic Anhydrase. *American Journal of Physiology*, 232:F291–F297.

Martonosi, A. N. 1984. Mechanisms of Calcium Release from the Sarcoplasmic Reticulum of Skeletal Muscle. *Physiology Revue*, 64:1240–1320.

Meldrum, N. U., and F. J. W. Roughton. 1933. Carbonic Anhydrase, its Preparation and Properties. *Journal of Physiology*, 80:113–142.

Mintz, G. R., S. De Francesco, and W. J. Lennarz. 1981. Spicule Formation by Cultured Embryonic Cells from the Sea Urchin. *Journal of Biological Chemistry*, 256:13105–13111.

Papahadjopoulos, D. 1972. Studies on the Mechanism of Action of Local Anesthetics with Phospholipid Model Membranes. *Biochimica et Biophysica Acta*, 265:169–186.

Perry, M. C. 1974. The Hormonal Control of Metabolism. Pages 587–607 in *Companion to Biochemistry. Selected Topics for Further Study, 1*, edited by A. T. Bull et al. London: Longmans Group.

Rozenfeld, F. 1980. Effects of Puromycin on the Differentiation of the Freshwater Sponge: *Ephydatia fluviatilis. Differentiation*, 17:193–198.

Simpson, T. L., and G. A. Rodan. 1976. Role of cAMP in the Release from Dormancy of Freshwater Sponge Gemmules. *Developmental Biology*, 49:544–547.

Weber, A., and R. Herz. 1968. The Relationship between Caffeine Contracture of Intact Muscle and the Effect of Caffeine on Reticulum. *Journal of General Physiology*, 52:750–759.

Wilkinson, C., and R. Garrone. 1980. Nutrition of Marine Sponges. Involvement of Symbiotic Bacteria in the Uptake of Dissolved Carbon. Pages 157–161 in *Nutrition in the Lower Metazoa*, edited by D. C. Smith and Y. Tiffon. Oxford: Pergamon Press.

Tissue Organization, Morphology, and Mechanics

FREDERICK W. HARRISON
Department of Biology
Western Carolina University
Cullowhee, North Carolina 28723

NANCY W. KAYE
GORDON W. KAYE
Department of Anatomy
Albany Medical College
Albany, New York 12208

The Dermal Membrane of *Eunapius fragilis*

Abstract

The dermal membrane of *Eunapius fragilis* is structurally complex. Some regions consist of numerous ostia and porocytes, which overlie vestibules, and others are covered almost entirely by exopinacocytes. The exopinacocyte and endopinacocyte layers are separated by a well-defined fibrillar mat that in many respects resembles the lamina reticularis zone of the vertebrate epithelial basal lamina. Exopinacocytes are in close contact with the fibrous meshwork but appear capable of movement on it. A population of ameboid cells is present on the free surface of the dermal membrane. These ameboid cells interact with foreign structures such as pollen grains and Protozoa. In many cases the ameboid cells deposit fibers on the foreign body in what appears to be a primary tissue response.

The external surface of many sponges consists of a dermal membrane. This complex structure has an outer layer of exopinacocytes, an inner layer of endopinacocytes, and a thin mesohyl zone positioned between the two pinacocyte layers. Regions of ostia overlying a dermal space or vestibular cavity alternate with areas composed entirely of pinacocytes. A number of functions, including the regulation of water flow, fibrilogenesis, removal of epizoic organisms, and nutrient processing have been attributed to the pinacocytes of the dermal membrane. However, surprisingly few studies have looked at the dermal membrane. As a result, little is known about ostial structure, the possibility of exopinacocyte motility, and the processes involved in epizoite removal.

In a attempt to provide further information on the morphology and function of this structure, we examined the dermal membrane of the freshwater sponge, *Eunapius fragilis* with light microscopic histochemical techniques, and scanning electron microscopy.

Materials and Methods

Specimens of the freshwater sponge, *Eunapius fragilis* (Leidy) were collected in mid-July, 1984 and 1985, from Ten-Mile Creek below Lake Myosotis on the E. N. Huyck

Preserve, Rensselaerville, New York. The sponges were immediately fixed in Bouin's solution or cold ethanol-glacial acetic acid (3:1) for light microscopy, or cold 0.35% glutaraldehyde in 0.025M cacodylate buffer (De Vos, 1979) for scanning electron microscopy (SEM). Tissues were processed routinely, embedded in Paraplast-Plus, and sectioned serially at 7 μm for the light microscope. Tissues to be examined by SEM were dehydrated through a graded ethanol series, dried in a Samdri–790 critical point drier and coated with gold-palladium in a Sam-sputter-EA sputter coater (both from Tousimis Research Corporation, Rockville, Maryland).

Results

The dermal membrane of *Eunapius fragilis* is structurally complex. Ostial regions of the sponge surface alternate with zones consisting only of pinacocytes (Figure 1*a*).

Ostia as known from light microscopy may be of two types: "intercellular" (Figure 1*a*), which are formed by several cells bordering the ostium, and "intracellular," which are formed by a single cell that apparently closes on itself in the manner of a doughnut or by several cells that enclose the space in the manner of a small capillary. Porocytes extend through the dermal membrane, their interior margin projecting veil-like beyond the endopinacocyte layer of the dermal membrane into the dermal space (Figure 1*b*).

A dense fibrillar meshwork occupies the mesohylar region between the two pinacocyte layers of the dermal membrane (Figure 1*c*). The fibers of this network (most likely collagen fibrils of varying size that do not exhibit periodicity in these SEM preparations) range in diameter from 35 to 70 nm. We were unable to demonstrate fibrilogenic activity in either pinacocyte layer, but observed no other cell types in the dermal membrane mesohyl zone. The fibrous zone is PAS-positive and stains strongly with the fluorochrome brilliant sulfoflavine, which indicates the presence of glycoprotein and ionizable amino groups.

The exopinacocytes are closely associated with the fibrillar meshwork, but appear capable of movement on it (Figure 1*d*). They exhibit numerous filopodial and lamellipodial extensions (Figure 2*a*) and may be observed extending filopodia into ostial openings (Figure 2*b*).

The peripheral margin of each porocyte "cups" the fibrillar mat. This relationship is particularly evident when shrinkage artifact has separated the porocyte boundary from the fibrous layer (Figure 2*c*).

A population of amebocytes is present on the free surface of the dermal membrane. These cells interact with foreign structures, such as pollen grains and Protozoa. In many cases, the amebocytes deposit fibers on the foreign body in what appears to be a primary recognition response to the foreign object (Figure 2*d*). Although ex-

opinacocytes exhibit numerous pseudopodia, these cells were not observed phagocytizing epizoic organisms.

Discussion

The gross morphology of the dermal membrane of *Eunapius fragilis* resembles that seen in another freshwater sponge, *Ephydatia fluviatilis* (De Vos, 1979; Weissenfels, 1980). In both sponges, ostia are restricted to local areas of the surface and overlie a dermal space or vestibule. The dermal space is lined with endopinacocytes that give rise to numerous incurrent canals and pores. Other regions of the dermal membrane of both sponges contain no ostia.

The nature of ostia in demosponges has been debated extensively. Several authors have reported porocytes with intracellular pores (Wilson, 1910; Brien, 1932; Harrison, 1972a,b; Reiswig, 1975; Weissenfels, 1980), while others (Simpson, 1968; Boury-Esnault, 1972; De Vos, 1979), describe intercellular ostia created by junctional separations of adjacent pinacocyte margins. *Eunapius fragilis* was thought to contain both types of ostia, porocytes with intracellular pores, and intercellular pores between pinacocytes, but the SEM micrographs in our study indicate that what were described as intracellular ostia are more properly either small unicellular ostia, formed by a doughnut-shaped cell, or small ostia formed by only two cells. In fact, the term intracellular is, by any modern definition of cell structure, a misnomer; an ostium, a through passage, is extracellular, even if the hole is surround by the processes of a single cell. Both unicellular and multicellular ostia are present in demosponges. The variability reported may reflect maturational differences in the particular specimens observed.

Porocytes of *Eunapius fragilis* extend through the dermal membrane as do those of *Ephydatia fluviatilis* (Weissenfels, 1980). Similarly, *E. fragilis* porocytes contact both exo- and endopinacocytes at their lateral surface, but in addition, cup the fibrous meshwork of the interior of the dermal membrane.

The fibrillar component of the *Eunapius fragilis* dermal membrane differs from that seen in other sponge species. Typically, the interpinacocyte region of the dermal membrane consists of a mesohyl containing scattered collagen fibrils (Garrone and Pottu, 1973; Pottu-Boumendil, 1975; Garrone and Rozenfeld, 1981). The fibers in the *E. fragilis* dermal membrane form a dense mat that in many respects resembles the lamina reticularis zone of the vertebrate epithelial basal lamina. This similarity may only be coincidental. Such a relationship cannot be confirmed without more information on the nature of collagen and other glycoproteins in the sponge's dermal membrane.

We have not observed exopinacocytes or endopinacocytes engaged in fibrilogenesis. However, considering the apparent absence of ameboid cells in the mesohylar region of the *Eunapius fragilis* dermal membrane, pinacocytes

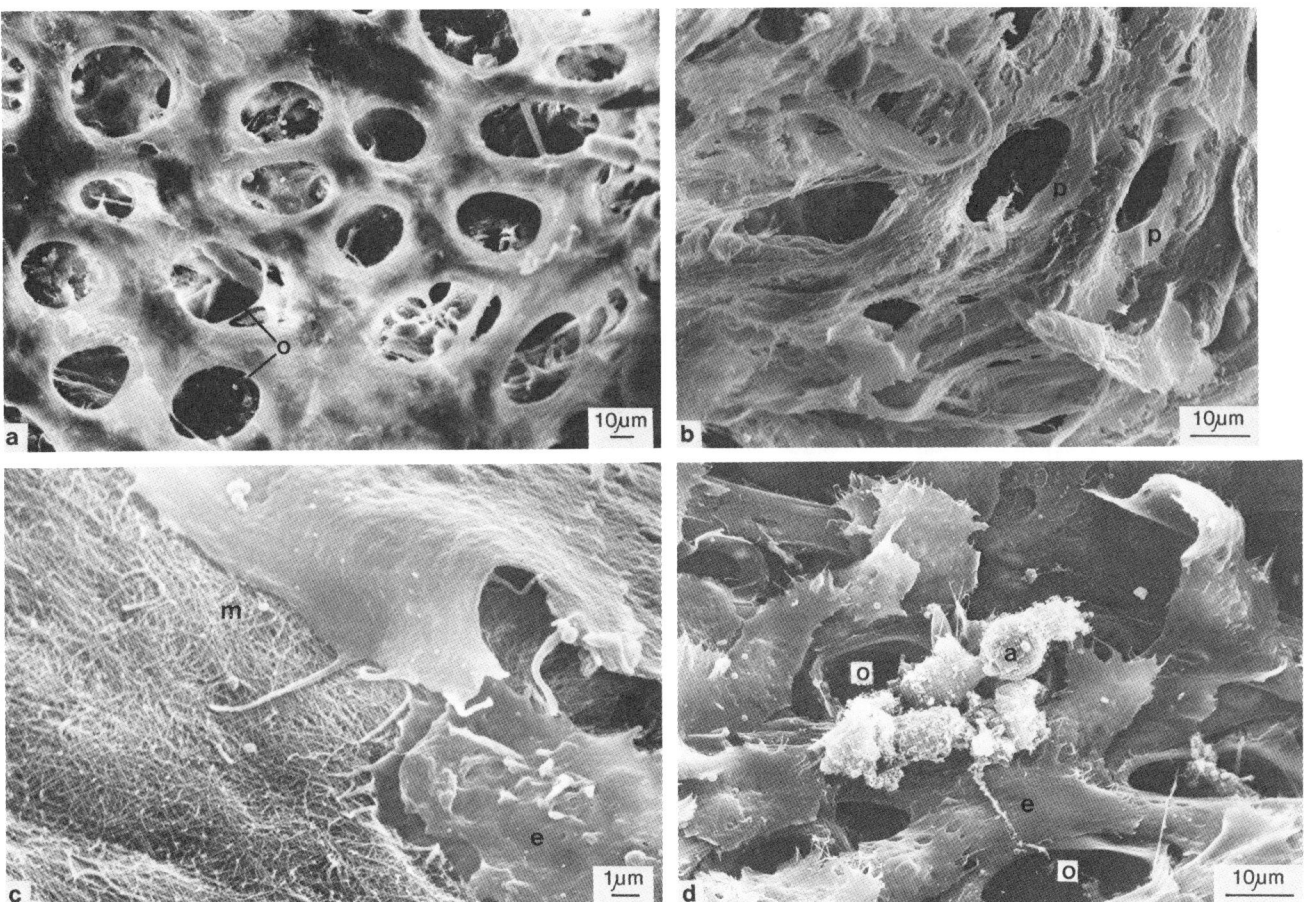

Figure 1. *Eunapius fragilis*, dermal membrane (SEM): *a*, ostia pierce dermal membrane and overlie a vestibular or "subdermal" space (average size of these ostia is 30 μm); *b*, viewed from underlying vestibular space, porocytes extend through dermal membrane, their interior margin projecting veillike beyond endopinacocyte layer of dermal membrane; *c*, exopinacocytes of dermal membrane are positioned upon a dense fibrillar mat (fibers of meshwork range in diameter from 35 to 70 nm); *d*, exopinacocytes of dermal membrane appear capable of movement upon sponge surface; a population of sponge amebocytes is present on sponge exterior and numerous ostia are visible. *a*, amebocyte; *e*, exopinacocyte; *m*, fibrillar mat; *o*, ostium; *p*, porocyte.

may well be involved in the synthesis of the mat fibrils. Pinacocytes are capable of elaborating cell surface and intercellular matrix components (see Garrone, 1978 for review). Exo- and endopinacocytes synthesize a surface glycocalyx in a number of sponges (Lévi and Porte, 1962; Bagby, 1970; Garrone et al., 1971; Thiney, 1972; Ledger, 1976; Evans, 1977; Willenz, 1983), while basopinacocytes are able to secrete such matrix components as basal spongin (Garrone, 1978) and fibronectin (Labat-Robert et al., 1981). In addition, a number of studies have indicated that exo- and endopinacocytes are involved in collagen synthesis. Much of the evidence revolves around the position of collagen fibrils with respect to dermal membrane pinacocytes (Boury-Esnault, 1973; Pottu-Boumendil, 1975; Garrone, 1978; Garrone and Rozenfeld, 1981). However, Pottu-Boumendil (1975) showed that in *Ephydatia muelleri*, only exopinacocytes incorporate ^{3}H-proline during collagen synthesis in the dermal membrane of that freshwater sponge. Therefore, the evidence

suggests that the fibrillar meshwork present in the *E. fragilis* dermal membrane is the product of synthetic activities of the pinacocytes.

Harrison (1972a, 1974a) observed that in the freshwater sponge *Corvomeyenia carolinensis*, a population of ameboid cells migrates on the external surface of the sponge and suggested (Harrison, 1974b) that these amebocytes cleanse the sponge surface and affect the sponge's tolerance to siltation. Frost (1976) reported that the superior surface of *Spongilla lacustris* is freer of detritus and epizoites than macrophyte surfaces in the same pond, but noted (Frost, 1978) that sediment may accumulate over gemmules to such an extent that successful hatching is prevented. It appears that growing sponges can cleanse their external surfaces of accumulated foreign materials primarily through the activities of an external ameboid cell population. These activities extend beyond exopinacocyte phagocytic processes described by Willenz (1980) and Willenz and Van de Vyver (1982). Their reports of the

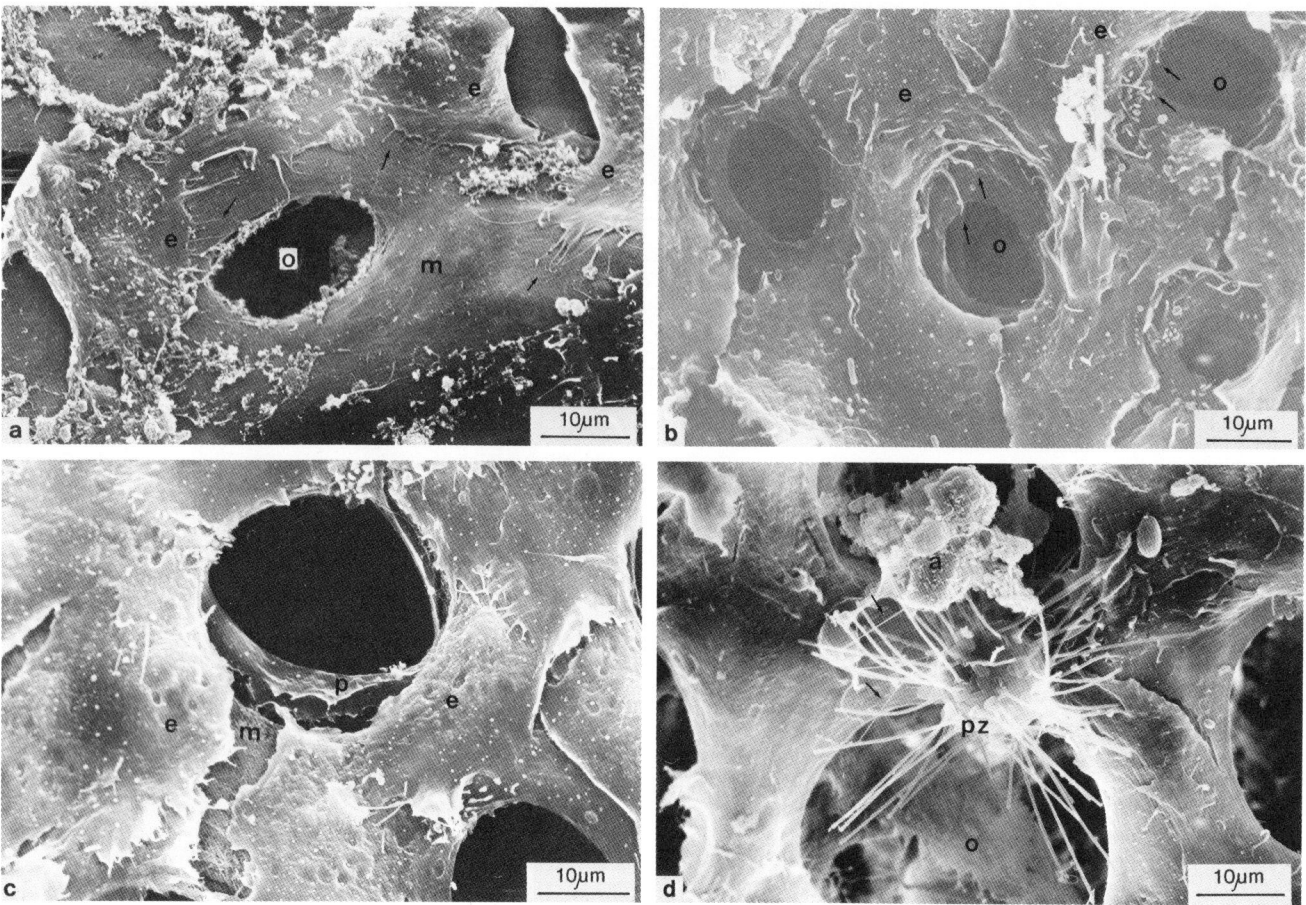

Figure 2. *Eunapius fragilis*, dermal membrane (SEM): *a*, exopinacocytes on fibrillar meshwork of dermal membrane extend filopodial processes *(arrows)* toward and into ostia; *b*, exopinacocyte filopodia *(arrows)* extend into ostia; *c*, a porocyte "cups" fibrillar meshwork of dermal membrane and is closely associated with several exopinacocytes; *d*, an amebocyte on sponge's superior surface deposits fibers *(arrows)* on a stalked protozoan that rests at an ostial margin. Vestibular space visible beneath dermal membrane. *a*, amebocyte; *e*, exopinacocyte; *m*, fibrillar mat; *o*, ostia; *p*, porocyte; *pz*, protozoan.

ingestion of latex beads suggest exopinacocyte assimilative capabilities that may not necessarily have anything to do with the cleaning of the sponge surface.

We did not observe phagocytic activity by the amebocyte population on the sponge surface. However, these cell were observed depositing fibers on foreign structures such as pollen grains and Protozoa. This foreign tissue reaction is identical to instances of sponge cell fibrogenesis in response to commensal or parasitic organisms (Connes, 1968; Connes et al., 1971) or to the experimental introduction of foreign materials (Cheng et al., 1968). In view of the phagocytic potential of sponge cells, it seems probable that the ameboid cell population observed on the external surface of *Eunapius fragilis* possesses phagocytic as well as fibrogenic capabilities in response to foreign objects.

It appears that the exopinacocytes of *Eunapius fragilis* are capable of movement on the underlying fibrous meshwork of the dermal membrane. Although Harrison (1974a) believes that the exopinacocytes of *Corvomeyenia*

carolinensis function as an integumentary or limiting surface, studies of wound healing in that species (Harrison, 1972a) reveal that basopinacocytes become ameboid under certain stimuli. In addition, Weissenfels (1978) has noted that exo- as well as basopinacocytes are capable of migration in the freshwater sponge *Ephydatia fluviatilis*, while Ledger (1976) has reported exopinacocyte migration into the mesohyl in calcareous sponges.

Although localized increases in membrane density that are sometimes associated with bundles of tonofilaments and that approach the structure of macula adherens have been reported in *Ephydatia fluviatilis* (Feige, 1969) and in *E. muelleri* (Pottu-Boumendil, 1975), true desmosomes have never been described in sponges. Thus, in the absence of desmosomal junctions and in view of other reports of pinacocyte mobility, it seems possible that the exopinacocytes of the *Eunapius fragilis* dermal membrane are capable of motility. However, De Vos (pers. comm.) warns that the apparent patterns of exopinacocyte motility in *E. fragilis* may reflect mechanical artifacts created by fixation or critical point drying action upon a pinacoderm under tension. We recognize the possibility of preparation artifact in such delicate tissues but suggest for the time that the exopinacocytes are capable of ameboid migration.

Acknowledgments

We gratefully acknowledge the fellowship provided by the Edmund Niles Huyck Preserve, Inc., Rensselaerville, New York, awarded to Frederick W. Harrison.

Literature Cited

Bagby, R. M. 1970. The Fine Structure of Pinacocytes in the Marine Sponge *Microciona prolifera* (Ellis and Solander). *Zeitscrift für Zellforschung,* 105:579–594.

Boury-Esnault, N. 1972. Une structure inhalente remarquable des spongiaires: le crible. Etude morphologique et cytologique. *Archives de Zoologie Expérimentale et Générale,* 113:7–23.

———. 1973. L'exopinacoderme des spongiaires. *Bulletin, Muséum National d'Histoire Naturelle* (3. Series), 178:1193–1206.

Brien, P. 1932. Contribution a l'étude de la régénération naturelle chez les Spongillidae, *Spongilla lacustris* (L.), *Ephydatia fluviatilis* (L.). *Archives de Zoologie Expérimentale et Générale,* 74:461–506.

Cheng, T. C., H. W. F. Yee, E. Rifkin, and M. Kramer. 1968. Cellular Reactions in *Terpios zeteki* to Implanted Heterologous Biological Materials. *Journal of Invertebrate Pathology,* 12:29–35.

Connes, R. 1968. Etude histologique, cytologique et expérimentale de la régénération et de la reproduction asexuée chez *Tethya lyncurium* Lamarck (= *T. aurantium* Pallas) (Demosponges). Thesis, Univ. Montpellier. 193 pp.

Connes, R., J. Paris, and J. Sube. 1971. Réactions tissulaire de quelques Démosponges vis-à-vis de leurs commensaux et parasites. *Naturaliste Canadien,* 98:923–935.

De Vos, L. 1979. Structure tridimensionelle de l'Eponge *Ephydatia fluviatilis.* Pages 159–164 in *Biologie des Spongiaires,* edited by C. Lévi and N. Boury-Esnault. Colloques Internationauxdu C.N.R.S. 291. Paris: Centre National de la Recherche SciÈntifique.

Evans, C. W. 1977. The Ultrastructure of Larvae from the Marine Sponge *Halichondria moorei* Bergquist (Porifera, Demospongiae). *Cahiers de Biologie Marine,* 18:427–433.

Feige, N. W. 1969. Die Feinstruktur der Epithelien von *Ephydatia fluviatilis. Zoologische Jahrbücher, Anatomie,* 86:177–237.

Frost, T. M. 1976. Investigations of the Aufwuchs of Freshwater Sponges. I. A Quantitative Comparison between the Surfaces of *Spongilla lacustris* and Three Aquatic Macrophytes. *Hydrobiologia,* 50:145–149.

———. 1978. The Ecology of the Freshwater Sponge *Spongilla lacustris.* Thesis, Dartmouth College. 196 pp.

Garrone, R. 1978. *‹IP0,7›Phylogenesis of Connective Tissue. Morphological Aspects and Biosynsthesis of Sponge Intercellular Matrix.* Basel: S. Karger. 250 pp.

Garrone, R., and J. Pottu. 1973. Collagen Biosynthesis in Sponges: Elaboration of Spongin by Spongocytes. *Journal of Submicroscopic Cytology,* 5:199–218.

Garrone, R., and F. Rozenfeld. 1981. Electron Microscope

Study of Cell Differentiation and Collagen Synthesis in Hydroxyurea-Treated Freshwater Sponges. *Journal of Submicroscopic Cytology,* 13:127–134.

Garrone, R., Y. Thiney, and M. Pavans de Ceccatty. 1971. Electron Microscopy of a Mucopolysaccharide Cell Coat in Sponges. *Experientia,* 27:1324–1329.

Harrison, F. W. 1972a. The Nature and Role of the Basal Pinacoderm of *Corvomeyenia carolinensis* Harrison (Porifera: Spongillidae). *Hydrobiologia,* 39:495–508.

———. 1972b. Phase Contrast Photomicrography of Cellular Behavior in Spongillid Porocytes (Porifera: Spongillidae). *Hydrobiologia,* 40:513–517.

———. 1974a. Porifera. Pages 29–66 in *Pollution Ecology of Freshwater Invertebrates,* edited by C. W. Hart, Jr. and S. L. H. Fuller. New York: Academic Press.

———. 1974b. Histology and Histochemistry of Developing Outgrowths of *Corvomeyenia carolinensis* Harrison (Porifera: Spongillidae). *Journal of Morphology,* 144:185–194.

Labat-Robert, J., L. Robert, C. Auger, C. Lethias, and R. Garrone. 1981. Fibronection-like Protein in Porifera: Its Role in Cell Aggregation. *Proceedings, National Academy of Sciences (U.S.A.),* 78:6261–6265.

Ledger, P. W. 1976. Aspects of the Secretion and Structure of Calcareous Sponge Spicules. Thesis, Univ. College North Wales. 125 pp.

Lévi, C., and A. Porte. 1962. Etude au microscope électronique de l'éponge *Oscarella lobularis* Schmidt et de sa larvae amphiblastula. *Cahiers de Biologie Marine,* 3:307–315.

Pottu-Boumendil, J. 1975. Ultrastructure, cytochimie, et comportements morphogénetiques des cellules de l'éponge *Ephydatia mülleri* (Lieb.) Thesis, Univ. Claude Bernard. 101 pp.

Reiswig, H. 1975. The Aquiferous Systems of Three Marine Demospongiae. *Journal of Morphology,* 145:493–502.

Simpson, T. L. 1968. The Structure and Function of Sponge Cells. *Bulletin, Peabody Museum of Natural History,* 25:1–141.

Thiney, Y. 1972. Morphologie et cytochimie ultrastructurale de l'oscule d'*Hippospongia communis* LMK et de sa régénération. Thesis, Univ. Claude Bernard. 63 pp.

Weissenfels, N. 1978. Bau und Funktion des Süßwasserschwamms *Ephydatia fluviatilis* L. (Porifera). V. Das nadelskelett und seine Entstehung. *Zoologische Jahrbücher, Anatomie,* 99:211–223.

———. 1980. Bau und Funktion des Süßwasserschwamms *Ephydatia fluviatilis* L. (Porifera). VII. Die Porocyten. *Zoomorphology,* 95:27–40.

Willenz, P. 1980. Kinetic and Morphological Aspects of Particle Ingestion by the Freshwater Sponge *Ephydatia fluviatilis* L. Pages 163–178 in *Nutrition in the Lower Metazoa,* edited by D. C. Smith and Y. Tiffon. Oxford: Permagon Press.

———. 1983. Aspects cinétiques quantitatifs et ultrastructuraux de l'endocytose, la digestion, et l'exocytose chez les Eponges. Thesis, Université Libre de Bruxelles. 107 pp.

Willenz, P., and G. Van de Vyver. 1982. Endocytosis of Latex Beads by the Exopinacoderm in the Freshwater Sponge *Ephydatia fluviatilis*: An in vitro and in situ Study in SEM and TEM. *Journal of Ultrastructure Research,* 79:294–306.

Wilson, H. V. 1910. A Study of Some Epithelioid Membranes in Monaxonid Sponges. *Journal of Experimental Zoology,* 9:537–577.

WILLARD D. HARTMAN
PHILIPPE WILLENZ*
Department of Biology and
Peabody Museum of Natural History
Yale University
New Haven, Connecticut 06511

Organization of the Choanosome of Three Caribbean Sclerosponges

Abstract

The fine structural organization of the living tissue of *Ceratoporella nicholsoni* (Hickson), *Goreauiella auriculata* Hartman (Ceratoporellidae), and *Calcifibrospongia actinostromarioides* Hartman (Calcifibrospongiidae) has been analyzed using transmission and scanning electron microscope techniques.

The choanocyte chambers of *Ceratoporella* are embedded in a thick mesohylar layer containing few sponge cells but large numbers of symbiotic bacteria. In contrast, the chambers of *Goreauiella* are surrounded by a much thinner mesohylar layer containing scarce, dispersed bacteria.

The diplodal choanocyte chambers of *Ceratoporella* are partly lined by endopinacocytes and are continuous with wide aphodi leading to the exhalant canal system. A remarkable feature of the aquiferous canals of *Ceratoporella* is the existence of valvules formed by pseudopodial processes of the endopinacocytes.

In *Calcifibrospongia*, the choanocyte chambers lie across the lumen of the terminal inhalant canals and open into exhalant canals of lesser diameter. The density of sponge cells in the mesohyl is higher than in *Ceratoporella*, and a dense bacterial population is also present. Central cells lie above the choanoderm of the chambers in this sponge.

Although Recent sponges with a compound skeleton consisting of siliceous spicules, collagen fibers, and a massive basal skeleton of calcium carbonate have been known for some time (*Astrosclera willeyana* Lister, the first to be named, was well described by Lister in 1900), the extent of their diversity on Recent coral reefs has become clear only recently (Hartman and Goreau, 1970). Although the question of how to classify these prodigal skeleton makers among the sponges is still open to dispute, two main views are being espoused. According to Vacelet (1981), the capacity to secrete a basal layer of calcium carbonate has

*Present address: Laboratoire de Biologie Animale et Cellulaire, Faculté des Sciences, Université Libre de Bruxelles, 50 av. F. D. Roosevelt, 1050 Bruxelles, Belgium.

developed several times during the course of the evolution of demosponges. In his view, several independent lines of demosponges have given rise to different lines of "hypercalcified" sponges. In contrast, Hartman and Goreau (1970) have argued that at some point in the past, a process of calcification was added to the general make-up of demosponges and that the resulting new line of sponges with a compound skeleton itself diverged into different groups.

Our interest in the evolutionary relationships between sclerosponges and other demospnge classes has led us to study the ultrastructure of the living tissue of these organisms. Now that a considerable body of information is available on the comparative micromorphology of the choanosome of demosponges and its potential phylogenetic significance has been discussed (Boury-Esnault et al., 1984, this volume; Langenbruch, 1983; Langenbruch et al., 1985), it seems appropriate to apply this approach to the sclerosponges.

Material and Methods

Ceratoporella nicholsoni and *Goreauiella auriculata* were collected in a reef tunnel at a depth of 28 m at Pear Tree Bottom, 5 km east of Discovery Bay, Jamaica, in July 1984 and February 1985. *Calcifibrospongia actinostromarioides* was collected under an overhang at a depth of 30 m on the forereef slope of Jamaica Bay, at the southern tip of Acklins Island, Bahamas, in August 1985.

Samples studied under the transmission electron microscope (TEM) consisted of fragments with attached skeleton that had been removed with hammer and cold chisel from the periphery of the sponges. They were fixed in situ according to a modification of the "low osmium pre-fixative" technique of Eisenman and Alfert (1981). A 5- to 10-min pre-fixation was performed in a mixture of solutions A and B (ratio 10:1) combined in situ immediately before use. Solution A consisted of 4% glutaraldehyde in 0.2 M sodium cacodylate buffer, pH 7.4, supplemented with 0.35 M sucrose and 0.1 M NaCl to obtain a final osmotic pressure of 1,700 mOsM. Solution B contained 1% osmium tetroxide in 0.2 M sodium cacodylate, supplemented with 0.3 M NaCl (final osmotic pressure = 1,040 mOsM). Fixation was then continued in the same fixative without the added OsO_4 for 24 hr at 4°C. Specimens were washed six times for 10 min in 0.2 M sodium cacodylate buffer, pH 7.4. To obtain a contrasting enhancement of the cell coat, ruthenium red was added to each solution in the proportion of 50 mg/100 ml (Luft, 1971a,b). Specimens were then decalcified for 2 to 6 weeks at 4°C in a 4.1% solution of disodium EDTA adjusted to pH 6.8 with NaOH, supplemented with 5% polyvinyl pyrrolidone (Fullmer, 1966) and 12% sucrose, to give a final osmotic pressure of 1,142 mOsM. Samples were then postfixed for 3 hr in 1% osmium tetroxide in 0.2 M sodium

cacodylate and 0.3 M NaCl, dehydrated through a graded ethanol series, and embedded in ERL 4206 according to Spurr (1969). Sections were obtained with a diamond knife on a Sorvall MT-2 ultramicrotome. Before sectioning, the siliceous spicules at the section plane were dissolved in 15–20% hydrofluoric acid in distilled water for 5 min. Thin sections double-stained with uranyl acetate and lead citrate (Reynolds, 1963) were examined in a Zeiss EM 10 electron microscope at 80 kV.

Scanning electron microscope (SEM) observations were carried out on specimens that had been fixed and decalcified as for TEM but without the addition of ruthenium red. After dehydration through a graded ethanol series, they were cryofractured in liquid nitrogen, thawed in 100% ethanol at ambient temperature, and dried by the critical point method from carbon dioxide. They were finally covered with gold, using a Balzer sputter coater. The preparations were examined in an ISI DS 130 scanning electron microscope at 20 and 30 kV.

Results

The choanocyte chambers of *Ceratoporella* are embedded in a thick layer of mesohyl containing few sponge cells but large numbers of symbiotic bacteria (Figure 1a,b). The chambers are small, (mean diameter 20.7 ± 1.8 μm, n = 50) and contain choanocytes on their inhalant side only. The exhalant half of the chamber is lined by endopinacocytes (Figures 1a,b). Each chamber is connected with the incurrent aquiferous canals by a fine canal, the prosodus, which leads to the side of the chamber lined by choanocytes (Figure 1a,b). The prosopyle appears to be a pore through an endopinacocyte. Water leaves the choanocyte chambers by way of a fine canal, the aphodus, which arises from the endopinacocyte-lined part of the choanocyte chamber and leads to an exhalant canal (Figure 1a). A pinacocytic apopyle, consisting of a single porocyte, allows water to enter the aphodus (see Figure 3a). A remarkable feature of the aquiferous canals of *Ceratoporella* is the presence of pseudopodial processes of the endopinacocytes lining them; these processes span the canals and form valvules (Figure 1a,b) that probably play a role in regulating water flow through the sponge.

The base of the choanocyte of *Ceratoporella* (5.6 ± 1.6 μm across, n = 20,) is greatly flattened, as is the nucleus itself (Figure 1c). Adjacent choanocytes are held together by lateral interdigitations. Choanocytes contain numerous phagosomes, mitochondria, and electron-dense spherical inclusions (Figure 1c).

The living tissue of *Goreauiella auriculata* is not organized into calicles, as is that of *Ceratoporella*, but instead fills the spaces between the arborescent processes of the aragonitic skeleton. In sections, the cells are loosely arranged; the mesohyl is not dense and lacks symbiotic bacteria. The choanocyte chambers are larger than in *Ceratoporella*

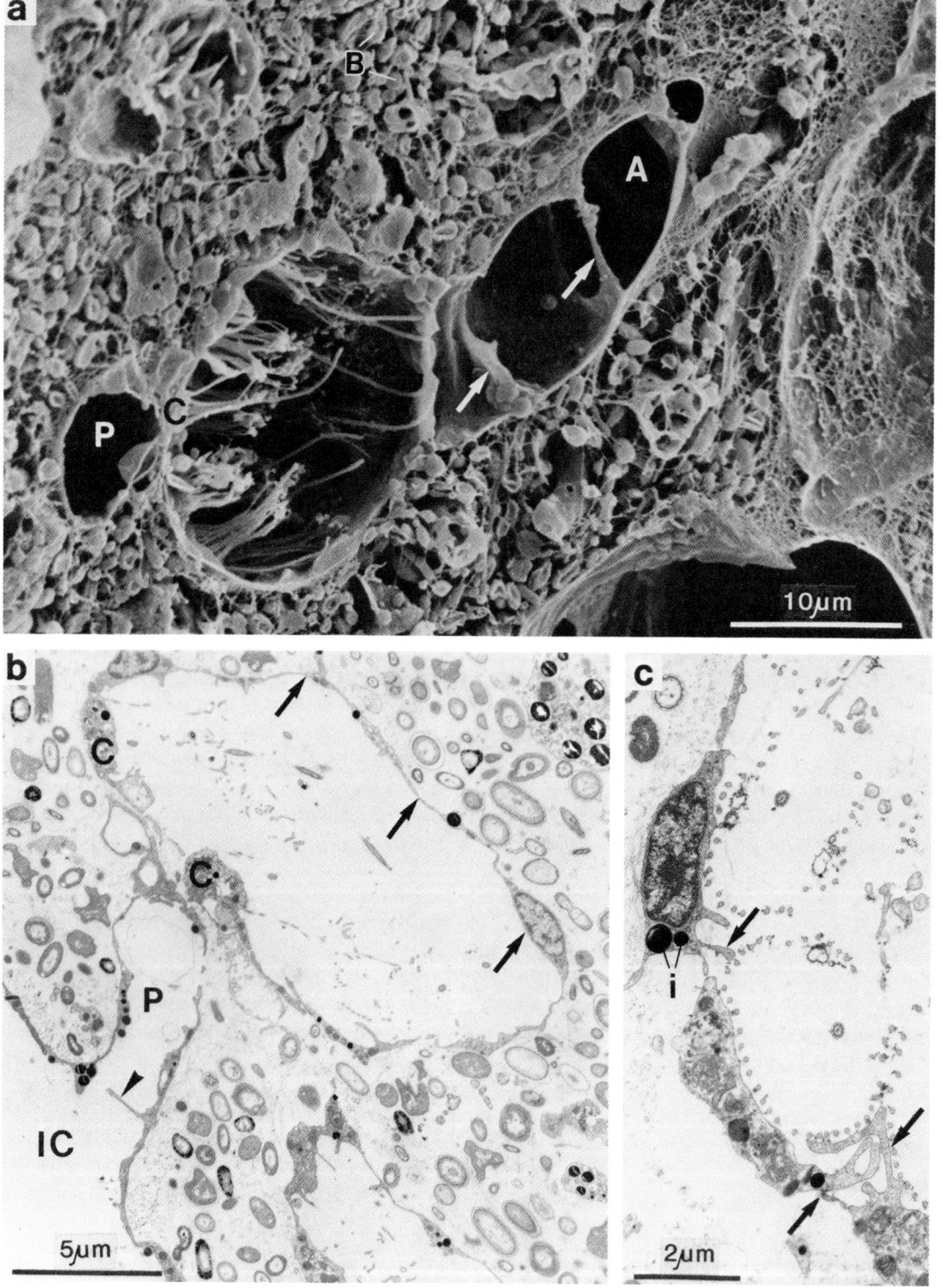

Figure 1. *Ceratoporella nicholsoni*: *a*, choanocyte chamber (SEM) connected with aquiferous system by a prosodus and an aphodus; arrows indicate pseudopodial processes of endopinacocytes of aquiferous system forming valvules; *b*, section through a choanocyte chamber (TEM) showing its connection with a prosodus; arrows indicate endopinacocytes lining a part of chamber, arrowhead indicates a pseudopodial process of an endopinacocyte forming a valvule; *c*, choanocytes (TEM) showing their flattened shape and unusual electron dense inclusions; arrows indicate lateral interdigitations connecting adjacent choanocytes. *A*, aphodus; *B*, symbiotic bacteria; *C*, choanocytes; *i*, electron-dense inclusions; *IC*, inhalant canal; *P*, prosodus.

(mean diameter 30.3 ± 2.0 μm, n = 50). They are formed by flattened choanocytes (mean diameter 8.3 ± 1.7 μm, n = 17) which are provided with flaring collars (Figure 2a,c). More than half of the chamber facing the exhalant canal system is lined by endopinacocytes (Figure 2a,b). Prosodi and aphodi are lacking in this species; the chambers are eurypylous. An apopyle consisting of a single porocyte opens into the exhalant canal system (Figure 3b). The choanocytes contain phagosomes, sparse mitochondria, and numerous electron-dense inclusions and are interconnected by thin glycocalyx bridges (Figure 2c).

In *Calcifibrospongia actinostromarioides*, the living tissue surrounds the reticulate aragonitic skeleton to a depth of 15–25 mm. The density of sponge cells in the mesohyl is greater than in *Ceratoporella*; as in *Ceratoporella*, a large population of extracellular symbiotic bacteria is also present (Figures 4a,b). The choanocyte chambers (mean diameter 28.8 ± 1.8 μm, n = 50) lie across the lumen of the inner ends of the inhalant canals (Figures 4a,b) and open into exhalant canals of lesser diameter by way of a pinacocytic apopyle. The choanocytes are therefore not in contact with the mesohyl. Two, three, or more endopinacocytes lie in contact with the choanocytes on the inhalant side of the chamber (Figure 4b). Pseudopodial processes run from the endopinacocytes on the basal sides of the choanocytes to the wall of the inhalant canal, thus helping to suspend the chamber in the canal. Pores apparent in some of these endopinacocytes may allow water from the inhalant canal to pass to the prosopyles of the choanocyte chambers. The choanocytes of *Calcifibrospongia* are small (mean diameter 4.2 ± 0.7 μm, n = 20). They are truncated on the upper surface (from which the flagellum arises) and bear phagosomes, mitochondria, Golgi complexes, and an abundance of small vacuoles (Figure 4c). A well-developed glycocalyx occurs at the surface of the choanocyte cell membrane inside the collar and extends between the microvilli of adjacent choanocytes (Figure 4c). The chambers open into an exhalant

canal by way of a pinacocytic apopyle apparently consisting of a single porocyte. Two or more central cells occur in the lumen of each choanocyte chamber (Figure 4b). These cells have pseudopodia that enclose choanocyte flagella and are generally located near the apoplye.

Discussion

The diversity of the micromorphology of the choanocyte chambers of the three sclerosponges described here (see Table 1 for summary) is greater than might be expected from what is known about the structure of the skeleton of these sponges (Hartman, 1969, 1979; Hartman and Goreau, 1972). This is especially true in comparing *Ceratoporella nicholsoni* and *Goreauiella auriculata*, which were initially placed in the same family, the Ceratoporellidae (Hartman and Goreau, 1972). The characteristic calicular configuration of the calcareous skeleton of *Ceratoporella* is related to the calcareous skeleton of *Goreauiella* by way of intermediate conditions induced by proximity to a scleractinian coral (Hartman and Goreau, 1972). The siliceous spicules of *Goreauiella* can be interpreted as truncated derivatives of those of *Ceratoporella*. The choanocyte chamber in the two animals are quite different, however. Those of *Ceratoporella* are small (20.7 μm in mean diameter), subspherical, and diplodal, whereas those of *Goreauiella* are larger (30.3 μm in mean diameter), of a more flattened shape (height much less than diameter), and eurypylous. Further, the mesohyl of *Ceratoporella* is packed with bacteria, but these are lacking in *Goreauiella*.

In the three orders of demosponges that have an exclusively fibrous intrinsic skeleton and for which the configurations of the choanocyte chambers are well known, these features tend to characterize family groups (Bergquist, 1980b). Thus, small, rounded diplodal chambers are found in the Spongiidae, Thorectidae, Aplysinidae, and Aplysinellidae, whereas large, sack-shaped to elongate and branched chambers occur in the Aplysillidae,

Table 1. Summary of microstructural characteristics in three sclerosponges

Characters	*Ceratoporella*	*Goreauiella*	*Calcifibrospongia*
Mesohyl	Thick; dense; symbiotic bacteria	Thin; no symbiotic bacteria	Thick; many canals; symbiotic bacteria
Shape of choanocytes	Flattened	Flattened	Hemispherical; truncated
Structural arrangements of chambers	Diplodal	Eurypylous	Intracanalicular with adherent cells
Apopyle	Pinacocytic	Pinacocytic	Pinacocytic
Prosopyle	Pinacocytic	Pinacocytic	Choanocytic + flattened cells
Central cells	No	No	1 or 2
Aquiferous canals	With valvules	Smooth	Smooth
Electron dense inclusions	In all cells	In all cells	None

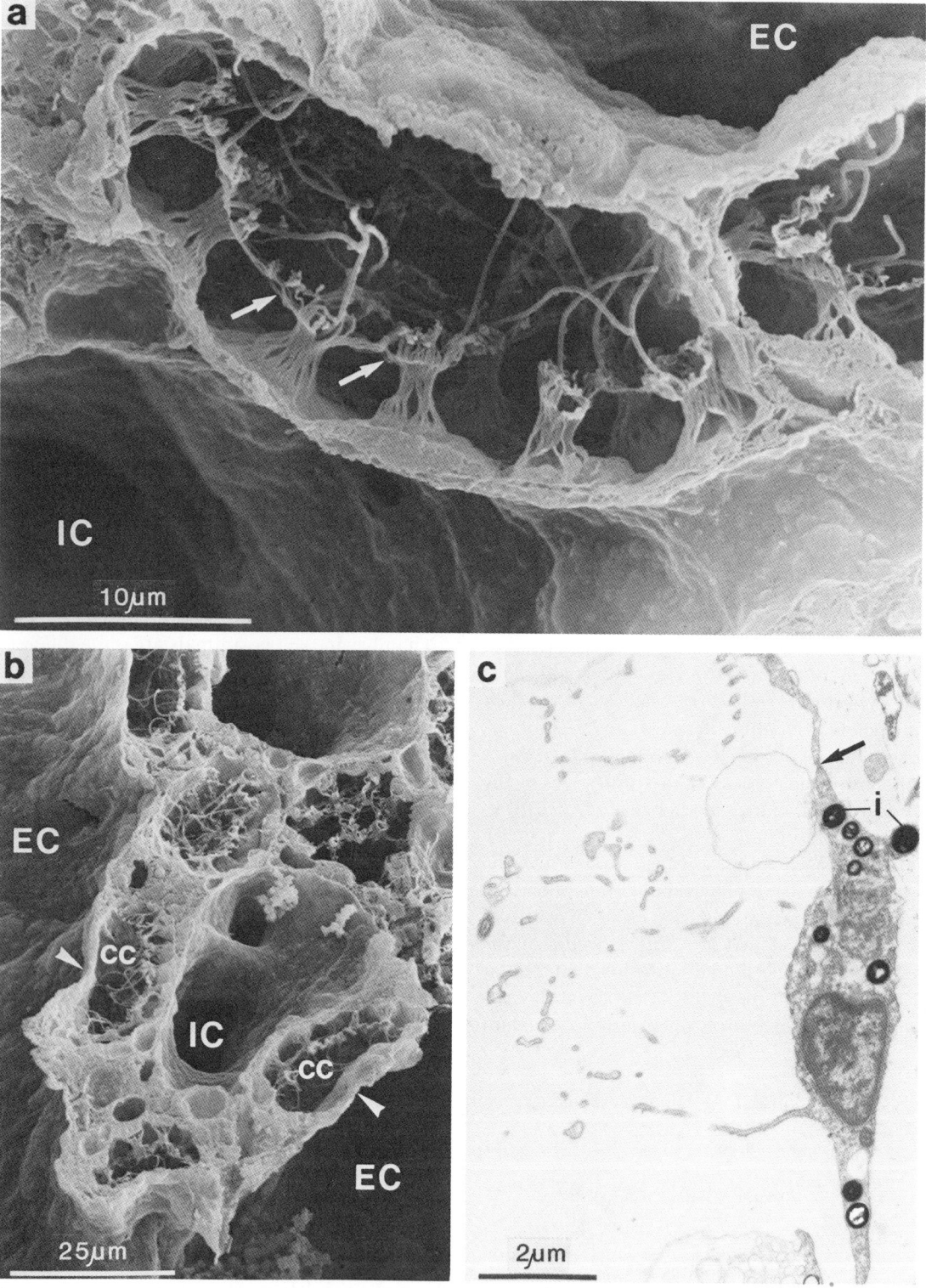

Figure 2. *Goreauiella auriculata: a,* SEM view of a cryofractured choanocyte chamber; note lack of symbiotic bacteria; flattened choanocytes have a flaring collar *(arrow); b,* SEM view of cryofractured specimen showing thinness of mesohyl; several eurypylous choanocyte chambers surround an inhalant canal; side of chambers facing exhalant canals is lined by endopinacocytes *(arrowheads); c,* choanocyte (TEM) showing its flattened shape and electron dense inclusions; arrow indicates a glycocalyx bridge interconnecting adjacent cells. *cc,* choanocyte chambers; *EC,* exhalant canals; *i,* electron dense inclusions; *IC,* inhalant canal.

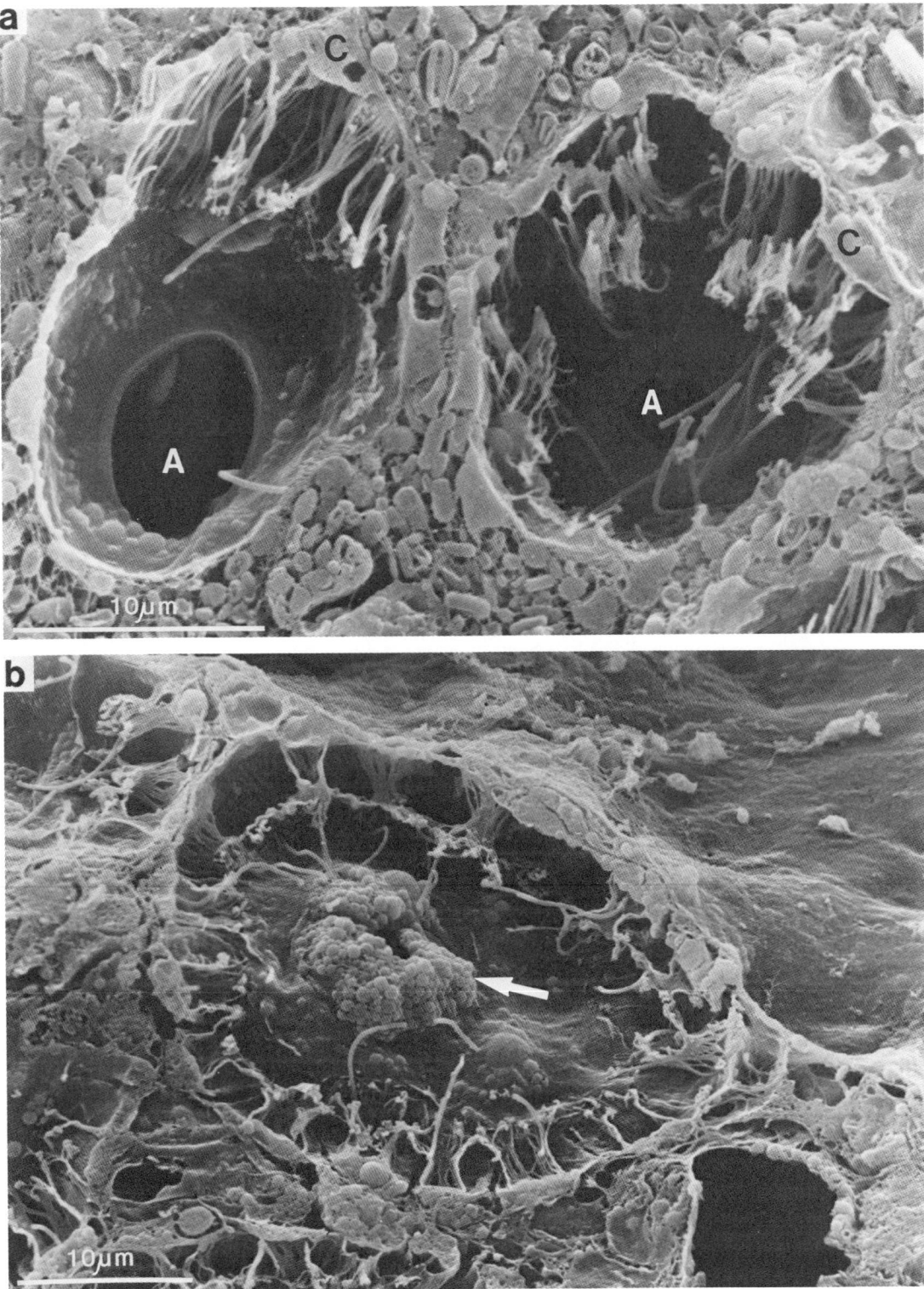

Figure 3. Fractured choanocyte chambers of sclerosponges (SEM): *a, Ceratoporella nicholsoni* chambers viewed toward aphodus; porocytes forming apopyles show degrees of openness; *b, Goreauiella auriculata,* chamber with highly contracted apopyle; arrow indicates an accumulation of dense inclusions around apopyle. *A*, apopyle; *C*, choanocyte.

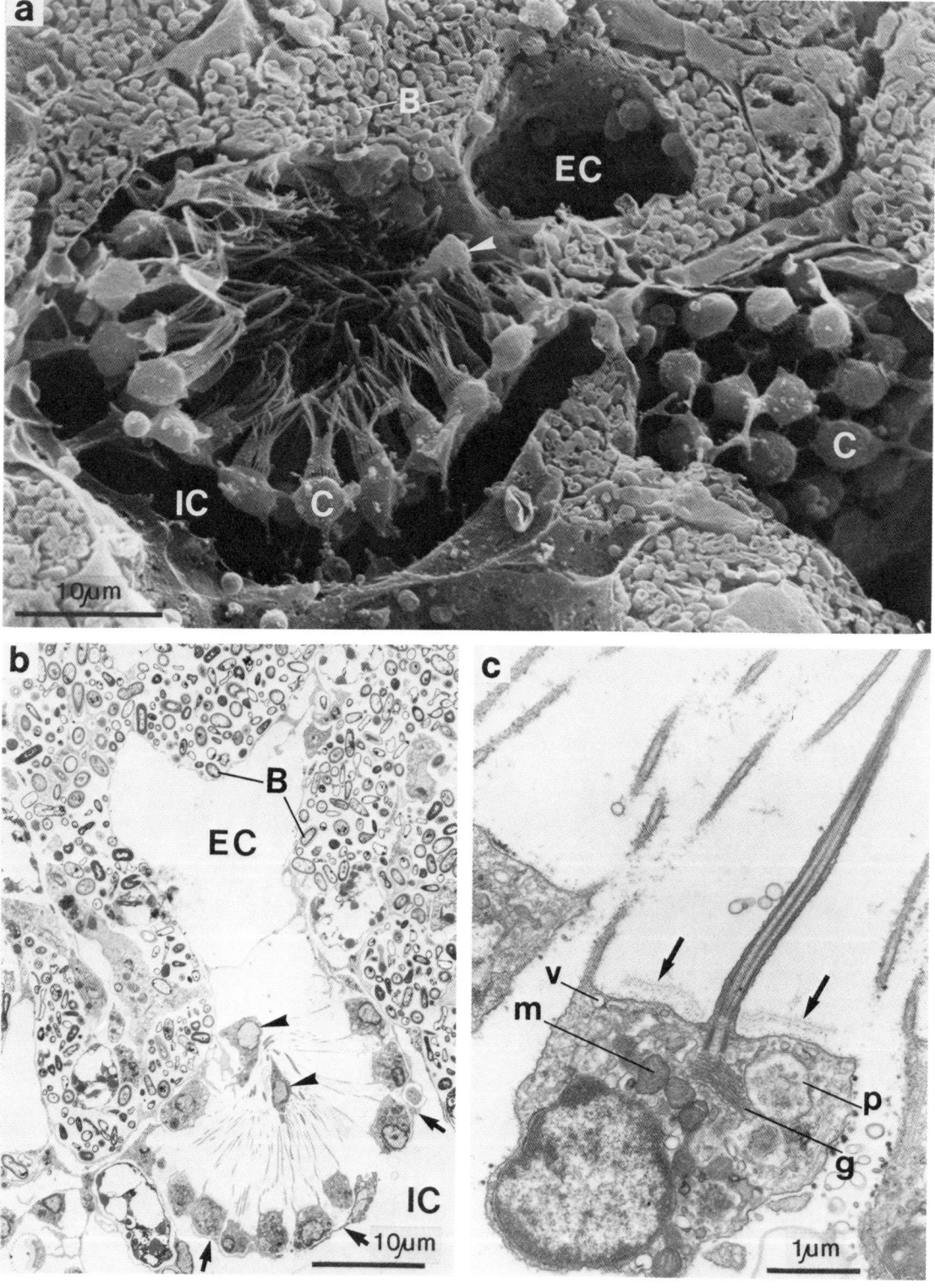

Figure 4. *Calcifibrospongia actinostromarioides*: *a,* choanocyte chamber (SEM) lying across lumen of a terminal inhalant canal and opening into an exhalant canal; arrowhead indicates a fragment of a central cell; *b,* similar view in TEM; note flattened cells adherent to inhalant side of chamber *(arrows)* and presence of central cells *(arrowheads);* *c,* choanocyte (TEM) showing its truncated shape; arrows indicate well-developed glycocalyx; note Golgi complex, abundance of mitochondria, large phagosome, and abundance of small vacuoles. *B,* symbiotic bacteria; *C,* choanocytes; *EC,* exhalant canal; *g,* Golgi complex; *IC,* inhalant canal; *m,* mitochondria; *p,* phagosome; *v,* vacuole.

Halisarcidae, and Ianthellidae. Among fibrous skeleton-bearing demosponges, only in the case of the Dendroceratida are the features of the choanocytes chambers of ordinal significance, according to the classification of Bergquist.

The strikingly different choanosomal features of *Ceratoporella* and *Goreauiella* suggest that these two sponges might belong to different family groups. However, additional characters are needed to confirm that supposition.

Calcifibrospongia actinostromarioides, with a reticulate calcareous skeleton enclosing thick tracts of siliceous strongyles, superficially resembles demosponges of the Order Petrosiida.[1] In section, the skeleton resembles certain Mesozoic stromatoporoids, and this led Hartman (1979) to include *Calcifibrospongia* in this large and mostly extinct order or subclass. However, the choanocyte chambers show a striking resemblance to the chambers of some species of the orders Haplosclerida and Petrosiida. The choanocyte chambers of *Calcifibrospongia* hang free within the terminus of the distal inhalant canals and open by way of pinacocytic apopyles into the exhalant system of canals. This configuration is shared by *Reniera* sp., a haploscleridan (Langenbruch, 1983), and *Petrosia ficiformis*, a petrosiidan (Langenbruch et al., 1985).

The three species in question differ in some respects. First, a cell ring stretches between the choanocyte layer and the pinacocytes of the exhalant canal wall in *Reniera*; in *Petrosia*, a ring of flagellum-bearing cone cells occupies the position of the cell ring; neither cone cells nor cell rings are present in *Calcifibrospongia*, according to our observations. Second, the choanocyte chambers of *Reniera* and *Petrosia* hang free in the inhalant canals, but those of *Petrosia* are covered basally by the pinacocyte epithelium of the inhalant canal system. The chambers of *Calcifibrospongia* are intermediate between these two conditions in that two to several isolated pinacocytes lie on the basal surface of the choanocytes, but there is no continuous pinacoderm separating the inhalant canal from the chamber itself.

Thus, *Calcifibrospongia* strongly resembles the genus *Petrosia*, both in regard to choanosomal and skeletal characteristics. Both also share an abundance of bacteria, but these are extracellular in the former genus and intracellular in the latter (Vacelet and Donadey, 1977; Langenbruch et al., 1985). The presence of central cells comparable in form to Type D of Reiswig and Brown (1977) and an aragonitic skeleton differentiate *Calcifibrospongia* most strongly. We have no information yet as to whether this sponge has sterols with a cyclopropene ring in the side chain and an oviparous developmental pattern (Bergquist, 1980a), as do genera of the order Petrosiida.

Although there are some suggestive similarities between *Calcifibrospongia* and the Petrosiida, we await the accumulation of additional data before drawing phylogenetic conclusions. Since studies of the micromorphology of the choanosome of sponges are only beginning, we do not yet know whether the peculiar spatial relationship between the choanocyte chambers and the distal inhalant canals found in *Reniera*, *Petrosia*, and *Calcifibrospongia* occurs widely or whether it is restricted to a group of related genera. Interestingly, the chambers in *Callyspongia* (until now regarded as a haplosclerid) do not hang freely in individual inhalant canals (Johnston and Hildemann, 1982), as they do in *Reniera*. It is quite probable that additional diversity will be revealed when we learn more about the ultrastructure of the choanosome of other groups of sponges.

Acknowledgments

We are indebted to Mark Mooseker, Department of Biology, Yale University, for permitting us to use the Zeiss EM 10 transmission electron microscope in his laboratory. We extend our gratitude to Louis De Vos and Paulette Van Gansen, Université Libre de Bruxelles, Laboratoire de Biologie animale et cellulaire and Laboratoire de Cytologie et d'Embryologie moléculaires, for their cordial reception in the scanning microscope facilities under their care. We thank Robin Hall for his assistance in SCUBA diving and express our gratitude to all members of the Discovery Bay Marine laboratory for optimal diving and laboratory conditions during our stays in Jamaica. Samples of *Calcifibrospongia* were collected during a cruise on the ORV *Cape Florida* to Acklins Island. We thank Rita Colwell and Deborah Santavy for their invitation to join this cruise, and we thank all scientists of the crew for SCUBA diving assistance. Britt Wheeler did the word processing for this typescript. This work was supported by NSF grant BSR-8317690 to Yale University and both a NATO science fellowship and a subvention of the Foundation AGATHON DE POTTER to one of us (PW). This is contribution no. 362 from the Discovery Bay Marine Laboratory, University of the West Indies, Discovery Bay, Jamaica, W.I.

Literature Cited

Bergquist, P. R. 1980a. The Ordinal and Subclass Classification of the Demospongiae (Porifera); Appraisal of the Present Arrangement, and Proposal of a New Order. *New Zealand Journal of Zoology*, 7:1–6.

———. 1980b. A Revision of the Supraspecific Classification of the Orders Dictyoceratida, Dendroceratida, and Verongida (Class Demospongiae). *New Zealand Journal of Zoology*, 7:443–503.

---

[1]This name was suggested (Hartman, 1982) to replace the order Nepheliospongiida (Bergquist, 1980a) since the fossil family Nepheliospongiidae is not especially close to the Recent petrosiids.

Boury-Esnault, N., L. De Vos, C. Donadey, and J. Vacelet. 1984. Comparative Study of the Choanosome of Porifera: 1. The Homoscleromorpha. *Journal of Morphology*, 180:3–17.

Eisenman, E. A., and M. Alfert. 1981. A New Fixation Procedure for Preserving the Ultrastructure of Marine Invertebrate Tissues. *Journal of Microscopy*, 125:117–120.

Fullmer, H. M. 1966. Histochemical Studies of Mineralized Tissues. *Annales d'Histochimie*, 11:369–374.

Hartman, W. D. 1969. New Genera and Species of Coralline Sponges (Porifera) from Jamaica. *Postilla*, 137:1–39.

———. 1979. A New Sclerosponge from the Bahamas and Its Relationship to Mesozoic Stromatoporoids. Pages 467–474 in *Biologie des Spongiaires*, edited by C. Lévi and N. Boury-Esnault. Colloques Internationaux du C.N.R.S. 291. Paris: Centre National de la Recherche Sciéntifique.

———. 1982. Porifera. Pages 640–666 in *Synopsis and Classification of Living Organisms*, 1, edited by S. P. Parker. New York: McGraw-Hill.

Hartman, W. D., and T. F. Goreau. 1970. Jamaican Coralline Sponges: Their Morphology, Ecology and Fossil Relatives. Pages 205–243 in *The Biology of the Porifera*, edited by W. G. Fry. Symposia of the Zoological Society of London 25. London: Academic Press.

———. 1972. *Ceratoporella* (Porifera: Sclerospongiae) and the Chaetetid "Corals." *Transactions of the Connecticut Academy of Arts and Sciences*, 44:133–148.

Johnston, I. S., and W. H. Hildemann. 1982. Cellular Organization in the Marine Demosponge *Callyspongia diffusa*. *Marine Biology*, 67:1–7.

Langenbruch, P.-F. 1983. Body Structure of Marine Sponges. 1. Arrangement of the Flagellated Chambers in the Canal System of *Reniera* sp. *Marine Biology*, 75:319–325.

Langenbruch, P.-F., T. L. Simpson, and L. Scalera-Liaci. 1985. Body Structure of Marine Sponges. III. The Structure of Choanocyte Chambers in *Petrosia ficiformis* (Porifera, Demospongiae). *Zoomorphology*, 105:383–387.

Lister, J. J. 1900. *Astrosclera willeyana*, the Type of a New Family of Sponges. Pages 459–482 in *Zoological Results, Based on Material Collected by A. Willey, Part 4*. Cambridge (England): University Press.

Luft, J. H. 1971a. Ruthenium Red and Violet. I. Chemistry, Purification, Methods of Use for Electron Microscopy and Mechanism of Action. *Anatomical Record*, 171:347–368.

———. 1971b. Ruthenium Red and Violet. II. Fine Structural Localization in Animal Tissues. *Anatomical Record*, 171:369–416.

Reiswig, H. M., and M. J. Brown. 1977. The Central Cells of Sponges: Their Distribution, Form and Function. *Zoomorphologie*, 88:81–94.

Reynolds, E. S. 1963. The Use of Lead Citrate at High pH as an Electron Opaque Stain in Electron Microscopy. *Journal of Cell Biology*, 17:208–212.

Spurr, A. R. 1969. A Low-Viscosity Resin Embedding Medium for Electron Microscopy. *Journal of Ultrastructural Research*, 26:31–43.

Vacelet, J. 1981. Eponges hypercalcifiées ("Pharétronides," "Sclérosponges") des cavités des récifs coralliens de Nouvelle-Calédonie. *Bulletin du Muséum National d'Histoire Naturelle*, 4ᵉ Série, 3:313–351.

Vacelet, J., and C. Donadey. 1976. Electron Microscope Study of the Association between some Sponges and Bacteria. *Journal of Experimental Marine Biology and Ecology*, 30:301–314.

NICOLE BOURY-ESNAULT
Centre d'Océanologie de Marseille
Station Marine d'Endoume
13007 Marseille, France

LOUIS DE VOS
Laboratoire de Biologie Animale et Cellulaire
Université Libre de Bruxelles
50 av. F. D. Roosevelt
1050 Bruxelles, Belgium

CLAUDE DONADEY
CERAM, Faculté des Sciences de St Jérôme
13397 Marseille Cedex 13, France

JEAN VACELET
Centre d'Océanologie de Marseille
Station Marine d'Endoume
13007 Marseille, France

Ultrastructure of Choanosome and Sponge Classification

Abstract

SEM and TEM comparisons of the anatomy of the aquiferous system and the ultrastructure of its cell components (choanocytes, endopinacocytes) reveal that the sclerosponges have close affinities with different orders of Demosponges and that the two heterogenous orders Dictyoceratida and Dendroceratida fall into a new classification.

The main criteria used to test the relevance of these cytological characters for systematic discrimination were volume and shape of choanocyte chambers, shape of choanocytes, number of choanocytes per chamber, presence of flagellum ornamentation, presence of a periflagellar sleeve, number of microvilli per collar, and presence of a flagellum on apopinacocytes.

According to the definition proposed by Bergquist (1978), sponges are sedentary filter feeding animals that utilize a single layer of flagellated elements—choanocytes within Calcarea and Demospongea, and collar bodies within Hexactinellida—to pump a unidirectional water current through their body. Classically, these choanocytes are described as cells with a characteristic collar of microvilli surrounding an apical flagellum. Among the early workers, Minchin (1896) stressed that calcareous homocoel sponges have two types of choanocytes, apinucleate and basinucleate. Bidder (1898) subsequently used these criteria to divide the Calcarea into Calcaronea and Calcinea. Next, Dendy and Row (1913) hypothesized that several distinct types of collared cells exist in the noncalcareous sponges. But the revolutionary suggestions of these authors did not gain attention for some time. In fact, it took more than 60 years for Bidder's classification to become widely accepted by of the spongological community. Even up to relatively recent times, the literature contained only a few works on the choanosome and choanocytes of noncalcareous sponges.

Ever since its ultrastructure was first described (Rasmont et al., 1957), the choanocyte has been considered a relatively homogenous cell type. As Lévi (1956) pointed out, the skeleton criteria emphasized by Schmidt and Bowerbank in the late 1800s still remain the only ones commonly used in taxonomy in spite of progress in biochemistry and cytology amd the introduction of scanning and transmission electron microscopy (SEM and TEM). Furthermore, sponges are the last phylum in which a consensus has not yet been reached at the ordinal and family level.

In a previous work (Boury-Esnault et al., 1984) we showed that in a homogenous group, the Homoscleromorpha, the general morphology of choanocytes, choanocyte chambers, and canals are homogenous, although it is possible to observe slight differences among species. To test the hypothesis that differences in the morphology of choanocytes, choanocyte chambers and canals, are phylogenetically significant, we had earlier undertaken a comparative study of the representatives of the entire phylum using TEM and SEM techniques (De Vos et al., 1981). In the present study, we set out to demonstrate the usefulness of such cytological criteria for systematics, using calcified sponges and Dictyoceratida and Dendroceratida as examples.

Material and Methods

All the species studied (see figure captions) were fixed in situ and observed both by TEM and SEM. For TEM studies, we used the fixative C described in our previous work (Boury-Esnault et al., 1984). For SEM studies, we used a fixative consisting of osmium tetroxide 2% and a saturated solution of mercuric chloride in a ratio of 6:1 (Johnston and Hildeman, 1982). Whenever necessary, specimens were desilicified in a solution of hydrofluoric acid 5%. The specimens were dehydrated in graded ethanol, and were fractured in liquid nitrogen when they were in ethanol 100%. Tissue fragments were critical point dried using liquid CO_2 as the transition medium, then were sputter-coated with gold. The specimens were observed with an ISI DS 130 scanning microscope at the Laboratoire de Biologie Animale et Cellulaire de l'Université Libre de Bruxelles.

Results
GENERAL CRITERIA

DIFFERENCES IN CHOANOCYTE CHAMBER SIZE. In order to make valid comparisons between the species, we paid particular attention to repeating the same fixation procedure. We eliminated from our estimation of size specimens in which the lumina of the choanocyte chambers and of the canals were in a contracted state (see Boury-Esnault et al., 1984). Still, the results presented here must be interpreted with caution as artifacts may nevertheless occur and the size of choanocyte chambers is difficult to measure accurately on sectioned or fractured specimens.

However, a relatively good concordance was found between measurements made either on SEM and TEM photographs or on semithin sections in light microscopy.

In demosponges, the volume of chambers within different species varies between 350 and 480,000 μm^3. Tetractinellida (Figure 1) and Agelasida (Figure 2) have the smallest chamber volume (350–500 μm^3), while Dysideidae (Figure 3), Aplysillidae (Figure 4), and Halisarcidae (Figure 5) have the largest chamber volume (40,000–480,000 μm^3). In most cases these chambers are spherical (Figures 6,7,8), while in Halisarcidae they are cylindrical (Figure 5).

It seems that the size of the chambers is correlated with the density of the mesohyl: when the mesohyl is filled with bacteria the chambers are small; and when the mesohyl has few bacteria, the chambers are large.

NUMBER OF CHOANOCYTES PER CHAMBER. This number was estimated by the indirect method used by Rasmont and Rozenfeld (1981). Tetractinellida and Agelasida contain 5–60 choanocytes per chamber; Dysideidae 800; Aplysillidae 1,100; and Halisarcidae 2,800.

SHAPE OF CHOANOCYTE CELL BODIES. As for the volume of the chambers, the conditions of fixation may affect the shape and size of choanocytes. An example of such artifacts is the hourglass-shaped choanocyte (Figure 9). Nevertheless, we were able to define two general shapes of choanocytes in Demospongea, cylindrical and flattened.

The cylindrical shape can be seen in *Dysidea* (Figure 10). Choanocytes lie on a collagenous layer and the junction between adjacent cells involves either the entire side wall or a large part of it. This type has been found in *Corticium*, *Oscarella*, *Ephydatia*, *Chelonaplysilla* (Figure 11), *Halisarca* (Figure 12), and *Axinella*, among others.

In the second type, the choanocyte cell bodies are flattened and in many cases penetrate into the mesohyl while lateral cytoplasmic extensions rest on the surface of the mesohyl. The junction between two neighboring cells occurs only at the tip of these extensions. We have found this form of choanocytes in *Geodia* (Figure 1), *Cinachyra*, *Agelas* (Figure 2), *Ceratoporella* (Figure 13), *Aplysina*, *Spongia* (Figures 14, 15), *Cacospongia* (Figure 16), and also in *Suberites* and *Acanthochaetetes*.

ULTRASTRUCTURAL DIFFERENCES WITHIN CHOANOCYTES. The choanocytes differ not only in shape and size but also in minute details: the collar is composed of a somewhat variable number of microvilli. We counted 22–27 microvilli per collar in Homoscleromorpha. We found 35–40 microvilli in the majority of Demosponges, with the exception of the Axinellidae, where we counted 50–55 microvilli per collar (Figure 17).

A periflagellar sleeve described by Connes et al. (1971) in *Suberites massa* Nardo has been found in other

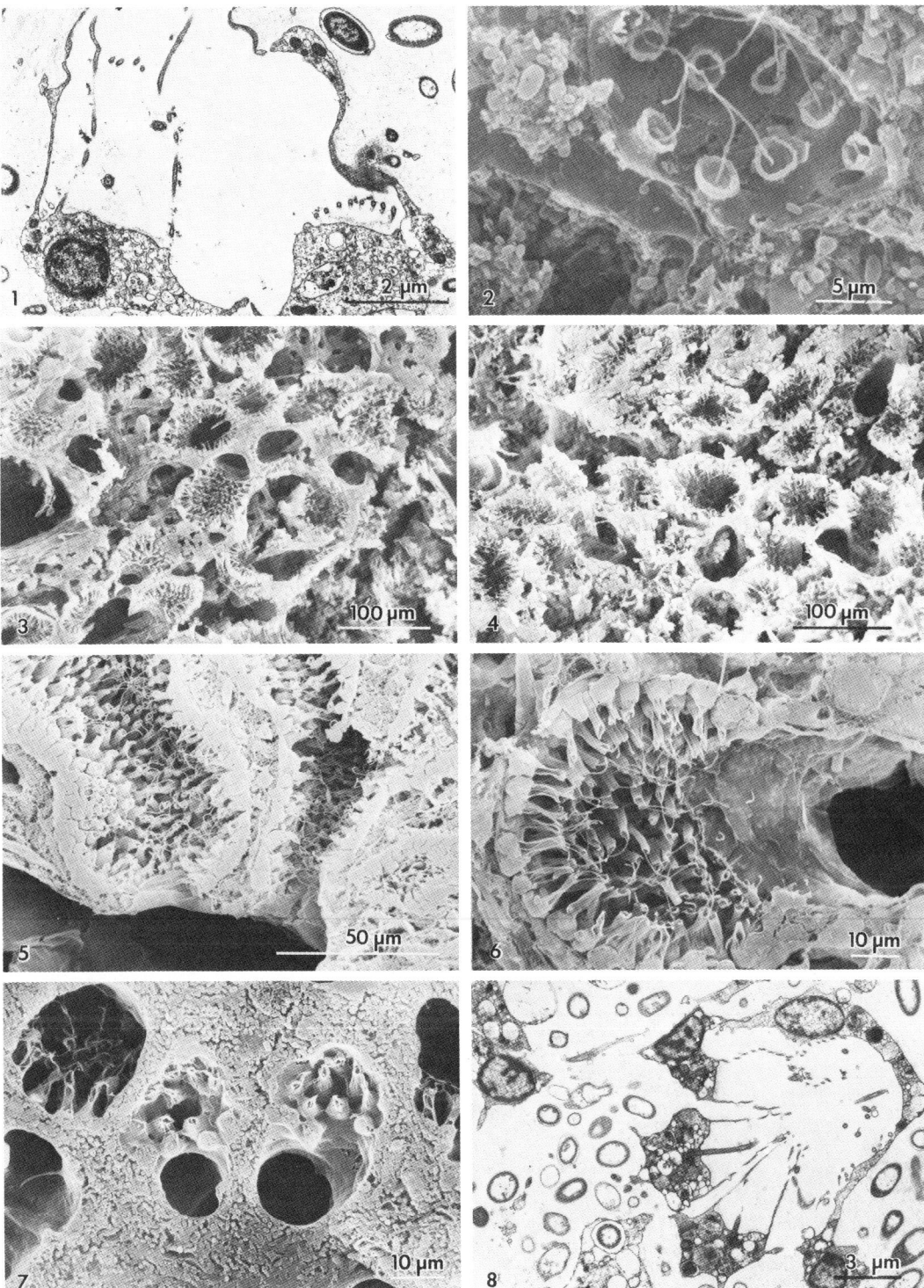

Figures 1–8. Views of choanosome and choanocyte chambers: *1*, choanocyte chamber of *Geodia cydonium* (Jameson), TEM (note the few numbers of choanocytes and their flattened shape); *2*, choanocyte chamber of *Agelas oroides* (Schmidt), SEM (note flattened shape of choanocyte and density of bacteria in mesohyl); *3*, choanosome of *Dysidea pallescens* (Schmidt), SEM; *4*, choanosome of *Chelonaplysilla noevus* (Carter), SEM; *5*, choanocyte chambers of *Halisarca dujardini* Johnson, SEM (note tubular branched form of these chambers); *6*, spherical chamber of *Dysidea pallescens*, SEM (note presence of prosopyles and a large apopyle); *7*, small spherical chambers of *Spongia nitens* (Schmidt), SEM (note high density of bacteria within mesohyl); *8*, choanocyte chamber of *Astroclera willeyana*, TEM (note density of bacteria within mesohyl).

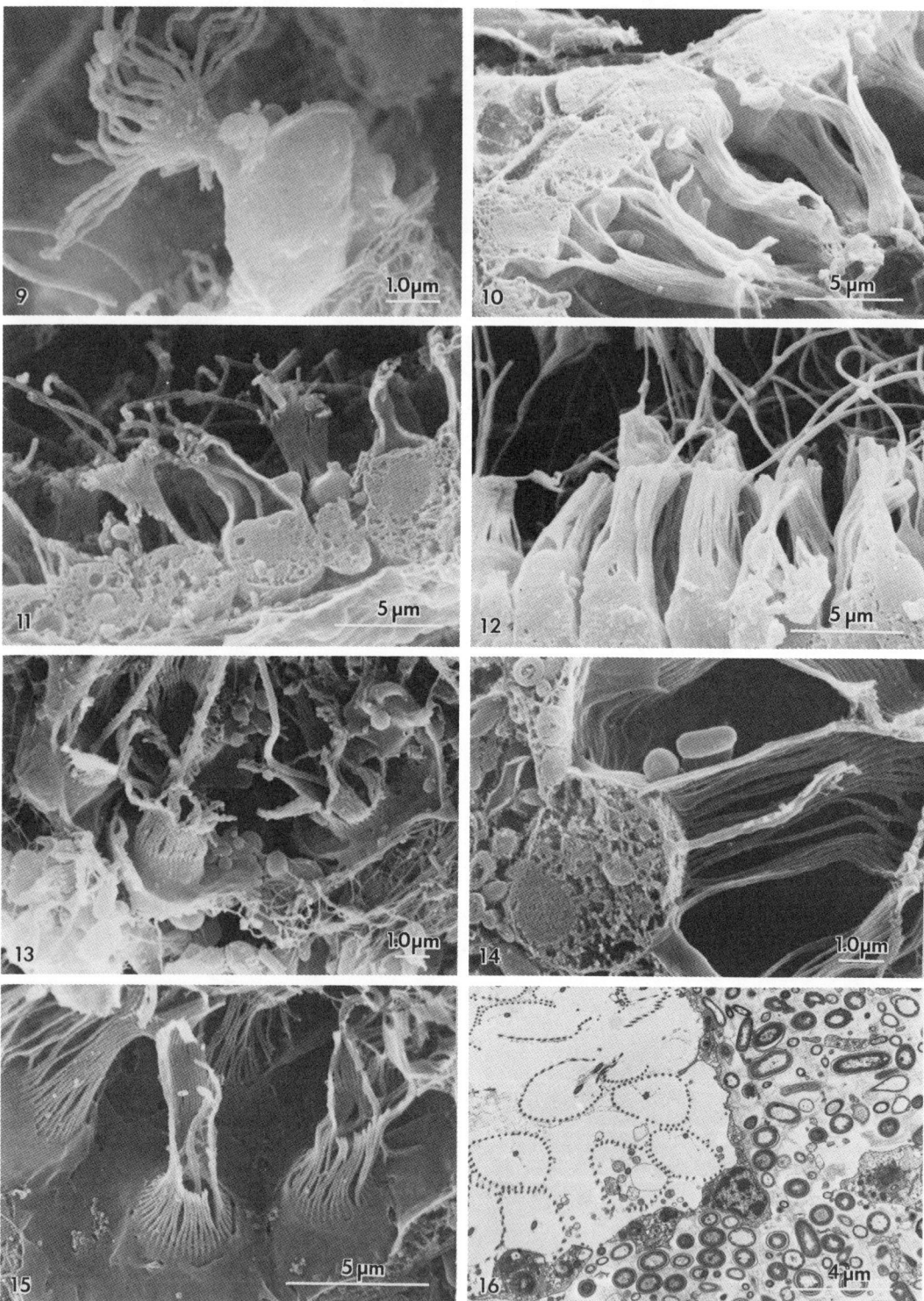

Figures 9–16. Views of choanocytes: *9*, hour-glass shaped choanocyte, SEM; *10*, cylindrical choanocytes of *Dysidea pallescens*, SEM; *11*, cylindrical choanocytes of *Chelonaplysilla noevus*, SEM; *12*, cylindrical choanocytes of *Halisarca dujardini*, SEM; *13*, flattened choanocytes of *Ceratoporella nicholsoni*, SEM (note bacteria in mesohyl); *14*, cross section of a choanocyte of *Spongia nitens*, SEM (note presence of glycocalyx at top of cell inside collar, lateral expansions of flagella, and small bridges of glycocalyx between microvilli); *15*, upper view of choanocytes of *Spongia nitens*, SEM (note well-developed lateral expansions of flagella, and presence of glycocalyx outside collar); *16*, choanocytes of *Cacospongia scalaris* (Schmidt), TEM (note flattened shape and glycocalyx between collar).

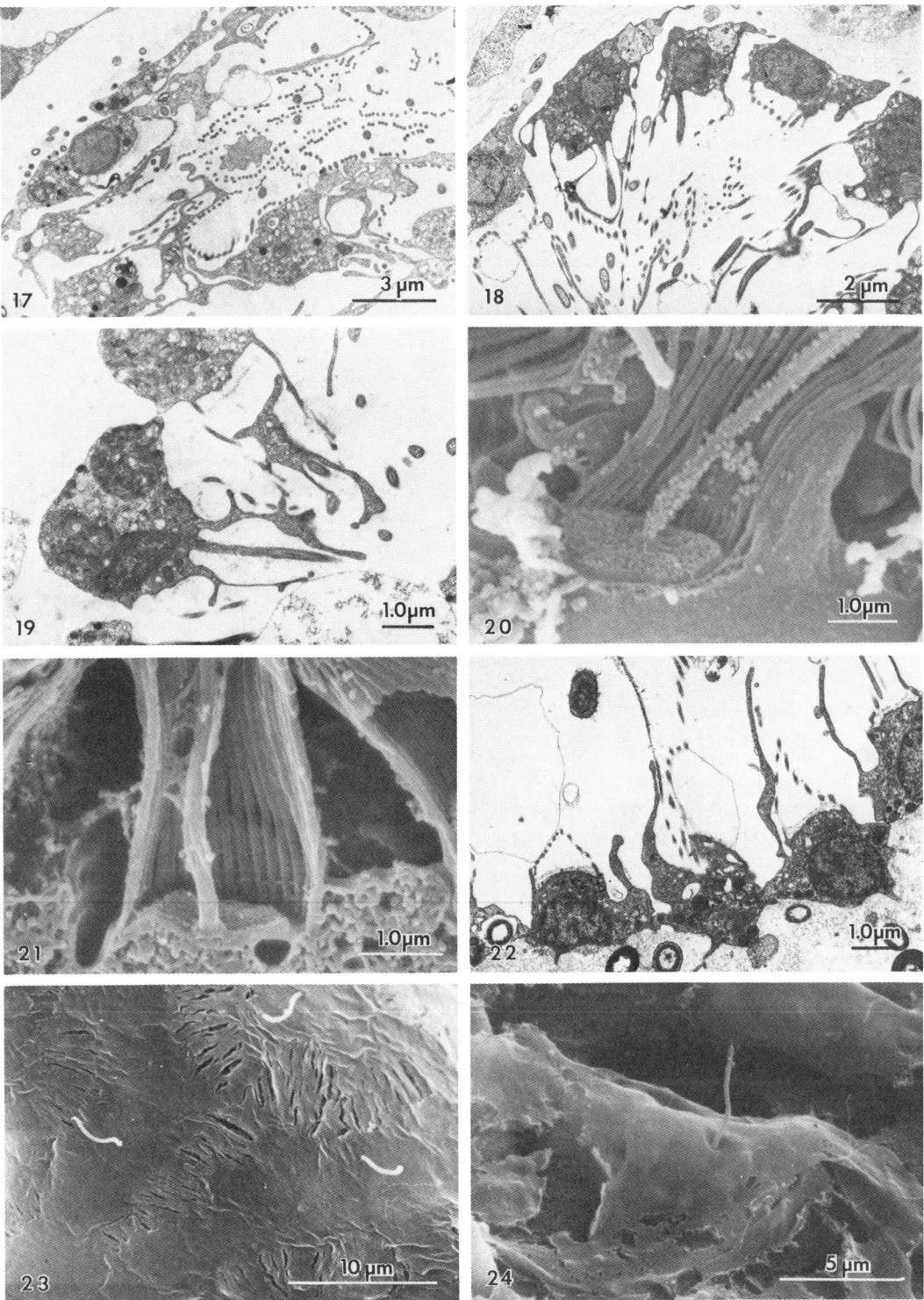

Figures 17–24. Views of choanoderm and pinacoderm: *17*, part of a choanocyte chamber of *Axinella polypoides* (Schmidt), TEM (note number of microvilli per collar); *18*, part of a choanocyte chamber of *Ficulina ficus*, TEM (note presence of a periflagellar sleeve); *19*, choanocyte of *Acanthochaetetes*, TEM (note periflagellar sleeve); *20*, details of a choanocyte of *Geodia cydonium*, SEM (note that top of cell and flagellum are covered by glycocalyx giving a roughened appearance); *21*, detail of a choanocyte of *Ephydatia fluviatilis* (L.), SEM (note lateral expansions of flagellum); *22*, choanocytes of *Aplysina aerophoba* (Schmidt), TEM (note presence of long apical pseudopods); *23*, flagellated apopinacoderm of *Spongia nitens*, SEM; *24*, flagellated apopinacocyte of *Spirastrella cunctatrix* Schmidt, SEM.

Hadromerida such as *Suberites domuncula* Olivi, *Ficulina ficus* Johnston (Figure 18), and a "Sclerosponge" such as *Acanthochaetetes* (Figure 19).

Details of flagellum ornamentation vary among species. The flagellum may be smooth *(Halisarca, Suberites, Acanthochaetetes)*, rough *(Polymastia, Geodia* [Figure 20], *Cinachyra)* and may have two lateral expansions (Dysideidae, Spongiidae, Spongillidae [Figure 21], Aplysillidae).

Willenz (1982) showed that phagocytosis in choanocytes involves apical or lateral pseudopods. We also found well-developed apical pseudopods in *Aplysina* (Figure 22) and in *Axinella,* where they are as long as the microvilli of the collars.

ENDOPINACOCYTES. In Homoscleromorpha, both prosendopinacocytes and apoendopinacocytes have a flagellum. In Ceractinomorpha, a more or less developed flagellum is present in Dictyoceratida (Figure 23) and Dendroceratida, with the exception of Halisarcidae. Flagella were not present in Tetractinomorpha, with the exception of *Spirastrella* (Figure 24), where a residual flagellum has been observed.

APOPYLAR CELLS AND CENTRAL CELLS. The presence of a particular cell type at the junction between the choanocyte chamber and the exhalant canal—that is, at the level of the apopyle—is reported by De Vos et al. (this volume).

RELEVANCE OF CYTOLOGICAL CHARACTERS TO SYSTEMATICS

CALCIFIED SPONGES. Of the calcified sponges ("sclerosponges") that are related to Demospongea (Vacelet, 1985), we studied four species: *Astrosclera willeyana* Lister, *Ceratoporella nicholsoni* (Hickson), *Acanthochaetetes wellsi* Hartman and Goreau, and *Vaceletia crypta* (Vacelet). The cytological characteristics we observed are summarized in Table 1.

A comparison of these results with those obtained from different orders of demosponges indicates that (1) *Astro-*sclera (Figure 8) and *Ceratoporella* (Figure 13) are close to Agelasida (Figure 2); (2) *Acanthochaetetes* (Figure 19) is close to Hadromerida; and (3) *Vaceletia* (Vacelet, 1979) is close to Ceractinomorpha (?Poecilosclerida).

KERATOSE SPONGES: DICTYOCERATIDA AND DENDROCERATIDA. The taxonomic criteria in keratose sponges are based on the reticulation of the spongin fibers. Spongiidae, Thorectidae, and Dysideidae, which have a reticulate skeleton, have been grouped together in the order Dictyoceratida, while the Aplysillidae, which have dendritic fibers, have been classified in the order Dendroceratida with Halisarcidae which have no skeleton. The cytological characteristics of these sponges are summarized in Table 2.

The organization of choanocyte chambers of Halisarcidae (Figure 5) resembles a syconoid structure (Lévi, 1956). The chambers are arranged radially around an exhalant canal. The tubular, digitate chambers may reach a volume of 480,000 μm^3, with 400–2,800 choanocytes per chamber. The cylindrical choanocytes have a smooth flagellum. The apopinacocytes are not flagellated.

In Aplysillidae the chamber organization is typically leuconoid and chamber volume is smaller (125,000–345,000 μm^3 with 380–1,100 choanocytes) (Figure 4). The chambers are saclike or tubular but never branched. The choanocytes are cylindrical and the flagellum has two lateral expansions of glycocalyx and the apopinacocytes are flagellated.

The organization of Dysideidae is very similar to that of Aplysillidae (Figure 3). The volume of the chambers varies between 44,000 and 325,000 μm^3 with 200–800 choanocytes per chamber (Figure 6). The choanocytes are cylindrical (Figure 10); the flagellum has lateral expansions and the apopinacocytes are flagellated. The mesohyl is poorly developed.

In the families Spongiidae and Thorectidae, the organization of chambers is also leuconoid (Figures 7, 16). The chambers are spheres that never exceed 40,000 μm^3. They have 25–200 choanocytes of the flattened type with the

Table 1. Characteristics of the choanosome of calcified sponges (C = cylindrical, F = flattened)

Species	Choanocyte chambers	Choanocytes				Mesohyl
	Volume (μm^3)	Number	Diameter[a]	Shape	Periflagellar sleeve	
Astrosclera	585–2,555	4–49	4.4–8.8	F	no	dense
Ceratoporella	585–1,553	5–17	6.0–9.1	F	no	dense
Acanthochaetetes	1,500–4,800	20–130	3.6–5.6	F	yes	loose
Vaceletia	1,400–2,800	12–91	3.2–6.0	C	no	medium

[a]basis of choanocytes in μm.

Table 2. Characteristics of the choanosome of Dictyoceratida and Dendroceratida (C = cylindrical, F = flattened, l = lateral glycocalyx expansions on the flagellum, s = smooth flagellum)

Family	Choanocyte chambers		Choanocytes				Mesohyl	Apopinacocyte
	Vol.×1000 (μm^3)	Shape	Number	Shape	Diameter[a]	Flagellum		
Spongiidae	12.3–37.8	sphere	37–200	F	4.5–10.0	l	dense	flagellated
Thorectidae	9.0–21.0	sphere	25–200	F	4–9	l	dense	flagellated
Dysideidae	44–325	sac–like	200–800	C	4.5–6.0	l	loose	flagellated
Aplysillidae	125–345	sac–like	380–1100	C	4.0–4.5	l	loose	flagellated
Halisarcidae	83–480	tubular, branched	400–2800	C	4–6	s	loose	not flagellated

[a]basis of choanocytes in μm.

cell body inside the mesohyl (Figures 14, 15, 16,). The mesohyl is densely filled with bacteria and the apopinacocytes are flagellated.

Discussion

The cytological criteria used here confirm that the so-called sclerosponges are Demospongea that are related to different orders (Vacelet, 1981, 1983, 1985). The arguments concerning skeletal features have been corroborated here by our new cytological criteria.

Hartman and Willenz (this volume) who used the same criteria on different calcified sponges confirm these affinities. In particular, Hartman and Willenz observed in *Calcifibrospongia* the same types of chambers and choanocytes that were described by Johnston and Hildeman (1982) in the haplosclerid *Callyspongia* and later by Langenbruch (1983) in the haplosclerid *Reniera*.

The so-called sclerosponges can thus be related to the different orders and subclasses of demosponges (Table 3).

Dendroceratida and Dictyoceratida are heterogenous orders: the organization of the choanosome as well as the ultrastructure of choanocytes and apopinacocytes are highly variable.

All the cytological criteria used here point to clear-cut differences between Halisarcidae and Aplysillidae. If the suggestions of Vacelet and Donadey (in press) are accepted, it seems that Halisarcidae must be rejected from Dendroceratida. For the time being, then, the family Halisarcidae should remain incertae sedis.

A comparison of Dysideidae with Aplysillidae on one hand and with Spongiidae and Thorectidae on the other hand shows clearly the close relationships between Dysideidae and Aplysillidae.

Bergquist et al. (this volume) come to the same conclusions after studying the terpenoid metabolites in Dendroceratida. Halisarcidae are far from Dendroceratida, while Dysideidae have a close relationship with Aplysillidae and must be put in the same order.

Heterogeneity showed up in both Dictyoceratida and Dendroceratida when cytological and chemical criteria were applied and thus cast doubt on the classification. We propose the following classification for the two orders:

Dictyoceratida: Spongiidae, Thorectidae
Dendroceratida: Dysideidae, Aplysillidae
Incertae sedis: Halisarcidae

Conclusions

Skeletal characteristics are without doubt important for the classification of sponges and they remain the most convenient way to identify many species in the field and in the laboratory. However, from a strictly systematic point of view, grouping a number of families into orders using these features alone remains problematic. We believe that the classification of sponges must be based on more than skeletal criteria. As illustrated in the present work, cytological, ultrastructural, and chemical criteria can help to clarify problems in sponge classification. However, the use of choanocyte characters in taxonomy is still hampered by the fact that a clear distinction has not yet been established between morphofunctional analogies and phylogenetic significance.

Acknowledgments

This work was supported by Fond de la Recherche Fondamentale Collective (FRFC) grant no. 2.4525.80 and funds from North Atlantic Treaty Organization (NATO) no. 85–0134. We aknowledge Chantal Bézac, Marie-Rose

Table 3. The relation of sclerosponges to orders and subclasses of demosponges

Family	Order	Subclass
Acanthochaetetidae	Hadromerida	Tetractinomorpha
Ceratoporellidae	Agelasida	Still in discussion
Astroscleridae	Agelasida	Still in discussion
Cryptocoelidae? (*Vaceletia*)	Poecilosclerida	Ceractinomorpha
Calcifibrospongiidae	Haplosclerida	Ceractinomorpha

Causi, Emile De Cock, and François Lambert for technical assistance.

Literature Cited

Bergquist P. R. 1978. *Sponges*. Berkeley and Los Angeles: University of California. 268 pp.

Bidder G. P. 1898. The Skeleton and Classification of Calcareous Sponges. *Proceedings, Royal Society of London*, 64:61–76.

Boury-Esnault, N., L. De Vos, C. Donadey, and J. Vacelet. 1984. Comparative Study of the Choanosome of Porifera. I. The Homoscleromorpha. *Journal of Morphology*, 180:3–17.

Connes R., J.-P. Diaz, and J. Paris. 1971. Choanocyte et cellule centrale chez la Démosponge *Suberites massa* Nardo. *Comptes-Rendus de l'Académie des Sciences, (Paris)*, 273(18):1590–1593.

Dendy, A., and R. W. H. Row. 1913. The Classification and Phylogeny of the Calcareous Sponges; with a Reference List of all Described Species, Systematically Arranged. *Proceedings, Zoological Society of London*, 1913:704–813.

De Vos, L., N. Boury-Esnault, J. Vacelet, and C. Donadey. 1981. Morphologie comparée des choanocytes des Spongiaires. Etudes aux microscopes électroniques à balayage et à transmission. Colloque, Société Française Microscopie électronique, Besançon. *Biology of the Cell*, 41:8a.

Johnston, I. S., and W. H. Hildemann. 1982. Cellular Organization in the Marine Demosponge *Callyspongia diffusa. Marine Biology*, 67:1–7.

Langenbruch, P. F. 1983. Body Structure of Marine Sponges. 1. Arrangement of the Flagellated Chambers in the Canal System of *Reniera* sp., *Marine Biology*, 75:319–325.

Lévi, C. 1956. Etude des *Halisarca* de Roscoff. Embryologie et systématique des Démosponges. *Archives de Zoologie expérimentale et générale*, 93:1–181.

Minchin, E. A. 1896. Suggestions for a Natural Classification of the Asconidae. *Annals and Magazine of Natural History*, 6(18): 349–362.

Rasmont, R., J. Bouillon, P. Castiaux, and G. Vandermeersche. 1957. Structure microscopique de la collerette des Choanocytes d'éponges. *Comptes-Rendus de l'Académie des Sciences (Paris)*, 245:1571–1574.

Rasmont, R., and F. Rozenfeld. 1981. Etude microcinématographique de la formation des chambres chez une éponge d'eau douce. *Annales de la Société Royale Zoologique de Belgique*, 111:33–44.

Vacelet, J. 1979. Description et affinité d'une éponge sphinctozoaire actuelle. Pages 483–493 in *Biologie des Spongiaires*, edited by C. Lévi and N. Boury-Esnault. Colloques Internationaux du C.N.R.S. 291. Paris: Centre National de la Recherche Scièntifique.

———. 1981. Eponges hypercalcifiées ("Pharétronides", "Sclérosponges") des cavités des récifs coralliens de Nouvelle Calédonie. *Bulletin du Muséum National d'Histoire Naturelle (Paris)*, 3A:313–351.

———. 1983. Les éponges calcifiées et les récifs anciens. *Pour la Science*, 68:14–22.

———. 1985. Coralline Sponges and the Evolution of the Porifera. Pages 1–13 in *The Origins and Relationships of Lower Invertebrates*, edited by S. Conway Morris, R. Gibson, and H. M. Platt. Systematics Association Special Volume 28. Oxford: Clarendon Press.

Vacelet, J., and C. Donadey. In press. A New Species of *Halisarca* (Porifera, Demospongiae) from the Caribbean with Remarks on the Cytology and Affinities of the Genus. In *European Contribution to the Taxonomy of Sponges*, edited by W. C. Jones. Sherkin Island Marine Station Publications.

Willenz, P. 1982. Aspects cinétiques, quantitatifs et ultrastructuraux de l'endocytose, la digestion et l'exocytose chez les éponges. Thesis, Université Libre de Bruxelles. 107 pp.

PAUL-FRIEDRICH LANGENBRUCH
Entwicklungsgeschichtliche Abteilung des
Zoologisches Institut der Universität Bonn
Poppelsdorfer Schloss, D–5300 Bonn 1
Federal Republic of Germany

LIDIA SCALERA-LIACI
Instituto di Zoologia ed Anatomia Comparata
Universitá di Bari, via Amendola 165/A
I–70126 Bari, Italy

Structure of Choanocyte Chambers in Haplosclerid Sponges

Abstract

A ring of cone cells has been found surrounding the apopyles of the choanocyte chambers of three marine haplosclerids: *Reniera* sp., *Haliclona elegans*, and Petrosia *ficiformis*. These cells mediate the contact between the choanocytes and the pinacocyte epithelia of the incurrent and excurrent canal systems. In *Haliclona elegans* and *Petrosia ficiformis* the cone cells bear a flagellum near the nucleus region. In all the marine haplosclerids investigated the choanocytes are separated from the mesenchyme by the pinacocyte epithelium of the incurrent canal wall. In *Reniera* sp. the curved outer surface of the chambers protrudes into the incurrent canals. Only some pinacocyte extensions are spread loosely over the chamber surface. In *Haliclona elegans* and *Petrosia ficiformis* the pinacocyte epithelium of the incurrent canal walls also covers the basal surface of the choanocytes, leaving only some prosopylar openings to the interior of the chambers. In contrast to the choanocyte chambers of the marine haplosclerids, those of the freshwater haplosclerid *Ephydatia fluviatilis* are in direct contact with the cells of the mesenchymal tissue. The relationship of the choanocyte chambers to the mesenchyme and the incurrent canal system is important for feeding in sponges and may be used as criteria for taxonomy.

Choanocyte chambers are pumping organs responsible for the water flow through sponges. They are also filters that are used in the capture and ingestion of food. The structure of these important organs has been investigated in detail in only a few Demospongiae (see, for example, Weissenfels, 1980; Langenbruch, 1983b; Boury-Esnault et al., 1984).

This paper not only presents the results of our investigation, but it also summarizes the recent work that has been done on the body structure of haplosclerid sponges (Langenbruch, 1983a; Langenbruch et al., 1985; Weissenfels 1975, 1980, 1982, 1983). In particular, these investigations have provided new information on the construction of the apopyles of choanocyte chambers and on the relationship of the choanocyte chambers to the mesenchyme and the incurrent canal system. With regard to the last point, the marine haplosclerids investigated by us show organizational types fundamentally different from other Demospongiae. These organizational differences appear to be important for feeding in sponges (Langenbruch, 1985). Because organizational characteristics are in general of high taxonomic value, these new findings may be useful for sponge systematics.

Material and Methods

Specimens of *Reniera* sp. were provided by the Biologische Anstalt Helgoland and kept in a 240 liter aquarium for nine months at 15°C. Specimens of *Haliclona elegans* (Bowerbank) and *Petrosia ficiformis* (Poiret) were collected by diving in the Mediterranean Sea near Bari, Italy. Immediately after collection the sponges were cut into small pieces with razor blades and fixed. All sponges were prepared for light and electron microscopy using the following protocol:

1. Fixation: 2% glutaraldehyde, 2% OsO_4, and 5.5% sucrose in 0.05 M or 0.2 M Na-cacodylate buffer solution, pH 7.0, 4°C, 1–2 h.
2. Washing: 6.9% sucrose in 0.05 M or 0.2 M Na-Cacodylate buffer, pH 7.0, 6 × 10 min.
3. Desilicification: 5% hydrofluoric acid (HF) in washing solution (2), 2 h.
4. Washing: Washing solution (2), 3 × 10 min.
5. Dehydration: Ethanol series ascending from 15% to absolute alcohol in 7 steps, 20°C, 3.5 h total.
6. Postcontrasting: 1% phosphotungstic acid and 1% uranyl acetate in the 70% ethanol of the dehydration series, 4°C, 2 h.
7. Embedding: styrol-methacrylate.

A method developed by Weissenfels (1982) was used for scanning electron microscopy. The embedded material was first sectioned on a microtome. The remaining sponge pieces were then freed from the embedding material by treatment with xylene (changed several times), transferred to pure amyl acetate through an ascending xylene-amyl acetate series, and dried in a critical-point dryer, Type CPD 010 (Balzers). The small dried sponge pieces were then attached to stubs for scanning electron microscopy with the cut surface upward and were coated with gold in a "cool" sputter coater (Polaron Equipment Ltd.). Phase-contrast microscopy was conducted using a Leitz Dialux microscope with Wild MPS 51 automatic photography attachment (film: Ilford PAN F). For transmission electron microscopy we used a Zeiss EM 9 S-2 (film: Agfa Scientia). Scanning electron microscopy was carried out on a Siemens Auto-scan (film: Agfa PAN 100 Professional).

Results
Reniera sp.

The transmission electron micrograph in Figure 1 shows a mediate section through a choanocyte chamber of *Reniera* sp. between an incurrent and an excurrent canal. The incurrent canal is lined by a pinacocyte epithelium, which extends on both sides of the chamber to the border of the apopyle and separates the choanocytes from the mesenchymal tissue. The basal surface of the choanocytes is uncovered in large areas. Only some pinacocyte processes spread loosely over the outer chamber surface (see also Figure 2).

A small piece of cell can be seen on both sides of the apopyle. The cone cells form a small ring around the apopyle. Weissenfels (1980) first found this type of cell in the apopyle region of choanocyte chambers in the freshwater sponge *Ephydatia fluviatilis* and called it a "cone cell" because of its characteristic shape. On the left side of Figure 1 the cone cell ring contacts the pinacocyte epithelia of the incurrent and of the excurrent canal wall as well as the choanocyte epithelium.

As already mentioned, the choanocyte chambers of *Reniera* sp. are nearly uncovered at the outer surface. As a result, in the incurrent canals the mulberrylike outer surface of the choanocyte chambers can be seen with the aid of SEM (Figure 2). The entire uncovered basal surface of the choanocytes is visible in Figure 2. The cells are separated by gaps that are bridged by small cell extensions (arrows). The water passes through the sievelike spaces en route from the incurrent canal to the interior of the chamber.

Normally the choanocyte chambers of *Reniera* sp. are partly covered by pinacocyte processes that spread loosely over the chamber surface (as in Figure 2). Transmission electron micrographs reveal that these pinacocyte extensions sometimes project into the chambers. They are ingesting particles from the chamber surface and even from the outer surface of the choanocyte collars (Langenbruch, 1985).

Haliclona elegans

The transmission electron micrograph in Figure 3 shows a section through a choanocyte chamber of *Haliclona elegans* that has been cut through the apopyle. The apopyle opens to an excurrent canal lined by a flat pinacocyte epithelium. The chamber consists of numerous choanocytes. Sectioned collars and flagella are seen in the interior of the chamber.

The choanocyte epithelium bulges out into an incurrent canal. On the basal surface the choanocytes are covered by the pinacocyte epithelium of the incurrent canal wall, leaving some gaps, prosopyles, between the adjacent cho-

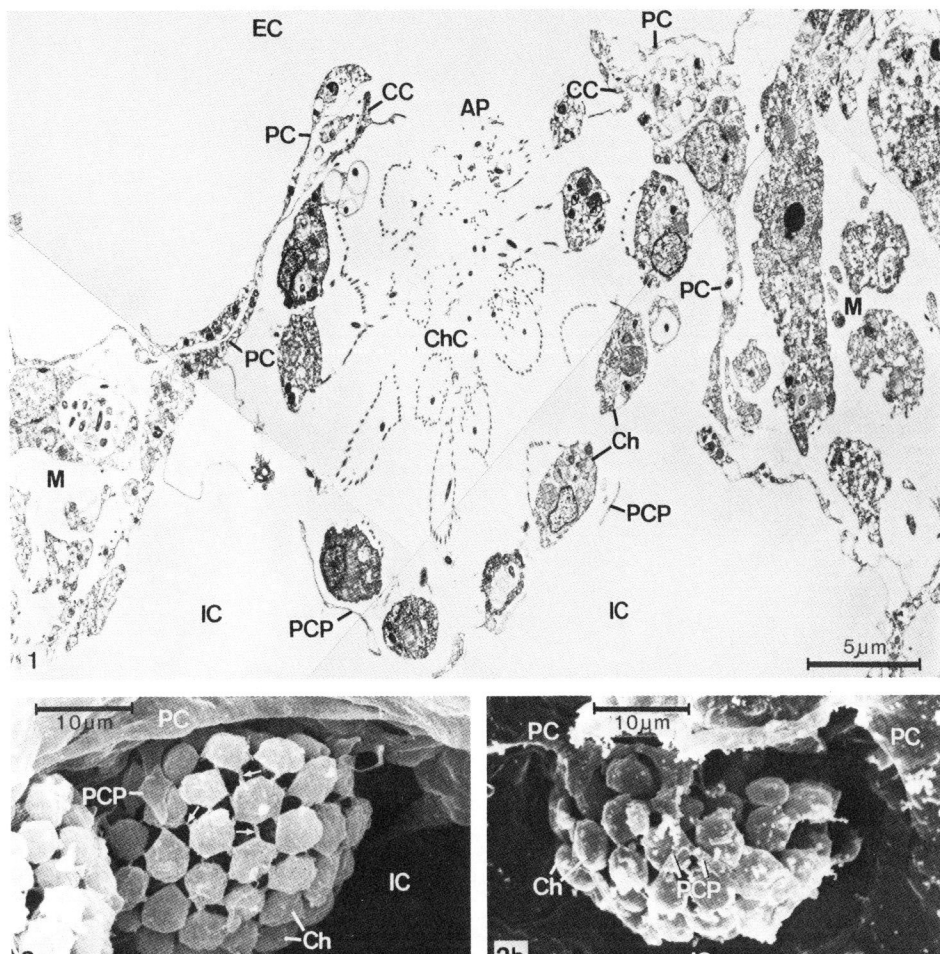

Figures 1–2. Choanocyte chambers of *Reniera* sp.: *1,* pinacocytes of incurrent and excurrent canal walls extending to cone cells at border of apopyle; outer chamber surface is only partly covered by pinacocyte processes (TEM); *2,* view from incurrent canal side on outer surface of two (*a, b*) choanocyte chambers (SEM); chambers only partly covered by small pinacocyte processes connecting choanocytes (*arrows*). *Ap,* apopyle; *CC,* cone cell; *Ch,* choanocyte; *ChC,* choanocyte chamber; *EC,* excurrent canal; *IC,* incurrent canal; *M,* mesenchyme; *PC,* pinacocyte; *PCP* pinacocyte process.

anocytes, through which the water flows into the chamber. The pinacocyte epithelium extends on both sides to a ring of cone cells at the border of the apopyle. On the right side it separates the choanocytes from a mesenchyme region with a spicule and spongine complex.

Figure 4 shows a cone cell of *Haliclona elegans* in higher magnification. Collar microvilli of the adjacent choanocyte are at the bottom of the picture. In the upper part a small cell piece (asterisks), probably of a central cell, lies at the border of the apopyle. A flagellum arises from the surface of the cone cell near the nucleus.

Figure 5 shows two choanocyte chambers lying in the plane of the sectioned tissue prepared for SEM (see "Material and Methods"). At the lower left is an open view into one of the chambers filled with collars and flagella. The chamber in the upper region is partly cut. The outer surface of this and another chamber can be seen in an incurrent canal in the center of the picture. The surface of the chambers is almost completely covered by pinacocytes. There are many prosopylar openings, which lead through gaps between the adjacent choanocytes into the chambers. Some of the choanocytes are partly uncovered (arrows).

Petrosia ficiformis

A transmission electron micrograph of a section through a choanocyte chamber of *Petrosia ficiformis* is shown in Figure 6*a*. The choanocyte chambers of *P. ficiformis* are of the aphodal type (Langenbruch et al., 1985), each being connected by a small aphodus (A in Figure 6*a*) to larger canals of the excurrent system. On the side opposite of the apopyle, the chamber in Figure 6*a* faces an incurrent canal.

In *Petrosia ficiformis,* as in *Haliclona elegans,* the outer surface of the chambers is covered by a pinacocyte epithelium. The pinacocyte cover (arrowheads in Figure 6), is in direct contact with the choanocytes and is therefore very difficult to distinguish from these cells. It continues on the left- and right-hand side in the pinacocyte epithelium of the incurrent canal wall. There the pinacocytes are in contact (arrows) with a ring of cone cells. Two overlapping pieces of cone cell are cut to the left of the apopyle.

A section through another cone cell of *Petrosia ficiformis* is shown in Figure 7 in higher magnification. In the basal part of the picture the cell is in contact with a choanocyte.

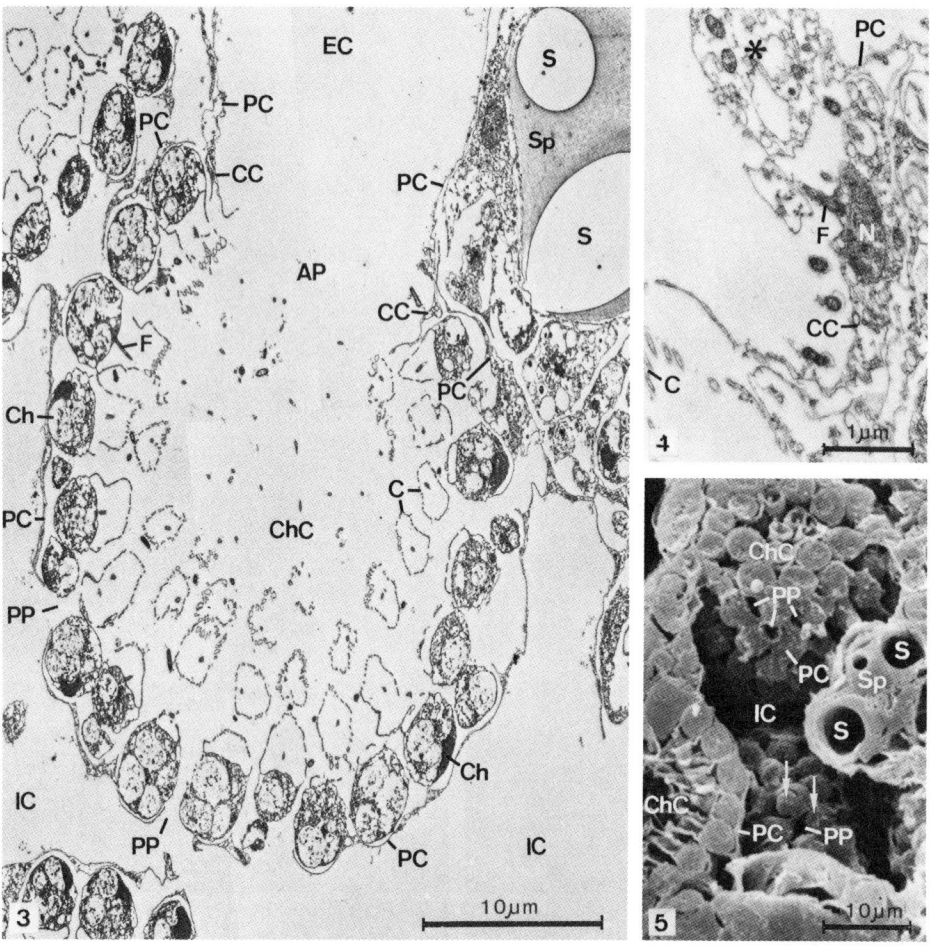

Figures 3–5. Choanocyte chambers of *Haliclona elegans: 3*, mediate section (TEM) shows outer surface of chamber covered by pinacocytes of incurrent canal wall; note cone cells at border of apopyle; *4*, cone cell cut through nucleus and basal part of a flagellum (TEM); asterisk indicates piece of central cell; *5*, view onto sectioned tissue surface and into an incurrent canal with choanocyte chambers covered by pinacocytes (SEM); note uncovered parts of choanocytes *(arrows)*. *AP*, apopyle; *C*, collar; *CC*, cone cell; *ChC*, choanocyte chamber; *EC*, excurrent canal; *F*, *flagellum; IC*, incurrent canal; *N*, nucleus; *PC*, *pinacocyte; PP*, prosopyle; *S*, spicule; *Sp*, spongin.

In the upper part, it contacts (arrows) the pinacocyte epithelium of an aphodus. As in *Haliclona elegans*, the cone cells of *P. ficiformis* bear a flagellum, the basal part of which is cut near the nucleus.

In the scanning electron micrograph of a section through *Petrosia ficiformis* tissue, bacteriocytes loosely filled with bacteria and a choanocyte chamber are seen in the cutting plane (Figure 8). The micrograph also provides a view into an incurrent canal. The pinacocyte epithelium of the canal wall covers a choanocyte chamber. The choanocytes (arrowheads) are bulging out under the pinacocyte cover. Some holes in the pinacocyte epithelium are prosopyles through which the water flows into the chamber.

Discussion and Conclusions

CHOANOCYTE CHAMBER MORPHOLOGY

The four semidiagrammatic drawings in Figure 9 summarize our findings in the choanocyte chambers of the three marine sponges examined here (Figures 9a–c), along with a choanocyte chamber of the freshwater haplosclerid *Ephydatia fluviatilis* (Figure 9d). Each diagram represents a section through a choanocyte chamber with the apopyle and the adjacent incurrent and excurrent canals.

All three marine species (Figure 9a–c) have a ring of cone cells in the apopyle region of the choanocyte chambers. In *Haliclona elegans* (Figure 9b) and *Petrosia ficiformis* (Figure 9c) the cone cells are flagellated. In *Reniera* sp. (Figure 9a) no flagella have yet been found; there are only some pinacocyte processes extending over the basal surface of the choanocytes (Figure 9a). Most of the chamber surface is uncovered and the water flows into the chambers via gaps between the choanocytes (small arrows).

In *Haliclona elegans* (Figure 9b) and *Petrosia ficiformis* (Figure 9c) most of the outer surface of the choanocyte chambers is covered by pinacocytes of the incurrent canal system, so that there are only some prosopylar openings (small arrows) to the interior of the chambers. Note that in all these marine haplosclerids the choanocyte chambers are separated from the mesenchymal tissue by the pinacocyte epithelium of the incurrent canal walls.

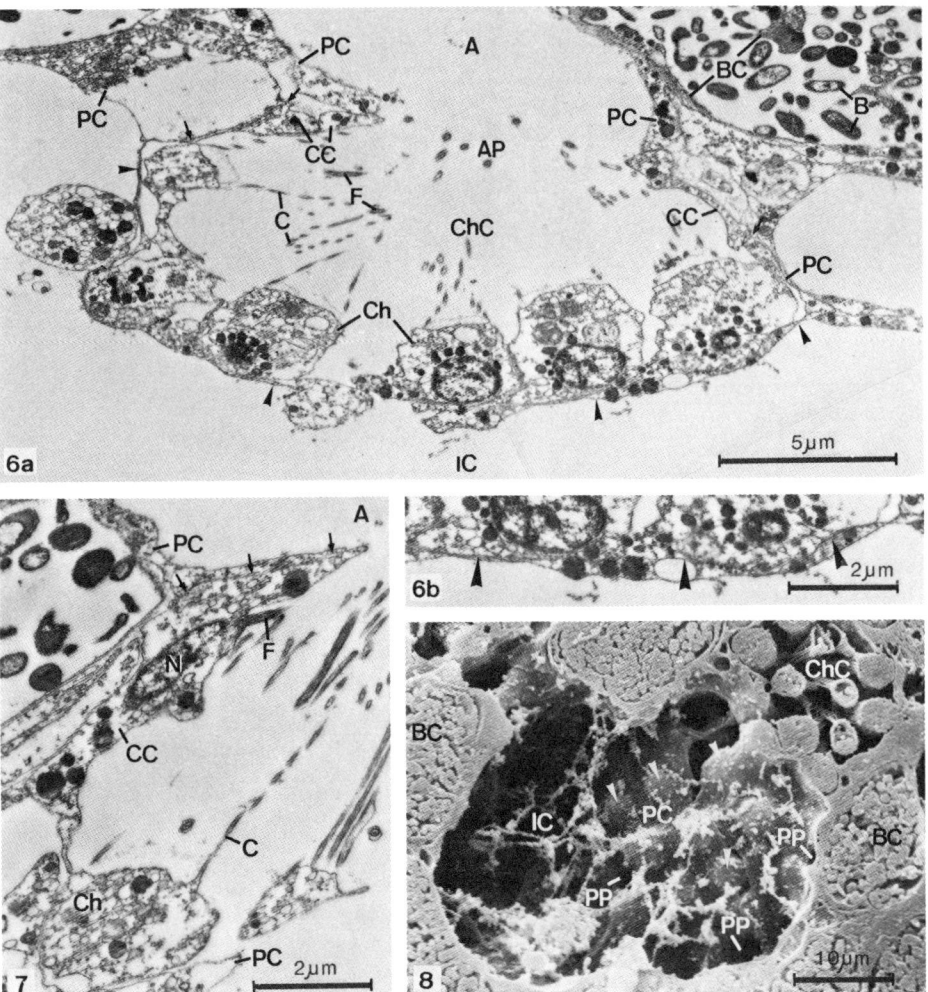

Figures 6–8. Choanocyte chambers of *Petrosia ficiformis*: *6a*, apopyle bordered by cone cells (TEM); note contact regions between cone cells and pinacocytes of incurrent canal *(arrows)* and pinacocyte cover on outer chamber surface *(arrowheads)*; *6b*, basal parts of two choanocytes in *a* with pinacocyte cover *(arrowheads)* (TEM); *7*, cone cell of *Petrosia ficiformis* bearing a flagellum near nucleus (TEM); note contact region between cone cell and a pinacocyte of aphodus *(arrows)*; *8*, tissue section (SEM) with bacteriocytes and a choanocyte chamber exposed; view into an incurrent canal with choanocytes *(arrowheads)* bulging out under pinacocyte epithelium of canal wall. *A*, aphodus; *AP*, apopyle; *B*, bacteria; *BC*, bacteriocyte; *C*, collar; *CC*, cone cell; *Ch*, choanocyte; *ChC*, choanocyte chamber; *F*, flagellum; *IC*, incurrent canal; *N*, nucleus; *PC*, pinacocyte; *PP*, prosopyle.

Figure 9*d* shows a section through a choanocyte chamber of the freshwater sponge *Ephydatia fluviatilis*. The drawing summarizes the results of recent investigations by Weissenfels (1975, 1980, 1982, 1983). The choanocyte chambers of *Ephydatia fluviatilis* also contain cone cells (CC) at the border of the apopyle. Unlike chambers of the marine haplosclerids investigated by us, those of *Ephydatia fluviatilis* and other spongillids are surrounded by the mesenchymal tissue.

CONE CELLS

Of particular interest in our study is the discovery of a ring of cone-shaped cells in the apopyle region of the choanocyte chambers of the investigated sponges. These cells, some of which are flagellated, appear to be structurally intermediate between the choanocytes and the pinacocytes of the excurrent canal walls. Cone cells may be stabilizing devices used to regulate the flow of water through the apopyles. A similar ring of cells has recently been reported by Boury-Esnault et al. (1984) in the Homo-

scleromorpha and by Boury-Esnault et al. (this volume) in many different Demospongiae.

FOOD UPTAKE

Investigations into the uptake of particulate food in *Reniera* sp. (Langenbruch, 1985) have recently shown that the pinacocytes of the incurrent canal walls ingest the greatest amount of particulate food. In sponges whose choanocyte chambers are separated from the mesenchyme, the pinacocytes of the incurrent canal system are particularly important for feeding because every food particle for the mesenchymal cells has to pass the walls of the incurrent canals.

In spongillids food taken up by the choanocytes may be passed on directly to the cells of the mesenchyme (Van Weel, 1948; Kilian, 1952; Schmidt, 1970; Weissenfels, 1976, 1983; Willenz, 1980). Furthermore, in *Ephydatia fluviatilis* all cells of the mesenchyme are able to ingest particles swept into the mesenchyme by the water current (Weissenfels, 1976, 1983).

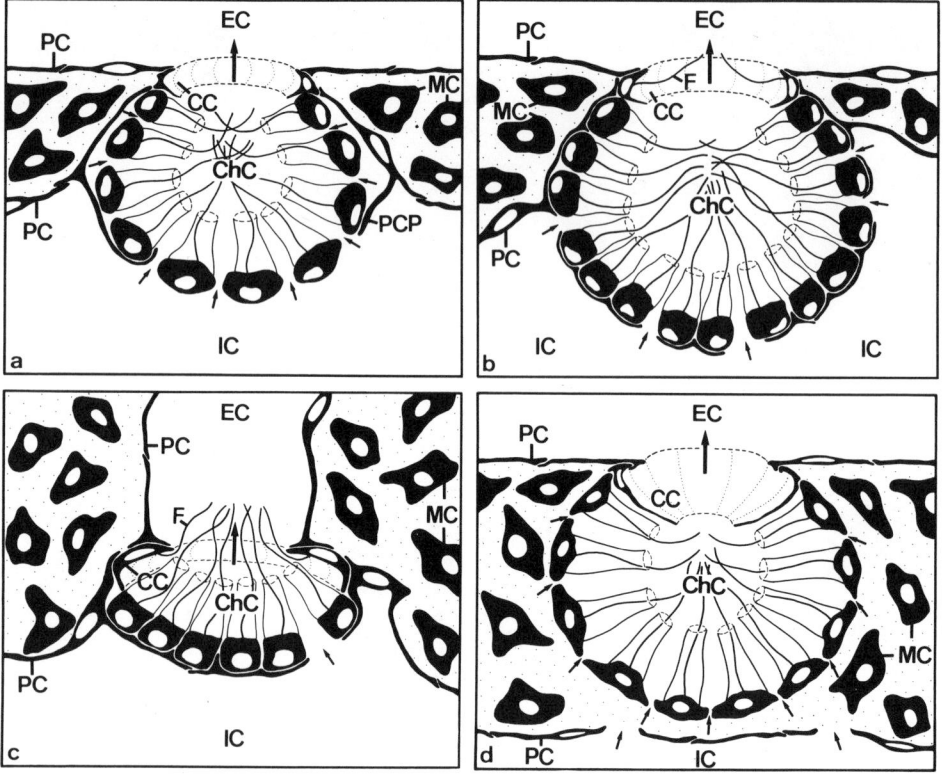

Figure 9. Semidiagrammatic drawings of sections through choanocyte chambers: *a*, *Reniera* sp.; *b*, *Haliclona elegans*; *c*, *Petrosia ficiformis*; *d*, *Ephydatia fluviatilis*. Three-dimensional arrangement of cone cell rings and edges of choanocyte collars indicated by dashed and dotted lines, directional water flow by arrows. Mesenchyme drawn schematically with "mesenchyme cells" and dotted in intercellular spaces without regard for real cellular organization. *CC*, cone cell; *ChC*, choanocyte chamber; *EC*, excurrent canal; *F*, flagellum; *IC*, incurrent canal; *MC*, mesenchyme cell; *PC*, pinacocyte; *PCP*, pinacocyte process.

TAXONOMY

Among the Demospongiae, choanocyte chambers separated from the mesenchyme have been reported only in the Haplosclerida. With the aid of SEM, Johnston and Hildemann (1982) recently showed that in the marine haplosclerid *Callyspongia diffusa* the outer surface of the choanocyte chambers can be seen from the incurrent canal side. The structure of these choanocyte chambers may be similar to those of *Reniera* sp. Our newest, unpublished results on *Adocia simulans* (Johnston) show that in this marine haplosclerid the choanocyte chambers are also separated from the mesenchyme.

Although we do not yet know whether the choanocyte chambers are separated from the mesenchyme in most marine haplosclerid sponges, it seems to be possible to define a group of sponges connected by this characteristic, which we consider to be a sign of a close relationship. Recently Bergquist (1980) separated the genus *Petrosia* from the Haplosclerida and transferred it, on the basis of skeletal, reproductive, and chemical criteria, into the new order Nepheliospongida (see also Bergquist and Warne, 1980; Bergquist et al., 1980). Actually, the choanocyte chambers of *P. ficiformis* are somewhat different from the choanocyte chambers of all other marine Haplosclerida investigated by us. They are of the aphodal type and consist of only a few choanocytes, which are arranged in a flat,

slightly outpocketed epithelium. This may be a sign of a certain systematic distance. At the same time, *P. ficiformis* shows the same separation of the choanocyte chambers from the mesenchyme as do the other marine Haplosclerida investigated by us; this common feature indicates a relationship between the two sponge groups in question (Wiedenmayer, 1977a,b).

A special problem arises for *Ephydatia* and related freshwater sponge genera. They may belong to a group of sponges in the Haplosclerida, characterized among other criteria by choanocyte chambers surrounded by the mesenchymal tissue. But if further investigations show that the separation of the choanocyte chambers from the mesenchyme turns out to be a common feature of marine haplosclerid sponges, it may be that *Ephydatia* and related freshwater sponges are not Haplosclerida.

Acknowledgments

We express our thanks to N. Weissenfels, Zoologisches Institut, Universität Bonn, for his helpful support throughout the project and to M. Geis, I. Nüssle, U. Müller, and B. Zarbock for technical assistance. The authors are also grateful to Ezio Amato, Agostino Cancellato, Nino Constantino, and Nicola Tedesco for their help in collecting the sponges. The scanning electron micrographs were prepared in the Max Planck Institut für Systemphysiologie,

Dortmund. We thank D. Lübbers, R. Kinne, D. Schäfer, and M. Seiffert. This work was supported by the Deutsche Forschungs-gemeinschaft.

Literature Cited

Bergquist, P. R. 1980. The Ordinal and Subclass Classification of the Demospongiae (Porifera); Appraisal of the Present Arrangement, and Proposal of a New Order. *New Zealand Journal of Zoology*, 7:1–6.

Bergquist, P. R., W. Hofheinz, and G. Oesterhelt. 1980. Sterol Composition and the Classification of the Demospongiae. *Biochemical Systematics and Ecology*, 8:423–435.

Bergquist, P. R., and K. P. Warne. 1980. The Marine Fauna of New Zealand: Porifera, Demospongiae, Part 3. Haplosclerida and Nepheliospongida. *New Zealand Oceanographic Institute Memoir*, 87:2–77.

Boury-Esnault, N., L. De Vos, C. Donadey, and J. Vacelet. 1984. Comparative Study of the Choanosome of Porifera: I. The Homoscleromorpha. *Journal of Morphology*, 180:3–17.

Johnston, I. S., and W. H. Hildemann. 1982. Cellular Organisation in the Marine Demosponge *Callyspongia diffusa*. *Marine Biology*, 67:1–7.

Kilian, E. F. 1952. Wasserströmung und Nahrungsaufnahme beim Süßwasserschwamm *Ephydatia fluviatilis*. *Zeitschrift für vergleichende Physiologie*, 34:407–447.

Langenbruch, P.-F. 1983a. Body Structure of Marine Sponges. I. Arrangement of the Flagellated Chambers in the Canal System of *Reniera* sp. *Marine Biology*, 75:319–325.

———. 1983b. Untersuchungen zum Körperbau von Meeresschwämmen. II. Das Wasserleitungssystem von *Halichondria panicea*. *Helgoländer wissenschaftliche Meeresuntersuchungen*, 36:337–346.

———..1985. Die Aufnahme partikulärer Nahrung bei *Reniera* sp. (Porifera). *Helgoländer wissenschaftliche Meeresuntersuchungen*, 39:263–272.

Langenbruch, P.-F., T. L. Simpson, and L. Scalera-Liaci. 1985.

Body Structure of Marine Sponges. III. The Structure of Choanocyte Chambers in *Petrosia ficiformis* (Porifera, Demospongiae). *Zoomorphology*, 105:383–387.

Schmidt, I. 1970. Phagocytose et pinocytose chez les Spongillidae. Etude in vivo de l'ingestion de bactéries et de protéines marquées a l'aide d'un colorant fluorescent en lumière ultra-violette. *Zeitschrift für vergleichende Physiologie*, 66:398–420.

Van Weel, P. B. 1949. On the Physiology of the Tropical Freshwater Sponge *Spongilla proliferens* Annand. I. Ingestion, Digestion and Excretion. *Physiologia Comparata et Oecologia*, 1:110–128.

Weissenfels, N. 1975. Bau und Funktion des Süßwasserschwamms *Ephydatia fluviatilis* L. (Porifera). II. Anmerkungen zum Körperbau. *Zeitschrift für Morphologie der Tiere*, 81:241–256.

———. 1976. Bau und Funktion des Süßwasserschwamms *Ephydatia fluviatilis* L. (Porifera). III. Nahrungsaufnahme, Verdauung und Defäkation. *Zoomorphology*, 85:73–88.

———. 1980. Bau und Funktion des Süßwasserschwamms *Ephydatia fluviatilis* L. (Porifera). VII. Die Porocyten. *Zoomorphology*, 95:27–40.

———. 1982. Bau und Funktion des Süßasserschwamms *Ephydatia fluviatilis* L. (Porifera). IX. Rasterelektronenmikroskopische Histologie und Cytologie. *Zoomorphology*, 100:75–87.

———. 1983. Bau und Funktion des Süßwasserschwamms *Ephydatia fluviatilis* (Porifera). X. Der Nachweis des offenen Mesenchyms durch Verfütterung von Bäckerhefe *(Saccharomyces cerevisiae)*. *Zoomorphology*, 103:15–23.

Wiedenmayer, F. 1977a. *Shallow-Water Sponges of the Western Bahamas*. Experientia Supplementum 28. Basel: Birkhäuser. 287 pp.

———. 1977b. The Nepheliospongiidae Clarke 1900 (Demospongea, Upper Devonian to Recent), an Ultraconservative, Chiefly Shallow-Marine Sponge Family. *Eclogae Geologicae Helvetiae*, 70:885–918.

Willenz, P. 1980. Kinetic and Morphological Aspects of Particle Ingestion by the Freshwater Sponge *Ephydatia fluviatilis* L. Pages 163–178 in *Nutrition in the Lower Metazoa*, edited by D. C. Smith and Y. Tiffon. Oxford: Pergamon Press.

CAROLYN K. TERAGAWA*
Department of Zoology
Duke University
Durham, North Carolina 27706

Mechanical Function and Regulation of the Skeletal Network in *Dysidea*

Abstract

Bending tests were used to study the functional morphology and mechanical properties of the skeletal fibers of *Dysidea etheria,* a dictyoceratid sponge that incorporates sand grains in its skeleton. Fibers with more particles were stiffer because they were able to compensate for the reduced stiffness of the fiber material through an increase in diameter. Manipulations of the living sponge demonstrated that tension was maintained in the dermal membrane. A model of the dermal membrane predicts the in situ stresses experienced by the dermal membrane and primary fibers and suggests that membrane tension controls particle incorporation into growing primary fibers.

The diverse skeletal structures of sponges make it possible to classify these organisms on the basis of their skeletal morphology. However, our understanding of the functional morphology of the sponge skeleton is still limited. Although the sharp spicules of some sponges may have a defensive role in repelling settling organisms or predators (Randall and Hartman, 1968; Wilkinson, 1979; Simpson, 1984), the main function of the sponge skeleton is to provide support for the sponge tissues and aquiferous system (Bergquist, 1978) and allow massive forms to develop. Three-dimensional growth can be advantageous to the sponge: taller sponges may pump water more efficiently by taking advantage of induced flow and by escaping the "boundary effects" of slow moving water near the substrate (Vogel, 1977, 1981); erect-growing sponges may escape substrate competition (Jackson, 1977); and larger volume sponges may increase their reproductive fitness as they have more biomass to devote to the production of gametes or embryos.

Dispersed spicular skeletons may provide the support

*Current address: Developmental Biology Center, University of California, Irvine, California 92717.

needed for three-dimensional growth by acting as stiffening fibers or filler particles in the sponge matrix, in the same way that hard particles act as stiffening agents when dispersed in man-made polymers: for example, carbon particles are dispersed in the rubber used to make tires (Nielson, 1962; Alexander, 1968; Wainwright et al., 1976). In a comparative study of sponges, cnidarians, and model tissues, Koehl (1982) demonstrated that spicules dispersed in biological matrices increased the tensile stiffness of the composite material and that greater volume fractions of spicules increased stiffness. Field studies on intertidal sponges suggest that such mechanisms may allow sponges to withstand physical forces in the environment (Palumbi, 1984, 1986). However, many sponges have discrete skeletons of fibrous networks or frameworks of fused spicules that may function to support the sponge more like scaffoldings (e.g., Reif and Robinson, 1976).

In the present study, the fibrous skeleton of the sponge *Dysidea etheria* (Dictyoceratida) was examined to determine the mechanical contribution of hard particles when constrained in fibers. Like many sponges, this species can incorporate foreign materials, such as sand grains or foreign spicules, in its skeleton. In contrast to some species that select specific types of particles, *D. etheria* incorporates a variety of particles into its radially arranged primary skeletal fibers; its cross-linking secondary fibers are relatively free of foreign material.

Shaw (1927) previously suggested that the particles could act as a stiffening material to compensate for the pliability of pure spongin. However, the mechanical behavior of particles aligned in fibers is little understood, and spicules or particles that tend to abut on one another, as do particles in fibers, will behave differently from dispersed particles (Wainwright et al., 1976, Koehl, 1982).

Static three-point bending tests were performed on isolated fibers of *Dysidea etheria* with variable amounts of particle contents to better define the mechanical contribution of particles in skeletal fibers. The mechanical behavior of the sponge dermal membrane was also examined to determine the type of loads applied to the fibrous skeletal network. These observations make it possible to construct a model in which the growth of primary fibers can be mechanically regulated.

Materials and Methods

OBSERVATION OF LIVE SPONGES. Sponges were carefully collected with their substrate intact from various locations in Bermuda and were maintained in the laboratory in a unidirectional flow-through tank. Samples were examined under a Wild M5 stereomicroscope.

MECHANICAL TESTS OF FIBERS. Three to six samples of both primary and secondary fibers were dissected from the skeletons of each of six sponges. The fibers selected contained varying amounts of particles and were as uniform in diameter and as straight as possible. Particles typically were concentrated in a central core in the fibers, and spongin was distributed in a layer around this core. The volume fraction of particles was calculated from the ratio of the volume of the inner cylinder of particles to the total volume of the cylindrical fiber. The ratio of the cylinder volumes was proportional to the ratio of the squared radii of the cylinders. An average volume fraction, hereafter called particle content, was calculated for each fiber from 20 paired measurements of the fiber radius and the particle core radius made with a compound microscope (250×, error ± 5 μm) at evenly spaced intervals along the fiber. Measurements of the particle core radius were corrected to exclude small areas of spongin between particles. The absence of particles was assigned a zero value in averages.

In static three-point bending tests, a fiber was treated as a simple beam with unfixed ends and with a concentrated load applied at its midpoint (Faupel, 1964). Tests were performed by supporting the ends of fibers on rigid plexiglas supports 2.6 to 2.8 mm apart and loading the fibers at their midpoints with small weights (ranging from 4.66–106.40 mg) made from bent pieces of wire. Fibers were kept moist during tests, and their ends were coated with petroleum jelly (Vaseline) to keep them in place on the supports without interfering with their freedom of movement in bending. The vertical deflections of loaded fibers were observed with a horizontally mounted Wild M5 stereomicroscope (50×, error ± 10 μm). Increasing weight was applied to each fiber until the fiber slipped off the supports. An average of six load deflections was measured for each fiber. Only small bending deflections of 12% or less of the fiber bending length (length between supports) were used for calculations since the equation used (see below) is applicable only at small deflections.

The stiffness of a material is represented by its Young's modulus or "E." Young's modulus is a property of materials expressed by the ratio of force per cross-sectional area (stress) to the resultant deformation, usually expressed as percent deformation (strain). E is the slope of the stress/strain curve for linearly elastic materials, and stiffer materials have higher values of E than do less stiff ones (Wainwright et al., 1976). Young's modulus was calculated from the three-point bending tests using the following equation: $y = (1/48)(WL^3/EI)$, where y is the vertical deflection measured at the midpoint of the fiber; W is the force applied (= weight); L is the bending length of the fiber, represented by the distance between the supports; and I is the second moment of area, a value dependent on the cross-sectional size and shape of the fiber (Vincent, 1978). The term EI is commonly known as the bending or flexural stiffness of a beam, and simultaneously considers the effects of material property *(E)* and beam shape *(I)* on bending behavior. The average value of W/y for each fiber was used to calculate E for each fiber. The value for *I* was estimated for each fiber from the following equation for a fiber of cylindrical cross section: $I = (\pi R^4/4)$, where *R* is the radius of the cylinder (Vincent, 1978). Twenty values of R^4 were obtained for each fiber; the average value of R^4 was used to calculate *I*.

The relationships between *E*, *I*, and *EI* to particle content were tested with linear regression models. In all tests a *p* value of less than 0.05 (95% confidence level) was considered significant. The values obtained for *E*, *I*, and *EI* were converted to natural logarithms to better fit assumptions about normal distributions and equality of variances required for linear regression models. Calculations and statistics were performed using mainframe SAS (SAS Institute, 1982) and Systat software for an IBM XT microcomputer.

Results

DERMAL MEMBRANE BEHAVIOR. The skeleton of *Dysidea etheria* is a loose meshwork formed by radially arranged primary fibers filled with particles. Smaller-diameter secondary fibers containing fewer particles link the primary fibers (Figure 1). Secondary fibers appear denser near basal portions of the sponge and often form evenly spaced, concentric layers in the skeleton. The apices of the primary fibers are not linked by secondary fibers (Figure 2) and meet the dermal membrane of the sponge to form conules (protrusions) on the surface (Figure 3). Between the conules, the dermal membrane adopts a curved, catenarylike shape.

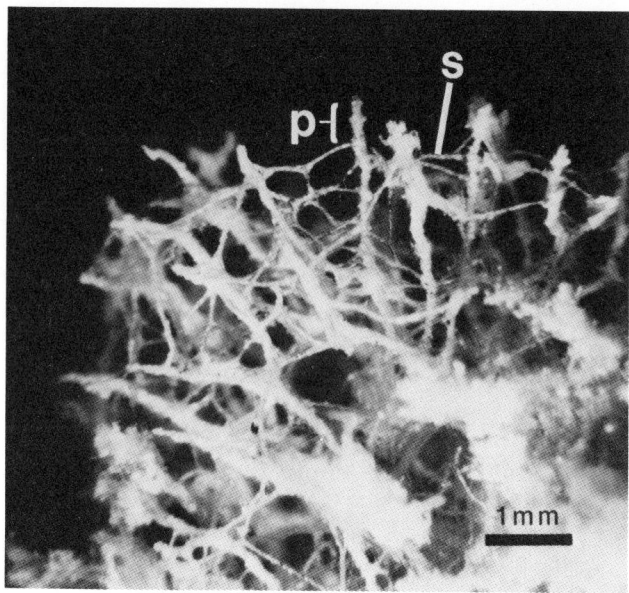

Figure 2. Apical region of *Dysidea etheria* skeleton showing free ends of primary fibers and cross-linking secondary fibers. *p*, primary; *s*, secondary.

The mechanical interaction of the primary fibers and the dermal membrane was explored by making small incisions in the dermal membrane with sharp razor blades. These small slits immediately opened to form convex, lens-shaped holes in the membrane (Figure 4*a*). The slits opened more rapidly than could be explained by the slow, contractile behavior observed in the dermal membrane. After an incision was made, the dermal membrane sagged with the release of tension. Measurements from camera lucida drawings of lateral views of incisions made between conule pairs showed that the vertical distance between the dermal membrane and the conule apices increased by

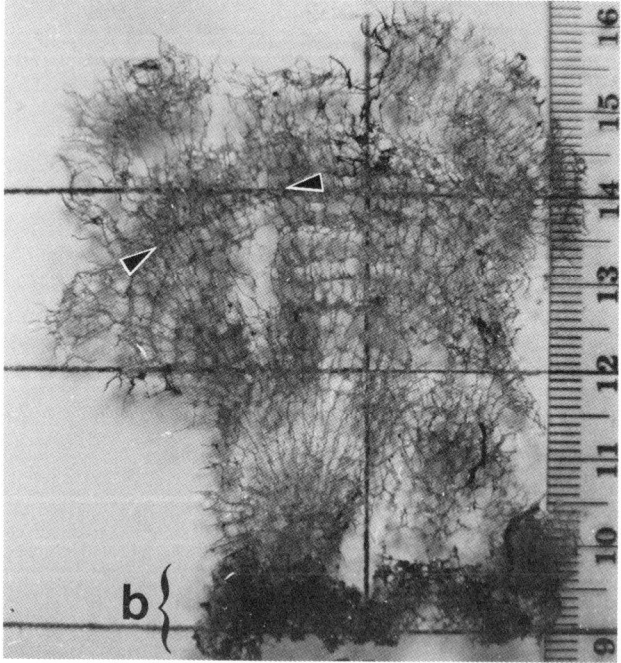

Figure 1. *Dysidea etheria*, cross section through whole isolated skeleton; basal portion is denser meshed; arrows indicate layer of secondary fibers. *b*, base; scale in mm.

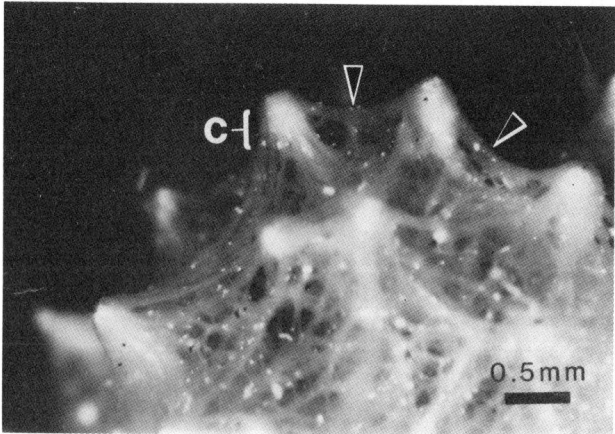

Figure 3. *Dysidea etheria*, surface of living sponge; white regions are primary fiber apices forming conules, arrows indicate curved dermal membrane between conules. *c*, conules.

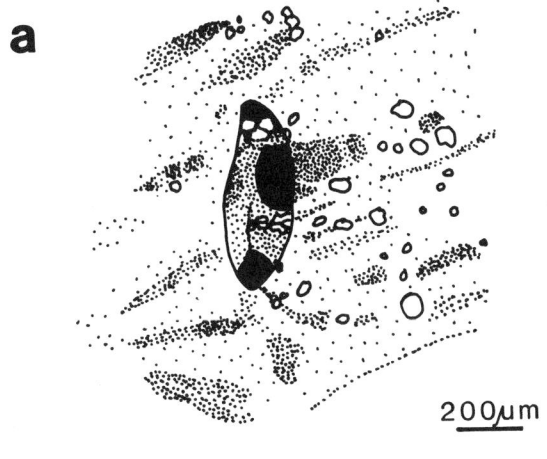

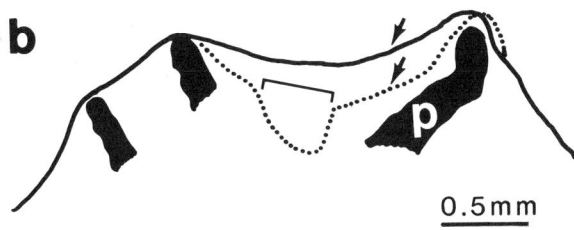

Figure 4. *Dysidea etheria*, dermal incision of a living sponge: *a*, tracing from photograph of lens-shaped hole in sponge surface, viewed from above immediately after cutting; *b*, camera lucida drawing of a dermal membrane incision, side view; original membrane position indicated by solid line, position after cutting by arrows; stippled line bracket indicates incision site. *p*, primary fiber.

86% after incision (midpoint distances, $n = 8$) (Figure 4*b*). No reliable measurement of primary fiber deflection could be obtained with this technique.

PARTICLE CONTENT AND FIBER STIFFNESS. Significant regressions were obtained for E ($p < 0.001$), I ($p < 0.001$) and and EI ($p < 0.05$) against particle content. The flexural stiffness *(EI)* of the skeletal fibers increased as the volume fraction of particles in fibers increased (Figure 5*a*). Examination of the individual behaviors of E and I as functions of particle content showed that E decreased (Figure 5*b*) whereas I increased (Figure 5*c*) with particle content to cause the overall increase in flexural stiffness.

Discussion

REGULATION OF FIBER STIFFNESS. An increase in the second moment of area of fibers containing more particles was large enough to offset the loss of material stiffness associated with increased particle content. As a result, there was an overall increase in the flexural stiffness of fibers with more particles. Since I is proportional to the fourth power of the fiber radius, only small increases in diameter are

necessary to produce large increases in I. Although the present data show that particles compromise material stiffness, previous studies have shown that particles enhance skeletal growth, probably by acting as a "filler material" occupying volume in the fibers (Teragawa, 1985, 1986). Particle incorporation thus allows *Dysidea etheria* to increase its fiber diameter to offset the decrease in material stiffness and still benefit from the filler role of particles. Maintenance of fiber stiffness is probably important to the sponge since its skeletal fibers support its tissues and aquiferous spaces. Individual skeletal fibers may need to be stiffer in sponges like *D. etheria*, which have a relatively loose-meshed skeletal network. Fiber stiffness may be especially critical at the uncross-linked apices of the primary fibers (Figure 2) which support the dermal membrane in the same way that tent poles support a tent. These fibers maintain the subdermal space, which may be critical for efficient flow through the sponge aquiferous system (Vogel, 1978).

Although the results of the bending test show that particles actually decrease the material stiffness *(E)* of fibers, the mechanics of sponge fibers should now be tested in pure compression and tension. Bending tests, although easy to perform, can be difficult to interpret for non-isotropic biological materials such as these skeletal fibers. Bending involves the application of both compressive and tensile stresses to the test material, and the microdistribution of particles in fibers can significantly affect the overall behavior of the fiber since the particles may behave differently, depending on whether they are subject to tension or compression. The maintenance of tension in the dermal membrane suggests that the primary fibers are loaded in compression at their apical ends (see Figure 6); compression tests of the primary fibers would most realistically measure the mechanical effects of particles in the sponge skeleton. In general, the hard sand particles should help the pliable spongin fibrils resist compressive loads.

DERMAL MEMBRANE TENSION. Tension in sponge tissues and its role in morphogenetic processes have been discussed previously (Lendenfeld, 1889; Brøndsted, 1953; Jones, 1957; Borojevic, 1971). In other systems, tension can cause growth of tissue culture neurites (Bray, 1984) and may organize developing connective tissues (Harris et al., 1981; Stopak and Harris, 1982). In *Dysidea etheria*, the tensed dermal membrane interacts directly with growing apices of primary fibers at the conules. Since the dermal membrane is sometimes associated with algal filaments or worm tubes growing up through the sponge and can form conulelike structures with these substitute primary fibers (Teragawa, 1985), mechanical rather than specific chemical interactions may govern growth and morphogenesis of the skeleton in *D. etheria*.

The mechanical forces of tension in the dermal membrane could affect two aspects of skeletogenesis: (a) der-

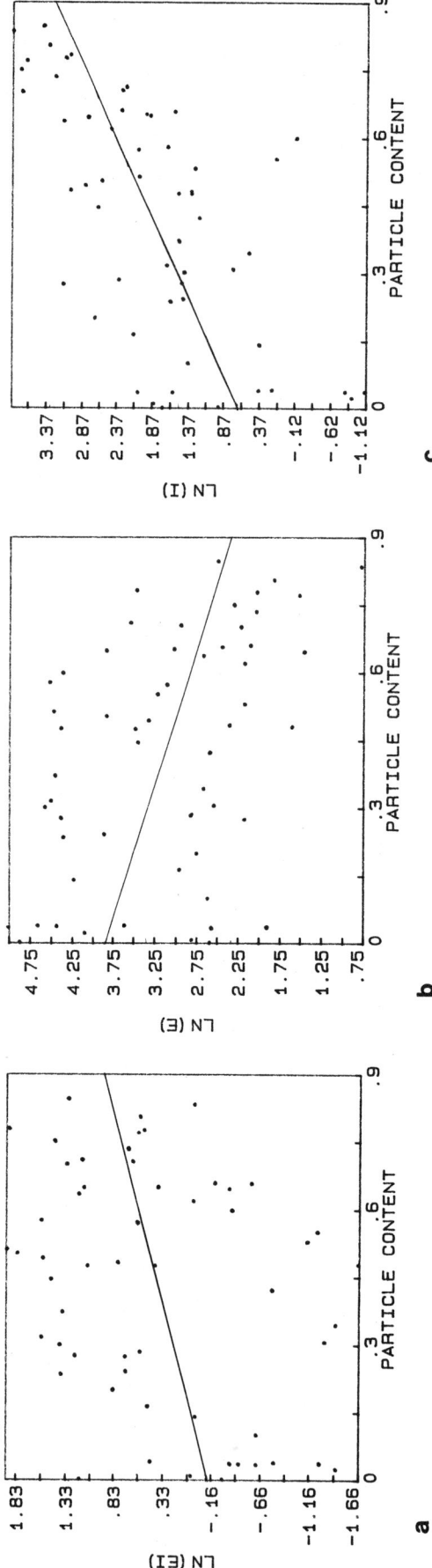

Figure 5. Mechanical properties of *Dysidea etheria* skeleton fiber: *a*, relationship of flexural stiffness (*EI*) to particle content (volume %); *EI* in units of $MN \times m^2 \times 10^{-15}$; slope = 1.174 ($p < 0.05$), intercept = -0.114; *b*, relationship of Young's modulus (*E*) to particle content (volume %); *E* in units of MN/m^2; slope = -1.693 ($p < 0.001$), intercept = 3.854; *c*, relationship of second moment of area (*I*) to particle content (volume %); *I* in units of $m^4 \times 10^{-17}$; slope = 2.872 ($p < 0.001$), intercept = 0.635.

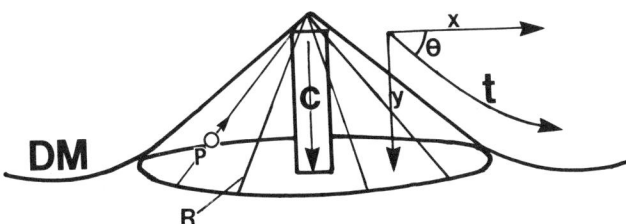

Figure 6. Model for mechanical interaction of dermal membrane and primary fibers. Shown is one cone-shaped conule formed by dermal membrane stretched over a primary fiber (rectangle). Tension in membrane will have a horizontal x component and a vertical y component. The latter compresses the primary fiber. Transport of sand particles to conule apices occurs along radially arranged lines of cells. C, compression; DM, dermal membrane; P, particle; R, radial cell line; t, tension.

mal membrane tension could directly affect the transport of sand particles to primary fibers, which occurs along lines of cells in the membrane that radiate from conule apices (Teragawa, 1986); and (b) membrane tension could also affect secretion of spongin at the tips of primary fibers forming the conule apices. If membrane tension could regulate primary fiber growth, it could also affect sponge morphogenesis, since primary fiber growth at conules determines the overall shape of the sponge.

One way to achieve tensile regulation of skeletogenesis would be through a dynamic balance between tensile stresses in the dermal membrane and compressive stresses in the primary fibers; the resulting self-regulating feedback system could control processes relevant to skeletal growth (Figure 6). Tension maintained in the dermal membrane will have a lateral, x component and a vertical, y component, and the magnitude of both components will depend on the angle of curvature (θ) of the dermal membrane. Since the membrane forms a curved surface, θ and the magnitude and direction of tension will be changing continuously in the membrane. If the sponge dermal membrane is similar to the tension distributions in the one-dimensional situation of a cable in a suspension bridge (Hibbeler, 1978), maximal tension will occur where θ is maximum, at the conule apex. As the sponge grows, the geometry of the conules and values of the angle θ can change, owing to divergent growth of the radially arranged primary fibers or nonproportional growth of skeletal fibers relative to sponge tissues (Lendenfeld, 1889). Large values of θ will thus increase membrane tension, which might slow particle transport; the y component of this tension will also create greater compression of primary fibers. Increased compression of primary fibers may halt processes of particle incorporation into fibers and spongin secretion; growth would not resume until stresses in the fibers and the membrane are relieved by tissue growth.

More information on in situ mechanical loading of

sponge fibers is needed to further develop models of mechanical regulation of skeletogenesis and to investigate the functional morphology of the sponge skeleton and the mechanical role of particles or spicules in fibers. Ideally, it would be desirable to examine the mechanical properties of whole skeletal networks and dermal membranes. Changes in dermal membrane shape could be measured after incisions are made and the results used to calculate the in situ strains sustained by the membrane. In addition, measurements of the tensile Young's modulus of the membrane could be used to calculate the magnitude of the in situ stresses in the membrane and skeletal fibers. Long-term growth experiments that alter tension in the dermal membrane while monitoring changes in skeletal growth and morphology would provide direct evidence for mechanical regulation of morphogenesis in sponges.

Conclusions

In the dictyoceratid sponge *Dysidea etheria* skeletal fibers with more particles were stiffer in bending because of the increase in their second moment of area. The sponge may increase the diameter of its fibers to compensate for the loss of material stiffness associated with increased particle content. Primary fiber stiffness may be important in maintaining the subdermal space and resisting compression generated by tensile stresses in the dermal membrane. In turn, membrane tension may regulate particle transport and the growth of primary fibers.

Acknowledgments

I thank P. C. Wainwright, S. A. Wainwright, S. Vogel, H. F. Nijhout, and M. LaBarbera for useful comments and assistance. I also thank the staff of the Bermuda Biological Station (BBS contribution number 1190) and the generous support of the Cocos Foundation, Exxon Corporation, Lerner-Gray Fund for Marine Research, Sigma-Xi, and the Duke University Graduate School.

Literature Cited

Alexander, R. McNeill. 1968. *Animal Mechanics*. Seattle: University of Washington Press. 346 pp.

Bergquist, P. R. 1978. *Sponges*. Berkeley: University of California Press. 268 pp.

Borojevic, R. 1971. Les comportement des cellules d'Eponge lors de processus morphogénétiques. *L'Année Biologique (series 4)*, 10:533–545.

Bray, D. 1984. Axonal Growth in Response to Experimentally Applied Mechanical Tension. *Developmental Biology*, 102:379–389.

Brøndsted, H. V. 1953. The Ability to Differentiate and the Size of Regenerated Cells after Repeated Regeneration in *Spongilla lacustris*. *Quarterly Journal of Microscopical Science*, 94:177–184.

Faupel, J. H. 1964. *Engineering Design, a Synthesis of Stress Analysis and Materials Engineering*. New York: John Wiley and Sons. 980 pp.

Harris, A. K., D. Stopak, and P. Wild. 1981. Fibroblast Traction as a Mechanism for Collagen Morphogenesis. *Nature*, 290: 249–251.

Hibbeler, R. C. 1978. *Engineering Mechanics: Statics*. (2d edition). New York: Macmillan. 483 pp.

Jackson, J. B. C. 1977. Competition on Marine Hard Substrata: The Adaptive Significance of Solitary and Colonial Strategies. *American Naturalist*, 111:743–767.

Jones, W. C. 1957. The Contractility and Healing Behavior of Pieces of *Leucosolenia complicata*. *Quarterly Journal of Microscopical Science*, 98:203–217.

Koehl, M. A. R. 1982. Mechanical Design of Spicule–Reinforced Connective Tissue: Stiffness. *Journal of Experimental Biology*, 98:239–267.

Lendenfeld, R. J. von. 1889. *A Monograph of the Horny Sponges*. London: Royal Society. 936 pp.

Nielsen, L. E. 1962. *Mechanical Properties of Polymers*. New York: Reinhold. 274 pp.

Palumbi, S. R. 1984. Tactics of Acclimation: Morphological Changes of Sponges in an Unpredictable Environment. *Science*, 225:1478–1480.

———. 1986. How Body Plans Limit Acclimation: Responses of a Demosponge to Wave Force. *Ecology*, 67:208–214.

Randall, J. E., and E. D. Hartman. 1968. Sponge Feeding Fishes of the West Indies. *Marine Biology*, 1:216–225.

Reif, W. E., and J. A. Robinson. 1976. On Functional Morphology of the Skeleton in Lychnisc Sponges (Porifera, Hexactinellida). *Paläontologische Zeitschrift*, 50:57–69.

SAS Insitute. 1982. SAS User's Guide: Basics and Statistics. Cary, N.C.: Stastical Analysis System Institute.

Shaw, M. E. 1927. Note on the Inclusion of Sand in Sponges. *Annals and Magazine of Natural History (series 9)*, 19:601–609.

Simpson, T. L. 1984. *The Cell Biology of Sponges*. New York: Springer. 662 pp.

Stopak, D., and A. K. Harris. 1982. Connective Tissue Morphogenesis by Fibroblast Traction. *Developmental Biology*, 90: 383–398.

Teragawa, C. K. 1985. The Biology of Particles in Skeletogenesis of a Keratose Sponge *(Dysidea etheria)*, Ph.D. Dissertation, Duke University, Durham, N.C. 237 pp.

———. 1986. Particle Transport and Incorporation During keleton Formation in a Keratose Sponge: *Dysidea etheria*. *Biological Bulletin*, 170:321–334.

Vincent, J. F. V. 1978. *Experiments with Biomaterials*. Nottingham, Great Britain: National Centre for School Technology. 107 pp.

Vogel, S. 1977. Current-induced Flow through Living Sponges in Nature. *Proceedings, National Academy of Sciences, (USA)*, 74:2069–2071.

———. 1978. Evidence for One-way Valves in the Water-flow System of Sponges. *Journal of Experimental Biology*, 76:137–148.

———. 1981. *Life in Moving Fluids, the Physical Biology of Flow*. Princeton, N.J.: Princeton University of Press. 352 pp.

Wainwright, S. A., W. D. Biggs, J. D. Currey, and J. M. Gosline. 1976. *Mechanical Design in Organisms*. New York: John Wiley. 423 pp.

Wilkinson, C. R. 1979. Skeletal Morphology of Coral Reef Sponges: A Scanning Electron Microscope Study. *Australian Journal of Freshwater Research*, 30:793–801.

CLAUDE DONADEY
CERAM, Faculté des Sciences de St Jérôme
13397 Marseille, Cedex 13, France

JEAN PARIS
Faculté des Sciences du Languedoc
34060 Montpellier Cedex, France

JEAN VACELET
Centre d'Océanologie de Marseille
Station Marine d'Endoume
13007 Marseille, France

Occurrence and Ultrastructure of Microraphides in *Axinella polypoides*

Abstract

Axinella polypoides Schmidt possesses siliceous raphides 20 μm in length, which, owing to their unusually small diameter (0.1 μm), are below the resolution of the light microscope and were overlooked in previous taxonomic studies. These "microraphides" are grouped in bundles within microsclerocytes; each cell contains about 1,000 parallel spicules. In section, these minute spicules do not display the characteristic fractures of the silica of mature spicules. However, their siliceous nature has been checked by EDS microanalysis, by dissolution with hydrofluoric acid on TEM preparations, and by observations in SEM after cleaning in nitric acid. They contain a relatively large organic axial filament, and their silicalemma is in direct contact with the cytoplasm of the microsclerocyte. The occurrence of these minute spicules in the type species of the genus *Axinella* confirms that raphides must be used with caution in taxonomic studies, especially as they may have been overlooked with the light microscope. Although this observation reduces the differences between *A. polypoides* and *A. dissimilis*, they are still considered distinct species.

The classification of siliceous sponges leans heavily on spicule description. However, the different kinds of spicules are not all given the same importance. Among microscleres, raphides, which are very thin spicules often forming bundles (trichodragmata), are generally regarded as being especially inconsistent with evolutionary lines, as they occur in widely separated groups of both Tetractinomorpha and Ceractinomorpha. However, they have been frequently used for specific and generic distinctions. For instance, in the Axinellidae, the distinction between the genera *Tragosia* Gray and *Phakellia* Bowerbank is based mainly on the presence of trichodragmata in the former (Dendy, 1922). In the same family, in the genus *Axinella* Schmidt, the genotype *A. polypoides* Schmidt, from the Mediterranean, was considered devoid of raphides, while a sibling species, *A. dissimilis* (Bowerbank) from the northeast Atlantic, which has sometimes been considered a synonym, possesses raphides. The taxonomic use of this microsclere has been criticized on the ground that it is inconstant in many species.

These thin spicules seem to be secreted in quite a different way from megascleres and most other microscleres: each spicule does not originate in a single cell, as usual, but many spicules form within a single sclerocyte (Sollas, 1888; Wilkinson and Garrone, 1980; Garrone et al., 1981). The ultrastructure of this special spiculogenesis is known only in one species (Wilkinson and Garrone, 1980).

The electron microscope observations presented here indicate that *Axinella polypoides* has raphides that are too small for resolution of the light microscope. Consequently, these spicules may have been overlooked in previous taxonomical studies of other sponges. It may be of interest to compare these microspicules to conventional ones, both for their morphology and for their mode of secretion, and to examine what may be their taxonomic significance.

Material and Methods

Specimens of *Axinella polypoides* were collected by SCUBA diving in the Mediterranean near Marseille and Banyuls. The sponges were fixed immediately after collection in 2.5% glutaraldehyde in seawater (20 h), postfixed in Osmium tetroxide 2% in seawater (2 h) and embedded in araldite. Ultrathin sections were cut using a diamond knife. They were contrasted with uranyl acetate and lead citrate, and examined with a Phillips EM 300 electron microscope. The silica was removed from some thin sections on grids by floating each grid face down on a 2.5% aqueous solution of hydrofluoric acid for 2 h, as described by Simpson and Vaccaro (1974).

For SEM observations, the sponge tissue was boiled in nitric acid. The spicule suspension was washed in several changes of distilled water and centrifuged in order to prevent loss of the tiny microscleres. The spicules were spread out on stubs, dried, sputter-coated with gold, and examined with a JEOL 35 CF electron microscope.

The silica content of the microspicules was checked by energy dispersion spectrometry (EDS) on ultrathin sections mounted on copper grids.

Specimens of *Axinella dissimilis* Bowerbank from Portugal and Ireland were provided through the courtesy of N. Boury-Esnault and B. E. Picton.

Results

TRANSMISSION ELECTRON MICROSCOPY. The choanosome of *Axinella polypoides* contains rare cells that exhibit a stacking of thin, parallel inclusions, which we interpret as microspicules. These cells (Figures 1, 6) are about 20 μm long and 6 μm wide and bear some short, irregular pseudo-

podia. The nucleus (4 μm in diameter) has no visible nucleolus. The cytoplasm contains mitochondria, rough-membraned vesicles, clear vesicles, and dark, homogeneous, spherical inclusions (0.6–0.8 μm in diameter) surrounded by a unit membrane. The Golgi complex is not apparent. Most of these inclusions and the nucleus are located at one side of the elongated cells. The remaining part of the cytoplasm contains numerous rods (0.1 μm in diameter) lying parallel to the great axis of the cell. The number of rods on a transverse section of a cell (Figure 6) is about 1,000. Each rod has a clear central area (20–40 nm in diameter) that very often contains a crystalline axial filament (Figures 2, 7). In cross sections (Figure 7), this axial area appears irregularly hexagonal or circular; its crystalline contents display a succession of dark lines 2 nm wide, separated by clear areas of 4 nm. In oblique or longitudinal sections (Figure 2), the axial filament shows a transverse banding pattern, with dark areas 4–5 nm wide and clear ones 20–40 nm wide. Around the axial filament, the rod is made of a dense, homogeneous material, which is not porous as described in the microscleres of *Neofibularia* (Wilkinson and Garrone, 1980) and does not display the usual shattered appearance of the silica of megascleres and larger microscleres in thin sections. There is no clear evidence of a distinction between dense and granular silica, as observed by Simpson and Vaccaro (1974) in freshwater sponges. The axial filament seems to have the same diameter up to the sharp extremity of the rod.

Each rod is tightly surrounded by an unit membrane 4.5 nm wide, which is interpreted here as a silicalemma (Figure 7). This membrane is in direct contact with the surrounding cytoplasm. Thus the spicule is not inside a vacuole, in contrast to what has been observed in other spicule genesis (Garrone et al., 1981) and in raphide genesis (Wilkinson and Garrone, 1980).

After a 2-h hydrofluoric acid treatment on ultrathin sections, only the axial filament remains (Figures 8, 9).

These spicules were not observed in the intercellular matrix, but were always intracellular.

EDS ANALYSIS. In EDS analysis performed on sections of the same cells (Figure 4), the ray Kα of silicon is found in the cytoplasm of the rod containing cells. The elemental map of Si Kα X-ray emission (Figure 3) of the same cell also shows the high silicon content of the cytoplasm.

SCANNING ELECTRON MICROSCOPY. After boiling in nitric acid, very thin raphides could be observed in SEM (Figure 5). Only a few microraphides have been found among hundreds of megascleres, but their scarcity may be due to the loss of these tiny microscleres during washing. They are straight, pointed at both ends, 12–21 μm in length and 0.10–0.16 μm in diameter. They could not be observed in light microscopy.

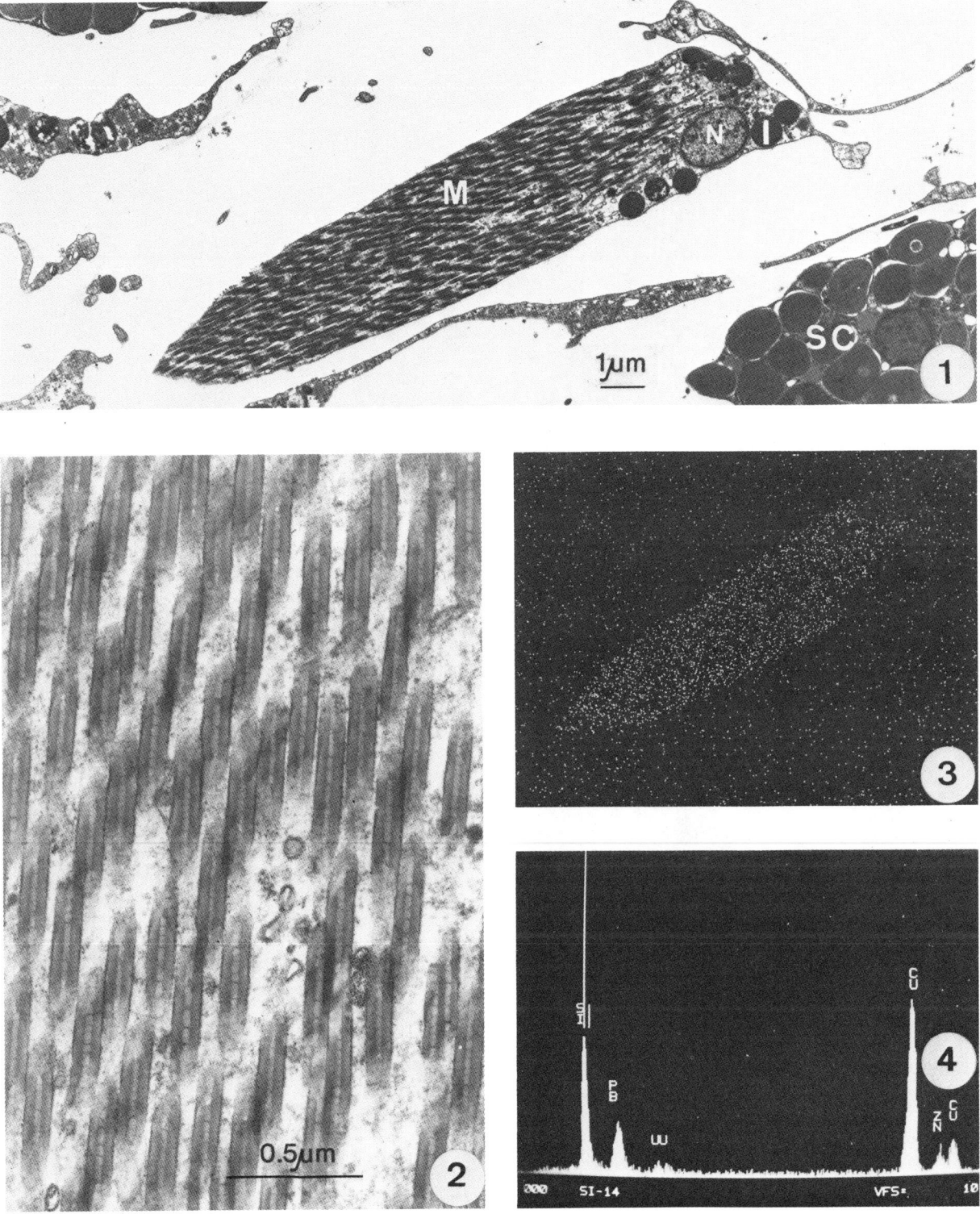

Figures 1–4. *Axinella polypoides*: *1*, microraphides in oblique section within a microsclerocyte; *2*, microraphides in oblique section, axial filament striated by dense, transverse bands; *3*, two-dimensional map of Si Kα X-ray emission from same field as *1*; *4*, EDS microanalysis from same field as *1* confirms siliceous nature of microraphides (Pb and U detection is due to contrasting agents used on section; Cu and Zn detection is due to grid components). *M*, microsclerocyte; *I*, dense inclusion; *N*, nucleus; *SC*, spherulous cell.

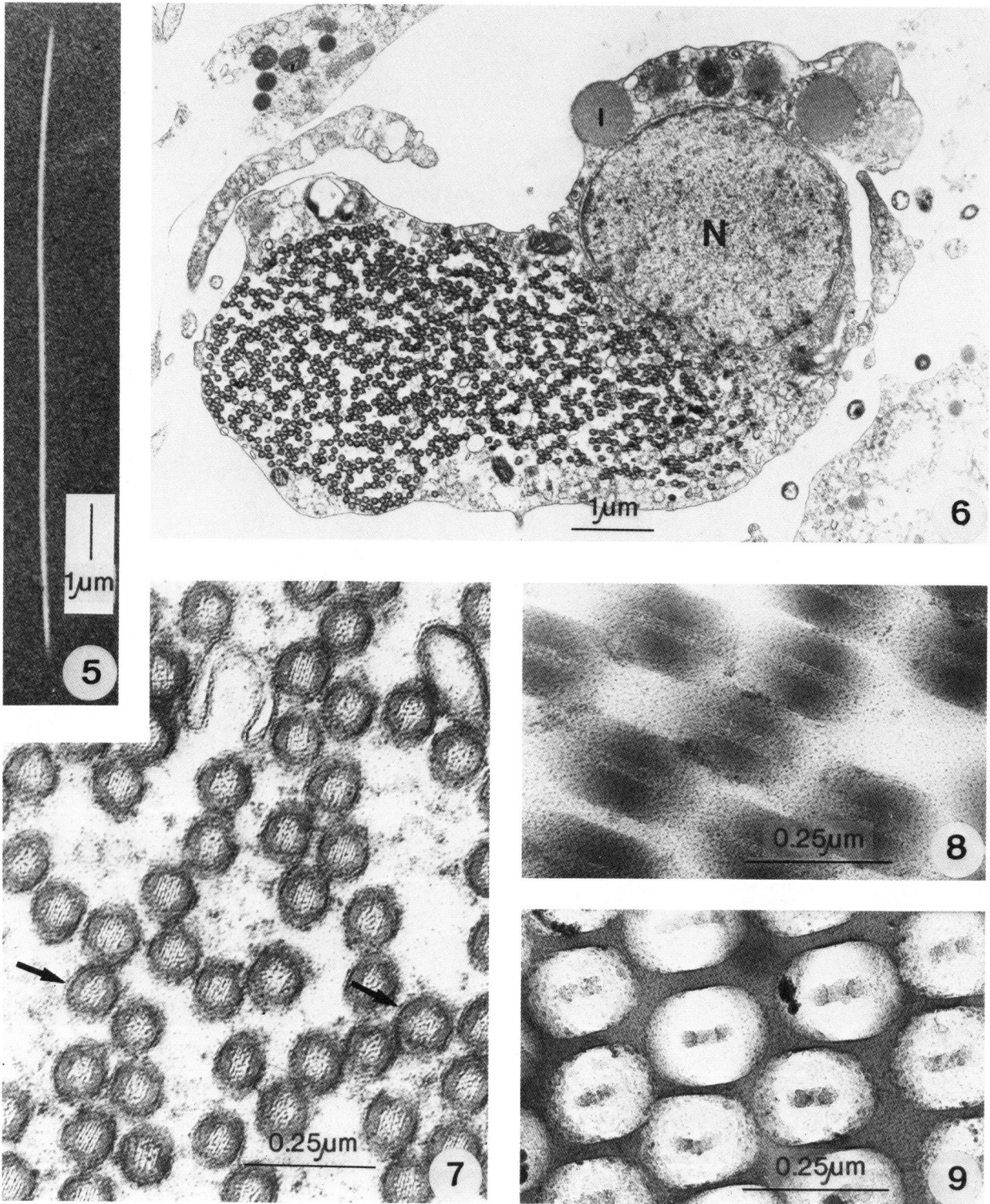

Figure 5–9. *Axinella polypoides: 5,* SEM picture of microraphide; *6,* microraphides within a microsclerocyte in cross section; *7,* microraphides in cross section (*arrows,* silicalemma); *8,* cross-section of microraphides before hydrofluoric acid treatment; *9,* cross-section of microraphides of same cell as in *8* after removal of silica by a hydrofluoric acid treatment. *I,* dense inclusions; *N,* nucleus.

Discussion

There is clear evidence that the intracellular rods described here are siliceous spicules: they have a high silicon content; they are dissolved by hydrofluoric acid; they display some morphological attributes of spicules, such as a crystalline axial filament; and spicules of similar size can be observed in SEM preparation after boiling in nitric acid. They appear to be similar to the raphides that are secreted together in the same sclerocyte and form trichodragmata.

However, these spicules, which can be called "microraphides," have some unusual features both in their morphology and in their mode of secretion. Their diameter is thinner than that of other previously known spicules and their size is below the resolution of the light microscope, which has been used in nearly all taxonomic studies and in most morphological studies. Their silica does not have the shattered appearance observed in larger spicules. Within their sclerocyte, the microraphides are surrounded by a silicalemma, as usual, but they are in direct contact with the cytoplasm and not within a vacuole.

It is noteworthy that these tiny spicules do have an axial organic filamant, and so it appears that this structure is always present in siliceous spicules, even in the thinnest ones. Sometimes the canal appears to be hexagonal, but most of the spicules observed in cross sections display a circular irregular axial canal. The ratio of filament diameter to spicule diameter has been calculated to be about 1:30 for both asters and triaenes in *Stelletta grubii*: this suggests a possible relationship between axial filament and spicule diameter (Simpson et al., 1985). This ratio is very different in the microraphides of *Axinella polypoides* (1:5 to 1:2.5) and does not confirm this suggestion.

It may be that the specimens examined here were immature spicules and that we observed early stages of secretion. According to a hypothesis of Wilkinson and Garrone (1980), the axial filament of the raphides of *Neofibularia* may each be secreted in different vacuoles, which connect during the course of silica deposition. It must be stressed that the microraphides of *Axinella polypoides* are not within a vacuole at this stage, although the silica is already present. The nonshattered appearance in thin sections may indicate incomplete mineralization. Therefore the spicules appear to be secreted directly within the cytoplasm.

SEM observations indicate that the diameter of the microraphides isolated by boiling in nitric acid is not significantly larger than that within sclerocytes. The slight difference observed may be due to the gold coating.

These observations may have further bearing on taxonomy. Raphides may be thin enough to be overlooked in light microscope observations. The genotype of *Axinella*, which has been considered to be devoid of microscleres, does in fact possess microraphides, which will perhaps be found in other species of the genus and in other sponges.

This also raises the problem of the distinction between the Mediterranean *Axinella polypoides* and *A. dissimilis* of the northeast Atlantic, which some have considered to be synonyms. These two species were previously separated mainly on the basis of the presence of raphides in *A. dissimilis*, but now are distinguished by the size of these raphides. The two sponges appear to be closely related in their morphological and skeletal features, and the lectin content confirms their close affinities (Bretting et al., 1981): unlike other *Axinella* species, both contain two lectins that show complete similarity except in inhibition studies with D N-Acetyl Galactosamin. However, according to Boury-Esnault (pers. comm.), the slight morphological differences and the change of color in alcohol (*A. dissimilis* turns grey whereas *A. polypoides* turns brown) are still evident in a mixing area, the Portugal (Sagres). We suggest that these two sponges be considered sibling species until a more thorough taxonomic comparison is made.

Literature Cited

Bretting H., C. Donadey, J. Vacelet, and G. Jacobs. 1981. Investigations on the Occurence of Lectins in Marine Sponges with Special Regards to Some Species of the Family Axinellidae. *Comparative Biochemistry and Physiology*, 70 B:69–76.

Dendy, A. 1922. Note on the Genus *Tragosia* Gray. *Annals and Magazine of Natural History (series 9)*, 11:169–174.

Garrone, R., T. L. Simpson, and J. Pottu. 1981. Ultrastructure and Deposition of Silica in Sponges. Pages 495–525 in *Silicon and Siliceous Structures in Biological Systems*, edited by T. L. Simpson and B. E. Volcani. New York: Springer.

Simpson, T. L., P. L. Langenbruch, and L. Scalera-Liaci. 1985. Silica Spicules and Axial Filament of the Marine Sponge *Stelletta grubii* (Porifera, Demospongiae). *Zoomorphology*, 105: 375–382.

Simpson, T. L., and C. A. Vaccaro. 1974. An Ultrastructural Study of Silica Deposition in the Fresh Water *Spongilla lacustris*. *Journal of Ultrastructure Research*, 47:296–309.

Sollas, W. J. 1888. Report on the Tetractinellida Collected by H.M.S. Challenger During the Years 1873–1876. *Report, H. M. S. Challenger, Zoology*, 25:1–458.

Wilkinson, C. R., and R. Garrone. 1980. Ultrastructure of Siliceous Spicules and Microsclerocytes in the Marine Sponge *Neofibularia irata* n. sp. *Journal of Morphology*, 166:51–54.

TRACY L. SIMPSON
Department of Biology and Health Sciences
University of Hartford
West Hartford, Connecticut 06117

Recent Data on Patterns of Silicification and the Origin of Monaxons from Tetraxons

Abstract

Classically, the four rays (axes) of tetraxon spicules are thought to give rise to diactinal and monactinal shapes by the loss of spicule rays. However, the direction of evolution (tetraxon to diactine and monactine or diactine and monactine to tetraxon) is unknown. The growth of diactinal and probably monactinal spicules involves silicic acid processing centers, as indicated by the accumulation of silica which can be induced by the incorporation of germanic acid into the structures. In diactinal spicules, pulsed exposure to germanic acid confirms the presence of such centers. These data support the classical view that the rays of tetraxons are related to those of diactines and monactines. The presumed silicic acid processing centers in birotules occur distally at each end of the spicule. Consequently, it is not clear how these structures are related to tetraxons, diactines, or monactines. However, they may indicate an unsuspected relationship to the asters of some tetractinomorphs. It may be that the pattern of growth of the organic axial filament does not always correspond to the pattern of silica deposition.

The process of silicification in sponges is important in a number of contexts. First, it represents an opportunity to study the complex biological processes involved in the polymerization of silicic acid and its deposition in the form of glass (silica) structures that constitute a portion of the skeleton of many living and fossil sponges. Siliceous skeletons, usually in the form of cell walls, are typically elaborated by lower (protistan) organisms and the use of silicic acid for biomineralization is consequently considered a process that harkens back to the early forms of eucaryotic life. The assumption underlying such a view is that the element silicon (Si) may have been important in evolution; indeed, some present-day organisms (diatoms) depend on Si for their perpetuation (Sullivan and Volcani, 1973).

Second, studies of biosilicification have the potential of providing new insights into the possible role of Si among higher multicellular organisms that lack the ability to

264

polymerize silicic acid and thus do not form siliceous skeletons. The classical view that Si is a biologically inert element in the latter groups has been discredited (Simpson and Volcani, 1981). Further, because sponge silica is deposited in very precise and predictable forms, knowledge of the sponge biosilicification system can increase our understanding of the determinants of these specific shapes. Such insight may clarify more general questions about the origin of shape and form in biological systems.

Finally, the shapes of sponge silica—that is, of spicules—have long been considered a kind of index of their evolutionary origin within the phylum (Reid, 1970; Hartman, 1981). More specifically, within some groups (Homoscleromorpha and Astrophorida) of the Demospongiae the tetraxon spicule shape with four converging axes has traditionally been considered ancestral to (1) the diactinal spicule with two pointed or similar ends and two presumed axes and (2) the monactinal spicule with one pointed end and one axis, both of which occur in other groups (Ceractinomorpha). This study presents the first experimental data that appear to support this traditional interpretation. However, as discussed later, our present knowledge of sponge silicification is far from complete, and the conclusions reached here must be considered in this light.

Materials and Methods

Newly developed spicules were obtained from hatching gemmules. In the freshwater species investigated (Spongilla lacustris, Ephydatia muelleri, E. fluviatilis), a synthetic freshwater medium (Rasmont, 1961) or 50% commercial mineral water (Evian) was the basal medium to which GeO_2 was added, routinely yielding a Ge:Si molar ratio of 0.1. Newly hatched sponges were unfed and raised at 20°C in moist chambers. Newly deposited spicules were cleaned in nitric acid, washed, and prepared either for light microscopy and autoradiography (Davie et al., 1983) or for scanning electron microscopy (SEM) and SEM-coupled, wavelength dispersive, X-ray analyses (Simpson et al., 1983). In one case, spicule (birotule) formation was induced by means of Theophylline (Rasmont, 1974; Simpson and Rodan, 1976). For the marine species examined (Suberites domuncula), filtered seawater acted as the basal medium to which GeO_2 was added. These techniques are described in detail elsewhere (Simpson et al., 1983; 1985a; Simpson and Langenbruch, 1984).

Results

CELLULAR BASIS OF SILICA DEPOSITION

Present data (see further Garrone et al., 1981; Simpson, 1984) indicate that silica spicules are elaborated intracellularly within a membrane (silicalemma)-limited cavity that contains a thin, organic axial filament. Deposited silica eventually surrounds and encloses the filament,

which thus comes to have the same overall shape as the spicule. Three distinct processes can be recognized: (1) filament assembly and growth, (2) membrane assembly and growth, and (3) silicic acid $[Si(OH)_4]$ accumulation and polymerization. Probably a fourth important process should also be considered—the arrest of the above three processes, which leads to a final, stable product.

Virtually all spicules examined to date (Simpson, 1984; Simpson et al., 1985b; Donadey et al., this volume) possess a central, axial filament or, if the filament has been lost, a central axial canal. Thus, it can be concluded that silica spicules all contain an organic core having the same overall shape as the spicule in which it occurs. Accordingly, all spicules can be said to appear similar, and it is only shape and size that distinguish one morphological type from another. That is to say, the pattern of assembly and growth of the silicalemma and filament coupled with the pattern of silicification determine specific shape and size.

GERMANIC ACID INCORPORATION INTO SILICA

Two independent methods—the isotope [68]Ge and SEM-coupled, wavelength dispersive X-ray analysis—were used to establish that when spicules are undergoing silicification, germanium is directly incorporated into the silica (Figure 1a,b). Although the chemical form in which germanium occurs is not known, its incorporation can result in a sensitivity to the preparative procedures for cleaning spicules. With successive cleaning in sodium hypochlorite and concentrated nitric acid, spicules in which germanium is incorporated can become noticeably etched and may even dissolve (Figure 1c). This occurrence may indicate the presence of a significant amount of organic material in the experimental spicules, which is consequently sensitive to oxidation. In any case, such etching acts as a marker of silica that contains incorporated Ge.

DIACTINES (OXEAS) OF FRESHWATER SPONGES

EFFECTS OF CONTINUOUS EXPOSURE TO $Ge(OH)_4$. The primary morphological effect of $Ge(OH)_4$ on developing spicules is the formation of a large central bulb (Figure 1c) and sometimes of smaller, secondary, more distal bulbs contrasting the remainder of the spicule which is very thin. Such an effect has now been demonstrated in Ephydatia muelleri, E. fluviatilis, Spongilla lacustris and other sponges (see Elvin, 1972). Other Ge effects on spicule dimensions and number also occur (Simpson et al., 1979) but are not pertinent to the present discussion. Note, however, that the bulbous spicules are much shorter (40%) and therefore that growth of the axial filament and the elongation of the silica must have been inhibited.

EFFECTS OF DELAYED ADDITION OF $Ge(OH)_4$. If newly developing spicules in control medium are exposed to

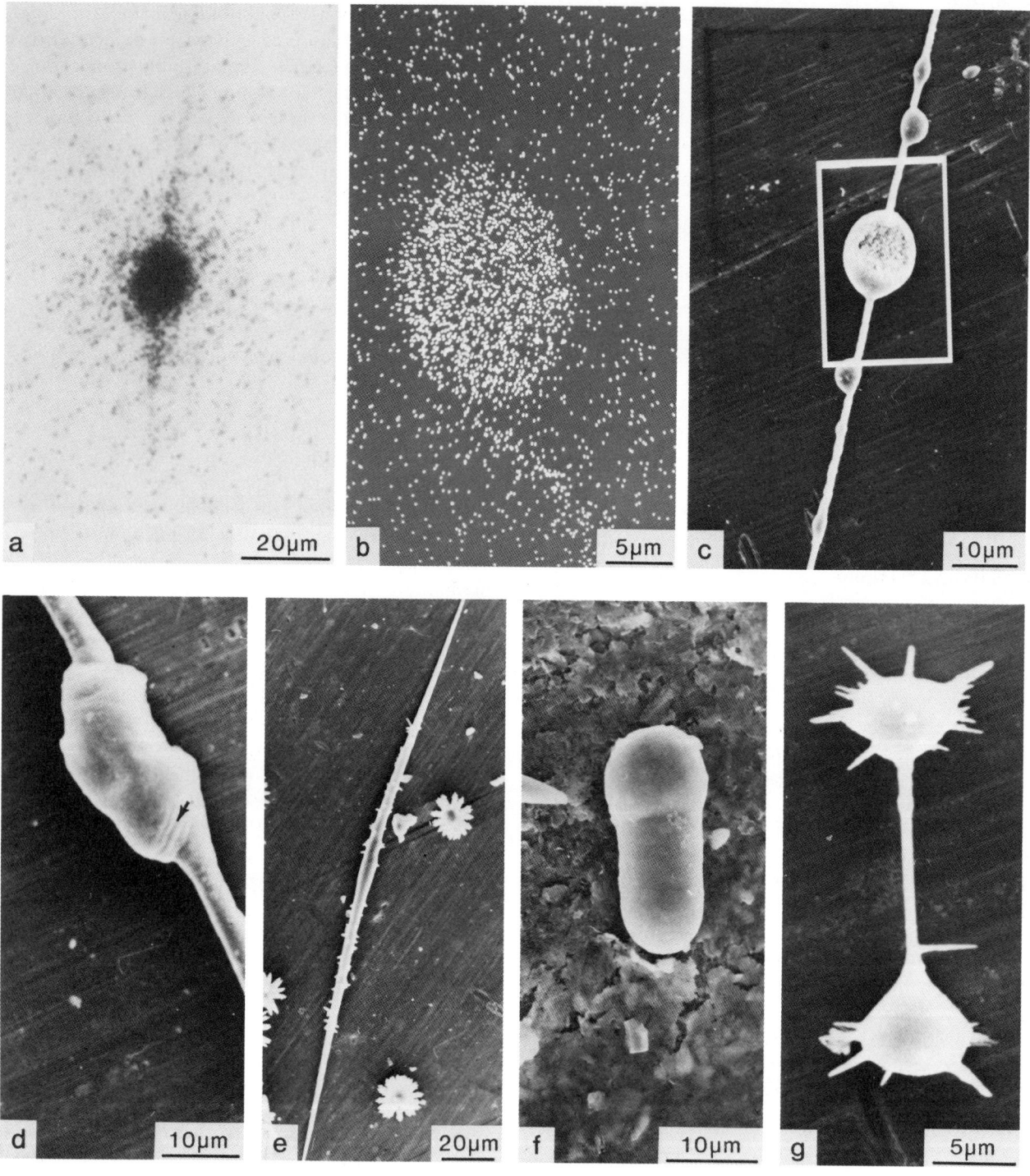

Ge(OH)$_4$ during their later development, etched areas occur in the central region of the spicules and sometimes just lateral to them; some spicules may develop unusual central bulblike structures, the edges of which have a distinct discontinuity in the silica (Figure 1d). Older, developing spicules similarly display central bulbs and discontinuities. The diversity of responses to pulsed Ge(OH)$_4$ may be partly due to the fact that spicules do not appear synchronously, so that different stages in spicule initiation and growth must be exposed to Ge. Further, the concentration of Ge(OH)$_4$ as it enters the tissue is likely to be variable from one region to the next and can thus introduce an additional source of nonuniformity of morphological response. These experiments clearly suggest that Ge(OH)$_4$, and thus Si(OH)$_4$, are transported or processed centrally in these spicules.

EFFECTS OF REMOVAL OF Ge(OH)$_4$. When Ge(OH)$_4$ is removed from young spicules that initially formed in its presence, the structures begin to recover normal morphology (Figure 1e) by becoming thicker and by developing short surface spines. The fact that both of these "improvements" occur in the central region of the spicule rather than at the thin tips clearly supports the conclusion that there is a central processing area in the silicification system.

MONACTINES (TYLOSTYLES) OF THE MARINE SPECIES Suberites domuncula

Like oxeas, Ge(OH)$_4$ reduces the length of tylostyles, but does not induce bulbs or reduce spicule width (Figure 1f). Despite these differences, germanium is still directly incorporated into the silica, as demonstrated by SEM-coupled, X-ray analyses. The very short structures still exhibit distinct swelling on the spicule head, as in normal tylostyles, and their widths are within the normal range. These results indicate that tylostyles grow outward from the head of the spicule, but they do not suggest per se the presence of a center for silicic acid processing as in diacti-

nal oxeas of freshwater sponges. Such a center may, nonetheless, be present, but may be relatively insensitive to Ge(OH)$_4$; the silicification system may be less selective (more premissive) and thus Ge may be processed in the same way as Si, without local accumulation. Alternatively, the biochemistry of the silicification events may be modified by the high salt content of seawater thus eliminating Ge accumulation and bulbs. Observations of young control spicules demonstrate that, in the very early stages of silicification the spicule head is silicified first and thus indeed appears as a center.

EFFECTS OF Ge(OH)$_4$ ON THE BIROTULES OF Ephydatia fluviatilis

An unexpected result of these investigations is the finding that Ge(OH)$_4$ induces the formation of two bulbs at the extremities of birotule microscleres; the intervening shaft is exceedingly thin (Figure 1g). The rotules may possess long, thin spines or may be relatively unspined. In the young stages of normal (control) spicules, the shaft appears first, followed by the rotules, but the two probably occur almost simultaneously since few young spicules can be found with only shafts; further, in the latter, the emerging rotule can always been seen. Taken together, the results indicate that there are two centers for Si(OH)$_4$ processing at the extremities of the spicule. Ge is similarly incorporated into the silica of birotules, as shown by SEM-coupled X-ray analyses.

Discussion

One of the basic problems in assuming that spicule morphology and the mechanism of silicification indicate taxonomic relatedness is that nothing is known of the selective advantages (if any) of a particular spicule shape or, for that matter, of siliceous spicules. Whether specific spicule morphology enhances other biological processes critical for survival remains unknown. However, it has been found that, in the absence of spicule formation, the

---

Figure 1. Observations on sponge spicule formation in presence of germanium: a, autoradiograph of an oxea (diactine) of Spongilla lacustris formed in a medium containing the isotope ^{68}Ge; it was subsequently acid cleaned and placed under photographic emulsion (note labeling of large central bulb and also of thinner shaft); b, SEM X-ray map of Ge localization in large bulb produced on oxea of Ephydatia muelleri grown in Ge-containing medium (this technique is not sufficiently sensitive to image the very thin spicule shaft); c, SEM of oxea of Ephydatia muelleri formed in Ge-containing medium; inside rectangle, a large central bulb has been partially eroded by preparative procedures; smaller secondary bulbs occur more distally; d, SEM of center of oxea of Ephydatia muelleri initially formed in the absence of Ge and then pulsed with Ge; accumulation of silica demonstrated at its borders by distinct discontinuities in silica (arrow) (note also asymmetry of bulb relative to shaft); e, SEM of oxea of Ephydatia muelleri initially formed in Ge-containing medium and then Ge withdrawn (note recovery of more normal morphology—thicker shaft and spines—in central region of spicule whereas spicule tips remain thin and unspined); f, SEM of tylostyle (monactine) of Suberites domuncula formed in Ge-containing medium; spicule extremely short, but enlarged spicule head still present and spicule thickness normal, indicating that these spicules grow unidirectionally outward from head; g, SEM of birotule of Ephydatia fluviatilis formed in Ge-containing medium; spicule possesses two terminal bulbs with rudimentary spines not fully grown outward, central shaft very thin.

overall development of young freshwater sponges is modified (Yourassowsky and Rasmont, 1983); thus siliceous spicules may be essential as support and connective tissue component for the normal modeling of other tissues. If this is the case for all sponges that form siliceous spicules, these products can be considered crucial for development, but the importance of their specific shape is still undetermined.

Recent data on the homoscleromorphs (Donadey, 1979; Boury-Esnault et al., 1984), many of which form tetraxonid spicules (calthrops), clearly demonstrate similar underlying tissue organization and thus suggest a coupling of the two. In hexactinellids, the hexactine is consistently associated with the unusual syncytial architecture of the animals (Mackie and Singula, 1983; Reiswig, this volume). Similarly, among astrophorids, triaenes (with four axes) are associated with a complex cortex development (Simpson et al., 1985c) and small choanocyte chambers (Boury-Esnault et al., this volume). Thus, in these cases specific spicule morphology clearly seems to be related to the tissue and cellular organization of the overall animal. In the absence of such data in other groups of demosponges, we find ourselves having to accept a priori, that specific spicule morphology is indicative of specific tissue structure (and of phylogeny).

The classical view of the evolution of demosponges via the reduction of spicule axes is thoroughly reviewed by Reid (1970) and little light has been shed on its validity since then (Figure 2). Indeed, as critically reviewed by Reid (1970) and more recently reiterated by Wiedenmayer (1980), the fossil evidence can be taken to suggest that the monaxonal-type spicule appeared first and thus is possibly the starting point for other demosponge spicule shapes (Finks, 1970). However, in Reid's (1970) view it is more likely that the phylogenetic sequence was from tetraxon to monaxon, not the reverse. Our studies focus on the more basic question of the possible existence of a physiological relationship between monaxons and tetraxons. That is to say, is there any observable relationship in the mode of formation of these two major types of skeletal structures?

Our preliminary results suggest that diactines (oxeas) possess a central silicic acid processing region and that monactines (tylostyles) posses a similar, terminal region. These conclusions coincide with the view that, developmentally, diactines possess two axes and that tylostyles possess one. Thus, the four axes of tetraxons can be developmentally related to both of these spicule, forms through reduction of axes (Figure 2). If such a relationship is evolutionarily valid, it may well be restricted to the tetraxons of the homoscleromorphs since those of the astrophorids probably belong to a different lineage. The latter are enormous structures in which the axial filaments appear significantly different (Simpson et al., 1985b). Interestingly, oxeas having two "axes" appear closer to tetraxons than

tylostyles which have a single axis. This similarity, albeit a narrow one, suggests that the haplosclerids are more closely related do the homoscleromorphs than do the hadromerids. However, such a suggestion raises several problems, both systematic and phylogenetic.

At present, we have no information on a critical aspect of spicule axis growth—the mechanism of growth of the organic, axial filament that cores each axis. The results reported here pertain to the pattern of silicification, that is, to mineralization, not to the assembly of this organic core. In the case of oxeas, the filament probably grows at the tips. If this is the case, the growth of the organic component is not altogether comparable to that of the mineral, although both do grow outward in two directions. Similarly, if the filament in tylostyles extends at the tip, its growth is unidirectional but not comparable to the probable pattern of silica desposition. It is not clear how to reconcile these differences which, however, do not significantly affect the overall pattern of growth of the whole spicule. Filament growth clearly needs to be studied in greater depth.

Spicule homologies and their evolutionary significance cannot be fully understood without precise knowledge of the biochemical and transport pathways involved in silica deposition. Although we are far from such an understanding, the effects of germanic acid incorporation into the deposited silica has established that in some cases (1) there are centers for silicic acid processing and (2) the transported species of silicic acid (dimers, trimers, etc.) are mobile and can be translocated along spicule axes. The new finding (Holvoet and Van de Vyver, this volume) that bipyridine can induce the same morphological abnormalities—namely, bulb formation—as does germanic acid suggests that the silicic acid processing center embodies a divalent cation that, according to our results, may be involved in the mobility of the transported silicic acid. Future investigations with bipyridine show considerable promise.

The effect of germanic acid incorporation into the birotules of *Ephydatia fluviatilis* suggests further complications in establishing spicule homologies. The presence of two distal silicic acid processing centers does not easily fit into interpretations of the relationship of tetraxons to monaxons. Because birotules are secreted by species that also elaborate diactines (oxeas), it is clear that a single animal can possess more than one type of silicification system, which doubtless also embodies differences in the assembly of the axial filament. Furthermore, it is not clear which of these two systems is likely to reflect a more "primitive" condition. If the multiple axes, each of which contains a filament branch (Drum, 1968; Garrone et al., 1981) at the ends of birotules, are considered to be similar to the multiple axes of asters—each of which in *Stelletta grubii* also contains a branch of the axial filament (Simpson et al., 1985b)—then some freshwater haplosclerids

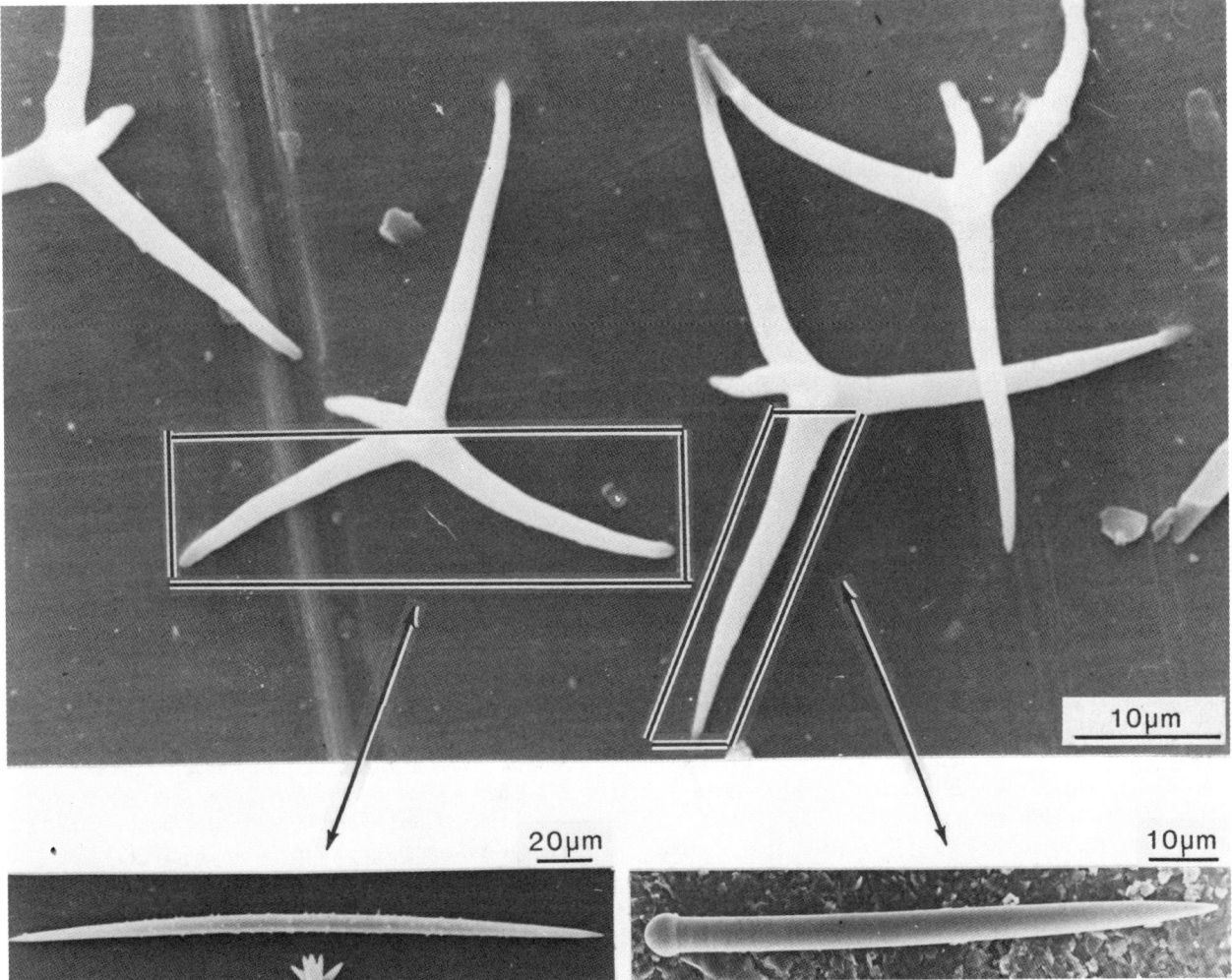

Figure 2. Representation of classical view that through loss of spicule rays tetraxon spicules are evolutionarily related to diactinal and monactinal spicules. Above, tetraxons (calthrops) of a homoscleromorph, *Corticium* sp. Boxed area to left contains two of four spicule rays depicted *(arrow)* as giving rise to a diactine (oxea of *Ephydatia muelleri*) through loss of remaining two rays. Boxed area to right contains one of four spicule rays depicted *(arrow)* as giving rise to a monactine (tylostyle of *Suberites domuncula*) through loss of remaining three rays. Arrows are double-headed to indicate that it is not known whether evolution has occurred from tetraxons to diactinal and monactinal forms or the reverse. All photos SEM micrographs.

may well be related to tetractinomorphs. The most that can be stated on the basis of the present results is that there is a good possibility that the axes of tetraxons are related to those of monactines and diactines. The precise manner in which they are related remains unclear.

Conclusions

Germanic acid is incorporated into siliceous spicules. At significant concentrations it induces central bulbs in the diactines (oxeas) of freshwater sponges. Pulsed exposure to germanium demonstrates that the spicules grow bidirectionally. The tylostyles of *Suberites domuncula* do not develop similar bulbs. Germanium inhibits their growth

in length which is unidirectional. Young, normal spicule stages indicate that the spicule head forms first and that the spicule grows outward from it. In the birotules of *Ephydatia fluviatilis*, germanium induces the formation of two terminal bulbs, which grow multidirectionally at right angles to the axis. These results indicate that there are three different patterns of silicification in these three spicule types. The patterns in oxeas and tylostyles can be related to the axes of the tetraxonid (calthrop) spicule.

Acknowledgments

Part of the research reported here was supported by a fellowship in the U.S.-France Cooperative Science Program funded jointly by National Science Foundation and Centre National de la Recherche Scientifique. Support and cooperation were also generously provided by N. Weissenfels and P.-F. Langenbruch, the Zoological Institute, University of Bonn, West Germany; J. Paris and R. Connes, the Laboratory of Animal Biology, University of Languedoc, Montpellier, France; and M. Pavans de Ceccatty and R. Garrone, the Laboratory of Experimental Histology, University Claude Bernard, Villeurbanne, France. Part of this work was conducted during a sabbatical and extended leave of absence granted by the University of Hartford.

Literature Cited

Boury-Esnault, N., L. DeVos, C. Donadey, and J. Vacelet. 1984. Comparative Study of the Choanosomes of Porifera: I. Homoscleromorpha. *Journal of Morphology*, 180:3–17.

Davie, E. I., T. L. Simpson, and R. Garrone. 1983. Experimental Germanium Incorporation into Siliceous Sponge Spicules. *Biology of the Cell*, 48:191–202.

Donadey, C. 1979.Contribution à l'étude cytologique de deux Démosponges Homosclerophorides: *Oscarella lobularis* (Schmidt) et *Plakina trilopha* (Schulze). Pages 165–172 in *Biologie des Spongiaires*, edited by C. Lévi and N. Boury-Esnault. Colloques Internationaux du C.N.R.S. 291, Paris: Centre National de la Recherche Scientifique.

Drum, R. 1968. Electron Microscopy of Siliceous Spicules from the Freshwater Sponge *Heteromyenia*. *Journal of Ultrastructure Research*, 22:12–21.

Elvin, D. 1972. Effect of Germanium upon Development of Siliceous Spicules of some Freshwater Sponges. *Experimental Cell Research*, 72:551–553.

Finks, R. 1970.The Evolution and Ecologic History of Sponges during Paleozoic Times. Pages 3–22 in *The Biology of the Porifera*, edited by W. G. Fry. Symposia of the Zoological Society of London 25. London: Academic Press.

Garrone, R., T. L. Simpson, and J. Pottu-Boumendil. 1981. Ultrastructure and Deposition of Silica in Sponges. Pages 495–525 in *Silicon and Siliceous Structures in Biological Systems*, edited by T. L. Simpson and B. E. Volcani. New York: Springer.

Hartman, W. D. 1981. Form and Distribution of Silica in Sponges. Pages 453–493 in *Silicon and Siliceous Structures in Biological Systems*, edited by T. L. Simpson and B. E. Volcani. New York: Springer.

Mackie, G. O., and C. L. Singula. 1983. Studies on Hexactinellid Sponges. I. Histology of *Rhabdocalyptus dawsoni* (Lambe, 1873). *Philosophical Transactions, Royal Society of London*, 301:365–400.

Rasmont, R. 1961. Une technique de culture des Eponges d'eau douce en milieu contrôlé. *Annales, Societé Royale Zoologique de Belgique*, 91:147–156.

———. 1974. Stimulation of Cell Aggregation by Theophylline in the Asexual Reproduction of Freshwater Sponges *(Ephydatia fluviatilis)*. *Experientia*, 30:792–794.

Reid, R. E. H. 1970. Tetraxons and Demosponge Phylogeny. Pages 63–89 in *The Biology of the Porifera*, edited by W. G. Fry. Symposia of the Zoological Society of London 25. London: Academic Press.

Simpson, T. L. 1984. *The Cell Biology of Sponges*. New York: Springer. 662 pp.

Simpson, T. L., R. Garrone, and M. Mazzorana. 1983. Interaction of Germanium (Ge) with Biosilicification in the Freshwater Sponge *Ephydatia mülleri*: Evidence of Localized Membrane Domains in the Silicalemma. *Journal of Ultrastructural Research*, 85:159–174.

Simpson, T. L., M. Gil, R. Connes, J-P. Diaz, and J. Paris. 1985a. Effects of Germanium (Ge) on the Silica Spicules of the Marine Sponge *Suberites domuncula*: Transformation of Spicule Type. *Journal of Morphology*, 183:117–128.

Simpson, T. L., and P.-F. Langenbruch. 1984. Effects of Germanium on the Morphogenesis of a Complex Silica Structure and on the Assembly of the Collageneous Gemmule Coat in a Freshwater Sponge. *Biology of the Cell*, 50:181–190.

Simpson, T. L., P-F. Langenbruch, and L. Scalera-Liaci. 1985b. Silica Spicules and Axial Filaments of the Marine Sponge *Stelletta grubii* (Porifera, Demospongiae). *Zoomorphology*, 105: 375–382.

———. 1985c. Cortical and Endosomal Structure of the Marine Sponge *Stelletta grubii*. *Marine Biology*, 86:37–45.

Simpson, T. L., L. M. Refolo, and M. Kaby. 1979. Effects of Germanium on the Morphology of Silica Deposition in a Freshwater Sponge. *Journal of Morphology*, 159:343–354.

Simpson, T. L., and G. A. Rodan. 1976. Recent Investigations of the Involvement of 3′, 5′ Cyclic AMP in the Developmental Physiology of Sponge Gemmules. Pages 83–97 in *Aspects of Sponge Biology*, edited by R. W. Harrison and R. R. Cowden. New York: Academic Press.

Simpson, T. L., and B. E. Volcani (eds.). 1981. *Silicon and Siliceous Structures in Biological Systems*. New York: Springer. 587 pp.

Sullivan, C. W., and B. E. Volcani. 1973. Role of Silicon in Diatom Metabolism. III. The Effects of Silicic Acid on DNA Synthesis in *Cylindrotheca fusisormis*. *Biochimica et Biophysica Acta*, 308:212–229.

Wiedenmayer, F. 1980. Siliceous Sponges, Development Through Time. Pages 55–86 in *Living and Fossil Sponges*. Notes for a short course, edited by W. D. Hartman, J. W. Wendt, and F. Wiedenmayer. Miami: University of Miami.

Yourassowsky, C., and R. Rasmont. 1983. The Differentiation of Sclerocytes in Freshwater Sponges Grown in a Silica-Poor Medium. *Differentiation*, 25:5–9.

Systematics, Biogeography, and Evolutionary Biology

JANE P. FROMONT*
PATRICIA R. BERGQUIST
Department of Zoology
University of Auckland
Private Bag, Auckland, New Zealand

Structural Characters and Their Use in Sponge Taxonomy: When Is a Sigma Not a Sigma?

Abstract

The new systematic techniques such as numerical taxonomy, morphometrics, and cladistic analysis rely heavily on structural characters as taxonomic discriminators and fail to recognize that many terms employed in sponge taxonomy to describe particular morphologies are broad descriptive terms and are not of themselves indicators of natural groups. An examination of cladistic analysis reveals that the characters employed in such analyses need to be adequately and precisely defined. Three specific cases serve to illustrate the points: the microsclere category "sigma," the megasclere form "acanthostyle," and erect surface structures termed "fistules."

Sponge classification systems have undergone many, frequently major modifications, particularly in recent years. Two of the main causes of this shifting and unstable classification are the paucity of characters available to the sponge taxonomist and the difficulty of specifying these accurately. Early workers such as Sollas, Schulze, Ridley, and Dendy relied on major spicule types and their distribution at suprageneric levels to categorize groups of species and genera (Lévi, 1957). Although structural characters are still regularly used, recent workers have recognized the need to incorporate a broader range of characters into taxonomic decisions. Thus ecological, biochemical, genetic, and reproductive data are now employed in addition to morphological characters.

Even though our knowledge of the biology of sponges has increased, many problems in systematics and phylogenetic relationships remain unresolved. An often-cited example is the grouping of the majority of species of the Demospongiae into two subclasses: the Ceractinomorpha, which have a viviparous reproductive mode and predominantly monaxonid spicules; and the Tetractinomorpha, which emphasize oviparous reproduction and have predominantly tetraxonid spicules. The small subclass Homoscleromorpha is considered to be quite separate.

The systematic disparity between ordinal groups of sponges is becoming increasingly evident, as is the great diversity at the species level. Consequently, an effort must be made to refine classification and to extend the criteria used in establishing relationships.

Cladistic Methods

Phylogenetic systematics or "cladistics," which first gained attention in recent times through the work of Hennig in 1950 (see Janvier, 1984), is based on the search for characters shared by two or more organisms. Since different characters probably evolved at different times, the

*Present address: Department of Marine Biology, James Cook University, Townsville, Q4811, Australia.

chronology of their occurrence in a group makes it possible to construct a phylogenetic tree in which ancestral species are inferred at branching points.

However, organisms can resemble each other with respect to advanced characters (synapomorphies) as well as primitive characters (symplesiomorphies). Hennig believed that synapomorphies alone indicate a relationship between organisms, and that all the organisms sharing a synapomorphy form a monophyletic group, which has its own history (Janvier, 1984). Synapomorphies must be homologous, that is, inherited from a common ancestor. This term is central in the ensuing discussion. Symplesiomorphies, or shared primitive characters, indicate a common ancestry that is not chronologically immediate, or exclusive, and hence such groups are termed paraphyletic.

A third group is recognized on the basis of shared nonhomologous characters that are convergent. Such groups are considered to be polyphyletic. For example, Vacelet (1985) and van Soest (1984a) have postulated that the class Sclerospongiae is polyphyletic.

Hennig also proposed criteria for determining whether characters are advanced or primitive, that is, whether they are apomorphous or plesiomorphous. His main criterion is embodied in the out-group rule, which states that if one takes a given character in members of a monophyletic group, and looks for its homologue in a sister group (i.e., a group more closely related to it than to any other group), then if the character occurs in both, it is plesiomorphous for the monophyletic group first considered. If it is absent in the sister group and unique to the group under study, it is an apomorphous character.

Cladists use a number of accessory criteria—such as ontogenetic character sequence, biogeographical succession, the paleontological record, and the correlation of character transformations—to determine character state

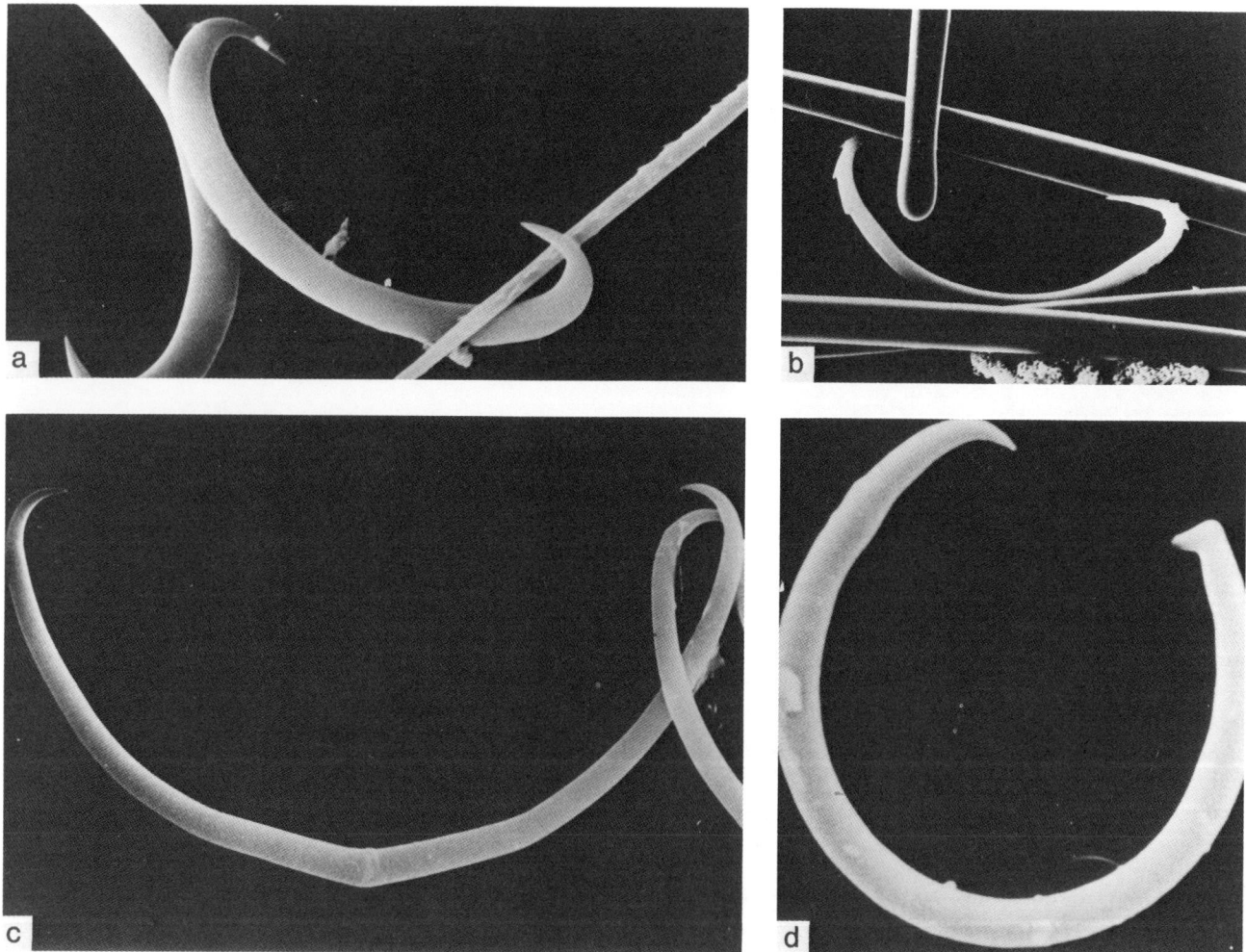

Figure 1. Examples of various forms of sigmas: *a*, C-shaped form of *Phorbas*; *b*, spined sigmas of *Paresperella*; *c*, centroangulate sigma of *Orina*; *d*, type of sigma that occurs in some mycalid species.

3d Int. Sponge Conf. 1985

but discussion of these is beyond the scope of this paper. The point here is that to establish a cladogram or a phylogenetic tree it is essential to use homologous characters.

Discussion

This point can be illustrated by using spicules as an example. The presence of the same general spicule type in a number of sponge entities cannot be held to imply homology. Consider those microsclere forms that are collectively titled "sigmas." Their common attribute is that they are C- or S-shaped: that is, they are curved but not reflexed microscleres without terminal expansions. "Sigma" usually refers to forms that are smooth, have sharp-pointed ends, and are shaped like a C or S. However, some have spines on the outside of the curve as in *Paresperella*, are centroangulate as in *Orina*, or are large and almost circular as in some mycalid species (Figure 1).

If the term "sigma" is interpreted as in Dendy (1921), it encompasses an even broader array of microsclere structures (Figure 2). Consider the profusely spined sigma of *Trachycladus*, which are frequently extremely contort and resemble a corkscrew; the small very faintly spined or "roughened" sigma that occurs in *Tetilla;* or the smooth contort sigmospires of *Rhabderemia* and *Rhabdosigma*. These few genera, which could be considered to have sigmoid microscleres, occur over four orders and span both subclasses of the Demospongiae.

In spicule nomenclature it has been the practice to apply one name to a general form as a broadly descriptive term that denotes the overall morphology of the spicule but does not indicate homology. Convergent forms, in a phylogenetic sense, that have the same broad morphological type can all justifiably be referred to by the same name. Hence, if the species mentioned earlier were grouped together because they contained "sigmas," the group would be polyphyletic in that the members share nonhomologous or convergent characters.

To be homologous, the character in question must to some extent exhibit similar morphogenetic processes dictated by a degree of genetic identity. The question then becomes, at what descriptive level is the homology recognizable? The genetic control of spicule geometry is not yet well understood, but the basic functional differences between sigmas packing a surface membrane and sigmas linked in tracts, accompanied by distinctive morphology, argue that the level of discrimination is far finer than "sigma" implies.

Acanthostyles provide a further example of the use of a broad descriptive term. Acanthostyles are megascleres that occur within the subclasses Tetractinomorpha, Ceractinomorpha, and the class Sclerospongiae. The term *acantho* refers to spining on the spicule, but this spining can vary considerably in pattern and morphology. Some acanthose spicules have faint spining localized to the

Figure 2. Sigmatose microscleres: *A*, two sigmas of *Tetilla baradensis*, ×870; *B*, sigma of *Gellius flagellifer*, ×320; *C*, sigma of *Axoniderma mirabile*, ×290; *D*, two sigmas of *Esperella porosa*, ×180; *E*, sigma of *Esperella simonis*, ×180; *F*, two sigmas of *Rhabderemia indica*, ×530; *G*, sigma of *Gellius angulatus* var. *canaliculata*, ×460; *H*, sigma of *Paresperella serratohamata*, ×350; *K*, spirula of *Trachycladus levispirulifer*, ×1000; *L*, chiastosigmas of *Myxilla pecqueryi*, ×160 (*L* after Topsent). (Reprinted from Dendy, 1921.)

head, as do the principal spicules of *Ectyodoryx*, or those of *Phorbas* which are more distinctive recurved spines (Figure 3). Acanthose spicules may be extensively spined over the whole shaft, either irregularly or in orderly whorls, a condition termed verticillate spining. Many authors have used the distribution of acanthostyles throughout a number of sponge families to express the relationship between these families. Using a cladistic mode, van Soest (1984b), for example, recently interpreted the presence of acanthostyles in the Poecilosclerida and certain axinellid groups as an indication of a close phylogenetic relationship, which earlier authors had suggested on other grounds (Bergquist et al., 1980).

Like "sigma," "acantho" is a broad descriptive term. It indicates the presence of spining on a spicule but not the degree, structure, or localization of the spining. Again, one name is applied to all spicules that express this condition and it does not selectively include those that are homologous. If incorporated into a cladistic analysis, the general character "acanthose megasclere" precludes the use of the specific details of structure that are expressions of differing morphogenetic processes and different genetic makeup.

A megasclere in a sponge skeleton is involved in maintaining the gross form of the sponge, and consequently megascleres have a function and a specific localization. This is not always true of microscleres. Thus acanthose spicules may occur as a dermal crust, echinating the primary skeletal tracts within the sponge, or in the skeletal tracts themselves. These patterns of tissue localization are almost as diverse as the variation in ornamentation of the

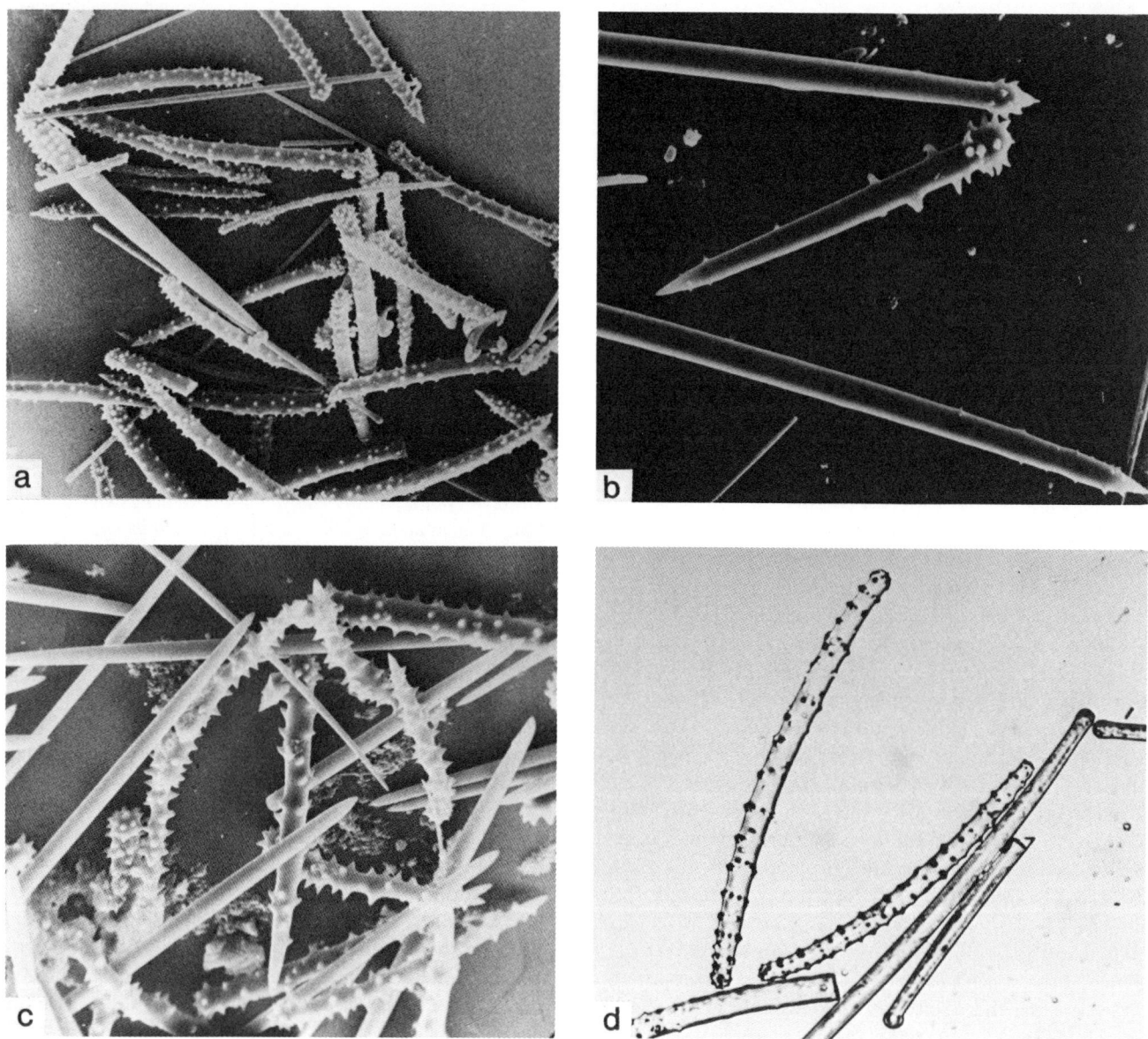

Figure 3. Various types of spining: *a*, spines on "head" of *Ectyodoryx* principal acanthostyle, and accessory acanthostyles; *b*, recurved spines on "head" of *Phorbas* acanthostyles; *c,* irregular spining on *Crella* acanthostyles; *d*, verticillate spined acanthostrongyles of *Zyzza massalis*.

spicules themselves. The fact that some are found either partly or completely embedded in spongin fiber whereas others occur interstitially or near the surface supports the idea that specific morphogenetic processes are associated with spicule localization. These factors must be taken into account when any "unit character" is used to analyze the relationship among forms, as occurs in cladistic techniques.

The "fistule" is another character that highlights the question of the descriptive level at which homology is recognizable. Its distribution has been used to express relationship, or lack of it within the Porifera. In its broad

descriptive sense, this term refers to elevated water conducting structures. These are erect tubes or papillae on the sponge surface that are generally associated with the conduction of water currents and that localize this function above the sponge surface. However, the morphology of these structures is extremely variable; ultrastructural work suggests that their function also varies.

For instance, Boury-Esnault (1974) has noted two types of papillae in *Polymastia*, those that have a terminal oscule and those that do not. However, both contain incurrent canals. Functionally there appear to be incurrent papillae and mixed papillae: in the former a central incurrent canal runs the length of the papillae and in the latter a single excurrent canal leads to a terminal oscule and the displaced incurrent canals are superficial and multiple.

Orina sagittara has a distinctive habit of a massive base with anastomosing fistules that subdivide terminally into a number of finer tubes, which expand into an apical oval structure. These structures may also occur on lateral branches and can become free-floating. Possibly they are involved in asexual reproduction (Figure 4).

Poecilosclerid genera such as *Hamigera* have highly organized, slightly raised "areolate" pore areas. Members of the families Hymedesmiidae and Phorbasidae have similar specialized pore regions which van Soest (1984b) has suggested indicate close relationship when considered in conjunction with skeletal and spicule similarities.

Van Soest (1984b) did not consider the tubular fistules of spherical coelosphaerids to be a good character on which to relate these genera because fistules of comparable morphology are associated with many sponge groups. If the skeletal and spicule characters of the Coelosphaeridae are also considered, these sponges do seem to constitute a close group. Although a fistule is thought to increase the surface area for water movement, its function may also be related to the ecology of the sponge; for example, *Siphonodictyon* has fistules that extend beyond the penetrated substrate, and sponges occurring in soft sediments (such as *Ciocalypta*) have long fistules, perhaps to prevent the sponge from being buried by silt. Similar surface structures may be active elements in asexual budding or may assist in dispersal of reproductive products.

Close attention to the basic morphology, detailed organization, and the function of "fistules" could shed some light on the affinity (or lack of it) among sponges that possess them. By interpreting the general descriptive term "fistule" as a unit character, cladistics precludes analysis of the individual form, function, and detailed morphology of these superficially similar structures.

Conclusions

A major obstacle for any worker in the field of sponge taxonomy or in general sponge biology is the huge number of terms used in classification. To complicate matters further, each term embraces an array of character states generated by incompletely known morphogenetic processes. Once this complexity is recognized, however, it becomes clear that attempts to group sponges by the presence of general morphological characters oversimplify this diversity. The use of "sigma," "acanthose," and "fistule" as homologous entities assumes it is not necessary to carry out the detailed morphogenetic studies that are essential if we are to identify the real genetic entities within these groups. Inevitably, relationships based only on simplistic

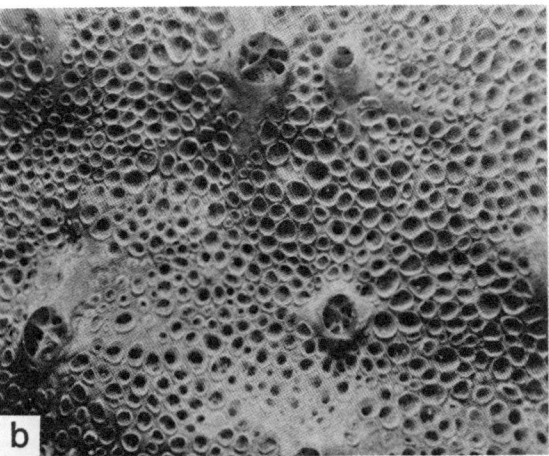

Figure 4. Examples of forms of fistules: *a, Polymastia granulosa*, papillae; *b, Hamigera* sp. (undescribed), "areolate" pore areas; *c*, coelosphaerid (undescribed), tubular fistules.

descriptive terms will be misleading and will deter sensible interpretations of sponge relationships.

Instead of oversimplifying, we must pay close attention to the processes responsible for the development of the sponge structures. Ultrastructual work on fistule types has already been published (Boury-Esnault, 1974). Although comparable work on microscleres and megascleres is still in its infancy, much is to be gained from this avenue of research. As the relatedness between groups becomes clearer with respect to truly homologous characters, we can expect to construct the cladograms and phylogenetic trees with a greater degree of probability. This will enable future workers to proceed without first having to determine when a sigma is not a sigma.

Literature Cited

Bergquist, P. R., W. Hofheinz, and G. Oesterhelt. 1980. Sterol Composition and the Classification of the Demospongiae. *Biochemical Sytematics and Ecology,* 8:423–435.

Boury-Esnault, N. 1974. Structure et ultrastructure des papilles d'éponges du genre *Polymastia* Bowerbank. *Archives de Zoologie Expérimentale et Générale,* 115:141–165.

Dendy, A. 1921. The Tetraxonid Sponge-Spicule: - A Study in Evolution. *Acta Zoologica,* 2:95–152.

Janvier, P. 1984. Cladistics: Theory, Purpose and Evolutionary Implications. Pages 39–75 in *Evolutionary Theory: Paths into the Future,* edited by J. W. Pollard. New York: John Wiley & Sons.

Lévi, C. 1957. Ontogeny and Systematics in Sponges. *Systematic Zoology,* 6:174–183.

Soest, R. W. M. van. 1984a. Deficient *Merlia normani* Kirkpatrick, 1908, from the Curaçao Reefs, with a Discussion on the Phylogenetic Interpretation of Sclerosponges. *Bijdragen tot de Dierkunde,* 54(2):211–219.

———. 1984b. Marine Sponges from Curaçao and other Caribbean Localities. Part III. Poecilosclerida. *Studies on the Fauna of Curaçao and Other Caribbean Islands,* 66(199):1–167.

Vacelet, J. 1985. Coralline Sponges and the Evolution of the Porifera. Pages 1–13 in *The Origin and Relationships of Lower Invertebrates,* edited by S. Conway Morris, J. D. George, R. Gibson, and H. M. Platt. Systematics Association Special Volume 28. Oxford: Clarendon Press.

RUTH DESQUEYROUX-FAUNDEZ
Muséum d'Histoire Naturelle
CH–1211 Genève 6, Switzerland

Silica Content of the New Caledonian Fauna of Haplosclerida and Petrosiida and Its Possible Taxonomic Significance

Abstract

Silica content, determined as total amount of spicules, in sponges was found to be a taxonomically important parameter in the New Caledonian reef fauna of Haplosclerida (families Niphatidae, Callyspongiidae, and Haliclonidae) and Petrosiida (families Petrosiidae and Oceanapiidae). Its relation to length and diameter of monaxon spicules was determined. We examined this measure as a systematic character in order to supplement more traditional taxonomic parameters employed in these groups.

The physiological mechanisms of silica deposition in marine sponges, the rate of deposition, and the conditions under which spicules are formed are poorly understood. Jones (1979, 1984), when studying spicule dimensions as taxonomic criteria in the identification of marine haplosclerid sponges, determined that a limited period of silica secretion produces changes in the correlation of thickness and length; thickness depending on the quantity of silica that can be supplied by the medium. According to Jones, the great variation in spicule length and width, even in a single species, may also be due to differences in the length of the axial organic filament at which silicification starts. In addition, factors such as temperature and environmental concentration of silica may play a role. Rützler and Macintyre (1978) have suggested that factors such as pH, temperature, SiO_2, and CO_2 concentration control the growth rate of spicules, although the relationship is not fully understood.

Experimental studies on freshwater sponges (Elvin, 1971) show that if *Ephydatia muelleri* grows in an environment of low silica concentration, still it produces spicules;

but if silica concentration is normal, the rate of growth is higher. Pé (1973) studied *Ephydatia fluviatilis* and observed that young sponges cultivated in water of low silica concentration (1.5 mg/l) produce a high number of thin and short spicules, whereas comparatively fewer and thicker spicules are produced in a silica-enriched medium (60 mg/l). Silica concentration requirements for these sponges are species dependent.

If we assume that physiological mechanisms involved in silica deposition in marine Haplosclerida and Petrosiida are the same as in freshwater sponges, we can also assume that the length and thickness of spicules should always be directly related to the variables mentioned above, especially to total soluble silica content of the habitat and the sponge's physiological needs and absorption potential.

Because the taxonomic characters used in identifying Haplosclerida and Petrosiida vary greatly and spicule dimensions alone are of little value in this regard, we tried to determine if the total amount of skeletal silica (spicule weight) could serve as a reliable character in identifying families in the group under investigation.

Materials and Methods

Between December 1976 and June 1978, J. Vacelet and C. Lévi collected specimens from depths of 35 m to 50 m in different localities of the lagoon on the southwest coast of New Caledonia. These were fixed in 90% alcohol or 4% formaldehyde immediately after collection; some were frozen. From these collections we examined 107 sample; 46 were identified to species, the remaining 61 were determined to genus but were specifically different from the ones above. In order to determine the amount of skeletal silica in relation to dry weight, we washed every sample in running water to eliminate foreign material. Pieces of sponge were chosen for testing without regard to the shape of the sample, one at the base or stem, and another at the end of tubes, branches, or other representative parts of the sponge. Samples were weighed after dehydration at 120°C for 24 h, or until the weight remained unchanged. Spongin

was eliminated by digesting with fuming nitric acid for 12 h at room temperature and for 24 h immersed in oil at 70°C. The spicule sample was centrifuged, washed in changes of distilled water and 100% alcohol, and dried at 120°C for 24 h before weighing two times at 0.01 mg precision.

Results

First, we measured the amount of spicules in relation to the dry weight of all the samples studied. These samples belonged to five families: Niphatidae, Callyspongiidae, and Haliclonidae (Haplosclerida), and Petrosiidae and Oceanapiidae (Petrosiida) (Table 1). We confirmed our first hypothesis—that mean silica content varies between groups in relation to dry weight. Regrettably, the differences are not significant enough to differentiate every family analyzed, as we had hoped they would be. The mean spicule content of the Haplosclerida families Niphatidae and Haliclonidae is very similar: 25.9% of the dry weight for the former and 29.6% for the latter ($t = 0.93$ at the 5% level). Members of the family Callyspongiidae in this order have a mean silica content of 4.5% of the dry weight. In contrast, the mean spicule content of the Petrosiida family Petrosiidae is 57.3% of the dry weight. The second family of this order, Oceanapiidae, has a mean spicule content of 33.7% but the standard deviation for this group is very high.

We have compared Haliclonidae and Oceanapiidae but with $t = 0.62$ at the 5% level differences are not significant. We can say the same for Niphatidae and Oceanapiidae with $t = 1.91$ at the 5% level. Thus, if we analyze the mean silica content at the order level, it seems possible to distinguish between orders (Table 2). However, we should determine the usefulness of the character "amount of silica" as a taxonomic measure in comparison with the traditional taxonomic characters used up to now within this group of sponges.

The second step of our analysis was to measure the length of the spicules in order to determine if maximum values of length could be used to delimit the different

Table 1. Spicular silica content in % of dry weight for 5 families of Haplosclerida and Petrosiida studied

Parameters	Family (no. of species)				
	Niphatidae (24)	Callyspongiidae (43)	Haliclonidae (22)	Petrosiidae (12)	Oceanapiidae (6)
Mean	25.9	4.5	29.6	57.3	33.7
SD	13.1	4.7	13.4	8.3	18.4
Minimum	8.6	0.8	12.8	45.6	12.8
Maximum	56.3	26.8	62.3	74.9	62.2
Range	47.7	26.0	49.5	29.3	49.4

Table 2. Spicular silica content in % of dry weight comparing the orders Haplosclerida and Petrosiida.

Parameters	Order (no. of species)	
	Haplosclerida (89)	Petrosiida (18)
Mean	16.5	49.5
SD	15.3	16.6
Minimum	0.8	12.8
Maximum	62.3	74.9
Range	61.5	62.1

groups (Table 3). If we compare Niphatidae and Ocean-apidae with $t = 0.34$ at the 5% level, no clear segregation of measurements into these families was observed, however, means of maximum lengths for the other three families of both orders are very different.

We also attempted to distinguish a regular interval between the mean of the maxima and the mean of the mini-ma of spicule length for each family (Tables 3, 4). The values for interval obtained for Haplosclerida families range from 12 μm to 33 μm, and for Petrosiida families they range from 52 μm to 57 μm. However, we do not know the taxonomic significance of this characteristic because of its great variability. If we analyze minimum lengths for the different families (Table 4), we cannot see a significant difference between Niphatidae and Oceanapi-idae ($t = 0.43$ at the 5% level); between Oceanapiidae and Haliclonidae ($t = 0.71$); and between Niphatidae and Haliclonidae ($t = 1.72$). Only the families Niphatidae, Callyspongiidae, and Petrosiidae exhibit a significant difference between them. The maximum diameter (thickness) and length of spicules were also examined in relation to silica content (Tables 5, 6) and found to have some connection in the different families analyzed. Generally it is possible, even if means are not significantly different, to establish a correlation between silica content, expressed as spicule content, and spicule size, but not spicule number.

Finally, we can see that correlations exist between minimum length and maximum diameter of spicules ($t = 0.96$)

Table 3. Maximum length (μm) of spicules of five families of Haplosclerida and Petrosiida

Parameters	Family (no. of species)				
	Niphatidae (24)	Callyspongiidae (43)	Haliclonidae (22)	Petrosiidae (12)	Oceanapiidae (6)
Mean	192.8	90.4	163.0	257.6	204.8
SD	80.4	23.0	48.5	48.8	71.3
Maximum	388.0	160.0	300.0	389.0	287.0
Range	308.0	105.0	200.0	179.0	182.0

Table 4. Minimum length (μm) of spicules of five families of Haplosclerida and Petrosiida

Parameters	Family (no. of species)				
	Niphatidae (24)	Callyspongiidae (43)	Haliclonidae (22)	Petrosiidae (12)	Oceanapiidae (6)
Mean	160.1	77.8	131.9	206.3	147.8
SD	63.7	19.7	46.4	43.4	57.4
Maximum	60.0	50.0	60.0	115.0	85.0
Range	220.0	90.0	215.0	179.0	149.0

Table 5. Maximum diameter of spicules (μm) compared to percentage of spicular silica content in five families of Haplosclerida and Petrosiida

Parameters	Niphatidae		Callyspongiidae		Haliclonidae		Petrosiidae		Oceanapiidae	
	Si %	ϕmax	Si %	ϕmax	Si %	ϕmax	Si %	ϕmax	Si %	ϕmax
Mean	25.9	7.0	4.5	1.8	29.6	6.6	57.3	10.2	33.7	8.0
SD	13.1	2.6	4.7	1.2	13.4	3.7	8.3	4.6	18.4	5.0
Maximum	56.3	10.0	26.8	4.0	62.3	20.0	74.9	16.0	62.2	16.0
Range	47.7	7.0	26.0	3.5	49.5	18.0	29.3	13.0	49.4	12.0

Table 6.   Maximum length (µm) of spicules compared to percentage of spicular silica in five families of Haplosclerida and Petrosiida

Parameters	Niphatidae Si %	lmax	Callyspongiidae Si %	lmax	Haliclonidae Si %	lmax	Petrosiidae Si %	lmax	Oceanapiidae Si %	lmax
Mean	25.9	192.8	4.5	90.4	29.6	163.0	57.3	257.6	33.7	204.8
SD	13.1	80.4	4.7	23.0	13.4	48.5	8.3	48.8	18.4	71.3
Maximum	56.3	388.0	26.0	160.0	62.3	300.0	74.9	389.0	62.2	287.0
Range	47.7	308.0	26.0	105.0	49.5	200.0	29.3	179.0	49.4	182.0

(Tables 4, 5); between maximum length and maximum diameter of spicules ($t = 0.99$) (Tables 5, 6); and between spicule content and maximum length of spicules ($t = 0.96$) (Table 6).

Discussion

If we assume that spicule length and width are correlated, as proposed by Jones (1984), we find that the longer the spicules the thicker they are in every species analyzed. We have also given the silica content values in relation to the dry weight, and if we observe the results for Haliclonidae

and Niphatidae we can see a clear inversion of the values for spicule dimensions and those for spicule content (Figure 1). However, we do not know if it is possible to use these relations as a diagnostic character in species discrimination, because spicule width and length vary widely in a single species and between species. The next step is to determine whether it is possible to use this relation as a diagnostic character in species discrimination.

Silica content pertains to fiber structure and thus some authors have tried to separate groups on the basis of the levels of spongin in relation to the spicule content. Dendy (1905) used this factor to determine the genera of the

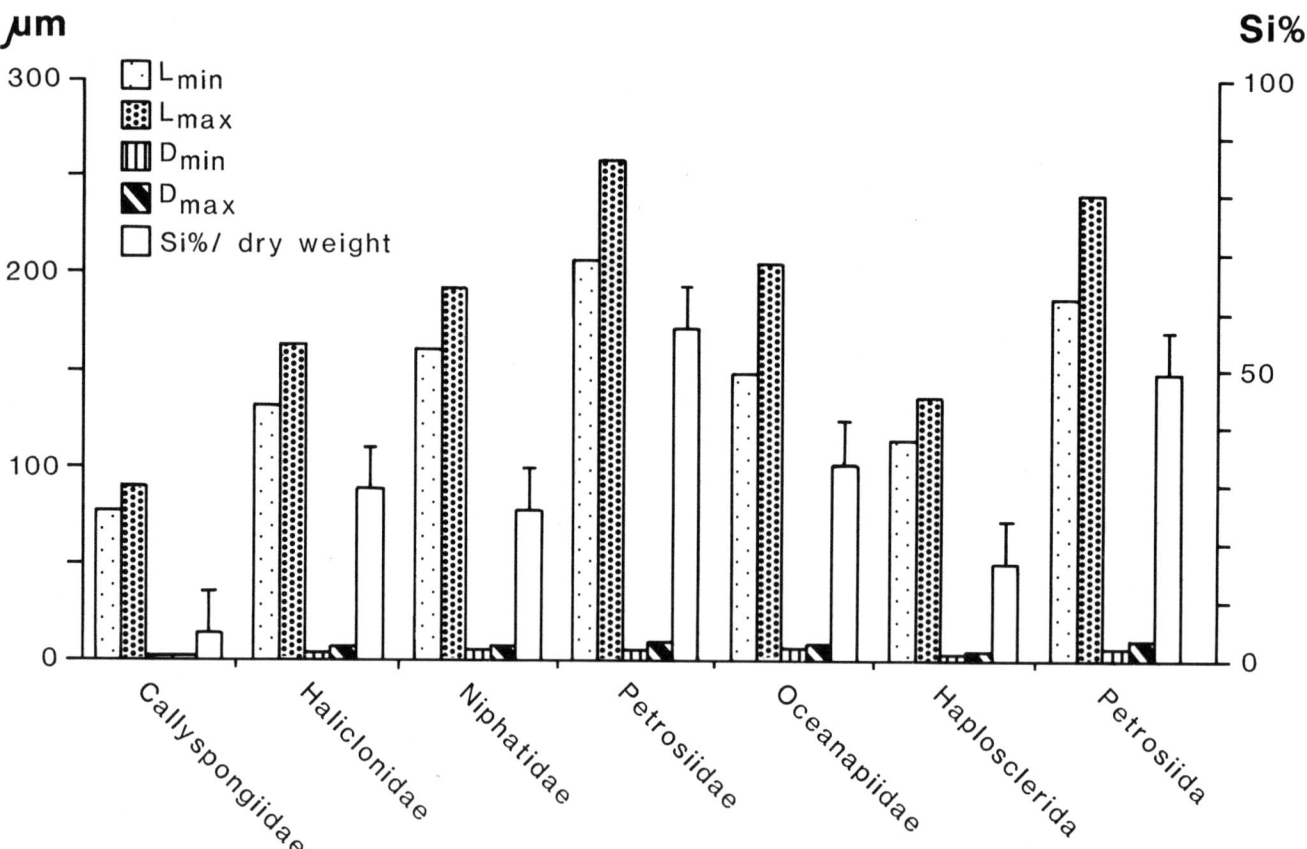

Figure 1. Mean minimum and maximum spicule length and diameter (µm) and mean spicular silica content in percentage of dry weight for three Haplosclerida and two Petrosiida families analyzed.

Gellininae and two other groups of Haplosclerida, Renierinae, and Chalininae. Pomponi (1976) considered the arrangement and abundance of spongin fibers as diagnostic character at the family level, even if she used only optical measurements. The coefficients of silica content calculated on the family level in this study, even if it does not reflect skeletal arrangement, express degrees of density which appear to be correlated with spicule number, length, and width. We already used the amount of silica, determined and expressed the same way as in this work, as a taxonomic character for families of Petrosiida (Desqueyroux-Faundez, 1987), but in combination with other morphological characters, such as surface skeleton and cortical structure. Subsequent studies should examine other parameters, such as dissolved silica concentration in the environment, dates of collection, variations in the rate of spicule growth, characteristics of sclerocytes secreting spicules, and any other factors that might influence spicule production and be useful in clarifying the mechanisms of silica deposition in marine sponges.

Acknowledgments

I am grateful to A. Rosa, who provided technical assistance in silica content analyses, and to D. Doumenc, who helped perform the computations.

Literature Cited

Dendy, A. 1905. Report on the Sponges Collected by Prof. Herdman, at Ceylon, in 1902. In *Ceylon Pearl Oyster Fisheries, Manaar Supplementary Reports*, 18:57–246.

Desqueyroux-Faundez, R. 1987. Description de la Faune de Petrosida (Porifera) de la Nouvelle-Calédonie. I. Petrosiidae–Oceanapiidae. *Revue suisse de Zoologie*, 94:177–243.

Elvin, D. 1971. Growth rates of the Siliceous Spicules of the Freshwater Sponge *Ephydatia muelleri* (Lieberkühn). *Transactions of the American Microscopical Society*, 90:219–224.

Jones, W. C. 1979. The Microstructure and Genesis of Sponge Biominerals. Pages 425–447 in *Biologie des Spongiaires*, edited by C. Lévi and N. Boury-Esnault. Colloques Internationaux du C.N.R.S. 291. Paris: Centre National de la Recherche Scientifique.

———. 1984. Spicule Dimensions as Taxonomic Criteria in the Identification of Haplosclerid Sponges from the Shores of Anglesey. *Zoological Journal of the Linnean Society*, 80:239–259.

Pé, J. 1973. Etude quantitative de la régulation du squelette chez une Eponge d'eau douce. *Archives de Biologie*, 84:147–173.

Pomponi, S. A. 1976. A Cytological Study of the Haliclonidae and Callyspongiidae (Porifera, Demospongiae, Haplosclerida). Pages 215–235 in *Aspects of Sponge Biology*, edited by F. W. Harrison and R. C. Cowden. New York: Academic Press.

Rützler, K., and I. G. Macintyre. 1978. Siliceous Sponge Spicules in Coral Reef Sediments. *Marine Biology*, 49:147–159.

JOHN N. A. HOOPER
Northern Territory Museum of Arts and Sciences
Darwin, NT, 5794, Australia

Character Stability, Systematics, and Affinities between Microcionidae (Poecilosclerida) and Axinellida

Abstract

Recent collections of marine demosponges from tropical northwest Australia yielded 65 species of Microcionidae (from 10 genera), and 108 species of Axinellida (34 genera, 7 families). The stability and importance of diagnostic morphological characteristics were examined through repetitive sampling of 32 common species (12 microcionids, 20 axinellids) from various habitats within the Northern Territory and the north of Western Australia (traversing 2300 km of coastline), over a two year period. Certain diagnostic characters were found to be consistent throughout the entire range of a species and can be used to separate taxa. Morphological differentiation of species is supported by general protein electrophoretic banding characteristics, which do not appear to vary significantly between specimens, irrespective of geographical locality, seasonality, or growth form. Examination of 90 other species and a detailed survey of the literature made it possible to evaluate 66 nominal Microcionidae genera, of which 11 are probably valid.

Shared morphological characters and statistical similarity of electrophoretic banding patterns suggest a closer affinity between Microcionidae and some Raspailiidae than is presently indicated by their current positions in different subclasses (Ceractinomorpha and Tetractinomorpha, respectively). However, ectosomal and other skeletal characteristics indicate that the two groups are distinct at the family level. Euryponidae and Desmacellidae are convergent with Microcionidae in some respects, but are less closely related to that group than is Raspailiidae. Axinellidae represents a diverse and probably unnatural assemblage of genera.

Morphological characteristics are of primary importance in separating sponge species and genera, whereas reproductive and biochemical characteristics are becoming increasingly significant in defining higher taxonomic categories (Lévi, 1957; Lévi, 1969; Bergquist, 1980; Bergquist and Wells, 1983; Lee and Gilchrist, 1985). In some cases, however, morphological evidence indicates that certain taxa are more closely related than otherwise suggested by their reproductive or biochemical characteristics (e.g., Hooper, 1984; van Soest, 1984). Since the reproductive and biochemical traits of the majority of species are not fully known, the placement of certain lower taxa in one or another higher grouping is often inferred through apparent morphological affinities (e.g., Lévi, 1957). This task is especially different in morphologically convergent groups for which little ontogenetic or biochemical data are available, notably, Desmacellidae and Biemnidae (e.g., Hooper, 1984); and Microcionidae (a senior synonym of Clathriidae; ICZN Article 40) and Raspailiidae (e.g., Bergquist, 1970). Although the use of reproductive and biochemical data and certain cytological characteristics (e.g., Simpson, 1968) may facilitate a clear separation of taxa at all levels, thus far few studies have found any correlation between "key" diagnostic characters and cytological, biochemical, or reproductive traits. In any case, those techniques are generally unavailable to most taxonomists in the field or cannot be applied to preserved museum collections (Simpson, 1968).

Affinities between species of Microcionidae and between genera of Microcionidae and various Axinellida were investigated as follows: (1) Morphological characteristics and character states were described from large samples of 32 species. These samples were collected from a variety of habitats at different times of the year to determine the stability of so-called diagnostic character states within species and genera. (2) Ninety other species of Microcionidae (mainly type specimens) were examined and the literature reviewed to obtain discriminatory generic diagnoses for seeking "key" diagnostic characters and for determining the validity of a number of genera, particularly those erected by Hallmann (1920). (3) Characters and character states were compared in genera of Microcionidae and Axinellida to determine whether shared traits merely represent the convergence of a few morphological characters. Alternatively, they may indicate a closer affinity than is suggested at present by their placement in the subclasses Ceractinomorpha and Tetractinomorpha, respectively. (4) Banding characteristics, which were derived from general protein electrophoresis, were examined for stability by comparing multiple samples, that incorporate both geographical and seasonal variability; these characteristics were compared statistically among species, genera, and families to supplement and compare morphological results.

Materials and Methods

Specimens were collected between 1982 and 1985 from an area covering approximately 2300 km of coastline in northwest Australia and extending from the Port Hedland region (Western Australia; 20° S, 118° E) to the Wessel Islands (Northern Territory; 10° S, 136° E) and Ashmore Reef (Territory of Australia; 12° S, 124° E). Most of the material was obtained from three localities: northeast of Port Hedland, Northwest Shelf (W.A.; 19°–20° S, 117°–119° E), here designated as NWS; Darwin region (N.T.; 12°–13° S, 130°–131° E), denoted as DAR; and the Cobourg Peninsula region (N.T.; 11°–12° S, 132°–133° E), denoted as CP. NWS material was collected by dredging and trawls (from depths of 32–360 m), while DAR and CP material was collected mainly by SCUBA (in the intertidal to 30 m). Collections for habitat comparisons were obtained from all three localities. Seasonal sampling was conducted over two years and was confined to the DAR region. All localities are subjected to large tidal ranges, which reach 8 m in some areas, and consequently most sponges studied are peculiar to areas of high sedimentation and turbidity. Specimen preparation and examination techniques have been described elsewhere (Hooper, 1984), and a discussion of characters used in cladistic analysis may be found in van Soest (1984) and Hooper (1987). Spicules were measured on 25 individuals of each category present in each species. Where available, type specimens were used to check species diagnoses. The following abbreviations are used in the text: AMU (Australian Museum, Sydney), BMNH (British Museum, Natural History, London), INM (National Museum of Ireland, Dublin), MMBS (Mukaishima Marine Biological Station, Onomichi, Japan), MNHN (Muséum National d'Histoire Naturelle, Paris), NMB (Natural History Museum, Basel), NMNZ (National Museum of New Zealand, Wellington), NMV (Museum of Victoria, Melbourne), NTM (Northern Territory Museum, Darwin), PMJ (Phyletisches Museum Jena), QMU (Queensland Museum, Brisbane), RSME (Royal Scottish Museum, Edinburgh), SAM (South Australian Museum, Adelaide), SMF (Senckenberg Museum, Frankfurt), USNM (United States National Museum of Natural History, Washington D.C.), WAM (Western Australian Museum, Perth), ZMB (Zoological Museum, Berlin).

Electrophoresis of frozen material was conducted by C. Keenan at the University of Queensland, Brisbane, using slab polyacrylamide gels (PAGE) with a discontinuous LiOH buffer system. The anodal portion of gels was stained using a modified silver technique (Keenan and Shaklee, 1985). Electrophoretic mobility bands were measured and compared using Lawson's index of similarity (Lawson et al., 1980).

Full results of sponge protein electrophoresis, including

methods and interpretation will be published elsewhere (Hooper et al., in press).

Results and Discussion
CHARACTER VARIABILITY

Of the 32 morphological characters examined in this preliminary analysis, most have been used routinely at one time or another for sponge classification. The expression of characters (character states) was based on an examination of approximately 800 specimens, collected in northwest Australia, which consisted of 12 species of Microcionidae (in 4 genera) and 20 species of Axinellida (in 16 genera and 6 families), as well as an additional 90 species of Microcionidae (255 specimens, mostly types). The stability of characters (a determination of whether or not character states varied between specimens of any given species over a geographical or seasonal range) was determined for those species examined from repetitive samples. A detailed analysis of morphological characters and character variability and a list of species examined will be presented elsewhere (Hooper, in press; in prep.).

For the Microcionidae, at the species level, several characters were shown to be important in differentiating allied species. Some characters were unstable, but observed variability occurred within only narrow limits (color, shape, surface sculpturing, basal spination of ectosomal and subectosomal megascleres, degree of development or density of a special ectosomal skeleton, and megasclere and microsclere dimensions). Some of those characters were highly unstable in certain species (e.g., shape of *Rhaphidophlus lendenfeldi* (Ridley and Dendy), color of *R. abietinus* (Lamarck)), but much more stable within other species (e.g., shape of *R. coppingeri* (Ridley), color of *Clathria tuberosa* (Bowerbank)). Those characters may be diagnostically useful for identifying species from restricted localities only. Other characters were consistently more stable within species. These include the distribution of oscula on the surface; basal spination of choanosomal megascleres; the presence or absence of a special ectosomal skeleton; the number of megasclere categories; the degree of development of the subectosomal extra-fiber skeleton; the fiber characteristics and distribution of megascleres within, on, or around those fibers or spicule tracts; the form of megascleres, particularly the spination pattern of echinating spicules; and the form and number of categories of microscleres. Those characters were found to be diagnostically important in identifying species, regardless of geographical or seasonal distribution of specimens. The use of spicule dimensions has been traditionally important in differentiating species, and in some cases species with close affinities have been separated on the basis of relatively small differences in spicule size. Until recently (e.g., Simpson, 1978; Jones, 1984), spicule variability has not been examined in detail, nor have the

effects of the environment on that variability. To date, only three species in this study have been examined statistically to determine variability in spicule size (*Rhaphidophlus abietinus*, *R. lendenfeldi*, and *Reniochalina stalagmitis* Lendenfeld), but all showed significant differences in size of all categories of spicules over seasonal and habitat distributions. Despite the apparent seasonal/habitat variability in the size of megascleres and microscleres, the use of those characters is valid in differentiating allied species, provided that sufficient material is examined and a range of variation is established. For example, an overlap in dimensions of choanosomal styles of *R. abietinus* (155–264 × 9–17 μm) and *R. lendenfeldi* (180–350 × 6–18 μm) is of less taxonomic significance in separating those species than is the combination of characters such as the distribution, form, and size of spicules.

EVALUATION OF CHARACTER IMPORTANCE

Several characters and character states were found to be useful in grouping species that are closely related morphologically. For example, the form and distribution of choanosomal or "principal" megascleres were similar between some species, notably, *Clathria clathrata* (Whitelegge), *C. costifera* (Hallmann), *Rhaphidophlus coppingeri* (Ridley), and *R. lendenfeldi* (Ridley and Dendy), which belong to Hallmann's (1920) "*spicata*" group; *C. tuberosa* (Bowerbank), *C. australiensis* (Carter), *C. reticulata* (Lendenfeld), and *C. mixta* Hentschel, which would fall into *Clathriopsamma*-like species; *C. curvichela* (Hallmann), *C. conectens* (Hallmann), *C. oxyphila* (Hallmann), *C. perforata* (Lendenfeld), *C. imperfecta* Dendy, and *C. myxilloides* Dendy, which are *Wilsonella*-like species. Another group of species may be characterized by the form, distribution, and spination of echinating spicules. This group includes *C. scotti* Dendy, *C. coccinea* (Bergquist), *C. parthena* (de Laubenfels), and *C. rubens* (Bergquist), all of which have characteristic radiating tufts of choanosomal and echinating spicules at fiber nodes. These characteristics, which combine similar species (e.g., *spicata*, *Wilsonella*, *Clathriopsamma* and *scotti* groups), occur elsewhere within the Microcionidae and outgroups (Myxillidae, Esperiopsidae, Raspailiidae) and thus are of limited value at the generic level of classification.

Similarly, the presence or absence of modified isochelae was found to be more common than previously reported (see van Soest, 1984), including twisted forms (15 species), cleistochelae (1 species), arcuate or anchorate isochelae (8 species), and isodictyalike chelae (2 species). Modified isochelae do not indicate natural groupings at the generic level, as suggested by various authors (e.g., de Laubenfels, 1936; Hallmann, 1920) because they occur throughout the family and sister groups and cut across a classification based on characters of more systematic value (see Figure 1).

In contrast, it has been confirmed elsewhere (Lévi, 1960; van Soest, 1984), as well as in this study that those characters are stable at the species level, and with few exceptions, they may be used to define species of Microcionidae. These exceptions include the following species, which represent new combinations or synonyms. (1) *Rhaphidophlus lendenfeldi* (BMNH 87.5.2.107) is the earliest available name for a group of species that Hallmann included in his *spicata* group. It would include as probable synonyms *R. bispinosus* Whitelegge (AMU type lost), *C. spicata* Hallmann (AMU type lost), *C. whiteleggei* Dendy, *C. coppingeri* var. *aculeata* Hentschel (SMF 1552, 1598, 1604, 1664, 1670), and *C. frondifera* var. *major* Hentschel (SMF 977). Other members of the *spicata* group (which spans the genera *Clathria* and *Rhaphidophlus*) are maintained as distinct species; *C. coppingeri* (BMNH 81.10. 21.246, and MNHN DT571 as *Spongia juniperina* var. *thuyaeformis* Lamarck) should be used for forms with a specialized reticulate planar flabelliform habit. That species is otherwise identical to *R. lendenfeldi* in architecture and spiculation and is placed in *Rhaphidophlus* on the basis of its ectosomal characteristics. Also in this group are *C. clathrata* (AMU G10530, 4355), which is a *Clathria* s.s. without a specialized ectosomal skeleton, and *C. inanchorata* (AMU E650), both of which have basally smooth choanosomal megascleres. In other morphological details, all species are very close. (2) *Rhaphidophlus vulpinus* (Lamarck) (MNHN DT639) is a senior synonym of *Clathria frondifera* (BMNH 82.70.17.42), which also includes var. *dichela* Hentschel (SMF 1673), var. *setotubulosa* Wilson (USNM 21257, 21256), and *R. seriatus* Thiele (NMB 16, 17), but excludes *C. nuda* Hentschel (SMF 1576) which was previously included as a junior synonym by various authors. Both *Rhaphidophlus vulpinus* and *R. lendenfeldi* are borderline cases between *Clathria* and *Rhaphidophlus*, because the development of ectosomal megascleres into brushes is highly variable. They are placed in *Rhaphidophlus* because ectosomal megascleres are of a distinctly smaller size category than those of the subectosome (e.g., van Soest, 1984). The existence of such species, with two distinct size classes of ectosomal-subectosomal megascleres (the composition of *Rhaphidophlus*) but with the ectosomal structure of *Clathria*, illustrates the tenuous nature of a division between genera based solely on that character. By comparison, *Dendrocia* has the ectosomal structure of *Raphidophlus* but the composition of *Clathria*. On the basis of their ectosomal characteristics, other species were transferred between *Clathria* and *Rhaphidophlus* (Hooper, in press; in prep.).

BIOCHEMICAL EVIDENCE

Support for the morphological diagnostic criteria used here is provided by banding patterns produced from general protein electrophoresis. Multiple specimens of each

species from different localities, different seasons, and in some cases, different growth forms were analyzed to determine (a) if proteins separated consistently for any given species, and (b) if consistent banding patterns were apparent between (i) morphologically allied species, and (ii) more morphologically distant species. Although only 20 species have been examined to date, each has distinct banding patterns (Hooper et al., in press). Analysis of multiple samples for those species shows that protein banding characteristics were consistent for each species regardless of locality, season, or growth form. In some samples only one or two extraneous bands were present, perhaps because of contamination by symbiotic organisms. Some characteristic bands were very faint in certain samples, possibly because of the differential concentration of homogenized samples. Although this technique is relatively insensitive and does not target specific enzyme systems, these analyses have shown that the method can be used to discriminate between morphologically similar species. Further investigations that are in progress are examining the carotenoid pigments using HPLC (Hooper et al., in press).

EVALUATION OF GENERA

Past efforts to establish clear morphological distinctions between genera have been based on skeletal architecture (e.g., Lévi, 1960), microsclere content (e.g., de Laubenfels, 1936), megasclere form and distribution (e.g., Hallmann, 1920), and ectosomal characteristics (e.g., van Soest, 1984). Although all those characters were found to be relatively stable within species regardless of geographic or seasonal variation, only a few characters retain the same expression throughout cogeneric groups. It is clear that interspecific variability of character states within genera of Microcionidae has in the past made the separation of these groups a difficult task, continues to be a problem in this study.

Of all Microcionidae examined so far (27 nominal genera), ectosomal characters were able to differentiate only 4 closely related groups, *Rhaphidophlus* and *Clathria*, and *Plocamilla* and *Antho*. However, the separation of those genera is not entirely clear from ectosomal structure alone, and it is widely believed that a valid distinction can be made only for *Clathria* and *Rhaphidophlus* on the basis of the presence of a special, smaller category of ectosomal megasclere in *Rhaphidophlus* (following van Soest, 1984). For example, species of *Clathria (Dendrocia)* have a continuous palisade of erect brushes on the surface, which are usually characteristic of *Rhaphidophlus* species, but in *Dendrocia* these brushes consist of a single category of undifferentiated choanosomal spicules (*C. alata* Dendy, *C. pyramida* Lendenfeld). In contrast, some *Rhaphidophlus* species have very light brushes of ectosomal spicules only occasional; sometimes they become paratangential to the ectosome but nevertheless consist of two size categories of spicules (e.g., *R. vulpinus, R. lendenfeldi, R. coppingeri, R. cratitius* [Lamarck]).

The fiber characteristics and the constituent coring and echinating spicules are undoubtedly important at the species level, as noted above, but they are of questionable significance at higher levels of classification. Hallmann (1920) erected, or resurrected, a number of genera using those characteristics. Of the genera that have obvious microcionid affinities (*Dendrocia* Hallmann, *Wilsonella* Carter, *Paradoryx* Hallmann, *Isociona* Hallmann, *Tenacia* Schmidt, *Clathriopsamma* Lendenfeld, *Tenaciella* Hallmann, *Axociella* Hallmann, *Isociella* Hallmann, and *Isopenectya* Hallmann), only *Isociella* is clearly a good genus. *Isopenectya* is unusual in having an axial condensation of spongin fibers, reminiscent of *Phakellia* and *Acanthella* (Axinellidae), but it has numerous close morphological affinities with *Antho*. Megasclere diversity and distribution in the genera *Wilsonella, Paradoryx, Clathriopsamma, Axociella,* and *Dendrocia* are depauperate as compared to the *Clathria/ Rhaphidophlus* states. The first three of these genera have obvious affinities with *Clathria* and are probably reduced forms of that genus. *Dendrocia* is enigmatic in that ectosomal development falls exactly between the *Clathria* and *Rhaphidophlus* states, but unlike those two genera, ectosomal and choanosomal spicules are undifferentiated. For this reason it is suggested that *Dendrocia* should be maintained as a separate genus (see Figure 1). Other genera of Hallmann (1920) with obvious microcionid affinities are junior synonyms of well-established genera: *Antho (Isociona, Plocamilla),* and *Rhaphidophlus (Tenacia, Tenaciella, Axociella).*

Van Soest (1984) lists other genera that were included in the Microcionidae at one time or another and suggests their probable affinities. Among the genera not mentioned are *Clathriella* Burton, which is close to *Isociella* but is maintained here tentatively owing to the presence of an oxeote subectosomal megasclere, that is much too large to be considered a microsclere. *Axocielita* de Laubenfels (type-species *M. similis* Stephens; INM 31.1914) is a synonym of *Clathria* rather than *Axociella (= Rhaphidophlus);* it lacks ectosomal specialization. The genus *Echinochalina* Thiele may be best placed elsewhere (e.g., Esperiopsidae, as suggested by van Soest, 1984), but it is included here in Microcionidae on a tentative basis only, because it has several morphological affinities with *Echinoclathria*. Conversely, it also represents a link with a Raspailiidae genus *(Echinodictyum)* in habit, fiber characteristics and spiculation. For example, *Echinodictyum ridleyi* Dendy and *E. spongiosum* Dendy are actually species of *Echinochalina* (here regarded as representing new combinations), but interpreted in a broad sense, both forms could be placed in either of those genera. The genus *Paratenaciella* Vacelet and Vasseur is most reasonably placed in *Clathria* s.l., since it has an ectosome consisting of a single category of poly-

tylote subtylostyles (cf. *C. ferrea* [de Laubenfels]), and also because the possession of a plumose, nonanastomosing choanosomal skeleton, characteristic of *Microciona (= Clathria),* occurs in other microcionid genera (e.g., *Clathria acanthotoxa* [Stephens], and *Rhaphidophlus* provisional species, 17, 32, 34). Similarly, the possession of microoxeas as microscleres is not unique in Microcionidae (e.g., *Artemisina archegona* Ristau). The genus *Plocamiopsis* Topsent appears to be a thinly encrusting *Plocamilla (= Antho)* with cleistochelae. From its ectosomal characteristics, *Melonchela* Koltun has obvious affinities with the Myxillidae. *Querciclona* de Laubenfels (type species *Antherochalina quercifolia* Keller; ZMB429) was included with microcionid taxa by de Laubenfels (1936), but is a synonym of *Phakellia* (Axinellidae).

SYSTEMATICS OF HIGHER TAXA

The cladistic analysis (Figure 1; Tables 1, 2) was based on a restricted set of morphological attributes (characters 1–19), and Myxillidae and Raspailiidae were used for outgroup comparisons. Results of that analysis must be interpreted with caution, however, because the microcionid genera are difficult to separate, and it is not clear whether apomorphy can be defined by single characters for this complex group. Of the 66 nominal genera erected for microcionid species, most are synonyms of other genera, differing only in minute details of spiculation, growth form, or architecture. Seven genera are differentiated relatively easily *(Antho, Isociella, Marleyia, Clathria, Rhaphidophlus, Artemisina,* and *Echinoclathria)* and four are closely allied to other genera *(Isopenectya, Clathriella, Dendrocia,* and *Qasimella).* Two genera are of doubtful Microcionidae affinity *(Echinochalina* and *Protophlitaspongia* Burton; included here as *Echinoclathria*-like taxa).

Cladistic analysis indicates that the following characters are important for the Microcionidae: (1) choanosomal architecture in conjunction with the distribution of choanosomal and subectosomal spicules, (2) ectosomal specialization, (3) spiculation; (4) the development of a supplementary choanosomal skeleton and subectosomal skeleton, and (5) a clear differentiation of an ectosomal and subectosomal/choanosomal region. Those characters were stable at the species level, and with the exception of *Dendrocia,* appear to be consistent at the generic level. A basic separation of two groups is implied from Figure 1: those with myxillid affinities *(Antho, Isociella, Isopenectya,* and possibly *Clathriella* and *Marleyia,* which are poorly known), and those that appear to be more closely related to the axinellids *(Clathria,* including *Dendrocia, Rhaphidophlus,* and *Echinoclathria).* The latter is intermediate. The affinities of *Artemisina* are not known, but its spiculation is similar to that of *Melonchela,* which is a Myxillidae. The genus *Qasimella* has *Clathria*-like spiculation and a tangen-

Table 1. Homologous characters used in the construction of Figure 1 (Note, characters were used for the apomorphic separation of well–established microcionid genera and outgroups Myxillidae and Raspailiidae)

Character	Plesiomorphy	Apomorphy
	Habit	
1. Specialized habit	a. No distinctive or peculiar gross morphology	b. Specialized tubular, pseudosynconoid construction
	Ectosomal Skeleton	
2. Architecture	a. Absent, or choanosomal megascleres present only, perpendicular or tangential to ectosome.	b. Ectosomal diactinal tylotes or tornotes in brushes or tangential c. Ectosomal subtylostyles in brushes or tangential d. Ectosomal oxeas, anisoxeas or quasimonacts in brushes or tangential
3. Megascleres	a. Basally spined or smooth tylotes	b. Tornotesor anisotornotes c. Ectosomal oxeas, anisoxeas
4. Specialization	a. Membraneous b. Tangentially disposed discrete choanosomal megascleres c. Paratangential tracts of choanosomal	d. Bundles or brushes of megascleres perpendicular or paratangential e. Bundles or brushes of megascleres surrounding protruding choanosomal megascleres. f. Tangential raphidiform microscleres
5. Diversification	a. Single category of ectosomal or subectosomal megascleres on surface	b. Two distinct size classes of ectosomal or subectosomal megascleres
6. Specialized fibers	a. No specialized ectosomal fibers	b. Distinctive tangential reticulation of ectosomal fibers
7. Megasclere ornamentation	a. Acanthose monactinal megascleres	b. Smooth monactinal megascleres c. Smooth diactinal megascleres d. Acanthose diactinal megascleres
	Subectosomal Skeleton	
8. Megasclere specialization	a. Only choanosomal or subectosomal megascleres subdermally	b. Special category of subdermal diactinal megascleres
9. Differentiation	a. No regional differentiation of skeletal, no ectosomal crust	b. Structural localization of undifferentiated choanosomal megascleres c. Architectural and geometric localization of choanosomal, subectosomal and/or ectosomal megascleres
	Choanosomal Skeleton	
10. Development	a. Fused skeleton of principal megascleres, or discrete spicules disposed without order	b. Discrete spicules forming skeletal columns, aggregated into reticulate plumose, or condensed architecture c. Skeleton reduced to a basally condensed hymedesmoid construction
11. Differentiation	a. No regional differentiation of axial or extra–axial components	b. Marginal condensation of axial skeleton, structurally distinct from peripheral extra–axial components
12. Architecture	a. Reticulate b. Plumose, without axial or extra–axial differentiation	c. Renieroid-subrenieroid (plumose primary, reticulate secondary lines) d. Axially reticulate, extra–axially plumose or plumo-reticulate e. Axially reticulate, extra–axially plumose f. Axially plumose, condensed, extra–axially plumose, radial g. Basally condensed spongin, uncored, extra–axially plumose h. Basally condensed spongin, uncored, extra–axially hymedesmoid
13. Secondary organization	a. Regularly or irregularly plumo-reticulate	b. Choanosomal organization only recognizable at periphery
14. Order	a. Halichondroid, or at least partly organized	b. No skeletal organization into tracts and fibers whatsoever
15. Skeletal condensation	a. Reticulate, without condensation	b. Axially condensed core fibers c. Basally condensed uncored fibers
16. Supplementary skeleton	a. Simple isodictyal reticulation of tracts or fibers	b. Development of multispicular primary and pauci- or aspicular secondary renieroid reticulation c. Plumose ascending tracts of subectosomal megascleres
17. Megasclere geometry (excl. echinators)	a. Basally spined monactinal spicules b. Smooth monacts (1 category)	c. Smooth monacts (2 categories) d. Smooth monacts (3 categories) e. Smooth styles, oxeas, and/or strongyles of at least 2 categories f. Acanthose rhabdostyles
18. Distribution of echinating megascleres	a. Evenly echinating on skeletal tracts	b. Clumped on basal spongin, absent from c. Clumped at bases of subectosomal megascleres, in contact with axial skeleton d. Dense, forming a rigid interlocking secondary skeleton e. Absent
19. Echinating spicule morphology	a. Microciona–like acanthostyles, evenly or vestigially spined b. Acanthostyles with aspinous bases c. Acanthostyles with large recurved spines	d. Acanthose rhabdostyles with vestigially spined aspinose bases e. Evenly acanthose rhabdostyles f. Clavulate acanthostyles on shaft g. Double-clad acanthostyles/strongyles h. Sagittal triacts, tetracts, diacts, or pentactinal megascleres i. Absent

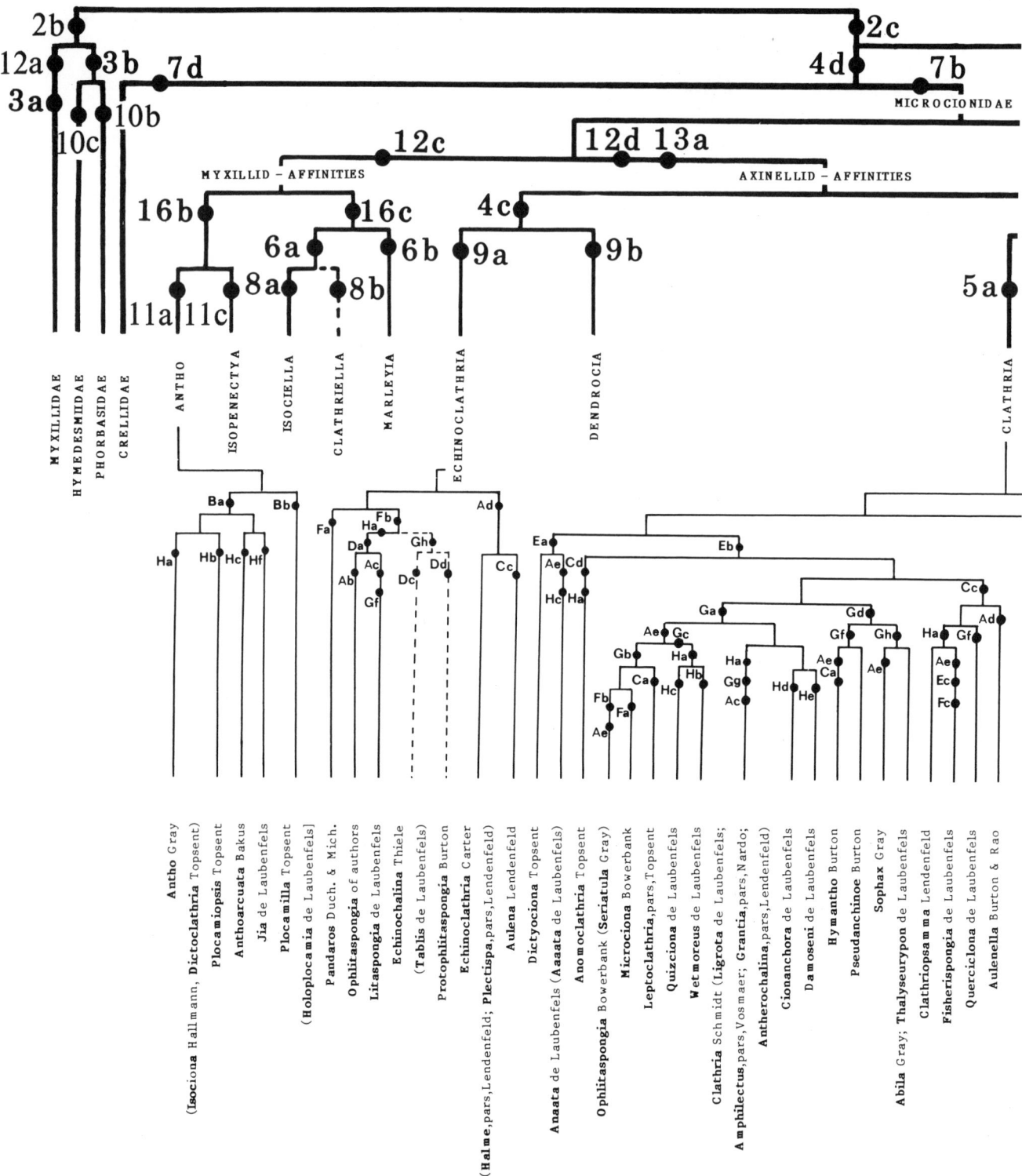

Figure 1. Cladistic analysis of Microcionidae genera: each number on cladogram (1–19) indicates an evolutionary change
nominal microcionid genera (characters A–H, Table 2), which represent probable synonyms of well-established genera.

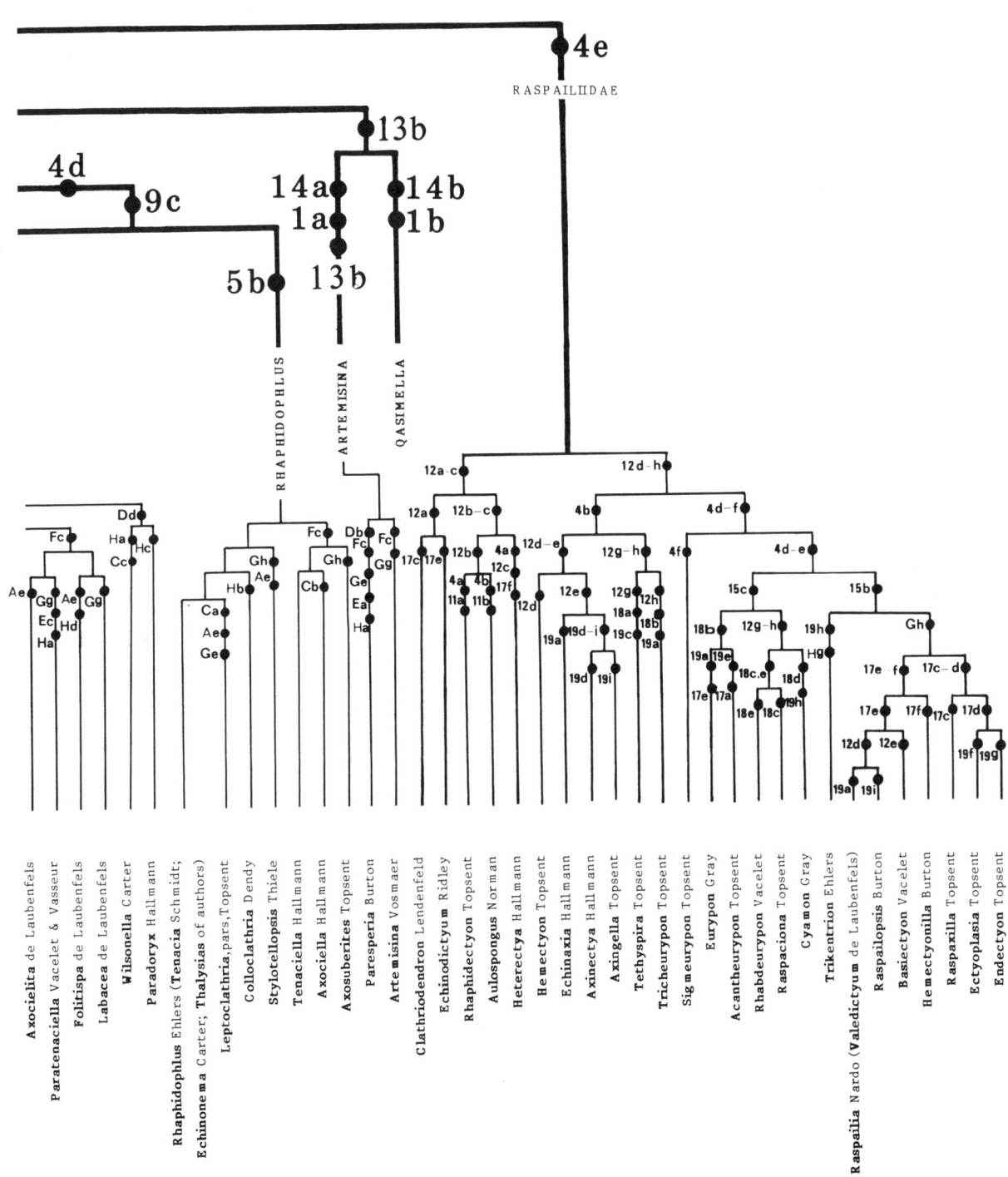

in corresponding character (see Table 1), from relatively plesiomorphic to relatively apomorphic condition; note affinities of all

Table 2. Nonhomologous characters used in the construction of Figure 1 (note, characters differentiate nominal microcionid genera, considered here to be possible synonyms of well–established genera)

Character	Plesiomorphy	Apomorphy
Habit		
A. Growth form	a. Massive	b. Flabelliform c. Erect, arborescent or ramose d. "Honeycombed" anastamosing form e. Incrusting
Ectosome		
B. Orientation	a. Tangential, undifferentiated	b. Erect brushes of undifferentiated megascleres
Choanosome		
C. Modification	a. Leptoclathriid–hymedesmoid construction of choanosomal styles erect on surface	b. Axial reticulation choanosomal styles erect on substrate c. Foreign particles incorporated into skeleton d. Algal filaments incorporated into skeleton
D. Spicule diversity	a. Megascleres consist of 2–3 differentiated	b. No special category of choansomal forms "principals" (replaced by auxiliaries) c. Auxiliary spicules monacts–quasi–diacts d. Hastate oxeas–quasimonacts
E. Megasclere spination	a. Choanosomal megascleres entirely acanthose	b. Megascleres smooth or spined basally c. Polytylote megascleres
F. Echinating spicules	a. Acanthose	b. Vestigial spination or smooth c. Absent
G. Microsclere diversity	a. Palmate isochelae and diverse forma of toxas b. Single category of palmate isochela	c. Palmate isochelae plus modified form d. Microsclere complement reduced e. Toxas absent, isochelae present f. Isochelae absent, toxas present g. Modified toxas (oxeote or spined) h. Microscleres absent
H. Microsclere form	a. Palmate isochelae g. Raphides, singly or in bundles h. Sigmas i. Microxeas	b. Cleistochelae c. Arcuate isochelae d. Anchorate isochelae e. Bidentate sigmoid isochelae f. J–shaped modified isochelae j. Diactinal acanthorhabds

tial subectosomal skeleton, but it has little else in common with other Microcionidae.

Although only few axinellid genera were available for examination, morphological comparison showed that some families of the Axinellida have character states similar to the Microcionidae. These include *Pseudaxinyssa, Phakellia, Reniochalina, Acanthella,* and *Teichaxinella* of the Axinellidae, and *Echinoclathria* (and to a lesser extent *Isopenectya*) of Microcionidae; *Echinodictyum* (Raspailiidae), and *Echinochalina* and *Echinoclathria* of Microcionidae; *Sigmaxinella* (Desmacellidae), *Trikentrion* (Euryponidae), *Ectyoplasia, Raspailopsis, Raspailia* and *Clathriodendron* (Raspailiidae), and *Rhaphidophlus* and *Clathria* (Microcionidae). The constituent megascleres, special ectosomal characteristics, subectosomal development, and axial condensation of the choanosome, all of which present in many of those axinellid species, are sufficiently different from the Microcionidae to show that the families are clearly separated. However, it is problematic whether those distinctions can be maintained over an ordinal and subclass classification. Characters such as the possession of acanthostyles, differentiation of axial and extra-axial components of the skeleton, the development of a subectosomal region, and ecto-

somal specialization, some of which are diagnostic for the microcionids, occur in species from both groups. For example, species such as *Ectyoplasia tabula* (Lamarck) and *Rhaphidophlus abietinus,* can be differentiated by the presence of a subectosomal category of megasclere, a regular choanosomal reticulation, morphology of echinating acanthostyles, and microsclere form of *R. abietinus,* but they scarsely differ in other characters with known diagnostic value. It is suggested that at least some members of the Raspailiidae and Microcionidae, in particular, have closer affinities than is at present recognized by existing classification schemes.

EVIDENCE FROM ELECTROPHORETIC PROTEIN BANDING

Electrophoretic protein banding characteristics support this argument (Figure 2). A higher similarity (33%) was obtained between species of Raspailiidae and Microcionidae than among the Axinellidae (22%) or other axinellid families. Protein banding pattern analysis also indicated that the Axinellidae do not appear to be homogeneous, at least for the four species examined thus far.

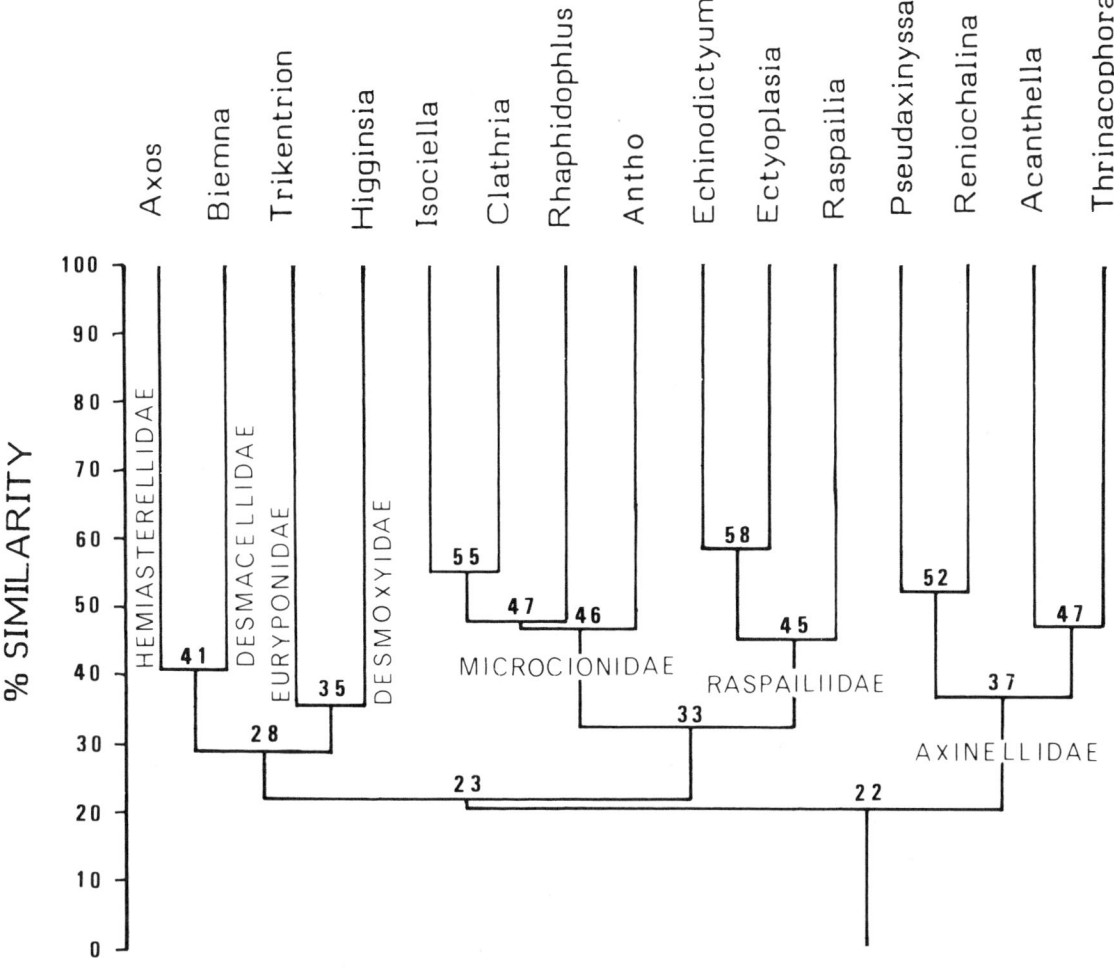

Figure 2. Comparison of genera of Microcionidae and Axinellida, based on electrophoretic protein mobility bands using Lawson's index of similarity.

They fall into two groups (*Pseudaxinyssa, Reniochalina* [52% similarity] and *Acanthella, Thrinacophora* [47%]), which were united by a similarity of only 37%. No morphological correlations are easily found to support this division, and it is probable that more than one such division occurs within this large family (51 nominal genera). It is clear from the literature that a major revision of the family Axinellidae is required; this study has shown that biochemical comparisons may be useful in aiding that revision.

Conclusions

Strong evidence for a classification of northwest Australian Microcionidae that is not substantially different from the scheme first proposed by Lévi (1960) comes from morphological investigations into character stability; the

suggested importance of those characters at the specific, generic, and family levels of classification; and limited protein electrophoretic results. That scheme was based on western European species and was also used by van Soest (1984) for West Indian species. The scheme, as interpreted here, emphasizes the importance of combining characters rather than relying on any single morphological character. These include choanosomal and subectosomal spicule distribution, fiber construction, and ectosomal structure. The presence or absence and form of certain spicule categories (microscleres and echinating megascleres) appear to be important mainly at the species level of classification, although the presence of acanthose megascleres in some Microcionidae, and in Raspaliidae and Euryponidae of the Axinellida, is indicative of closer affinities between the two groups. This relationship is also supported by other morphological characters. It is possi-

ble that echinating spicules can be given greater taxonomic importance at higher levels as indicators or "key"
characters, but apparently not at the generic level of classification (Simpson, 1968; van Soest, 1984), because they
represent the retention of an ancestral character state.
These data suggest that at least some Raspailiidae are
more closely related to the microcionids than to other
axinellid groups, and thus support the older concept of
Microcioniformes proposed by de Laubenfels (1936). The
close relationship between the two groups is supported by
protein electrophoresis. This study indicates that Microcionidae and Raspailiidae, in particular, should not be
separated at the subclass level, and possibly not at the
ordinal level of classification. Nevertheless, it would be
retrogressive to advocate a return to the Microcioniformes
scheme as proposed by de Laubenfels, since many other
relationships covered by that system are known to be invalid. The data presented here are insufficient at this stage
to propose an alternative system for higher classification.

Acknowledgments

I thank C. Keenan (University of Queensland) for con-
ducting electrophoretic analyses, S. Stone (BMNH), K.
Ruetzler (USNM), F. Rowe (AMU), C. Lu (NMV), W.
Zeidler (SAM), L. Cannon (QMU), L. Marsh (WAM),
C. Lévi (MNHN), U. Rahm (NMB), M. Grasshoff
(SMF), D. Kühlmann (ZMB), F. von Knorre (PMJ, via
F. Wiedenmayer), F. Climo (NMNZ), C. O'Riordan
(INM), S. Chambers (RSME), and T. Hoshino (MMBS)
for access to type specimens and other material. I am
grateful to T. Ward and the CSIRO Division of Fisheries
and Oceanography (Hobart) for seatime and assistance
with collections from the Northwest Shelf of Australia,
and to the Northern Territory University Planning Au-
thority (Darwin) for financial assistance for biochemical
analyses. This investigation was suggested by P. Bergquist (University of Auckland), and I thank her for her
encouragement and comments; I am also grateful to R.
van Soest (Zoological Museum Amsterdam) for productive discussions on the systematics of poecilosclerids.

Literature Cited

Bergquist, P. R. 1970. The Marine Fauna of New Zealand: Por-
    ifera, Demospongiae, Part 2 (Axinellida and Halichondrida).
    *Bulletin, New Zealand Department of Scientific and Industrial Re-
    search*, 197 (Memoir 51, NZ Oceanographic Institute): 1–85.
———. 1980. The Ordinal and Subclass Classification of the
    Demospongiae (Porifera); Appraisal of the Present Arrange-
    ment, and Proposal of a New Order. *New Zealand Journal of
    Zoology*, 7:1–6.
Bergquist, P. R., and R. J. Wells. 1983. Chemotaxonomy of the
    Porifera: The Development and Current Status of the Field.

Pages 1–50 in *Marine Natural Products*, 5, edited by P. Scheuer.
    New York: Academic Press.
Hallmann, E. R. 1920. New Genera of Monaxonid Sponges
    Related to the Genus *Clathria*. *Proceedings, Linnean Society of New
    South Wales*, 44:767–792.
Hooper, J. N. A. 1984. *Sigmaxinella soelae and Desmacella ithystela*,
    Two New Desmacellid Sponges (Porifera, Axinellida, Des-
    macellidae) from the Northwest Shelf of Western Australia,
    with a Revision of the Family Desmacellidae. *Northern Territory
    Museum of Arts and Sciences, Monograph Series*, 2:1–58.
———. 1987. New Records of *Acarnus* Gray (Porifera: De-
    mospongiae: Poecilosclerida) from Australia, with a Synopsis
    of the Genus. *Memoirs of the Queensland Museum*, 25:71–
    105.
———. In press. Revision of the Australasian Raspailiidae (Por-
    ifera: Demospongiae). *Invertebrate Taxonomy*.
———. In prep. Revision of the Australasian Microcionidae
    (Porifera: Demospongiae). *Invertebrate Taxonomy*.
Hooper, J. N. A., R. J. Capon, C. P. Keenan, and D. L. Parry. In
    press. Biochemical and Morphometric Differentiation of Two
    Sympatric Sibling Species of *Clathria* (Porifera: Demo-
    spongiae) from Northern Australia. *Invertebrate Taxonomy*, 4(1).
Jones, W. C. 1984. Spicule Dimensions as Taxonomic Criteria in
    the Identification of Haplosclerid Sponges from the Shores of
    Anglesey. *Zoological Journal of the Linnean Society*, 80:239–259.
Keenan, C. P., and J. B. Shaklee. 1985. Electrophoretic Identi-
    fication of Raw and Cooked Fish Fillets and Other Marine
    Products. *Food Techology in Australia*, 37:117–128.
Laubenfels, M. W. de. 1936. A Discussion of the Sponge Fauna
    of the Dry Tortugas in Particular, and the West Indies in
    General, with Material for a Revision of the Families and
    Orders of the Porifera. *Papers of the Tortugas Laboratory*, 30:1–
    225.
Lawson, R., D. H. Davies, J. Casey, and S. A. Mohammad. 1980.
    The Interpretation of Protein Separations on Polyacrylamide
    Gels. *Laboratory Practice*, 29:1065–1066.
Lee, W. L., and B. M. Gilchrist. 1985. Carotenoid Patterns in
    Twenty-Nine Species of Sponges in the Order Poecilosclerida
    (Porifera:Demospongiae): A Possible Tool for Chemosystem-
    atics. *Marine Biology*, 86:21–35.
Lévi, C. 1957. Ontogeny and Systematics in Sponges, *Systematic
    Zoology*, 6(4):174–183.
———. 1960. Les Demosponges des côtes de France. I. Les
    Clathriidae. *Cahiers de Biologie Marine*, 1:47–87.
———. 1969. Remarques sur la Taxonomie des Demospongea.
    Pages 497–502 in *Biologie des Spongiaires*, edited by C. Lévi and
    N. Boury-Esnault. Colloques Internationaux du C.N.R.S.
    291, Paris: Centre National de la Recherche Scièntifique.
Simpson, T. L. 1968. The Structure and Function of Sponge
    Cells: New Criteria for the Taxonomy of Poecilosclerid
    Sponges (Demospongiae). *Peabody Museum of Natural History,
    Bulletin*, 25:1–141.
———. 1978. The Biology of the Marine Sponge *Microciona
    prolifera* (Ellis and Solander). 3. Spicule Secretion and the
    Effect of Temperature on Spicule Size. *Journal of Experimental
    Marine Biology and Ecology*, 35(1):31–42.
Soest, R. W. M. van. 1984. Marine Sponges, from Curaçao and
    other Caribbean Localities. Part III. Poecilosclerida. *Studies
    on the Fauna of Curaçao and Other Caribbean Islands*, 66(112):1–
    167.

TAKAHARU HOSHINO*
Mukaishima Marine Biological Station
Faculty of Science
Hiroshima University
Onomichi P. O., Hiroshima 722, Japan

Merlia tenuis n. sp. Encrusting Shell Surfaces of Gastropods, *Chicoreus,* from Japan

Abstract

Merlia tenuis new species was collected from depths of less than 20 m along the coast of the South West Archipelago and Kii Peninsula, Japan. It thinly encrusts (ca. 200 μm) grooves between surface granules and varices of muricid gastropods *(Chicoreus carneolus, C. saulii,* and *C. microphyllus).* The sponge appears to be associated with about 15% of *Chicoreus* spp. shells, whether host gastropods are living or not. Like *Merlia deficiens, M. tenuis* lacks a calcareous basal skeleton, and therefore differs from *M. normani.* The spiculation consists of tylostyles (146–*168.6*–180 μm × 1–2 μm); clavidiscs (65–*90.5*–104 μm × 33–*43.9*–52 μm), and raphides (65–100 μm × 1–1.5 μm); commata are absent. The coloration is spectrum red or carmine. *M. tenuis* can be identified by its spiculation, coloration, and habitat.

Observation of the aberrant clavidiscs shows that the spicules are closely related to the diancistra of *Hamacantha.* This and other details of the spiculation indicate that the genus *Merlia* belongs to the family Bieminidae (Poecilosclerida).

In July 1984, a muricid gastropod *(Chicoreus carneolus)* associated with a sponge was collected from the shallow coastal waters (20 m deep) off the Kerama Islands in the South West Archipelago, Japan. Gastropods of the genus *Chicoreus* are generally light brown, hence, the association with a sponge was readily recognized by orange or red color. Examination of the sponge spicules in the laboratory showed that the sponge was a species of the genus *Merlia* (Poecilosclerida). Suspecting that other species of the genus *Chicoreus* from different localities could also be associated with *Merlia,* several Japanese species of the gastropod were examined.

*Deceased; direct correspondence to the Director, Mukaishima Marine Biological Station.

Materials and Methods

Several species of *Chicoreus* were collected from Zamami Island, the Kerama Islands in the South West Archipelago, Nada on the Kii Peninsula of Middle Japan (Figure 1), and other areas. *Merlia* was found on three species; the numbers observed are shown in Table 1. *Chicoreus carneolus* (Figure 2a) is the most abundant species of the genus in the subtidal region around Western Japan; *C. saulii* and *C. microphyllus* are rare. All the gastropods with *Merlia* from the South West Archipelago were alive, those from Nada, which were caught by shrimp nets, consisted of shells inhabited by the hermit crab, *Dardanus aspersus*. Immediately after collection, sponge color was recorded and the sample fixed in alcohol. Subsequently, the spicules, skeleton structure, and other parts of the sponge encrusted on the shell surface were studied under light and scanning electron microscopes.

Results

Merlia occurred as a thin encrustation over the entire surface of a few shells, but in most specimens covered only a small portion of the shell surface (Figure 2b). *Merlia* seems to prefer the concave portion between spiral granules or between varices. The sponge is carmine when alive, but after fixation by alcohol it turns to gray or white.

A vertical section of shell associated with the sponge shows that the sponge encrustation is only a few hundred micrometers thick and has no calcareous basal skeleton (Figure 3a). The spicules, tylostyles, clavidiscs, and raphides appear to have a confused arrangement.

Scanning electron microscopy of the sponge surface (Figure 3b) reveals that the spicules, clavidiscs, tylostyles, and raphides are arranged randomly on the shell surface, especially between the granules or between the spiral lines of the shell. (The raphides are not distinct, but they can be identified under a light microscope.)

Table 1. Muricid gastropods, *Chicoreus* spp., associated with *Merlia tenuis*

Location and species	No. specimens collected	No. specimens associated with *Merlia*
South West Archipelago (< 20 m in depth)		
Chicoreus saulii	2	1
C. microphyllus	2	1
C. carneolus	2	1
Nada, Kii Peninsula (depth unkown, caught in shrimp net)		
C. carneolus	62	9

The spicule dimensions of each specimen are given in Table 2. Note that the clavidiscs from Nada are larger than those from the South West Archipelago.

Numerous aberrant clavidiscs were observed on the spicule mount of one specimen (Figure 4a). The aberration is was an incomplete fusion between terminal ends of C-formed spicules, otherwise a common feature only of early stage spicules.

Diagnosis of *Merlia tenuis* New Species

HOLOTYPE: Jap-197, Zamami Island, Kerama Islands, 20 m in depth.

DEPOSITION: Mukaishima Marine Biological Station.

MORPHOLOGY: Extremely thin (about 200 μm) crust on shell surface of *Chicoreus* spp. (Muricidae, Gastropoda). Color carmine. Skeleton, confused arrangement of tylostyles, clavidiscs, and raphides. Commata absent. Calcareous basal skeleton deficient. Tylostyles, smooth, nearly straight; base spherically swollen, diameter 1.5–2 times as thick as shaft, opposite end sharply pointed. Clavidisc (Figure 4b), oval, with spindle-shaped hole in the central part. Raphides, rough, nearly straight, tapering from middle to each pointed end.

Table 2. Spicule dimensions (μm) of specimens of *Merlia tenuis* (minimum–mean–maximum) for length × thickness (or width) based on ten spicules for each specimen; n.d. = no data

Locality and gastropod	Tylostyle	Clavidiscs	Raphides
South West Archipelago			
Chicoreus saulii	140–155 × 1	50–64.0–74 × 25–36.9–50	ca. 70 × 1
C.microphyllus	n.d	46–54.0–65 × 24–27.1–30	n.d.
C. carneolus	140–*157.9*–170 × 1–2	50–*68.3*–83 × 29–32.2–36	60–100 × 1
Nada, Kii peninsula			
C. carneolus	110–*156.7*–180 × 1–2	70–*86.2*–104 × 38–40.1–43	60–70 × 1–1.5
	146–*168.6*–180 × 1–2	65–*90.5*–104 × 33–43.9–52	65–100 × 1–1.5
	148–*162.9*–178 × 1–2	75–*90.1*–110 × 35–41.8–46	
	151–*164.6*–171 × 1–2	60–*87.6*–91 × 52–37.3–43	65–100 × 1
	145–*163.3*–175 × 1–2	58–*79.0*–100 × 33–37.5–43	65–90 × 1
	138–*160.1*–175 × 1–1.5	70–*80.3*–98 × 33–39.1–45	
	139–*153.7*–163 × 1–2	55–*67.2*–98 × 29–34.6–47	85–102 × 1
	158–*168.8*–176 × 2–3	63–*87.3*–102 × 32–38.8–46	70–110 × 1
	150–*160.0*–183 × 1–2.5	65–*79.8*–91 × 33–40.7–46	55–110 × 1–2

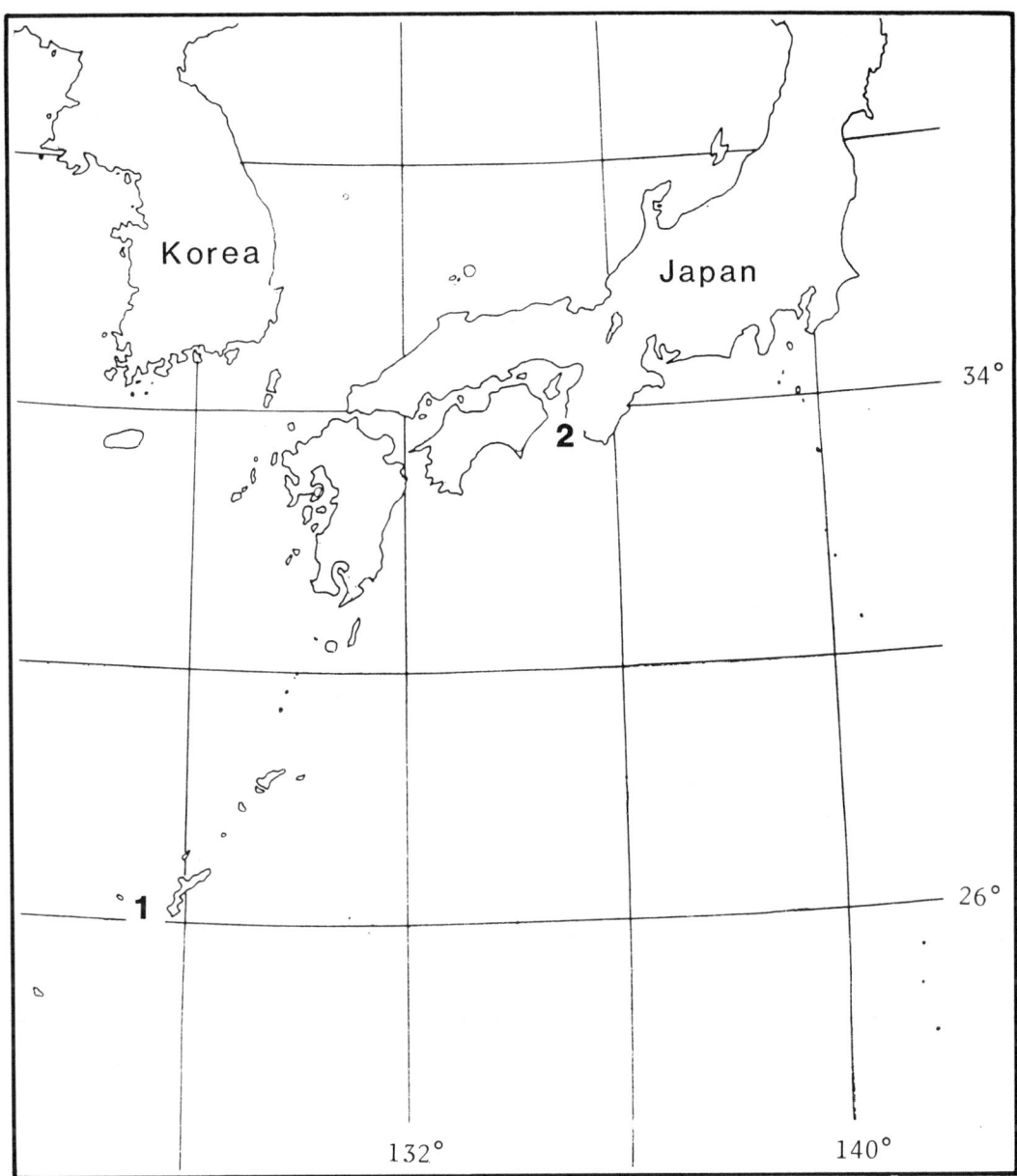

Figure 1. Map showing study sites: Zamami Island, South West Archipelago (1) and Nada, Kii Peninsula (2).

SPICULE MEASUREMENTS OF HOLOTYPE SPECIMEN, (RANGES AND MEANS): Tylostyles, 140–*157.9*–170 μm × 1–2 μm; Clavidiscs, 50–*68.3*–83 μm × 29–*32.2*–36 μm; Raphides, 60–*100* μm × 1 μm.

Discussion

Since Kirkpatrick (1908) first described *Merlia normani* from Porto Santo, Madeira Archipelago, this species has been found in several locations: Solomon and Mauritius (Dendy, 1922); Madeira and Mediterranean (Vacelet, 1967), Jamaica (Hartman and Goreau, 1970); Bay of Na-

ples (Pulitzer-Finali, 1970); and Tulear (Vacelet and Vasseur, 1971).

Another *Merlia* species, *M. deficient* Vacelet (1980) is distinguished by the absence of a basal skeleton. It was found in Marseille (Pouliquen, 1972–as *M. normani*); Marseille, Jamaica, and New Caledonia (Vacelet, 1980, 1981); Curaçao and Porto Santo (van Soest, 1984–as *M. normani*); and Bermuda (Vacelet, pers. comm. [ed.]).

Merlia tenuis differs from the other two species in several respects. Like *M. deficiens*, it lacks a calcareous basal mass present in *M. normani*. It differs from *M. deficiens* in coloration, in the lack of commata, and in habitat. Several au-

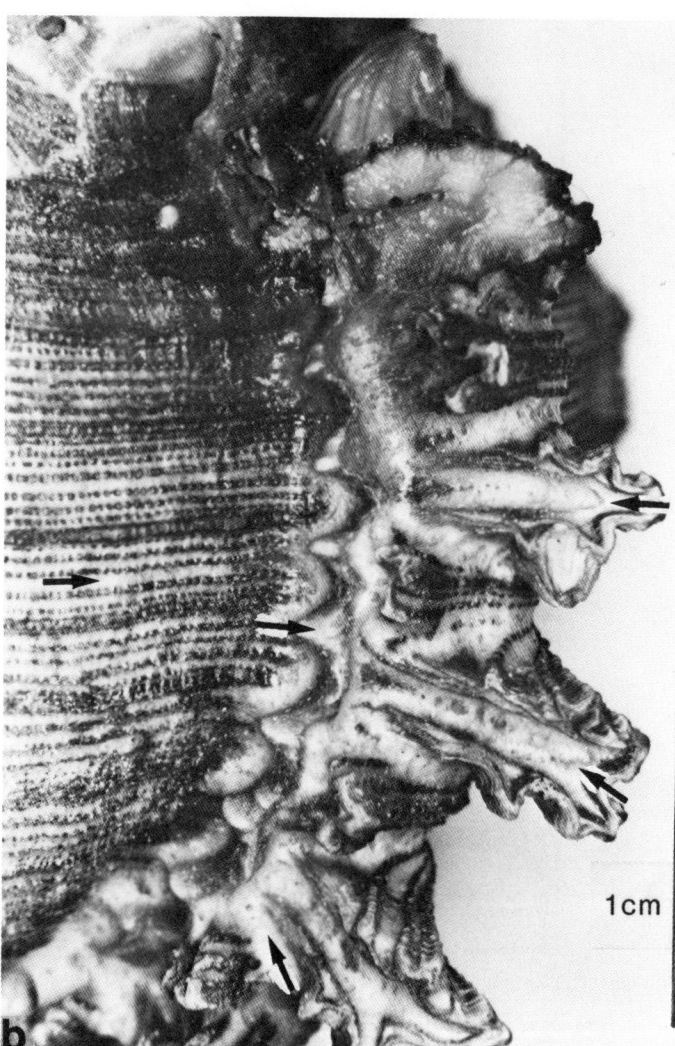

Figure 2. *Chicoreus carneolus* collected from Nada, inhabited by a hermit crab, *Dardanus aspersus*: *a*, total view (shell length, 10.5 cm); *b*, detail of shell surface coated by *Merlia tenuis (arrows)*.

thors have suggested that *M. deficiens* is a shallow-water form of *M. normani* in which the development of a calcareous basal mass has been arrested, but no intermediate form between *M. normani* and *M. deficiens* has yet been found. When distribution is taken into account, a strong case can be made for separating the species: the ecological habitat of *M. tenuis*, encrustation on *Chicoreus*, seems to be specific.

The lack of a calcareous basal skeleton, the aberrant clavidiscs, and the similarity between clavidiscs of the genus *Merlia* and diancistra of the genus *Hamacantha* indicate that the genus *Merlia* belongs to the order Poecilosclerida. The author agrees with van Soest (1984) that both clavidiscs and diancistras are sigmatose microscleres. The observation of aberrant clavidiscs supports this view. Furthermore, many species in the genus *Hamacantha* have macroscleres similar to those of the genus *Mer-*

lia. Therefore, *Merlia* and *Hamacantha* should be considered closely related genera, a view already expressed by Kirkpatrick (1909). The systematic position assigned to the genus *Merlia* by subsequent authors is as follows:

Monaxonellida (Kirkpatrick, 1910)
Merlinae, Haploscleridae, Monaxonellida (Kirkpatrick, 1911)
Merlinae, Haplosclerida (Dendy, 1922)
Desmacidonidae, Poecilosclerida (Topsent, 1928)
Ophlitaspongiidae, Poecilosclerida (de Laubenfels, 1936)
Sigmaxinellidae, Clavaxinellida (Vacelet, 1964)
Sclerospongiae (Hartman and Goreau, 1970)
Sclerosponges (Vacelet and Vasseur, 1971)
? Sclerospongiae (Lévi, 1973)
Ceratoporellida, Sclerospongiae (Bergquist, 1978)

3d Int. Sponge Conf. 1985

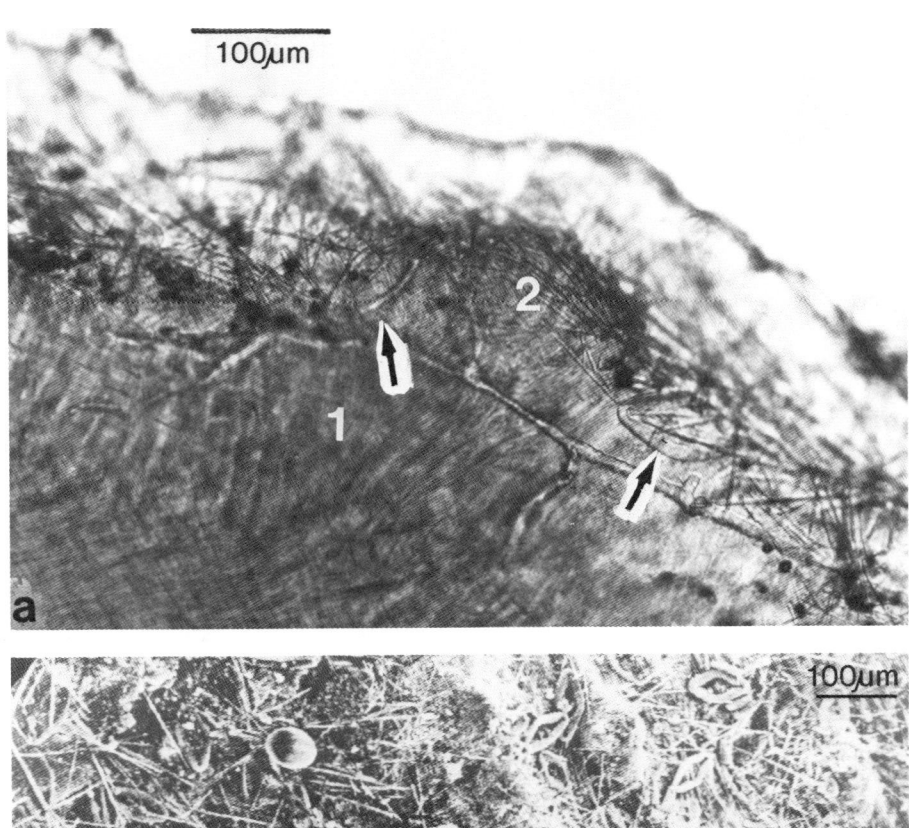

Figure 3. *Merlia tenuis* on *Chicoreus* shell: *a*, vertical section; *b*, scanning electron micrograph of sponge surface. *1*, shell; *2*, sponge; arrows point to clavidisc spicules.

Merlida, Tetractinomorpha (Vacelet, 1981, 1983)
Biemnidae, Poecilosclerida (van Soest, 1984)

Today there are three predominant views on the systematic position of the *Merlia*: some believe the genus belongs to the class Sclerospongiae (e.g., Hartman and Goreau, 1970; Bergquist, 1978); others put it in the order Merlida, subclass Tetractinomorpha (Vacelet, 1981, 1983); and still others place it in the order Poecilosclerida, Ceractinomorpha (van Soest, 1984).

This study indicates that *Merlia* is not a sclerosponge, but offers no clear evidence as to whether *Merlia* belongs to the Tetractinomorpha or to the Ceractinomorpha. However the results point to a similarity between the spicules of the genus *Merlia* and those of the order Poecilosclerida. Hence, we agree with van Soest (1984) that *Merlia* should be assigned to the family Biemnidae.

If the three species discussed here belong to the same genus, the presence of a calcareous basal skeleton in *Merlia normani* should be considered a species character. That

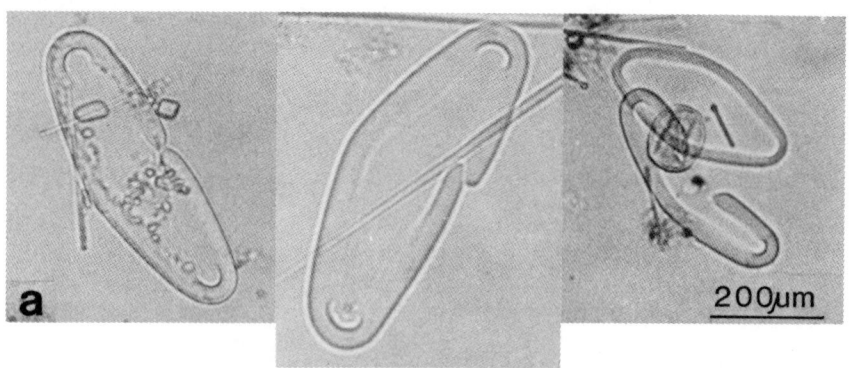

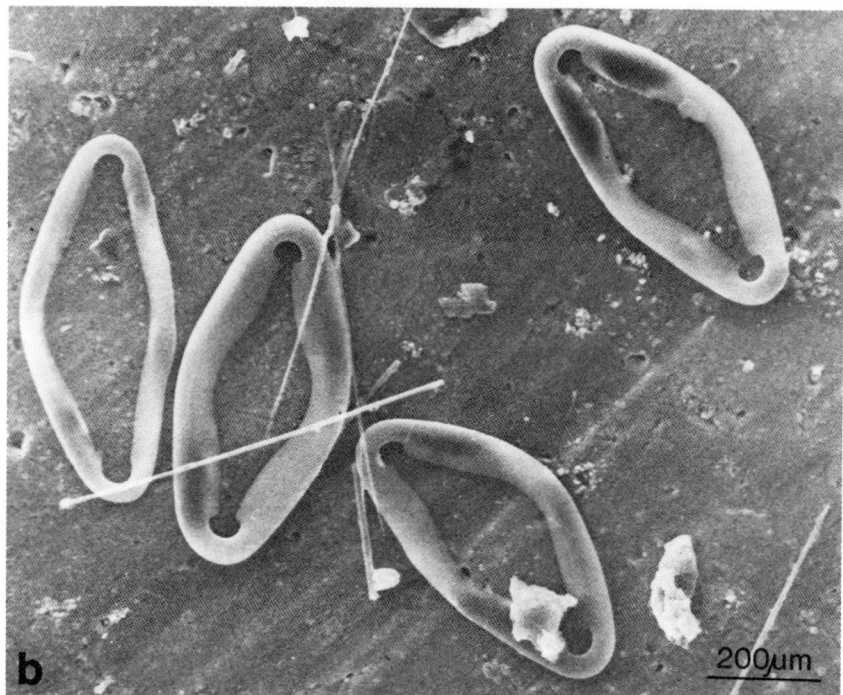

Figure 4. *Merlia tenuis* clavidiscs: *a*, light micrograph showing incomplete fusion; *b*, scanning electron micrograph (broken tylostyles also seen).

is to say, it can be considered unstable in the genus but stable in the species *Merlia normani*.

Acknowledgments

I am grateful to Ryo Katashima, Director and Professor of the Mukaishima Marine Biological Station, Faculty of Science, Hiroshima University, and Akihiko Inaba, Professor Emeritus of the Hiroshima University, for their encouragement throughout this study. Contribution of the Mukaishima Marine Biological Station, no. 270.

Literature Cited

Bergquist, P. R. 1978. *Sponges*. London: Hutchinson. 268 pp.

Dendy, A. 1922. Report on the Sigmatotetraxonida Collected by H. M. S. "Sealark" in the Indian Ocean. *Transactions, Linnaean Society, London* (series 2, Zoology.), 18:1–164.

Hartman, W. D., and T. F. Goreau. 1970. Jamaican Coralline Sponges: Their Morphology, Ecology and Fossil Relatives. Pages 205–243 in *The Biology of the Porifera*, edited by W. G. Fry. Symposia of the Zoological Society of London 25. London: Academic Press.

Kirkpatrick, R. 1908. On Two New Genera of Recent Pharetronid Sponges *Annals and Magazine, Natural History* (series 8), 2:503–514.

———. 1909. Notes on *Merlia normani* Kirkpatrick. *Annals and Magazine, Natural History* (series 8), 4:228–291.

———. 1910. Further note on *Merlia normani* Kirkpatrick. *Annals and Magazine, Natural History* (series 8), 5:287–291.

———. 1911. On *Merlia normani*, a Sponge with a Silicious and Calcareous Skeleton. *Quarterly Journal for Microscopical Science*, 56:657–702.

Laubenfels, M. W. de. 1936. A Discussion of the Sponge Fauna of the Dry Tortugas in Particular, and the West Indies in General, with Material for a Revision of the Families and Orders of the Porifera. *Papers of the Tortugas Laboratory*, 3:467:1–225.

Lévi, C. 1973. Systematique de la classe des Demospongiaria (Démosponges). Pages 577–631 in *Spongiaires, Traité de Zoologie*, edited by P. P. Grassé. Paris: Masson.

Pouliquen, L. 1972. Les Spongiaires des grottes sousmarines de la region de Marseille: Ecologie et systematique. *Tethys*, 3:717–758.

Pulitzer-Finali, G. 1970. Report on a Collection of Sponges from the Bay of Naples, I. Sclerospongiae, Lithistida, Tetractinellida, Epipolasida. *Publicazioni Stazione Zoologica, Napoli*, 38:328–354.

Soest, R. W. M. van. 1984. Deficient *Merlia normani* Kirkpatrick, 1908, from the Curaçao Reefs, with a Discussion on the Phylogenetic Interpretation of Sclerosponges. *Bijdragen tot de Dierkunde*, 54(2):211–219.

Topsent, E. 1928. Spongiaires de l'Atlantique et de la Méditerranée provenant des croisières du Prince Albert Ier de Monaco. *Résultats des Campagnes Scientifiques par Albert Ier Monaco*, 74:1–376.

Vacelet, J. 1964. Etude monographioque de l'éponge calcaire Pharetronide de Mediterranée, *Petrobiona massiliana* Vacelet et Lévi, les Pharetronides actuelles et fossiles. *Recueil des Traveaux de la Station Marseille d'Endoume*, 50(34):1–125.

———. 1967. Quelques Eponges Pharentronides et silicocalcaires de grottes sous-marines obscures. *Recueil des Traveaux de la Station Marseille d'Endoume*, 58(42):121–132.

———. 1980. Squelette Calcaire Facultatif et corps de regeneration dans genre *Merlia*, éponges apparentees aux chaetetides fossiles. *Comptes Rendus, Academie des Sciences, (Paris)*, 290:227–230.

———. 1981. Eponges hypercalcifiées ("pharetronides", "sclerosponges") des cavités des recifs coralliens de Nouvelle-Caledonie. *Bulletin, Muséum National Histoire Naturelle, (Paris), (series 4)*, 3:313–351.

———. 1983. Les Eponges hypercalcifies, reliques des organisms constructeurs de Récifes du Paleozoique et du Mesozoique. *Bulletin, Societé Zoologique de France*, 108(4):547–557.

Vacelet, J., and Vasseur, P. 1971. Eponges des recifs coralliens de Tulear, Madagascar. *Tethys*, supplement 1:51–126.

ROB W. M. VAN SOEST
Instituut voor Taxonomische Zoologie
Zoölogisch Museum Amsterdam, Mauritskade 57
P.O. Box 4766, Amsterdam, The Netherlands

Shallow-Water Reef Sponges of Eastern Indonesia

Abstract

In a preliminary study of coral reef sponges collected during the Indonesian-Dutch Snellius II Expedition (1984–1985) in the eastern part of the Indonesian Archipelago, 56 species (out of 300–400 collected) were found to be common in this area. Other findings include a novel symbiotic association between the octocoral *Tubipora* and a *Mycale* and the first sclerosponge ever to be recorded from Indonesia (i.e., *Astrosclera willeyana*). Literature reveals that some of the common reef sponges of Indonesia are also common in other parts of the Indo-Pacific area: 26 appear to be common in Northeast Australian reefs, 37 in Central Pacific reefs, and 24 in Western Indian Ocean reefs. From the overall distribution of sponges in these four areas it is tentatively concluded that Indonesia is the distributional center of Indo-Pacific reef sponges. At the same time, species vary locally, and only 10 species (out of an estimated total of 204) may be common throughout the Indo-Pacific area. In addition, Indo-Pacific reefs seem highly dissimilar to those of the tropical Atlantic with respect to the generic composition of the common species of sponges.

Surprisingly little is known about the sponge fauna of Indonesia, supposedly the richest of the world's oceans. The most recent publication containing taxonomic descriptions of original material dates back to before World War II (Brøndsted, 1934). There are only a few earlier major publications on the Indonesian sponge fauna: Topsent (1897), Kieschnick (1896, 1900), Thiele (1899, 1903), and Hentschel (1912). The collections on which these works were based have not been revised, with the exception of Topsent's (1897; see Desqueyroux, 1981). One reason for this dearth of information is that the sponges in the large collections made by major expeditions in Indonesian waters (namely, the Siboga Expedition, 1898–1899, and the Snellius Expedition, 1925) were never described, with the exception of the Siboga Calcarea (Burton, 1930) and the genera *Placospongia* (Vosmaer and Vernhout, 1902) and *Spirastrella* (Vosmaer, 1911).

The Indonesian-Dutch Snellius II Expedition 1984–

1985 was organized around five principle projects in the fields of physical oceanography, marine geology, and marine biology, the last of which comprised a pelagic and a coral reef program. I joined the team interested in in exploring the coral reefs aboard the R/V *Tyro*, which traveled from Ambon (Amboina) to various locations around the Band and Savu seas, ending in Sulawesi (Figure 1). The eight localities visited were chosen for their divergent physical environment, namely, sheltered and exposed lagoons, deep lagoons, sea grass beds, coastal reef, exposed reef flats, drowned atolls, and exposed outer reefs.

Methods

During the cruise, these eight areas were systematically sampled by diving, snorkeling, and wading. Sponges were removed from the substrate, taken aboard, characterized briefly, labeled, and preserved in ethanol. Many specimens were photographed. Whenever possible, sampling in each location progressed as follows: dive 10–15 m, dive 4–10 m, snorkeling 1–4 m, intertidal wading. Now and then material was collected from depths exceeding 15 m. In addition, many specimens were obtained from deep-reef habitats by dredging and trawling.

Results
DEPTH AND HABITAT DISTRIBUTION

Approximately 1,200 specimens belonging to 300–400 species at 67 stations spread over Eastern Indonesia were collected. Fifty-six species were found to be common in more than three of the eight localities visited; many of these were common throughout the area. These common species and their depth distribution are listed in Table 1. Some identification problems still exist, and all identifications have yet to be checked against the type specimens.

On the basis of rough estimation of relative sediment cover, prevailing wave action, and current strength, the sampled stations were categorized as "sheltered" and "exposed." Although this is admittedly an oversimplification, it nevertheless provides some indication of the ecological preferences of many of the common sponges. The characteristic species found at different depths in sheltered and exposed habitats are listed in Table 2.

NEW RECORDS AND OBSERVATIONS

FIRST INDONESIAN SCLEROSPONGE. In view of other reports of *Astrosclera willeyana* Lister from Polynesia, Christ-

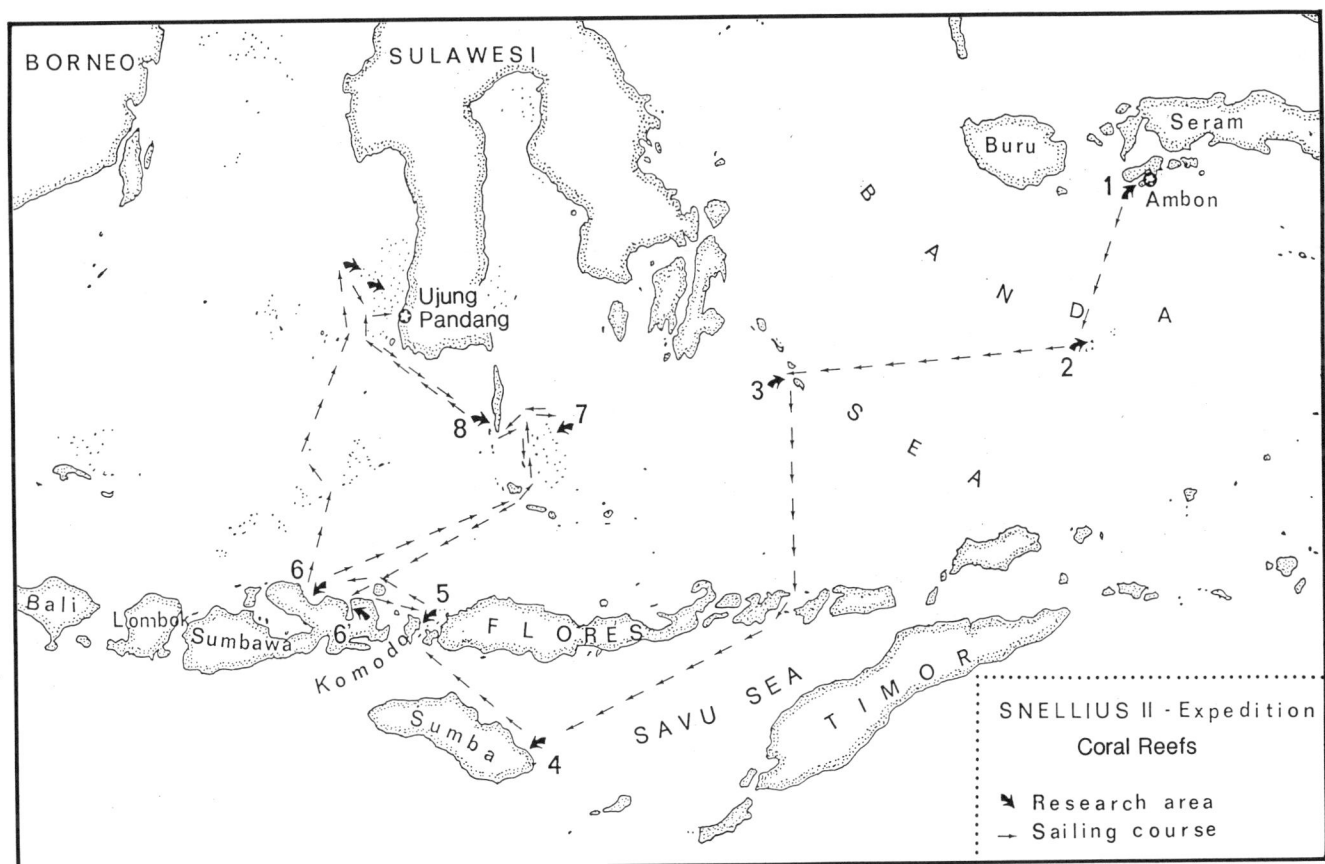

Figure 1. Map of eastern Indonesia showing cruise track of R/V *Tyro* and coral reef sampling stations of Snellius II expedition: *1*, Ambon; *2*, Pulau 2 Maisel; *3*, Tukangbesi; *4*, Sumba; *5*, Komodo; *6*, Sumbawa; *7*, Taka Bone Rate; *8*, Selayar; *9*, Ujung Pandang.

Table 1.   Depth distribution of common shallow–water reef sponges of eastern Indonesia; abundance estimates are based on number of observations (long bar = abundant in three or more localities; medium bar = occasional in three or more localities; short bar = rare in three or more localities)

*(In the table below, abundance is coded from the bar lengths: **A** = long bar (abundant), **O** = medium bar (occasional), **R** = short bar (rare).)*

Species	0–1 m	1–4 m	4–10 m	10–15 m	> 15 m
Coelocarteria singaporense	A	R		R	
Haliclona cymaeformis	A	R			
Gelliodes pumilus	A	A			
Haliclona viola	A	A			
Hymenacidon conulosa	A	A			
Ircinia ramosa	A	A			
Cinachyra australiensis	A	A	R		
Xestospongia exigua	A	A	R	R	
Stelletta clavosa	A	A		R	
Clathria reinwardti	A	A	O		
Dysidea herbacea	A	A	A	O	
Spirastrella vagabunda	A	R			
Dactylospongia elegans		O	O		
Psammaplysilla purpurea		O	R	R	R
Xestospongia cf. *carbonaria*	R	R			
Liosina paradoxa	R	O	A		
Haliclona cf. *nematifera*	R	O	A		
Tethya robusta	R	O	A	R	
Phyllospongia foliascens	R	R	O	O	
Phyllospongia papyracea	R	O	A		
Spirastrella solida	R		O		
Haliclona sp. red	R				
Aaptos cf. *suberitoides*		O	R		
Halichondria cartilaginea		O	R		
Gelliodes fibulatus		A	A	R	
Hyrtios erectus		A	A	O	
Haliclona amboinensis		A	A	O	
Dysidea granulosa		A	A	O	
Stelletta globostellata		O	A	R	
Axinella carteri		O	A		R
Plakortis cf. *nigra*		O	A	R	R
Petrosia contignata		O	O	A	A
Theonella swinhoei		A	A	A	O
Oceanapia amboinensis		O	O	A	O
Agelas mauritiana		O	O		
Asteropus sarassinorum		O	A	R	
Placospongia melobesioides		O	O		
Acervochalina confusa		O	O		R
Haliclona cf. *turqoisia*			O		
Ircinia cf. *irregularis*			O	R	
Spirastrella decumbens			A		
Petrosia nigricans			A	O	
Petrosia testudinaria			A	O	
Fascaplysinopsis reticulata			O	O	
Niphates olemda			O	O	R
Callyspongia joubini			O	A	
Acanthella cavernosa			O	A	O
Gelliodes petrosioides			O	A	R
Callyspongia confoederata			R	A	
Dysidea cinerea			A	O	
Clathria basilana			O	O	
Liosina arenosa			A	O	
Myrmekioderma granulata			O	O	
Cliona sp. orange			O	O	
Theonella conica			R	O	
Agelas ceylonica				O	R

Table 2. Vertical distribution of characteristic species of sheltered and exposed habitats in eastern Indonesia

Depth	Habitat type	
	Sheltered	Exposed
0–1 m	*Coelocarteria singaporense* *Gelliodes pumilis* *Haliclona cymaeformis* *Haliclona* cf. *viola* *Spirastrella vagabunda*	*Hymeniacidon conulosa* *Dysidea herbacea* *Stelleta clavosa* *Cinachyra australiensis* *Xestospongia exigua*
1–4 m	*Ircinia ramosa* *Clathria reinwardti* *Dactylospongia elegans* *Psammaplysilla purpurea* *Gelliodes fibulatus*	*Asteropus sarassinorum* *Phyllospongia foliascen* *Phyllospongia papyracea* *Hyrtios erectus* *Halichondria cartilaginea*
4–10 m	*Petrosia contignata* *Tethya robusta* *Spirastrella solida* *Axinella carteri* *Petrosia nigricans* *Spirastrella decumbens* *Myrmekioderma granulata*	*Liosina paradoxa* *Dysidea granulosa* *Stelleta globostellata* *Agelas mauritiana* *Petrosia testudinaria* *Theonella swinhoei* *Haliclona* cf. *turquoisia* *Fascaplysinopsis reticulata*
10–15 m	*Acervochalina confusa* *Clathria basilana* *Niphates olemda*	*Oceanapia amboinensis* *Gelliodes petrosioides* *Acanthella cavernosa* *Callyspongia confoederata* *Theonella conica*

mas Island, Madagascar, New Caledonia and Northeast Australia (Ayling, 1982), it is hardly surprising to find this species at depths of 10 m near Taka Karlarang (Taka Bone Rate).

"SOFT" LITHISTIDS. The lithistid *Theonella swinhoei* Gray is a common sponge in shallow, rather exposed reef habitats. It is compressible owing to the weakly developed lithistid skeleton, especially near the surface. A similar "soft" lithistid is found in West Indian coral reefs, namely, *Discodermia dissoluta* Schmidt.

SYMBIOTIC ASSOCIATIONS. Symbiotic relationships are generally more numerous and complex in Indo-Pacific reefs than in West Indian reefs. It is not at all unusual to find sponges in association with macroalgae. Two examples of such "compound" organisms are *Haliclona cymaeformis* and *Halichondria cartilaginea* (Esper, 1794). The former is relatively well known (Vacelet, 1981); it is intimately associated with the red alga *Ceratodictyon spongiosum*, and is known throughout the Indo-Pacific: particularly in North Australia (Bergquist and Tizard, 1967, as *Sigmadocia symbiotica*), Hong Kong (van Soest, 1980, as *S. symbiotica*) Madagascar (Vacelet and Vasseur, 1971, as *Gellius cymiformis*), Mozambique Channel (Thomas, 1979, as *Sigmadocia fibulata*), and New Caledonia (Vacelet, 1981, as *Gellius cymiformis*). It was earlier reported from Indonesia by Weber and Weber (1889, as *Reniera fibulata*), and the Siboga collections also hold several specimens.

Less well known is *Halichondria cartilaginea* (senior syn-

onym of *Halichondria symbiotica* Lévi, 1961). It lives in intimate association with the green alga *Cladophoropsis vaucheriaeformis*, forming corrugated green mats that have branching projections. Weber and Weber (1889) reported this form from Indonesia as *Halichondria* spec. *Struvea delicatula* (see their Plate V, Figures 3–5).

An association that apparently has not yet been described is that of the red alga *Amphiroa* and a species of *Mycale* showing affinities with *M. cockburniana* Hentschel (1911). The association consists of yellow, tubelike forms; the alga is visible as red "veins" in the walls of these forms. It was found in three separate localities.

Much more spectacular is a novel *Mycale-Tubipora* association found exclusively around the island of Komodo. It forms quite characteristic groups of tubes, which are apparently the result of the sponge acting upon the growth form of the normally hemispherical or lobate octocoral. The compound organism was found to be quite common in shallow water around Komodo, but repeated attempts to find it elsewhere (for instance around nearby Sumbawa) have failed. The fact that in all discovered instances the same *Mycale* species was found growing over and between the red polyp tubes indicates that this may well be an obligatory relationship.

A further remarkable association is the only discovered case of a *Hyattella* species completely interwoven with a hydroid of the thecate family Haleciidae. The few hydroids reported to be living in association with sponges have been Athecata.

In contrast to the situation in the West Indian region, only a few zoanthids were found encrusting sponges; most of these examples are from caves or deep water.

Discussion
BIOGEOGRAPHIC DISTRIBUTIONAL CENTERS IN THE INDO-WEST PACIFIC

A major biogeographic issue in the Indo-Pacific is the alleged presence of a Western Indian Ocean coral reef distributional center which is thought to reflect the separate geological history of this area (see Rosen, 1971, 1975; Briggs, 1974). Following Rosen (1971), I attempted to establish whether such a distributional center in the Western Indian Ocean is also evident for reef sponges by comparing the lists of genera known from different coral reef areas of the western region of the Indo-Pacific. The numbers of genera of four separate areas (Indonesia, the Central Pacific, Northeast Australia, and Madagascar) are presented in Table 3. This information was gathered from recent articles dealing solely with reef sponges; all those containing data on dredged material were omitted. These reports were as follows: on Western Indian Ocean reefs, Vacelet and Vasseur (1965, 1971, 1977), Vacelet et al. (1976), Rützler (1972); on Central Pacific reefs, de Laubenfels (1954), Bergquist (1965), Bergquist et al.

Table 3. Comparisons of coral reef sponge faunas of four Indo–West Pacific areas and the West Indies region, using numbers of genera reported for each area and the numbers of species reported as "common" in each area. (– = not applicable; for further explanation, see text)

Compared areas	Genera				"Common" species			
A – B	Total A	Shared (W)	Total B	Cz %	Total A	Shared (W)	Total B	Cz %
Indonesia – Northeast Australia	189	66	79	49	98	26	54	34
Indonesia – Central Pacific	189	66	83	49	98	37	61	47
Central Pacific – Northeast Australia	83	31	79	38	61	16	54	28
W. Indian Ocean – Indonesia	121	85	189	55	79	24	98	27
W. Indian Ocean – Northeast Australia	121	32	79	32	79	15	54	24
W. Indian Ocean – Central Pacific	121	43	83	42	79	11	61	16
Indonesia – West Indies	189	82	117	53	–	–	–	–
Total Indo–Pacific – West Indies	244	94	117	52	–	–	–	–
"Common" genera: Total Indo–Pacific – West Indies	48	10	24	28	–	–	–	–

(1971); and on Northeast Australian reefs, Burton (1934), Bergquist (1969), Bergquist and Tizard (1967), Pulitzer-Finali (1982). Data on Indonesian reefs were obtained from both the Snellius II and the Siboga collections.

The lists of genera have been compared using Czekanowski's coefficient:

$$Cz = \frac{2W}{A+B} \times 100$$

in which W is the number of shared genera of area A and B, A is the number of genera in area A, B is the number of genera in area B (after Dauer and Simon, 1975). Czekanowski's values for the different areas are given in Table 3. Other coefficients used in biogeographic comparisons (e.g., Jaccard's or Simpson's, see Udvardy, 1969) yield comparable results. If Rosen's data on reef coral genera concur with those on reef sponge genera, the genera of Indonesia, the Central Pacific, and Northeast Australia can be expected to show a high degree of similarity and few differences compared to those of the Western Indian Ocean. Although generic endemism appears to be higher (25%) in the Western Indian Ocean than anywhere else (Indonesia 20%, Central Pacific 18%, and Northeast Australia 15%), the Cz values for the respective areas are quite similar, so no conclusion may be drawn concerning the existence of a separate Indian Ocean center. Also, note that closely studied cryptic habitats in Madagascar reefs have yielded several unusual genera, whereas this habitat has not been investigated or has been studied only cursorily in other areas. Moreover, the generic composition of all four areas is by no means equally well known.

If we extend our comparison to West Indies coral reef sponge genera described in the recent literature (Hechtel, 1965; Wiedenmayer, 1977; van Soest, 1981; Gómez López and Green, 1984; Zea, 1987; Alvarez et al., this volume), we find this fauna shows the same high similarity with Indonesia as the Western Indian Ocean. Compared to the Western Indo-Pacific as a whole, the Cz value remains equal. From these comparisons, crude and preliminary as they may be, it appears that reef sponge genera tend to have large, often cosmotropical, distributions, and, thus, that any differences between the areas under consideration must be based on a comparison of species. The present state of our knowledge makes such a comparison impossible. Very few regional studies have included comparisons of type specimens, so that many species identifications will obviously prove to be wrong. Furthermore, the different areas have by no means been studied with equal intensity, so the absence of species from the regional lists does not necessarily mean they do not occur in a particular area. At most, we are able to compare the distribution of the more common, relatively well-known species in the various parts of this huge area.

DISTRIBUTION OF "COMMON" REEF SPONGES

Reef sponges were judged to be common in the four areas studied on the basis of assessments and actual numbers reported in the literature. The numbers of species concerned and their comparison values (Cz values) are presented in Table 3. Although this procedure can only yield very preliminary results, some trends are already apparent: Local variations of sponge faunas are pronounced in different parts of the Indo-Pacific. Even closely adjoining areas such as Indonesia and Northeast Australia have a Cz value of only 34% (based on 26 shared species out of a total of 152 species considered common in both areas). The highest Cz value was for Indonesia and the Central Pacific (47% based on 37 species shared out of a total of 159 considered common in both areas); the lowest Cz value was for Madagascar and the Central Pacific (16% based on 11 species shared out of a total of 140 considered common in both areas). Only 10 species were found to be common in all four areas: *Spirastrella vagabunda, Tethya robusta, Myrmekioderma granulata, Iotrochota baculifera, Clathria reinwardti, Haliclona cymaeformis, Ircinia ramosa, Phyllospongia*

Table 4. Examples of morphologically similar, vicariant species in Indo–Pacific and West Indies reefs

Indo–Pacific	West Indies
Monanchora unguiculata	M. unguifera
Mycale sp.	M. laevis
Mycale euplectelloides	M. laxissima
Iotrochota baculifera	I. birotulata
Petrosia testudinaria	Xestospongia muta
Callyspongia confoederata	C. vaginalis
Niphates sp.	N. digitalis

foliascens, *Phyllospongia papyracea*, and *Psammaplysilla purpurea*. Although species lists for the separate areas may eventually turn out to be rather homogeneous (as early evidence suggests), it is still likely that the faunas are quite dissimilar owing to the different local dominance of species.

Indonesia appears to be the main distributional center for the Indo-Pacific sponge fauna since Cz values for Madagascar, Northeast Australia, and the Central Pacific (including Indonesia) are consistently higher than the values for these areas compared among themselves. This coincides with most biogeographic observations (cf. Briggs, 1974) and concurs with the fact that Indonesia is situated in the center of the four considered areas.

West Indian coral reef sponge faunas are dissimilar to Indo-Pacific sponge faunas. The generic composition of the common sponges of the Indo-Pacific area was based on the same data garnered from the literature noted above. Genera containing one or more species cited in three or more instances as common were considered "common." In the same way, a list of common genera was assembled for the West Indian region using recent faunal surveys. A comparison of these lists yielded a Cz value of 28% (ten shared common genera out of total of 72; see Table 3). The species diversity of these ten shared genera is much higher in the West Indian reefs (half of the common species in the West Indies belong to these 10 shared genera) than in the Indo-Pacific (only about one-fifth).

In contrast to the observed dissimilarity, there are some strikingly similar species in both areas; a list of examples is given in Table 4. These forms are considered to be evidence of slow evolutionary change in some sponge genera.

Acknowledgments

The Dutch Council for Research (NRZ) and the Indonesian Institute of Oceanography (LIPI) supported the Snellius II sponge project. I am grateful to J. Van der Land for inviting me to join in on the exploration of the Coral Reef Theme (4) on board R/V *Tyro*. W. Prud'homme van Reijne identified the algae. J. Vermeulen provided technical assistance in the preliminary identifications.

Literature Cited

Ayling, A. L. 1982. A Redescription of *Astrosclera willeyana* Lister (1900) (Ceratoporellida, Demospongiae), a New Record from the Great Barrier Reef. *Memoirs of the National Museum of Victoria*, 43:99–103.

Bergquist, P. R. 1965. The Sponges of Micronesia, Part I. The Palau Archipelago. *Pacific Science*, 19:123–204.

———. 1969. Shallow Water Demospongiae from Heron Island. *University of Queensland Papers*, 1:63–72.

Bergquist, P. R., J. E. Morton, and C. A. Tizard. 1971. Some Demospongiae from the Solomon Islands with Descriptive Notes on the Major Sponge Habitats. *Micronesica*, 7:99–121.

Bergquist, P. R., and C. A. Tizard. 1967. Australian Interidal Sponges from the Darwin area. *Micronesica*, 3:175–202.

Briggs, J. C. 1974. *Marine Zoogeography*. New York: McGraw-Hill. 475 pp.

Brøndsted, H. V. 1934. Sponges. *Résultats scientifiques du Voyage aux Indes Orientales Néerlandaises du Prince Léopold de Belgique*, 2:1–27.

Burton, M. 1930. The Porifera of the Siboga Expedition III. Calcarea. *Monographe Siboga Expeditie*, 6a2:1–18.

———. 1934. Sponges. *Great Barrier Reef Expedition, Scientific Reports*, 44:513–621, 2 pls.

Dauer, D. M., and J. L. Simon. 1975. Repopulation of the Polychaete Fauna of an Intertidal Habitat Following Natural Defaunation Species Equilibrium. *Oecologia (Berlin)*, 22:99–117.

Desqueyroux, R. 1981. Révision de la collection d'éponges d'Amboine Moluques, Indonésie) constituée par Bedot et Pictet et conservée au Muséum d'Histoire Naturelle de Genève. *Revue Suisse de Zoologie*, 88:723–764.

Esper, E. J. C. 1794. *Die Pflanzenthiere in Abbildungen nach der Natur mit Farben erleuchtet nebst Beschreibungen. 2. Theil.* Nürnberg: Raspe. 303 pp.

Gómez López, P., and G. Green. 1984. Sistematica de las Esponjas marinas de Puerto Morelos, Quintana Roo, México. *Annales, Instituto de Ciencias del Mar y Limnologia (Mexico)*, 11:65–90.

Hechtel, G. J. 1965. A Systematic Study of the Demospongiae of Port Royal, Jamaica. *Bulletin, Peabody Museum of Natural History*, 12:i–xii, 1–155.

Hentschel, E. 1911. Tetraxonida, 2.Teil. *Fauna Südwest-Australiens*, 3(10):277–393.

———. 1912. Kiesel- und Hornschwämme der Aru- und Kei-Inseln. *Abhandlungen der Senckenbergischen Naturforschenden Gesellschaft*, 34:293–448.

Kieschnick, O. 1896. Silicispongiae von Ternate. *Der Zoologische Anzeiger*, 19:526–534.

———. 1900. Kieselschwämme von Amboina. *Denkschriften der Medizinischen und Naturwissenschaftlichen Gesellschaft Jena*, 8:545–582.

Laubenfels, M. W., de. 1954. The Sponges of the West-Central Pacific. *Oregon State Monographs Studies in Zoology*, 7:1–320.

Lévi, C. 1961. Eponges intercotidales de Nha Trang (Viet Nam). *Archives de Zoologie expérimentale et générale*, 100:127–148.

Pulitzer-Finali, G. 1982. Some New or Little-known Sponges from the Great Barrier Reef of Australia. *Bolletino, Museo e Istituto di Biologia dell'Universitá di Genova*, 48–49(1980–1981):87–141.

Rosen, B. R. 1971. The Distribution of Reef Coral Genera in the

Indian Ocean. Pages 263–299 in *Regional Variation in Indian Ocean Coral Reefs*, edited by D. R. Stoddart and C. M. Young. Symposia of the Zoological Society of London, 28. London: Academic Press.

————. 1975. The Distribution of Reef Corals. *Reports of the Underwater Association* (new series), 1(NS):1–16.

Rützler, K. 1972. Principles of Sponge Distribution in Indo-Pacific Coral Reefs: Results of the Austrian Indo-West Pacific Expedition 1959/60. Pages 315–332 in *Proceedings of the Symposium on Corals and Coral Reefs*, edited by C. Mukundan and C. S. Gopinadha Pillai. Cochin: Marine Biological Association, India.

Soest, R. W. M. van. 1980. A Small Collection of Sponges (Porifera) from Hong Kong. Pages 85–95 in *The Marine Flora and Fauna of Hong Kong and Southern China 2*, edited by B. Morton and C. K. Tseng. Hong Kong: University Press.

————. 1981. A Checklist of the Curaçao Sponges (Porifera, Demospongiae) Including a Pictorial Key to the More Common Reef Forms. *Verslagen en Technische Gegevens, Instituut voor Taxonomische Zoologie, Universiteit van Amsterdam*, 32:1–39 (mimeographed).

Thiele, M. E. 1899. Uber einige Spongien von Celebes. *Zoologica*, 24, II(2):1–33.

————. 1903. Kieselschwämme von Ternate I und II. *Abhandlungen der Senckenbergischen Naturforschenden Gesellschaft*, 25: 19–80, 933–968.

Thomas, P. A. 1979. Studies on Sponges of the Mozambique Channel. *Annalen van het Koninklijk Museum voor Midden-Afrika Tervueren, Zoölogische Wetenschappen*, 227:1–73.

Topsent, E. 1897. Spongiaires de la baie d'Amboine. *Revue Suisse de Zoologie*, 4:421–487.

Udvardy, M. D. F. 1969. *Dynamic Zoogeography*. New York: Van Nostrand Reinhold. 445 pp.

Vacelet, J. 1981. Algal-Sponge Symbioses in the Coral Reefs of New Caledonia: A Morphological Study. Pages 713–719 in *Proceedings of the 4th International Coral Reef Symposium 2*, edited by E. D. Gomez et al. Manila: Marine Sciences Center, University of the Philippines.

Vacelet, J., and P. Vasseur. 1965. Spongiaires des grottes et surplombs des récifs de Tulear (Madagascar). *Travaux de la Station Marine de Tuléar* (supplement 4):71–123.

————. 1971. Eponges des récifs coralliens de Tuléar (Madagascar). *Téthys* (supplement 1):51–126.

————. 1977. Sponge Distribution in Coral Reefs and Related Areas in the Vicinity of Tuléar (Madagascar). Pages 113–117 in *Proceedings of the 3d International Coral Reef Symposium*, edited by D. L. Taylor. Miami: Rosenstiel School of Marine and Atmospheric Sciences.

Vacelet, J., P. Vasseur, and C. Lévi. 1976. Spongiaires de la pente externe des récifs coralliens de Tuléar (sud-ouest de Madagascar). *Memoires, Museum National d'Histoire Naturelle (Paris)*, (séries A, Zoology), 99:1–116.

Vosmaer, G. C. J. 1911. The Porifera of the Siboga Expedition II. The Genus *Spirastrella*. *Monographe Siboga Expeditie*, 6a1:1–69.

Vosmaer, G. C., and J. H. Vernhout. 1902. The Porifera of the Siboga Expedition I. The Genus *Placospongia*. *Monographe Siboga Expeditie*, 6a:1–17.

Weber, M., and A. Weber. 1889. Quelques nouveaux cas de symbiose. Pages 48–71 in *Zoologische Ergebnisse einer Reise in Niederländisch Ost-Indien I*, edited by M. Weber. Leiden: E. J. Brill.

Wiedenmayer, F. 1977. *A Monograph of the Shallow-Water Sponges of the Western Bahamas*. Basel und Stuttgart: Birkhäuser Verlag. 287 pp.

Zea, S. 1987. *Eponjas del Caribe Colombiano*. Bogotá: Catálago Cientifico. 486 pp.

MARIA-JESUS URIZ
Centro de Estudios Avanzados
Camí de Santa Bárbara
Blanes (Gerona), Spain

Possible Influence of Trawl Fishery on Recent Expansion in the Range of *Suberites tylobtusa* in the Southeast Atlantic

Abstract

Suberites tylobtusa, a species previously thought to be restricted to the Red Sea, has been found in the Southeast Atlantic off Namibia, South-West Africa, where it thrives on sandy and muddy bottoms at depths of 100 to 500 m. The Fishing activities of an important trawl fishery operating in this region since 1960 do not seem to adversely affect the sponge, but rather seem to have helped it spread, since it was not reported in the region in biological surveys carried out prior to 1960. A quantitative study of the bathymetric and latitudinal distribution of this species between parallels 23° and 29° S is presented. Taxonomic affinities with other species, *S. ficus* in particular, are discussed, and hypotheses as to the causes that have allowed it to flourish in the region are advanced.

The activities of humans may inadvertently lead to a variety of changes in the distribution of marine species (introduction of species associated with other cultured species, fouling, construction of canals facilitating species migration, etc.) and have been studied by numerous authors (Zibrowius, 1983). Normally, however, only the negative effects of trawl fisheries receive attention because the stress they create, which limits the growth of the macrobenthos and may even bring about the disappearance of certain species.

The sponge *Suberites tylobtusa* (Demospongiae, Hadromerida), described by Lévi (1958) from specimens from the Red Sea, is extremely abundant on the continental shelf off Namibia and South Africa. It dwells at depths of 100 to 500 m in an area visited by trawlers of a sizable international fishery.

The sponge population does not seem to be adversely affected by the fishing activity, which commenced in 1960.

On the contrary, trawling seems to play an important role in the spread of the species, which prior to 1960 was not found in the region or in nearby areas (Ridley and Dendy, 1887; Kirkpatrick, 1902, 1903; Stephens, 1915; Burton, 1932, 1933) and which today is, quantitatively speaking, one of the most important noncommercial macroinvertebrates present in this fishing area (Figure 1).

Not many species of sponges inhabit soft bottoms, or at least they do not usually generate a biomass of any appreciable size, particularly if the bottoms are subject to the constant action of trawl fishing and the concomitant instability, changes at the water-sediment interface (A. Lobo, unpublished manuscript), and increased sedimentation rate. Few filter feeders are able to survive in extreme conditions of the kind in which *Suberites tylobtusa* thrives.

Material, Methods, and Study Area

The samples used in the present study were collected on the Benguela 6 (January–February 1984), Benguela 7 (July–August 1984), and Benguela 8 (July 1985) surveys conducted by the fishing vessel Chicha-Touza using a bottom trawl with a codend mesh size of 21 mm.

Sampling lasted for 30 minutes of effective towing time, at a speed of 3 knots/hour.

A sampling grid of 245 stations was set up (Figure 2a) using semirandom stratified techniques (Macpherson et al., 1985), and of these stations 195 were accepted as valid

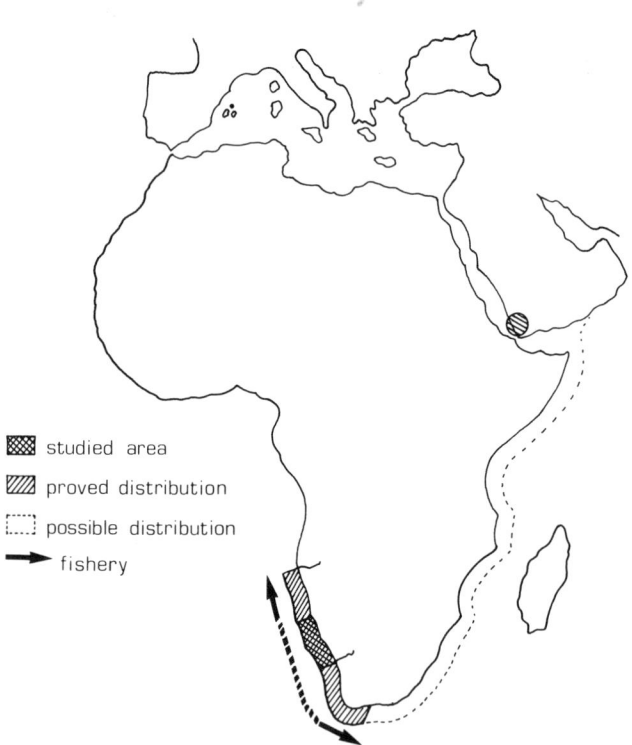

studied area
proved distribution
possible distribution
fishery

Figure 1. Geographic distribution of *Suberites tylobtusa*, Red Sea and South-West Africa.

in that the bottoms exhibited the desired characteristics. Stations in rocky or abiotic regions, with completely different characteristics, are covered in a separate study; *Suberites tylobtusa* was not present in any samples collected at those stations.

To calculate the area swept by each trawl, the average net opening size was taken to be $\bar{x} = 5.7$ m, on the basis of the angle of inclination of the warp lines (Macpherson et al., 1985). Biomass data are given in g drained wet weight, and the number of individuals is also provided. When the catch taken in any given tow was too large to be examined in its entirety, the biomass was calculated proportionally by multiplying the results of a portion of the sample by a factor estimated in each instance.

The study area is a littoral zone slightly more than 80 km in width, with bottom depths ranging from 60 to 550 m, running from Walvis Bay (23° S) to the mouth of the Orange River (29° S). The cold Benguela current passes through the region, and upwelling takes place.

Bottoms are relatively uniform, that is muddy or a mixture of mud and fine sand, with isolated rocky areas. Productivity is high, and the region is characterized by vertical homeothermy caused by the upwelling. Bottom temperature ranges from 9.5° to 12°C in winter (August).

Results

The density of distribution of *Suberites tylobtusa* in the study area is shown in Figure 2b. Before describing the distributional parameters, some data on the morphological properties of the population should be presented.

MORPHOLOGY OF *Suberites tylobtusa*

Suberites tylobtusa is massive with rounded edges. Two morphological types are distinguishable: (a) globulous, widest at the base, with one large apical osculum, and (b) elongate, with compressed base that gives rise to one or more globular lobes which are widest in the middle and end in small oscula (Figure 3).

Although occasionally attached to a gastropod shell, *Suberites tylobtusa* usually grows free, unattached to any substrate (in no case was a small shell found inside a specimen). Its surface is smooth and uniform, soft to the touch, with short, even bristles visible only under a microscope. The color is an even straw yellow, but mud-encrusted specimens are grayish. The consistency of the species varies, depending on the degree of contraction of the specimen. The ectosome is indistinguishable to the naked eye. The water current system is arranged around a central atrium longitudinally traversing each lobe and opening in an osculum measuring 0.3 to 3.0 cm in diameter.

Suberites tylobtusa has three distinct types of megascleres, which occur in varying proportions depending on the

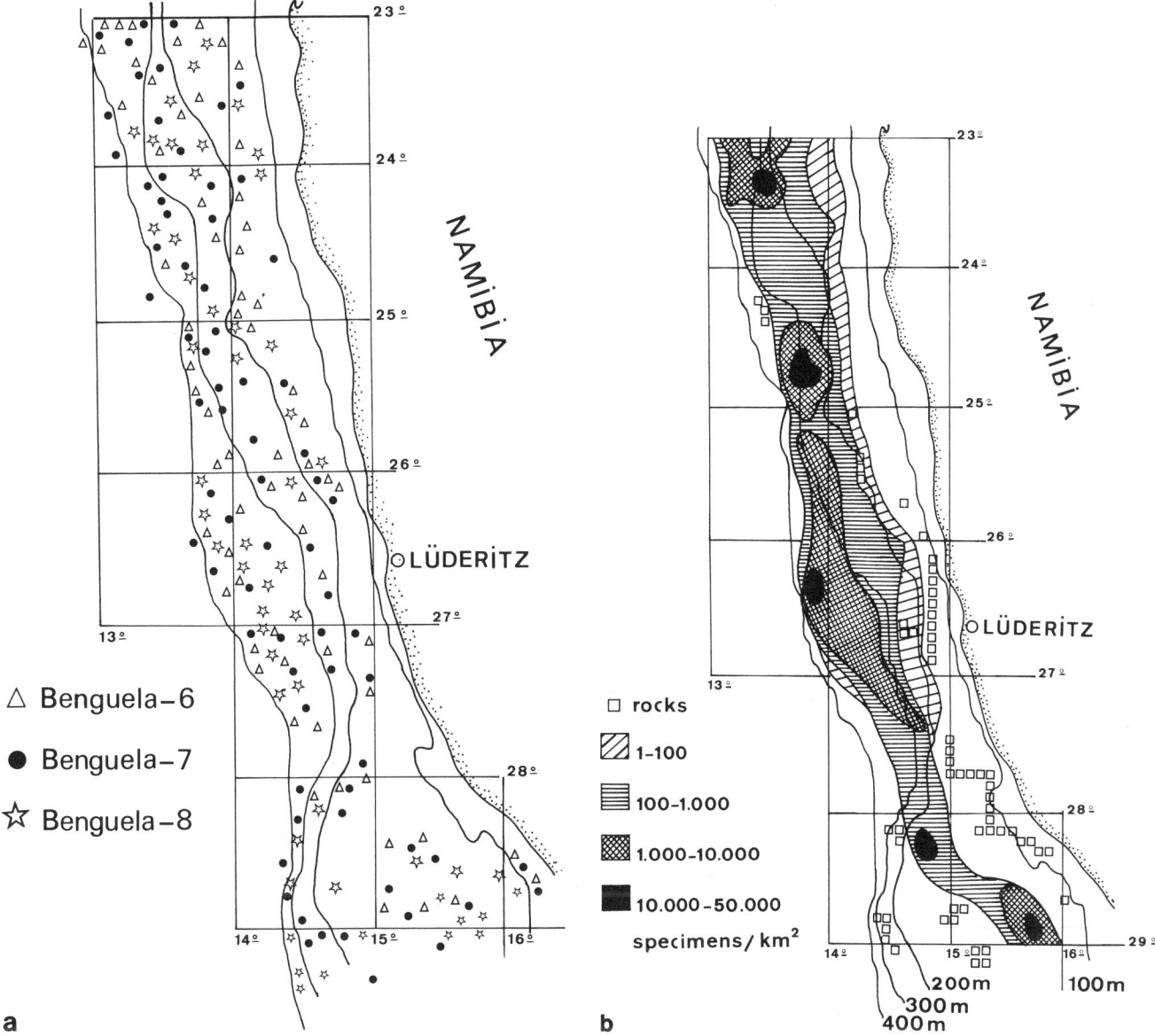

a

△ Benguela-6

● Benguela-7

☆ Benguela-8

b

□ rocks

▨ 1–100

▤ 100–1.000

▨ 1.000–10.000

■ 10.000–50.000

specimens / km²

200 m
300 m
400 m

100 m

Figure 2. Study area off Namibia: *a*, trawling survey stations and sampling grid; *b*, isobaths and plot of density of *Suberites tylobtusa*.

specimen and area (Figure 4*a*). Tylostyle I is straight, curved, and even sinous, with a gradually tapering, sharp point and an uneven, poorly differentiated base, measuring 280–620 μm × 8–15 μm; located throughout the sponge but more abundant in the distal portion and periphery, forming a ring around the oscula in all specimens (Figure 5). Tylostyle II is relatively robust, somewhat curved, wider in the apical section (looks somewhat like a scimitar) with a rounded, clearly differentiated base and blunt or obtuse point, measuring 290–420 μm × 15–23 μm at its widest point (9–16 μm under the base); it is most abundant in the base of the sponge and in the periphery. Tylostrongyles are very thick and variable in length, with

a rounded, undifferentiated base and rounded point often wider than the base, measuring 105–320 μm × 18–25 μm; abnormal shapes, also common in other *Suberites* species are abundant. The morphology of the centrotylote microstrongyles is similar to that of *S. ficus*, with a smooth appearance under an optical microscope but a slightly rough appearance under the electron microscope (Figure 4*b*); it is present in variable proportions in all the specimens examined.

The skeletal structure consists of irregularly arranged polyspicular bundles mixed in with scattered individual spicules in the choanosome, oriented perpendicular to the surface in the periphery of the sponge, where they can be

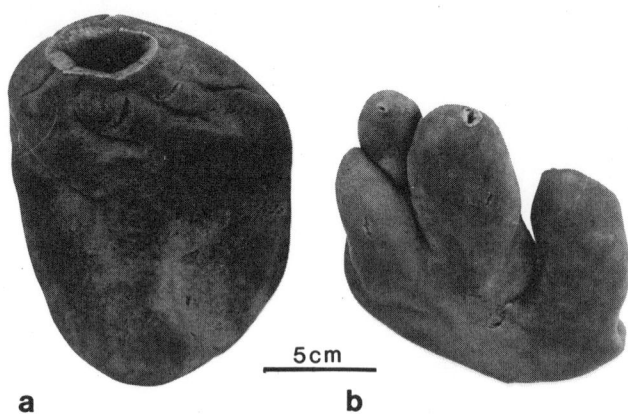

Figure 3. Different morphological aspects of *Suberites tylobtusa*: *a*, globulous with large osculum; *b*, elongate with projections and small oscula.

intermixed with a dense layer of spicules forming a palisadelike structure interrupted only by the incurrent channels (Figure 6).

DISTRIBUTION IN THE STUDY AREA

The known distribution of this species is currently restricted to the Red Sea (Lévi, 1958) and to the littoral zone off Namibia and South Africa, from latitude 35° S in the Indian Ocean (B. Morales and B. Roel, pers. comm.) to latitude 17° S in the South Atlantic (E. Macpherson and R. Allue, pers. comm.). The environmental conditions in the Red Sea and off Namibia are so different as to suggest that the species has a broad ecological adaptability, which in turn would suggest a more extensive geographical distribution for this species than that known to date.

Figure 7 shows three sponge population parameters on on depth and latitude in the study area: (1) abundance (mean number of individuals/sampling station, all stations), (2) density (mean number of individuals/sampling station where the species was found), and (3) dispersion (percentage of sampling stations where the sponge was found).

On the basis of these figures it can be concluded that (a) the sponge is most abundant in the 300-m depth stratum (from 250 to 350 m), where it is also most uniformly distributed although population density is lowest; and (b) in contrast, in the 200- and 400-m strata, where apparently the impact of trawl fishing is less severe, density is very high, abundance average and dispersion low.

Dispersion and density are inversely related at 23° and 28° S. The percentage of sponge producing stations is highest at 23° and lowest at 28° where the mean number of individuals/sponge-producing station was extremely high, indicating that the species distribution is limited to specific areas. The values of all three parameters were low at 27° S, with average to high values at 26° S. There is a

direct relationship between the parameters at both these latitudes.

Figure 8 depicts mean weight/individual as an index of favorable environmental conditions. The largest mean size (with the appropriate reservations, in view of the high variance) occurs at latitude 28° S, coinciding with very high densities and a less disperse population, all of which suggests a more stable environment, perhaps less affected by the fishing activity.

Discussion
TAXONOMIC ASPECTS

When a large number of *Suberites tylobtusa* specimens is examined, one finds considerable variation in the percentage of different spicule types present. This is also the case in other species in this genus, chiefly *S. ficus*, with which the species from Namibia exhibits a number of affinities.

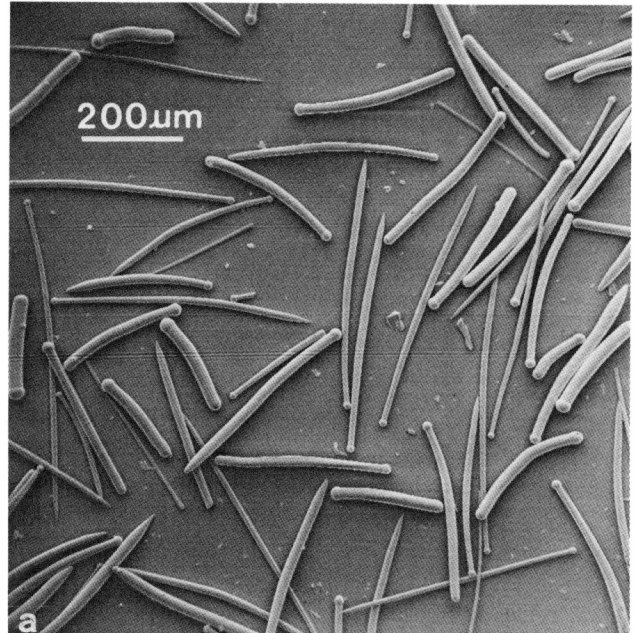

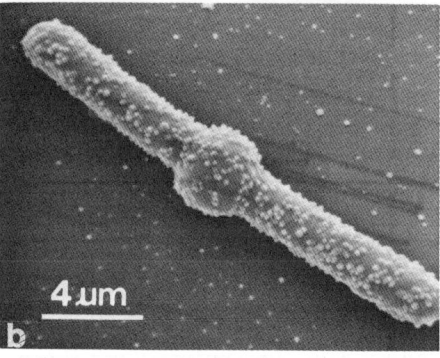

Figure 4. *Suberites tylobtusa*, scanning electron micrographs of spicules: *a*, different megasclere types; *b*, centrotylote microstrongyle.

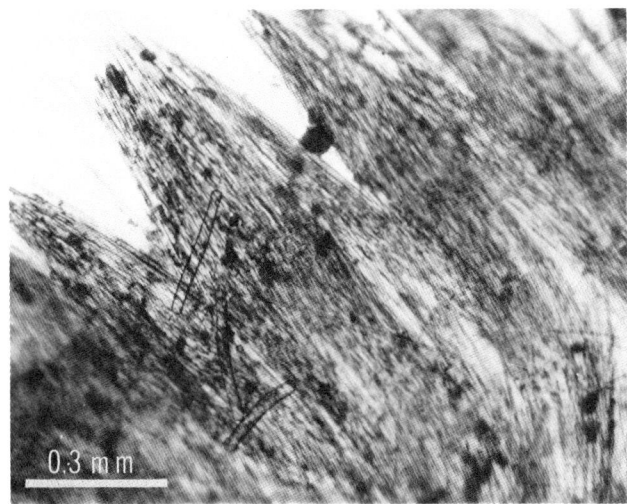

Figure 5. Spicule arrangement around osculum in *Suberites tylobtusa*.

Externally, *Suberites tylobtusa* could be mistaken for *S. compacta* Verrill. I was able to examine specimens of the latter species from Martha's Vineyard and Alaska in the Smithsonian Institution's collection. Hartman (1958) apparently considered them to be a different species from *S. ficus*.

The characteristic spicules of *Suberites tylobtusa*, scimitar-shaped tylostyles and tylostrongyles, have also been described on occasion for *S. suberea* specimens from the Pacific

Figure 6. *Suberites tylobtusa*, outer skeletal arrangement: *a*, perpendicular spicular bundles; *b*, palisade arrangement.

(Thiele, 1889) and for the base of *S. domuncula* specimens from the Mediterranean (Hartman, 1958). It might be thought that this type of spicule, more robust with blunter ends, is a result of constant mechanical chafing they are more frequent in the outer portion and in the base of the sponge) produced by the free-living, bottom-dwelling life-style of the specimens from the area off Namibia. However, these modifications do not occur in *S. ficus* specimens, which are also free-living bottom dwellers on the fishing grounds in Block Island Sound (Hartman, 1958), whereas they do occur in the specimens from the Red Sea that live attached to rocks (Lévi, 1958).

As Hartman (1958) pointed out, it seems quite likely that several *Suberites* species may have been grouped under the name *ficus*, in which case this variability would not be as great as currently thought. Recent genetic studies (Solé-Cava and Thorpe, this volume) would appear to corroborate this hypothesis. This, together with the fact that specimens with typical spicule and morphological characteristics have been found in the same area as *Suberites tylobtusa* (Uriz, 1985), suggests the conclusion that the latter is a separate species.

IMPACT OF THE FISHERY

The quantity of organic matter reaching the sediment over the continental shelf off Namibia is exceptionally high (L. Cros, pers. comm.) owing to the high level of plankton productivity in the region. This rain of particulate organic matter (POM) might be a factor favoring the development of filter feeders and suspension feeders in general, provided that they have mechanisms to prevent the high amount of sedimentation from blocking their filters.

The already high natural sedimentation rate in the region is augmented considerably by the action of trawlers, which, by stirring up the topmost layer of sediment, returns to suspension both the POM, which again becomes available to suspension feeders, as well as the mineralized matter in the sediment.

Another, more acute effect of fishing on the macrobenthos is the progressive disappearance of species unable to withstand decompression, physical injury, and the period out of the water accompanying capture. At the same time, some species may proliferate if they are able to adapt and it they encounter optimum feeding conditions and reduced competition.

Four separate species of nonarthropod macroinvertebrates dominate the benthic environment on the fishing grounds off Namibia and may even become characteristic of certain facies: (1) an unidentified solitary ascidian; (2) the gastropod *Fusitrion magellanicus*; (3) an actinian of the genus *Actinauge*; and (4) the sponge *Suberites tylobtusa*.

Two factors may enable fishing to exert a positive effect

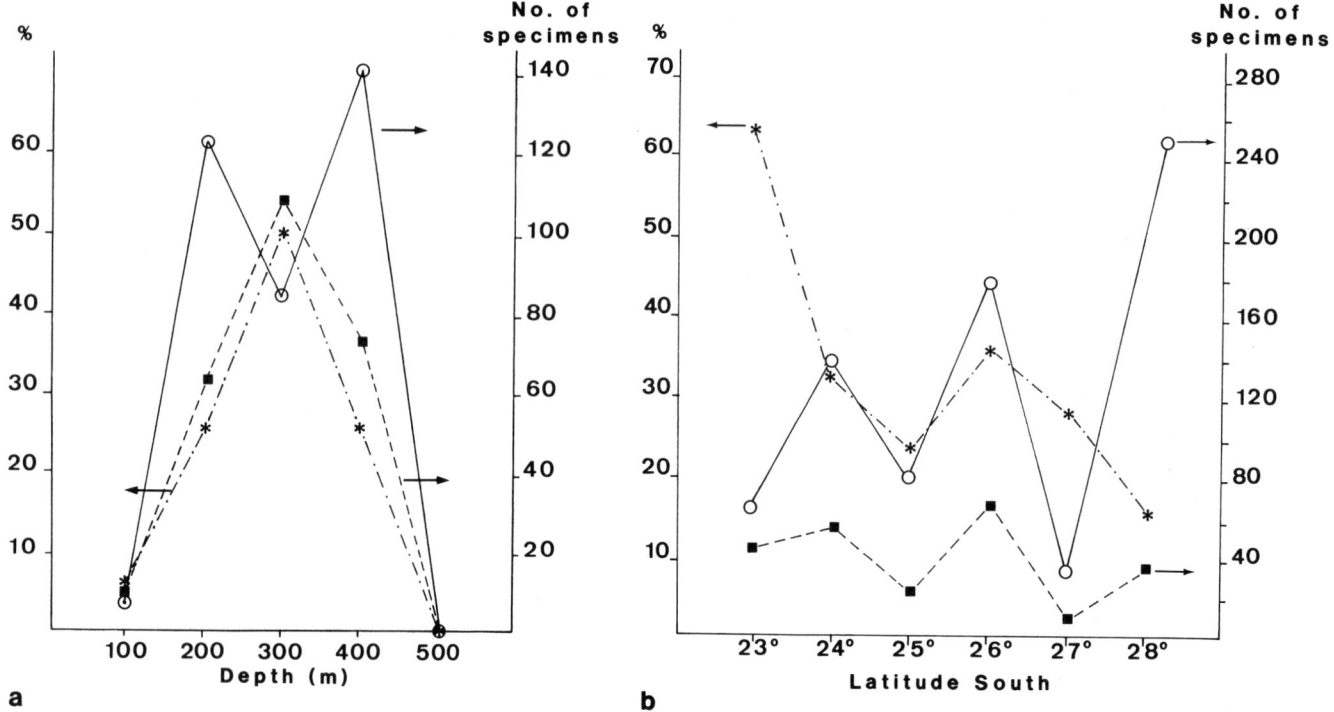

Figure 7. *Suberites tylobtusa*, quantitative parameters: *a*, sponge abundance (*black squares*, mean number of individuals/sampling station, all stations), density (*circles*, mean number of individuals/sampling station where sponge individuals were present), and dispersion (*asterisks*, sampling stations where sponges were collected, in %) on depth; *b*, same parameters as in *a* but on latitude (arrows point to scales).

on the distribution of these species. The first of these is the nature of fishing operations in the Atlantic. The trawlers operate in all directions over large areas for months at a time without stopping in port, making haul after haul, and discarding the noncommercial species caught, returning them to the sea in a region of similar environmental conditions. These organisms, if they can survive the trauma of capture and a relatively short period out of water, may continue living at a distance several km from the spot where they were caught. This would not greatly affect distribution of species whose larvae are planktonic, but it might help to spread certain species of sponges whose larvae are not very mobile, such as those of the order Hadromerida, to which *Suberites tylobtusa* belongs.

Second, the dominant species in the study area share several characteristics: First, they are capable of surviving relatively short periods out of water by storing water inside—in its tissue in the case of the sponge, in spongy tissue by closing the operculum in the case of the gastropod, and by contracting the siphon or the pharynx in the case of the ascidian and the actinian, respectively. Second, they are capable of surviving the mechanical chafing of the catching process, by means of a thick protective layer (ascidian and actinian), a shell (gastropod),

and high overall contractibility (contracted specimens having the consistency of rubber) and a dense outer layer of spicules (sponge). Third, they are capable of landing in such a position such that they may continue to feed after they have been returned to the sea following capture, because of a globulous shape and low center of gravity, thanks to high spicule density in the base, which does not bear any water current channels (sponge), a glob of mud held within (actinian) or adhering to (ascidian) the base or the ability to turn around on its axis (gastropod). And fourth, they are all capable of a free-living life-style on the bottom, with the result that the organism does not suffer irreparable damage if torn from its substrate.

How the sponge withstands sedimentation cannot readily be explained although its contractibility is high (an atrial chamber 1 cm in diameter in a fully distended specimen can shrink in size to less than 2 mm on contraction), and contractions could help free it of sediment. Furthermore, the capacity of certain sponges to reverse the water current flow, thus expelling particles that obstruct its pores, has been referred to by certain authors. Although this phenomenon is not well studied, its importance should not be treated lightly in the case of *Suberites tylobtusa*.

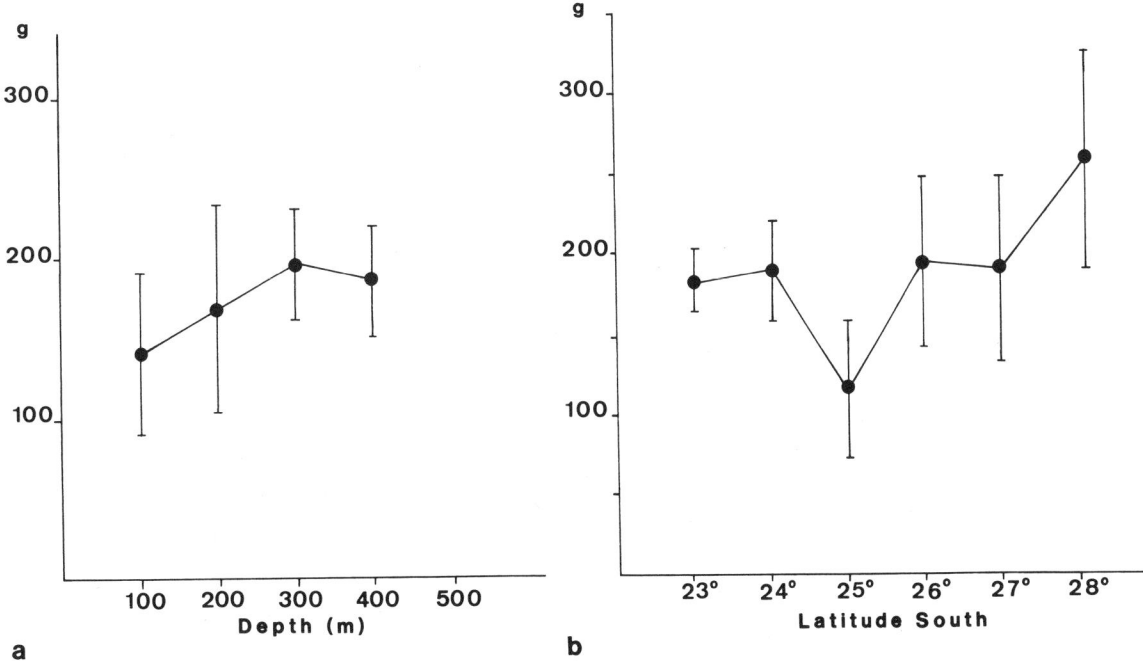

Figure 8. Mean specimen weights: *a*, on depth; *b*, on latitude (bars indicate standard deviation).

Acknowledgments

The author wishes to thank the crew of the Chicha-Touza, who successfully turned a fishing trawler into an efficient oceanographic vessel; E. Macpherson, survey director, and R. Allue and his team for their assistance and companionship during the months spent on board ship. Thanks also go to K. Rützler for kindly enabling the author to examine the Smithsonian Institution's sponge collection.

Literature Cited

Burton, M. 1932. Sponges. *Discovery Reports*, 6:237–392.
———. 1933. Report on a Small Collection of Sponges from Stil Bay, S. Africa. *Annals and Magazine, Natural History*, (series, 10), 11:235–244.
Hartman, W. 1958. Natural History of the Marine Sponges of Southern New England. *Bulletin, Peabody Museum of Natural History* (Yale University), 12:1–155.
Kirkpatrick, F. Z. S. 1902. Description of South African Sponges. Part II. *Marine Investigations in South Africa*, 2:171–180.
———. 1903. Description of South African Sponges. Part III. *Marine Investestigations in South Africa*, 2:233–264.
Lévi, C. 1958. Campagne 1951–1952 de la Calypso en Mer Rouge. 11, Spongiaires de Mer Rouge. *Annales, Institut Oceanographique*, 34:1–46.
Macpherson, E., B. Roel, and B. Morales. 1985. Reclutamiento de la merluza y abundancia y distribucion de diferentes especies comerciales en las Divisiones 1.4 y 1.5 durante 1983–1984. *Collection of Scientific Papers, International Commission of SE Atlantic Fisheries*, 12:1–61.
Ridley, S. O., and A. Dendy. 1887. Report on the Monaxonida Collected by H.M.S. "Challenger". *Reports on the Scientific Results Challenger, Zoolology*, 20:1–275.
Stephens, J. 1915. Atlantic Sponges Collected by the Scottish National Antarctic Expedition. *Transactions, Royal Society of Edinburgh*, 50 II(15):423–467.
Thiele, J. 1898. Studien über pazifische Spongien. *Zoologica*, 24:1–72.
Uriz, M. J. 1985. Primera aportacíon al conocimiento de la fauna espongícola de la plataforma de Namibia. Demosponjas de la Benguela VI y de la Valdivia I. *International Symposium on Upwelling Areas of Western Africa*. Barcelona: Instituto de Investigacious Pesqueras. 2:639–649.
Zibrowius H. 1983. Extension de l'aire de répartition favorisée par l'homme chez les invertébrés marins. *Oceanis*, 9:289–293.

CHUNG JA SIM
Department of Biology
Han Nam University
Daejeon, Republic of Korea

Distribution of the Tetractinomorpha in South Korea

Abstract

This study of the Korean Tectractinomorpha is based on material collected from 29 localities, including many islands and coastal seas of the East (Sea of Japan), the West (Yellow Sea), and the South of Korea. Species compositions of Tetractinomorpha in the three coastal seas were compared. One among the species collected, *Suberites ficus*, is widely distributed in the South of Korea and shows local variation in its morphology.

Tetractinomorpha were studied in coastal localities of the Republic of Korea including many islands. It is the purpose of this paper to list the species and their distribution in the three coastal seas—the Japan, the Yellow, and the South seas.

Among the common species collected is *Suberites ficus* which is widely distributed off the southern coast of Korea. Its morphological variation in different localities will be examined.

Material and Methods

The specimens examined in this study were collected in 31 localitites (Figure 1) of Korean waters during 1964–1985 by B. J. Rho, C. J. Sim, and others. The specimens were sampled with fishing net or scuba, or by wading during low tide. The material is deposited in the Department of Biology, Ewha Womans University, and the Department of Biology, Han Nam University.

Clorox (commercial bleach) preparations were made for examination of spicule types. Thick sections were made by hand with a razor blade after the specimen was hardened in alcohol so that skeleton structure could be examined.

Results

The Korean Tetractinomorpha identified in this study (Table 1) belong to the orders Choristida (19 species, 9 genera, 5 families), Hadromerida (19 species, 7 genera, 6 families), Axinellida (16 species, 6 genera, 2 families),

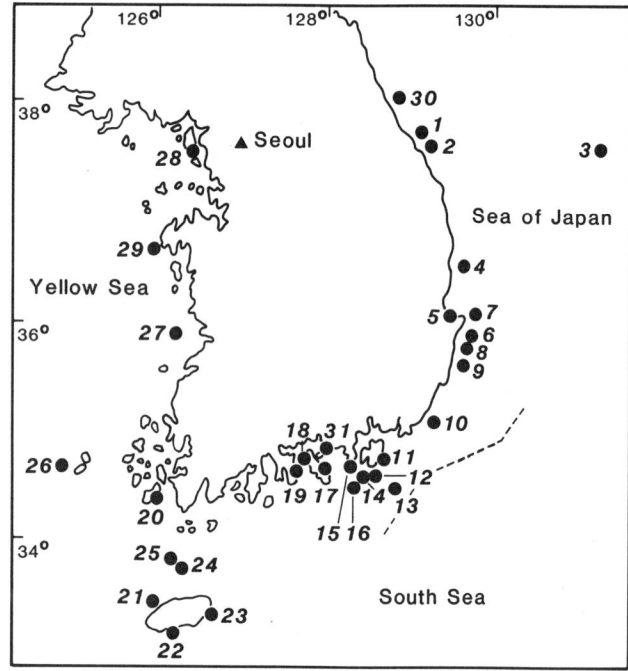

Figure 1. Map of South Korea showing localities for sponges listed in Table 1. Sea of Japan: *1*, Aninjin; *2*, Mugho; *3*, Ulreung I.; *4*, Chugsan; *5*, Pohang; *6*, Yeong-il Bay; *7*, Guman; *8*, Guryongpo; *9*, Gampo; *30*, Jumunjin. South Sea, mainland coast: *10*, Mipo; *11*, Geoje I.; *12*, Yundol I.; *13*, Galdo I.; *14*, Bijin I.; *15*, Chungmu; *16*, Indae I.; *17*, Mijo Ri; *18*, Namhae I.; *19*, Nodo I.; *20*, Jindo I; *31*, Samchonpo. South Sea, Jeju I.: *21*, Hanrim; *22*, Seogwipo; *23*, Seongsanpo; *24*, Sasu I.; *25*, Hoeng-gan I. Yellow Sea: *26*, Hongdo I.; *27*, Bieung I.; *28*, Incheon; *29*, Anhung.

316

Table 1. Distribution of Tectractinomorpha species in Korean seas (for localities see Figure 1)

Species	Japan Sea										South Sea — Mainland coast												Jeju I.					Yellow Sea			
	1	2	3	4	5	6	7	8	9	30	10	11	12	13	14	15	16	17	18	19	20	31	21	22	23	24	25	26	27	28	29
Stelleta crassispiculata																											+				
Stelletta misakiensis																	+														
Stelletta gigantea				+																											
Penares incrustans													+					+									+				
Papyrula hilgendorfi																								+							
Papyrula metastrosa																								+							
Geodia variospiculosa																	+								+						
Geodia japonica																	+							+							
Geodia reniformis																								+							
Geodinella cylindrica				+																											
Geodinella hyotania					+																										
Erylus koreanus																								+							
Pachastrella tenuilaminaris																								+							
Pachastrella doederleini																								+							
Pachastrella cribrum																								+							
Pachastrella sollasi																								+							
Pachastrella japonica																								+							
Asteropus simplex																								+							
Plakortis simplex														+													+				
Tetilla ovata																								+							
Discodermia japonica																								+							
Discodermia calyx																								+							
Discodermia kiiensis																							+	+							
Theonella swinhoei																								+							
Suberites ficus	+	+		+	+	+		+	+				+											+							
Suberites virgultosa				+																				+						+	
Suberites caminatus												+																			
Suberites excellence						+						+																			
Suberites microstomus																					+					+					
Suberites sericeus																		+		+											
Suberites japonicus											+									+									+		
Rhyzaxinella clavata																												+			
Cliona celata				+			+	+	+	+			+											+							
Cliona lobata																	+		+		+			+					+	+	
Spirastrella panis																								+							
Tethya amanensis																								+							
Tethya japonica				+								+													+						
Tethya seychellensis																															
Tethya aurantium				+			+													+											
Tethya koreana																										+					
Latrunculia ikematsui				+								+												+							
Chondrilla australiensis																											+				
Chondrilla mixta																								+							
Axinella copiosa			+																												
Axinella tenuis																							+	+			+				
Auletta halichondroides																								+							
Auletta consimilis																								+							
Acanthella vulgata												+																			
Acanthella simplex																															
Phakellia elegans												+												+							
Phakellia paupera																								+		+	+				
Phakellia ventilabrum																								+							
Dactyella hilgendorfi				+								+												+							
Ceratopsis clavata																											+				
Ceratopsis ramosa																								+							
Raspailia hirsuta												+												+							
Raspailia folium																								+							
Raspailia villosa				+																				+							
Raspailia koreana																								+							

Lithistida (4 species, 2 genera, 1 family), and Spirophorida (2 species, 1 genus, 1 family). The distribution of Korea's Tetractinomorpha appears to vary with the conditions in the three coastal seas surrounding the Korean peninsula (Table 2). Of the 60 species found, 55 were collected in the South Sea, 13 in the Japan Sea, and 9 in the Yellow Sea.

3d Int. Sponge Conf. 1985

Several authors have discussed the validity of differences between *Suberites ficus* and *S. domunculus* (Lendenfeld, 1897; de Laubenfels, 1949; Burton, 1953; Hartman, 1958; Tanita, 1965). Our material is assigned to *S. ficus*. Specimens have both megascleres and microscleres. The megascleres are chiefly tylostyles, the microscleres centrotylote microstrongyles (Figure 2a). Microscleres seldom occur

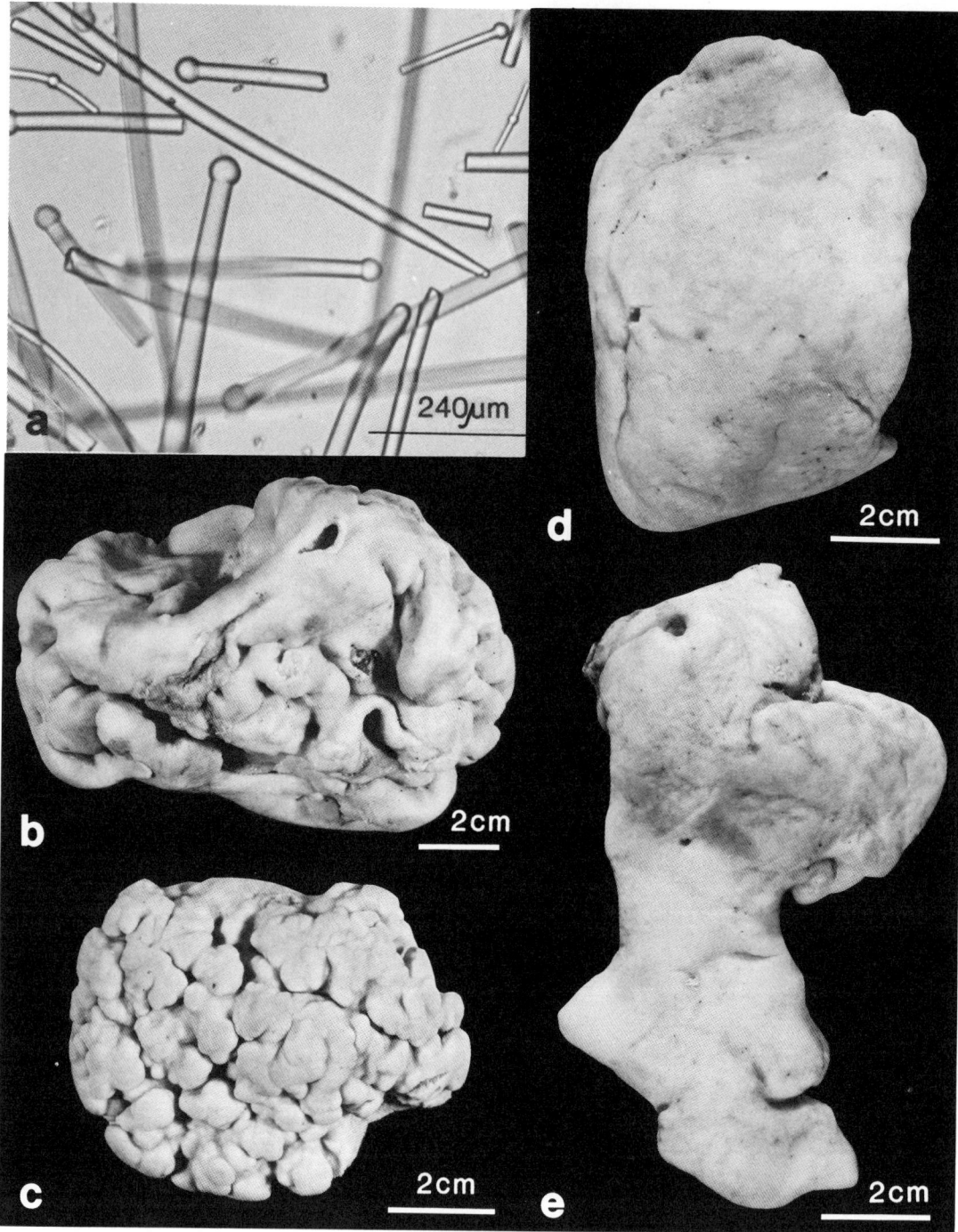

Figure 2. *Suberites ficus*, spicules and morphological variation: *a*, tylostyle and centrotylote microstrongyle; *b*, large-lobate form from Sea of Japan; *c*, small-lobate form from South Sea, mainland coast; *d*, solid form from South Sea, Jeju Island; *e*, irregular, hostless form from Yellow Sea.

on top of the sponge, but are common inside the hole formed by the hermit crab host. Morphological differences in specimens from different localities are listed in Table 3 and illustrated in Figure 2*b–e*.

Discussion and Conclusions

Thirteen species of Tetractinomorpha were found in the Japan Sea, 55 species in the South Sea, and 9 species in the

Table 2. Average oceanographic conditions in Korea's coastal seas

| Principal parameters | Sea of Japan | South Sea | | Yellow Sea |
		Mainland coast	Jeju I.	
Geomorphology and depth (m)	Steep slope 1,000–3,000	Many bays and islands 100	Isolated island shelf 100	Wide continental 45
Tide (m), predominant tidal current	0.2–0.4 north–south	1.4–3.7 east–west	3	9.7 (Incheon, location no. 28)
Temperature (°C)	Summer 27–29 Winter 3–10	Summer 26–30 Winter 5–11	Summer 26–30 Winter 14	Summer 25–28 Winter 2–8
Salinity (‰)	Summer 33–30 Winter 34.5–35.0	Summer 33–30 Winter 34.5–35.0	Summer 34 Winter 34.6	32 (North) 25 (center)
Predominant	Tsushima warm current; Liman cold current	Warm current	Warm current all year	Very weak warm current (Kuroshio current)

Table 3. Morphological variation in *Suberites ficus* from different locations

Location	Growth form	Color	Microscleres (% of all spicules)	Host
Japan Sea	Rounded, lobate; few large lobes	red	1	*Pagurus pectinatus*
South Sea mainland coast	Rounded, lobate; many small lobes	purple	1	*Pagurus pectinatus*
South Sea Jeju Island	Rounded, solid; no projections	yellow	50	*Dardanus arrosor* (in shell)
Yellow Sea	Irregular, elongate; no projections	yellow	50	No host

Yellow Sea; only six occur in all three areas. The coast of the Yellow Sea does not appear to be suitable for sponge growth, presumably because of its adverse oceanographic conditions, such as great tidal range, high turbidity, and lack of hard substrates. Six of the nine species collected there came from Hongdo Island which is closest to and probably affected by the South Sea. The richest location is Jeju Island, at the southern limit of the South Sea, where 44 species were collected. Three species common throughout the three coastal seas are *Suberites ficus, Cliona celata,* and *Suberites excellence.*

When classified by preferred water temperature conditions, the samples were found to contain three cold-temperate species (5%), forty-eight temperate water species (80%), four tropical water species (6.7%) and five cosmopolitan species (8.3%). Fifty-three of the species found in Korea also occur in nearby Japanese water. *Suberites ficus,* which is widely distributed in South Korea, shows morphological variation with locality. Tanita (1965) observed that specimens occupied by hermit crabs are very rare; however, most of our specimens contained live hermit crabs.

Acknowledgments

I am especially grateful to Boon Jo Rho, Department of Biology, Ewha Womans University, Korea, and G. Bakus, Department of Biological Sciences, University of Southern California, for their valuable suggestions during the present work. This work was supported by a grant from the Korea Science and Engineering Foundation.

Literature Cited

Burton, M. 1953. *Suberites domunculus* (Olivi): Its Synonymy, Distribution and Ecology. *Bulletin, British Museum (Natural History), Zoology,* 1:358–378.

Hartman, W. D. 1958. Natural History of the Marine Sponges of Southern New England. *Peabody Museum of Natural History, Bulletin,* 12:1–55.

Laubenfels, M. W. de. 1949. The Sponges of Wood Hole and Adjacent Waters. *Bulletin, Museum of Comparative Zoology,* 103(1):1–55.

Lendenfeld, R. 1897. Die Clavulina der Adria, *Nova Acta, Abhandlungen der Deutschen Akademie der Naturforscher* (Halle), 69(1):1–251.

Tanita, S. 1965. A Sponge and a Hermit-crab. *Bulletin, Japan Sea Regional Fisheries Research Laboratory,* 14:95–97.

ANNA TRAVESET
Biology Department
University of Pennsylvania
Philadelphia, Pennsylvania 19104
and
Departament d'Ecologia
Universitat de Barcelona
Barcelona 08028, Catalunya, Spain

Notes on Iberian Freshwater Sponges

Abstract

Freshwater sponges have received very little attention in the Iberian Peninsula. Some preliminary taxonomical, ecological, and biogeographical findings suggest that freshwater sponges (like other limnic taxa) are a much more diverse group than previously thought. In a relatively ancient lake (Banyoles), sponges have lost the ability to gemmulate and they probably represent an endemic taxon, convergent with other sponges from lakes such as Baikal, Malawi, Ochrid, and Tiberias.

Little is known about the freshwater Porifera in the Iberian Peninsula. In most cases, the available information consists of long lists of local freshwater faunas. The purpose of this study was to review all published reports and available material in order to compile some preliminary taxonomic, ecological, and biogeographical data on the freshwater sponges of this region. This material, collected by several limnologists including the author, comes from an extensive survey throughout Iberia (1974–1984) (Figure 1) and the intensive study of a few sites.

Freshwater sponges have been largely ignored by limnologists even though these organisms may contribute considerably to the biomass and primary and secondary production in some limnic ecosystems and provide a refuge for animals (mainly insect larvae) and food for some specialists. Many species of freshwater sponges contain a large number of mutualistic *Chlorella*-like algae, and therefore function as autotrophs.

The Iberian freshwater sponge fauna is a poor subset of the European fauna, which consists of 6 genera and 14 species. The species present in the Iberian Peninsula are *Spongilla lacustris*, *Ephydatia fluviatilis*, *Ephydatia muelleri*, and *Heteromeyenia baileyi*. These extant species plus *Trochospongilla horrida* were found in a Quaternary fossil deposit in central Portugal (Moura, 1958).

Ecological Considerations

Freshwater sponges live under a wide range of environmental conditions, both physicochemical and biotic. In

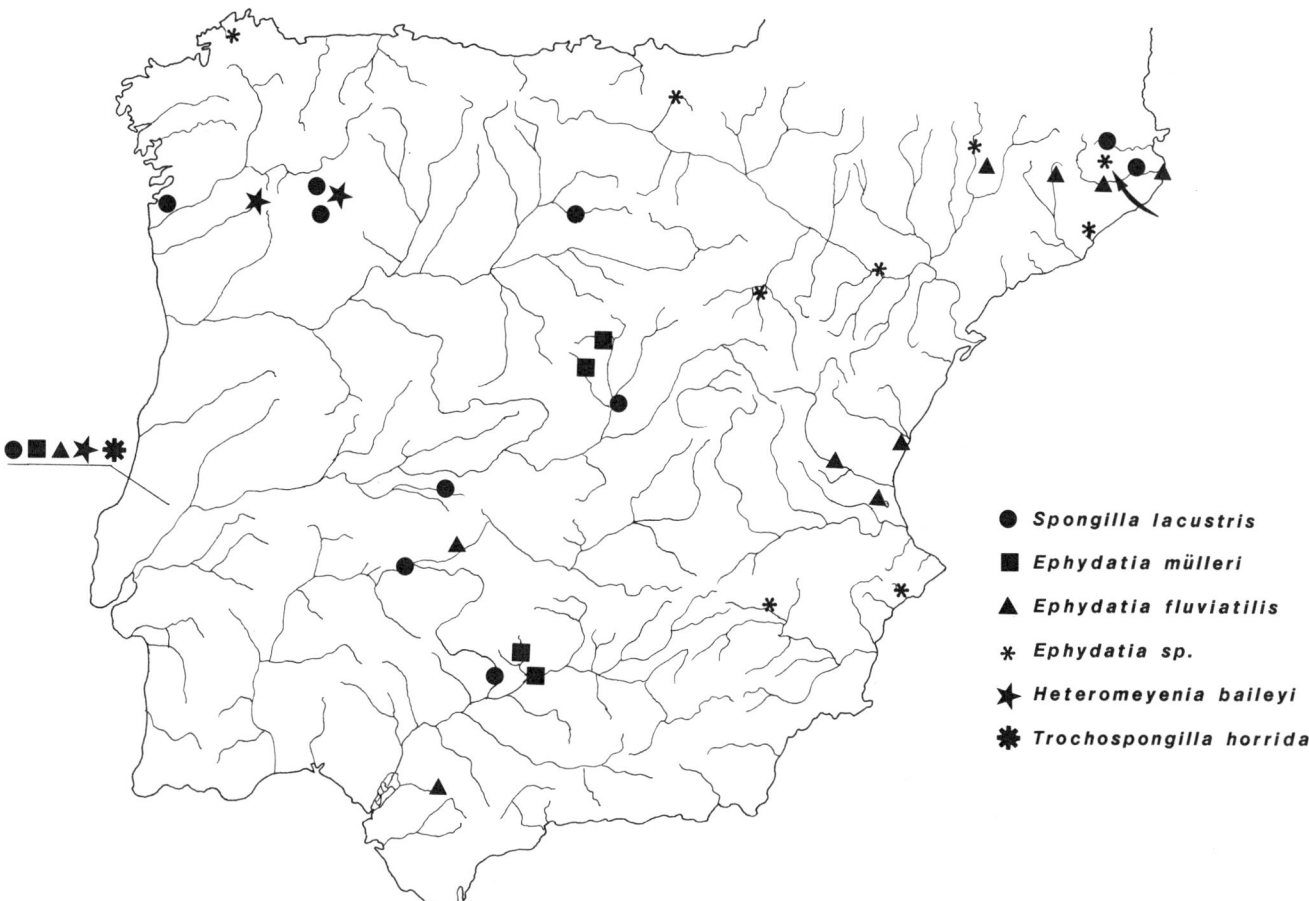

Figure 1. Map of Iberian peninsula showing collecting sites and distribution of freshwater sponges. *Ephydatia* sp. refers to samples with no gemmules and therefore unidentifiable at species level. Arrow points to Lake Banyoles.

the Iberian Peninsula, *Ephydatia* is the only one genus found in reservoirs. *E. muelleri* seems restricted to these environments, while *E. fluviatilis* also lives in streams. *Heteromeyenia baileyi* has been found only in two lakes, one of which was recently drained (Antela Lake). *Spongilla lacustris*, on the other hand, lives only in rivers. All are green when growing under sunlight owing to the large load of mutualistic algae in their tissues. The green sponges always grow faster and attain a larger size than white ones; this affects positively the occupation of the substrate and the production of gemmules.

In permanent waters, freshwater sponges may lose the ability to gemmulate. This seems to have happened in a relatively ancient karstic lake (Banyoles). Intensive collecting in this lake every month throughout the year showed no production of gemmules, but abundant larvae in the spring (I. Olivella, pers. comm.). Hence this population must be maintained by sexual reproduction. These sponges are difficult to identify owing to their small size and the absence of gemmoscleres. Other characters assign these sponges to the genus *Ephydatia*. It is quite possible that they represent an endemic taxon, convergent with other lacustrine Porifera (*Baikalospongia*, *Ochridaspongia*, *Malawispongia*, etc.) that have lost gemmulation as an adaptation to permanent waters.

Freshwater sponges may harbor a high and diverse fauna in their tissues ($H = 2.68$; Matteson and Jacobi, 1980). Different groups of Insecta (Plecoptera, Ephemeroptera, Trichoptera, Neuroptera, Diptera, Coleoptera), Mollusca (Gastropoda, Bivalvia), Crustacea (Isopoda, Amphipoda), Oligochaeta, Hirudinea, Nematoda, and Hydracarina have been found on or within freshwater sponges, although few of these animals have a direct relationship with the sponges.

Three specialized parasites inhabit Iberian freshwater sponges: *Sisyra iridipensis* (Neuroptera), *Xenochironomus xenolabis* (Diptera: Chironomidae) and *Unionicola crassipes* (Hydracarina: Unionicolidae). They are also host to a large variety of transient animals (Traveset, 1985).

In contrast with marine sponges, the mechanisms of defense that freshwater sponges possess against parasites and the tissue reactions produced in contact with commensals have not been studied. In one sample from Iberia, a dark layer of tissue in an otherwise white sponge surrounds the body of a chironomid larva. This may suggest that the sponge is using some type of chemical defense against the foreign body.

Biogeographical Considerations

Spongilla lacustris, Ephydatia fluviatilis, and *E. muelleri* are ubiquitous whereas *Heteromeyenia baileyi* is restricted to the westernmost part of the peninsula. *Trochospongilla horrida* is known only as a Pleistocene fossil from central Portugal.

Spongilla lacustris was first considered a cosmopolitan species but after the revision of Penney and Racek (1968), it appears to be restricted to the Northern Hemisphere.

Ephydatia fluviatilis, apparently cosmopolitan, has been divided into two subspecies by Ezcurra de Drago (1975): *E. fluviatilis fluviatilis* (Northern Hemisphere) and *E. fluviatilis ramsayi* (Southern Hemisphere). This subdivision is artificial, as it is based on just four highly variable, ecomorphic parameters of the gemmoscleres. Besides, it would create a very unlikely disjunctive pattern of distribution.

Ephydatia muelleri is also considered a typically Holarctic species, although Ezcurra de Drago (1975) reports it from Zimbabwe.

Heteromeyenia baileyi occurs along eastern North America, southern Brazil, and northeastern Argentina. In Europe it is reported from Germany and Poland; in Iberia, it has a restricted distribution, which probably reflects isolation during the last glaciation.

Trochospongilla horrida has a discontinuous boreal distribution. It is most probably extinct in Iberia.

Taxonomic Considerations

Almost all Iberian specimens are easily identified according to the generally accepted taxonomy of freshwater sponges (Penney and Racek, 1968). Nevertheless, environmental variables are usually not measured making it impossible to assess their influence on ecophenotypic plasticity. Unfortunately, the number of characters and character-states used in freshwater sponge systematics are too few to detect possible convergences and closely related but distinct species.

Therefore, it is hypothesized that Iberian freshwater sponges are a much more diverse group than previously thought. However, this suggestion cannot be substantiated until detailed morphological analyses have been conducted, as has been done with ciliates (Corliss, 1979); immunological and electrophoretic techniques should also prove helpful. Thorough phylogenetic systematics are needed to elucidate the ecology, life history, colonization, and radiation of freshwater sponges in this area.

Acknowledgments

I want to thank C. R. Altaba, R. Margalef, and C. Volkmer-Ribeiro for their help and valuable comments on this paper. I am also grateful to N. Prat, I. Olivella, and V. Montserrat for providing some of the material.

Literature Cited

Corliss, J. O. 1979. The Impact of Electron Microscopy on Ciliate Systematics. *American Zoologist,* 19:573–587.

Ezcurra de Drago, I. 1975. El género *Ephydatia* Lamoroux (Porifera, Spongillidae). Sistemática y distribución. *Physis,* 34(89): 157–174.

Matteson, J. D., and G. Z. Jacobi. 1980. Benthic Macroinvertebrates Found on the Freshwater sponge *Spongilla lacustris. Great Lakes Entomologist,* 13:169–172.

Moura, A. 1958. Espongílidos fósseis no diatomito de Azenhas do Vale de Atela (Alpiarça). *Memorias e Notícias,* (University of Coimbra), 46:1–13.

Penney, J. T., and A. A. Racek. 1968. Comprehensive Revision of a Worldwide Collection of Freshwater Sponges (Porifera, Spongillidae). *United States National Museum, Bulletin,* 272:1–184.

Traveset, A. 1985. Contribució al coneixement de les esponges d'aigua dolça a la Península Ibèrica. Tesi de Llicenciatura. University of Barcelona.

CECILIA VOLKMER-RIBEIRO
Departamento de Biologia
Pontifícia Universidade
Católica do Rio Grande do Sul,
and
Fundaçao Zoobotânica do Rio Grande do Sul
Porto Alegre, R. S., Brazil.

A New Insight into the Systematics, Evolution, and Taxonomy of Freshwater Sponges

Abstract

The evolutionary transformational series of characters revealed in a revisive study of several neotropical species of freshwater sponges sheds new light on specific and generic definitions. Evidence of the retention or slow modification in the freshwater habitat of some characters considered to be of important diagnostic value for marine sponges provides a means of detecting symplesiomorphies and of excluding convergent evolutionary patterns.

Hennig's principles for a phylogenetic systematics can thus be applied to such revised species. When this was done for the neotropical species of the genus *Metania* Gray, a series of new characteristics emerged that demonstrate its relationship to the marine poecilosclerid genus *Acarnus* Gray.

Sixteen evolving characters were recognized in the four neotropical species of *Metania*, and the direction of their change was determined by using the genus *Acarnus* as the taxonomical outgroup. In addition, a cladogram was constructed to illustrate how *Metania* may have evolved in the neotropical region.

Although the interrelationship among the genera of the gemmule-producing freshwater sponges is said to be "fairly well established" according to Penney and Racek (1968:6), the Spongillidae are still considered by them as a poorly defined family, owing to the lack of information on the nongemmuliferous freshwater sponges. Whereas de Laubenfels (1936) defined the Spongillidae as those sponges normally occurring in freshwater, Penney and Racek restricted the family to gemmule-producing freshwater sponges. Penney and Racek's (1968) new genus *Radiospongilla* would close the gap between the freshwater sponges with rodlike gemmoscleres (Carter's subfamily Spongillinae) and the freshwater sponges with birotulate gemmoscleres (Vejdovsky's subfamily Meyeninae) and thus render that subdivision unnecessary.

The idea of a monophyletic origin for the Spongillidae as proposed by Penney and Racek was reinforced by

Racek and Harrison (1975), who suggested that the family evolved from a *Radiospongilla* stock. Penney and Racek did not mention what particular marine group might be the source of that stock.

Earlier, Marshall (1883), had proposed a polyphyletic origin for freshwater sponges, conceived as successive invasions originating from the Renieridae.

Brien (1970), who proposed the new family Potamolepidae for the Ethiopian genera *Potamolepis* Marshall and *Potamophloios* Brien, both gemmule-producing freshwater sponges, hypothesized a broader polyphyletism involving estuarine invasions by more than one marine family in different periods. He set the Spongillidae as defined by Penney and Racek apart from the Potamolepidae because of the differences in their parenchymula larvae as well as their skeletal and gemmular structures. Brien suggested that the gemmule production is a convergent phenomenon owing to its highly adaptive significance.

Volkmer-Ribeiro and De Rosa-Barbosa (1979) reported the occurrence of the Potamolepidae in the neotropical region with the aggregation of their genus *Sterrastrolepis*, genus *Oncosclera* Volkmer-Ribeiro, and genus *Uruguaya* Carter, which was redefined and transferred from the Spongillidae. These authors supported Brien's (1970) suggestion of a hadromerid origin for the Potamolepidae (because the gemmoscleres of potamolepid genera retain sterrastrose characteristics) and they adopted Brien's conception of polyphyletism for the Spongillidae.

Volkmer-Ribeiro and De Rosa-Barbosa (1979) register the fact that endemic genera of freshwater sponges are recorded from lakes relict of past marine transgressions to support their proposition of a passive mechanism of invasion by marine sponges instead of the active one advanced by Brien. No endemic genus of freshwater sponges has ever been recorded from estuarine areas.

In establishing the genus *Sanidastra* and describing *S. yokotonensis*, Volkmer-Ribeiro and Watanabe (1983) found further evidence of a hadromerid stock whose characteristics are retained in a gemmule-producing freshwater sponge, among them that of a sanidaster transformed into a gemmosclere.

The work of Volkmer-Ribeiro and De Rosa-Barbosa (1979) and of Volkmer-Ribeiro and Watanabe (1983) confirms a polyphyletic origin for freshwater sponges and indicates that it may have been more widespread than previously thought. It has long been recognized that the evolution of the Porifera has been slow, as evident from the occurrence of "living fossils" (Lévi, 1973; Bergquist, 1978; Hartman et al., 1980). The fact that spicular components and the structure of some marine sponges may pass to the freshwater branches of that stock with little or no modification is good reason to discount convergent evolution in phylogenetic propositions for some freshwater sponges.

Recent work (Volkmer-Ribeiro, 1979, 1984) on the neo-

tropical species of the genus *Metania* Gray has disclosed new characteristics that have made it necessary to redefine the genus and propose new intergeneric relationships (Volkmer-Ribeiro, 1986) vis á vis the demonstration of its affinities to the marine poecilosclerid genus *Acarnus* Gray. The purpose of this study is to test the applicability of Hennig's principles (Hennig, 1966) in tracing phylogenetic relationships in the neotropical branch of *Metania*.

Methods

To construct a cladogram according to the principles of phylogenetics expounded by Hennig (1966) and now followed by an ever-increasing number of systematists, the derived characters of each group must be uniquely shared by the members of that group. In this study, sister-group relationship was postulated and tested by the out-group comparison method (see Watrous and Wheeler, 1981). The ancestral or derived condition for each character or character state was also established.

The gondwanic (Figure 1) genus *Metania* has four neotropical species (Volkmer-Ribeiro, 1979, 1984) for which an array of new characteristics have been described. Of particular interest here is a boletiform gemmosclere—which has a spiny shaft, a collar of spines under the expanded lower rotule, and a knobbed upper rotule—that is common to all four species (Figures 2, 3) but is not present in other sponge genera. Studies in progress by the author indicate that this gemmosclere is also produced in *Metania* species from Asia, Ethiopia, and Australia.

Fifteen other characters of *Metania reticulata* (Bowerbank), which is the type species of the genus, and of *M. spinata* (Carter), *M. fittkaui* Volkmer-Ribeiro, and *M. subtilis* Volkmer-Ribeiro were found to form transformational evolution series. Among these characters are the skeletal reticulum, the growth pattern, the shape and hardness of the sponges (Figure 4), the number of the spicular categories, the shape of the spicules, and the covering on them (Figures 2, 3).

Before the evolution of the neotropical branch of *Metania* could be examined, however, it was necessary to determine the direction of the evolution along those transformational series of characters.

Because of the growing evidence of a polyphyletic origin for freshwater sponges, added to the peculiarity of the gemmosclere in *Metania* and the demonstration of a complex spicular set, it seemed appropriate to compare this genus with marine genera of the order Poecilosclerida. The comparison (Volkmer-Ribeiro, 1986) revealed that the spicular components and skeletal structure in genera *Metania* and *Acarnus* are strikingly similar.

If *Metania* and *Acarnus* had a common ancestor, as is inferred from their overall similarity (Volkmer-Ribeiro, 1986), and if *Acarnus* is the marine branch, it appears that

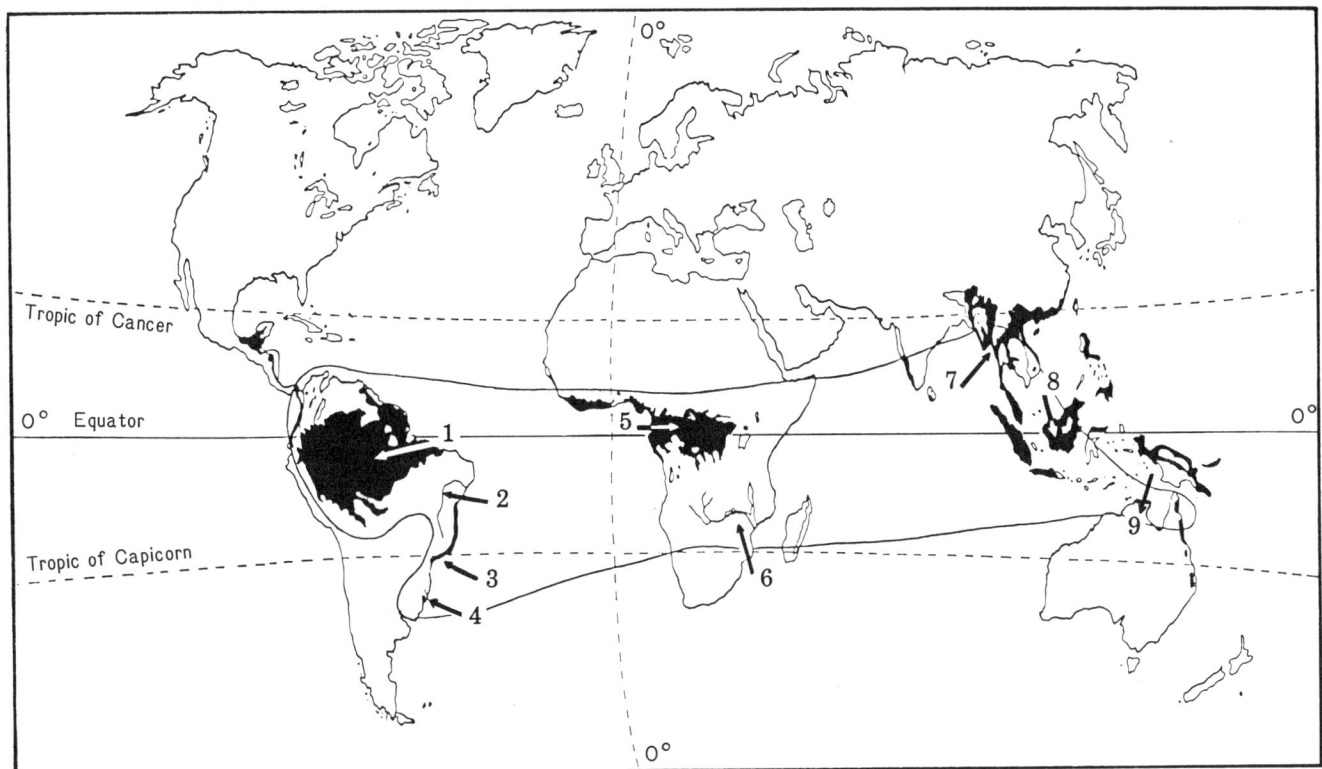

Figure 1. Gondwanic distribution of genus *Metania* Gray. Areas 1–4 after Volkmer-Ribeiro (1979, 1984); areas 5–8 after Penney and Racek (1968); area 9 after Stanisic (1979).

less evolutionary pressure was experienced by the branch that remained in the original habitat, that is, *Acarnus*. Thus, for each homologous character found in a different state in the two genera, the state found in *Acarnus* would be the primitive one and that in *Metania* would be the derived or apomorphic one. A homologous character found in the same condition both in *Metania* and in *Acarnus* would be a common primitive character or a symplesiomorphy as defined by Hennig (1966).

Results

The sixteen evolving characters identified in the four neotropical species of *Metania* are listed in Table 1 according to their ancestral or derived condition. A cladogram (Figure 5) was constructed on the basis of the synapomorphies and autapomorphies determined in Table 1.

The directionality of transformation series for the listed characters was determined by immediate out-group comparison (i.e., with *Acarnus* for characters 1, 2, 5, 9, 13, 14, and 15, and with *Metania spinata* for the other characters up to 16). *M. spinata* was the species with the largest number of primitive characters or character states. Thus, for each character not found in *Acarnus* but found in *M. spinata*, the state in *M. spinata* was considered the primitive one.

Discussion

THE CHARACTERS

1. The particular transformation of an acanthocladotylote into a boletiform gemmosclere with a collar of spines under the lower rotule is a character shared exclusively by the *Metania* species and is thus considered the common derived character, or the synapomorphy that places those species into a group. Traces of the ancestral cladiform terminations of the acanthocladotylote are still present on the inner face of the expanded rotule of some gemmoscleres in *Metania* (Figure 3b), where lateral wings departing from the shaft meet the border of the rotule and determine its polygonal profile.

2. Some *Acarnus* species have a third class of megasclere: a short acanthostyle with spines concentrated at the extremities. Some *Metania* species also have a second class of megasclere that is shorter and spined. The ancestral condition of this megasclere is particularly conspicuous in *Metania fittkaui*, in which some of the beta megascleres have one extremity rounded and the other pointed (Figure 2c). In *M. spinata*, the beta megasclere has smooth extremities (Figures 2a, 3a). The condition in *M. spinata* is thus considered derived and an autapomorphy for that species.

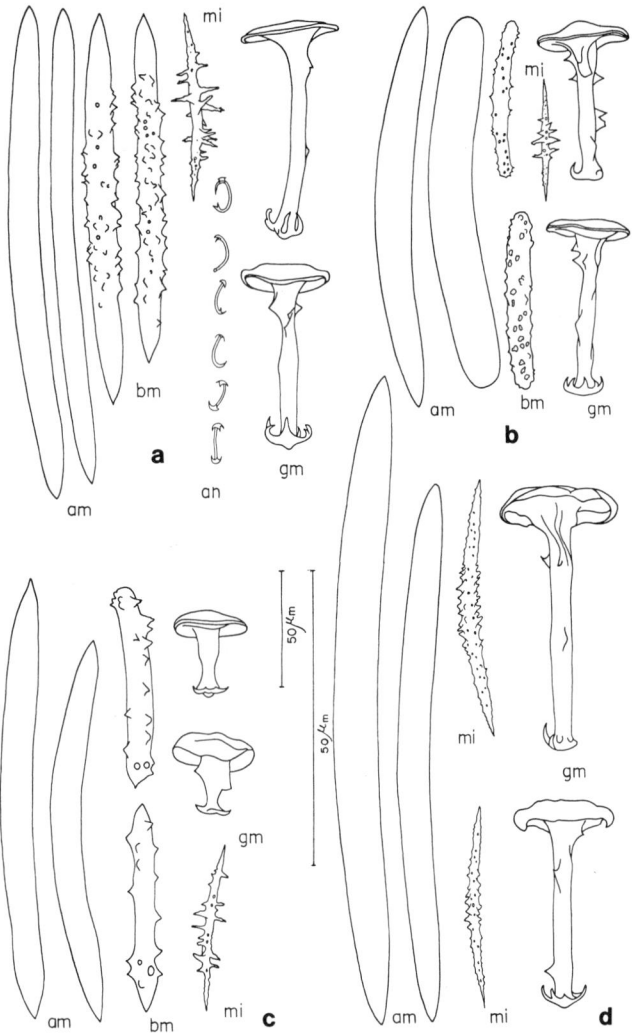

Figure 2. Spicule complement of neotropical species of *Metania*: a, *M. spinata* (Carter); b, *M. reticulata* (Bowerbank); c, *M. fittkaui* Volkmer-Ribeiro; d, *M. subtilis* Volkmer-Ribeiro. am, alpha megascleres; an, anchors (chelas); bm, beta megascleres; gm, gemmoscleres; mi, micro-acanthoxea.

3. The acanthocladotylote in all *Acarnus* species occurs as an echinating spicule along the ascending fibers. Since in *Metania* this spicule was transformed into the boletiform gemmosclere, it may be inferred that gemmular localization along the main fibers is a primitive condition (plesiomorphy) and that the gemmules in *M. spinata* not linked to the skeletal fibers and in a basal position are another autapomorphy for that species.

4. The lower rotule of the gemmosclere in *Metania spinata* does not project much farther beyond the shaft (Figure 3a) and thus resembles a primitive condition of the transformation of the cladiform extremity of the acanthocladotylote into an expanded boletiform rotule. The fact that the lower rotule extends quite far from the shaft in *M. reticulata*, *M. fittkaui* and *M. subtilis* (Figures 3b–d) indicates that the character is derived and suggests a fourth synapomorphy for those species.

5. *Metania spinata* is the only neotropical species of *Metania* to have retained the chela series of microscleres present in *Acarnus* (Figures 2a, 3a). The absence of the character in *M. reticulata*, *M. fittkaui* and *M. subtilis* is thus another synapomorphy for these three species.

6 and 7. The pneumatic layer of the gemmule is thick in *Metania spinata* and contains two layers of gemmoscleres around the foraminal tube (Volkmer-Ribeiro, 1984: figure 7). In *M. reticulata* (Volkmer-Ribeiro, 1984: figure 4), *M. fittkaui* and *M. subtilis* the pneumatic layer is thin and the gemmoscleres coat the gemmule in one layer. Out-group comparison reveals that the conditions of the two characters found in these three species are derived and constitute two more synapomorphies for them.

8 and 9. The acanthostyle in *Acarnus* is located at the basal part of the sponge. This primitive condition was retained in *Metania spinata*. The condition found in *M. reticulata* and *M. fittkaui* is thus derived, the beta megasclere being used in the building of the cages that link the gemmules to the primary skeletal fibers. The derived state of the character is therefore a synapomorphy for *M. reticulata* and *M. fittkaui*. It is interesting to note that *M. reticulata* is the most abundant species of *Metania* in the Amazonian flooded tropical forest (Volkmer-Ribeiro, 1984). This successful colonization must be attributed to the production of the gemmular cages, which retain the gemmules in the body of the mother sponges when the level of the Amazon falls. When flooding next occurs the gemmules initiate a new growth upon the dried skeleton of the parent sponge.

10, 11, and 12. Gemmosclere shafts with a few spines and a fragile consistency of the sponges (characters 10 and 11) are present in the out-group *(Metania spinata)* and in the other neotropical species, with the exception of *M. reticulata*, which has several spines on the gemmosclere shaft (Figures 2b, 3d) and usually produces quite hard specimens (Figure 4b). Thus the states found in *M. reticulata* are autapomorphies for that species. The same reasoning applies to the length of the gemmosclere, which is extremely reduced only in *M. fittkaui* (Figures 2c, 3c) and thus must be considered an autapomorphy for that species.

13, 14, 15, and 16. The beta megascleres (character 13), the reticulate skeleton (14), an irregular surface (15) and long spines on the middle portion of the anfioxeote microscleres (16) are present in the out-group *(Acarnus* or *Metania spinata)* and in the other neotropical species, with the exception of *M. subtilis* (Figures 2d, 3b, 4d). The conditions

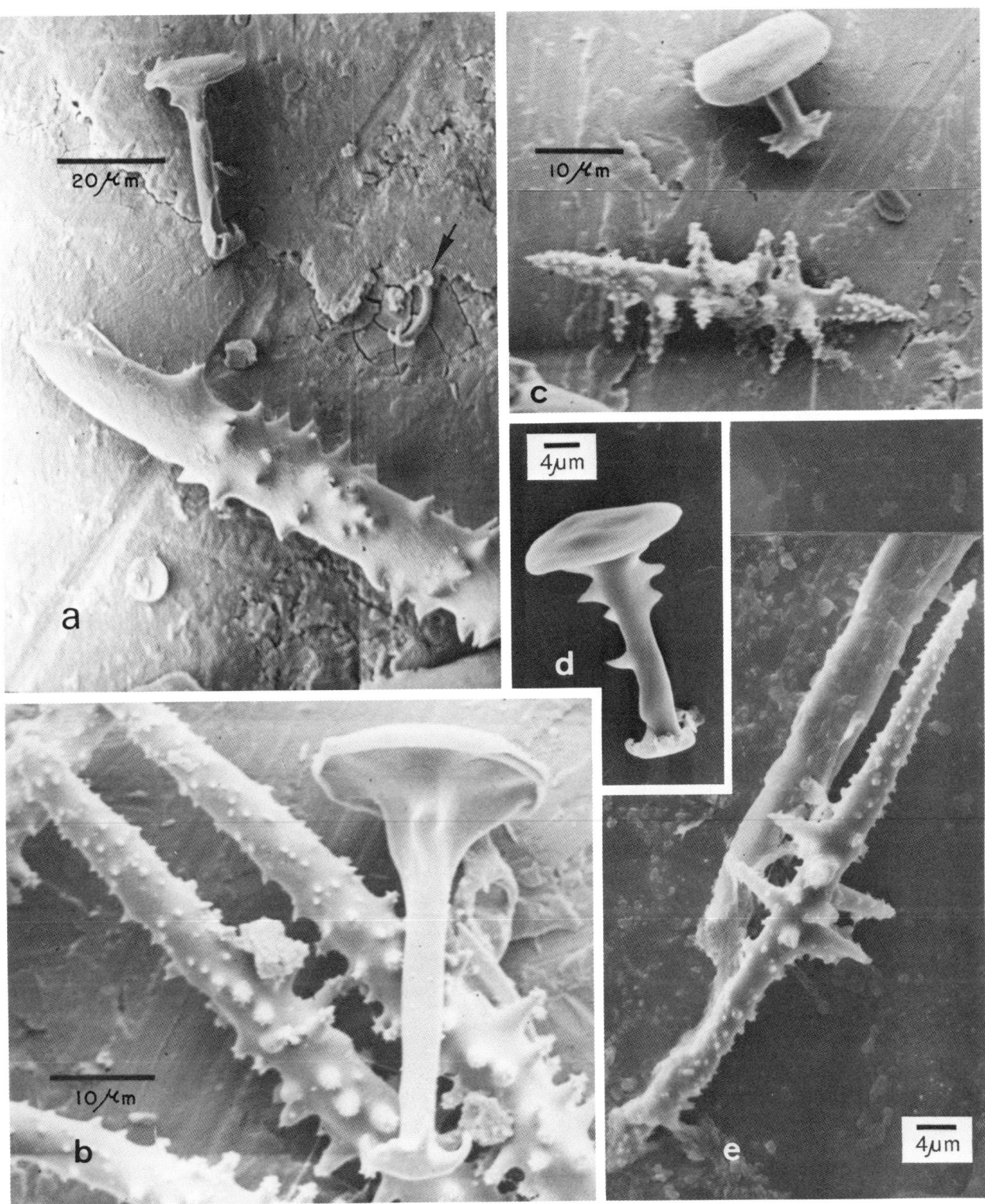

Figure 3. Scanning electron micrographs of spicules of neotropical species of *Metania* illustrating ancestral and autapomorphic states of spicular evolution (listed in Table 1): *a*, gemmosclere, extremity of beta megasclere, and chela microsclere of *M. spinata*; *b*, gemmosclere and microscleres of *M. subtilis*; *c*, gemmosclere and microsclere of *M. fittkaui*; *d*, gemmosclere of *M. reticulata*; *e*, part of megasclere and entire microsclere of *M. reticulata*.

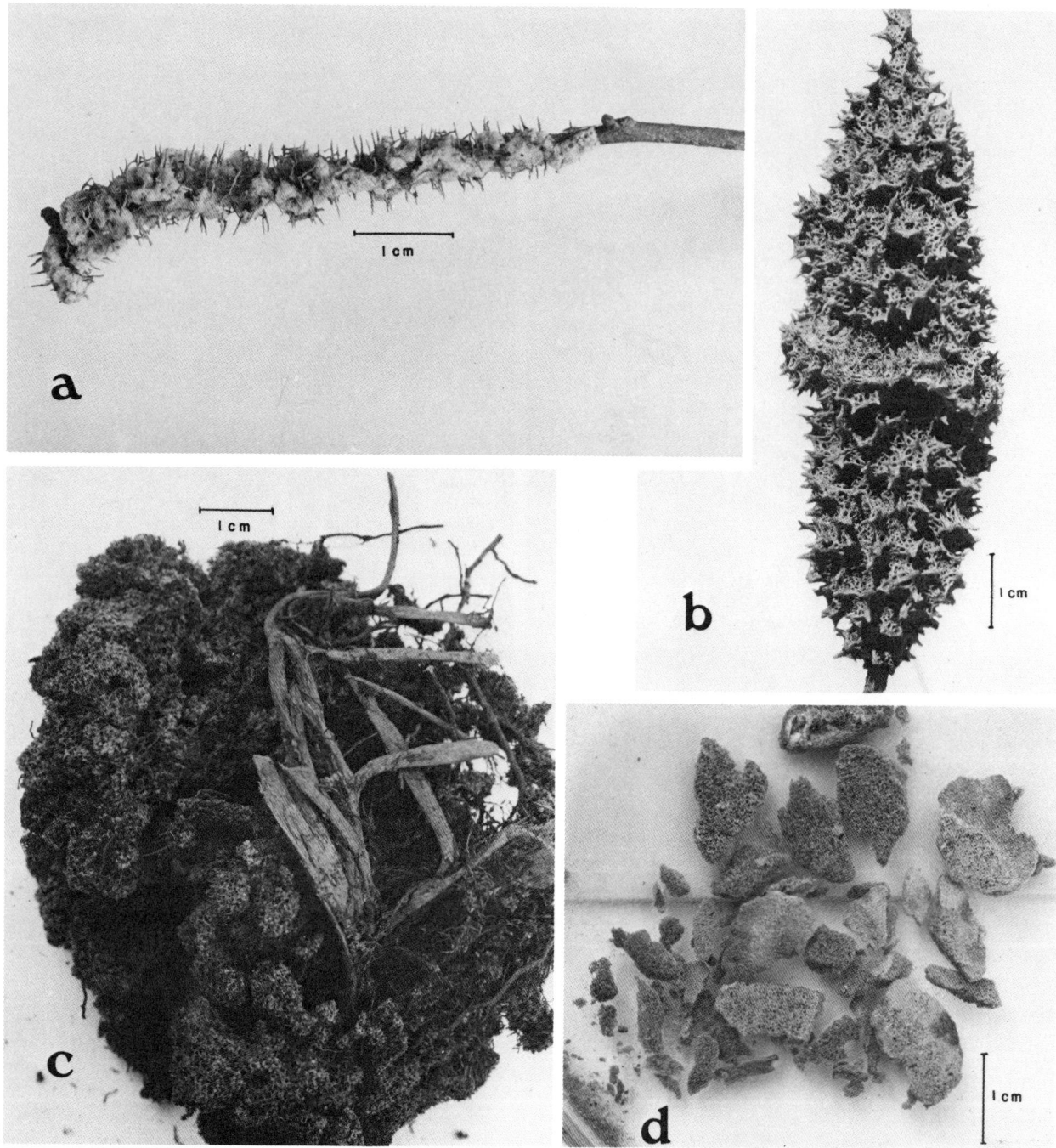

Figure 4. Specimens of neotropical species of genus *Metania: a, M. spinata* (Carter); *b, M. reticulata* (Bowerbank), with a "nonmerging front" in center, which indicates that this is not one sponge but two growing side by side and competing for the same substrate; *c, M. fittkaui* Volkmer-Ribeiro; *d, M. subtilis* Volkmer-Ribeiro.

found in *M. subtilis* for those characters are derived and autapomorphies for that species.

THE CLADOGRAM

The cladogram (Figure 5) obtained using the synapomorphies to establish the sister-group (Hennig, 1966) rela-

tionships and the autapomorphies to distinguish between the species indicates that evolution has progressed primarily through reduction of the characters (i.e., the chelate microscleres, the beta megascleres, the reticulate skeleton, and the size and consistency of the sponges) and to a lesser extent through modification of the character state (i.e., the number and pattern of the spine covering, length

Table 1. The sixteen evolving characters analyzed in a phylogenetic study of the neotropical branch of the genus Metania [* indicates derived (apomorphic) condition of the character; N.A. = not applicable]

Character	Outgroup (Acarnus)	M. spinata	M. reticulata	M. fittkaui	M. subtilis
1. Acanthocladotylote spicule	Cladiform extremity not expanded. No particular distribution of shaft	*Cladiform extremity expanded in a boletiform manner. Some shaft spines forming a collar under the expanded extremity	*Cladiform extremity expanded in a boletiform manner. Some shaft spines forming a collar under the expanded extremity	*Cladiform extremity expanded in a boletiform manner. Some shaft spines forming a collar under the expanded extremity	*Cladiform extremity expanded in a boletiform manner. Some shaft spines forming a collar under the expanded extremity
2. Extremities of the beta megascleres	Spined	*Smooth	Spined	Spined	N.A.
3. Gemmules fibers	N.A.	*Not linked to skeletal	Linked to skeletal fibers	Linked to skeletal fibers	Linked to skeletal fibers
4. Border of the lower rotule of the gemmoscleres	N.A.	*Poorly developed	*Well developed	*Well developed	*Well developed
5. Chelate	Present	Present	*Absent	*Absent	*Absent microscleres
6. Pneumatic layer of gemmules	N.A.	Thick	*Thin	*Thin	*Thin
7. Gemmoscleres	N.A.	In a double layer	*In a single layer	*In a single layer	*In a single layer
8. Gemmules	N.A.	Not in capsules	*In capsules of beta megascleres	*In capsules of beta megascleres	Not in capsules
9. Position of beta megascleres	At the base	At the base	*Not basal	*Not basal	N.A.
10. Shaft of gemmoscleres	N.A.	With a few spines	*With several spines	With a few spines	With a few spines
11. Consistency of sponge	N.A.	Fragile	*Hard	Fragile	Fragile
12. Shaft of gemmoscleres	N.A.	Long	Long	*Short	Long
13. Presence of beta megascleres	Present	Present	Present	Present	*Absent
14. Skeleton	Reticulate	Reticulate	Reticulate	Reticulate	*Not reticulate
15. Surface	Irregular	Irregular	Irregular	Irregular	*Smooth
16. Spines at middle portion of oxeote microscleres	N.A.	Long	Long	Long	*Short

3d Int. Sponge Conf. 1985

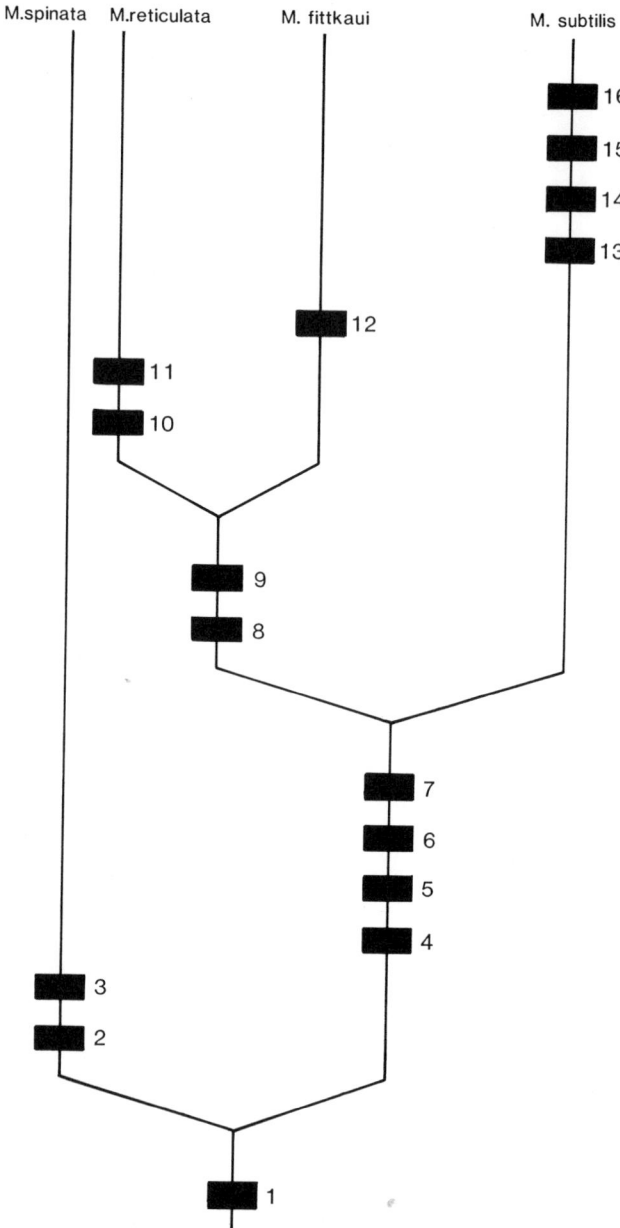

Figure 5. Cladogram of proposed phylogenetic relationship of four known neotropical species of *Metania*. Numbers correspond to characters listed in Table 1. Black boxes indicate derived condition of each character.

efforts in South America (not to mention Africa and Australia) may be considered scant and quite discontinuous; (2) in the first adaptive steps to the new habitat, the missing spicules were the first to be lost; (3) the *Metania* species studied arrived from an *Acarnus* ancestor already deprived of the tylotes. The toxa, but not the tylotes are present in all extant *Acarnus* species (Hooper, 1987). Another possibility is that in the South American species of *Metania* those spicular categories may yet be found, since few specimens apart from *M. reticulata* have been examined up to now. A characteristic of the *Acarnus* species is that some spicular categories are absent from certain specimens of the same species (Hooper, 1987). The prevailing pathway in the evolution of *Acarnus*, as inferred by Hooper from the study of extant species, is via the reduction of spicular categories.

Conclusions

When Hennig's principles were used to trace phylogenetic relationships, the neotropical branch of the genus *Metania* was found to be composed of four species of which *M. spinata* and *M. subtilis* are the least related, and *M. reticulata* and *M. fittkaui* are the closest.

If it is assumed that the genus *Metania* and the poecilosclerid genus *Acarnus* have a common ancestor, *M. spinata* has retained the largest number of ancestral characteristics, whereas *M. subtilis* has retained the least number of such characteristics. The *M. reticulata-M. fittkaui* intermediate pair is close to the first specie in that it exhibits strengthened skeletal structure and the gemmules are bound to the skeletal fibers by gemmular cages; these characteristics led to the successful invasion of a new habitat, the flooded tropical forest along the Amazon.

This study has demonstrated speciation events along which the ancestral characteristics of a cladotylote spicule were transformed and approached those of a birotulate spicule. The results strongly suggest that the genus *Metania* does not belong to the "birotulate" genera of freshwater sponges as previously supposed. The present birotulate condition of the gemmosclere, as in *Metania fittkaui* (Figure 3c) is the result of a convergent process of evolution.

Acknowledgments

This study was supported by fellowship no. 30.6134/76 and grant no. 40.5893/85 of Conselho Nacional de Desenvolvimento Científico e Tecnológico (CNPq) Brazil. My sincere thanks go to Klaus Rützler, National Museum of Natural History, Smithsonian Institution for the support and encouragement. I am indebted to Francisco Kees, Faculdade de Engenharia of Universidade Federal do Rio Grande do Sul, and to Arno A. Lise from Museu de Ciências Naturais of Fundaçao Zoobotânica do Rio Grande do

of the gemmosclere shaft, transformation of the inner and outer rotules). The out-group comparison of *Metania spinata* and *M. subtilis* clearly shows that the former has retained the largest number of ancestral characteristics and the latter the least. In the four species studied, some of the *Acarnus* characters, such as the sigma microscleres and the ectosomal tylotes, do not appear at all. Three hypotheses can be formulated to explain this absence: (1) It may be that the *Metania* species containing those spicules have not yet been found, since the geographic coverage of collecting

Sul, for the SEM photos in Figures 3a–c, and to George M. Davis, Malacology Deptartment of the Academy of Natural Sciences of Philadelphia, for the SEM photos in Figures 3d,e made possible by Grant TMP–11373 of the U.S. National Institutes of Health to Dr. Davis. I also thank John N. A. Hooper, Division of Natural Sciences of the Northern Territory Museum of Arts and Sciences, Australia, for making accessible his manuscript in press.

Literature Cited

Bergquist, P. R. 1978. *Sponges*. London: Hutchinson. 268 pp.

Brien, P. 1970. Les Potamolepides Africaines nouvelles du Luapula et du Lac Moëro. Pages 163–187 in *The Biology of the Porifera*, edited by W. G. Fry. Symposia of the Zoological Society of London 25. London: Academic Press.

Laubenfels, M. W. de. 1936. A Discussion of the Sponge Fauna of the Dry Tortugas in Particular, and the West Indies in General, with Material for a Revision of the Families and Orders of the Porifera. *Papers of the Tortugas Laboratory*, 30:1–225.

Hartman, W. D., J. W. Wendt, and F. Wiedenmayer, eds. 1980. *Living and Fossil Sponges*. Notes for a short course. Sedimenta 8. Miami: University of Miami. 274 pp.

Hennig, W. 1966. *Phylogenetic Systematics*. Urbana: University of Illinois Press. 263 pp.

Hooper, J. N. A. 1987. New Records of *Acarnus* Gray (Porifera: Demospongiae: Poecilosclerida) from Australia, with a Synopsis of the Genus. *Memoirs of the Queensland Museum*, 25:71–105.

Lévi, C. 1973. Systématique de la classe des Demospongiaria (Démosponges). Pages 577–631 in *Spongiaires, Traité de Zoologie* 3(1), edited by P. P. Grassé. Paris: Masson.

Marshall, W. 1883. On Some New Siliceous Sponges collected by M. Pechuël-Lösche in the Congo. *Annals and Magazine of Natural History*, 12:391–412.

Penney, J. T., and A. A. Racek. 1968. Comprehensive Revision of a Worldwide Collection of Freshwater Sponges (Porifera: Spongillidae). *U. S. National Museum Bulletin*, 272:1–184.

Racek, A. A., and F. W. Harrison 1975. The Systematic and Phylogenetic Position of *Paleospongilla chubutensis* (Porifera: Spongillidae). *Proceedings, Linnean Society of New South Wales*, 99:157–165.

Stanisic, J. 1979. Freshwater Sponges from the Northern Territory (Porifera: Spongillidae). *Proceedings, Linnean Society of New South Wales*, 103:123–130.

Volkmer-Ribeiro, C. 1979. Evolutionary Study of Genus *Metania* Gray, 1867 (Porifera: Spongillidae): 1. The New Species. *Amazoniana*, 6:639–649.

———. 1984. Evolutionary Study of the Genus *Metania* Gray, 1867 (Porifera: Spongillidae): II. Redescription of Two Neotropical Species. *Amazoniana*, 8:541–553.

———. 1986. Evolutionary Study of the Freshwater Sponge Genus *Metania* Gray, 1867: III. Metaniidae, New Family. *Amazoniana*, 9:493–509.

Volkmer-Ribeiro, C., and R. De Rosa-Barbosa. 1979. Neotropical Freshwater Sponges of the Family Potamolepidae Brien, 1967. Pages 503–511 in *Biologie des Spongiaires*, edited by C. Lévi and N. Boury-Esnault. Colloques Internationaux du C.N.R.S. 291. Paris: Centre National de la Recherche Scientifique.

Volkmer-Ribeiro, C., and Y. Watanabe. 1983. *Sanidastra yokotonensis*, n. gen. and n. sp. of Freshwater Sponge from Japan. *Bulletin, National Science Museum, Zoology*, 9:151–159.

Watrous, L. E., and Q. D. Wheeler. 1981. The Out-Group Comparison Method of Character Analysis. *Systematic Zoology*, 30:1–11.

ANTONIO M. SOLE-CAVA*
JOHN P. THORPE
University of Liverpool
Marine Biological Station
Port Erin, Isle of Man, United Kingdom

High Levels of Genetic Variation in Marine Sponges

Abstract

Levels of genetic polymorphism are estimated for natural populations of the sponges *Halichondria panicea, Mycale macilenta, Suberites subereus,* and two species of *Suberites "ficus"* from the Irish sea. The animals were analyzed by horizontal starch gel electrophoresis and staining for specific enzymes. A total of 86 enzyme loci covering 18 different enzymes were detected over the five populations studied. Between 50 and 75% of the enzyme loci were polymorphic in each species, and levels of mean heterozygosity varied from 0.168 *(S. ficus)* to 0.335 *(S. subereus)*. These levels are high when compared with values normally obtained for most animals or plants, but compare well with heterozygosities found in marine cnidarians and some mollusks. Possible relationships between levels of genetic variation and random genetic processes or environmental factors are discussed.

A basic premise of the theory of evolution is that genetic variation exists within natural populations. Without this variation there would be little scope for the action of natural selection, genetic drift, or other evolutionary processes. However, the real magnitude of that variation was not fully appreciated until researchers began to use electrophoretic techniques to estimate genetic polymorphism. The main advantage of electrophoresis over other (non-biochemical) techniques in this application is that electrophoresis makes it possible to directly detect variation in products of single chosen gene loci in individual animals or plants (Lewontin, 1974; Ayala, 1976; Ferguson, 1980). Furthermore, samples of populations or groups of individuals can be readily investigated over a substantial number of loci; the estimates obtained are statistically fairly robust and not much affected by the sample size (Nei and Roychoudbury, 1974; Nei, 1978; Gorman and Renzi, 1979). Useful results can therefore be obtained even from uncommon species for which only small sample sizes may be available.

*Present address: Departamento de Biologia Geral, Universidade Federal Fluminense, C. P. 183, 24000 Niteroi, R. J., Brazil.

Data on the levels of genetic variability in natural populations of a wide variety of organisms have been available for many years. A notable feature of findings to date is that, irrespective of the species involved, levels of genetic variability are generally similar, within a limited range, in the vast majority of populations (see reviews, e.g., of Powell, 1975; Selander, 1976; Nevo, 1978; Burton, 1983). Among both animal and plant populations, the proportion of loci polymorphic *(P)* generally falls within the range 0.10–0.50 while mean heterozygosity per locus *(H̄)* varies from about 0.02 to 0.15. The comprehensive survey by Nevo (1978: table 1) covering 277 populations spread over 243 species of vertebrates, invertebrates, and plants records only four heterozygosity estimates greater than 0.25, the highest being 0.309 (from the bisexual parthenogenic weevil *Otiorrhynchus scaber*, Soumalainen and Saura, 1973). Very low *H̄* values are uncommon and are (predictably) frequently linked to very low population size (e.g., Selander and Kaufman, 1975; Bonnell and Selander, 1974). However, *H̄* values greater than 0.3 are considerably rarer, and we know only of three published studies giving such data: the recent work of Ritte and Pashtan (1982) on Red Sea populations of two species of the marine gastropod *Cerithium* (in which *H̄* values are around 0.6), of Beaumont and Beveridge (1984) on the bivalve *Chlamys distorta* (*H̄* = 0.321), and of Solé-Cava et al. (1985) on sea anemones of the genus *Urticina* (*H̄* = 0.41–0.44).

No heterozygosity estimates are available for Porifera; the few electrophoretic studies that have been carried out on sponges to date either do not include a genetic interpretation of the results (e.g., Connes et al., 1974; Urbaneja and Lin, 1981; Hooper, this volume) or use too few loci for meaningful heterozygosity estimates, although the numbers are adequate for biochemical taxonomic comparison (e.g., Sará, this volume).

In this paper we provide estimates of levels of genetic variation in five common species of marine sponge.

Materials and Methods

Halichondria panicea (Pallas) was collected from low in the intertidal zone at Port St. Mary Ledges, Port St. Mary, Isle of Man. *Suberites* cf. *ficus* (Johnston) species *A*, *S*. cf. *ficus* (Johnston) species *B*, *S. subereus* (Johnston), and *Mycale macilenta* (Bowerbank) were collected by dredging at 10–30 m off Port Erin, Isle of Man (northern Irish Sea) by the Liverpool University research vessel *Cuma* (for further information on the three *Suberites* species, see Solé-Cava and Thorpe, 1986).

Samples were analyzed immediately after collection. To avoid diluting the enzymes during extraction, we homogenized the samples without any buffer. Homogenized samples were analyzed by horizontal starch (12.5%) gel electrophoresis. Several buffer systems were tried, but the best

results in terms of enzyme activity and resolution were all obtained using a tris-citric acid pH 8.0 system from Ward and Beardmore (1977). (Composition of the electrode buffer was 30.3 g tris + 12.0 g citric acid + 1000 ml distilled water; the gel buffer consisted of 38.5 ml of electrode buffer diluted to 1000 ml with distilled water.)

The staining of the gels followed standard procedures (Shaw and Prasad, 1970; Harris and Hopkinson, 1978). Enzyme nomenclature follows that of Harris and Hopkinson (1978).

All samples were analyzed for 23 enzymes, but no activity could be obtained for octanol dehydrogenase (E.C. 1.1.1.1), aldolase (E.C. 4.1.2.13), leucine aminopeptidase (E.C. 3.4.1.1), superoxide dismutase (E.C. 1.15.1.1), or aconitase (E.C. 4.2.1.3). The other 18 enzymes (Table 1) each gave useful results for at least one of the populations studied.

Observed heterozygosity for each locus was calculated as the proportion of individuals in the sample heterozygous for that locus (i.e., number of heterozygotes/total number of genotypes). Expected heterozygosity per locus was calculated as $1 - \Sigma x^2_{ij}$, where x is the frequency of the ith allele in the locus j. Heterozygosities for each species

Table 1. Calculated heterozygosities for each of the studied loci in five sponge species (na = no activity observed, He = the mean calculated heterozygosity for each species)

Locus	Halichondria panicaea	Mycale macilenta	Suberites ficus A	Suberites ficus B	Suberites subereus
ak	0.210	na	na	na	na
aldox	0	0.444	0	0.069	0
cat	0.198	0	0.375	0.180	0.451
est.1	na	0	0.219	0.291	0.720
est.2	na	0	0.180	0.204	na
fum	0	0.245	0.219	0	0.117
gdh	na	na	na	na	0.561
got	0	0	0	0	na
gpd.1	0.375	0.245	0.117	0	na
gpd.2	0	na	na	na	na
hk.1	0.401	0.278	0	0.124	0.627
hk.2	0.475	na	na	na	na
idh	na	0	0.219	0.444	0.531
mdh	0	na	0	0	0.225
me	0.153	0.459	na	na	na
mpi	na	0	0.305	0.293	0.461
pep.1	0.500	0.245	0	0.180	0
pep.2	0.153	na	na	na	na
pgd	0.424	0	0.111	0.305	0.492
pgi.1	na	0.459	0	0	0
pgi.2	na	0	0.420	0.451	0.490
pgm.1	0.647	0.520	0.570	0.420	0.061
pgm.2	na	0.500	0.170	0.549	0.398
xod	0.204	0	0.117	0	0.227
He	0.234	0.189	0.168	0.190	0.335

Enzymes analyzed are: adenylate kinase (ak, E.C. 2.7.4.3), aldehyde oxidase (aldox, E.C. 1.2.1.3), catalase (cat, E.C.1.11.1.6), esterases (est, E.C. 3.1.1.1), fumarase (fum, E.C. 4.2.1.2), glutamate dehydrogenase (gdh, E.C. 1.4.1.3), glutamate–oxaloacetate transaminase (got, E.C. 2.6.1.1), glycerophosphate dehydrogenase (gpd, E.C. 1.1.1.8), hexokinase (hk, E.C. 2.7.1.1), isocitrate dehydrogenase (idh, E.C. 1.1.1.42), malate dehydrogenase (mdh, E.C. 1.1.1.37), malic enzyme (me, E.C 1.1.1.40), manosephosphate isomerase (mpi, E.C. 5.3.1.8), peptidase (pep, E.C. 3.4.11–17.–.), 6–phosphogluconate dehydrogenase (pgd, E.C. 1.1.1 43), phosphoglucose isomerase (pgi, E.C 5.3.1.9), and phosphoglucomutase (pgm, E.C. 2.7.5.1), and xanthine oxidase (xod, –E.C. 1.2.3.2).

are presented as the arithmetic means of the values obtained over all loci studied for that species (see Nei, 1975).

Another measure of genetic variation used here is the proportion of loci polymorphic, P, expressed as a percentage of loci for which the frequency of the commonest allele was lower than 95% ($P_{0.95}$) or lower than 99% ($P_{0.99}$).

Results

A total of 86 enzyme loci as detected over the five species studied (Table 1). At least 16 loci were analyzed for each species, and the mean number of individuals studied per locus for each species ranged from 7 to 18 (Table 2). The numbers of loci and individuals were based on the numbers recommended by Gorman and Renzi (1979) for the estimation of genetic variation in natural populations.

Discussion

The heterozygosity levels (Table 2) observed for the five species of sponges studied here were very high in comparison with values normally observed for other animals and plants (see, e.g., Nevo, 1978). Interestingly, all the sponges studied—even the more polymorphic species of *Cerithium, Chlamys,* and *Urticina* (Ritte and Pashtan, 1982; Beaumont and Beveridge, 1984; Solé-Cava et al., 1985)—are marine invertebrates. Nevo (1978) lists 14 (out of 243) studies in which heterozygosity values exceeded 0.2: one of these is on a plant, seven are on insects (mainly *Drosophila*), and six on marine invertebrates (Ayala et al., 1973, 1975; Manwell, 1975; Campbell et al., 1975; Valentine and Ayala, 1976). If our data and those of Ritte and Pashtan (1982), Beaumont and Beveridge (1984), and Solé-Cava et al., (1985) are included, most species known to have heterozygosities greater than 0.2 and all those considerably higher than 0.3 appear to be marine invertebrates. This is surely remarkable in that only a minority (29/243 from table 1 in Nevo, 1978) of the vast number of

published estimates of heterozygosity are for marine invertebrate species.

Note, too, that among the marine invertebrates crustaceans on average appear to account for the lowest levels of heterozygosity (see Nelson and Hedgecock, 1980), whereas more primitive organisms with limited mobility—such as actiniarians (Bucklin, 1985; Solé-Cava et al., 1985), brachiopods (Hammond and Poiner, 1984), and mollusks (Ritte and Pashtan, 1982; Beaumont and Beveridge, 1984)—account for the highest heterozygosities.

Selander and Kaufman (1973) have suggested that heterozygosity may be inversely correlated with both the mobility and the degree of complexity of an organism. They note that more mobile species are better able to avoid environmental fluctuations while more complex organisms may be expected to have a greater capacity for homeostatic control. Therefore the enzymes of such species will generally have less need for genetic variability, which would enable the enzymes to function successfully under more widely fluctuating internal environmental conditions. This idea is supported, for example, by a negative correlation between vagility and heterozygosity in decapod crustaceans (Nelson and Hedgecock, 1980). Moreover, there is evidence among teleosts, that demersal, specialized, and less mobile (i.e., narrow niche) species generally show higher levels of genetic variation than do pelagic and more generalist species (Smith and Fujio, 1982).

Thus, heterozygosity could be related to environmental grain (i.e., the way one species "sees" the environment; Levins, 1968), as suggested by Valentine (1976) and supported, with slight modifications, by Nevo (1978), Nelson and Hedgecock (1980), and Smith and Fujio (1982). Many benthic invertebrates seem to have adopted a "coarse grain" strategy; that is, they experience the environment as a mosaic of niche opportunities rather than a homogeneous entity. For example, marine invertebrates can be very specialized when it comes to the microhabitat selected by the larva during settlement (Campbell, 1974; Shroeder and Hermans, 1975). In groups where the larval settlement is less specific, as observed in the Porifera (Fell, 1974; Sará and Vacelet, 1973; Bergquist, 1978), differential larval mortality is another way of producing differential microhabitat colonization (Fell, 1974). In a heterogeneous environment this differential mortality could produce a selective pressure for diversification of genotypes specialized to particular microclimatic conditions and thus could lead to higher levels of gene polymorphism in the population as a whole.

Nonselectionist "neutralist" hypotheses are also available to explain the levels of heterozygosity in natural populations (Kimura, 1968, 1983). According to neutralist concepts, population size, genetic drift, and divergence time are expected to be the main determinants of levels of

Table 2. Genetic variation in five sponge species (ni = mean number of individuals analyzed per locus, nl = number of loci analyzed, $P_{(0.95)}$ = proportion of polymorphic loci in the sample, Ho = mean observed heterozygosity, He = mean expected heterozygosity; $\bar{x}$ and σ = mean and standard deviation of $P_{(0.95)}$, Ho and He over all species studied)

Species	ni	nl	$P_{(0.95)}$	Ho	He
Halichondria panicea	18	16	0.688	0.227	0.234
Mycale macilenta	7	18	0.500	0.246	0.189
Suberites ficus A	13	18	0.667	0.179	0.168
Suberites ficus B	11	18	0.611	0.215	0.190
Suberites subereus	16	16	0.750	0.365	0.335
$\bar{x}$			0.643	0.265	0.223
σ			0.094	0.071	0.067

gene polymorphism. Precise estimates of population size are not available for the sponges studied. *Halichondria panicea* is very common under a wide range of ecological conditions in all temperate seas and probably represents large populations (Vethaak et al., 1982); *Suberites* cf. *ficus A* and *B* are normally epizoic upon the shells of *Chlamys* spp. (Bloom, 1975; Solé-Cava et al., 1985). *Chlamys* are fished commercially and have large populations in the Irish Sea and elsewhere, and therefore population sizes for these two *Suberites* species may also be large (note that the species of *Chlamys* that act as hosts to *Suberites* are among the molluscan species with highest heterozygosities; Beaumont and Beveridge, 1984). *S. subereus* and *Mycale macilenta*, however, seem to have far smaller populations, at least in the Irish Sea (Solé-Cava and Thorpe, unpublished results), yet their levels of heterozygosity are as high as those observed for the other sponge species (Table 2).

According to the neutralist hypothesis, levels of heterozygosity may be related to the phylogenetic "age" (Soule, 1972) and to the "conservativeness" (Gorman and Kim,

1977) of the taxonomic group–old and conservative groups being expected to show higher levels of genetic variation than younger or more speciose ones. In fact, if the average expected heterozygosities are plotted against levels of polymorphism for all the groups of organisms cited by Nevo (1978), we find that the more "primitive" groups tend to show the highest levels of gene variation (Figure 1). However, we should not overlook the possibility that heterozygosity is related to other factors as well, such as differences in the size or homeostatic efficiency of each group.

Conclusions

As can be seen from the above discussion and some recent reviews (Nevo, 1978; Smith and Fujio, 1982; Burton, 1983; Kimura, 1983), various, often conflicting, hypotheses, have been proposed to explain observed variations in the levels of naturally occurring genetic polymorphism. Much can be said for and against each of these hypoth-

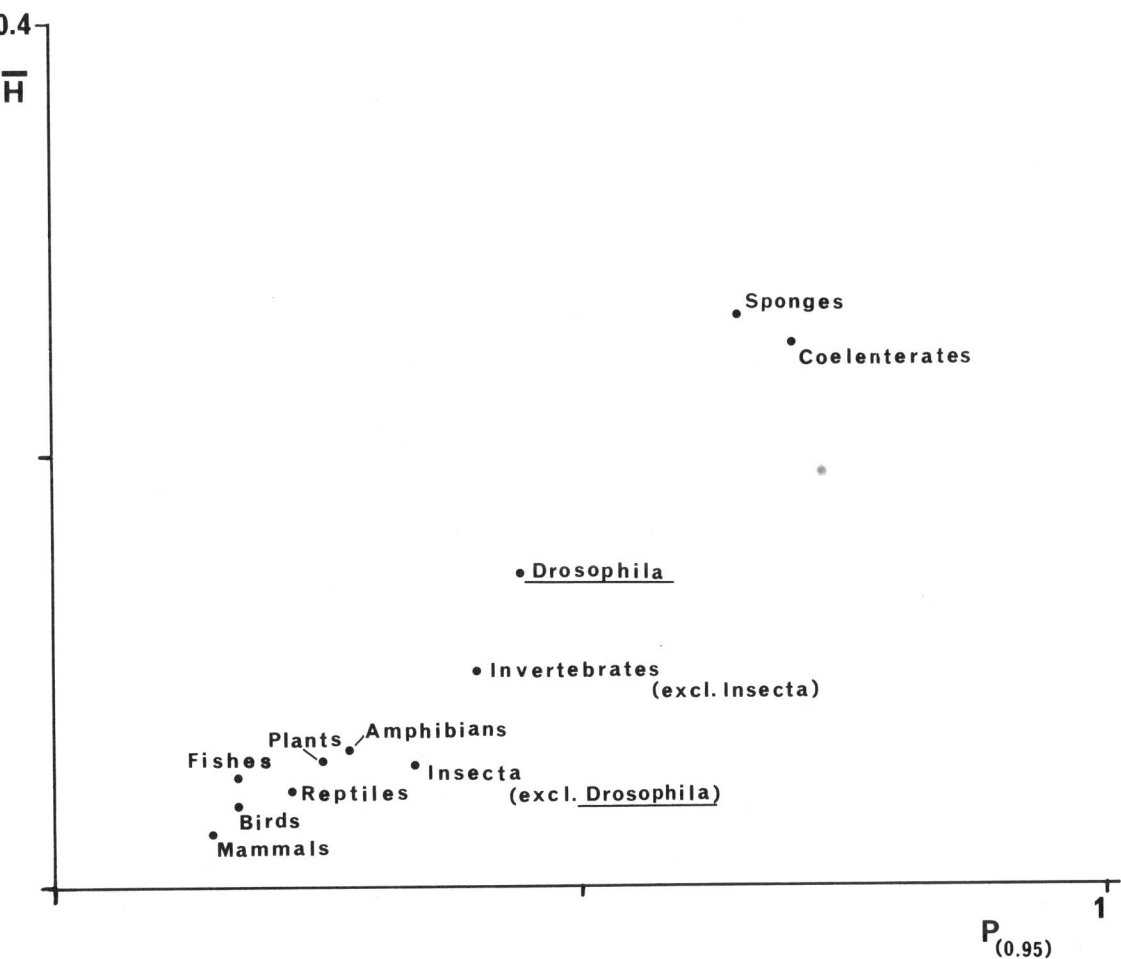

Figure 1. Levels of genetic variation, expressed as a percentage of polymorphism *(P)* and average heterozygosity *(H̄)* in different groups of organisms. Data obtained from Nevo (1978) and Solé-Cava et al. (1985).

eses, but it is very difficult, because of the nature of the problem, to test each suggestion objectively. In the case of marine sponges, a great deal of work still needs to be done, particularly with regard to population dynamics and genetics, before any conclusions can be drawn concerning the likely reasons for the observed high levels of genetic variation.

Acknowledgments

The authors are most grateful to S. M. K. Stone, C. Volkmer-Ribeiro, M. Sarà, and several other participants, for their stimulating discussions during this symposium. We also thank T. A. Norton for providing facilities, and the Brazilian government for financial assistance through grant CAPES 4334/82 to A. M. Solé-Cava.

Literature Cited

Ayala, F. J. 1976. *Molecular Evolution.* Sunderland, Massachusetts: Sinauer Associates. 278 pp.

Ayala, F. J., D. Hedgecock, G. S. Zumwalt, and J. C. Ehrenfeld. 1973. Genetic Variation in *Tridacna maxima*, an Ecological Analog of some Unsuccessful Evolutionary Lineages. *Evolution*, 27:177–191.

Ayala, F. J., J. W. Valentine, D. Hedgecock, and L. G. Barr. 1975. Deep-sea Asteroids: High Genetic Variability in a Stable Environment. *Evolution*, 29:203–212.

Beaumont, A. R., and C. M. Beveridge. 1984. Electrophoretic Survey of Genetic Variation in *Pecten maximus, Chlamys opercularis, C. varia* and *C. distorta* from the Irish Sea. *Marine Biology*, 81:299–306.

Bergquist, P. R. 1984. *Sponges.* London: Hutchinson. 268 pp.

Bloom, S. A. 1975. The Motile Escape Response of a Sessile Prey: A Sponge-Scallop Mutualism. *Journal of Experimental Marine Biology and Ecology*, 17:311–321.

Bonnell, M. L., and Selander. 1974. Elephant Seals: Genetic Variation and Near Extinction. *Science*, 184:908–909.

Bucklin, A. 1985. Biochemical Genetic Variation, Growth and Regeneration of the Sea Anemone, *Metridium*, of the British Shores. *Journal of the Marine Biological Association of the United Kingdom*, 65:141–157.

Burton, R. S. 1983. Protein Polymorphism and Genetic Differentiation of Marine Invertebrate Populations. *Marine Biology Letters*, 4:193–206.

Campbell, C. A., J. W. Valentine, and F. J. Ayala. 1975. High Genetic Variability in a Population of *Tridacna maxima* from the Great Barrier Reef. *Marine Biology*, 33:341–345.

Campbell, R. D. 1974. Cnidaria. Pages 133–200 in *Reproduction of Marine Invertebrates I*, edited by A. C. Giese and J. S. Pearse. London: Academic Press.

Connes, R., J. P. Diaz, G. Negre, and J. Paris. 1974. Etude morphologique, cytologique et sérologique de deux formes de *Suberites massa* de l'étang de Thau. *Vie et Millieu*, 24:213–224.

Fell, P. E. 1974. Porifera. Pages 1–131 in *Reproduction of Marine Invertebrates I*, edited by A. C. Giese and J. S. Pearse. London: Academic Press.

Ferguson, A. *Biochemical Systematics and Evolution.* Glasgow: Blackie. 194 pp.

Gorman, G. C., and Y. J. Kim. 1977. Genotypic Evolution in tje Face of Phenotypic Conservativeness. *Abudefduf* (Pomacentridae) from the Atlantic and Pacific Sides of Panama. *Copeia*, 694–697.

Gorman, G. C., and J. Renzi. 1979. Genetic Distance and Heterozygosity Estimates in Electrophoretic Studies: Effects of Sample Size. *Copeia*, 242–249.

Hammond, L. S., and I. R. Poiner. 1984. Genetic Structure of Three Populations of the "Living Fossil" Brachiopod *Lingula* from Queensland, Australia. *Lethaia*, 17:139–143.

Harris, H., and D. A. Hopkinson. 1978. *Handbook of Enzyme Electrophoresis in Human Genetics.* Amsterdam: North Holland.

Kimura, M. 1968. Evolutionary Rate at the Molecular Level. *Nature*, 217:624–626.

———. 1983. *The Neutral Theory of Molecular Evolution.* London: Cambridge University Press. 367 pp.

Levins, R. 1968. *Evolution in Changing Environments.* Princeton: Princeton University Press. 120 pp.

Lewontin, R. C. 1974. *The Genetic Basis of Evolutionary Change.* New York: Columbia University Press. 346 pp.

Manwell, C. 1975. Enzyme Variability in the Protochordate *Amphioxus*. *Nature*, 258:606–608.

Nei, M. 1975. *Molecular Population Genetics and Evolution.* Amsterdam: North Holland. 288 pp.

———. 1978. Estimation of Average Heterozygosity and Genetic Distance from a Small Number of Individuals. *Genetics*, 89:583–590.

Nei, M., and A. K. Roychoudbury. 1974. Sampling Variances of Heterozygosity and Genetic Distances. *Genetics*, 76:379–390.

Nelson, K., and Hedgecock, D. 1980. Enzyme Polymorphism and Adaptive Strategy in the Decapod Crustacea. *Anerican Naturalist*, 116:238–280.

Nevo, E. 1978. Genetic Variation in Natural Populations: Patterns and Theory. *Theoretical Population Biology*, 13:121–177.

Powell, J. R. 1975. Protein Variation in Natural Populations of Animals. *Evolutionary Biology*, 8:79–119.

Ritte, U., and A. Pashtan. 1982. Extreme Levels of Genetic Variability in Two Red Sea *Cerithium* Species (Gastropoda: Cerithidae). *Evolution*, 36:403–407.

Sarà, M., and J. Vacelet. 1973. Ecologie des démosponges. Pages 462–576 in *Biologie des Spongiaires*, edited by C. Lévi and N. Boury-Esnault. Colloques Internationaux du C.N.R.S. 291. Paris: Centre National de la Recherche Scièntifique.

———. 1975. Annelida: Polychaeta. Pages 1–214 in *Reproduction of Marine Invertebrates III*, edited by A. C. Giese and J. S. Pearse. London: Academic Press.

Selander, R. K. 1976. Genetic Variation in Natural Populations. Pages 21–45 in *Molecular Evolution*, edited by F. J. Ayala. Sunderland, Mass.: Sinauer Associates.

Selander, R. K., and D. W. Kaufman. 1975. Genetic Structure of Populations of the Brown Snail *(Helix aspersa)*, I—Micrographic Variation. *Evolution*, 29:385–401.

Shaw, P. R., and R. Prasad. 1970. Starch Gel Electrophoresis of Enzymes—A Compilation of Recipes. *Biochemical Genetics*, 4:297–320.

Smith, P. J., and Y. Fujio. 1982. Genetic Variation in Marine Teleosts: High Variability in Habitat Specialists and

Low Variability in Habitat Generalists. *Marine Biology*, 69: 7–20.

Solé-Cava, A. M., and J. P. Thorpe. 1986. Genetic Differentiation between Morphotypes of the Marine Sponge *Suberites ficus* (Demospongia: Hadromerida). *Marine Biology*, 93:247–254.

Solé-Cava, A. M., J. P. Thorpe, and J. G. Kaye. 1985. Reproductive Isolation with Little Genetic Divergence between *Urticina* (= *Tealia) felina* and *U. eques* (Anthozoa: Actiniaria). *Marine Biology*, 85:279–284.

Soule, M. 1972. Phenetics of Natural Populations III. Variations in Insular Populations of a Lizard. *American Naturalist*, 106: 429–446.

Soumalainen, E., and A. Saura. 1973. Genetic Polymorphism and Evolution in Parthenogenetic Animals. I. Polyploid Curculionidae. *Genetics*, 74:489–508.

Urbaneja, M., and A. L. Lin. 1981. A Preliminary Study on the Isozyme Patterns and Taxonomy of Tropical Sponges. *Comparative Biochemistry and Physiology*, 70B:367–373.

Valentine, J. W. 1976. Genetic Strategies of Adaptation. Pages 78–94 in *Molecular Evolution*, edited by F. J. Ayala. Sunderland, Mass.: Sinauer Associates.

Valentine, J. W., and F. J. Ayala. 1976. Genetic Variability in Krill. *Proceedings of the National Academy of Sciences (USA)*, 73:658–660.

Vethaak, A. D., R. J. A. Cronie, and R. W. M. van Soest. 1982. Ecology and Distribution of Two Sympatric, Closely Related Sponge Species, *Halichondria panicea* (Pallas, 1766) and *H. bowerbanki* Burton, 1930 (Porifera, Demospongiae), with Remarks on Their Speciation. *Bijdragen tot de Dierkunde*, 52(2):83–102.

Ward, R. D., and J. A. Beardmore. 1977. Protein Variation in the Plaice *(Pleuronectes platessa). Genetic Researcch*, 30:45–62.

MICHELE SARA
Istituto di Zoologia
Universitá di Genova
via Balbi 5, 16126 Genova, Italy

Divergence between the Sympatric Species *Tethya aurantium* and *Tethya citrina* and Speciation in Sponges

Abstract

Tethya aurantium (Pallas) and *Tethya citrina* Sará and Melone (Demospongiae: Hadromerida) live sympatrically at many locations along the Mediterranean and North European coasts. The two species are clearly distinguished by their external features—notably color (orange-red in *aurantium*, pale yellow in *citrina*), surface, and consistency—and show a high degree of morphological divergence in the structure of the cortex and in its spiculation. The two species differ also in their ecological distribution and requirements. The divergence in cortical structure and spiculation may be readily explained by their adaptation to different ecological requirements. No sign of hybridization between the sympatric populations of the two species has been found. At Palese (Bari), *T. citrina* shows a slight but highly significant delay in the maturation time in comparison with *T. aurantium*. A preliminary electrophoretic allozyme analysis shows considerable genetic distance between the species and therefore a high divergence time (at least 7–8 million years).

The parapatric ecologically distinct distribution of the two species and the likely adaptive significance of the main morphological divergence suggests that parapatric speciation occurred in the distant past. It is hypothesized that *Tethya citrina* evolved from *Tethya aurantium* through mutational changes in the regulatory gene system controlling the morphogenesis of the cortex and the contemporaneous opening of new suitable niches for *citrina*. Subsequently, reproductive barriers arose and genetic divergence was accumulated, as indicated by allozyme analysis.

About 6000 living species are known in the phylum Porifera and their taxonomic status is based mainly on minute spicular details. However, difficulties in delimiting species on this basis have led some researchers—such as

Vosmaer (1933–1935), who worked on Demospongiae, and Burton (1963), who worked on Calcarea—to lump species together excessively. Therefore it has become necessary to add new features (Griessinger, 1971), including histological and cytological ones (Simpson, 1968). Additional characters may also be found in the chemical composition of species (Lee and Gilchrist, 1985) although the chemotaxonomy of sponges is mainly directed at elucidating their general phylogeny and higher taxa (Bergquist and Wells, 1983). At the heart of the species problem in sponges, however, as in other groups, is the need not only to identify sponges correctly, but also to clarify the significance and value of species from a biological point of view. Spongologists are now beginning to take the first steps in this direction through the study of life histories, population dynamics, and ecological behaviour, along with immunological research and transplantations. The more direct study of genetic structure and divergence through allozyme electrophoretic patterns has just begun. The biological characterization of species is directly connected with the speciation problem. Unfortunately, up to now only a few researchers have looked at the evolutionary processes involved (Sarà, 1956a,b, 1959; Hartman, 1957; Wiedenmayer, 1974; Chen, 1976; Vethaak et al., 1982). New data on the morphological, biological, and ecological divergence between *Tethya aurantium* and *T. citrina*, species that live sympatrically along the Mediterranean and other coasts of Europe, may shed some light on these species and speciation problems in sponges.

Material and Methods

The cortical morphology and spiculation of *Tethya aurantium* and *T. citrina* were studied in slides and sections from my collection and G. Pulitzer-Finali's collection of Mediterranean specimens along with specimens in the British Museum of Natural History (BMNH) from northern European and other waters. Micraster shape and spinulation were studied with the aid of a scanning electron microscope. SCUBA was used to investigate the ecological distribution of the two species and data on depth, exposure to light, hydrodynamism, sediment, and epiphyte cover were obtained for both species at the Adriatic station of Palese (Bari), in the Ionian lagoon of Porto Cesareo (Lecce), and in the South Tyrrhenian lagoon of Stagnone (Marsala, Sicily). At this station, temperature and salinity data were also collected. Previously published data (Scalera-Liaci et al., 1971) on the reproductive cycles of the two species were reexamined and submitted to a X^2 test in order to detect statistically significant differences in their maturation times. The allozyme pattern at twelve gene loci of two geographically different populations of each species was analyzed by means of cellulose acetate and starch gel electrophoresis.

Results

Tethya aurantium (Pallas) and *T. citrina* Sarà and Melone are two clearly different species of Demosponges (Sarà and Melone, 1965) but they were lumped together under *T. aurantium* (and its synonyms) for more than a century and were not separated until recently. In view of their wide divergence and the absence of true intermediates, this confusion is indeed astonishing, even though it may be partly explained by the lack of field studies in the past.

Their external appearance, as judged by color, dimensions, consistency and surface structure, is very different. *Tethya aurantium* is orange-red, papillate with rounded papillae, and generally larger, and firmer than *T. citrina*, which is yellowish or greenish, and very often drab due to sediment covering. Furthermore, the surface of *T. citrina* lacks clear papillae and is smooth or conulose. These as well as the spicular, reproductive, ecological, and other differences are summarized in Table 1, which compares Mediterranean populations of the two species living sympatrically in shallow waters near Bari.

Spicular divergence is evident from megaster and micraster shape and distribution in the sponge. The megaster of *Tethya aurantium* has the shape of a spheraster in which the ratio of ray length to centrum diameter is < 1, while that of *T. citrina* has the shape of an oxyspheraster with a ratio > 1. Variability for this character is not overlapping in all the Mediterranean populations hitherto analyzed, and the distinction is always correlated with sponge color. This constant relationship has been the main diagnostic feature used to identify the two species (Pulitzer-Finali, 1983). In addition, the megasters have a different pattern of distribution, a function of the development of the cortex in the two species, which is thicker and more differentiated in *T. aurantium*, and thinner and less distinct in *T. citrina*. The spherasters of *T. aurantium* are arranged in many layers, building a strong armor in the deeper cortical region, whereas the oxyspherasters of *T. citrina* are distributed in a single layer at the boundary between the cortex and the choanosome. The two species also have different micrasters. In *T. aurantium* a cortical variable (oxyaster–chiaster–tylaster) micraster with a mean diameter of 12 μm may be distinguished from a choanosomal oxyaster with a mean diameter of 20 μm, wheras in *T. citrina* cortical and choanosomal micrasters are alike and more similar to the cortical type of *T. aurantium*.

Further, the species differ in their sexual and asexual phases. When the sexual cycles described by Scalera-Liaci et al. (1971) are reexamined for sympatric Apulian populations, it is evident that despite some overlapping, the two species are distinct, at least in terms of oogenesis, with *Tethya aurantium* maturing a little earlier. A probability test is highly significant ($p < 0.05$) in this regard. The existence of efficient reproductive barriers between

Table 1.   Differential characters in *Tethya aurantium* and *Tethya citrina* from sympatric shallow-water populations, Adriatic Sea, Bari

Characters	*T. aurantium*	*T. citrina*
Morphological characters		
Sponge diameter	10.0–45.5 mm (30.5 mm)	5.5–26.0 mm (15 mm)
Cortex thickness	1.5–3.5 mm (2.6 mm)	1.0–1.5 mm (1.2 mm)
Cortex structure	With a thick collagenous layer	With a thin collagenous layer
Surface	Papillate	Smooth or conulose
Color	Red orange	Yellowish or greenish
Megasters		
Shape	Spherasters	Oxyspherasters
Diameter	16.2–93.6 μm 53.6 μm)	18.0–64.8 μm (42.9 μm)
Ratio, ray length: centrum diameter	0.4–0.8 (0.5)	0.6–1.4 (1.1)
Rays, number	21–30 (25)	10–20 (15)
Distribution in the cortex	Multilayered	Monolayered
Micrasters	Two categories: (1) Cortical, variable between oxyaster, chiaster, and tylaster shape, 10–15 μm (12 μm); (2) choanosomal, oxyaster, 18–21 μm (20 μm)	One category:In cortex and choanosome the same type, variable between oxyaster, chiaster, and tylaster, but more oxyaster–like than in aurantium, 10–15 μm (12 μm)
Pigments	Echinenone (4–cheto β–carotene)	β–carotene
Reproduction		
Oogenesis (timing)	Months VII–X	Months VIII–XI
Bud dimensions	1.5–4.0 mm	0.25–1.0 mm
Bud number (maximum)	55	250
Ecology		
Depth	1–4 m	0–1 m
Distribution	Various, also sunny positions in the upper infralittoral	Sheltered locations at the mediolittoral–infralittoral boundary
Sediment cover	None or light	Heavy
Symbionts	*Phormidium spongeliae* (cyanobacteria, in the cortex)	No algal symbiont

Note: Measurements taken from 166 specimens of *T. aurantium* and 215 specimens of *T. citrina;* revised from Sarà and Melone, 1965

the species is indicated by the fact that even in the three locations (Apulian shore and the lagoons of P. Cesareo and of Marsala) where sympatric populations with some overlapping distribution have been studied, no intermediates suggesting hybridization have been found among hundreds of specimens. Buds are smaller and more numerous in *T. citrina* but the biological consequences of this difference and whether there is also a difference in timing is presently unknown.

Significant differences exist in the ecological distribution and requirements of the two species. Up to now these differences have been studied only in the Mediterranean populations. In this area *Tethya aurantium* is commonly found only between 1 m and 40 m, while *T. citrina* is common between 0 m and 1 m, but it is also the only *Tethya* recorded between 40 m and 135 m. In shallow waters both species may occasionally occur at the same sites, demonstrating the genotypic basis of their divergences in external appearance. In general, however, the two species are distinguished by their microhabitats, which reflect differences in their ecological niches. The ecological distribution of the two species has been studied in detail for

sympatric populations living in depths of 0–3 m along the shore near Bari (on the Adriatic coast) and in a lagoon near Marsala (on the Sicily Channel). *T. aurantium* occurs in a slightly deeper zone but is more exposed to light and to hydrodynamism than *T. citrina*, which lives in sheltered regions of the upper zone. In the lagoon of Marsala it seems more tolerant of higher temperatures and salinities than *T. aurantium*. *T. citrina*'s preference for sheltered locations is reflected in the fact that this species inhabits the deeper circalittoral stations. Furthermore, unlike *T. aurantium* it is frequently covered by sediment and epiphytes in the shallow water locations. The uneven bathymetric distribution of *T. citrina* may be interpreted by considering *T. aurantium* to be competitively dominant in the infralittoral stations. *T. citrina* is therefore excluded from these more favorable areas by *T. aurantium* and confined to refuge areas at sheltered locations in the lower mediolittoral to upper infralittoral fringe in addition to the circalittoral zone.

Examination of the BMNH slides and specimens indicates that both *Tethya citrina* and *T. aurantium* are also common in the North European seas. It is also probable that

T. norvegica (Bowerbank), up to now attributed to *T. aurantium*, would instead be a form, perhaps differentiated, of *T. citrina*, but this suggestion needs to be corroborated by further inquiry. *T. citrina* also occurs in the Canary Islands. *T. aurantium* has a wider distribution, which extends into tropical seas. Direct examination of this species confirms its occurrence in the Caribbean Sea, the northwestern coasts of Africa and the Cape Verde Islands, Red Sea, Indian Ocean (Sri-Lanka and Seychelles). Further work needs to be done at the Australian and New Zealand locations. South African, Californian, and Japanese reports probably refer to other *Tethya*.

Preliminary data (Sarà et al., 1989) on the allozyme pattern have been obtained for twelve different loci in 14 specimens of *Tethya aurantium* and 12 specimens of *T. citrina* in both stations from the two different populations of Palese (Bari) and Marsala (Sicily). The overall genetic distance between the species, is $I = 1.400$, that between the two *T. citrina* populations of Palese and Marsala is $I = 0.244$, while that between the two populations of *T. aurantium* is $I = 0$. These distances indicate a very high divergence time, particularly if one considers that the loci analyzed were mainly of the slow-evolving, one-substrate type. According to Nei (1972), divergence times should be at least 7,345,000 years between the two species and at least 1,450,000 years between the two populations of *T. citrina*. The two populations of *T. aurantium* show no divergence in this respect.

Discussion

The data on the different cortical morphology of *Tethya aurantium* and *T. citrina*, spiculation, ecological distribution and requirements, and high divergence time, as determined by allozyme electrophoretic analysis, suggest that parapatric speciation is responsible for their separation. It may be that a speciation event in ancient times caused *T. citrina* to be separated from *T. aurantium*. However, it seems unlikely that *T. aurantium* originated from *T. citrina*, since the latter may be considered more ecologically specialized and less tolerant than *aurantium*. Moreover, *T. aurantium* has a wider geographical distribution and is more similar to the majority of species in *Tethya*.

The hypothesis of parapatric speciation is supported by the geographical and ecological distribution of the two species. They live sympatrically at various Mediterranean locations and in the North European seas. At the Mediterranean shallow-water locations for which field data are available, this sympatry is mainly of the parapatric type, and the two species are distributed along contiguous bands that are bathymetrically and ecologically different. Moreover, within these bands they live in different positions and consequently show a different tolerance toward some ecological factors: *T. aurantium* is more tolerant of light and hydrodynamism, and *T. citrina* of sedimen-

tation. The clear ecological differences between the species make it seem unlikely that secondary parapatric distribution is due to migration after speciation in allopatry. Furthermore, the hypothesis of parapatric speciation is consistent with the occurrence in *Tethya* of asexual reproduction by budding besides the sexual reproduction. Through the establishment of clones the new genotypes may rapidly spread, favoring their diffusion in the new available niche before the onset of genetic isolating barriers. This diffusion may also be favored by some other biological features proper to shallow-water sponges, such as the short-distance propagation of larvae and propagula, and by the ability of habitat selection during the colonization of substrata which is well known for many marine benthic larvae.

A sympatric or microallopatric speciation has been suggested for *Leucosolenia* at Roscoff and *Clathrina* at Naples, (Sarà, 1956a,b; 1959) and for the clionids at Rovigno (Hartman, 1957; Wiedenmayer, 1974). A sympatric speciation, likely of the stasipatric type through poliploidy, has also been considered for some morphologically similar but ecologically different species of *Halisarca* (Chen, 1976). On the other hand, Vethaak et al. (1982) suggest for the species pair *Halichondria panicea* and *H. bowerbanki* along the European and North American coasts, which have sympatric distribution but are not clearly ecologically different, an allopatric type of speciation with secondary sympatry due to reciprocal migration. In the parapatric type of speciation (White, 1978) in which the distributional ranges of the species are contiguous, some spatial barriers, even if slight, may be operating to prevent gene flow in a different degree according to the diffusion capabilities of the different species. Parapatric speciation has been considered by Murray and Clarke (1980) for scarcely movable animals such as the terrestrial pulmonates. There is little evidence for the diffusion of parapatric ways of speciation in the sessile and sedentary benthos mainly because little work has been done on speciation in these groups.

A second important point to note with respect to the parapatric hypothesis of speciation discussed here is the adaptive value of the main morphological divergence between *Tethya aurantium* and *T. citrina*. The two species can be distinguished by the different development of the cortex in relation to the choanosome. This divergence is accompanied by a morphofunctional correlation of many distinctive characters that are likely to be adaptively related to the ecological requirements of each species. The more developed cortex of *T. aurantium* correlated with the spheraster shape and pattern and the micraster differentiation seems a good adaptation toward greater values of light and hydrodynamism. On the other hand, the thinner and less structured cortex of *T. citrina* gives the sponge a greater plasticity with greater possibility for expanding and contracting the body, which favors its pumping activity, and this may be considered an advantage for a species that lives in a sheltered environment with low hydrodynamism and abundant sedimentation. It is interesting to note that *T. crypta* (Laubenfels), a tropical species that lives in sites comparable in certain respects to those of *T. citrina,* shows a series of parallel modifications such as the oxyspheraster shape of its megasters, an indistinct cortex, great plasticity, and a covering with sediment and epiphytes (Wiedenmayer, 1977).

The adaptive significance of the divergence between *Tethya aurantium* and *T. citrina* suggests that this divergence was affirmed and diffused under the influence of selective pressures. At the same time, the relationship between the distinctive morphological characters—which may be largely connected with the different development of the cortex in the two species—could suggest a saltational, not gradualistic origin of this divergence. This may be due to mutations in major regulatory genes that have had an effect on morphogenesis (Arthur, 1984), bringing about underdevelopment of the cortex in *T. citrina,* perhaps through heterochronic changes, as indicated in the theory of punctuated equilibria (Gould, 1977). Finally, the high genetic distance and the following very high divergence time of less than 7,350,000 years between the species indicate that the speciation event is very old. It is important to conceptually separate, even if they may eventually coincide, the morphological adaptive divergence from the electrophoretically detected allozyme divergence on which the divergence time is based (Carson, 1975; Nei et al., 1983), because the latter is largely stochastic and selectively neutral.

In conclusion, an evolutionary scenario can be envisaged in which *Tethya aurantium* (or a *T. aurantium*-like ancestor) with a cortex like that of most *Tethya* species gave rise to a *T. citrina* (or a *T. citrina*-like form) through a genotypic change favoring the reduction of the cortex. The latter could occupy suitable ecological niches, newly opened in relation to climatic changes, for example. The morphological, ecological, and reproductive study of the North European populations of *T. citrina* and *T. aurantium,* in comparison with the Mediterranean ones, could clarify whether these speciation events occurred, as suggested by the more northern diffusion of *T. citrina*. In this case, the colonization of the Mediterranean by *citrina* could be considered a successive event related to Plio-Pleistocenic climatic changes, and the present parapatric distribution of the two species in this sea, with *citrina* populations disjoint at the periphery of the *T. aurantium* zone, could perhaps be considered a relict one. This parapatric speciation hypothesis would explain the different ecological distribution and adaptive morphological divergence between *T. aurantium* and *T. citrina*. It does not exclude the possibility that these species, in view of their remote origin, could have been successively differentiated in an allopatric way that gave rise to different races and perhaps species (in

which case *T. aurantium* and *T. citrina* should be considered groups of species or superspecies). An indication of this possibility is the high genetic distance between the two geographically separated populations of *T. citrina* at Palese and Marsala with a divergence time of at least 1,450,000 years. In this regard it would be interesting to test the populations of *T. citrina* inhabiting bottoms at depths below 40 m. For *T. aurantium,* there is no genetic distance between the populations of Palese and Marsala, but these locations are relatively close. On the other hand, *T. aurantium* shows a very wide range, including tropical seas. Within this range—for example in the Red Sea populations—considerable spicular differences have been observed.

Conclusions

Tethya aurantium and *T. citrina* show a sympatric distribution in various Mediterranean and North European locations. They differ widely in their external features such as color (*T. aurantium* is orange-red; *T. citrina* is yellowish), surface (*T. citrina* is smoother), and consistency (*T. citrina* is more plastic). The main structural difference is that the cortex is more developed in *T. aurantium* and spiculation is different.

In the surveyed shallow-water Mediterranean stations, the two species inhabit bathymetrically distinct habitats (*Tethya aurantium* occurs more frequently between 1 and 40 m, *T. citrina* between 0 and 1 m, and also beyond 40 m) with different conditions of light, hydrodynamism, and sedimentation (*T. aurantium* lives in more exposed sites for light and hydrodynamism, *T. citrina* in more sheltered ones or is covered by sediment).

The different development of the cortex and the related spicular characters seem good adaptations of the two species to their respective niches.

The two species show no intermediates, even when their populations occasionally overlap, as at Palese. An analysis of the reproductive cycle of the two species in this locality shows that a slight but statistically highly significant difference exists in their maturation times.

Preliminary data of an electrophoretic study on the allozyme patterns of two populations of each species show that the two species are separated by a very high divergence time (at least 7,350,000 years). Furthermore, two geographically separated populations of *Tethya citrina* are well separated genetically, with a divergence time of at least 1,450,000 years.

A parapatric hypothesis of speciation by which *Tethya aurantium* and *T. citrina* separated in the remote past has been put forth. *T. citrina* originated from *T. aurantium* through mutations affecting the development of the cortex and the characters of the relative spiculation. This change in the architecture of the sponge allowed *T. citrina* to colonize a new ecological niche to which it was better adapted than *T. aurantium*. Isolating mechanisms then arose, even in the parapatric situation, favored also by the peculiar reproductive biology of the two species. Subsequently, considerable genetic divergence occurred, as indicated by the allozyme patterns. Successive intraspecific differentiation by allopatric separation cannot be discounted, but requires further documentation.

Literature Cited

Arthur, W. 1984. *Mechanisms of Morphological Evolution*. Chichester: J. Wiley & Sons. 275 pp.

Bergquist, P. R., and R. J. Wells. 1983. Chemotaxonomy of the Porifera. The Development and Current Status of the Field. Pages 1–50 in *Marine Natural Products*, edited by P. J. Scheuer. New York: Academic Press.

Burton, M. 1963. *A Revision of the Classification of the Calcareous Sponges*. London: British Museum (Natural History). 693 pp.

Carson, H. L. 1975. The Genetics of Speciation at the Diploid Level. *American Naturalist*, 109:83–92.

Chen, W. T. 1976. Reproduction and Speciation in *Halisarca*. Pages 113–137 in *Aspects of Sponge Biology*, edited by F. W. Harrison and R. R. Cowden. New York: Academic Press.

Griessinger, J. M. 1971. Etude des réniérides de Mediterranée (Démosponges, Haplosclérides). *Bulletin, Museum National d'Histoire Naturelle (Paris)*, 3:98–180.

Gould, S. J. 1977. *Ontogeny and Phylogeny*. Cambridge, Massachusetts: Harvard University Press. 501 pp.

Hartman, W. D. 1957. Ecological Niche Differentiation in the Boring Sponges (Clionidae). *Evolution*, 11:294–297.

Lee, W. L., and Q. M. Gilchrist. 1985. Carotenoid Patterns in 29 Species of Sponges in the Order Poecilosclerida (Porifera: Demospongiae): A Possible Tool for Chemosystematics. *Marine Biology*, 86:21–36.

Murray, J., and B. Clarke 1980. The Genus *Partula* in Moorea: Speciation in Progress. *Proceedings, Royal Society of London B*, 211:83–117.

Nei, M. 1972. Genetic Distance Between Populations. *American Naturalist*, 106:283–292.

Nei, M., T. Maruyama, and I. Wu Chung. 1983. Models of Evolution of Reproductive Isolation. *Genetics*, 103:557–559.

Pulitzer-Finali, G. 1983. A Collection of Mediterranean De-mospongiae (Porifera) With, in Appendix, a List of the Demospongiae Hitherto Recorded for the Mediterranean Sea. *Annali, Museo Civico di Storia Naturale di Genova*, 84:445–621.

Sarà, M. 1956a. Aspetti genetici ed ecologici dell'ibridazione naturale fra differenti specie di *Leucosolenia* a Roscoff. *Bollettino di Zoologica*, 23:149–162.

———. 1956b. Variabilitá della specie ed ecologia nei Poriferi. *Bollettino di Zoologica*, 23:65–78.

———. 1959. Sulla coesistenza di specie strettamente affini di Poriferi litorali. *Bollettino di Zoologia*, 26:1–8.

Sarà, M., and N. Melone. 1965. Una nuova specie del Genere *Tethya*, *T. citrina* sp. n. del Mediterraneo (Porifera, Demospongiae). *Atti Societá Peloritana di Scienze Fisiche, Matematiche, e Naturali*, Supplement, 11:123–138.

Sarà, M., P. Mensi, R. Mancone, G. Bavestrello, and E. Balletto. 1989. Genetic Variability in Mediterranean Populations of *Tethya* (Porifera: Demospongiae). Pages 293–298 in *Reproduction, Genetics and Distributions of Marine Organisms*, edited by J. S. Ryland and P. A. Tyler. Fredensborg: Olsen and Olsen.

Scalera-Liaci, L., M. Sciscioli, O. Papa, and E. Lepore. 1971. Raffronto fra i cicli sessuali di *Tethya aurantium* (Pallas) Gray e *Tethya citrina* Sará e Melone (Porifera, Hadromerina); Analisi Statistica. *Atti, Societá Peloritana di Scienze Fisiche, Matematiche e Naturali*, 17:287–298.

Simpson, T. L. 1968. The Structure and Function of Sponge Cells: New Criteria for the Taxonomy of Poecilosclerid Sponges (Demospongiae). *Peabody Museum of Natural History, Bulletin*, 25:5–141.

Vethaak, A. D., R. J. A. Cronie, and R. W. M. van Soest. 1982. Ecology and Distribution of Two Sympatric, Closely Related Sponge Species, *Halichondria panicea* (Pallas, 1766) and *H. bowerbanki* Burton, 1930 (Porifera, Demospongiae), with Remarks on Their Speciation. *Bijdragen tot de Dierkunde*, 52(2): 83–102.

Vosmaer, G. C. J. 1933–1935. *Sponges of the Bay of Naples. Porifera Incalcaria*. The Hague: Martinus Nijhoff.

White, M. J. D. 1978. *Modes of Speciation*. San Francisco: W. H. Freeman. 455 pp.

Wiedenmayer, F. 1974. Recent Marine Shallow-Water Sponges of the West Indies and the Problem of Speciation. *Verhandlungen der Naturforschenden Gesellschaft in Basel*, 84:361–375.

———. 1977. *Shallow-Water Sponges of the Western Bahamas*. Basel: Birkhäuser Verlag. 336 pp.

Toward a Phylogenetic Classification of Sponges

ROB W. M. VAN SOEST
Instituut voor Taxonomische Zoologie
Zoölogisch Museum Amsterdam, Mauritskade 57
P.O. Box 4766, Amsterdam, The Netherlands

Abstract

Basic principles of phylogenetic classification are defined and their use explained with examples from orders and families of sponges presently under study. Many genera and families, as well as higher categories currently in use are found to lack a phylogenetic basis. The cladistic approach is favored because the resulting classifications not only provide considerable information but also have predictive value.

In sponge classification (as in any other biological classification) there seem to be two schools of classifiers. One school is primarily concerned with making the organisms known and accessible to as large a public as possible, using preferably simple character differences to differentiate between the organisms. Its adherents argue that a system of classification should be easy to use and should be easily converted into identification keys. This school of classifiers has made use of so-called phenetic methods, classifying organisms into groups with overall morphological similarity. Among spongologists who represent this school are von Lendenfeld, Topsent, and de Laubenfels, although they may not be self-confessed.

The other school is primarily concerned with unraveling the presumed order in nature, in this case, in sponges. These classifiers try to analyze relationships among organisms, through which they hope to reconstruct the evolutionary process that gave rise to present-day diversity and distributions of organisms. Often this classification concurs with that of the phenetic school; however, in many important cases, overall similarities are not decisive. There are apparently two different kinds of similarity, namely, "genuine" and "false", the latter caused by processes of parallel or convergent development. Thus, the two systems of classification may disagree on certain important points. Representatives of the phylogenetic approach among spongologists appear to be Dendy, Hentschel, and Burton, and, I hope, most of the recent taxonomists.

The main advantage of the phylogenetic classification is that its systems hold more information than the phenetic systems. If two species bear the same generic name, it not only means that they show important morphological similarities, but it also implies that the two share a common ancestor in the near past and evolved by a fairly recent process of speciation. To many people, this is the only scientific approach to classification, whereas phenetic classification is considered akin to a clerk's job, to stamp collecting, or to pigeon holing. I am not inclined to go so far in my judgment, because I believe the primary aim of both systems is to come to grips with the dazzling diversity of organisms.

However, sponge systematics has long been dominated by the pheneticists, with the result that hypotheses about phylogenetic relationships are poorly developed or nonexistent. Only since the mid-1950s has higher systematics been seriously discussed by sponge biologists, although progress has been limited to attempts to corroborate the ruling Topsent–de Laubenfels system with phylogenetic theories. Notwithstanding many admirable attempts at phylogenetic reconstruction, there is still no coherent phylogenetic sponge system, either at the higher taxa levels or at the lower ones.

Recently, there has been a movement within the phylogenetic school toward applying more rigorous methods. In fact, two subschools have developed, the Evolutionary Systematists and the Cladists. I will not go into the finer distinctions between the two (which are amply described in many recent issues of *Systematic Zoology*), but suffice it to say that where both construct phylogenetic trees, the cladists do so by strict application of logical methods, making their trees highly objective and certainly more reproducible (for a useful review of the cladistic approach, see Eldredge and Cracraft, 1980). Also, the cladistic approach focuses the attention on problematic parts of the classification. It is my firm belief that sponge classification can benefit enormously from close application of the cladistic principles. If I cannot convince readers of the advantage of these methods, at least I can explain the basis for my past and future taxonomic decisions.

Definitions and Points of View
TAXA

The basis of the phylogenetic system is the monophyly of its taxa. A taxon comprises all the descendants of a common ancestor, but only these (no others). Taxa of a higher level consist of monophyletic groups of taxa of a lower level; the latter are called sister taxa.

CHARACTERS

The characters or character states (conditions) of a taxon are either inherited unchanged from its ancestor (so-called primitive or plesiomorphous characters/character states) or they are newly developed or modified (so called derived or apomorphous characters/character states). A

taxon can only be defined by its derived characters/character states. If lower taxa are to be united into monophyletic groups (higher taxa), they must share the same derived characters (which are thus primitive characters for the respective sister taxa). These shared derived characters are called synapomorphies, and they are likewise the only characters on which higher taxa may be defined. Shared primitive characters (so-called symplesiomorphies) cannot be used to group taxa into monophyletic higher taxa, although they point to common ancestry at some higher level. Two examples from sponge taxonomy will illustrate these points:

> Example 1, Cladogram of the Spirophorida–Astrophorida–Hadromerida (Figure 1): The radiate architecture is a synapomorphy for the three. This character cannot be used to differentiate between families or genera in either of the three groups. The triaene spicule is a synapomorphy for the Astrophorida + Spirophorida; it cannot be used to differentiate between families and genera (e.g., *Jaspis* and *Stelletta*).

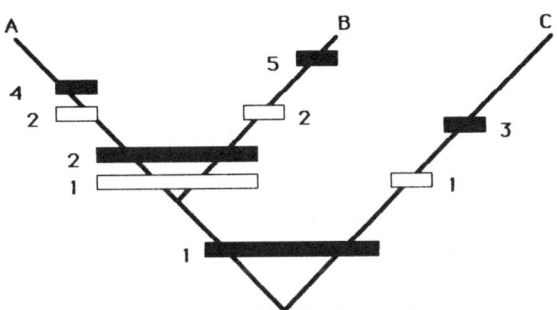

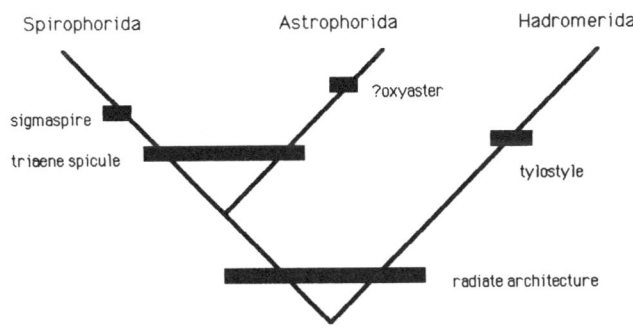

Figure 1. Demonstration of cladograms: *top*, cladogram of hypothetical higher taxon ABC, characterized by synapomorphy 1; sister taxa AB and C are characterized by synapomorphy 2 and apomorphy 3, respectively; sister taxa A and B are characterized by apomorphies 4 and 5; open rectangles symbolize (sym-) plesiomorphous conditions of characters; *bottom*, cladogram of Spirophorida-Astrophorida-Hadromerida to illustrate use of synapomorphies (all problematic sister taxa, such as Homoscleromorpha and Lithistida, have been ignored).

Example 2, *Crella* and *Pytheas* (Figure 2): Definition of the genus *Crella*: Crellidae with ectosomal acanthoxea or acanthostyles; with basal acanthostyles; without chelae. Definition of the genus *Pytheas*: Crellidae with ectosomal acanthoxea or acanthostyles; with basal acanthostyles; with chelae.

The loss of chelae could very well be a derived character, as most other Crellidae, and indeed most Poecilosclerida possess chelae. *Pytheas* is defined as is *Crella*, except that it retained the ancestral chelae. Thus, *Pytheas* cannot be defined as a monophyletic group. It is highly doubtful whether species referred to *Pytheas* share a common ancestor, that they do not also share with species of *Crella*. Numerous similar cases may be found among Poriferan genera classified by de Laubenfels and Topsent.

This is an example of the so-called "*A*/not-*A* classification" applied already by Aristotle. Not-*A* is defined as lacking the characters by which *A* can be recognized. On theoretical grounds, it can be argued that Not-*A* comprises the whole universe excepting *A*. If we want our classifications to reflect evolutionary descent, it is necessary to define each taxon with positive characters (apomorphous character).

HOW TO DETERMINE THE PRIMITIVE OR DERIVED STATE OF A CHARACTER

First, note that such a "determination" always remains a hypothesis—as is proper for a scientific effort—which may have to be abandoned if it is shown to be untenable. Aside from the useful but often dangerous reasonings, such as adaptability of certain characters, functional morphology, biogenetic laws, and the like, there are two general principles or methods that can serve to unify attempts at classification:

1. Out-group comparison. The character or its condition must be compared to that in related groups, particularly sister groups.

   Example 3, microscleres in the Haplosclerida (Figure 3): Microscleres have been widely used to define Haplosclerid genera and even families. The family Gelliidae unites all Haplosclerida with microscleres. There is also a completely artificial system of genera with and without certain microscleres. This is a typical phenetic system that can readily be used in identification keys. However, if we examine the distribution of these microscleres (sigmata, toxa, and rhaphides) within the Haplosclerida and in other ceractinomorph groups, we discover that they are widespread and thus must be considered primitive characters. For this reason, we have proposed that the use of genus

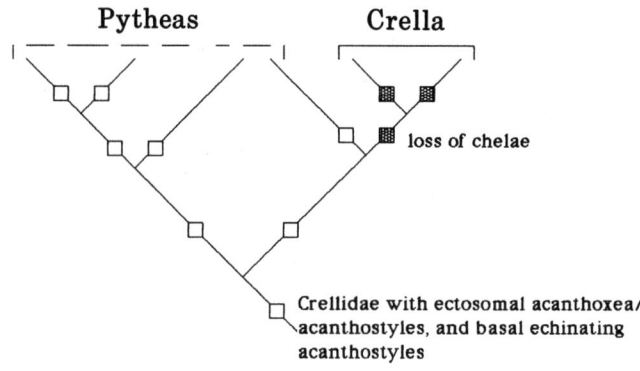

Figure 2. Hypothetical cladogram of a group of species belonging to the *Crella-Pytheas* group of the Crellidae, to illustrate the "emptiness" of the "genus" *Pytheas*. This is an example of the so-called "*A*/Not-*A* classification".

names such as *Sigmadocia*, *Orina*, and *Gellius* be abandoned until the family Haliclonidae can be revised on a cladistic basis.

2. Principle of parsimony. When in doubt, the least complicated transformation of a character is presumed to have taken place.

   Example 4, the class Sclerospongiae: This example is elaborated in an earlier paper (van Soest, 1984) in which I support Vacelet's (1985) conclusion that sclerosponges are polyphyletic and should be incorporated in the Demospongiae, within existing orders. The parsimony principle forces us to admit that it is highly unlikely that such elaborate and narrowly distributed spicules as are found in several sclerosponges (clavidiscs in *Merlia*, spirasters in *Acanthochaetetes*, and verticillate acanthostyles in ceratoporellids) have been developed in a parallel way in two separate classes; by the same token, it is unlikely that these spicules represent the primitive types that were already in existence at the time the two alleged classes developed from an unknown ancestor.

A PRIORI ASSUMPTIONS

Paleontological evidence has traditionally been given high (a priori) priority in determining the postulated evolution of a group. Without doubt, fossils provide the evidence needed to support the theory behind our phylogenetic reconstructions, which would otherwise be completely abstract formulations. However, it is extremely unlikely that the fossil specimens that have been found represent the ancestral species precisely as many earlier paleontologists have led us to believe. Fossils must be treated just like any other recent species (except that they have the added dis-

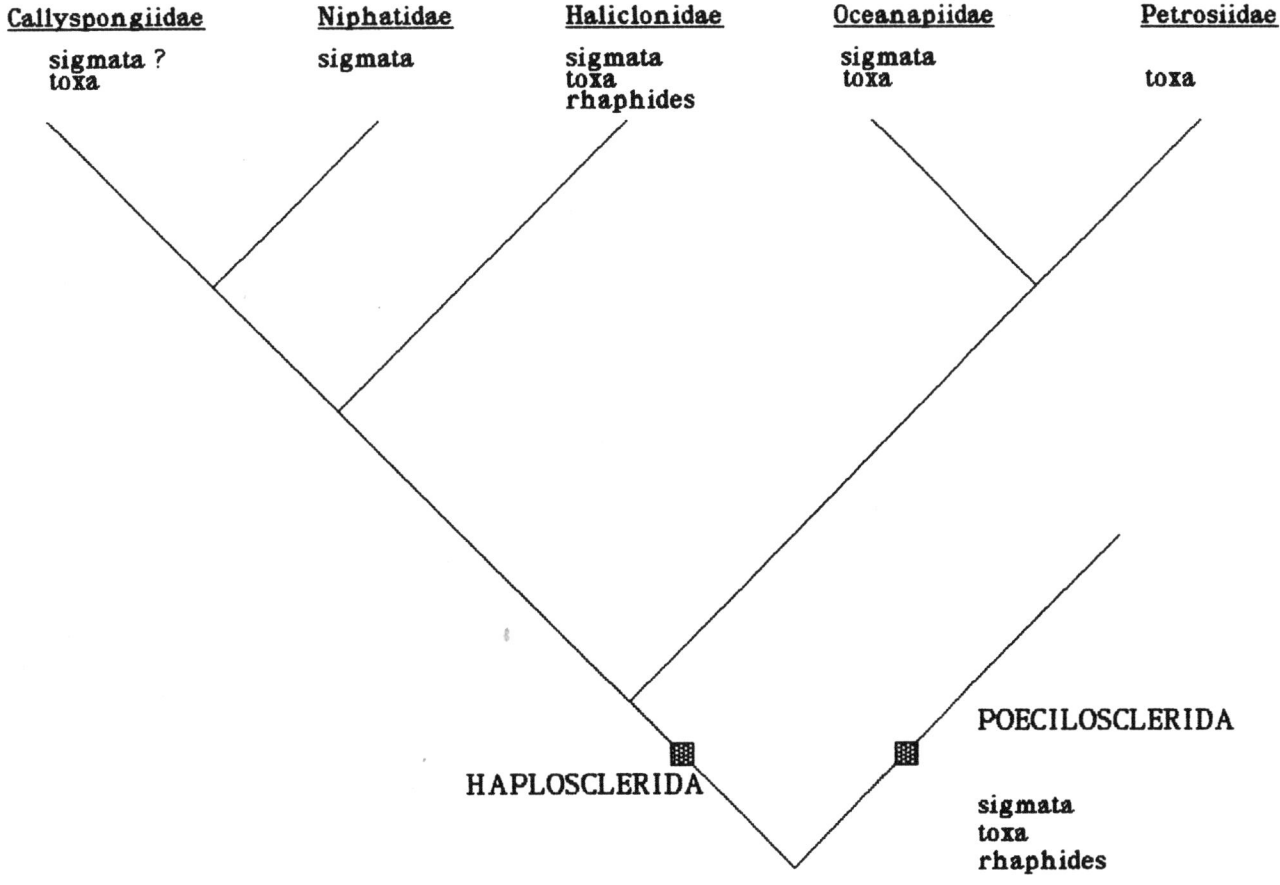

Figure 3. Cladogram of families of the order Haplosclerida s.l. and distribution of microscleres to illustrate that, according to the principle of parsimony, the presence of microscleres in haplosclerid taxa is a primitive (plesiomorphous) character.

advantage of an incompletely known morphology) and should be classified, if possible, into one and the same system.

Certain characters are said to be endowed with particular significance; they determine singly (often against the weight of combined morphological evidence) how relationships should be viewed. Such characters are biochemical compounds, serological reactions, isoenzymes, and so on. It is argued that these are nonadaptive, objectively determined characters that are not subject to interpretation. This type of a priori assumption is unjustified; the fact that we do not understand how many of these characters have evolved makes them difficult to use. They represent characters that can be used as (syn)apomorphies, however, if their distribution within the studied groups is sufficiently well known.

RANKING

The rank of a taxon is not a matter of random choice; still, to a large extent it may be subject to change, and in general it remains somewhat academic. There seems to be no

absolute way of determining rank. People tend to use morphological distance for this purpose, with the result that *Homo sapiens,* for instance, is relegated to a separate family of its own by some systematists, whereas others place man in a subfamily together with chimpanzees and gorillas. The solution to such problems is to look exclusively at synapomorphies; as soon as one of the groups has been ranked, then it is relatively simple to assign remaining higher and lower taxa to their corresponding ranks. Examples 5 and 6 will illustrate my reasoning:

Example 5, ordinal status of the Nepheliospongiida (Petrosiida): In Reevaluating ranks of higher taxa such as orders within the Porifera, Bergquist (1980, 1985) stressed certain apomorph characters. Although such emphasis is necessary to determine the monophyly of a group, it does not provide firm evidence for separate ordinal rank. The Nepheliospongiida share quite a few characters with the Haplosclerida s.s (the original taxon to which they were assigned), which they do not share with any other order of the Porifera. For this reason, I think there is ample ground for retaining the Nepheliospongiida with the Haplosclerida, whether as a suborder or a superfamily.

Example 6, phylum status of the Hexactinellida: The same reasoning applies to Bergquist's (1985) proposal that the Hexactinellida be raised to phylum rank. Here too, her evidence has to do with autapomorphies, which can only serve to establish the monophyly of the Hexactinellids. The fact remains that the Hexactinellids share characters with the Cellularia that they do not share with any other phylum. To me, the obvious conclusion is that Symplasma and Cellularia should be recognized at the subphylum level, as Reiswig and Mackie (1983) suggest.

We should be strict in ranking such crucial taxa levels as genera, orders and classes because these are also widely used by nonsystematists.

Conclusions

I do not intend to come forward with a "new" system of sponge classification (at least not in the foreseeable future) and to place it alongside already existing ones. Rather, it is my intention to apply the above principles, which are the basic elements of the cladistic approach, to any smaller or larger part of the sponge system I am (or will be) familiar with. I can only hope that others will feel impelled to do the same.

The following is a list of higher sponge taxa in which classification "problems" are apparent. The list is based on recent summaries given by Bergquist (1978, 1985), Hartman et al. (1980) and Möhn (1984). Problems refer here to violations of cladistic principles, such as the use of plesiomorphous characters and nonparsimonious theories of evolution:

STATUS OF THE SCLEROSPONGIAE
Ceractinomorpha—Tetractinomorpha
Lithistida—Tetractinellida
Axinellida—Halichondrida
Haplosclerida—Nepheliospongida
Haplosclerida—Poecilosclerida
Calcarea—Homoscleromorpha
Calcinea—Calcaronea.

Literature Cited

Bergquist, P. R. 1978. *Sponges*. London: Hutchinson. 268 pp.
———. 1980. Ordinal and Subclass Classification of Demospongiae (Porifera). *New Zealand Journal of Zoology*, 7:1–6.
———. 1985. Poriferan Relationships. Pages 14–27 in *On the Origin and Relationships of Lower Invertebrates*, edited by S. C. Morris, J. D. George, R. Gibson and H. M. Platt. Systematics Association Special Volume 28. Oxford: Clarendon Press.
Eldredge N., and J. Cracraft. 1980. *Phylogenetic Patterns and the Evolutionary Process*. New York: Columbia University Press. 349 pp.
Hartman, W. D., J. Wendt, and F. Wiedenmayer, eds. 1980. *Living and Fossil Sponges*. Notes for a short course. Sedimenta 8. Miami: University of Miami. 274 pp.
Möhn, E. 1984. *System und Phylogenie der Lebewesen I*. Stuttgart: Schweizerbart. 884 pp.
Reiswig, H., and G. O. Mackie. 1983. Studies on Hexactinellid Sponges. III. The Taxonomic Status of Hexactinellida within the Porifera. *Philosophical Transactions, Royal Society of London, B. Biological Sciences*, 301:419–428.
Soest, R. W. M. van. 1984. Deficient *Merlia normani* Kirkpatrick, 1908, from the Curaçao Reefs, With a Discussion on the Phylogenetic Interpretation of Sclerosponges. *Bijdragen tot de Dierkunde*, 54(2):211–219.
Vacelet J. 1985. Coralline Sponges and the Evolution of the Porifera. Pages 1–3 in *On the Origin and Relationships of Lower Invertebrates*, edited by S. C. Morris, J. D. George, R. Gibson and H. M. Platt. Systematics Association Special Volume 28. Oxford: Clarendon Press. Oxford: Clarendon Press.

Community Structure
and Ecology

PEDRO M. ALCOLADO
Instituto de Oceanologia
Academia de Ciencas de Cuba
Ciudad de la Habana, Cuba

General Features of Cuban Sponge Communities

Abstract

Recent investigations have produced considerable information on the general species composition, abundance, diversity, and population density of Cuba's sponge communities, whose characteristics have hitherto been poorly known. Sponge communities from different biotopes (reefs, mangroves, grass beds, inshore rock pavements, soft bottoms, and the upper bathyal zone) show distinct patterns of community structure which depend on evironmental stress factors. Greatest diversity occurs in reefs at depths of 10 to 30 m and on inshore rock pavements where salinity does not vary much and siltation is not excessive. Diversities are moderately high in some mangrove systems near reefs.

Although the sponge communities of Cuba have received little attention until recently information about their general features in different biotopes of the shelf and in the upper bathyal zone (up to 600 m depths)is now rapidly accumulating (see Alcolado, 1976, 1978, 1979a,b, 1980, 1981a, 1985a,b; Alcolado and Gotera, 1985; Alcolado and Herrera, in press). This paper presents an overview of what is known about species composition with respect to abundance, frequency, diversity, and population density in these communities.

Material and Methods

This analysis is based on information from earlier investigations as well as firsthand observations. Sampling was done by SCUBA diving on the Cuban shelf, in depths to 30 m, and observations were made from the minisubmersible Argus, in the upper bathyal zone (to depths of about 600 m). A description of the numerous sampling sites is omitted here for the sake of brevity, but these details can be found in the author's earlier publications on this area (cited above).

Diversity H' (Shannon and Weaver, 1963) and equitability J' (Pielou, 1966) were calculated using natural logarithms (nat). The results are not homogeneous for all sampled biotopes, because the design, purpose, and time spent on the investigation varied from site to site.

The index of water agitation (see Alcolado, 1981b, 1984), which is based on the summed percentages of individuals belonging to surf-resistant gorgonian species, was used as a comparative indicator of hydrodynamic stress.

Results and Discussion
REEF SPONGE COMMUNITIES

About 130 sponge species have been collected from Cuban reefs. Of these, about 110 have been identified to the species level.

The most common species at depths of less than 10 m are *Aplysina fistularis* f. *aggregata*, *Clathria juniperina*, *Ulosa ruetzleri*, *Spirastrella coccinea*, *Cliona aprica*, *C. varians*, *Tethya crypta*, and *Chondrilla nucula*. Any one of these species may be predominant in a given location, depending on ecological conditions, as determined by the interaction and intensities of various stressors, such as surf beating, sedimentation, abrasion, light intensity, substrate type, or substrate structure.

Between 10 m and about 30 m, the most abundant species, more or less in decreasing order, are *Iotrochota birotulata* f. *typica*, *Smenospongia aurea*, *Niphates areolata*, *N. digitalis*, *Ulosa ruetzleri*, *Mycale laevis*, *Aplysina fistularis*, *Callyspongia vaginalis*, *Ectyoplasia ferox*, *Spirastrella coccinea*, *Erylus formosus*, and *Cliona vesparium*. In areas of the reef where the bottom is soft (i.e., the forereef slope), *Tethya crypta*, *Keratylum rubrum* and *Oceanapia oleracea* are also abundant. Some other species that are common here and there are *A. cauliformis*, *Pseudoceratina crassa*, *Holopsamma helwigi*, *Ircinia felix*, *I. strobilina*, *N. erecta*, *Cliona schmidti*, *M. laxissima*, and *Agelas dispar*.

Thus far, it is very difficult to predict which of these species will constitute the dominant assemblage in any given location with a benign environment, which is not the case in more physically controlled places (at depths of less than 3 m). However, in areas affected by organic pollution, *Clathria venosa* and *Iotrochota birotulata* f. *musciformis* (a soft, wine-colored encrusting form) are likely to occur in the dominant species assemblage, as demonstrated by Alcolado and Herrera (in press).

The diversity of sponge species in reefs reflects the degree of environmental stress to which communities are submitted. Thus, in physically controlled communities of Cuba, diversity values can be as low as 0.35 nat (Figure 1), whereas in more benign environments, diversity values can reach 3.53 nat. Diversity values decline to 0.35–1.60 nat in the shallow zones of the reefs, mainly in depths below 5 m, but sometimes down to 10 m in areas of very high energy (e.g., the reef lagoon, back-reef zone, reef flat, surf zone, barren zone, and rock pavement flats).

Low values (<0.7) were obtained for the equitability index in sites where stressors act in an unpredictable manner, higher values were obtained in more constant or predictable environments.

At depths of more than 10 m, diversity tends to exceed 2.5 nat, owing to the more benign conditions, and equitability is greater than 0.7 as a result of high environmental predictability (in the mixed zone, *Acropora cervicornis* zone, spur zone, forereef, patch reefs on the forereef slope, and the deep forereef). At these depths, low diversities occur only in those parts of the reef where the bottom is soft (surf channels and forereef slope).

Alcolado and Herrera (in press) report that at depths of 10 m, diversity is low (1.03–2.24 nat), and *Clathria venosa* and *Iotrochota birotulata* f. *musciformis* are present, as noted above. Thus it is suggested that both these features can be used to detect the effect of organic pollution on reefs at depths of 10–15 m.

At depths of less than 10 m, population density may vary from almost zero to about 7–9 individuals/m². In these areas, sponges are generally small (commonly less than fist size) although some boring sponges (e. g., *Cliona varians* and *C. aprica*) may cover large areas. At greater depths, population density can attain 8–15 individuals/m², and sponges occur in a wide range of size. (Alcolado and Gotera, 1985).

The number of species observed on the bottom in depths greater than 10 m ranged from 21 to 56. In the shallower parts of the reefs, the numbers ranged from 2 to 19.

MANGROVE SPONGE COMMUNITIES

Thirty-seven species, of which 30 have been identified to species rank, were collected in Cuban mangroves, The most abundant species, which also constitute the dominant species assemblages are, *Reniera manglaris*, *Ulosa ruetzleri*, *Dysidea etheria*, *Mycale microsigmatosa*, *Lissodendoryx isodictyalis*, *Tedania ignis*, and *Hyrtios proteus*. Also common species are *Adocia implexiformis*, *Ircinia felix*, *Clathrina coriacea*, *Halichondria melanodocia*, *Myriastra kallitetilla*, and *Spongia* sp. From the population density and average size of sponges, it appears that *T. ignis*, *D. etheria*, *L. isodictyalis*, and *H. proteus* contribute most to the highest biomass in mangrove communities.

The sponge fauna in the mangroves of Cuba is very heterogeneous and depends on location. Thus, for example, in mangroves growing along the coasts (far from the reefs), near rivers, and in estuaries, sponge species are few, and in many areas sponges do not exist at all. The situation is similar in shallow areas where silting or intense illumination inhibit the establishment of a sponge community. Closer to the reefs, in mangroves located along the shore in depths of about 1 m or more, where foliage offers protection against light, the sponge communities are rich and diverse (2.00–2.37 nat), as is the case in some of the mangroves at Punta del Este, Island of Youth (Alcolado, 1985a).

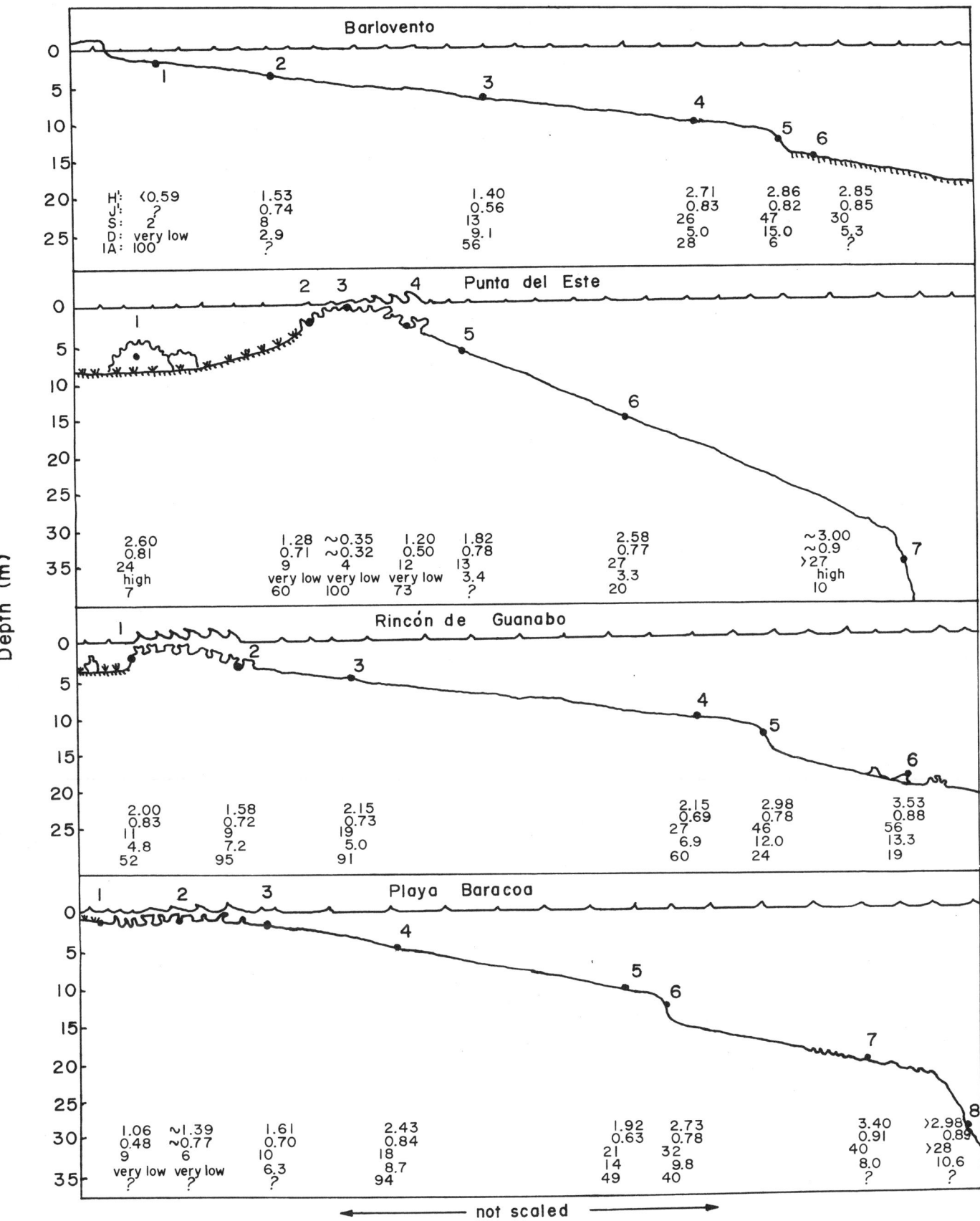

Figure 1. Sponge community features in four Cuban reef profiles; numbered dots are sampling stations. H', index of species diversity; J', equitability index; S, number of species; D, population density; IA, index of water agitation.

3d Int. Sponge Conf. 1985

In some of these mangroves, diversity is stratified with depth along the submerged parts of the roots. At Punta del Este, for example, sponge samples from mangrove roots growing to depths of 2–3 m were more diverse in the upper half than in the lower half of this zone (Figure 2). Apparently the stress of siltation produced by waves and currents is less strong in the upper half. The fact that equitability is fairly high in both strata indicates that stressors act in a rather constant way over the life span of mangrove sponges. These values are not higher probably because of some unrecorded storm effect that was not strong enough to be catastrophic.

Population density at Punta del Este was estimated to be about 50–80 individuals per meter of shore. The number of species in the different mangrove systems sampled varied from zero to 20.

SPONGE COMMUNITIES OF INSHORE SHELF MACROLAGOONS

The sea floor of Cuba's shelf macrolagoons varies greatly, and this part of the shelf is submitted to different degrees of water exchange with the open ocean and to terrigenous influence (through rivers and run-off). Thus it is difficult to generalize about the features of the sponge community there, and the sponge fauna can only be described in relation to particular sites.

The area that has received the greatest attention is the Gulf of Batabanó (Alcolado, 1985b). Of the 90 species collected there, 76 have been identified to the species level. In order of priority, the most important species with respect to biomass are *Aplysina fistularis* f. *fulva*, *Cliona vesparium*, and *C. varians*; the most important with respect to population density are *Chondrilla nucula*, *A. fistularis* f. *fulva*, and *C. varians*. Among the most common species, in addition to those above, are *Hyrtios violacea*, *Tedania ignis*, *Hyrtios proteus*, *Clathria schoenus*, *Ulosa ruetzleri*, *Ircinia felix*, *Amphi-*

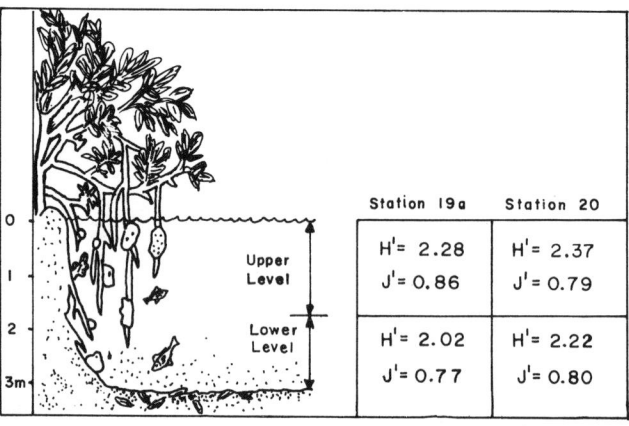

Figure 2. Sponge diversity *(H')* and equitability *(J')* at two depths in mangrove stations at Punta del Este, Island of Youth (Alcolado, 1984).

medon viridis, *Geodia gibberosa*, *Spongia obscura*, and *Dysidea etheria* (in order of decreasing importance). These species grow in grass beds of *Thalassia testudinum*, on inshore rocky pavements, and on sandy bottoms devoid of macrophytes. In one location (Ensenada de la Broa, at the northeast end of the gulf), on a muddy bottom without macrophytes, the dominant species was *Halichondria melanodocia*, which is a eurytopic species that inhabits sites characterized by poikilohaline conditions and muddy bottoms (Wiedenmayer, 1977).

Species composition in Cuba appears to be similar to that found in the Bimini lagoon (Wiedenmayer, 1977). The differences that exist among the most common or abundant species can be explained more by ecological peculiarities than by zoogeography, but in any case are beyond the scope of this paper. Although it is not yet possible to say to what extent chaos may contribute to such differences even when similar ecological conditions prevail, chaos appears to play an important role in the unpredictable environments characteristic of lagoons (inshore waters).

In the Gulf of Batabanó, higher diversities (H' = 2.19–3.40 nat) were found at inshore rocky pavements, and the same was true of equitability (J' = 0.72–0.78). This distribution is attributable to the benign and predictable or quasi-constant environmental conditions (little hydrodynamical and sediment stress) in that area. However, in a rock pavement located south of Ensenada de la Broa, where salinity drops drastically (in some cases to less than 29‰) and siltation is intense, diversity was only 1.77 nat, and equitability 0.71, which indicates an almost constant effect of such stressors (Figure 3).

Similarly, Wiedenmayer's (1977) data show that the diversity index in a rock pavement in the Bimini lagoon ranged from 1.00 to 2.42 nat. Where a sand veneer covered the rock pavement, diversity did not exceed 1.96 nat, and where this veneer was very thick, Wiedenmayer found only one or two species (H' < 1 nat).

In grass beds of the Gulf of Batabanó, diversity varies from almost zero to 1.85 nat, owing to a combination of factors such as silting, sediment instability, salinity fluctuation, light intensity, lack of stable substrate, and so on.

On soft bottoms with different vegetation densities, Wiedenmayer's (1977) diversity values (0.42–1.76 nat) coincide with my results. Higher values of biomass (21–70 g dry weight/m²) were calculated for areas that are devoid of strong terrigenous influences, and where ocean water flows in freely (Alcolado, 1985b) (Figure 4).

The distribution of population density has a more complicated pattern. Some areas affected by terrigenous influence exhibit comparatively high densities, but individuals are generally small in size and are members of opportunistic species.

Another type of macrolagoon occurs in the Gulf of Ana María-Guacanayabo (southeast of Cuba); it is character-

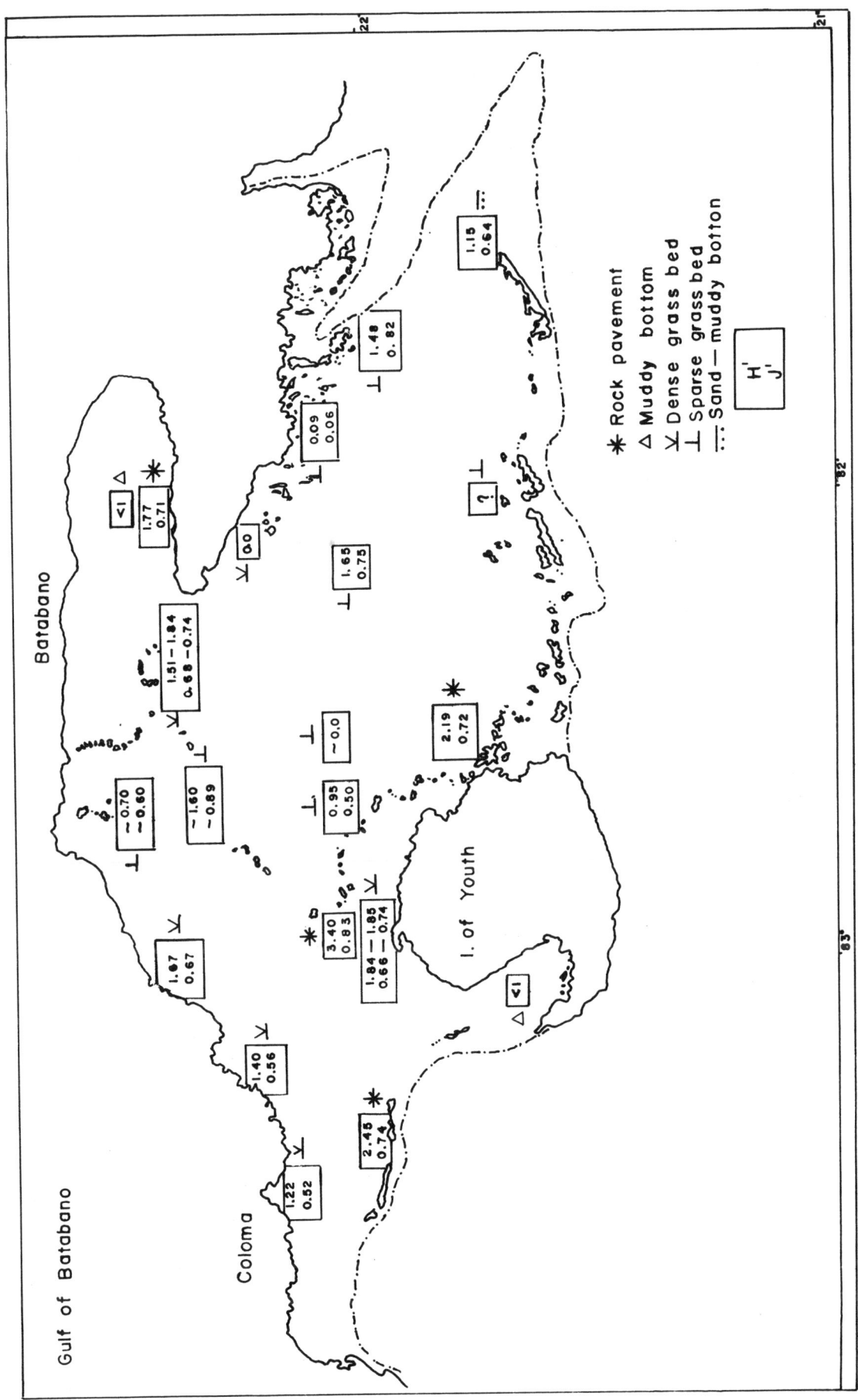

Figure 3. Sponge species diversity (H') and equitability (J') in the Gulf of Batabanó (Alcolado, 1985b).

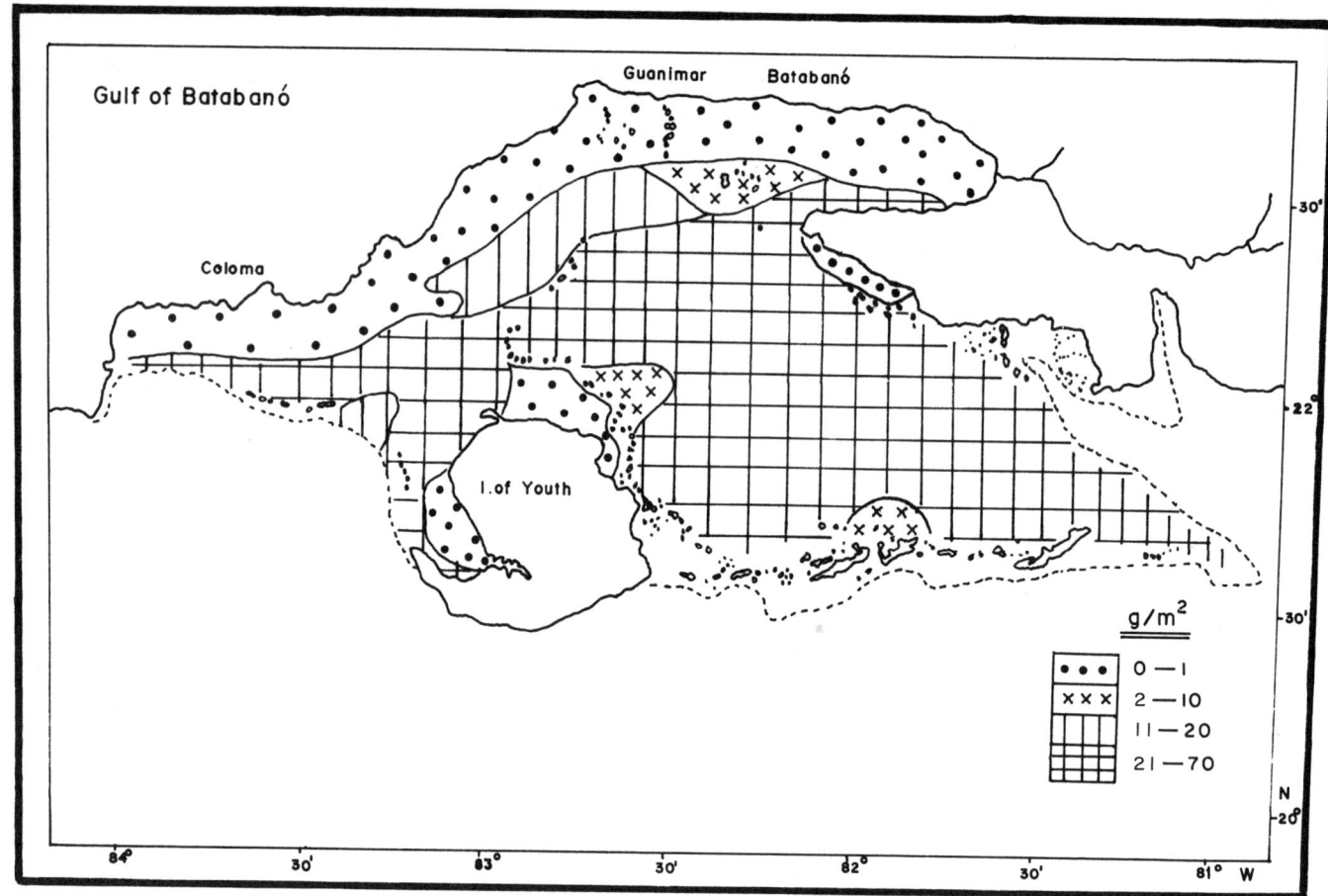

Figure 4. Distribution of sponge maximum dry biomass (g/m²) in the Gulf of Batabanó (Alcolado, 1985b).

ized by soft muddy bottoms with little or no macrophytes, rich phytoplankton, and turbid water. This area has not yet been thoroughly investigated, but it appears that the most important species with respect to number and frequency are *Mycale angulosa*, followed by *Niphates ramosa*, which are anchored in the mud with their dead base parts. Other species found on this shelf are *Terpios zeteki* and *Timea squamata*. Diversity appears to be quite low as no more than 4 species in any one location were collected.

Another area recently investigated is the Bay of Cardenas (on the northwest coast of Cuba). Here, the bottom is predominantly muddy and the density of macrophytes (mainly seagrasses) varies greatly. In terms of biomass, number, and frequency of occurrence, the sponge fauna is dominated by *Halichondria melanodocia*, followed by *Aplysina fistularis*, *Amphimedon viridis*, and *Chondrilla nucula*. Other fairly common species are *Terpios zeteki*, *Foliolina peltata*, *Tedania ignis*, and *Dysidea etheria*.

The number of species collected ranged from 6 to 24 in grass beds, from 21 to 48 on inshore rock pavements, and from zero to 8 on muddy bottoms.

BATHYAL SPONGE COMMUNITIES

About 20 species were observed or collected between depths of 150–608 m in the upper bathyal zone northwest and southwest of Cuba. Of these, 7 could be identified to species level. The dominant species in that depth range, is *Dactylocalyx pumiceus* (150–470 m), followed by *Pachastrella lithistina* and *P. monilifera* (290–600 m). Some patches of *Pheronema annae* were also locally dominant at depths of 370–560 m. A few other sponges were collected, but they were not abundant; these include *Myliusia conica* (185–550 m), *Margaritella coeloptychioides* (245 m), *Neothenea enae* (456 m), *Oceanapia* sp. (290–450 m), *Xestospongia* sp. (120–255 m), and a species of Jaspidae (202 m). A vivid yellow, thinly encrusting sponge was common on rocks.

Sponges occur sporadically in these depths and are related to the presence of a hard rocky bottom emerging from the sediment layer; this hard-bottom is sometimes covered with a thin veneer of sediment.

At lesser depths (100–150 m), the number of species increases significantly. Of these, *Ceratoporella nicholsoni*,

Stromatospongia vermicola, and *Discodermia polydiscus* were identified and collected.

Moving up the slope, at a depth of about 60–100 m, which is already out of the bathyal zone, we find a diverse sponge fauna with gigantic individuals freely occupying the available substrate without having to compete with scleractinian corals. This sponge "paradise" calls for future investigations, which will undoubtedly provide science with many new species and interesting ecological findings concerning community structure.

Literature Cited

Alcolado, P. M. 1976. Lista de nuevos registros de poríferos para Cuba. *Academia de Ciencias de Cuba, Series Oceanología*, 36:1–11.

———. 1978. Ecological Structure of the Sponge Fauna in a Reef Profile of Cuba. Pages 297–302 in *Biologie des Spongiaires*, edited by C. Lévi and N. Boury-Esnault. Colloques Internationaux du C.N.R.S. 291. Paris: Centre National de la Recherche Scientifique.

———. 1979a. Estructura ecológica de la comunidad de esponjas en un perfil costero de Cuba. *Ciencias Biológicas*, 3:105–127.

———. 1979b. Nueva especie de porífero (género *Strongylophora*) encontrada en Cuba. *Poeyana*, 196:1–5.

———. 1980. Esponjas de Cuba: Nuevos registros. *Poeyana*, 197:1–10.

———. 1981a. Guía para la identificación de algunos poríferos Cubanos (clase Demospongiae). *Academia de Ciencias de Cuba, Informe Científico-Técnico*, 184:1–42.

———. 1981b. Zonación de los gorgonaceos someros de Cuba y su posible uso como indicadores de tensión hydrodinamica sobre los organismos del bentos. *Academia de Ciencias de Cuba, Informe Científico-Técnico*, 187:1–43.

———. 1984. Utilidad de algunos índices ecológicos estructurales en el estudio de comunidades marinas de Cuba. *Ciencias Biológicas*, 11:61–77.

———. 1985a. Estructura ecológica de las comunidades de esponjas en Punta del Este. *Academia de Ciencias (Cuba), Reporte de Investigación*, 38:1–63.

———. 1985b. Distribución de la abundancia y composición de las comunidades de esponjas, en la macrolaguna del Golfo de Batabanó. Pages 5–10 in *Symposium on Marine Sciences and 7th Scientific Meeting of the Institute of Oceanology (20th Anniversary)*. Havana: Academy of Sciences of Cuba.

Alcolado, P. M., and G. G. Gotera. 1985. Estructura de las comunidades de esponjas en arrecifes cubanos. Pages 11–15 in *Symposium on Marine Sciences and 7th Scientific Meeting of the Institute of Oceanology (20th Anniversary)*. Havana: Academy of Sciences of Cuba.

Alcolado, P. M., and A. Herrera. In press. Efectos de la contaminación sobre las comunidades de esponjas al Oeste de la Bahía de la Habana. Havana: Academy of Sciences of Cuba.

Pielou, E. C. 1966. The Measurement of Diversity in Different Types of Biological Collections. *Journal of Theoretical Biology*, 13:131–141.

Shannon, C. E., and W. Weaver. 1963. *The Mathematical Theory of Communication*. Urbana, Illinois: University of Illinois Press. 117 pp.

Wiedenmayer, F. 1977. *A Monograph of the Shallow-Water Sponges of the Western Bahamas*. Basel and Stuttgart: Birkhäuser. 287 pp.

BELINDA ALVAREZ*
M. CRISTINA DIAZ**
ROGER A. LAUGHLIN
Fundación Científica Los Roques
Apartado 1
Caracas 1010-A, Venezuela

The Sponge Fauna on a Fringing Coral Reef in Venezuela, I: Composition, Distribution, and Abundance

Abstract

Niphates erecta, Ulosa ruetzleri, Cliona delitrix, and *Ectyoplasia ferox* are the most frequent and abundant species in the sponge community of the fringing coral reef in the Archipiélago de los Roques National Park, Venezuela. Of the 60 species found in this area, 25 constitute 90 to 95% of the total values obtained for frequency of occurrence, area covered, and number of individuals. Some species occur along the entire depth gradient from 1 to 35 m *(N. erecta, U. ruetzleri, C. delitrix)*; others are restricted to the deeper zones, below 18 m *(Pseudaxinella lunaecharta, Cinachyra alloclada, Teichaxinella morchella)*; and one occurs only in the shallow areas, above 18 m *(Mycale laevis),* always growing on the coral *Montastrea annularis.* Area coverage and density vary among the species and along the transects but are not necessarily related to depth. For two species, *E. ferox* and *U. ruetzleri,* these parameters are high below a depth of 20 m. Only *C. delitrix* has high values in depths of 9 to 20 m, where large colonies of *M. annularis* and other corals provide the proper substrates.

The coral reef is one of the most diverse and productive communities of the sea (Odum and Odum, 1955), with stony corals, soft corals, and sponges among its most common compounds. Research in this community has focused primarily on a few conspicuous groups, mainly the stony corals, fishes, and algae, whereas sponges have been ne-

*Present address: Department of Invertebrate Zoology, National Museum of Natural History, Smithsonian Institution, Washington, D.C. 20560.
**Present address: Marine Sciences Offices, 273 Applied Sciences Building, University of California, Santa Cruz, California 95064.

glected in spite of their important role in the dynamic ecology of reef environments (Rützler, 1978).

Systematic difficulties, and uncertainty about the physiology and reproductive biology (Bergquist, 1978), and sampling problems in the reef environment are among the reasons for the lack of ecological studies of sponges. However, many of the systematic reports (Laubenfels, 1936, 1950; Hechtel, 1965) provide at least data on habitat and depth for many species. These data have helped to establish some patterns of the ecological and geographical distribution and have provided clues to some of the physical factors that control sponge occurence (Sarà and Vacelet, 1973; Alcolado, 1979). Recently, much effort has been put into determining the role of sponges in coral reefs (Reiswig, 1971, 1973, 1974, 1981; Wilkinson, 1981, 1983; Wulff, 1984). However, quantitative information on the structure of the reef sponge communities is still lacking.

Sponges are very important faunal components of Venezuelan coral reefs. A few studies provide taxonomic descriptions of several sponges inhabiting mangrove roots in the Morrocoy National Park, on the western-central coast (Díaz et al., 1985), and in Mochima Bay in eastern Venezuela (Olivares, 1976), but little is known of the ecology of sponges in Venezuelan reefs. This study was undertaken with a view to defining the principal ecological parameters of the sponge community here. The results of the study are presented in two parts, of which this is Part I. It is concerned with the composition, distribution, and abundance of sponge species along the depth profile in the study area. Part II (Díaz et al., this volume) looks at the ecology of this community.

Materials and Methods

The study area is located in a fringing coral reef at the southwestern end of the Dos Mosquises Sur Cay (11°47′91″ N; 66°53′75″ W) in the Archipiélago de Los Roques National Park (Figure 1). Los Roques is an insular complex formed by 42 islands and more than 200 sandbanks and reefs irregularly distributed around a shallow lagoon, 1 to 5 m deep. The archipelago is about 150 km north of the north-central coast of Venezuela, and rests on a submarine igneous plateau that rises sharply from depths of 900 to 1800 m.

Sampling was conducted from January 1983 to August 1984 using Scuba equipment. Five transects were placed perpendicularly to the shoreline and extended to a depth of 28 to 35 m (Figure 2). The number of sponge species and specimens and their area coverage were determined in a 1-m² quadrat frame that was successively moved along each of the five transects. The habitat quality (light exposure, substrate type and complexity) was also examined. For the massive and incrusting species, the area coverage was determinated using a plastic grid with squares measuring 13.7 cm². For the tubular and ramose

species, the perimeter of basal area was determined with a plastic measuring tape. The sponge species were codified and characterized in situ by their shape, consistency, color, surface structure, and position and diameter of the oscules. The substrate type where each specimen occurred was also recorded. Sponges were photographed using a Nikonos IV underwater camera with a close-up unit and flash. For taxonomic study, pieces of each specimen were fixed in 10% formalin-seawater and then preserved in 70% alcohol.

Frequency of occurence of species is defined as the ratio of the total number of specimens found per total quadrats sampled (1290), density is the number of individuals per m², and abundance is the quantitative importance based on three criteria: frequency of occurence, total number of individuals sampled, and area covered.

Results

Coral reef morphology in the study area changes with depth (Figure 3). From the shore to a depth of 2 m the substrate is an extensive, almost flat sand-bank having a ridge of dead *Acropora palmata* on the seaward edge. Some live colonies of *A. palmata*, *Millepora alcicornis*, *Madracis decactis*, *Porites porites*, *A. cervicornis*, *Montastrea annularis*, and fragments of dead coral and sand are also present. Between depths of 2 m and 6 m, large domeshaped colonies of *M. annularis* are common and are interspersed with colonies of *M. decactis*, *P. porites*, *A. cervicornis*, and *Diploria* spp. From 6 m to 9 m, large colonies (up to 3 m in height) of *M. annularis*, *Colpophyllia natans* and *Diploria* spp. are dominant, but *M. annularis* is by far the most abundant. Below depths of 9 m, *M. annularis* remains the most common coral species, but in this area the colonies are morphologically heterogeneous, and feature large crevices and cavities. Between 12 m and 20 m, the coral cover thins out and many coral boulders appear; these are varied in shape and are isolated by small patches of sand. Below 20 m, sand becomes the main substrate, but small colonies of several coral species occur here and there along with small coral rocks. Below 28 m, the sand slope is covered by numerous sponges and octocorals; beyond this point, it drops off sharply to oceanic depths.

Sixty different demosponges were encountered in the study area but only 35 species were identified to species level (Table 1). These belong to 22 genera, 18 families, and 9 orders of the Demospongea. Three species (*Eurypon laughlini*, *Topsentia roquensis*, *Epipolapsis reiswigi*) were recently described (Díaz et al., 1987). One sponge type, encoded "incrusting red", was later identified as a mixture of three species (genera *Terpios*, *Acarnus*, and *Cliona*).

The species-area curve (Figure 4), which was fitted using data from all sampled quadrats (1290), reflects the diversity of sponge species in the study area. A minimal area of 600 m² is required to register most (93%) of the

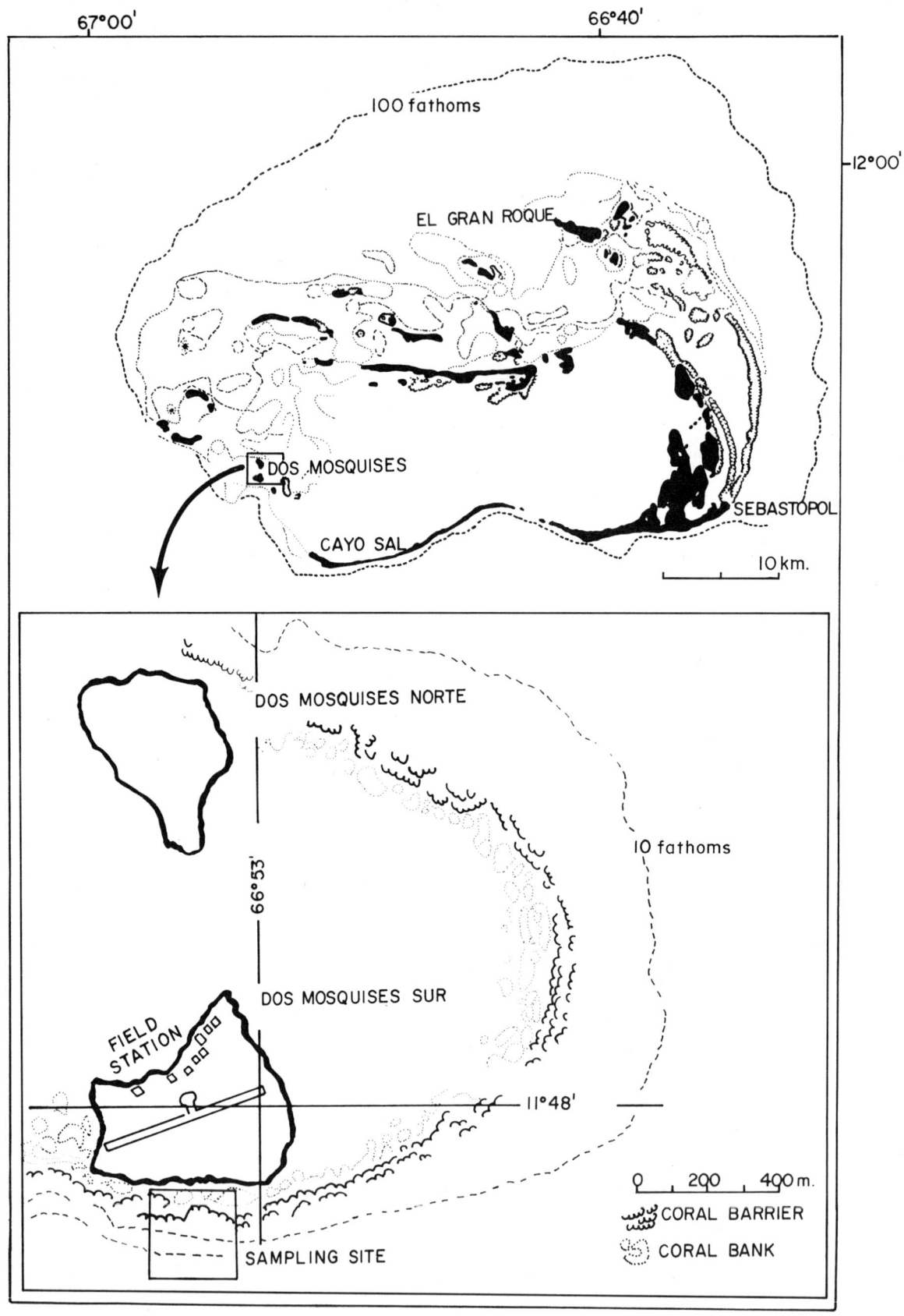

Figure 1. Archipiélago Los Roques National Park *(top)*, with location of sampling site enlarged *(bottom)*.

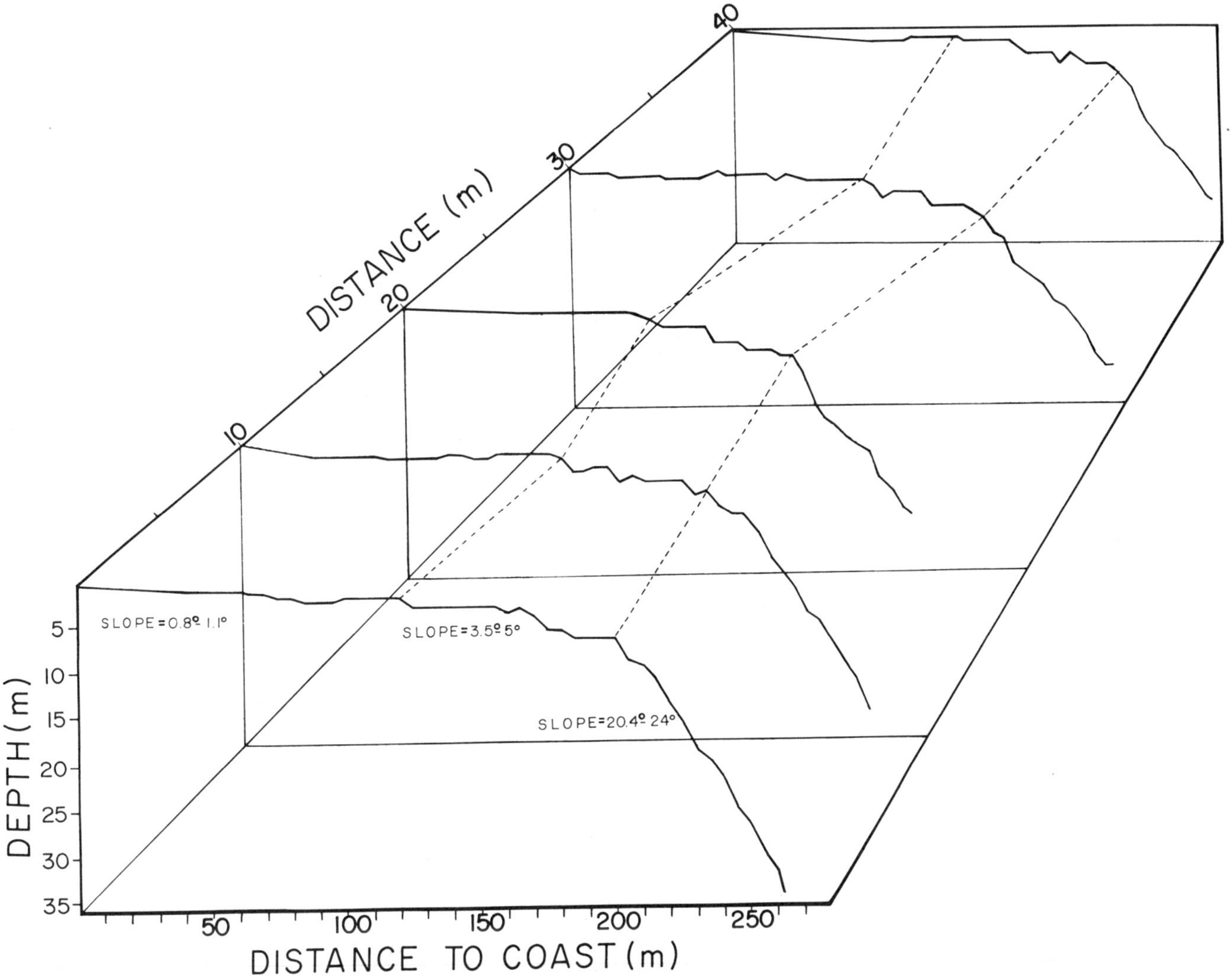

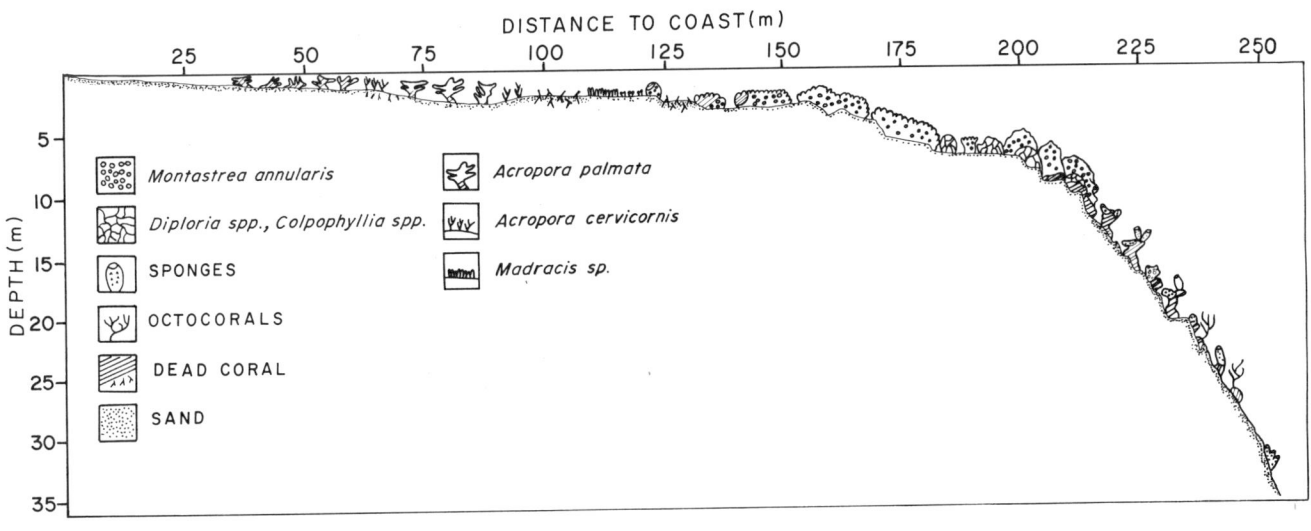

Figure 2. Three-dimensional view of study area with location of five transects oriented perpendicular to shoreline.

Figure 3. Morphological features of reef at study site.

Table 1. Species list and depth distribution of sponges samples along the 5 study transects. Undetermined taxa are represented by codes; species are arranged in order of decreasing depth range.

Species	Depth (m)						
	1–5	6–10	11–15	16–20	21–25	26–30	31–35
Niphates erecta Duch. & Mich.	X	X	X	X	X	X	X
Cliona delitrix Pang	X	X	X	X	X	X	X
Ectyoplasia ferox (Duch. & Mich.)	X	X	X	X	X	X	X
Eurypon sp.	X	X	X	X	X	X	X
Ulosa ruetzler Wiedenmayer	X	X	X	X	X	X	X
Callyspongia fallax Duch. & Mich.	X	X	X	X	X	X	X
Aplysina archeri (Higgin)	X			X	X	X	
Ircinia felix (Duch. & Mich.)		X		X	X	X	X
Agelas sceptrum (Lamarck)	X	X	X		X	X	X
Callyspongia vaginalis (Lamarck)	X	X	X	X	X	X	
Spheciospongia vesparium (Lamarck)		X		X		X	X
Incrusting red[a]		X	X	X	X	X	X
Agelas dispar Duch. & Mich.		X	X	X	X	X	X
Plakortis angulospiculatus (Carter)		X	X	X	X	X	X
Pseudoceratina crassa (Hyatt)		X	X	X	X	X	
Verongula rigida (Esper)		X	X	X	X		
Agelas conifer (Schmidt)		X	X	X	X		
Ircinia strobilina (Lamarck)		X	X	X	X		
Epipolasis reiswigi (Diaz et al.)			X	X	X	X	
Aplysina lacunosa (Pallas)			X	X	X	X	
Agelas clathrodes (Schmidt)				X	X		X
Stelletta sp.		X		X	X		
Xestospongia muta (Schmidt)				X	X	X	X
Iotrochata birotulata (Higgin)				X	X	X	X
Cinachyra alloclada Uliczka				X	X	X	X
Mycale laevis (Carter)	X	X	X	X			
Aplysina cauliformis (Carter)			X	X	X		
Petrosia weinbergi van Soest				X	X	X	X
Ircinia campana (Lamarck)		X			X		
Halichondria sp.					X	X	X
Pseudaxinella lunaecharta (Ridley & Dendy)					X	X	X
Myxillidae		X		X			
Axinellida 1				X		X	
Axinellida 2				X		X	
Neofibularia sp.					X	X	X
Halichondrid				X			
Teichaxinella morchella Wiedenmayer					X	X	X
Agelas sp.				X		X	
Unidentified			X			X	
Topsentia sp.				X	X	X	
Geodia neptuni (Sollas)					X	X	
Ircinia strobilina ? (Lamarck)					X	X	
Anthosigmella varians (Duch. & Mich.)					X	X	
Pandaros acanthifolium (Duch. & Mich.)				X	X		
Plakortis sp.						X	X
Spirastrella sp.					X	X	
Esperiopsidae				X	X		
Xestospongia subtriangularis (Duch. & Mich.)					X	X	
Aplysina fistularis (Esper)				X	X		
Axinellida 4					X	X	
Calcarea 1						X	
Calcarea 2						X	
Axinellida 3							X
Spheciospongia sp.							X
Xestospongia sp.							X
Spongidae					X		
Petrosia sp.					X		
Ircinia sp.				X			
Haplosclerida				X			
Erylus formosus Sollas		X					

[a] Later identified as *Terpios* sp., *Acarnus sourieri* (Lévi), *Cliona* sp.

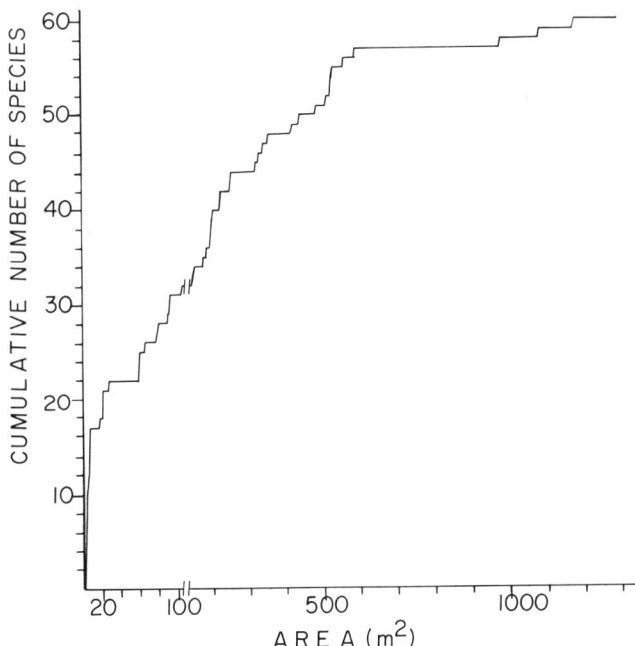

Figure 4. Species-area curve, fitted with all quadrats (1290) sampled on reef at study site.

sponge species. *Niphates erecta, Cliona delitrix,* and *Ectyoplasia ferox* occur along the entire depth profile. Other species, such as *Callyspongia vaginalis, Spheciospongia vesparium,* and *Agelas sceptrum,* occur discontinuously along the gradient. Still others, such as *Agelas conifera, A. dispar,*

Plakortis angulospiculatus, and *Aplysina lacunosa,* are restricted to depths of 8 to 35 m, and some species, such as *Petrosia weinbergi, Pseudaxinella lunaecharta,* and *Epipolasis reiswigi* are found only in the deeper zones of the reef. *Mycale laevis* is the only species limited to the shallow area, and always occurs between the domes of *Montastrea annularis* (Table 1).

Of the 1290 m² sampled, only 338 m² contained sponges. Of the 60 species found there, only 19 have a frequency of occurrence greater than 1% (Table 2). The most frequent species are *Niphates erecta, Ulosa ruetzleri, Cliona delitrix* and *Ectyoplasia ferox;* they occur in 30 to 50% of the quadrats that contain sponges. When ranked according to the total number of individuals sampled (2609), few species are noticeably dominant (Table 3). Only 4 species (*U. ruetzleri, N. erecta, C. delitrix* and *E. ferox*) constitute 61.1% of the total abundance. In this calculation the rank is somewhat different from that obtained above, although the same species share the highest hierarchical positions.

Similar trends were observed when the species were ranked according to the total area covered (Table 4). Three species alone account for more than 50% of the total area (12.8 m²) covered by sponges. However, the hierarchical order of species obtained with this criterion differs sharply from that given earlier. *Cliona delitrix* is now ranked first, with 25.6% of the total area coverage, *Ectyoplasia ferox* second with 20.7%, and *Ulosa ruetzleri* third with 6.0%. Many massive and encrusting species (e.g., *Agelas dispar, Pseudoceratina crassa,* and *Ircinia strobilina*) with

Table 2. Frequency of occurrence for the most common species in the study area

Rank	Species	Frequency of occurrence (%)
		1 2 3 4 5 6 8 10 12 14
1	*Niphates erecta*	
2	*Ulosa ruetzleri*	
3	*Cliona delitrix*	
4	*Ectyoplasia ferox*	
5	*Callyspongia fallax*	
6	*Eurypon* sp.	
7	Incrusting red[a]	
8	*Ircinia felix*	
9	*Agelas dispar*	
10	*Pseudoceratina crassa*	
11	*Iotrochata birotulata*	
12	*Cinachyra alloclada*	
13	*Plakortis angulospiculata*	
14	*Callyspongia vaginalis*	
15	*Mycale laevis*	
16	*Teichaxinella morchella*	
17	*Aplysina lacunosa*	
17	*Halichondria* sp.	
17	*Aplysina archeri*	
18	*Verongula rigida*	
19	*Agelas conifera*	
	Others (30 spp.)	

[a]Later identified as *Terpios* sp., *Acarnus sourieri* (Lévi), *Cliona* sp.

Table 3.   Number of individuals of the most common species in the study area (total sample = 2609)

Rank	Species	Total number of individuals (%)	Cumulative
		1 2 3 4 5 8 10 12 14 16	
1	Ulosa ruetzleri		16.37
2	Niphates erecta		31.66
3	Cliona delitrix		46.84
4	Ectyoplasia ferox		61.14
5	Callyspongia fallax		66.54
6	Eurypon sp.		70.53
7	Incrusting red[a]		74.09
8	Ircinia felix		76.85
9	Agelas dispar		79.23
10	Iotrochata birotulata		81.15
11	Pseudoceratina crassa		82.91
12	Cinachyra alloclada		84.60
13	Plakortis angulospiculatus		85.83
14	Mycale laevis		87.02
15	Callyspongia vaginalis		88.13
16	Teichaxinella morchella		89.13
	Others (44 spp.)		100.00

[a]Later identified as *Terpios* sp., *Acarnus sourieri* (Lévi), *Cliona* sp.

low values of frequency of occurrence and abundance were ranked high according to this parameter.

Area coverage and density vary along all five transects sampled but do not seem to be related to depth (see Alvarez and Díaz, 1985). Only two species, *Ectyoplasia ferox* and *Ulosa ruetzleri*, show high values for these parameters in the deeper zones of the reef. One species, *Cliona delitrix*, is dominant between 9 m and 20 m; it occurs in areas where large colonies of *Montastrea annularis* and other corals abound. Most common species show low density values (1–4 individuals per m²) along the entire depth profile; only a few, such as *Niphates erecta, Ulosa ruetzleri, Cliona*

delitrix, and *Ectyoplasia ferox*, are found to have high densities.

Discussion

All identified sponges in the study area, with the exception of the new ones, have also been found in other Caribbean regions (Hechtel, 1965, 1969,1976; Van Soest, 1978,1980, 1984; Wiedenmayer, 1977; Alcolado, 1981). However, the sponge fauna of Los Roques cannot be fully compared with that of other regions until the remaining species are described.

Table 4.   Area coverage of the most common species in the study area (total area covered by sponges = 12.8 m²)

Rank	Species	Total area coverage (%)	Cumulative
		1 2 3 4 5 10 15 20 25	
1	Cliona delitrix		25.62
2	Ectyoplasia ferox		46.34
3	Ulosa ruetzleri		52.34
4	Agelas dispar		56.97
5	Niphates erecta		61.32
6	Pseudoceratina crassa		65.65
7	Callyspongia fallax		68.94
8	Incrusting red[a]		71.77
9	Callyspongia vaginalis		74.35
10	Ircinia felix		76.78
11	Ircinia strobilina		78.75
12	Eurypon sp.		80.62
13	Petrosia weinbergi		82.47
14	Verongula rigida		84.17
15	Agelas conifera		85.64
16	Anthosigmella varians		86.99
	Others (44 spp.)		100.00

[a]Later identified as *Terpios* sp., *Acarnus sourieri* (Lévi), *Cliona* sp.

Several criteria (frequency of occurrence, area coverage, and total number of individuals) were used to determine the quantitative importance of the sponge species at Los Roques because any one by itself may not accurately reflect the situation. For example, area coverage by itself is imprecise as a measure of abundance for ramose, tubular, or very massive species which are better quantified by tissue volume (Rützler, 1978). Furthermore, area coverage should not be a basis of comparison between sponges and other reef taxa, such as corals, because it underestimates the ecological importance of the filter feeders. Again, volume is a more appropriate measure but was not determined in this study, because it requires destructive sampling, and was not considered essential to the objectives.

The distribution of sponge species in the study area is similar to the patterns registered by Hechtel (1965), Wiedenmayer (1977), Van Soest (1978,1980, 1984), and Alcolado (1979). Exceptions are *Iotrochota birotulata*, *Pseudaxinella lunaecharta*, *Cinachyra alloclada*, and *Teichaxinella morchella* which elsewhere occur at depths of 1 to 3 m, whereas at Los Roques they are restricted to the deeper reef.

The distribution of sponge species along the depth profile may be indicative of their ecological tolerance. Species that are widely distributed are more tolerant (eurytopic generalist) than those restricted to certain depths. For instance, *Niphates erecta*, *Ulosa ruetzleri*, *Cliona delitrix*, and *Ectyoplasia ferox*, are distributed along the entire depth gradient, show high values of density in many of the sampled quadrats, and are therefore considered to be generalists with respect to their requirements. On the other hand, wide depth range may merely reflect the distribution of particular microhabitats along the depth profile. *Callyspongia fallax*, for example, is widely distributed in respect to depth because the shaded crevices or cavities it inhabits occur at all depths. Another example is *Mycale laevis*, which is restricted to the shallow areas of the Dos Mosquises reef but has a very different distribution in other Caribbean reefs. According to Goreau and Hartman (1966), it can be found between 1 and 80 m, and is particularly abundant below 40 m. Thus, the ecological tolerance of the species cannot be judged on the basis of local distribution. Further studies similar to those of Reiswig (1971, 1973, 1974, 1981) are required to properly assess these relationships.

The variation in area coverage and density of species along the transect may be due to several physical and biological factors (Reiswig, 1973; Sarà and Vacelet, 1973), but the methodology used here does not permit us to explain the reasons for these differences. However, one suggestion has been put forth to explain variations in abundance of the burrowing species *Cliona delitrix*. According to MacGeachy (1977), the dead basal area of scleractinian corals is highly suceptible to burrowing by these sponges.

C. delitrix, in the study area, shows high values of area coverage and number of individuals wherever there are large coral colonies with dead basal areas. This suggests a direct relationship of substrate availability and abundance of *C. delitrix*.

Most common species show low density values in many of the sampled quadrats. This may be an indication of the way in which the local populations are distributed. Species with low density values along the entire depth profile are homogeneously distributed. Those with high density values in some of the sampled quadrats have clumped distributions. Further studies of species life histories and environmental requirements are required to explain many of these results.

Conclusions

Of the 60 demosponges encountered in the study area, the 35 most frequent species were identified at species level.

As in other communities, the sponge fauna consists of a few abundant species and many uncommon ones. The most common are *Niphates erecta*, *Ulosa ruetzleri*, *Cliona delitrix*, and *Ectyoplasia ferox*.

The most frequent species are also the most widely distributed along the depth gradient. They also show high densities (more than two individuals/m^2) in many of the quadrats sampled. The other species are restricted to the deeper zones of the reef. Only *Mycale laevis* is restricted to the shallow areas; the reason for this limited local distribution has not been determined.

For all species, the values of density and area coverage varies along the transects but seems independent from depth. The abundance of *Cliona delitrix* is correlated with availability of dead coral substrate.

Acknowledgments

This work was sponsored in part by Consejo Nacional de Investigaciones Científicas y Tecnológicas (CONICIT). The participation of two of us in the Third International Conference on the Biology of Sponges was made possible by grants from the Smithsonian Institution. We wish to thank all those who have helped us with this effort, especially our friend Ernesto Weil for his important help with the field work. We are deeply indebted to R. W. M. van Soest and K. Rützler for their assistance in the taxonomic work. We are also grateful to Clive Wilkinson and Hugh Caffey for their comments on the manuscript, and to the staff and members of the Fundación Científica Los Roques, whose cooperation made this survey possible. The Laboratory of Photography of the Universidad Central de Venezuela photographed the figures, and Pablo Rodríguez helped us with the drawings. This is Scientific Contribution No. 21 of the Fundación Científica Los Roques.

Literature Cited

Alcolado, P. M. 1979. Estructura ecológica de la comunidad de esponjas en un perfil costero de Cuba. *Ciencias Biológicas,* 3:105–127.

———. 1981. Guía para la identificación de algunos Poríferos cubanos (Clase Demospongiae). *Academia de Ciencias de Cuba, Informe Científico-Técnico,* 184:1–42.

Alvarez, B., and Díaz, M. C. 1985. Las esponjas de un arrecife coralino en el Parque Nacional Archipiélago de Los Roques. I.- Taxonomia. II.- Estructura ecológica. Licenciature thesis. Universidad Central de Venezuela. 216 pp.

Bergquist P. R. 1978. *Sponges.* Berkeley and Los Angeles: University of California Press. 268 pp.

Díaz, M. C., B. Alvarez, and R. W. M. van Soest. 1987. New species of Demospongiae (Porifera) from the National Park "Archipiélago de Los Roques", Venezuela. *Bijdragen tot de Dierkunde,* 57:31–41.

Díaz H., M. Bevilacqua, and D. Bone. 1985. Esponjas en manglares del Parque Nacional Morrocoy. Caracas: Fondo Editorial. Acta Científica Venezolana. 64 pp.

Goreau, T. F., and W. D. Hartman. 1966. Sponge: Effect on the Form of Reef Corals. *Science,* 151:343–344.

Hechtel, G. J. 1965. A Systematic Study of the Demospongiae of Port Royal, Jamaica. *Peabody Museum of Natural History Bulletin,* 20:1–103.

———. 1969. New Species and Records of Shallow Water Demospongiae from Barbados, West Indies. *Postilla,* 132:1–38.

———. 1976. Zoogeography of Brazilian Marine Demospongiae. Pages 237–260 in *Aspects of Sponge Biology,* edited by F. W. Harrison and R. R. Cowden. New York, San Francisco and London: Academic Press.

Laubenfels, M. W. de. 1936. A Discussion of the Sponge Fauna of the Dry Tortugas, in particular and the West Indies in General, with Material for a Revision of the Families and Orders of the Porifera. *Papers from the Tortugas Laboratory,* 30:1–225.

———. 1950. The Porifera of the Bermuda Archipelago. *Transactions, Zoological Society of London,* 27:1–153.

MacGeachy, J. K. 1977. Factors Controlling Sponge Boring in Barbados Reefs Corals. Pages 477–483 in *Proceedings, Third International Coral Reef Symposium 2, Miami, Florida.* Rosenstiel School of Marine and Atmospheric Science, University of Miami.

Odum, H. T., and E. P. Odum. 1955. Trophic Structure and Productivity of Windward Coral Reef Community on Eniwetok Atoll. *Ecological Monographs,* 25:291–320.

Olivares, M. A. 1976. Estudio taxonómico de algunas Demosponjas (Porífera) de la Bahia de Mochima, Sucre, Venezuela. Trabajo de Ascenso, Universidad de Oriente, Escuela de Ciencias, Dpto. de Biología. 77 pp.

Reiswig, H. M. 1971. Particle Feeding in Natural Populations of Three Marine Demosponges. *Biological Bulletin,* 141:568–591.

———. 1973. Population Dynamics of Three Jamaican Demospongiae. *Bulletin of Marine Science,* 23(2):191–226.

———. 1974. Water Transport, Respiration and Energetics of Three Tropical Marine Sponges. *Journal of Experimental Marine Biology and Ecology,* 14:231–249.

———. 1981. Partial Carbon and Energy Budgets of the Bacteriosponge *Verongia fistularis* (Porifera: Demospongiae) in Barbados. *Marine Ecology,* 2:273–293.

Rützler, K. 1978. Sponges in Coral Reefs. Pages 299–314 in *Coral Reefs: Research Methods,* edited by D. R. Stoddart and R. E. Johannes. Monographs on Oceanographic Methodology 5. Paris: Unesco.

Sarà, M., and M. Vacelet. 1973. Ecologie des Démosponges. Pages 462–526 in *Traité de Zoologie 3(1), Spongiaires,* edited by P. P. Grassé. Paris: Masson.

Soest, R. W. M. van. 1978. Marine Sponges from Curaçao and Other Caribbean Localities. Part I. Keratosa. *Studies on the Fauna of Curaçao and other Caribbean Islands,* 56(94):1–94.

———. 1980. Marine Sponges from Curaçao and Other Caribbean Localities. Part II. Haplosclerida. *Studies on the Fauna of Curaçao and other Caribbean Islands,* 62(104):1–173.

———. 1984. Marine Sponges from Curaçao and Other Caribbean Localities. Part III. Poecilosclerida. *Studies on the Fauna of Curaçao and other Caribbean Islands,* 66(112):1–167.

Wiedenmayer, F. 1977. *The Shallow-Water Sponges of the Western Bahamas.* Basel und Stuttgart: Birkhäuser Verlag. 287 pp.

Wilkinson, C. R. 1981. Significance of Sponges with Cyanobacterial Symbionts on Davis Reef, Great Barrier Reef. Pages 705–712 in *The Reef and Man,* edited by E. D. Gomez et al. Proceedings, Fourth International Coral Reef Symposium 2. Quezon City, Philippines, University of the Philippines.

———. 1983. Role of Sponges in Coral Reef Structural Processes. Pages 263–276 in *Perspectives on Coral Reefs: Reviews Arising from a Workshop Held at the Australian Institute of Marine Science,* edited by D. J. Barnes. Townsville: Australian Institute of Marine Science.

Wulff, J. 1984. Sponge-Mediated Coral Reef Growth and Rejuvenation. *Coral Reefs,* 3:157–164.

M. CRISTINA DIAZ*
BELINDA ALVAREZ**
ROGER A. LAUGHLIN
Fundación Científica Los Roques
Apartado 1
Caracas 1010-A, Venezuela

The Sponge Fauna on a Fringing Coral Reef in Venezuela, II: Community Structure

Abstract

The community structure of Porifera in open reef habitats at Archipiélago de los Roques National Park, Venezuela, was analyzed in terms of species number, area coverage, density, species diversity, and evenness. No sponges were found in depths of 0 to 2 m range (80 to 100 m from shore). Below 2 m, species richness and abundance increase with depth. Species diversity (Shannon-Weaver) is highest in the deeper zones of the reef. Evenness values are high and constant along the depth profile (to 35 m), and indicate that no single sponge species is dominant. Although all the values obtained appear to increase with depth, the data vary greatly, even in contiguous areas of the reef. From field observations it appears that this variation is related to changes in substrate features. Seven reef zones were differentiated long the depth profile on the basis of type, extent, and structural complexity of the substrate. Species richness and abundance were found to increase in zones with high structural complexity and high substrate availability. Five sponge growth forms were differentiated on the basis of area coverage and volume. Of these, encrusting species were the most abundant in five of the seven reef zones.

Sponges are important organisms in coral reef environments. Their biomass, ecological tolerances, and diversity frequently exceed those of the reef-building corals (Rützler, 1970; Reiswig, 1973; Suchaneck et al., 1983). Moreover, their role as efficient filter feeders (Reiswig, 1971), bioerosive agents (Goreau and Hartman, 1963; Rützler, 1975; Wilkinson, 1983), coral reef consolidators (Wulff and Buss, 1979), and providers of food and shelter for many invertebrates and fishes (Randall and Hartman,

*Present address: Marine Sciences Offices, 273 Applied Sciences Building, University of California, Santa Cruz, California 95064.
**Present address: Department of Invertebrate Zoology, National Museum of Natural History, Smithsonian Institution, Washington, D.C. 20560.

1968; Rützler, 1976; Villamizar, 1985) underscore their important ecological role in coral reefs.

Ecological studies and field observations have shown that physicochemical factors (sedimentation, light, temperature, hydrodynamics, nutrients, extent and composition of the substrate; see Sarà and Vacelet, 1973; Alcolado, 1979), biological disturbances (competition and predation; see Jackson, 1983), and life history patterns (larvae, growth rate, mortality; see Reiswig, 1973) determine the distribution and abundance of this group, both locally and geographically. Even so, few quantitative studies have been conducted on the distribution of sponges in tropical environments, particularly in coral reef habitats. The present study constitutes a preliminary examination of the structural parameters of Porifera growing on open reef habitats at the Archipiélago de los Roques National Park, Venezuela.

Variations in species richness, density, area coverage, and species diversity and evenness along the depth profile were studied in order to evaluate the changes in the ecological structure of the group with respect to depth-dependent factors. And, since size, composition, and substrate type are important factors in determining structure of benthic communities (Jackson, 1979), the possible relationships between community parameters and substrate characteristics were also investigated. In view of the importance of growth form in sessile organisms in relation to the physical and biological environment, we also examined growth forms of the sponge species in the field.

Material and Methods

The data analyzed in this study were collected along five transects perpendicular to the coast extending from 0.1 m to 35 m depth in a fringing coral reef at Dos Mosquises Sur Cay, Archipiélago de Los Roques National Park (Alvarez et al., this volume). Species richness (total number of sponge species), density (total number of specimens), and area coverage (cm²) were determined inside a 1-m² quadrat which was successively moved along each transect. Diversity (Shannon-Weaver) and evenness (Pielou, 1966) indexes were calculated for each quadrat. The formulas used to calculate these indexes are:

$$\text{Diversity } H' = - \sum_{i=1}^{n} p_i \cdot \log_2 p_i,$$

where $p_i = n_i/n$, n_i = number of individuals of the species i or the area coverage of species i, n = total number of individuals or total area coverage;

$$\text{Evenness } V' = H'/H_{max},$$

where $H_{max} = \log S$, S = total number of species.

Species zonation was investigated through cluster analysis using the 2PM program of B.M.D.P. (Dixon and Brown, 1979). Two types of analyses were performed, one on species number and density and the other on species number and area coverage. The apparent relationship between sponge community structure and habitat characteristics was studied by analysing seven reef zones differentiated on the basis of depth, extent, structural complexity of the substrate, and environmental conditions (Table 1). In situ drawings to scale of the substrate profile made along each transect on a plexiglass slate were used to prepare a representative diagram of each zone (Figure 1). The depth range, length, and some characteristics of environmental conditions for each zone are given in Table 1. The same ecological parameters studied for each quadrat were calculated for each reef zone, pooling together the data from the five studied transects.

The relationship between the shape and the abundance of sponge species was examined for five growth forms differentiated by the relationship between substrate area covered and volume occupied in the water column (Table 2). Relative density values for each growth form were calculated for each zone.

Results

GRADIENT ANALYSIS. Sponges were practically absent between 0 m and 2 m in open reef habitats. Species richness, area coverage, and density increase with depth to about 25 to 30 m, then a decrease in these parameters occurs. The predominance of sand at these depths might cause this situation. It is interesting to note the great variability of the parameters between adjacent points along the depth profile. This mainly the consequence of sudden changes that occur in hard substrate availability (Figure 2).

To study the relationship between species richness and abundance, we calculated the ratio of area coverage to species number and of density to species number for each quadrat (Figure 2). These ratios, particularly that of density to species number, show a relatively narrow range of variation. For the ratio of coverage to species number, high values were found at intermediate depth levels; these correspond with quadrats in which species with high coverage (such as *Cliona delitrix*) or massive size (such as *Ircinia felix, I. campana,* and *Spheciospongia vesparium*) are present. These results appear to indicate a direct relationship between species richness and abundance parameters.

Diversity values increase with depth whereas evenness remains high and almost constant along the depth profile (Figure 2) thus no species is dominant at any point along the depth profile.

The results of the cluster analysis confirm that the sponges in the study area lack a well defined pattern of zonation. All quadrats without sponges were grouped at a dissimilarity value of zero; the remaining quadrats were

Table 1. General characteristics of reef zone types indentified in Dos Mosquises Sur fringing coral reef

Zone	Depth range (m)	Length (m)	Structural complexity	Environmental conditions	Comments
1	0(shore)–2	105–143	High	Extreme (high light intensity; wave stress; high turbidity; low tide exposure)	Shoreward: Dead and live *Acropora palmata.* Seaward: Coral rubble, sand, and massive corals.
2	2–9	63–74	Medium	Moderate	Large colonies of dome-shaped *Montastrea annularis* few dead corals
3	9–12	7–27	Low	Moderate	Large (1–3m high) coral heads (*M. annularis*); dead coral scarce
4	12–18	9–19	Medium	Moderate	Coral cover similar to zone 3 but less compact, more dead coral
5	18–23	15–21	High	Moderate	Coral heads and rocks small (0.5–2 m high)
6	23–28	13–15	Low	Moderate	Mainly sand with small (0.5m high) corals and rock
7	28–35	1–15	Very Low	Moderate	Sandy bottom with scattered coral rubble

grouped at higher levels of dissimilarity, and no relationship was found between them and their position on the depth profile.

REEF ZONE ANALYSIS. From the species–area curve for each zone (Figure 3), it appears that the species number and the slope of the curve increase with depth. Most of the curves reach a plateau except for those representing the deeper reef zones, 5, 6, and 7. More area has to be studied to estimate the total number of species occurring in these zones.

The values of the species number, mean area coverage, mean density, species diversity, and evenness for each zone are shown in Figure 4. The values for each parameter increase from zone 1 to 5, and remain high and almost constant in reef zones 6 and 7. The largest increases in species richness and abundance occur in zone 2 (2–9 m in depth); a smaller increase in these values occurs in zone 5, where values double.

GROWTH FORM ANALYSIS. Density values (individuals/m^2) for each growth form (Table 2) in each zone show that all

Table 2. Sponges grouped into five major growth forms, characterized by the relationship between covered substrate area and volume

Shape A — crust	Shape B — cushion	Shape C — large massive	Shape D — erect	Shape E — small massive
Niphates erecta *Ulose ruetzleri* *Cliona delitrix* *Anthosigmella varians* *Eurypon* sp. 3 unidentified species	*Ectyoplasia ferox* *Callyspongia fallax* *Plakortis angulospiculatus* *Mycale laevis* *Halichondria* sp. *Pseudaxinella lunaecharta* *Agelas conifera* *Agelas sceptrum* *Agelas* sp. *Petrosia weinbergi* *Epipolasis* sp. *Topsentia* sp. 8 unidentified species	*Ircinia felix* *Ircinia strobilina* *Agelas dispar* *Pseudoceratina crassa* *Verongula rigida* *Xestospongia muta* *Spheciospongia vesparia* *Agelas clathrodes* 3 unidentified species	*Niphate erecta* *Ircinia campana* *Iotrochata birotulata* *Callyspongia vaginalis* *Aplysina cauliformis* *Aplysina archeri* *Aplysina lacunosa* 2 unidentified species	*Cinachyra alloclada* *Teichaxinella morchella* *Geodia neptuni* 9 unidentified species

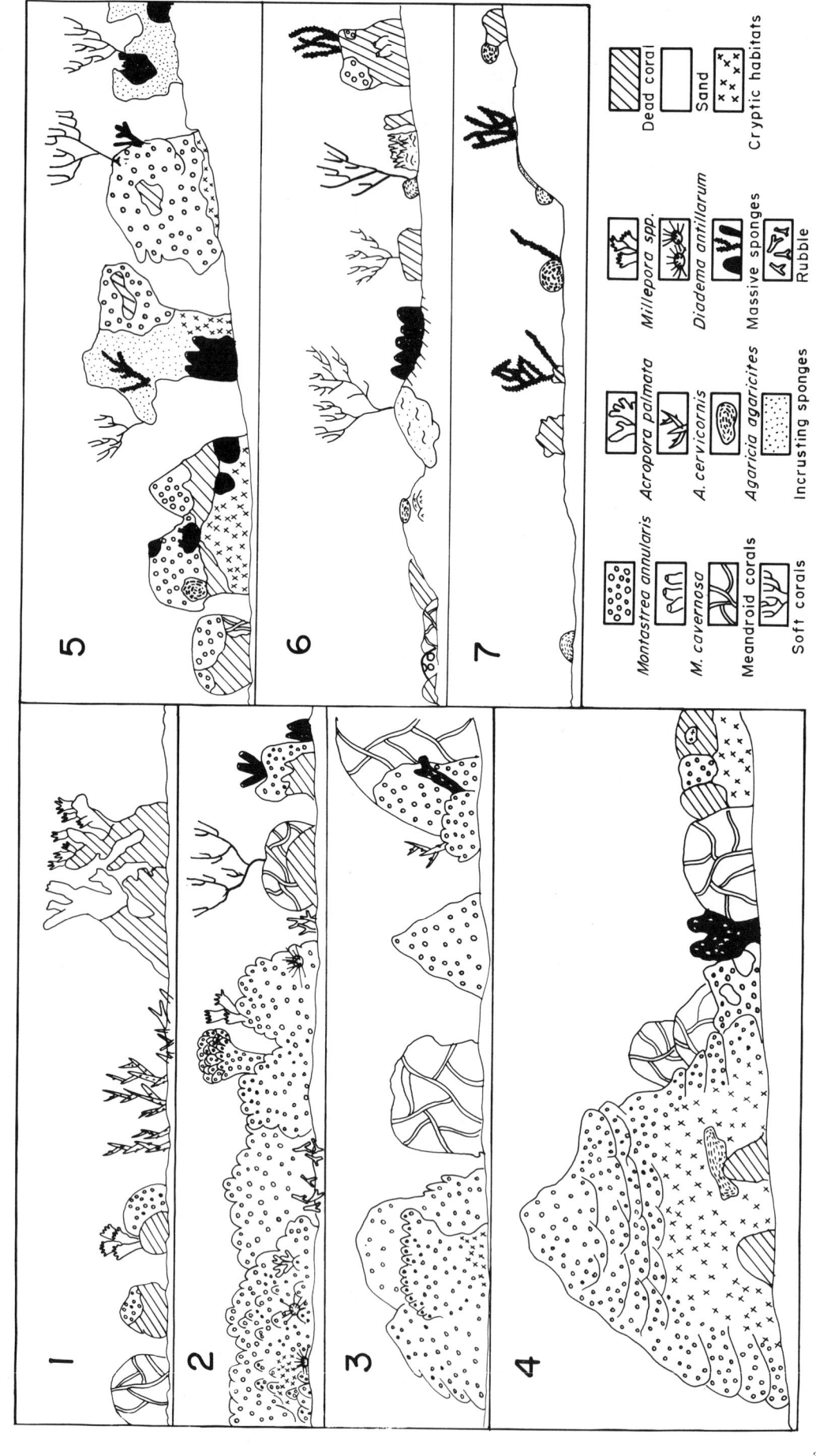

Figure 1. Generalized diagram of zone types identified in a fringing coral reef at Dos Mosquises Sur cay, Archipiélago de Los Roques National Park. Deeper areas of each zone are to left; zone number indicated at top left of each diagram (see Table 1 for more details).

3d Int. Sponge Conf. 1985

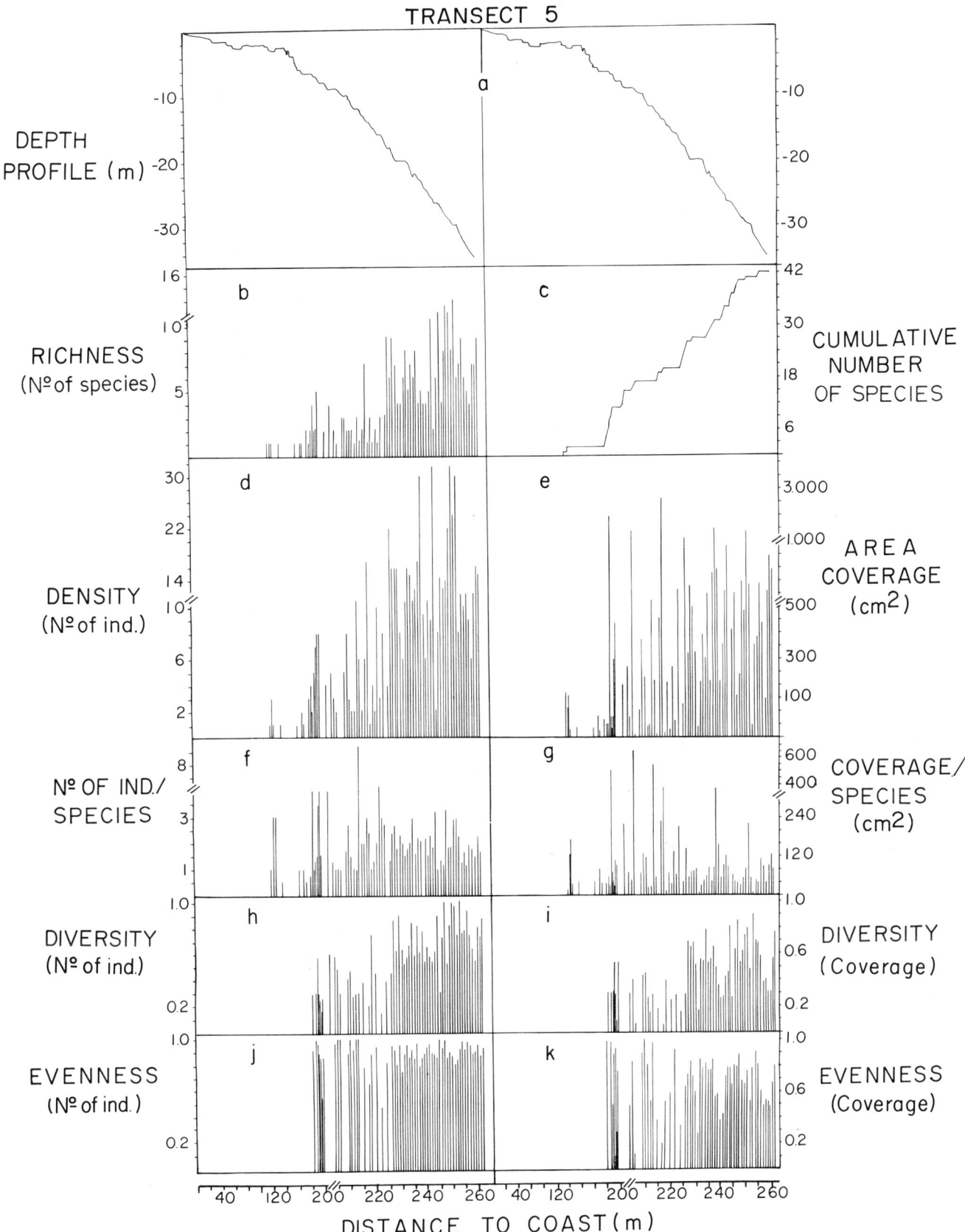

Figure 2. Variation in ecological parameters examined in study area along transect 5, perpendicular to coast.

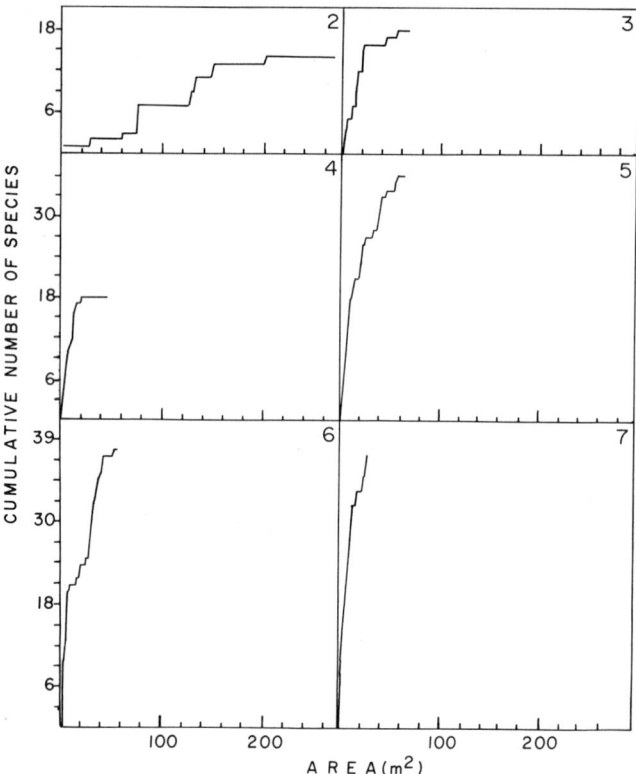

Figure 3. Species-area curves for each reef zone *(2–7)* differentiated in study area (zone 1 revealed only traces of sponges).

shapes except E (small, massive) were found from reef zones 2–7 and densities increased toward the deeper zone. Encrusting species (group A) constitute the most abundant group in many reef zones (3–7), followed in abundance by the massive encrusting species (group B). Group E was only represented in the deep reef zones (6, 7). The massive sponges were the most poorly represented. Interestingly, the species belonging to group A which are low in number (8) have the highest abundance in values all reef zones.

Discussion

Among the factors affecting sponge distribution in the study area, light and wave action may play an important role. High light intensities can cause the death of some sponge species through ultraviolet radiation (Jokiel, 1980). Wave stress may limit the colonization and growth of Porifera by generating substrate unstability, high turbidity, and turbulence. Therefore, the variations in these factors along the depth profile may in part account for the absence of sponges at shallow open areas and the increased abundance of sponges with depth—a pattern that exists in other tropical reefs as well (Hechtel, 1969; Reiswig, 1973; Alcolado, 1979; Wilkinson, 1981; Schmahl, this volume). The lack of dominance, although evident among

the sponges in the study area, is not apparent among the other sessile groups here (e.g., stony corals, Hung, 1985), nor is it a general feature of reef sponges elsewhere. Sarà (1970), however, reported similar findings in a cave biotope and established a direct relationship between species diversity and density of sponges. Sarà suggests that the causative factor is the low incidence of competition between sponge species in this biotope and a possible cooperation phenomenon in the phylum. Under certain conditions, however, some species can monopolize the substrate, as in the case of *Anthosigmella varians* at the study site, where it covers small areas of substrate in depths of 25–29 m. On another reef of the Archipiélago (Cayo Sal, located 4 km southeast of the study site), the species was found covering large areas up to 10 m² in depths of 8–10 m. This extensive growth could be a consequence of the strong directional currents that prevail in that area. Another example of dominance among reef sponges is reported by Vicente (this volume), who found that *Chondrilla nucula* Schmidt is the most abundant sessile component in a coral reef at Cayo Enrique at the southwest coast of Puerto Rico and is the principal and most persistent aggressor over many reef taxa.

The absence of a pattern of zonation among sponge species in the study area could be a result of the fact that sponges are not specialized with regard to feeding and substrate competition, as are organisms in other phyla (see Reiswig, 1973). At the same time, the existence in the phylum of competitive mechanisms such as epibiosis (Rützler, 1970; Sarà, 1970) and the use of allelochemical substances (Jackson and Buss, 1975; Green, 1977) may promote the coexistence of a high number of sponge species, and thus inhibit monopolization of the substrate. Furthermore, the characteric structural complexity of the substrate in reef environments and the relatively high density of predators and grazers in the study area may also contribute to maintain a high species diversity there. The grazing activity of herbivores, especially fishes, could affect the structure of sponge populations: first, by continuously generating free substrate of various sizes, and, second, by limiting the abundance of superior space competitors. In the study area, however, we saw few signs of predation among adult sponges, although predation might be more common during the early stages of the life cycle. Field experiments are required to evaluate the effect of predation on sponge populations in coral reef environments.

As indicated earlier, the low abundance of sponges in zone 1 could be the result of the environmental conditions in these shallow depths. Species number and abundance of sponges show the highest increase in zone 2, even though the availability and complexity of the substrate are low; this could indicate that environmental conditions at this depth are more favorable. The second highest increase in these parameters takes place in zone 5, where the

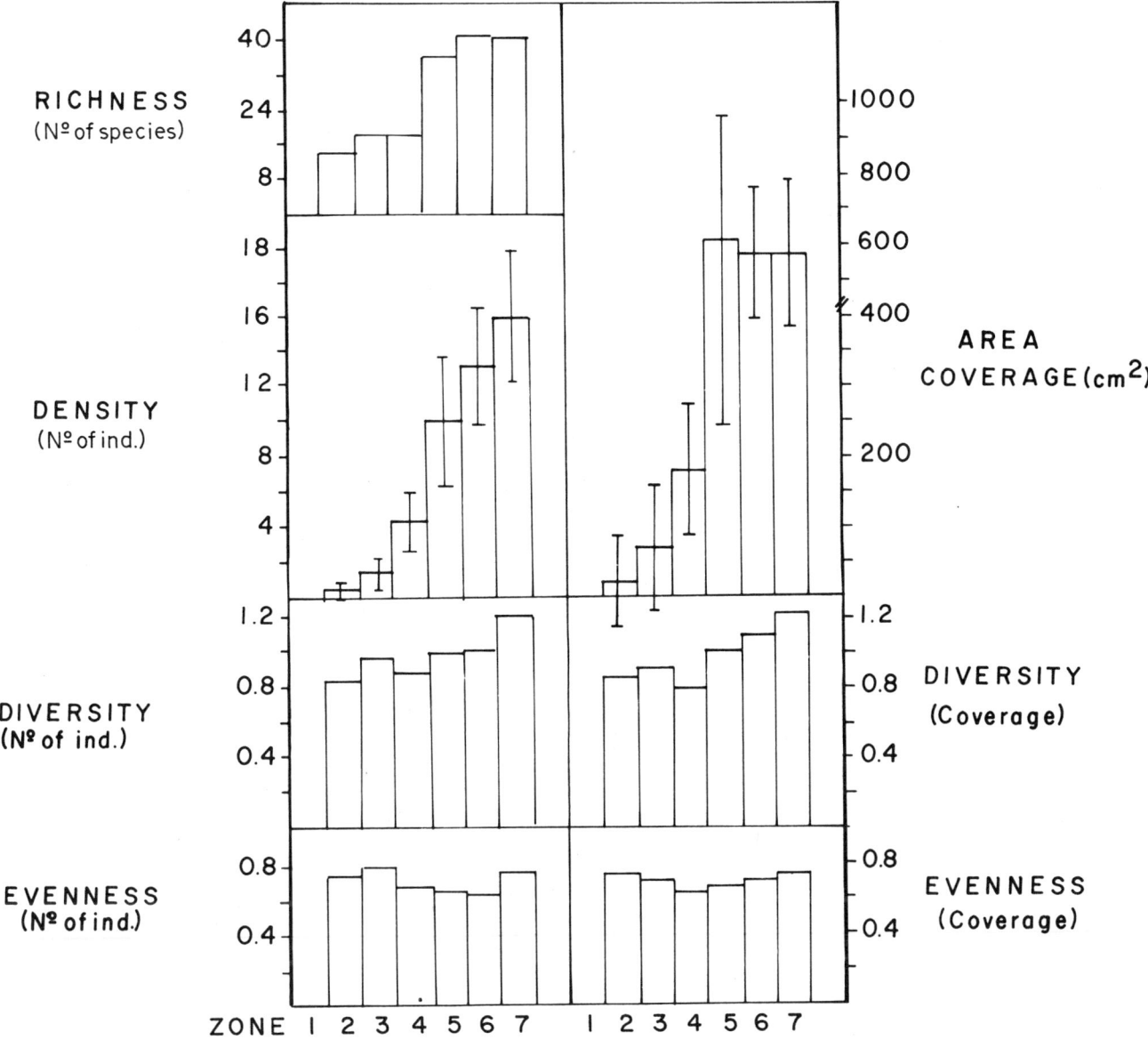

Figure 4. Variation in ecological parameters examined in seven reef zones identified at study site. Mean and standard deviation values shown for both density and area coverage. Vertical bar represents one standard deviation.

substrate is fairly complex and is densely occupied by a great diversity of sponges, stony corals, soft corals, and encrusting algae.

Species number and density also increase in reef zones 6 and 7, but morphological complexity of the substrate decrease. This finding appears to contradict the expected relationship between species number and morphological complexity of the substrate. Other environmental features in these reef zones, however, could explain this situation, at these depths low light intensity and high sedimentation rates limit the development of important space competitors such as stony corals (Wells, 1957; Porter, 1972; Loya, 1972). Thus more substrate is available for sponge colonization and growth. In these reef zones, however,

sponges are small and are frequently buried by sand; this burying could possibly limit sponge growth and may contribute to mortality. These factors in combination may affect the structure of sponge communities in these deep reef areas, where diversity is higher and evenness is fairly constant.

This analysis suggests that the structure of a sponge community is governed by biological and physical disturbances that vary in frequency and intensity in a nonlinear gradient along the depth profile. High levels of disturbance, such as those found in zone 1, limit community development. On the other hand, intermediate levels of disturbance (biological, as in zone 5, or physical, as in reef zones 6 and 7), could favor relatively high diversities.

These ideas need further testing, however, to determine the precise effect of biological and physical disturbances on the growth rate of sponge populations at different depths in the study area.

One of the more interesting findings of this study is the predominance of encrusting species at almost all depths, although they do not constitute a diverse group. Two hypothesis can be put forth to explain this situation in view of the fact that the study area is subjected to water currents coming from the inner lagoon of the archipelago. These turbid currents, which are frequent at certain times of the year, may carry sediments and organic material from other communities wich are very abundant the east of the study reef.

In discussing the importance of the growth form of sessile organisms in relation to their interspecific interactions (competition and predation), Jackson (1979) suggested that in environments where disturbance levels and substrate availability are low, as in the case of coral reefs, the relative frequency of each shape would depend on food abundance. In areas where food is scarce, erect growth forms would have an advantage over encrusting ones, owing to their greater accessibility to the water column. But in areas where food is abundant, the erect forms would lose this advantage, and the encrusting species, by means of their lateral growth, could prevent the colonization of the substrate by other forms. The large supply of food available for sponges in the study area could therefore favor high abundance of encrusting species.

The second explanation finds support in Reiswig's (1971) suggestion that the amebocyte cycle, one of the processes for capturing particles in Porifera, must be slow and easily saturable for massive sponges. Hence at the study site, where turbid waters are relatively common, massive sponges could have problems with this system restraining their growth rate and survival. This could also explain the high abundance of tiny specimens, group E, on the sandy substrate, and the relatively small size of massive species in the study area as compared with specimens in other regions. For example, at Cayo Sal reef, which is subjected to strong directional and clear oceanic currents, massive sponges attain extremely large sizes.

Conclusions

Species richness, density, area coverage, and species diversity of Porifera tend to increase toward the deeper zones of the reef at Dos Mosquises Cay, although these parameters vary greatly between adjacent points along the depth profile.

High and constant evenness along the depth profile and some field observations suggest an absence of local substrate monopolization by any sponge species at the study site.

The structure of sponge populations at the study site

appears to be determined both by depth-dependent factors and by parameters, such as substrate morphology, space competitors, and predators, whose intensities do not vary linearly with depth.

Encrusting sponges display the highest density of all form types at almost all reef zones along the depth profile in the study area.

Acknowledgments

This work was sponsored in part by Consejo Nacional de Investigaciones Científicas y Tecnológicas (CONICIT). The participation of two of us, in the Third International Conference on the Biology of Sponges was made possible by grants from the Smithsonian Institution.

We wish to thank all those who have helped us with this effort, particularly our friend Ernesto Weil, whose assistance in the field was invaluable. We are deeply indebted to R. W. M. van Soest and K. Rützler for their assistance in the taxonomic work. We are also grateful to Clive Wilkinson for his comments on the manuscript, and to the staff and members of the Fundación Científica Los Roques, whose cooperation made this survey possible. The Laboratory of Photography of the Universidad Central de Venezuela photographed the figures, and Pablo Rodríguez helped us with the drawings. This is Scientific Contribution No. 22 of the Fundación Científica Los Roques.

Literature Cited

Alcolado, P. M. 1979. Estructura ecológica de la comunidad de Esponjas en un perfil costero de Cuba. *Ciencias Biológicas*, 3:105–127.

Dixon, W. J., and M. W. Brown. 1979. *Biomedical Computer Programs, P–Series*. Berkeley: University of California Press. 877 pp.

Goreau, T. F., and Hartman, W. D. 1963. Boring Sponges as Controlling Factors in the Formation and Maintenance of Coral Reefs. Pages 24–25 in *Mechanisms of Hard Tissue Destruction*, edited by R. F. Sognnaes. Washington, D.C.: American Association for the Advancement of Science, Publication 75.

Green, G. 1977. Ecology of Toxicity in Marine Sponges. *Marine Biology*, 40:207–215.

Hechtel, G. J. 1969. New Species and Records of Shallow Water Demospongiae from Barbados West Indies. *Postilla*, 132:1–38.

Hung, M. 1985. Los corales Pétreos del Parque Nacional Archipiélago de Los Roques. Licenciature Thesis. Universidad Central de Venezuela. 204 pp.

Jackson, J. B. C. 1979. Morphological Strategies of Sessile Animals. Pages 499–555 in *Biology and Systematics of Colonial Organisms*, edited by G. P. Larwood and B. R. Rosen, London: Academic Press.

———. 1983. Biological Determinants of Present and Past Sessile Animal Distributions. Pages 39–107 in *Biotic Interactions in Recent and Fossil Benthic Communities 3*, edited by Michael J. S. Tevesz and P. L. McCall. New York: Plenum Press.

Jackson, J. B. C., and L. W. Buss. 1975. Allelopathy and Spatial Competition Among Coral Reef Invertebrates. *Proceedings, National Academy of Science, (USA),* 72:5160–5163.

Jokiel, P. L. 1980. Solar Ultraviolet Radiation and Coral Reef Epifauna. *Science,* 207:1069–1071.

Loya, Y. 1972. Community Structure and Species Diversity of Hermatypic Corals at Eilat, Red Sea. *Marine Biology,* 13:100–123.

Pielou, E. C. 1966. The Measurement of Diversity in Different Types of Biological Collections. *Journal of Theoretical Biology,* 13:131–144.

Porter, J. W. 1972. Patterns of Species Diversity in Caribbean Reef Corals. *Ecology,* 53:745–748.

Randall, J. E., and W. D. Hartman. 1968. Sponge Feeding Fisches of the West-Indies. *Marine Biology,* 1:216–225.

Reiswig, H. M. 1971. Particle Feeding in Natural Populations of Three Marine Demosponges. *Biological Bulletin,* 141:568–591.

———. 1973. Population Dynamics of Three Jamaican Demospongiae. *Bulletin of Marine Science,* 23:191–226.

Rützler, K. 1970. Spatial Competition among Porifera: Solution by Epizoism. *Oecologia,* 5:85–95.

———. 1975. The Role of Burrowing Sponges in Bioerosion. *Oecologia,* 19:203–216.

———. 1976. Ecology of Tunisian Commercial Sponges. *Tethys,* 7:249–264.

Sarà, M. 1970. Competition and Cooperation in Sponge Populations. Pages 273–284 in *The Biology of the Porifera,* edited by W. G. Fry. Symposia of the Zoological Society of London 25. London and New York: Academic Press.

Sarà, M., and J. Vacelet. 1973. Ecologie des Démosponges. Pages 462–526 in *Traité de Zoologie III. Spongiaires,* edited by P. P. Grassé. Paris: Masson.

Suchanek, T. H., R. C. Carpenter, J. D. Witman, and C. D. Harvell. 1985. Sponges as Important Space Competitors in Deep Caribbean Coral Reef Communities. Pages 55–59, in *The Ecology of Deep and Shallow Coral Reefs,* edited by M. L. Reaka. Symposia Series for Undersea Research 3(1). Rockville, Maryland: NOAA Undersea Research Program.

Villamizar, E. 1985. Fauna asociada a las esponjas *Aplysina lacunosa* y *Aplysina archeri* en el Parque Nacional Archipiélago de Los Roques, Venezuela. Licenciature Thesis. Universidad Central de Venezuela. 200 pp.

Wells, J. W. 1957. Coral Reefs. *Memoirs of the Geological Society of America,* 67:609–631.

Wilkinson, C. R. 1981. Significance of Sponges with Cyanobacterial Symbionts on Davis Reef, Great Barrier Reef. Pages 705–712 in *The Reef and Man,* edited by E. D. Gomez et al. Proceedings, Fourth International Coral Reef Symposium 2. Quezon City, Philippines, University of the Philippines.

———. 1983. Role of Sponges in Coral Reef Structural Processes. Pages 263–276 in *Perspectives on Coral Reefs,* edited by B. J. Barnes. Townsville: Australian Institute of Marine Science.

Wulff, J. L., and L. W. Buss. 1979. Do Sponges Help Hold Coral Reefs Together? *Nature,* 281:474–475.

GEORGE P. SCHMAHL[*]
Department of Zoology
University of Georgia
Athens, Georgia 30602

Community Structure and Ecology of Sponges Associated with Four Southern Florida Coral Reefs

Abstract

The distribution and abundance of sponges associated with four shelf-edge reefs located off southern Florida in the upper Florida Keys were investigated at three depth zones between the surface and 20 m. Sponges were enumerated within 1-m^2 quadrats at 5-m intervals along 100 m transects running parallel to representative depth contours. More than 80 species of sponges were found in abundance at these locations, whereas substrate coverage by large reef-building corals is low. Species composition and abundance with depth are similar in all four reefs. Species diversity and abundance appear to increase with depth, the deep (20 m) stations having significantly higher densities than the shallower sites. In spite of differences in abundance, the mid-depth and deep reef sites have more similar species compositions than the shallow locations. Intraspecies abundance also varies significantly with depth, and most species increase in density at the deeper stations. Turbulence and wave action due to frequent winter cold fronts and occasional hurricanes appear to limit the occurrence of many species in the shallow zones. Sponges common at the shallow sites are adapted to high turbulence and are usually encrusting or boring species. Because these reefs are near the northern extent of reef growth in South Florida, temperature may also be a limiting factor. The distributional patterns of sponges may be used to identify the ecological factors that influence the communities in this area.

Sponges are an important component of the benthic community of the coral reefs of southern Florida and the Florida Keys. Although work has been done on the systematics and species composition of the sponge fauna of this area, little is known about their distribution and abundance. Sponge communities are partitioned along habitat

---

[*]Present address: Office of Coastal Programs, Department of Community Affairs, 2740 Centerview Drive, Tallahassee, Florida 32399.

and depth gradients on coral reefs and are influenced by a variety of environmental factors (Laubenfels, 1950; Sarà and Vacelet, 1973). The purpose of this study is to document the occurrence and distribution of sponges associated with four shelf-edge reefs in Biscayne National Park, Florida, and to determine the variability of those assemblages with depth and location. Knowledge of the distributional patterns may provide insight into the ecological factors that influence species occurrence and community structure.

Study Area

Biscayne National Park is located near the southeastern tip of Florida (25°22′ N to 25°31′ N) and includes the waters of southern Biscayne Bay, the uppermost islands of the Florida Keys (primarily Elliott Key), and offshore waters to the 10 fathom (18 m) depth contour. Reefs within the park represent the approximate northern limit of an extensive coral reef system, known as the Florida Reef Tract, which extends 360 km along a southwestern arc from Miami to the Dry Tortugas. Biscayne National Park is the site of four elongated shelf-edge reefs—known as Triumph, Long, Ajax and Pacific—located approximately 7 km east of Elliott Key. These reefs lie in a general north-south orientation (Figure 1) and are separated from Elliott Key by a large, shallow lagoonal area in which thousands of patch reefs occur (Marszalek et al., 1977). The subject reefs are poorly developed, as evidenced by the relative scarcity of *Acropora palmata* and the low abundance of other corals (Burns, 1985). Vertical relief with depth is minimal (Figure 2), and vigorous reef growth ceases between 20 m and 30 m, at the edge of a broad sandy plain. Low water temperatures are usually considered the primary reason for the lack of reef growth north of this area (Vaughn, 1918).

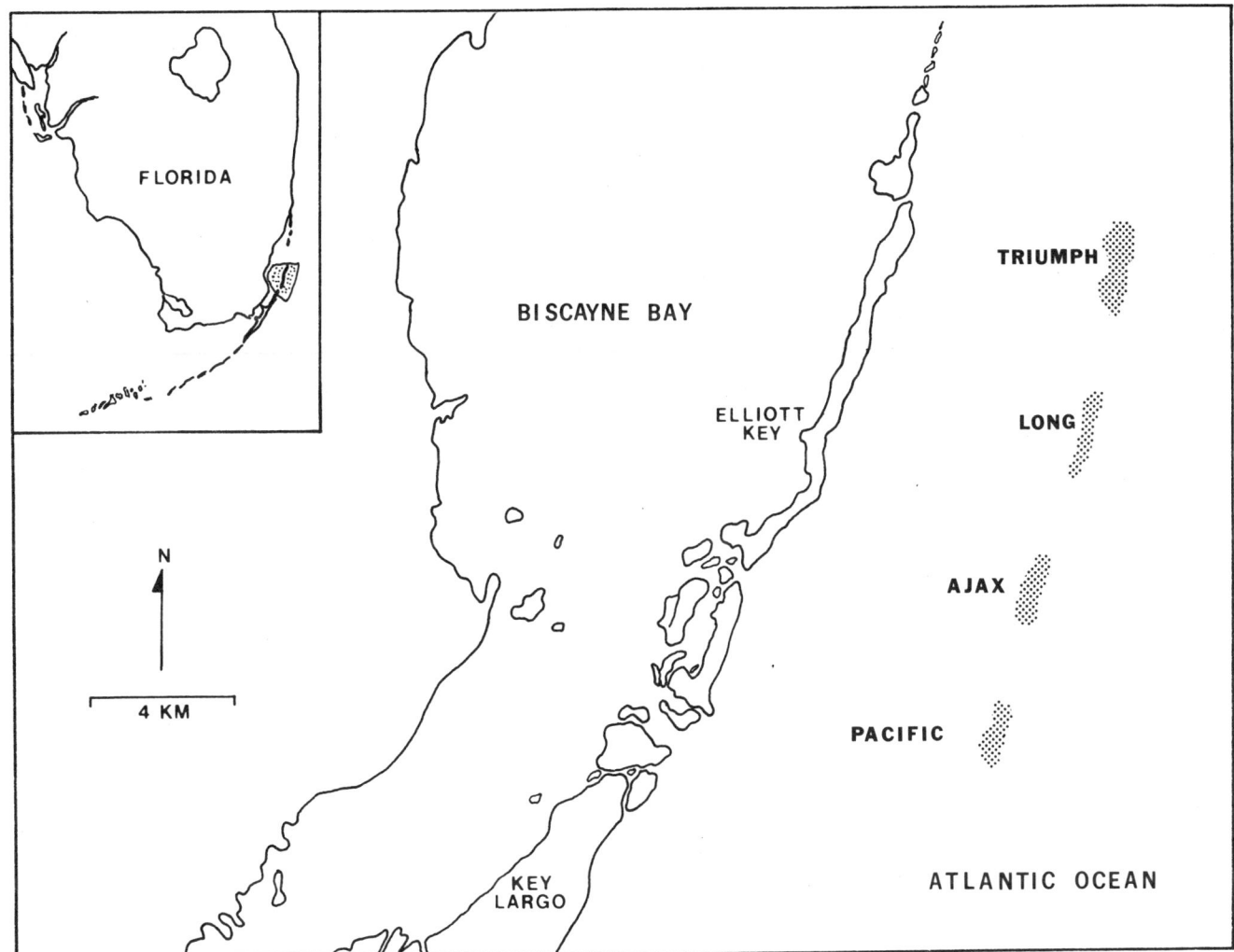

Figure 1. Map showing Biscayne National Park (shaded area of insert) and study reef locations.

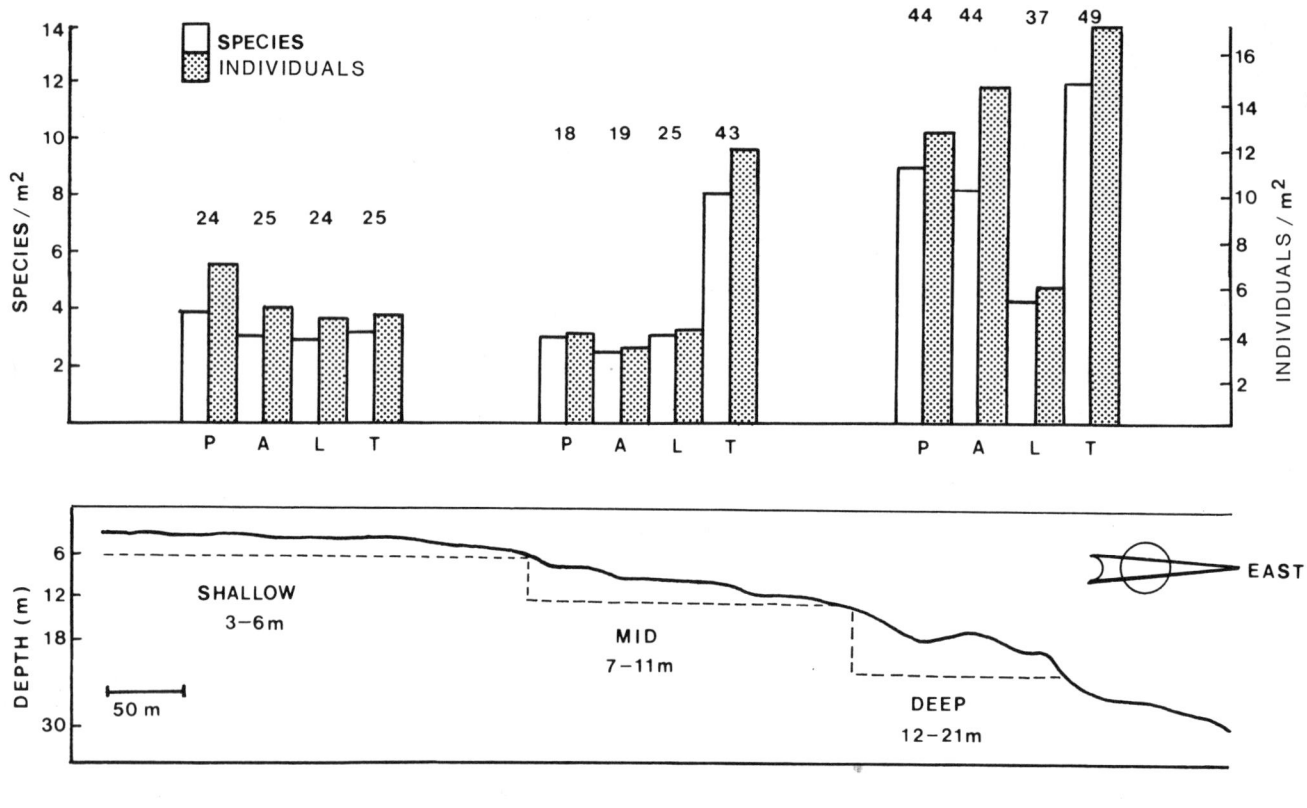

Figure 2. Reef profile of Ajax reef indicating depth-zone sample areas, and histogram of average number of species and individuals per square meter for all study reefs by zone. *A,* Ajax; *L,* Long; *P,* Pacific; *T,* Triumph; numbers above histograms indicate total number of species found.

Materials and Methods

On the basis of large-scale visual surveys and fathometer tracings of the study area, the reefs were divided into three depth zones for quantitative sampling: 1–5 m = shallow (reef flat); 6–12 m = mid (reef slope); and 13–18 m = deep (forereef) (Figure 2). At similar depths at each reef, two 100 m transect lines were randomly placed parallel to the specified depth contour. A 1-m² quadrat was placed at 5-m intervals along each transect, and all epibenthic sponges were recorded and enumerated within quadrats, for a total of 40 m² sampled per station. Identifications were made in the field whenever possible, and voucher specimens were collected for verification. Sponge identifications were based on the available literature. This study was carried out during the summer of 1981.

The size of representative samples was established on the basis of species/area curves derived from randomly assorted quadrats at each station. Typically, between 20 m² and 35 m² were required to adequately sample the community. Mean number of species and individuals were calculated per square meter for each reef zone. Diversity was measured using the Shannon-Weaver index. The statistical significance of differences between reef zones

was tested by ANOVA on mean numbers per quadrat and by nonparametric STP (Sokal and Rohlf, 1981) using totals of randomly allocated groups of five quadrats as the sample statistic. Nonparametric STP was also used to test differences in abundance within species. In addition, mean values from eight previously studied lagoonal patch reefs were compared with mean values obtained from eight randomly allocated groups of quadrats in the shallow and deep zones, using nonparametric STP.

Pairs from every combination of location and depth were compared using the percentage similarity index (Gauch, 1982). Group average classification analysis was carried out on the similarity matrix using a computer program given in Orloci (1975), which was adapted for use with nonmetric indices.

Results

A total of 84 species of sponges was found in the sampled quadrats. Of these, 14 species had extremely low population levels (one or two individuals) and could not be identified. The remaining 70 species were considered in the analysis (Table 1). In general, sponge abundance and diversity increase with depth at all four reef locations (Fig-

Table 1. Species list and abundance of sponges (numbers of individuals) on four south Florida reefs (S = shallow zone, M = mid zone, D = deep zone)

Species	Triumph			Long			Ajax			Pacific			Total
	S	M	D	S	M	D	S	M	D	S	M	D	
Ircinia felix	6	3	8	2		1			5	2	2		38
Ircinia campana			2										2
Ircinia strobilina	2	4	20			2		1	5		2		36
Ircinia ramosa (?)	2	2	2		1	1	12	3	1	11			3
Dysidea janiae				1									1
Dysidea etheria		10	12		1	2	1		4			3	33*
Dysidea fragilis			14						3			4	21
Dysidea variabilis		7	10									1	18
Aplysina fistularis		3		3	3	5		1				6	21
Aplysina archeri									1				1
Aplysina lacunosa		1	2		1	2			5			5	16*
Aplysina cauliformis		41	44		3	11		2	17		4	8	130*
Pseudoceratina crassa	4	8	12			5	2		8			6	45*
Verongula rigida			4			5						3	12
Smenospongia aurea			2			1			22			2	27*
Aplysilla sulfurea				1					1	1			3
Adocia implexiformis (?)			16										16
Amphimedon compressa	4	13	20	2	7	6	3		11	2	11	16	95*
Niphates erecta	4	25	42		3	17	3	1	56		7	62	220*
Niphates digitalis	4	27	22	5	20	25	8	23	30	5	25	28	222*
Niphates amorpha		45	20	6	33	28	11	36	29	5	31	38	282*
Siphonodictyon coralliphagum						6							6
Callyspongia vaginalis	1	19	10	9	15	15	10	17	14		14	15	139*
Callyspongia armigera	2	15	22						9	1	10	6	65
Callyspongia fallax		1	1	1					1			2	6
Callyspongia plicifera		3	12			1			5		1	1	23
Xestospongia muta		18	22		2	1		1	16		1	16	77*
Mycale laevis		17	2		1	8	2	2	27	2	10	15	86*
Mycale laxissima					1					4	5		10
Holopsamma helwigi		11	60			6			19				96*
Iotrochota birotulata		14	36			1			12		1	14	78*
Monanchora unguifera	4		2			1							7
Merriamium tortugasensis		7	10			12			22			29	80*
Tedania ignis				5	3	7	7	3		14		1	40
Clathria spinosa	2	1	2									2	7
Rhaphidophlus juniperinus	6	1		12			13		1	26			59*
Pandaros acanthifolium		3	4			1	2						10
Ulosa ruetzleri	34	42	74	10	2	20	14	11	117	53	7	30	414
Homaxinella rudis		2	4			4	1		4	1		1	17*
Teichaxinella morchella		4	12						2			1	19
Pseudaxinella lunaecharta	1	14	6	2	1	2				1	3	4	34
Ectyoplasia ferox	1	2	24	3	4	5	8	4	28	18	11	13	121*
Agelas schmidti	1	1								4			6
Agelas conifera		3	4	1									8
Agelas clathrodes				2	1		1						4
Spirastrella cunctatrix		23	10	6	1		2	1				2	45
Spirastrella coccinea	8	3	2	5	18	2	10	4	2	4	10	10	78
Anthosigmella varians		1	20	1	2	1	7	3	6	1	3	9	54
Spheciospongia vesparium		5	6			3		2	4			3	23*
Timea mixta	12	2							1	31		6	52
Tethya crypta	18	11	8	39	25		13						114*
Cliona delitrix	2	5	6	11		5	5		23	7	1	20	85*
Cliona caribbaea			2						1		2		5
Cliona schmidti	16												16
Cliona vermifera		1					1		2	2		3	9
Cliona sp. (brown)	30			2			27						59
Cliona laticavicola (?)	12			4						5			21
Cinachyra alloclada			2										2
Cinachyra kuekenthali						1							1
Chondrilla nucula	1						1					3	5
Chondrosia collectrix												1	1
Clathrina coriacea									1				1
Unidentified:													
Orange axinellid		2						1	3			1	7
Red bushlike									4				4
Orange encrusting	2	2	8	15	11	5	4	4		6			57
Black encrusting			8						4				12
Black endolithic				18			32		1	64			115
Red encrusting		30	12									19	61
Black massive		3	2	5		1			1	3		1	16
Orange massive			12		1				16			1	30
Totals	179	454	658	169	161	220	198	120	546	269	154	426	

*Significant difference in abundance among depth zones (all reefs combined)

ure 2, Table 2). Sponge abundance tends to be low in the shallow zones (4.6–6.9 individuals per quadrat) and does not increase significantly in the mid-depth regions. In fact, abundance and diversity between the shallow and the mid zones decrease slightly at three of the four reefs, although these differences are not significant. A dramatic exception to this pattern occurs in the mid-depth zone at Triumph reef, where species and individuals per quadrat increase significantly over the shallow zone. At all locations except Long reef, significant increases in abundance and diversity were noted at the deep reef stations. At Long reef, no significant differences were noted among the three depth zones, although there was an increase in the total number of species at the deep sample site. The highest abundance values in this study (averaging 17.4 individuals per quadrat) were recorded at the deep reef zone on Triumph reef. This, however, was not a significant increase over the unusually abundant mid-zone fauna at the same location.

Abundance patterns were found to be similar at corresponding depth contours at all four reefs. There were no significant differences in numbers of species or individuals per square meter in any of the shallow or mid-depth zones, except in the Triumph mid zone, as noted above. Abundance values in that zone and in all of the deep zones, except for Long reef, were also not significantly different. The Long reef deep zone was found to differ only slightly from the mid zones of the other reefs. In general, two abundance patterns are represented in the study area. The first is a high abundance group, which includes the mid-depth station at Triumph and all deep reef stations except for Long reef. Significantly lower abundance values were recorded at the shallow and remaining mid-depth stations. This group includes the deep station at Long reef, although some significant differences were noted between this site and some shallow transects. All in all, however, the density of species and individuals are similar over the depth range studied at each reef.

The occurrence and relative abundance of species within the sample sites were compared between all reef zones using pairwise calculations of the percentage similarity index and classification analysis. The sample sites generally grouped together according to depth zones (Figure 3). At first glance, two major divisions seem to emerge. The first consists solely of the four shallow zones, which are associated with each other at relatively low similarity levels. The second major division includes both the mid and deep reef zones. The latter group can be subdivided into what are essentially mid-depth and deep reef clusters. The mid-depth zone at Triumph reef is grouped with the deep stations, and the deep reef zone at Long reef is grouped with the mid-depth cluster. Interestingly, although the shallow and mid zones do not differ significantly in terms of abundance measures, the species composition of the mid zone is much more similar to that of the deep reef. The highest similarity values were recorded between Ajax and Pacific reefs.

Results of the classification analysis suggest that assemblages of species may account for the distinct clustering of locations by depth. Among the species in the study area, abundance patterns vary with depth and location (Table 1). Some species are particularly abundant; 21 species accounted for more than 75% of the total individuals observed. Every species was tested for significant differences between depth zones. For the purposes of this analysis, all reefs were combined. A total of 21 species exhibited signif-

Table 2. Abundance and diversity of sponges in four South Florida reefs (Sp. = species; Ind. = individuals; SD = standard deviation; H' ln = Shannon–Weaver diversity index, natural log base; J = index of evenness)

Reef	Zone	Zone totals		Transect means (SD)		Diversity	
		Sp.	Ind.	Sp./m²	Ind./m²	H'ln	J
Triumph	Shallow	25	179	3.25 (1.6)	4.75 (2.5)	2.66	0.83
	Mid	43	454	8.08 (2.2)	12.05 (3.9)	3.23	0.86
	Deep	49	658	10.40 (3.8)	17.45 (7.3)	3.39	0.87
Long	Shallow	24	169	2.98 (1.2)	4.58 (2.1)	2.73	0.86
	Mid	25	161	3.12 (1.1)	4.15 (1.9)	2.52	0.78
	Deep	37	220	4.32 (2.4)	5.90 (3.9)	3.07	0.85
Ajax	Shallow	25	198	3.02 (1.7)	5.08 (3.5)	2.80	0.87
	Mid	19	120	2.50 (1.4)	3.28 (2.3)	2.23	0.76
	Deep	44	546	8.38 (3.6)	14.79 (8.5)	3.04	0.80
Pacific	Shallow	24	269	3.98 (1.4)	6.92 (3.1)	2.43	0.76
	Mid	19	154	3.02 (1.8)	3.95 (2.6)	2.50	0.85
	Deep	44	426	8.97 (3.0)	12.89 (5.3)	3.19	0.84

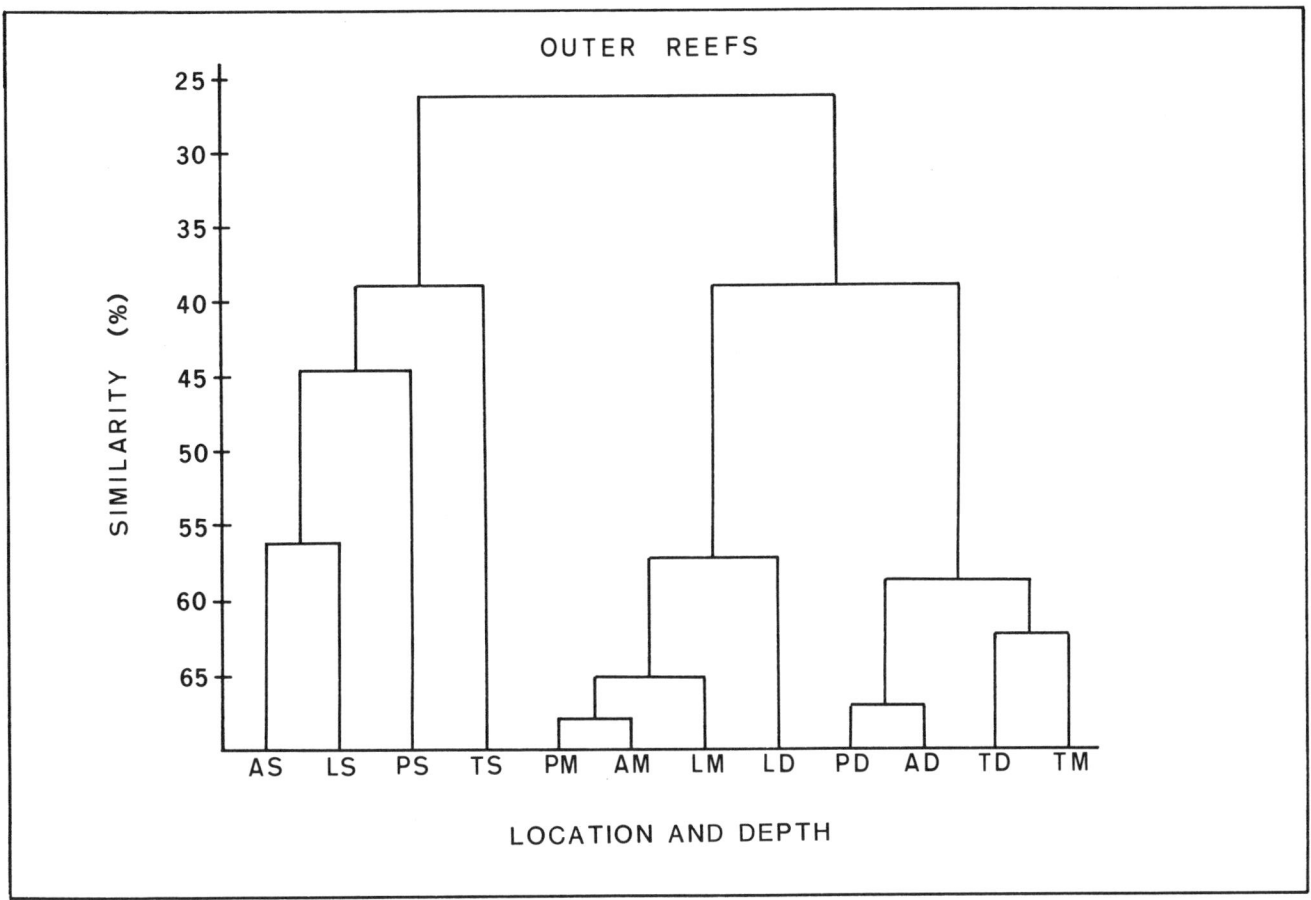

Figure 3. Cluster analysis of reef zones using percentage similarity index and group-average classification. *A*, Ajax; *L*, Long; *P*, Pacific; *T*, Triumph; *D*, deep; *M*, mid; *S*, shallow.

icant differences over depth. Two species, *Rhaphidophlus juniperinus* and *Tethya crypta*, were found to be significantly more abundant in the shallow zones. *Cliona delitrix* was found to be significantly less abundant in the mid-depth region but had similar abundance in the shallow and deep zones. Among the remaining species, abundances increase significantly with increasing depth. Five species were found to be significantly more abundant only in the deep zones: *Pseudoceratina crassa, Smenospongia aurea, Niphates erecta, Phorbas amaranthus,* and *Homaxinella rudis*. The following species increase significantly in abundance over depth, with the mid zone showing intermediate values: *Aplysina lacunosa, Dysidea etheria, Amphimedon compressa, Holopsamma helwigi, Mycale laevis, Iotrochota birotulata, Spheciospongia vesparium* and *Ectyoplasia ferox*. Those species in which the deep and mid zones showed significantly greater abundance than the shallow zone include: *Aplysina cauliformis, Niphates amorpha, Niphates digitalis, Callyspongia vaginalis,* and *Xestospongia muta*. Common species that showed no significant difference with depth are *Ulosa ruetzleri, Spirastrella coccinea, Callyspongia armigera,* and *Anthosigmella varians*.

Discussion

The results of this study are generally consistent with those reported by Alcolado (1979) and Wilkinson (1982), who documented increases in sponge abundance over a similar depth range. Ecological factors that vary with depth appear to be responsible for the observed distributions. One such factor is turbulence, or physical disturbance due to wave action, which decreases substantially with increasing depth. Wave action is thought to play an important role in sponge distributions on other Caribbean coral reefs (Reiswig, 1973; Alcolado, 1979). A study conducted along other transects across the Florida shelf in this area (Enos, 1977) has demonstrated that the outer bank reefs buffer the large lagoonal area from the wave action related to prevailing easterly winds and storm events, which commonly accompany winter cold fronts. Therefore, information on the abundance of sponges associated with the lagoonal patch reefs may provide evidence of how sponge occurrence is affected by exposure to wave stress. An earlier study of eight representative patch reefs has documented the abundance of sponges in square-meter quadrats (Schmahl, 1987). These patch reefs occur in depths less than 6 m, which correspond with the depth of the shallow zone in the present study. Sponge abundance on the patch reefs is substantially higher than in the shallow zones of the outer reefs (averaging 12.7 individuals per square meter) and is generally similar to abundance values observed in the deep reef zones. This pattern is consistent with results obtained for some scleractinian coral communities in the Caribbean, which are reported to be significantly more abundant on the leeward areas of exposed reefs (Storr, 1964; Jordan et al., 1982).

The abundance of species found to be significantly more abundant in the deeper zones of the outer reefs was compared with that of the same species on the patch reefs. Eight of 18 such species were found to have similar abundances on the patch reefs and on the deep reefs. Six of these species were found to be significantly more abundant on the patch reefs than at similar depths on the shallow outer reef. Therefore these species seem able to live in shallow water and it may be that protection from wave action by the outer reef is an important factor influencing their distribution. Four of these species—*Aplysina cauliformis*, *Niphates erecta*, *Amphimedon compressa*, and *Iotrochota birotulata*—are branching forms that would probably be unable to withstand the mechanical stress caused by wave action. Another species in this group, *Dysidea etheria*, although semiencrusting, is quite fragile and easily torn. Sponges that do occur in high-impact wave areas usually exhibit morphological adaptations, such as an increase in wall thickness (tubular forms) or reinforcement of support tissues, that are useful in resisting damage (Reiswig, 1973; Palumbi, 1984). Most of the common species in the shallow zone of the study area are encrusting or boring sponges. *Rhaphidophlus juniperinus*, which can exhibit both encrusting and branching morphs, occurs on the shallow outer reef exclusively in the encrusting form. Our observations of the patch reef populations indicate that this species occurs primarily in shallow areas, since patch reef abundance did not differ significantly from that on the shallow outer reefs, while both areas had significantly higher abundances than the deep reef zones. The fact that *Tethya crypta* was significantly more abundant only on the shallow zone of the outer reef suggests that this species may be adapted to the high-energy zone under the influence of waves.

The distribution of the other species does not correspond solely with exposure to wave surge. For example, eight of the remaining ten species that were found to increase with depth on the outer reefs were also significantly less abundant or absent on the patch reefs. These species appear to be restricted mainly to deeper reef locations (13 m) within this study area and include *Pseudoceratina crassa*, *Aplysina lacunosa*, *Niphates digitalis*, *Xestospongia muta*, *Holopsamma helwigi*, *Phorbas amaranthus*, *Homaxinella rudis*, and *Spheciospongia vesparium*. Although wave stress may have

some affect on the distribution of these species, the fact that they are not common on the shallow patch reefs indicates that other factors are also involved.

The harmful effects of ultraviolet light in very shallow water, for example, may restrict the occurrence of some species to deeper water (Wilkinson, 1982). The high abundance of sponges on the shallow patch reefs demonstrates that this phenomenon is not a limiting factor for all species. This suggestion can only be confirmed through experimental transplants. Note that the outer and patch reefs differ in other respects beside protection from wave surge. The lagoonal reefs are subjected to reduction in water circulation and currents, increased turbidity, and greater fluctuations in temperature (Enos, 1977). The currents associated with the Gulf Stream, which flows north adjacent to the outer reefs, are strong and relatively consistent. Some sponges, especially large forms such as *Xestospongia muta*, may require strong currents to assist in water transport and reduce energetic constraints (Vogel, 1977). Other species, such as *Aplysina lacunosa*, appear to be adversely affected by increased sedimentation (Gerrodette and Flechsig, 1979). Therefore, the reduced currents and increased turbidity in the patch reef area may influence the abundance of such species.

Increased fluctuations in temperature may restrict the occurrence of some sponges in the lagoonal area and variably with depth. Water temperatures within this segment of the Florida Reef Tract vary from a minimum of 18.5°C at Carysfort reef, located just south of the study area, to a minimum of 15.6°C at Fowey Rocks, which is immediately north of Triumph reef (Vaughn, 1918). The low temperatures recorded at Fowey Rocks are most probably due to the influence of the cooled water from Biscayne Bay, which enters the reef tract through the passes north of Elliott Key (known as the "safety valve"); this water can reach very low levels after periods of cold weather (Marszalek et al., 1977). A minimum of 16.7°C has been recorded within the lagoonal area at one of the patch reefs used in these comparisons (Jaap, 1984). Burns (1985) suggests that the influence of cold water from tidal passes is the most important factor in restricting coral growth over depth in the study area. Although such temperatures do not seem to restrict total abundance of sponges in the lagoonal area, the occurrence of certain species may be affected by wide temperature fluctuations. Some species are apparently able to withstand lower water temperatures, as suggested by the presence of a number of West Indian sponges off North Carolina (Wells et al., 1960), including *Callyspongia vaginalis*, *Homaxinella rudis*, and *Spheciospongia vesparium*.

The distributional patterns on the four study reefs indicate that similar environmental factors influence sponge occurrence throughout this geographical area. This is in contrast to other studies, however, where community structure, even within a limited area and at similar depths, can vary greatly. For example, whereas measures of abundance were similar at the same depth contour at three sites in Salt River Canyon, St. Croix, U.S. Virgin Islands, community composition was substantially different (Targett and Schmahl, 1984). Factors unrelated to depth, such as substrate orientation and degree of sedimentation, appear to be responsible for the observed distributions at that location.

Conclusions

The distribution and abundance of sponges on four adjacent shelf-edge reefs in South Florida exhibit similar community patterns. In general, sponge abundance increases with depth; we recorded the greatest values in water deeper than 13 m. Wave action and turbulence appear to be a primary factor in the observed distributions, for a number of the species that increase with depth on the outer reefs are found in equal abundances on shallow-water lagoonal patch reefs that are relatively protected from wave stress. However, the occurrence of other species cannot be explained by this factor alone. Other factors that probably are involved include light, temperature, current regimes, and turbidity.

Literature Cited

Alcolado, P. M. 1979. Ecological Structure of the Sponge Fauna in a Reef Profile of Cuba. Pages 297–302 in *Biologie des Spongiaires*, edited by C. Lévi and N. Boury-Esnault. Colloques Internationaux du C.N.R.S. 291. Paris: Centre National de la Recherche Scientifique.

Burns, T. P. 1985. Hard-Coral Distribution and Cold-Water Disturbances in South Florida: Variation with Depth and Location. *Coral Reefs*, 4:117–124.

Enos, P. 1977. Quarternary Sedimentation in South Florida. Part I. Holocene Sediment Accumulations of the South Florida Shelf Margin. *Geological Society of America, Memoirs*, 147:1–130.

Gauch, H. G., Jr. 1982. *Multivariate Analysis in Community Ecology*. Cambridge: Cambridge University Press. 298 pp.

Gerrodette, T., and A. O. Flechsig. 1979. Sediment Induced Reduction in Pumping Rate of the Tropical Sponge *Verongia lacunosa. Marine Biology*, 55:103–110.

Jaap, W. 1984. The Ecology of South Florida Coral Reefs: A Community Profile. Washington, D.C.: U.S. Fish and Wildlife Service, FWS/OBS–82/08. 138 pp.

Jordan, E., M. Merino, O. Moreno, and E. Martin. 1982. Community Structure of Coral Reefs in the Mexican Caribbean. Pages 303–308 in *Proceedings, Fourth International Coral Reef Symposium* 2. Quezon City, Philippines: Marine Science Center, University of the Philippines.

Laubenfels, M. W. de. 1950. An Ecological Discussion of the Sponges of Bermuda. *Transactions, Zoological Society of London*, 27:154–201.

Marszalek, D. S., G. Babashoff, Jr., M. R. Noel, and D. R. Worley. 1977. Reef Distribution in South Florida. Pages 223–229 in *Proceedings, Third International Coral Reef Symposium* 2, Miami, Florida: Rosenstiel School of Marine and Atmospheric Science, University of Miami.

Orloci, L. 1975. *Multivariate Analysis in Vegetation Research*. The Hague: Junk. 167 pp.

Palumbi, S. R. 1984. Tactics of Acclimation: Morphological Changes of Sponges in an Unpredictable Environment. *Science*, 225:1478–1480.

Reiswig, H. M. 1973. Population Dynamics of Three Jamaican Demospongiae. *Bulletin of Marine Science*, 23:191–226.

Sarà, M., and J. Vacelet. 1973. Ecologie des Démosponges. Pages 462–526 in *Traité de Zoologie 3(1), Spongiaires*, edited by P. P. Grassé. Paris: Masson.

Schmahl, G. P. 1987. Aspects of the Ecology of Sponges (Porifera) Associated with Coral Reefs of Biscayne National Park, Upper Florida Keys. Masters Thesis, University of Georgia, Athens. 49 pp.

Sokal, R. R., and F. J. Rohlf. 1981. *Biometry*. 2d Edition. San Francisco: W. H. Freeman. 859 pp.

Storr, J. F. 1964. Ecology and Oceanography of the Coral Reef Tract, Abaco Island, Bahamas. *Geological Society of America, Special Paper*, 79:1–98.

Targett, N. M., and G. P. Schmahl. 1984. Chemical Ecology and Distribution of Sponges in the Salt River Canyon, St. Croix, U.S.V.I. NOAA Technical Memoir, OAR NURP–1. Rockville, Maryland: National Oceanographic and Atmospheric Administration. 30 pp.

Vaughn, T. W. 1918. The Temperature of the Florida Reef Tract. *Carnegie Institute of Washington, Publication*, 213:319–339.

Vogel, S. 1977. Current-Induced Flow Through Living Sponges in Nature. *Proceedings, National Academy of Science*, (USA), 74:2069–2071.

Wells, H. W., M. J. Wells, and I. E. Gray. 1960. Marine Sponges of North Carolina. *Journal, Elisha Mitchell Scientific Society*, 76:200–245.

Wilkinson, C. R. 1982. Significance of Sponges with Cyanobacterial Symbionts on Davies Reef, Great Barrier Reef. Pages 705–712 in *The Reef and Man*, edited by E. D. Gomez et al. Proceedings, Fourth International Coral Reef Symposium. Quezon City, Philippines: Marine Science Center, University of the Philippines.

SHIRLEY A. POMPONI*
DONALD W. MERITT
University of Maryland
Horn Point Laboratories
P.O. Box 775
Cambridge, Maryland 21613

Distribution and Life History of the Boring Sponge *Cliona truitti* in the Upper Chesapeake Bay

Abstract

Cliona truitti bores into oyster shells on 90% of the oyster bars under investigation in the upper Chesapeake Bay. Boring may affect the growth rates of oysters or cause early mortality since weakened shells become more susceptible to predation. *Cliona truitti* reproduces both sexually and asexually by gemmules, which begin to hatch in early spring when the water temperature rises above 15°C (mid-March). Gemmule hatching is completed and rapid cell proliferation is initiated after the water temperature rises above 20°C (in late May to early June). Gametogenesis begins as early as March and continues throughout June. Larval settlement and metamorphosis occur during June and July and coincide with oyster spat set and consequent availability of fresh calcium carbonate substrate. The sponge gemmulates when water temperature drops below 20°C in late summer and early fall. Adult tissue regresses during the winter. Approximately 50% of the oyster shell is affected by borings, regardless of age, and highest rates occur during the period of rapid somatic growth in both adult and recently settled sponges. It is likely that *Cliona truitti* affects oyster growth and productivity at certain sites on the upper Chesapeake Bay.

Cliona truitti Old (1941) is an estuarine sponge that bores into the calcium carbonate shell of the commercially important American oyster *Crassostrea virginica*. In 1941, Old estimated that 25–75% of all oysters and cultch in low salinity waters of Chesapeake Bay were affected by this species. Although *Cliona truitti* can be found in empty shells, it is most commonly associated with live oysters, including recently settled spat. The oyster responds to boring organisms by depositing more shell to wall off the penetration (Bailey-Brock and Ringwood, 1982). In shells

---

*Present address: Harbor Branch Oceanographic Institution, 5600 Old Dixie Highway, Fort Pierce, Florida 34946.

that are heavily bored, the oyster tissue appears thin and watery (Fasten, 1931), and the shell becomes weakened, with the result that the oyster is more susceptible to predation (Bailey-Brock and Ringwood, 1982). It has been estimated that shell deposition may require as much as one-third of the total energy of growth (for a review, see Wilbur and Saleuddin, 1983). Shell repair in response to boring could result in reduced growth rates and stunting (Cole and Waugh, 1959; Kennedy and Breisch, 1981) and, consequently, in reduced oyster productivity and marketability. Stunting has been observed in at least two oyster bars in the upper Chesapeake Bay and has led to intensive studies of one subestuary, Broad Creek, to determine possible causes. This research was part of those studies.

Our objectives were to determine the distribution and abundance of *Cliona truitti* in the Chesapeake Bay, to study the life history of *C. truitti* and establish relationships between the annual cycle of this species and that of *Crassostrea virginica* and to begin to assess the impact of sponge boring on the oyster by measuring extent of boring.

Methods

DISTRIBUTION AND ABUNDANCE. We surveyed by dredging 53 commercially important oyster bars in the upper Chesapeake Bay in October 1981. Abundance of boring sponge species was recorded for each site. Observations on selected bars were made using scuba.

LIFE HISTORY. *Cliona truitti* populations at Deep Neck Bar in Broad Creek, a well-studied subestuary of the Choptank River, were monitored for abundance, growth, sexual reproductive activity, gemmule production, and dormancy from 1981 through 1983. Populations were sampled biweekly from October 1982 through April 1983 and weekly from May through September 1983, and were examined histologically. Tissue was fixed in Bouin's, 10% buffered formalin, or 2% glutaraldehyde/1% osmium tetroxide (cacodylate buffered). Bouin's- and formalin-fixed tissue was decalcified prior to embedding. Paraffin sections were stained with hematoxylin and eosin; semi-thin epoxy sections were stained with toluidine blue.

EXTENT OF BORING. Oysters from two year classes (1980, n = 328; 1981, n = 62) were randomly sampled from populations in Broad Creek, in September 1981 and biweekly from late April through August 1982. Extent of boring was calculated from measurements of bored areas seen on X-radiographs of shells (Figure 1). Total shell area, as well as area containing borings (bored area), were measured with a digitizer tablet. The sponge chamber area was calculated by tracing borings from X-rays to an

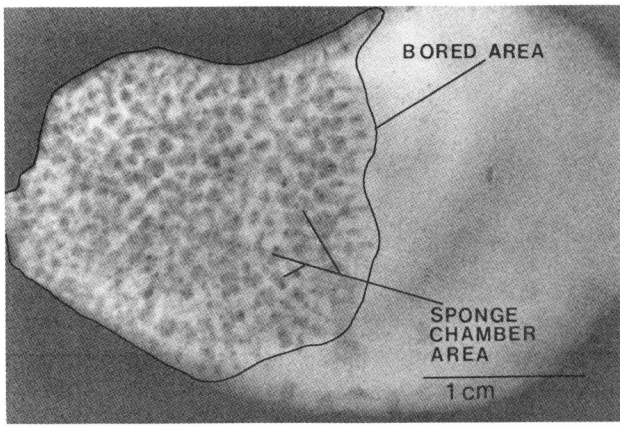

Figure 1. X-ray of oyster shell bored by *Cliona truitti. Bored area*, surface area of shell affected by borings; *sponge chamber area*, surface area of bored chambers.

acetate overlay and measuring the tracing with a Li-Cor Model 3100 Area Meter.

Results and Discussion
DISTRIBUTION AND ABUNDANCE

The distribution of *Cliona truitti* in the upper Chesapeake Bay is illustrated in Figure 2. Live *C. truitti* was observed in living oyster shells on 90% of the bars surveyed. The most abundant populations occurred on three oyster seed-producing sites: Broad Creek, the Little Choptank River, and the lower Potomac River. An oyster seed-producing bar is characterized by good cultch (settlement substrate) and good spat set. In two of these sites, Broad Creek and the Little Choptank, many oysters are transplanted to other areas where they can grow to a more desirable size.

When oysters are commercially transplanted from seed areas, any boring sponge present can be inadvertently killed as the dredged oysters lie exposed to air during transport to the transplantation site. The sponge will not survive desiccation unless gemmules are present in the tissue and the site to which the oysters are transplanted will support sponge growth. If gemmules are present, the potential exists for spreading the sponge via plantings to sites where it previously did not occur. There is evidence to suggest that this has occurred at locations in the Potomac, Patuxent, and upper Choptank rivers.

Our hypothesis that larger pieces of the shells that we collected by dredging were recently broken by blue crab predation was confirmed by observations using scuba. Spat settle gregariously (Hidu, 1969) and form clumps of several individuals attached to each other as they grow. Once the oysters reach a certain size, however, they break apart from one another. Our observations indicate that while clumped, the oysters are protected from blue crab predation by their overall size. Once they break apart,

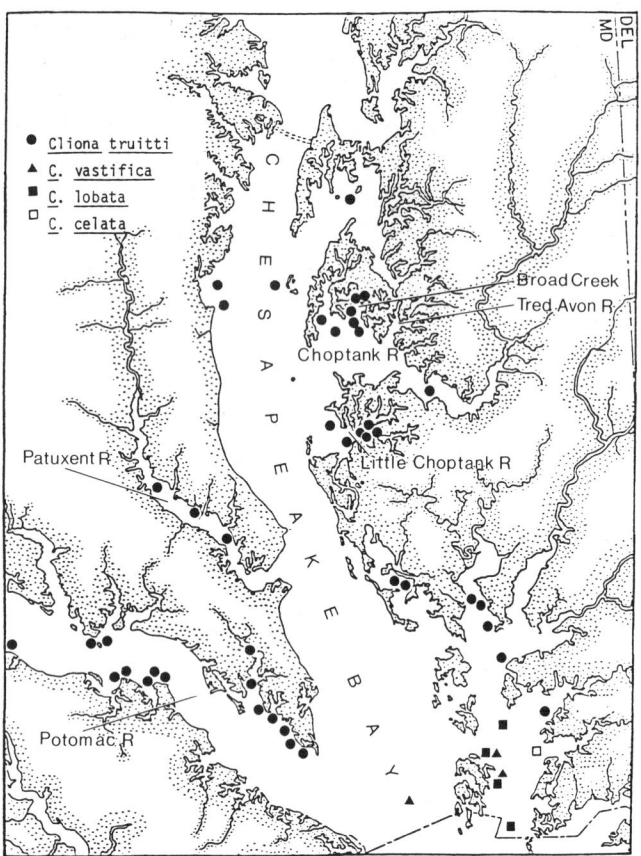

Figure 2. Distribution of boring sponges in upper Chesapeake Bay, October 1981.

however, they can be grasped by blue crab claws. If weakened by borings, the shell is easily snapped or crushed.

Some researchers have hypothesized that the stunted growth of oysters observed in Broad Creek is the result of environmental factors. Comparisons of nutrients, chlorophyll levels, and phytoplankton size, biochemistry, and abundance in Broad Creek and the Tred Avon, an adjacent subestuary of the Choptank River in which oysters are not stunted, have, to date, revealed no significant differences between the two systems (T. Jones, University of Maryland, Horn Point Laboratories, pers. comm.). Berg and Newell (1986) report no differences in food quality between the two systems, although "food quantity" was higher in the Tred Avon River during the two summers of their study. They suggest that higher food quantity in the Tred Avon River accounts for the larger size of the oysters.

One noticeable difference between the two systems is that *Cliona truitti* occurs on the bars in Broad Creek where oysters are stunted and does not occur in the Tred Avon River where oysters grow to normal size. "Stunting" could be the result of reduced growth rates or an artifact of mortality. Early mortality as a consequence of increased susceptibility of bored oysters to predation has been discussed. Somatic growth rates could decline if energy is required for shell repair (Cole and Waugh, 1959; Kennedy and Breisch, 1981), particularly since boring occurs at the same time as oyster gametogenesis (Table 1), which imposes an additional energy demand on many bivalves (Bayne and Newell, 1983).

LIFE HISTORY

Cliona truitti reproduces both sexually and asexually. Oviparity seems to be the rule for sexual reproduction in *Cliona* species (Lévi, 1956) except for a single report of viviparity in *C. lobata* (Topsent, 1900). We have not observed either release of gametes or larval development for *C. truitti*, so we cannot report which mode of reproduction this species follows.

Oogenesis occurs as early as March and continues through the middle of June. Oocytes are formed both prior to gemmule hatching by cells that have overwintered

Table 1. Annual cycle of *Cliona truitti* in the Chesapeake Bay and its temporal relationship with oyster gamete release, spat settlement, and shell growth

Species and parameters	Mar	Apr	May	Jun	Jul	Aug	Sep	Oct	Nov	Dec	Jan	Feb
Cliona truitti												
Sexual reproduction												
Oogenesis	—	—	—	—								
Spermatogenesis		—	—	—								
Asexual reproduction												
Gemmule hatching	—	—	—									
Gemmule formation				—	—	—	—					
Somatic growth												
Boring	—	—	—	—	—	—	—	—				
Regression								—	—	—	—	—
Crassostrea virginica												
Gamete release			—	—	—							
Spat settlement			—	—	—							
Increase in shell length*			—	—								
Increase in shell width*			—									

*From Loosanoff and Nomejko, 1949.

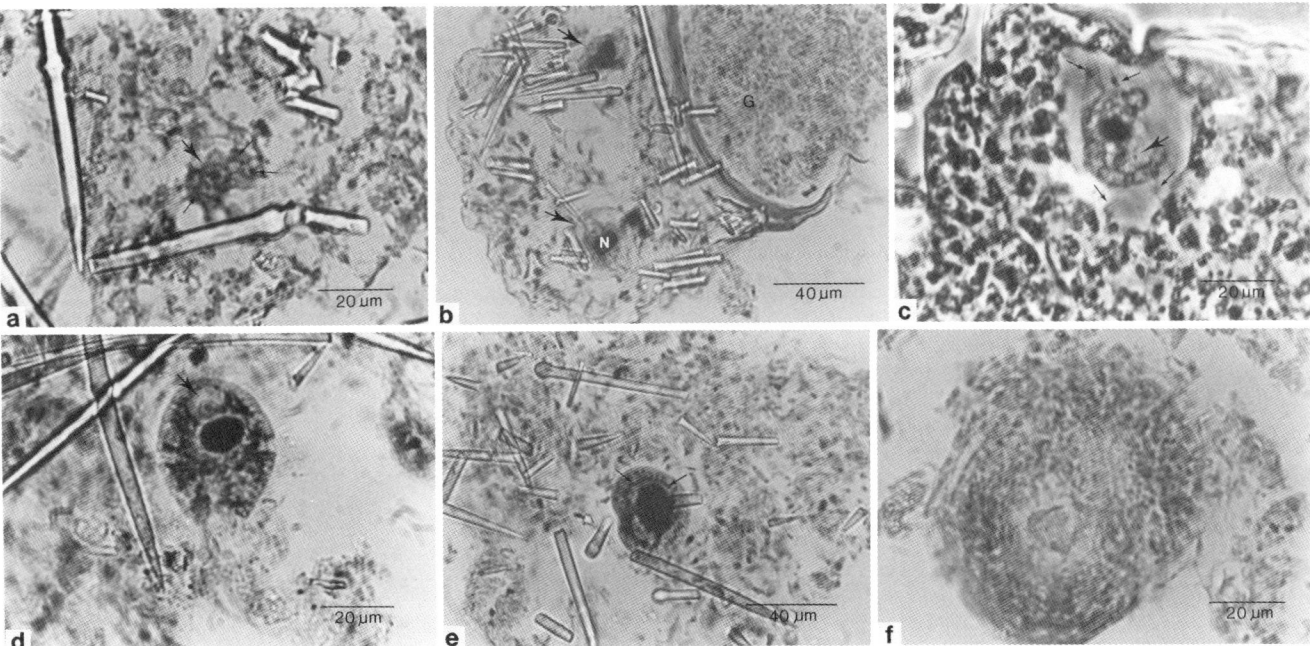

Figure 3. Gametogenesis in *Cliona truitti: a*, young oocyte *(large arrow)*, formed prior to gemmule hatching, in contact with nurse cells (sensu Simpson, 1984) *(small arrows); b*, young oocytes *(arrows)* with vesicular cytoplasm and large, basophilic nucleus formed by differentiated gemmular cells; hatched gemmule; *c*, oocyte with cytoplasmic bridges *(small arrows)* and phagocytosis of nurse cell *(large arrow); d*, phagocytosis of archaeocyte *(arrow)* by oocyte; *e*, cytoplasm of mature oocyte organized into two zones, separated by basophilic inclusions *(arrows); f*, spermatogenic cyst. *N*, nucleus; *G*, gemmule.

(Figure 3*a*) and subsequent to gemmule hatching by differentiated gemmular cells (Figure 3*b*). Precise cellular origin of oocytes is not known, but is presumed to be from archaeocytes because the oocytes are located in the mesohyl. The oocytes are vesicular and contain a large, basophilic nucleus (Figure 3*b*). Cytoplasmic bridges (Figure 3*c*) and phagocytosis of surrounding cells (Figures 3*c,d*) have been observed under the light microscope. The cytoplasm of mature oocytes appears to be organized into two zones (Figure 3*e*), as described by Diaz (1979) for *Suberites massa*.

Spermatogenesis has been observed only rarely during May and June. The spermatozoa are grouped in large cysts in the mesohyl (Figure 3*f*). Details of spermatogenesis within the cysts could not be discerned in our samples.

We have no field or histological evidence for larval development or metamorphosis, although new borings were observed in July in one-month-old, hatchery-reared oyster spat (Figure 4) that were not in contact with bored adult oysters. This indicates that larvae were present in the water at that time. Field and laboratory observations suggest, therefore, that *Cliona truitti* larval settlement coincides with oyster spat settlement (Table 1).

Cliona truitti reproduces asexually by the production of gemmules. The adult tissue of freshwater and marine sponges that produce gemmules may completely degenerate. In certain of these species, gemmules are an obligate stage in their life history (Simpson, 1984). In other marine sponges that do not produce gemmules, adult tissue may regress during the winter to a smaller mass of cells lacking choanocytes (Simpson, 1984). *Cliona truitti* has characteristics of both groups. During the winter, few incurrent and excurrent papillae are present. Tissue is reduced but not completely absent. Only during one extremely cold winter

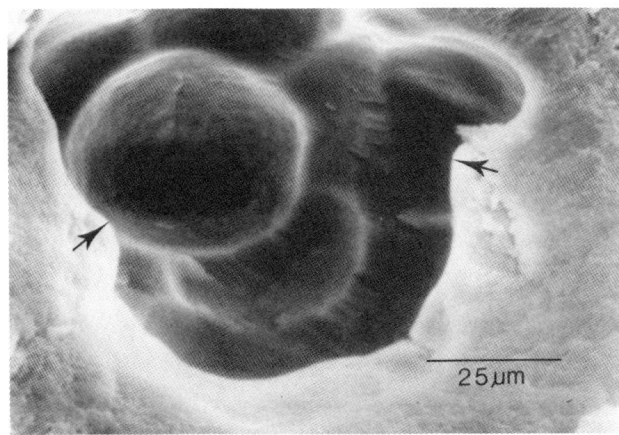

Figure 4. Scanning electron micrograph of one-month-old oyster spat shell bored by *Cliona truitti*. Pits *(arrows)* represent areas from which chips of shell were excavated by the sponge.

during our study did some adult tissue completely degenerate.

Gemmules are 1 to 2 mm in diameter. The gemmule coat is single-layered, aspicular, and collagenous. Gemmules were often found close together, but rarely fused (Figure 5a).

When the water temperature rises above 15°C (mid-March), the gemmules begin to hatch. Cytological evidence suggests that as the gemmules develop, archaeocytes differentiate into etching cells and begin boring immediately into the oyster shell (Figure 5b). Complete gemmule hatching and rapid cell proliferation occur after the water temperature rises above 20°C (in late May to early June). Empty gemmule coats are found within the sponge tissue, but cells generally develop within broken capsules or coats (Figure 5a). Intact gemmules are rarely observed in the sponge tissue during the summer. As the water temperature drops below 20°C in the early fall, sponge cells again regroup to form gemmules.

Boring resumes in early April and continues through September or October. Spherulous cells observed in cavities within the matrix of decalcified oyster shell (Figure 5c) are believed to be accessory etching cells (Pomponi, 1979).

Adult tissue begins regressing in mid-October, after gemmules have formed.

EXTENT OF BORING

The impact of *Cliona truitti* boring on oysters was assessed by calculating the ratio of bored area to total area of the shell (Figure 1, Table 2). This measure indicates the percent of shell weakened by boring and thus susceptibility of the oyster to predation. The ratio was about the same for both year classes: 46% in two-year-old oysters and 45% in one-year-old oysters, or nearly half of the shell weakened by boring.

An estimate of the percent of shell removed was made by calculating the ratio of sponge chamber area to total shell area (Figure 1, Table 2). Volume measurements were difficult to make owing to differences in shell thickness, so estimates were based on area instead of volume. Again, there was no significant difference between the two year

Table 2. Boring activity of *Cliona truitti* in *Crassostrea virginica* in Broad Creek. Mean values per shell calculated for two year-classes sampled during 1982.

Parameter	Year Class	
	1980 (n = 328)	1981 (n = 62)
Total area (cm²)	7.38	3.27
Bored area (cm²)	3.36	1.48
Sponge chamber area (cm²)	0.95	0.40
Bored area/total area (%)	46	45
Sponge chamber area/total area (%)	13	12

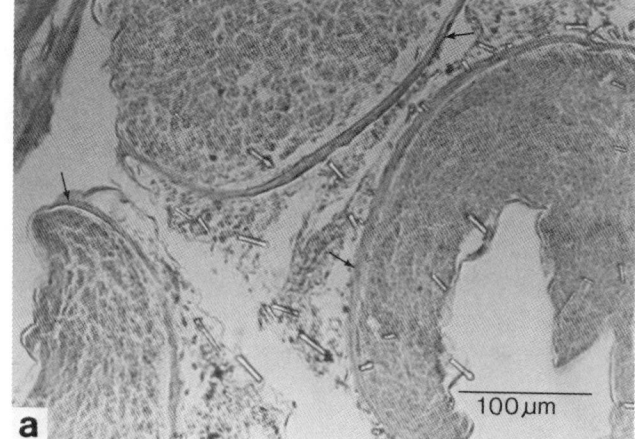

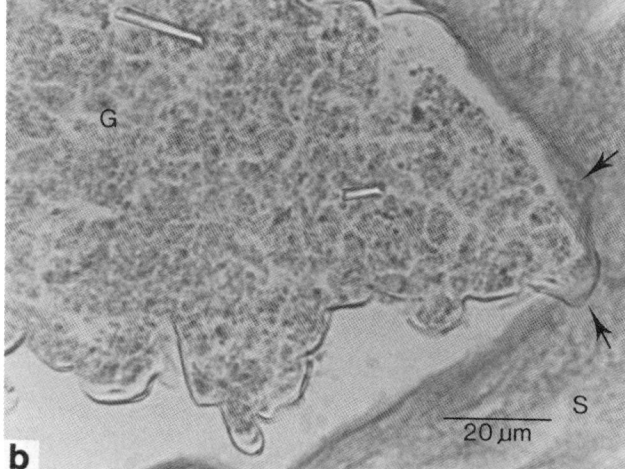

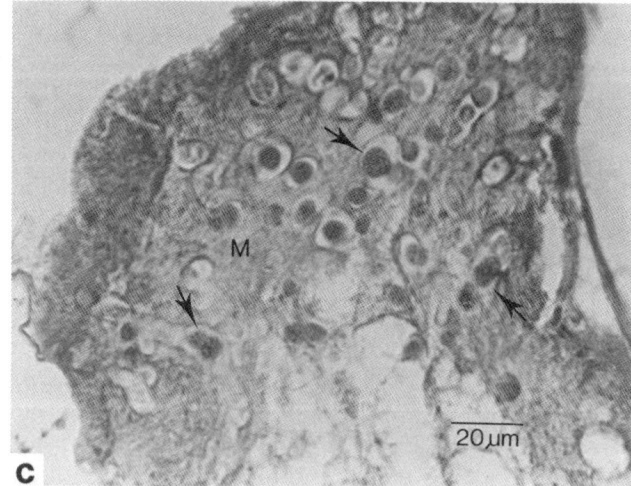

Figure 5. *Cliona truitti*, histology. *a*, hatched gemmules with cells developing within broken gemmule coats *(arrows); b*, gemmular cells differentiate into etching cells *(arrows)* and bore into oyster shell; *c*, spherulous cells *(arrows)* are accessory etching cells in cavities remaining in oyster shell matrix after chips of shell are excavated. *G*, gemmular cells; *M*, shell matrix; *S*, shell.

classes: 13% for two-year-olds and 12% for one-year-olds. These figures are similar to those reported for clionids boring into coral skeletons (Hein and Risk, 1975; Mac-Geachy, 1977; Moore and Shedd, 1977).

To determine if there were variations in the extent of boring during the annual growth period of both *Cliona truitti* and *Crassostrea virginica* (Table 1), we calculated the ratios of sponge chamber area to total shell area and bored area to total area in subsamples of the oyster population at Broad Creek over a ten-month period (Figure 6). We selected random samples of two year classes of oysters at approximately biweekly intervals. The ratio of sponge chamber area (i.e., the expansion of existing chambers) to total shell area provides an estimate of changes in chamber size relative to changes in area of oyster shell. The ratio of bored area to total shell area is an estimate of pioneering spreading of new sponge chambers relative to changes in area of the oyster shell.

From late April through June, the relative area affected by sponge borings decreased or remained the same. Mac-Geachy (1977) observed differences in boring rates of sponges in corals and suggested that they were due, in part, to differences in rates of calcium carbonate deposition by the host organisms. Decreases or no changes in relative surface area of sponge borings during the months of May and June may be explained by the high rates of oyster shell growth at this time (Loosanoff and Nomejko, 1949). Therefore, even though boring may be intense during this period, it may be offset by simultaneous increases in rates of skeleton formation (and perhaps repair) by the oysters. Our data suggest that sponge growth during this period was devoted more to enlargement of existing chambers by both adult cells and hatched gemmules rather than to spreading of new chambers.

During July, there was a rapid increase in both chamber are and spreading of *Cliona truitti*. This is most likely due to a combination of the diversion of energy from reproduction to somatic growth in adult sponges during this period (Table 1), as well as rapid initial penetration rates of recently settled sponge larvae (Rützler, 1975). Our measurements of the extent of boring in one-year-old oysters (Figure 6b) suggest a greater increase than that in two-year-old oysters (Figure 6a). This could be because younger oysters have shells which are not as heavily fouled as older oysters, so there is more space available for settlement and survival of sponge larvae. Shells of younger oysters are thinner than older oysters, so enlargement of chambers and pioneering spreading would be more evident in younger oysters with the methods we used to estimate extent of boring. In X-rays of older oysters with thicker shells, chamber images are likely to be superimposed and our measurements of area would underestimate the extent of boring.

Indeed, our field observations of older oysters indicate that borings penetrate through the entire shell thickness.

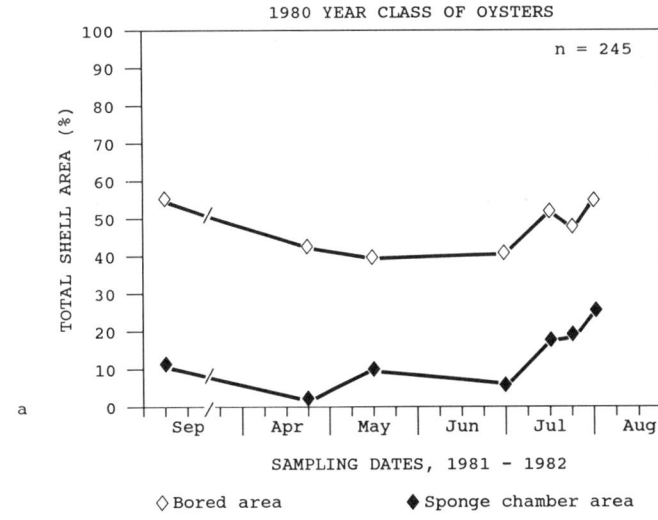

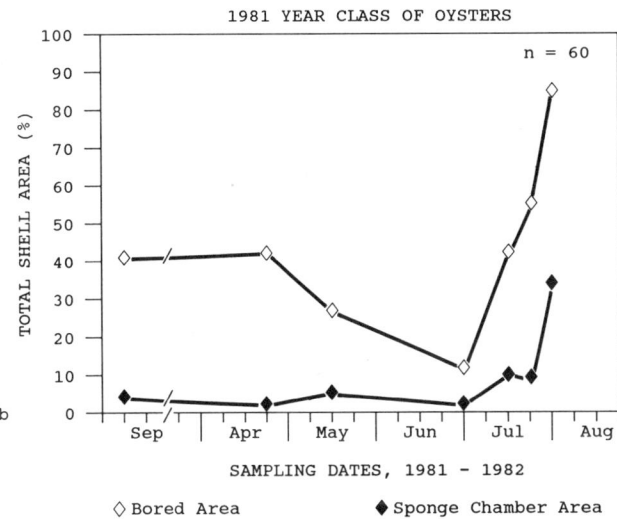

Figure 6. Extent of boring during annual growth period of *Cliona truitti* into oysters: *a*, one- to two-year-old oysters (1980 year class); *b*, oysters less than one year old (1981 year class).

In heavily bored shells, the sponge can penetrate through to the inside of the shell. The oyster responds by depositing more shell, often forming a "blister" which becomes filled with sponge. Frequently, "blisters" are observed in the area of muscle attachment to the shell. Similar observations of weakened muscle attachment sites in *Chama macerophylla* bored by *Cliona lampa* were reported by Rützler (pers. comm.).

Conclusions

The annual growth cycle of *Cliona truitti* correlates with that of the American oyster *Crassostrea virginica*, into which it bores (Table 1). Gemmule hatching and somatic growth coincide with periods of oyster shell deposition. *Cliona truitti* larval settlement occurs at the same time that new substrates (oyster spat shells) are available for settlement.

Measurements of extent of boring indicate that in areas of stunted oyster growth (Broad Creek and Little Choptank River), approximately 50% of the shell is weakened by sponge borings, and therefore is susceptible to predation.

It is most likely that *Cliona truitti* has an effect on oyster growth and productivity. This suggestion is supported by concurrent studies that indicate there is no difference in nutrient levels and phytoplankton biochemistry between two subestuaries of Chesapeake Bay, one in which *Cliona truitti* occurs and the oysters are stunted (Broad Creek) and one in which *C. truitti* does not occur and the oysters are not stunted (Tred Avon River). Furthermore, when the sponge tissue is killed and stunted oysters are transplanted to sites which do not support sponge growth, the previously affected oysters grow to normal size.

Acknowledgments

We thank Jay Harper and his staff for providing X-ray facilities, supplies, and expertise. Eva Potter and Amy Malkus assisted with histology and measurements of oyster shells. John Hochheimer contributed valuable suggestions regarding analysis of boring rates. This research was supported by the Maryland Department of Natural Resources and by a grant from the National Sea Grant College Program, NOAA, Department of Commerce, to the University of Maryland.

Literature Cited

Bayley-Brock, J. H., and A. Ringwood. 1982. Methods for Control of the Mud Blister Worm, *Polydora websteri*, in Hawaiian Oyster Culture. *Sea Grant Quarterly*, 4:1–6.

Bayne, B. L., and R. C. Newell. 1983. Physiological Energetics of Marine Molluscs. Pages 407–515 in *The Mollusca 4, Physiology 1*, edited by A. S. M. Saleuddin and K. M. Wilbur. New York: Academic Press.

Berg, J. A., and R. I. E. Newell. 1986. Temporal and Spatial Variations in the Composition of Seston Available to the Suspension Feeder *Crassostrea virginica*. *Estuarine Coastal Shelf Science*, 23:375–386.

Cole, H. A., and G. D. Waugh. 1959. The Problem of Stunted Growth in Oysters. *Journal du Conseil International pour l'Exploration de la Mer*, 24:355–365.

Diaz, J. P. 1979. Variations, différenciations et fonctions des categories cellulaires de la démosponge d'eaux saumâtres, *Suberites massa*, Nardo, au cours du cycle biologique annuel et dans des conditions expérimentales. Thése. Université des Sciences et Techniques du Languedoc. 332 pp.

Fasten, N. 1931. The Yaquina Oyster Beds of Oregon. *American Naturalist*, 65:434–468.

Hein, F. J., and M. J. Risk. 1975. Bioerosion of Coral Heads: Inner Patch Reefs, Florida Reef Tract. *Bulletin of Marine Science*, 25:133–138.

Hidu, H. 1969. Gregarious Settling in the American Oyster *Crassostrea virginica* (Gmelin). *Chesapeake Science*, 10:85–92.

Kennedy, V. S., and L. L. Breisch. 1981. Maryland's Oysters: Research and Management. College Park, Maryland: University of Maryland Sea Grant Program. 286 pp.

Lévi, C. 1956. Etude des *Halisarca* de Roscoff: Embryologie et systématique des Démosponges. *Archives de Zoologie Expérimentale et Générale*, 93:1–181.

Loosanoff, V. L., and C. A. Nomejko. 1949. Growth of Oysters, *Crassostrea virginica*, during Different Months. *Biological Bulletin*, 97:82–84.

MacGeachy, J. K. 1977. Factors Controlling Sponge Boring in Barbados Reef Corals. Pages 477–483 in *Proceedings, Third International Coral Reef Symposium, 2*. Miami, Florida: Rosenstiel School of Marine and Atmospheric Sciences, University of Miami.

Moore, C. H., Jr., and W. W. Shedd. 1977. Effective Rates of Sponge Bioerosion as a Function of Carbonate Production. Pages 499–505 in *Proceedings, Third International Coral Reef Symposium*, edited by D. L. Taylor. Miami, Florida: Rosenstiel School of Marine and Atmospheric Sciences, University of Miami.

Old, M. C. 1941. The Taxonomy and Distribution of the Boring Sponges (Clionidae) Along the Atlantic Coast of North America. Solomons Island, Maryland: Chesapeake Biological Laboratory Publication 44. 30 pp.

Pomponi, S. A. 1979. Ultrastructure of Cells Associated with Excavation of Calcium Carbonate Substrates by Boring Sponges. *Journal, Marine Biological Association of the United Kingdom*, 59:777–784.

Rützler, K. 1975. The Role of Burrowing Sponges in Bioerosion. *Oecologia*, 19:203–216.

Simpson, T. L. 1984. The Cell Biology of Sponges. New York: Springer. 662 pp.

Topsent, E. 1900. Etude monographique des spongiaires de France: III. Monaxonides Hadromerina. *Archives de Zoologie expérimentale et générale*, 8:1–331.

Wilbur, K. M., and A. S. M. Saleuddin. 1983. Shell Formation. Pages 236–287 in *The Mollusca 4, Physiology*, edited by A. S. M. Saleuddin and K. M. Wilbur. New York: Academic Press.

JON D. WITMAN
KENNETH P. SEBENS
Marine Science Center
Northeastern University
Nahant, Massachusetts 01908

Distribution and Ecology of Sponges at a Subtidal Rock Ledge in the Central Gulf of Maine

Abstract

The distribution and abundance of sponges at four depth zones between 30 and 60 m on Ammen Rock pinnacle were investigated by quantitative photographic techniques. Sponges were found to be an important component of benthic communities along the depth gradient. The percentage of sponge cover and species composition of sponge populations differed between depths, and between habitats (horizontal vs. vertical rock surfaces) at the same depth. The percentage of cover increased with depth to 45 m. Sponges with an encrusting growth form were significantly more abundant (by percentage of cover) than sponges with an upright growth form at 30, 50, and 60 m. Upright sponges were more abundant than encrusting sponges at 45 m, owing to the dominance of *Mycale lingua* and *Artemisina* sp. at this depth. Four poecilosclerids found at Ammen Rock pinnacle have not been previously reported from the western North Atlantic Ocean.

Sponges play important roles in Antartic (Dayton et al., 1974), Mediterranean (Sarà, 1970), and coral-reef benthic communities (Reiswig, 1973; Hartman, 1977; Wulff and Buss, 1980; Wilkinson, 1981). Although sponges in the Gulf of Maine, in the eastern United States, construct assemblages of sessile invertebrates on hard substrata, little is known about their distribution and ecology.

The most extensive study of sponges from the New England region was conducted by Hartman (1958) who described the systematics, distribution, and ecology of 12 species of demosponges from Long Island and Block Island sounds. In the Gulf of Maine, *Polymastia infrapilosa* has been reported as the most abundant sponge species on horizontal and sloping rock surfaces at a depth of 30 m at a site 10 km offshore (Witman 1985). The sponges *Halichondria panicea*, *Leucosolenia* sp., and *Halisarca dujardini* are important space occupiers on vertical rock walls at a depth of 8 m at coastal sites in Massachusetts (Sebens, 1986).

The objective of this research was to assess the abundance and distribution of sponge populations in deep rocky habitat at Cashes Ledge in the central Gulf of Maine. This chapter represents the first report of depth variation in sponge populations from offshore regions of the Gulf of Maine.

Study Area and Methods

Sponge populations were studied at an unnamed pinnacle (hereafter Ammen Rock pinnacle: 42°51′25″ N latitude, 68°57′11″ W longitude) 2 km south of Ammen Rock on Cashes Ledge (Figure 1). The peak of Ammen Rock pinnacle is situated at a depth of 27 m below mean low water. The substratum consists of sloping granite bedrock to a depth of approximately 65 m. A pavement of cobbles and gravel extends from 65 m to approximately 90 m, where the substratum changes to fine sand. Large boulders (2 m high) are scattered along the deep flanks of the pinnacle.

Two photographic sampling methods were employed to quantify the abundance and distribution of sponges from 30 to 60 m depths. Haphazardly placed 0.25-m² quadrats were photographed on vertical rock walls ($n = 11$ quadrats) and on horizontal rock surfaces ($n = 23$ quadrats) at a 30 m depth with a Nikonos IV camera mounted on a fixed camera frame (quadrapod). Photographic transects were made with a Benthos 35 mm camera mounted on the bow of the Johnson Sea Link submersible. Sponge abundance was assessed in 25 photographs (area = 7.0 m²) from each of the following depths: 30, 45, 50, and 60 m. Scuba and submersible dives were conducted during July 1985. A limited number of observations and collections were also made in August 1984 from the Mermaid II submersible.

The percentage of sponge cover in the photographs was analyzed by projecting the color transparency onto a grid of 100 randomly placed dots. The organism in the clear center of each 2-mm-diameter circle was identified and counted (method of Sebens, 1986). Individual sponges were photographed in situ, and if possible the same specimen was collected for identification. Sponge identifications were provided by Willard D. Hartman. Species were classified by growth form as either encrusting or upright. Encrusting sponges were those species that spread laterally across the substratum and were less than 1 cm thick. Species with an upright growth form included mound-shaped and ramose sponges which were characterized by substantial vertical relief (> 1–30 cm tall).

Results

A preliminary list of species of demosponges at Ammen Rock pinnacle is presented in Table 1. Eight sponges have been positively identified to species, and eight more have been assigned to genera. The two unidentified sponges were recognized in photographs and videotapes, but they have not been collected. Four Poecilosclerida—*Leptosia* sp., *Hymedesmia* sp. 1, *Hymedesmia* sp. 2, *Hymedesmia* sp. 3—have not been previously recorded from the western North Atlantic.

The mean percentages of sponge cover on vertical rock walls and horizontal and sloping rock surfaces are compared in Figures 2 and 3. Mann Whitney U tests (Sokal

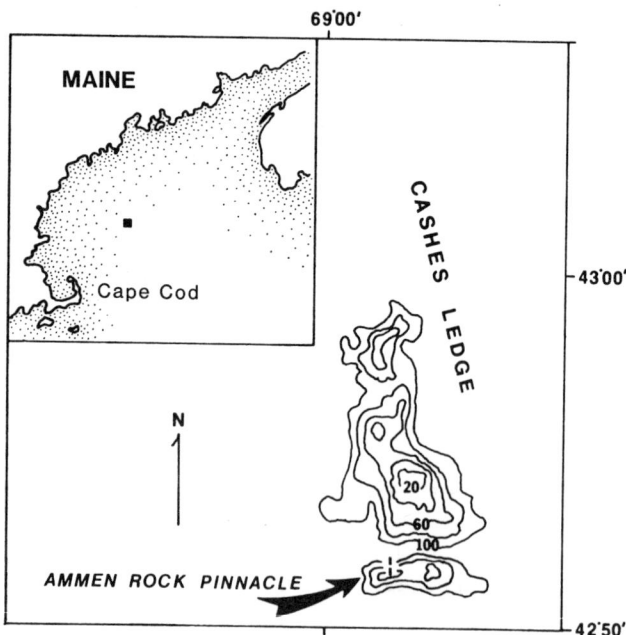

Figure 1. Map of study sites on Cashes Ledge.

Table 1.   Species list of sponges at Ammen Rock pinnacle. (E = encrusting, U = upright)

Taxon	Growth form
Order Hadromerida	
Polymastia infrapilosa Topsent	U
Polymastia sp.	U
Order Halichondrida	
Halichondria panicea (Pallas)	E
Order Poecilosclerida	
Myxilla fimbriata (Bowerbank)	U
Iophon pattersoni (Bowerbank)	E
Hymedesmia sp. 1	E
Hymedesmia sp. 2	E
Hymedesmia sp. 3	E
Leptosia sp.	E
Artemesina sp.	U
Mycale lingua (Bowerbank)	U
Isodictya sp.	U
Order Haplosclerida	
Haliclona urceola (Rathke and Vahl)	U
Gellius arcoferus Vosmaer	U
Order Dendroceratida	
Halisarca dujardini Johnston	E
Aplysilla sp.	E
Unidentified sponges	
orange encrusting	E
tan encrusting	E

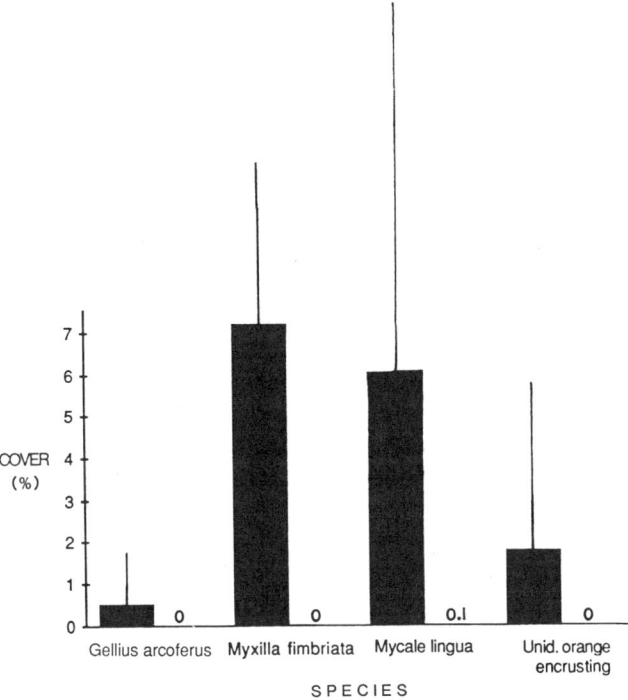

Figure 2. Mean percentage of sponge cover on vertical *(black blocks)* and horizontal rock surfaces at depth of 30 m. Category *Mycale lingua* in this and later figures includes a small amount of *Artemisina* sp.; these two species were difficult to separate in photographs. Error bars signify 95% confidence intervals.

and Rohlf, 1969) indicated that the percentage of *Myxilla fimbriata* cover and the reddish-orange encrusting sponge *Hymedesmia* sp. 3 was significantly greater on vertical rock walls than on horizontal rock surfaces ($U = 230$, p < 0.001; both species). Although *Gellius arcoferus, Mycale lingua, Artemisina* sp., and the unidentified orange encrusting

sponge commonly occurred on rock walls, there was no significant difference in the mean percentage of such species cover between rock walls and horizontal rock surface ($U = 126.5$, *G. arcoferus*; $U = 134.0$, *M. lingua*; $U = 126.5$, orange encrusting sponge). In contrast, the mean percentage of *Halichondria panicea* and *Isodictya* sp. cover was significantly greater on horizontal rock surfaces than on vertical rock walls ($U = 200.0$, p < 0.001; both species). There was no significant difference in the mean percentage of *Halisarca dujardini* cover and the unidentified tan encrusting sponge between habitats ($U = 24.5$, *H. dujardini*; $U = 20.0$, tan encrusting sponge).

There was a striking twofold increase in the total percentage of sponge cover (all cover data pooled by depth) with depth from 30 to 45 m, where the maximum of 20.8% cover occurred (Figure 4). Sponge abundance decreased with depth from 45 m to 60 m.

Figure 5 shows the variation in the mean percentage of cover of 13 common species of demosponge along the 30 to 60 m depth gradient at Ammen Rock pinnacle, based on the analysis of photographs taken form the submersible. The most abundant species at 30 m is *Hymedesmia* sp. 3, which covers an average of 4.9% of the substratum ($SD = 4.3$, $n = 25$). *Halichondria panicea* is also common at 30 m, with a mean cover of 3.4% ($SD = 3.5$, $n = 25$), yet it was not present deeper than 40 m. Sponge populations at 45 m were overwhelmingly dominated by a large (40 cm maximum diameter), yellow, mound-shaped sponge, *Mycale lingua*, which attained an average of 10.8% cover ($SD = 6.9$, $n = 25$). Two encrusting species, *Hymedesmia* sp. 2 and *Leptosia* sp., an orange sponge with a distinctive stellate pattern of oscular canals, covered an average of 2.8% and 2.5% of the bottom at 45 m ($SD = 3.0$ and 2.8, respectively, $n = 25$). *Leptosia* sp. was the most abundant species in the 50 m sponge assemblage. *M. lingua* and *Artemisina* sp. were comparatively less abundant at 50 m (mean = 2.8%, $SD = 5.9$, $n = 25$) than at 45 m. The rare blue encrusting sponge, *Hymedesmia* sp. 1, occurred at 50 m.

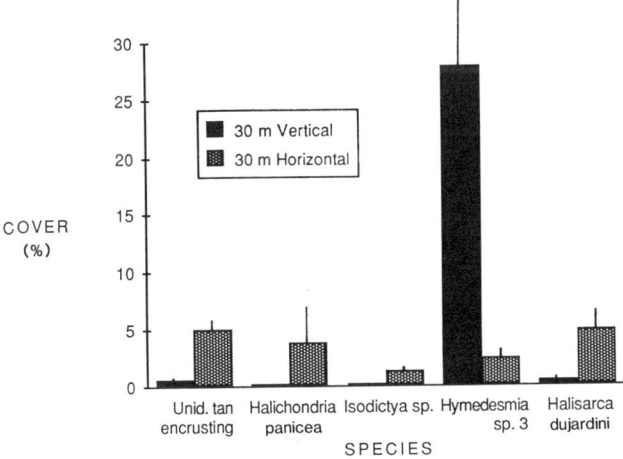

Figure 3. Mean percentage of sponge cover on vertical and horizontal rock surfaces at depth of 30 m. Error bars signify 95% confidence intervals.

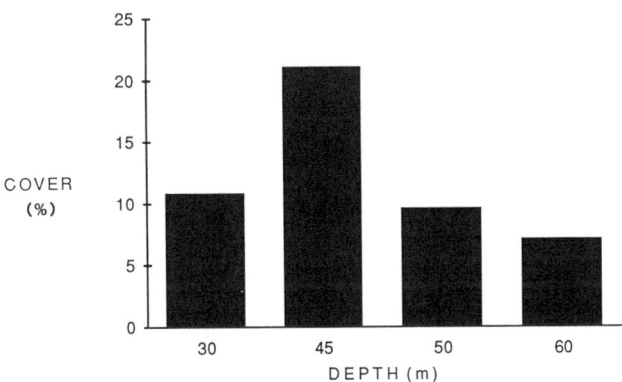

Figure 4. Total percentage of sponge cover along 30–60 m depth profile at Ammen Rock pinnacle.

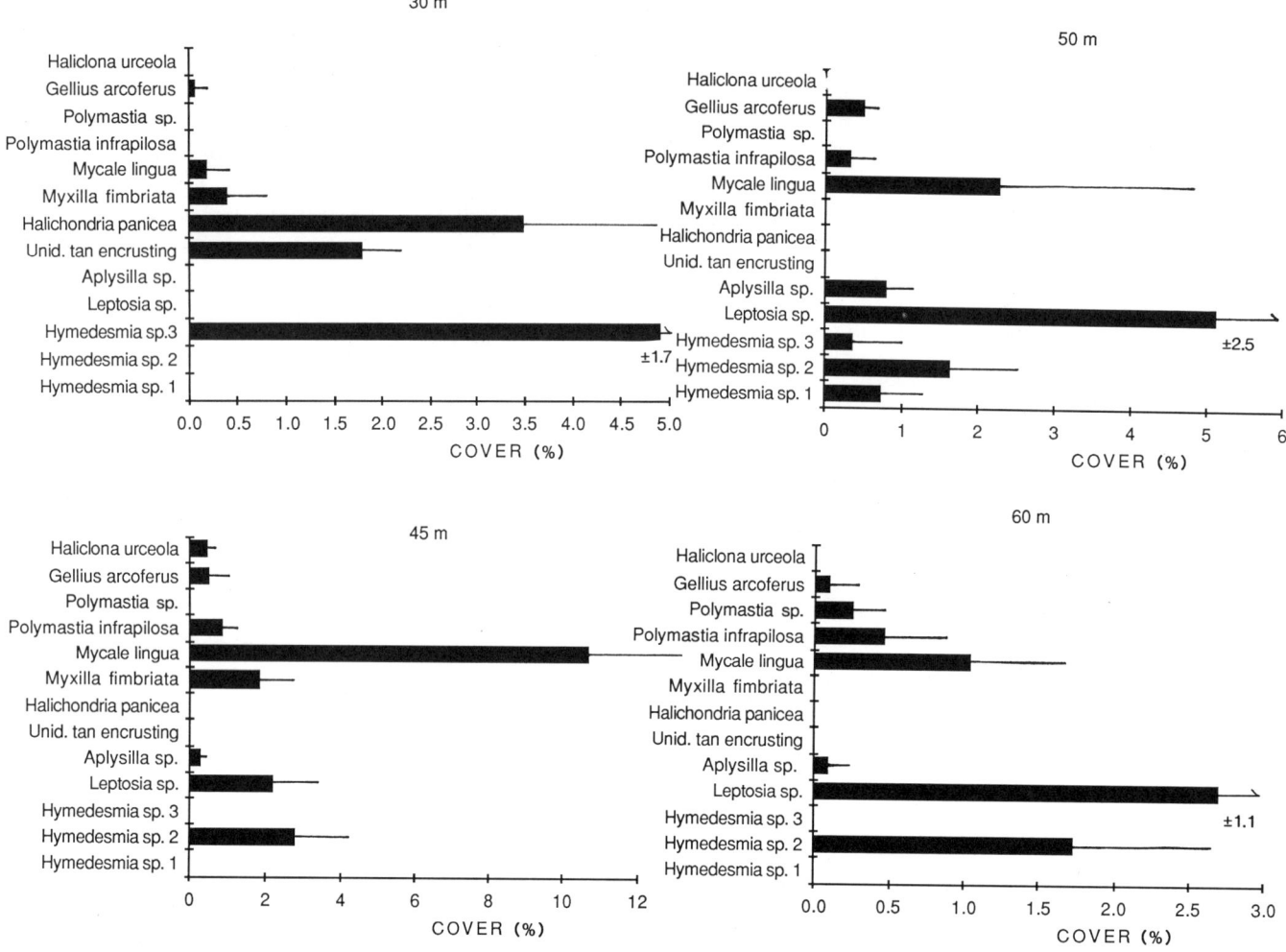

Figure 5. Mean percentage of cover consisting of 13 common species of sponges at Ammen Rock pinnacle.

Leptosia sp. continued to dominate deep sponge populations at 60 m, although the total percentage of sponge cover at this depth was low (7.0%; Figure 4). Several species (e.g., *Polymastia infrapilosa* and *Aplysilla* sp.) showed no trend in abundance with increasing depth (Figure 5). *Iophon pattersoni* was common on brachiopod shells at depths below 30 m, and *Haliclona urceola* also appeared at those depths, although it was rare.

Data on percentage of cover data were arcsine-transformed before comparisons were made of mean percentage cover of upright and encrusting sponge cover by a student's t test (Sokal and Rohlf, 1969). Encrusting species were significantly more abundant (by percentage of cover) than sponges with an upright growth form at all depths except 45 m (Figure 6; $t = 6.7, p < 0.001, 30$ m; $t = 9.4, p < 0.001, 50$ m; $t = 3.1, p < 0.01, 60$ m; all tests with 1 and 24 df). The mean percentage of upright sponge cover was 2.5 times greater than the mean percentage cover of encrusting sponges at 45 m ($t = 4.1, p < 0.001, 1$ and 24 df) owing to the dominance of mound-shaped *Mycale lingua* and *Artemsina* sp. at 45 m. Given the great differences in

shape and degree of erectness, the relationship between biomass and percentage of cover may vary greatly.

Discussion

Sponges are clearly an important component of hard substratum benthic communities in the central Gulf of Maine; quantitative photo transects taken from the Johnson Sea Link submersible revealed that sponges occupied 21% of the substratum at a depth of 45 m. There was considerable zonation in both species composition and abundance of sponge populations along the 30 to 60 m depth profile at Ammen Rock pinnacle. Species zonation patterns included a *Halichondria panicea–Halisarca dujardini* zone from 30 to 40 m, a *Mycale lingua–Artemisina* sp. zone centered around 45 m, and a deep *Leptosia* sp. zone from 50 to 60 m. Few comparisons can be made with other temperate rocky subtidal habitats because data on sponge populations from similar depths are not presently available. Sponge assemblages at Ammen Rock pinnacle have affinities with the boreal sponge fauna of northern Europe.

3d Int. Sponge Conf. 1985

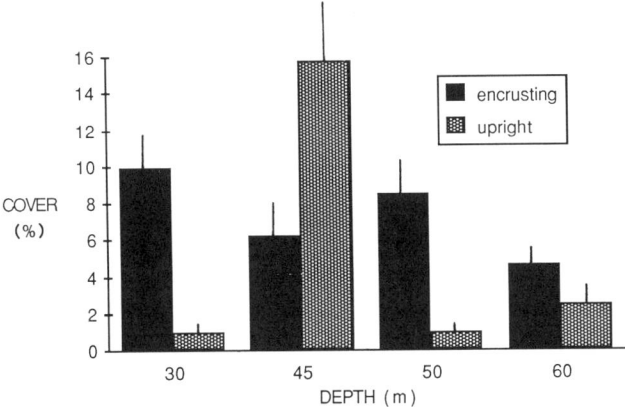

Figure 6. Comparison of mean percentage of sponge cover composed of upright and encrusting growth forms. Error bars signify 95% confidence intervals.

For example, *Myxilla fimbriata*, *Polymastia infrapilosa*, *Leptosia* sp., and several species in the genus *Hymedesmia* were reported from Iceland and Greenland by Lundbeck (1905). The distribution of *Halichondria panicea* is also Amphi-Atlantic; it is common in subtidal habitats off Britain (Firth, 1976) and Wales (Hoare and Peattie, 1979).

Large and small-scale patterns of sponge distribution and abundance at Ammen Rock pinnacle may be related to competition for space with macroalgae, and to sedimentation, water movement, and light. For example, sponge cover is low on macroalgal-dominated horizontal-sloping rock surfaces at 30 m (mean cover of macroalgae = 74.0%, $SD = 17.2$, $n = 23$ photo quadrats). In contrast, sponges are abundant on vertical rock walls at that depth (mean cover of macroalgae = 4.6%, $SD = 4.7$, $n = 10$ photo quadrats). Additional evidence of negative association between sponges and macroalgae comes from the inverse relationship between increasing sponge abundance from 30 to 45 m and the reduction in algal abundance at the 40-m extinction depth of laminarian algae (Vadas and Steneck, 1988). Thus, the prevalence of sponges at 45 m may represent a release from competition for attachment space with kelp on horizontal and sloping rock surfaces above a depth of 40 m. At shallower depths in the Gulf of Maine, the upward extension of the horse mussel *Modiolus modiolus* is limited by interference competition with kelp (Witman, 1987). Kelp "whiplash" effects limit the small-scale distribution of sessile invertebrates in rocky subtidal habitats of South Africa (Velimirov, 1983). Similar processes may explain the low percentage of sponge cover on horizontal rock surfaces at 30 m at Ammen Rock pinnacle.

Although sedimentation rates have not been measured, our observations suggest that sediment covers larger areas of the bottom on deep (60 m) rock surfaces at Cashes Ledge than in shallower habitats, which are characterized by strong currents (> 50 cm·s^{-1} mainstream velocity, un-

published data). It is possible that the low percentage of sponge cover at 60 m is due to adverse effects of sedimentation on the growth and survival of sponges. *Halichondria panicea* occurred only on shallow, well illuminated rock surfaces. All specimens of *Halichondria* were pale green, presumably owing to the presence of green algal symbionts within the mesohyl (Long, 1968). Thus, it is possible that the distribution of *H. panicea* is influenced by light, although little is known about the importance of algal symbionts to the nutrition of *Halichondria*.

Echinoderm and nudibranch predators influence the distribution of sponges between 30 and 60 m depths in Antarctica (Dayton et al., 1974). The sea star *Henricia sanguinolenta* and the nudibranch *Cadlina laevis* consume sponges in the Gulf of Maine (Sebens, 1986; J. Witman, personal observations). Predation on *Halichondria panicea* by *H. sanguinolenta* was evident in 17.4% of the 0.25-m^2 quadrat photographs (4 of 23) taken on horizontal rock surfaces at a depth of 30 m at Ammen Rock pinnacle. *Henricia* predation produced lesions ranging from < 1.0 to 4.0 cm^2 on the surface of *H. panicea*. *C. laevis* is often seen feeding on *Halisarca dujardini*. Although predators may influence the small-scale distribution of sponges in rocky subtidal habitats, our qualitative impression is that predation cannot explain large-scale depth zonation patters of sponges in the Gulf of Maine. Clearly, this topic deserves further attention.

This preliminary study has demonstrated that sponges form a considerable proportion of deep, rocky subtidal communities in the Gulf of Maine. The role that sponges play in modifying benthic community structure at Ammen Rock pinnacle remains unknown. Experimental manipulations are planned for the next stage of this research to evaluate the effects of biotic and abiotic processes on the observed patterns of sponge distribution and abundance.

Acknowledgments

This study was funded by grants from the National Undersea Research Program at the University of Connecticut, Avery Point. The crew of the R/V Edwin Link greatly facilitated our research from the Johnson Sea Link submersible. We appreciate field assistance given by Ammen Rock Research Group (ARRG) team colleagues R. S. Steneck, R. L. Vadas, and M. R. Patterson. We are especially grateful for the taxonomic assistance provided by Willard D. Hartman. This is contribution no. 144 of the Marine Science Center, Northeastern University, and contribution no. 1 of the NURP-UCAP Gulf of Maine program.

Literature Cited

Dayton, P. K., G. A. Robilliard, R. T. Paine, and L. B. Dayton. 1974. Biological Accommodation in the Benthic Community at McMurdo Sound, Antarctica. *Ecological Monographs*, 44:105–128.

Firth, D. W. 1976. Animals Associated with Sponges at North Hatling, Hampshire. *Zoological Journal of the Linnean Society*, 58:353–362.

Hartman, W. D. 1958. Natural History of the Marine Sponges of Southern New England. *Peabody Museum of Natural History, Bulletin* 12:1–155.

———. 1977. Sponges as Reef Builders and Shapers. Pages 127–134 in *Reefs and Related Carbonates–Ecology and Sedimentology*, edited by S. H. Frost, M. P. Weiss, and J. B. Saunders. Tulsa, Oklahoma: American Association of Petroleum Geologists.

Hoare, R., and M. E. Peattie. 1979. The Sublittoral Ecology of the Menai Strait. I. Temporal and Spatial Variation in the Fauna and Flora along a Transect. *Estuarine and Coastal Marine Science*, 9:663–675.

Long, E. 1968. The Associates of Four Species of Marine Sponges of Oregon and Washington. *Pacific Science*, 22:347–351.

Lundbeck, W. 1905. Porifera (Part II) Desmacionidae (Pars). *Danish Ingolf-Expedition*, 6A(2):1–219.

Reiswig, H. M. 1973. Population Dynamics of Three Jamaican Demspongiae. *Bulletin of Marine Science*, 23:191–226.

Sarà, M. 1970. Competition and Cooperation in Sponge Populations. Pages 273–284 in *The Biology of Sponges*, edited by W. G. Fry. Symposia of the Zoological Society of London 25. London: Academic Press.

Sebens, K. P. 1986. Community Ecology of Vertical Walls in the Gulf of Maine, USA: Small Scale Processes and Alternative Community States. Pages 346–371 in *The Ecology of Rocky Coasts*, edited by P. G. Moore and R. Seed. Kent, England: Hodder and Stoughton.

Sokal, R. R., and F. J. Rohlf. 1969. *Biometry*. San Francisco, California: W. H. Freeman.

Vadas, R. L., and R. S. Steneck. 1988. Zonation of Deep Water Benthic Algae in the Gulf of Maine. *Journal of Phycology*, 24:338–346.

Velimirov, B. 1983. Succession in a Kelp Bed Ecosystem: Clearing of Primary Substrate by Wave-Induced Kelp Sweeping. *Oceanologica Acta*, 15:201–206.

Wilkinson, C. R. 1981. Significance of Sponges with Cyanobacterial Symbionts on Davies Reef, Great Barrier Reef. Pages 705–712 in *The Reef and Man*, edited by E. D. Gomez et al. Proceedings of the Fourth International Coral Reef Symposium, Quezon City, Philippines: University of the Philippines.

Witman, J. D. 1985. Refuges, Biological Disturbance, and Rocky Subtidal Community Structure in New England. *Ecological Monographs*, 55:421–445.

———. 1987. Subtidal Coexistence: Storms, Grazing, Mutualism and the Zonation of Kelps and Mussels. *Ecological Monographs*, 57:167–187.

Wulff, J. L., and L. W. Buss. 1979. Do Sponges Help Hold Coral Reefs Together? *Nature*, 281:474–475.

CHRISTOPHER N. BATTERSHILL*
Department of Chemistry
University of Canterbury
Christchurch 1, New Zealand

PATRICIA R. BERGQUIST
Department of Zoology
University of Auckland
Private Bag, Auckland, New Zealand

The Influence of Storms on Asexual Reproduction, Recruitment, and Survivorship of Sponges

Abstract

Spatial patterns among sponge individuals on a deep reef flat in northeastern New Zealand reveal that species distribution and abundance are closely correlated with sediment depth and quality in this area. Significant clumping occurred in only one plane, perpendicular to the swell direction, and a number of bispecies associations were evident. The community was stable over time and recruitment appeared to be precluded by the sediment overlayer and thin encrusting floral and faunal components on the rock surface. The unusual predominance of asexual reproduction, the subsequent peculiar development of the products, and the physical factors mediating recruitment (such as storms) together suggest a mechanism for the distribution of adult sponges.

Information on the reproductive biology of sponges suggests a remarkable plasticity of reproductive mode (Fell, 1974; Bergquist, 1978; Ayling, 1980; Simpson, 1984). However, there are almost no data on the ecology of larvae and asexual propagules in their natural habitats (Hartman, 1958; Fell, 1974; Ayling, 1980; Simpson, 1980, 1984). The relative importance of sexual versus asexual modes of reproduction in maintaining sponge populations in different habitats is also unknown. Can sponge reproductives select settlement sites? What is the role, if any, of the physical and biological environment in mediating reproduction and recruitment? The few published studies on the ecology of reproductives in situ suggest that "distribution of sponges is not merely the result of selective mortality following unselective settlement" (Bergquist et al., 1970), and that asexual modes of reproduction may be important (Bergquist et al., 1970; Wulff, 1985; 1986).

The purpose of this study was to evaluate the relative importance of sexual and asexual modes of reproduction

*Present address: Australian Institute of Marine Science, PMB3, Townsville, Queensland 4810, Australia.

by assessing recruitment into the adult population. The behavior of reproductives was investigated in situ, and the role of physical and biological disturbance in generating primary space and mediating recruitment was evaluated. The problem was approached on two levels. The spatial distribution of adult sponges was assessed and hypotheses generated as to which environmental factors limited distribution. On a fine scale, the ecology of reproductives and their behavior in relation to their microhabitats were investigated.

The study was carried out on a deep reef flat where sponges (predominantly Hadromerida and Axinellida) reproduce both sexually and asexually. This chapter presents an overview of this study.

Material and Methods

DISTRIBUTION AND ABUNDANCE OF SPONGES. The distribution and abundance of sponges occurring on a deep reef flat were investigated by sampling the community on a grid format. The reef (which had the appearance of a "sponge garden") consisted of a homogeneous conglomerate rock platform covered by a layer of sediment up to 5 cm deep. The reef is located in 20 m of water within a marine reserve (36°16′ S, 174°48′ E), 100 km north of Auckland, New Zealand (Figure 1). Sponge density and population size were estimated by sampling the 22,500 m^2 of reef that took in the entire "sponge garden" habitat as well as adjacent areas characterized by other faunal and floral assemblages, including kelp forest and open sediment flats. All sponges found within 1-m^2 quadrats placed on a 10-m × 5-m grid were identified, counted by species, and placed into size categories. Sediment types and depth over the rock platform were measured at each of the 450 quadrats.

PHYSICAL AND BIOLOGICAL DISTURBANCE. Mode and intensity of disturbance of sponges and substratum were quantified on a monthly basis. This involved sampling five randomly placed 1-m^2 quadrats within four 10-m × 10-m permanent sites (marked by plastic dowling stakes). Two such sites were set up in each of two adjacent habitats that had been defined from observations made during the survey described above. In the sponge garden habitat, the sponge population was relatively dense, and sediments were 0.5 cm to 1.0 cm deep. In the adjacent reef habitat, sponges were significantly less abundant and sediments were 1.0 cm to 2.0 cm deep (Figure 1).

REPRODUCTIVE ACTIVITY. Reproductive activity was monitored on a monthly basis through meticulous in situ sampling of five replicate quadrats randomly placed within the 10-m × 10-m sites (located in both habitats), as described above. The surface and matrix of the sediment were searched for reproductive products (buds, frag-

ments, and larvae). Reproductives over 0.5 mm in diameter could be reliably sampled, and those over 1.0 mm could be identified to species in most cases. Reproductive identity and field estimates of abundance were verified in the laboratory. Additional sampling was carried out immediately after storms.

REPRODUCTIVE BIOLOGY. The behavior of reproductive products was followed through time in the field. This was achieved by sampling 1-m^2 quadrats (located by tagged nails) twice a day for as long as weather conditions permitted, or until the reproductives were lost. The small size of reproductives precluded tagging. However, individual buds (fragments, recruits) could be recognized by carefully describing their features and mapping their location. Bidaily sampling proved sufficient to track those reproductives influenced by water movement. Further observations and experiments were carried out in the laboratory. Briefly, the laboratory methodology consisted of placing freshly collected buds of two species (*Polymastia granulosa* and *Polymastia* sp., yellow, undescribed) in Petri dishes with three substratum treatments (gravel, broken shell, and no sediment). These were then placed in a water bath that permitted a gentle flow of filtered seawater over the petri dishes. The reproductives were not disturbed, and each of the ten buds in each of the three replicate dishes (for each species and treatment) were followed separately through time.

Results

DISTRIBUTION AND ABUNDANCE

The distribution of sponges was strongly correlated to sediment type and depth. Figure 2 summarizes the results for four of the most common sponges. In all cases, sponge presence or absence was determined within narrow limits of sediment depth. Sponge density was also influenced by sediment grain size. Very few sponges were found in sediment depths greater than 4 cm. In all species encountered, a medium grain (0.50–0.25) size appeared optimal. Several species (e.g., *Aaptos aaptos* and *Cinachyra* sp., undescribed) were commonly found in slightly deeper sediments. These species were more abundant on the periphery of the sponge garden habitat. Most of the other sponges in this habitat were found in sediment depths of 0.5 cm to 1.0 cm. Mats of turfing coralline algae are also present in this area of relatively dense sponge growth. Immediately adjacent to the sponge garden, sediment depths increased to over 3 cm, total sponge density dropped by an order of magnitude. There was little turfing algal cover. Sediment depth increased with distance from the sponge garden. Uncolonized basal rock approached 100%, but tube-building amphipods within sediments became more abundant. In sediment depths

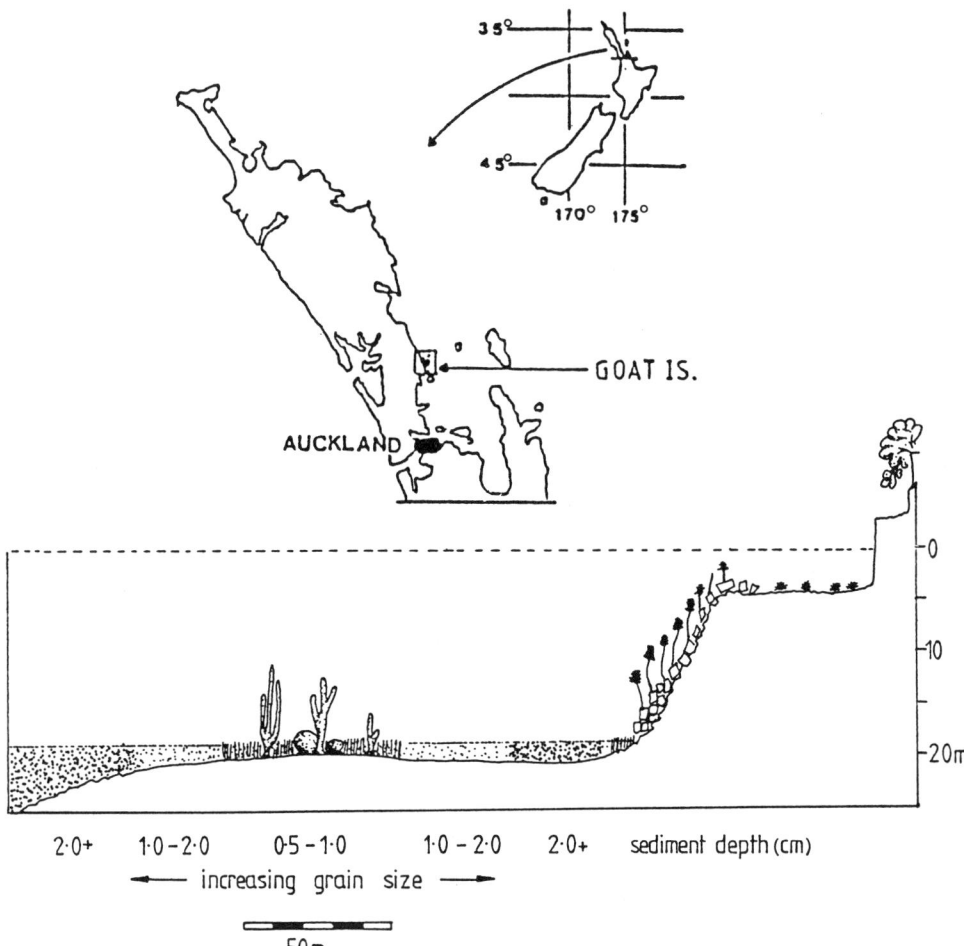

Figure 1. Location of study area *(top)* and profile view of sponge garden habitat with adjacent kelp forest and sediment flats.

exceeding 5 cm, amphipod tubes formed a dense and extensive mat layer.

Close association was common among a number of species. (Approximately 50% of all massive sponges were found within 5 mm of another massive or erect individual). Sponges were also highly clumped at higher scales. Clumps were significantly orientated east-west ($p <$ 0.001), such that sponge individuals were grouped in oblong clusters.

PHYSICAL AND BIOLOGICAL DISTURBANCE

Biological modes of disturbance were negligible in terms of restructuring the community. Sponge damage was the result of fungal attack and the grazing activities of the monacanthid *Parika scaber* and the opisthobranchs *Umbraculum sinicum* and *Archidoris wellingtonensis*. Grazers were present in relatively low numbers and seldom removed an entire sponge or created primary space (Ayling, 1978).

Physical disturbance occurred during storms when swells from the northeast rolled into the bay. Water movement caused some damage to sponges, but its greatest effect could be seen in the formation of ripples, which

induced a restructuring of sediment regimes. Basal rock was exposed along the edges of the ripples aligned east-west (perpendicular to the swell direction). These areas formed oblong patches orientated east-west which remained clear for up to a week, after which time sediment levels gradually returned to normal.

Surge movement also sorted sediments, so that fine material was found within the ripple and toward the ripple crest and coarse-grained particles were left at the ripple troughs. (Similar observations have been made by Cook and Gorshine, 1972).

Sediment ripples (also called megaripples or sand waves; see Cook and Goshine, 1972), did not readily form in areas of high sponge density, or where turfing organisms were present. The sponge garden area and regions of amphipod infestation were disturbed only by the most severe storms. Thus the sponge-dominated area was surrounded by belt of reef that was regularly disturbed. In intermediate areas of low sponge density and reduced turf cover, sediment was scoured around the bases of erect features (Figure 3).

No sponges were seen to be completely removed by storms. Storm damage appeared to be limited to breakage

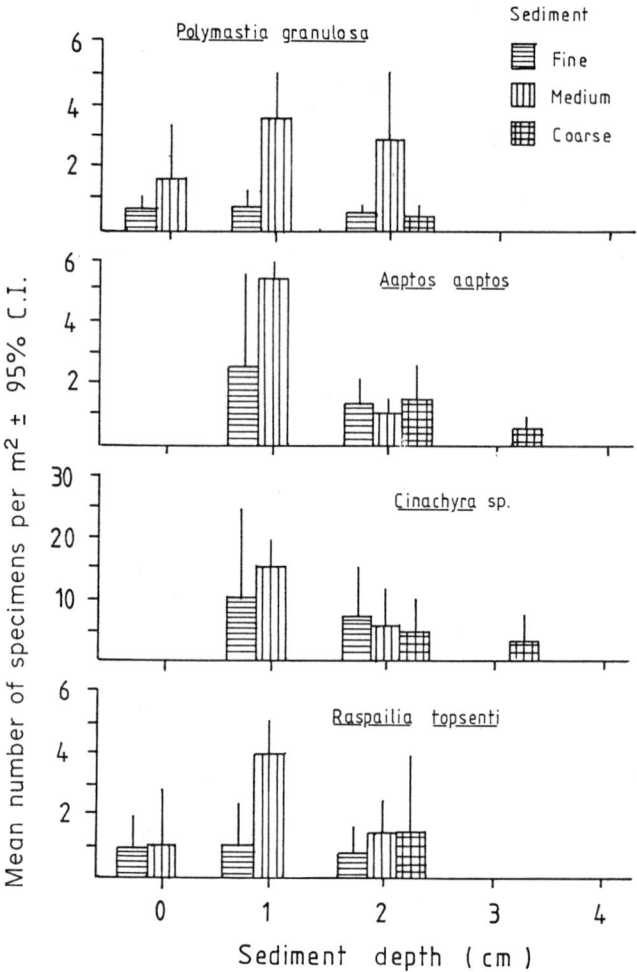

Figure 2. Relationship of sediment depth and grain size to abundance of four representative sponges. *Fine*, 0.250–0.125 mm; *medium*, 0.50–0.25 mm; *coarse*, 1.0–0.5 mm.

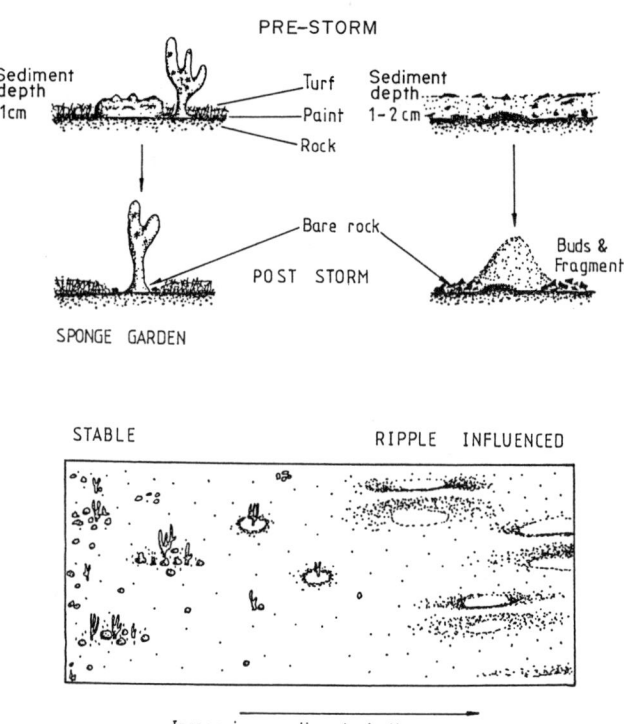

Figure 3. Diagrammatic representation of sediment restructuring during storms comparing sponge garden *(left)* with adjacent sediment flat *(right)*. Plan view at bottom; *turf* and *paint* refer respectively to erect branching and thin encrusting growth forms of a number of coralline algae species.

in some cases (*Raspailia topsenti* and *Axinella* sp.), liberation of buds in several instances (*Polymastia granulosa, Polymastia* sp. and *Aaptos aaptos*), and fragmentation (*Cinachyra* sp.).

REPRODUCTIVE BIOLOGY AND ECOLOGY

Reproductives were found rolling free on the bottom throughout the year, but became more abundant following storms. From laboratory experiments and field studies, it appears that propagules are derived predominantly by asexual means. Figure 4 summarizes the developmental stages in *Polymastia* sp. and *Polymastia granulosa*. Buds from a number of species exhibited developmental sequences similar to those shown for *Polymastia* sp. (Figure 4a). Buds were 1 to 5 mm in diameter and possessed a developed ectosome. Buds could remain in this spherical state for over a month. During storms or in surge conditions, buds would adhere to large fragments of rock and

shell (5 mm diameter) or to other buds (irrespective of bud genotypes). The increased size and weight, as well as the change in shape, altered the reproductive's behavior in turbid conditions such that the buds gravitated to ripple troughs or to depressions scoured around erect features (Figure 3, bottom right profile). Final attachment to cleared areas of basal rock was thus possible. Sexually produced reproductives were seldom encountered in monthly surveys, although they have been observed being released from parent sponges (Ayling, 1978). Observations during our field experiments suggest that larval recruitment and survival are rare in this habitat.

Figure 4*b* illustrates the development and peculiar behavior of *Polymastia granulosa* buds. Buds are "pinched off" the parent sponge and immediately round up. They remain spherical for approximately one week, then become slightly elongated (5 mm × 2 mm on average). During this period a central cavity forms. Severe water movement (or some other as yet unknown stimulus) causes the bud to elongate rapidly. Buds have been observed to elongate 10-fold in 48 hours (to over 60 mm in length and 1 mm diameter). During the following 24 hours, the bud knots at intervals along its length. The knots then break apart as the joining cell separate. This process has been called

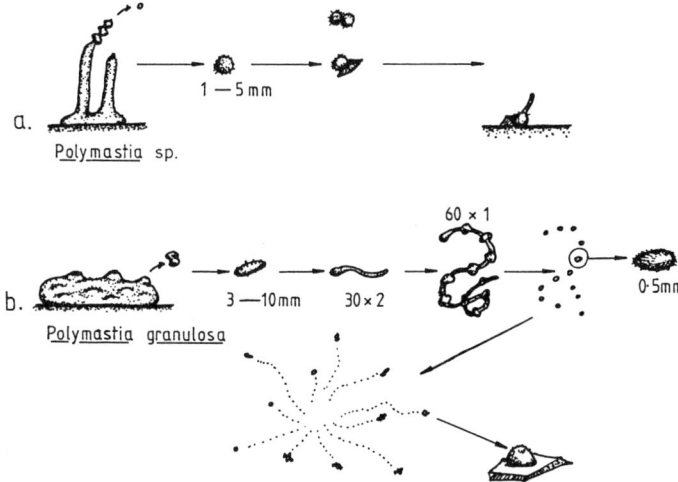

Figure 4. Developmental sequence of *Polymastia* buds: *a,* *Polymastia* sp.; *b, P. granulosa.*

"beading." The smaller beads (less than 1 mm in diameter) can creep short distances (up to 1 cm in 24 hours) and eventually settle onto the basal rock or onto shell and rock fragments. An oscule and other cell types develop rapidly. A functional sponge is complete in a further three days. The oscule may become extremely long in *Polymastia* species (over 20 mm on a recruit 3 mm in diameter) if sediments recover the sponge.

It is interesting to note that the rate of development varies with substratum type. Buds develop relatively quickly when placed on gravel. Development is significantly slower on shell-chip sediments.

Discussion

The predominance of asexual reproduction and the complex developmental phenomenon in the study area are unique. This life history pattern is well adapted to the mode of disturbance that liberates primary space in this habitat and improves the reproductives' chances of finding suitable settlement surfaces. Sediment levels over 0.5 cm preclude settlement. Turfing components stabilize sediment and resist disturbance. In control areas within the sponge garden habitat, there was no significant recruitment in over three years (Ayling, 1978). Under the present sediment regime, the sponge population is stable in both space and time. The region adjacent to the sponge garden is regularly disturbed by storm surge. Sediments are contoured into ripples that are always aligned east-west, as are patches of basal rock exposed by ripple formation. In intermediate areas, the reduced cover of turfing flora permits sediment disruption around the base of erect features, including sponges. Sponge reproductives tend to collect in these areas by virtue of their physical behavior in turbid conditions, during which they rapidly adhere to other propagules, or to rock and shell fragments of relatively large size. Settlement onto basal rock is thus possi-

ble, and the relatively large size of recruits enhances survivorship.

Long-term survival is related, species specifically, to ambient sediment depth and grain size. Recruits that survive longest are those that can tolerate being covered by sediment (e.g., *Cinachyra* sp. and *Aaptos aaptos*) or that can produce a long oscule (e.g., *Polymastia* spp.). Sexually produced reproductives of these species do not survive such conditions. These findings are consistent with the observed distribution of sponges in the sponge garden. The origin and maintenance of the distribution patterns in this habitat can thus be explained. The hypothesis that recruitment is mediated by storm disturbance of sediments and that subsequent survival is related to sediment regimes was tested experimentally in the field. Significant recruitment occurred in experimentally cleared patches that mimicked sediment scour and return. Longevity of recruits was related to sediment depth and grain size. The hypotheses suggested by the findings outlined above were supported, and assumptions that undisturbed sediment precluded settlement were validated.

The intimate association of physical disturbance and asexual reproduction, as well as the complex development of buds of a number of species of sponge, is intriguing. The importance and frequent occurrence of asexual modes of reproduction have been well documented for a number of animal and plant groups, perhaps most notably the corals (see Wallace, 1985). The biology of asexual modes of reproduction in sponges has also been well documented (see Bergquist, 1978; Simpson, 1984), although it is generally assumed that asexual reproduction is relatively uncommon in natural habitats (see "Discussion" in Harrison and Cowden, 1976; Simpson, 1973). However, little information is available on the ecology of sponge reproduction in general, let alone the significance of asexual modes (Fell, 1974; Simpson, 1980). Evidence presented here and in other studies (Burton, 1949; Bergquist et al., 1970;

Stone, 1970; Wulff, 1985), suggests that in certain environments asexual modes of reproduction are important.

Some evidence also indicates that a certain degree of passive selection of settlement site is possible even with nonmotile asexual propagules. Reproductives tend to accumulate in depressions and on basal rock that has been exposed by sediment scour. This passive movement is due to the physical dynamics of large, irregularly shaped particles in surge conditions. The subsequent rapid development of recruits enhances survival. In *Polymastia granulosa*, complex development of buds and the ability of bud fragments to creep on the rock surface suggest a level of selective settlement not previously considered. This phenomenon is obviously well suited to environmental conditions. Parallel observations have been made by Highsmith (1982) and Wulff (1985), who describe "adaptations" of branching sponges facilitating fragmentation and fragment survival during storms.

A good deal more work needs to be carried out on the ecology of sponge reproductives in their natural habitat before any generalizations can be made concerning the reproductive strategies of sponges in different environments. Since much of the literature pertains to selective settlement in benthic invertebrates other than sponge (see Meadows and Campbell, 1972; Crisp, 1974), little is known of the factors influencing variability in sponge recruitment (Dayton, 1979; Underwood and Denley, 1982; Keough, 1984; Connell, 1985). The work presented here indicates that microenvironmental conditions and the behavior of reproductives in situ (be it active or passive) play important roles in the recruitment of benthic invertebrates.

Conclusions

Sponge distribution and abundance on a subtidal sponge-dominated reef flat covered by a shallow sediment layer were found to be related, species specifically, to sediment depth and grain size. This distribution can be directly attributed to patterns in recruitment that do not appear to be random. Successful sponge recruitment was achieved predominantly by asexual modes of reproduction and was mediated by storm disturbance of sediments. During storms, the buds of many species exhibited similar developmental behavior. Buds became attached to large shell or rock fragments or to other buds. These particles gravitated to the edges of sediment ripples or were caught in depressions and against erect objects. Settlement onto the basal rock was thus possible. The large size of asexually derived recruits appears to enhance survival. Sexually produced larvae were seldom encountered and do not survive when re-covered by sediment. Subsequent long-term survival of recruits is species specific and depends on sediment depth and grain size.

Acknowledgments

This work was supported by grants from the University of New Zealand Grants Committee, University of Auckland Research Committee, and the Roche Research Institute of Marine Pharmacology. We thank K. Pritchard, B. Hickey, G. McIvor, and N. Moltschaniwskyj for their practical assistance with various aspects of this study.

Literature Cited

Ayling, A. L. 1978. Population Biology and Competitive Interactions in Subtidal Sponge Dominated Communities of Temperate Waters. Ph.D. Dissertation, University of Auckland. 113 pp.

———. 1980. Patterns of Sexuality, Asexual Reproduction and Recruitment in Some Subtidal Marine Demospongiae. *Biological Bulletin*, 158:271–282.

Bergquist, P. R. 1978. *Sponges*. Los Angeles: University of California Press. 268 pp.

Bergquist, P. R., M. E. Sinclair, and J. J. Hogg. 1970. Adaptation to Intertidal Existence: Reproductive Cycles and Larval Behaviour in Demospongiae. Pages 247–271 in *The Biology of the Porifera*, edited by W. G. Fry. Symposia of the Zoological Society of London 25. London: Academic Press.

Burton, M. 1949. Non-sexual Reproduction in Sponges, with Special Reference to a Collection of Young *Geodia*. *Proceedings, Linnean Society of London*, 160:163–178.

Connell, J. H. 1985. The Consequences of Variation in Initial Settlement vs. Post-Settlement Mortality in Rocky Intertidal Communities. *Journal of Experimental Marine Biology and Ecology*, 93:11–45.

Cook, D. L., and D. S. Gorshine. 1972. Field Observations of Sand Transport by Shoaling Waves. *Marine Geology*, 13:31–55.

Crisp, D. L. 1974. Factors Influencing the Settlement of Marine Invertebrate Larvae. Pages 177–265 in *Chemoreception in Marine Organisms*, edited by P. T. Grant and A. M. Mackie. London: Academic Press.

Dayton, P. L. 1979. Ecology, a Science and a Religion. Pages 3–18 in *Ecological Processes in Coastal and Marine Systems*, edited by R. J. Livingston. New York: Plenum Press.

Fell, P. E. 1974. Porifera. Page 51–132 in *Reproduction of Marine Invertebrates*, edited by A. C. Giese and J. S. Pearse. New York: Academic Press.

Harrison, F. W., and R. R. Cowden. 1976. Discussion Topics 1–9. Pages 20–47 in *Aspects of Sponge Biology*, edited by F. W. Harrison and R. R. Cowden. New York: Academic Press.

Hartman, W. D. 1958. Natural History of the Marine Sponges of Southern New England. *Bulletin, Peabody Museum of Natural History*, 12:1–155.

Highsmith, R. C. 1982. Reproduction by Fragmentation in Corals. *Marine Ecology Progress Series*, 7:207–226.

Keough, M. J. 1984. Dynamics of the Epifauna of the Bivalve *Pinna bicolor:* Interactions among Recruitment, Predation, and Competition. *Ecology*, 65:677–688.

Meadows, P. S., and J. I. Campbell. 1972. Habitat Selection by Aquatic Invertebrates. *Advances in Marine Biology*, 10:271–382.

Simpson, T. L. 1973. Coloniality Among the Porifera. Pages 549–565 in *Animal Colonies*, edited by R. S. Boardman, A. H. Cheetham, and W. A. Oliver. Stroudsburg: Dowden, Hutchison and Ross.

———. 1980. Reproductive Processes in Sponges: A Critical Evaluation of Current Data and Views. *International Journal of Invertebrate Reproduction*, 2:251–269.

———. 1984. *The Cell Biology of Sponges*. New York: Springer. 625 pp.

Stone, A. R. 1970. Growth and Reproduction of *Hymeniacidon perleve* (Montagu) (Porifera) in Langstone Harbour, Hampshire. *Journal of Zoology, London*, 161:443–459.

Underwood, A. J., and E. J. Denley. 1982. Paradigms, Explanations, and Generalizations in Models for the Structure of Intertidal Commuities on Rocky Shores. Pages 151–180 in *Ecological Communities: Conceptual Issues and the Evidence*, edited by D. R. Strong, L. G. Abele, D. Simberloff and A. B. Thistle. Princeton University Monograph. Princeton: University Press.

Wallace, C. C. 1985. Reproduction, Recruitment and Fragmentation in Nine Symaptric Species of the Coral Genus *Acropora*. *Marine Biology*, 88:217–233.

Wulff, J. L. 1985. Dispersal and Survival of Fragments of Coral Reef Sponges. Pages 119–124 in *Proceedings, Fifth International Coral Reef Congress, Tahiti*, 5. Moorea, French Polynesia: Antenne Museum-EPHE.

———. 1986. Variation in Clone Structure of Fragmenting Coral Reef Sponges. *Biological Journal, Linnean Society*, 27:311–330.

MAURIZIO PANSINI
ROBERTO PRONZATO
Istituto di Zoologia
Universitá di Genova
via Balbi 5, 16126 Genova, Italy

Observations on the Dynamics of a Mediterranean Sponge Community

Abstract

The development, dynamics, and evolution of an undisturbed population of Porifera living on a rocky bottom in the northwestern Mediterranean at depths of 12 to 28 m were examined by underwater photography over a period of six years. Plastic frames, divided into quadrats by nylon threads and nailed to the bottom, provided a size scale for the photographs and for the maps subsequently drawn. The recorded growth rates of the thirteen monitored species suggest that their life span probably extends considerably beyond the length of the observation period. Growth is slow in most species and often discontinuous. Some morphological variations, such as dilation and contraction of the body or color changes, were observed, but they do not appear to be seasonal.

Some newly recruited larva-derived specimens were occasionally observed, while fragmentation and fusion of adult specimens occurred more frequently. The relationships among neighboring sponges develop through continuous changes, but competition for free space was observed only after the death of encrusting organisms. Predation effects are negligible, whereas epibiosis may have lethal effects on erect species over the long term.

The composition of the population was taken to be almost constant during the period of observation since no new species appeared to enter the community and species disappeared from the station area only twice. The evolution and development of such a community is extremely slow and probably influenced by factors governed by water depth, being less pronounced at the deeper levels than in the shallow zone.

The long-term study of the sciaphilous hard-bottom communities often populated by sponges may be an important source of information on sponge recruitment, competition for substrate, growth, predation, and mortality, the causes of which are still poorly understood. Seasonal variations and population dynamics have usually been studied in intertidal or shallow-water environments (Burton, 1949;

404

Hartman, 1958; Stone, 1970; Reiswig, 1973; Elvin, 1976; Fell and Lewandrowski, 1981), and only a few studies have investigated subtidal or deep bottoms (Dayton, 1974, 1979; A. L. Ayling, 1980, 1983). As a result, little information is available on these environments, and much of it refers to the southern hemisphere.

Temporal variations of sponges on natural substrata have been studied for short periods of time (less then one year) in Mediterranean shallow-water caves (Labate, 1968) and a rocky littoral zone (Sarà, 1966). Other works on the subject (Harmelin, 1980) indicate that in deeper waters or in dark caves the evolution of the benthic community appears to be so slow as to require long-term investigations.

Researchers have used different field techniques at different depths to monitor the development of and variations in benthic organisms. Photography has been used at deeper levels, where staying time is necessarily short (Dayton, 1979; A. L. Ayling, 1983), and drawing has been used more often in intertidal or shallow subtidal zones (Burton, 1949; Labate, 1968; Sarà, 1970).

The present research took place at three stations on a steep rocky coast, which were monitored for six years. The purpose of this work was to collect information on the development, dynamics, and evolution of the undisturbed sponge populations in this area.

Material and Methods

Beginning in July 1979, observations were made on a sedimentary rocky bottom below the lighthouse of Punta di Portofino (eastern Ligurian coast). This point faces southeast and is therefore exposed to southern winds and a coastal current streaming out of Golfo Marconi. The combined action of the terrestrial runoff, wave mixing of bottom detritus, and flow of the current generates in high amounts of sediment particles, especially in winter. A coralligenous circalittoral biocoenose (sensu Pérès and Picard, 1964) thrives around depths of 30 m, but is also present in sciaphilous spots in somewhat shallower depths. Marine life covers at least 100% of the substrate and even more owing to the numerous epibioses.

Three square areas on vertical or subvertical positions measuring 30 × 30 cm were marked off by plastic frames nailed to the bottom. Station A was placed at a depth of 12 m, B at 22 m, and C at 26 m. Other frames, divided into quadrats measuring 5 cm on the side by nylon threads, were fitted to the fixed frames before each survey in order to provide a scale for the photographs (Balduzzi et al., 1981). A Hasselblad camera in a waterproof case was used for the photographic surveys, which were performed monthly at first and than at longer intervals later. A series of photographs was taken daily. The photographs were mapped with the aid of stereoscopic microscope, which was used to observe the transparencies, and a camera

lucida, which was used to draw the outlines of each sponge. Only area increases in the plane of the sponge were considered in evaluating growth, equally scoring the encrusting and massive species and excluding the few erect and branching ones as well as cryptic and boring forms. The areas covered by each sponge were measured on the maps with a Salmoiraghi Planimeter Model 236. Only very small samples of some monitored species were taken for identification purposes.

Results

The total number of sponge species living at each station is fairly high (20–25, according to our unpublished data from same area). However, only sponges of a certain size—those that were easily and consistently recognizable—were monitored (Table 1). The observations are reported by species, except for *Diplastrella bistellata*, which is included with *Pleraplysilla*.

Oscarella lobularis. This species, which is present at all the stations, is the only one for which we observed definite recruitment and successful development in two specimens in June and October 1982. Growth was rapid but discontinuous, and alternated with periods of regression (Figure 1). Fragmentation and successive coalescence of the fragments occurred often and may have caused the original specimen to be displaced. Shape varied considerably and seemed to have little to do with the stage of growth. On addition, individual specimens varied in color from violet to green to light brown and yellow, and depigmentation was often detectable in winter. Mortality was recorded once in summer and once in November. Four monitored specimens lived for more than four years, while another one is still living after six years.

Axinella damicornis. Three specimens of different sizes were monitored at station B. The largest one had rather

Table 1. Species of sponges monitored at each station.

Species	Stations and depth		
	A 12 m	B 22 m	C 26 m
1. *Oscarella lobularis* (Schmidt)	*	*	*
2. *Diplastrella bistellata* (Schmidt)	*	*	*
3. *Axinella damicornis* (Esper)		*	
4. *Acanthella acuta* Schmidt		*	
5. *Eurypon vescicularis* Sar & Siribelli	*	*	
6. *Anchinoe tenacior* Topsent	*	*	
7. *Crella mollior* Topsent	*		
8. *Reniera fulva* Topsent	*		
9. *Petrosia ficiformis* (Poiret)	*		
10. *Cacospongia scalaris* Schmidt	*	*	*
11. *Ircinia variabilis* Schmidt		*	*
12. *Dysidea avara* (Schmidt)	*	*	
13. *Pleraplysilla spinifera* (Schulze)	*	*	

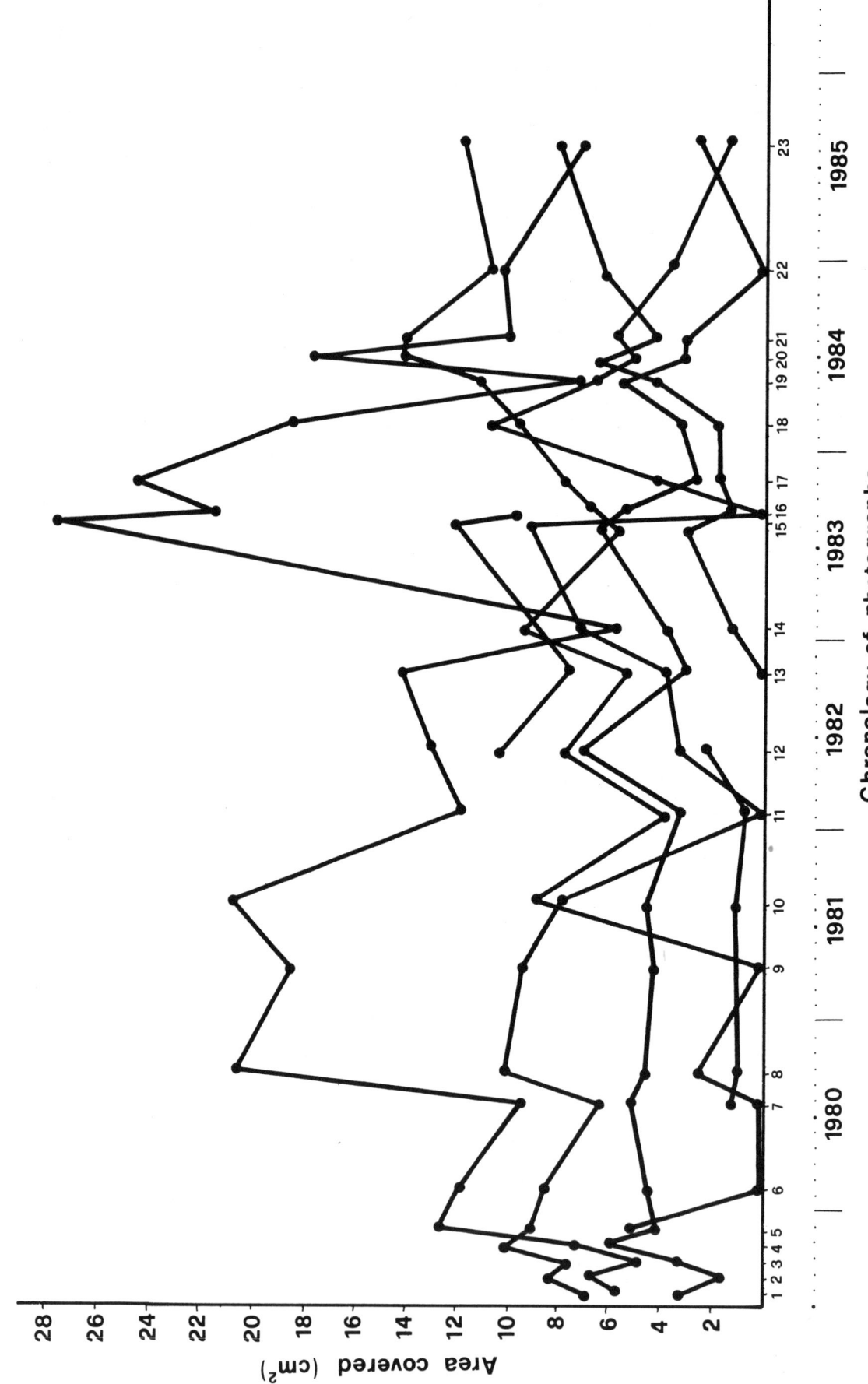

Figure 1. *Oscarella lobularis*: growth rates of monitored specimens.

flattened branches that hosted a colony of *Parazoanthus axinellae*. The demise of this specimen between September and November 1983 (Figure 2*a*) was probably related to the excessive weight of the epizoic cnidarians, which at first caused some sponge branches to become detached (arrows in Figure 2*a*). A smaller specimen, with two flattened lobes, was contiguous to a massive *Ircinia variabilis*, the development of which progressively enclosed *Axinella* and probably caused its death in the spring of 1983. Because *Ircinia* developed in this position, its tissues were left with a deep groove (Figure 2*b*). A third specimen, even smaller than the others, lived four years without showing any distinct signs of growth. In November 1983 it showed signs of senescence, being slightly contracted and densely covered by epizoic organisms, and it finally disappeared between May and June of the following year (Figure 2*b*). The verified longevity of the two smaller specimens is close to five years, but the largest one, which was fully developed at the beginning of the research, must obviously be much older. Small "buds" (Boury-Esnault, 1970) were observed on the oscular membranes of *Axinella* in summer and autumn, but nothing is known of their role, if any, in asexual reproduction.

Acanthella acuta. A single specimen of this species was monitored at station B for six years. As in other erect species, the growth of this specimen was difficult to quantify, but was evident by the end of six years. The sponge shows two aspects: it may be dilated (Figure 3*a*), with all the branch tips rounded, or contracted (Figure 3*b*), bristling with conules bearing bunches of styles. Four periods of contraction were observed during the study. Such periods do not show a precise seasonality, and their duration is still undetermined, but they probably last only a short time. Further monitoring at close intervals is needed to ascertain their periodicity.

Eurypon vesicularis. Four specimens monitored at stations A and B showed scanty morphological variations, discontinuous and slow growth, and remarkable stability. Specimens on open substrate did not seem to grow at a faster rate than those confined by neighboring sponges. The observed longevity exceeded the six years of observation.

Anchinoe tenacior. At station A this very common species showed an encrusting habit, although it can be found erect elsewhere. Numerous patches changed continuously in outline, but growth appeared to be extremely slow. The disappearance, in August 1981, of a broad colony of the encrusting bryozoan *Parasmittina* sp. provided an opportunity to investigate whether the availability of free space may stimulate growth. First hydroids settled on the free spot over the winter, then *Anchinoe* and a neighboring *Crella* several months afterward, limiting one another and

partly regaining the space lost by *Parasmittina*. Such recolonization by encrusting sponges remained incomplete, however, since other bryozoan species succeeded in occupying the area over the following months.

Crella mollior. Several small specimens of this species existed at station A from the beginning of the study and are still viable. During the course of the study, their surfaces expanded slowly but steadily, whereas their number increased at a faster pace. The pattern of colonization shown by the new recruits to the right of the parent specimen (Figure 4) may be due either to fragmentation or to the settlement of a swarm of larvae. *Crella* dilated and contracted periodically, but remained bright orange in color. Border outlines with neighboring encrusting species such as *Anchinoe* and *Eurypon* remained constant for long periods, but growth seemed to be stimulated by the death of other invertebrates (especially bryozoans).

Reniera fulva. The single specimen observed filled a cavity near the border of station B at a depth of 22 m. Variations in shape were also observed in this species, but they were not measured because of the awkward position of the sample. *Reniera* disappeared without apparent cause between September and November 1983 after having been monitored for more than four years.

Petrosia ficiformis. A single specimen monitored at station B showed no change in morphology, color (without seasonal light-induced variations as in *Ircinia,*) and outlines. Growth was not noticeable with the technique used, although small variations in the diameter of oscula and a slight increase in volume were recorded. The longevity of *Petrosia* must be much longer than the six years monitored.

Cacospongia scalaris. Several *Cacospongia* specimens of various sizes were monitored at stations B and C. The small and medium ones grew slowly and regularly (Figure 5), whereas the large ones underwent noticeable degeneration throughout the entire thickness of the sponge. Just before this occurred, the specimen became swollen and its color faded. Immediately after the degeneration the sponge began to rebuild the damaged region, restoring it with minor changes in outlines in about three years. Subsequently, new degeneration occured, followed by fragmentation and merging of contiguous specimens (Figure 6). Overgrowth by other sponges was not observed, whereas epibiosis by other invertebrates was common and especially intense just before the onset of degeneration. Since none of the specimens disappeared, the longevity of the species may well exceed six years.

Ircinia variabilis. Two specimens were monitored at station B and one at station C. The latter is occasionally visible owing to the continual presence of epibiotic organ-

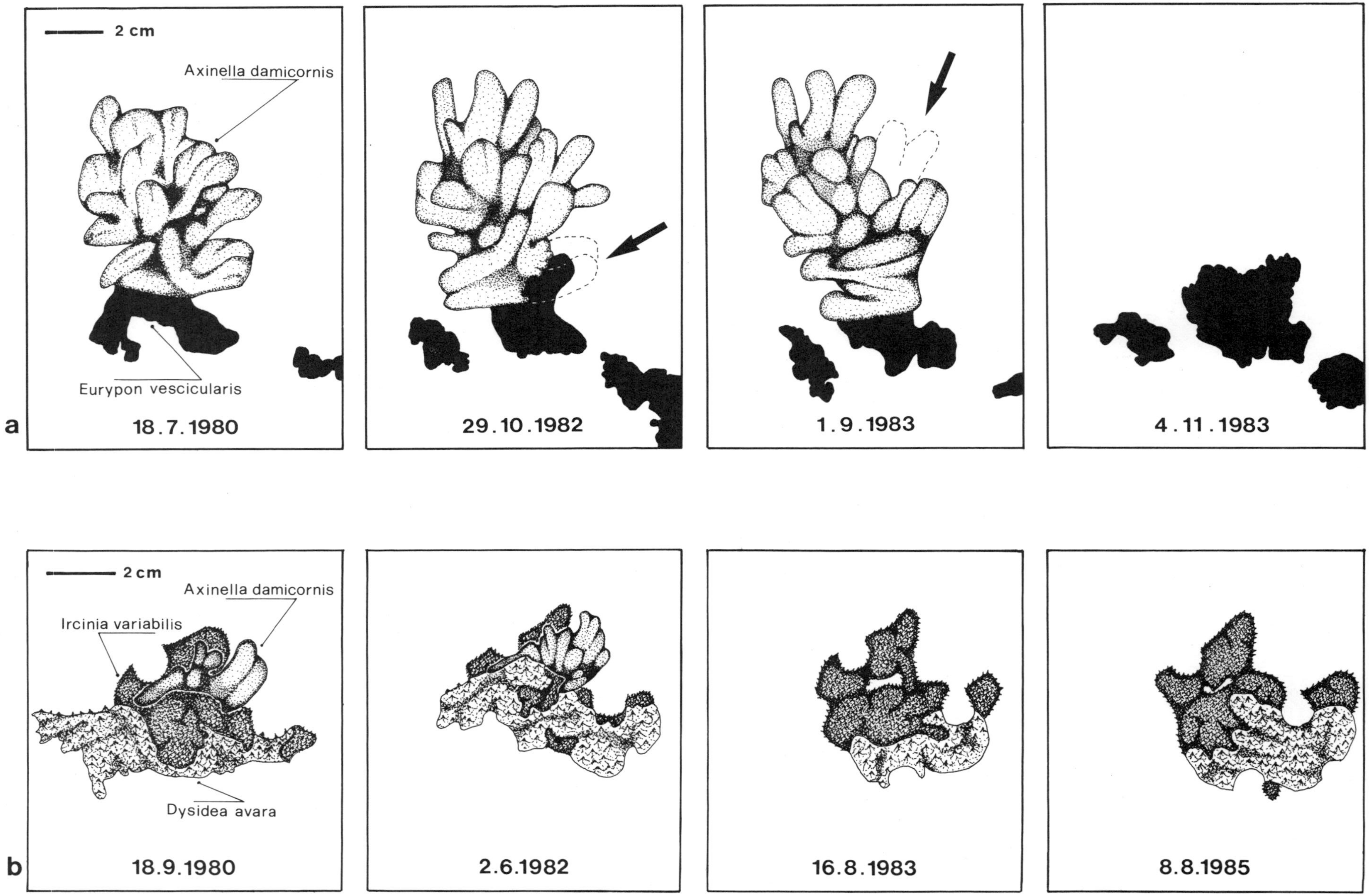

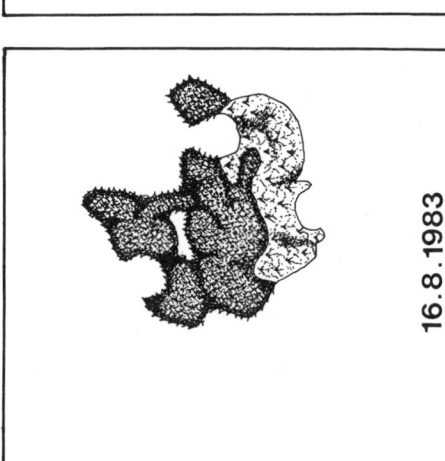

2 cm

Axinella damicornis

Eurypon vescicularis

18 . 7 . 1980

29 . 10 . 1982

1 . 9 . 1983

4 . 11 . 1983

a

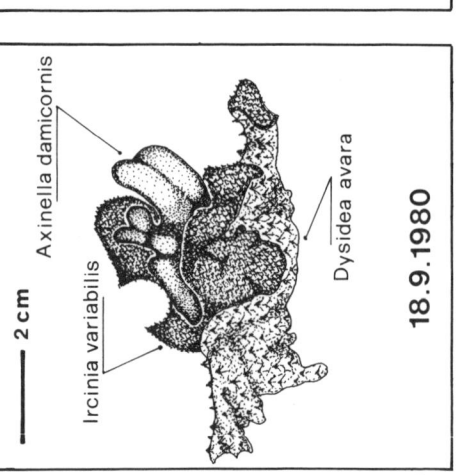

2 cm

Axinella damicornis

Ircinia variabilis

Dysidea avara

18 . 9 . 1980

2 . 6 . 1982

16 . 8 . 1983

8 . 8 . 1985

b

Figure 2. Map of individual changes through time: *a*, growth, loss of some branches, and death of *Axinella damicornis* specimen overgrown by colony of *Parazoanthus axinellae* at station B; *b*, development of three contiguous species at station B; *Ircinia variabilis* overgrown by *Dysidea avara* was not harmed, while *Axinella damicornis*, enclosed by *Ircinia*, disappeared in the spring of 1983.

3d Int. Sponge Conf. 1985

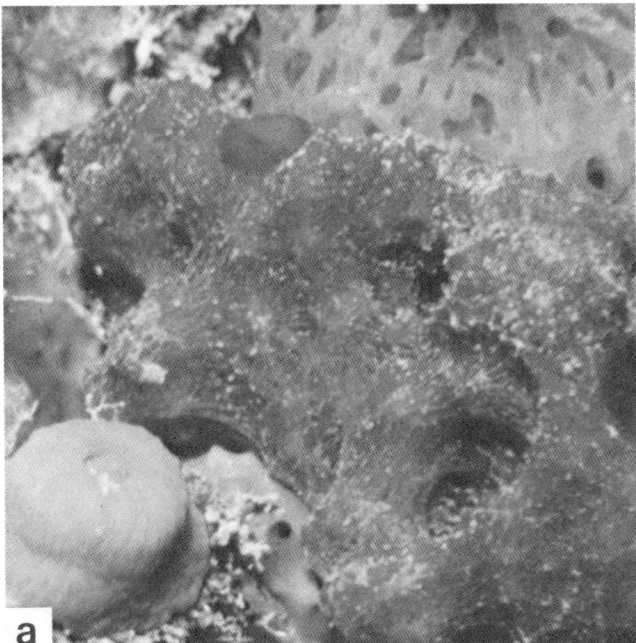 

Figure 3. Different morphological aspects in a specimen of *Acanthella acuta*.

isms. These specimens varied in size and their growth appeared discontinuous and was not seasonal. Overgrowth by sponges or other organisms seems to be well tolerated by this species: a specimen at station C showed dense coverings of epizoic fauna for long periods, and a specimen at station B appeared unaffected after being overgrown for six years by a *Dysidea avara* (Figure 2b), except for the development of funneled oscula, which are unusual in *Ircinia variabilis*. Massive neighboring species, such as *Cacospongia scalaris*, may have limited the growth of *Ircinia* which expanded toward the side where free space was available. Development was always very slow, however, since the groove left in the *Ircinia* specimen by the death of *Axinella* remained unfilled after two years. The violet color of *Ircinia variabilis*, which is due to the presence of symbiotic microorganisms (Sará, 1971), faded in winter, probably because less light reached the surface of the sponge during that period. The life span of the monitored specimens exceeded six years.

Dysidea avara. From July 1979 three patches of *Dysidea* were monitored at station A in addition to the rather static specimens overgrowing *Ircinia* (Figure 2b). In November 1979 these patches merged into a single specimen, which, after two years of growth, underwent fragmentation. The smaller fragments fused again in about a year, whereas the larger ones retained their individuality. The recorded life span at station C was over six years, whereas at station A the fragmentation makes it difficult to evaluate the longevity.

Pleraplysilla spinifera. Several specimens were evenly distributed throughout the area around station B. Their morphology was not constant as periodical contraction induced an atypical cribrous aspect. These variations did not seem to be seasonal. The species appeared to be more sensitive than others to the competition for space as some specimens disappeared from a densely populated area within two years. Their degeneration began with fragmentation. We observed a case of direct competition between *Pleraplysilla spinifera* and *Diplastrella bistellata*. After being dominated by *Diplastrella* for two years, *Pleraplysilla* made a remarkable recovery, growing at a rapid rate and spreading over the area previously occupied by *Diplastrella*. Then, at different times, both species disappeared. At the same time, several adjacent specimens of *Pleraplysilla* died on the right side of station A (Figure 4), whereas on the opposite side another one was monitored for more than six years.

Discussion

Most of the monitored species appear to grow very slowly and their development is marked by periods of standstill or regression here and there within an individual or in the entire organisms. The fastest growth rate was observed in *Oscarella lobularis*, in which we also observed the recruitment and development of young specimens. The slowest growing species seems to be *Petrosia ficiformis*. The growth process does not appear to be seasonal, as single con-

410

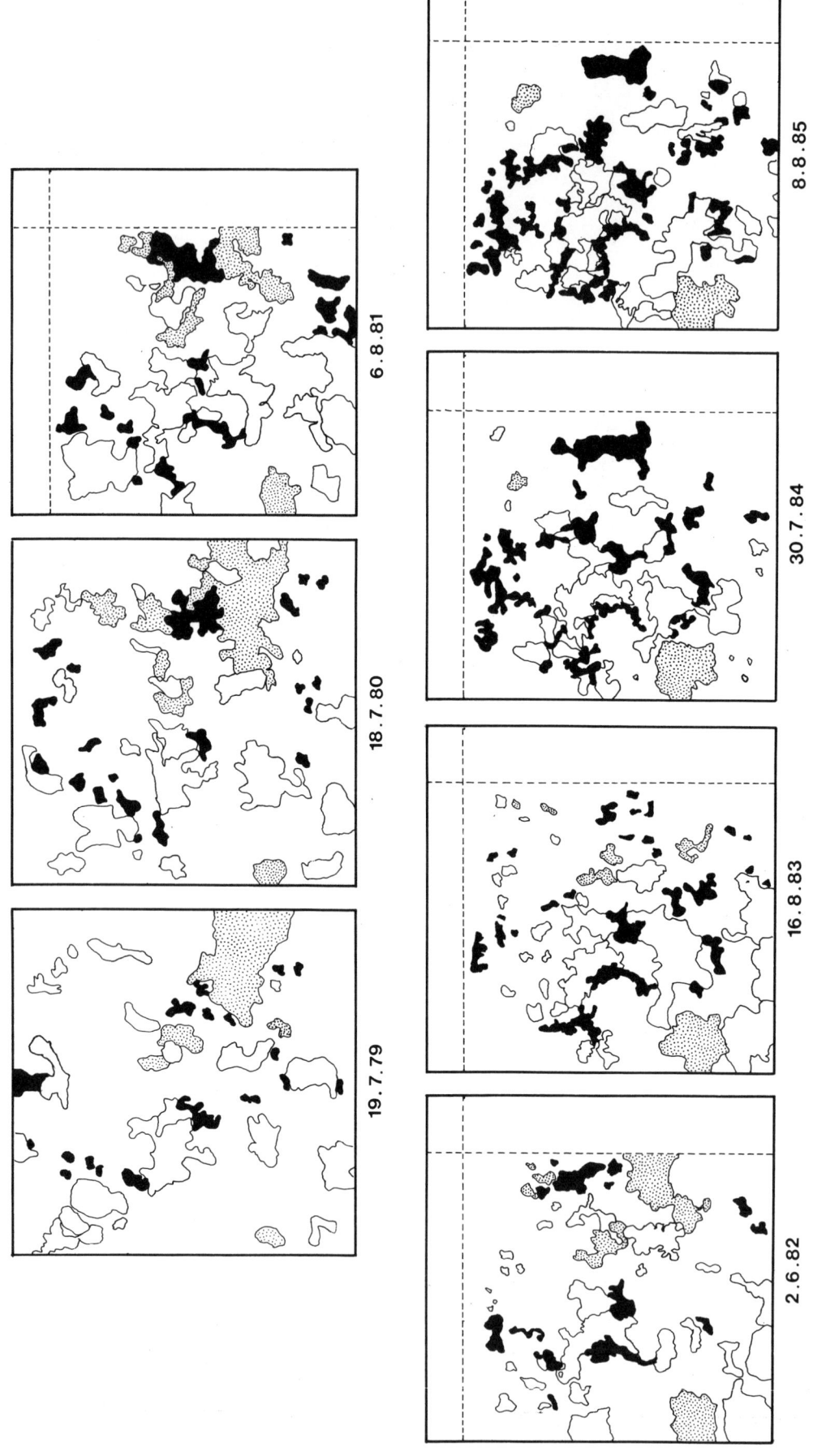

6.8.81

18.7.80

19.7.79

8.8.85

30.7.84

16.8.83

2.6.82

Figure 4. Remarkable mutability of two sponge species at station A at depth of 12 m: *Crella mollior* (*black*); and *Pleraplysilla spinifera* (*stipled*); other sponge species outlined only. Dotted lines indicate accidental displacement of frame from its original position.

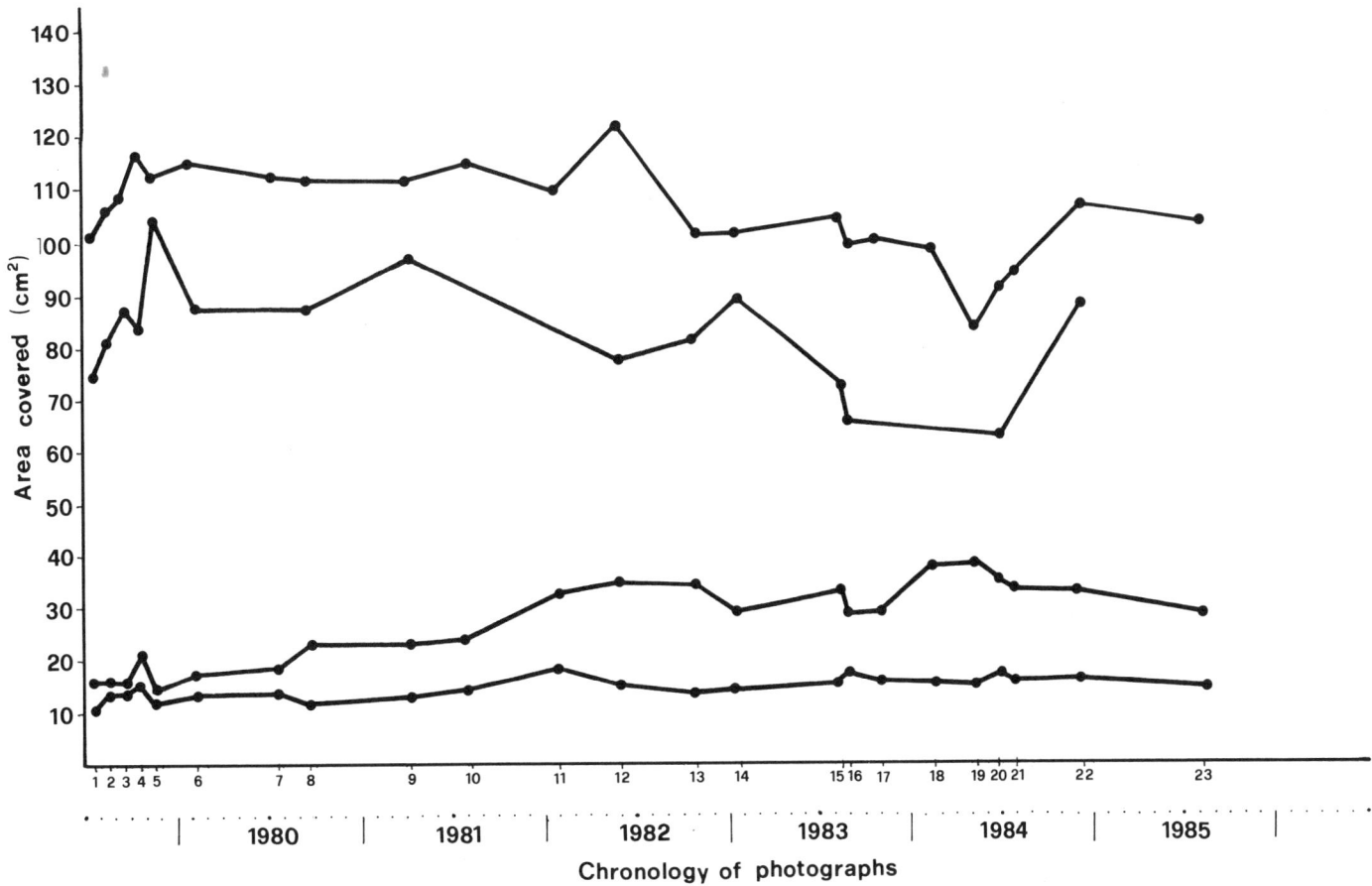

Figure 5. *Cacospongia scalaris:* growth rates of four monitored specimens.

specific specimens develop independently and asynchronously. Therefore, as A. L. Ayling (1983) has noted, the size variation in these species does not seem to be related to their reproductive cycles. Growth rates, which are very difficult to quantify owing to the different thickness of the sponges and to their size fluctuations, seem to be closer to those recorded by A. L. Ayling (1983) in the subtidal zone (12 and 18 m) than those recorded in the shallow subtidal or intertidal zones (Labate, 1968; Elvin, 1976; Fell and Lewandrowski, 1981), where development is faster and often related to the season. In fact, shallow-water sponge populations are known to be subjected to great temporal variations (Burton, 1949; Sarà, 1970; Stone, 1970; Johnson, 1979) as a result of the constant changes taking place in their environments.

Although the populations at the two depths studied were very similar, owing to the sciaphilous position of the 12 m station, some significant differences were also noted. For example, mutability was greater at the shallow subtidal station (with increased fragmentation, higher reproduction and mortality rates, and intensified competition for substrate), possibly because the variability of both biotic and abiotic factors decreases with depth. The season-

al alga, hydroid, and bryozoan populations are actually much more abundant here than at deeper levels.

Some species, such as *Cacospongia scalaris*, may even undergo extensive degeneration, although this appears to occur only in larger specimens. Regressions are common in sponges (Fell and Lewandrowski, 1981 and A. L. Ayling, 1983), whereas partial or total degeneration has been observed after exposure to the direct rays of the sun (Burton, 1949). The regeneration rate of the damaged sponge tissue is more rapid than the normal growth rate. A. M. Ayling (1981), in a simulated urchin-grazing experiment, found that *Stylopus* sp. regained lost space by spreading a thin layer of tissue over it at a rate 200 times higher than the normal growth rate of the species. Ayling reported regeneration rates 22 to 2900 times the normal growth rate. However, degeneration occured in a specimen of *Cacospongia scalaris* that was not encrusting and that was more than 2 cm thick; it took about three years to heal its grooved scars, and it subsequently even resumed growth.

The death of other encrusting invertebrates, which makes spaces available to sponges, seems to stimulate the expansion of neighboring species, but probably does not affect their normal growth rate. When A. M. Ayling

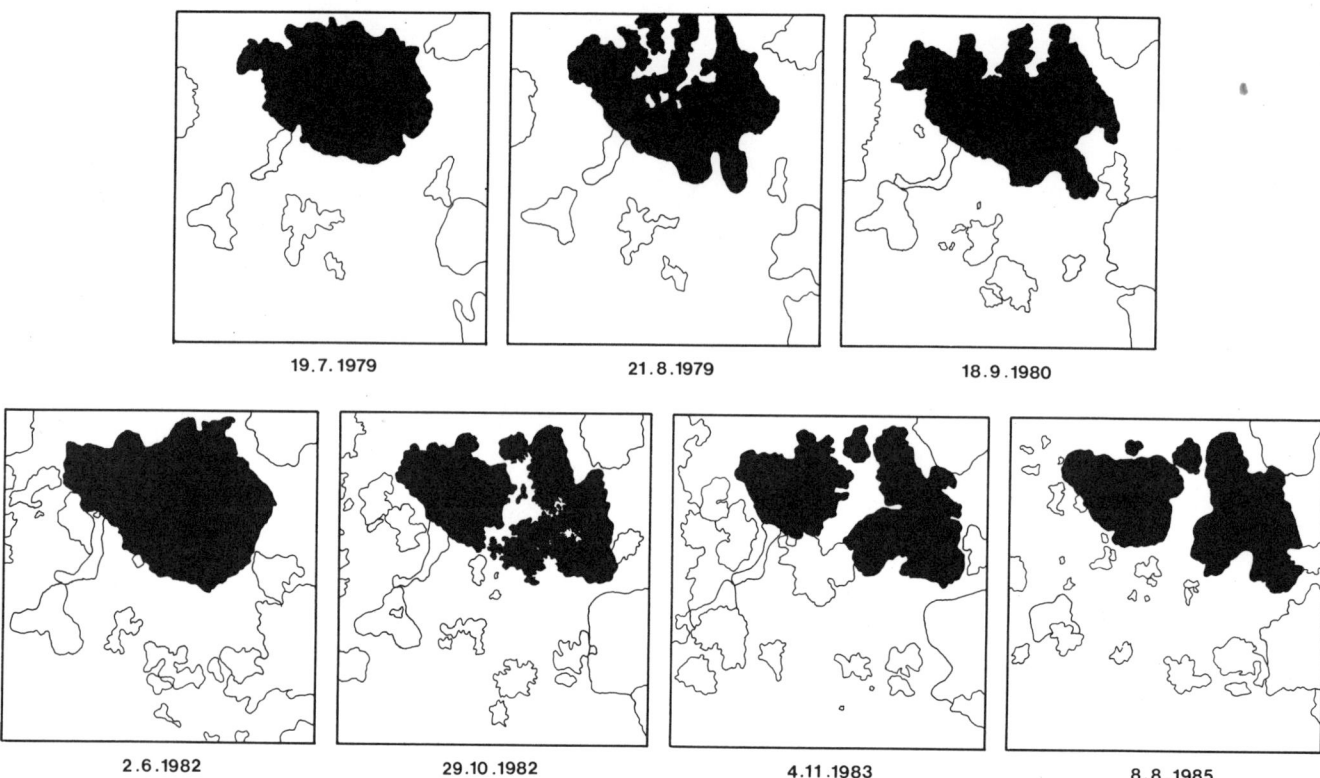

Figure 6. Development, degeneration, and regeneration of *Cacospongia scalaris (black)* at station C. Other sponge species outlined only.

(1981) attempted the experimental removal of invertebrates surrounding sponges, the results were unsatisfactory because some of the growth may have been stimulated by accidental sponge damage.

Fragmentation and fusion of sponge individuals have been reported for both Calcispongiae (Johnson, 1979) and Demospongiae (Burton, 1949; Elvin, 1976; Fell and Lewandrowski, 1981). We, too, observed these processes in *Cacospongia, Dysidea, Oscarella,* and probably *Anchinoe* and *Crella* among our monitored species. Fusion appears to occur among fragments coming from a single parent specimen or, as recently demonstrated (Evans and Curtis, 1979; Hildemann et al., 1979; Jokiel et al., 1982), even from specimens belonging to the same strain, on other words showing genetic identity. When a sponge breaks and some of the fragments succeed in developing and then coalesce, the original specimen may be deplaced, and eventually species distribution may be affected.

Many species show high plasticity and continual variations in size and shape, whereas others—such as *Petrosia, Diplastrella,* and *Eurypon*—appear to be more constant. Dilation and contraction, which show short periodicity and follow still undetermined rhythms, are very common but poorly understood in marine Demospongiae. Modifications of the type that we observed, because of their short duration, might be related to the physiological sta-

tus of a sponge, such as variable pumping rates (Reiswig, 1971; Gerrodette and Flechsig, 1979) and may cause temporary changes in volume. The contractions do not seem to be directly related to the increased turbidity resulting from storms or terrestrial runoff.

The color of a specimen, although generally constant, may change in some species. Such changes are rarely seasonal when the color is pigment-induced, as in *Oscarella lobularis* (Donadey, 1979), but may be periodical when it is related to the presence of symbiotic cyanobacteria, as in *Ircinia variabilis* (Sarà, 1971). However, such variations were not observed in *Petrosia*. This is rather surprising because *Petrosia* cyanobacteria should be highly sensitive to variations in light intensity, to judge by the existence of decolored cave-dwelling specimens and by the different shades observed in various parts of the same specimen living outside the caves (Sarà and Vacelet, 1973).

Since the signs of senescence, which in some species precede the degeneration and death of a sponge (Burton, 1949), were never observed in our photographs, it may be that some monitored specimens disappeared rapidly. On the other hand, the presence of numerous epizoic organisms, at least in those species that are normally devoid of them, may be symptomatic of an unhealthy or dying specimen. By and large, however, the causes of death remained unknown except for an *Axinella damicornis* specimen that broke off and disappeared owing to the excessive weight of epizoic *Parazoanthus*. Species proximity generally produces balance; however, in two adjacent *Ircinia* and *Axinella*, the development of the massive species seems to have been responsible for the death of the latter. The competition for substrate does not seem to be harmful to sponges, although both specimens of *Pleraplysilla* and *Diplastrella*, in this study disappeared at different times after a period of dominance by one and then the other specimens.

Mortality in most specimens occurs independently of that of other individuals, although some *Pleraplysilla* specimens living in close proximity on one side of station A died at the same time. When other pathological phenomena are lacking, mortality rates may be influenced by very high temperatures (Johnson, 1979; Fell and Lewandrowski, 1981). Reiswig (1973) attributed the high mortality rates he observed for *Mycale* sp. in Jamaica to the effect of winter storms. Mortality is generally higher for young postlarval specimens (Hartman, 1958; Reiswig, 1973; A. L. Ayling, 1980; Fell and Lewandrowski, 1981). Since temperature varies less with depth, as does wave action, death is more likely due to intrinsic factors rather than environmental ones. The degeneration and death of large horny sponge specimens *(Spongia* and *Cacospongia)* that we observed outside the monitored stations was a consequence of bacterial epidemics. This phenomenon has already been recorded for commercial sponges (Storr, 1964).

Seasonal epizoic fauna prefer to settle on certain species (such as the large horny sponges *Cacospongia* and *Ircinia*) and usually do not damage the host sponge. Overgrowth among different sponges is generally well tolerated by the underlying species and may last for a long time (Rützler, 1970; Sarà, 1970). Overgrowth is not common in our stations, however, even though the high density of benthic organisms greatly restricts the space available for settlement.

Neighboring species that do not occupy the same layer, for instance a massive and erect one, do not interfere with one another as they develop, although they may expand toward areas free of other sponges during the growth process. If the sponges are contiguous, however, as might be the case with two encrusting species, a slow and continuous variation in outline may be seen, and each sponge may in turn predominate for a period of time. The scant information on competition for substrate among Porifera (Labate, 1965; Rützler, 1970; Sarà, 1970) gives no indication that a sponge might destroy another one by overwhelming it (Sarà and Vacelet, 1973). Furthermore, A. L. Ayling (1983), whose data are comparable to ours, has recorded something of a balance, with outlines being more or less static among neighboring species. In the monitored area, we observed a dynamic balance among long-lasting species, and the other invertebrates present (such as hydroids

and bryozoans) had a much shorter life cycle and occupied the free areas only temporarily. The mortality of these organisms, primarily encrusting bryozoans, may from time to time free new areas and thus allow the neighboring species to expand. Nevertheless, competition between sponges and bryozoans, as reported for coral reef communities (Jackson, 1977, 1981; Jackson and Buss, 1975), was observed only for *Crella mollior* and *Anchinoe tenacior*.

During our study, the joint action of benthic invertebrates caused a visible modification of some substrate characteristics such as the filling of even quite deep grooves.

The effects of predation on this community seem to be negligible. Only a few nudibranch species were frequently observed at the stations, but some of these feed on hydroids *(Flabellina affinis, Coryphella pedata)*, and only *Hypselodoris tricolor* is a known sponge eater.

The data from this study confirm the remarkable longevity of many Demospongiae species in subtidal environments. We were fortunate to be able to monitor one or more specimens of most of the sponges living at each station—with the exception of *Axinella* and *Reniera*—for the entire period of the study. However, since a sponge may undergo fragmentation, fusion, and even regression, it is difficult to estimate longevity from growth rate measurements (A. L. Ayling, 1983). Therefore, only the observed ages are reported, although the size and the slow growth rate of many specimen suggest that some specimens live much longer than the six monitored years. A completely different situation was observed by Burton (1949), Stone (1970), Elvin (1976), Fell et al. (1979) in intertidal environments, where many short-lived (2 to 3 years) species are present and the dynamics of population is more accelerated. Some species live much longer; for example, in deeper zones Reiswig (1973) has estimated some specimens of *Verongia gigantea* as old as 50 to 100 years, and A. L. Ayling (1983) has estimated the largest specimen of *Stylopus* sp. she observed to be 78 years old, according to its mean growth rate.

The structure of the community we studied appears to be governed by the rather stable conditions of its environment. That is to say, periodic disturbances due to physical factors are almost absent and the influence of biotic activities such as predation seems to be negligible. In the absence of biological control, the sponge community appears to impose internal controls not unlike the cooperative phenomena said to arise in dense, continuous sponge populations (Sarà, 1970). Opportunistic species with high dispersal ability are scant; among the monitored Demospongiae only *Oscarella, Anchinoe* and *Crella* may be included in this group. Some connections probably exist among the different species which maintain the population equilibrium over a long period of time with minimum competition, although such connections are difficult to observe. The community may be defined as persistent (Sarà, 1985), because it maintains its structure according to species composition and abundance; in six years only two species, *Axinella* and *Reniera*, disappeared, and none immigrated to the study site. This persistence is the result of a specialized strategy in the dynamics of the community (Ott, 1981), which is characterized by a high biomass, a low dispersal ability, and development in a highly predictable environment where catastrophic events are normally absent.

Conclusions

The population of Porifera in the study area does not show yearly cycles or a precise seasonality. Growth in most species is slow and often discontinuous.

Some monitored sponges undergo fragmentation, fusion, regression and regeneration, which may cause displacement of individuals on the substrate.

Many species show morphological variation, such as dilation and contraction of the body, of short periodicity, but its significance is still undetermined.

Recruitment and mortality were seldom observed, since most sponge species have a long life span.

The relationships among neighboring sponges develop through continuous changes, but no single organism seems to prevail in the competition for the substrate.

Predation, being limited to a few nudibranch species, does not seem to influence the equilibrium of the population. However, in the long run, epibiosis may have lethal effects on the erect species.

Although the effect of physical disturbances related to depth appears to be minor, population activity decreased as depth increased.

The composition of the population remained more or less constant during the six years of the study since no new species immigrated to the study site and only two species disappeared.

The persistence of the community is assumed to be due to a specialized strategy of community dynamics.

Literature Cited

Ayling, A. L. 1980. Patterns of Sexuality, Asexual Reproduction and Recruitment in Some Subtidal Marine Demospongiae. *Biological Bulletin*, 158:271–282.
———. 1983. Growth and Regeneration Rates in Thinly Encrusting Demospongiae from Temperate Waters. *Biological Bulletin*, 165:343–352.
Ayling, A. M. 1981. The Role of Biological Disturbance in Temperate Subtidal Encrusting Communities. *Ecology*, 62:830–847.
Balduzzi, A., F. Boero, D. Pessani, M. Pansini, and R. Pronzato. 1981. Emploi des rélevements photographiques dans l'étude de l'évolution des biocoenoses de substrat dur naturel. *Rap-*

port, *Commission International pour la Mer Méditerranée (France)*, 27(9):249–250.
Boury-Esnault, N. 1970. Un phénomène de bourgeonnement externe chez l'Eponge *Axinella damicornis* (Esper). *Cahiers de Biologie Marine*, 11:491–496.
Burton, M. 1949. Observations on Littoral Sponges, including the Supposed Swarming of Larvae, Movement and Coalescence in Mature Individuals, Longevity and Death. *Proceedings, Zoological Society of London*, 118:893–915.
Dayton, P. K. 1979. Observations of Growth, Dispersal and Population Dynamics of Some Sponges in McMurdo Sound, Antarctica. Pages 271–282 in *Biologie des Spongiaires*, edited by C. Lévi and N. Boury-Esnault. Colloques internationaux du C.N.R.S. 291. Paris: Centre National de la Recherche Scientifique.
Dayton, P. K., G. A. Robilliard, R. T. Paine, and L. B. Paine. 1974. Biological Accomodation in the Benthic Community at McMurdo Sound, Antarctica. *Ecological Monographs*, 44:105–128.
Donadey, C. 1979. Contribution á l'étude cytologique de deux démosponges Homosclérophorides: *Oscarella lobularis* (Schmidt) et *Plakina trilopha* Schulze. Pages 165–172 in *Biologie des Spongiaires*, edited by C. Lévi and N. Boury-Esnault. Colloques internationaux du C.N.R.S. 291. Paris: Centre National de la Recherche Scientifique.
Elvin, D. W. 1976. Seasonal Growth and Reproduction of an Intertidal Sponge, *Haliclona permollis* (Bowerbank). *Biological Bulletin*, 151:108–125.
Evans, C. W., and A. S. G. Curtis. 1979. Graft Rejection in Sponges: Its Relation to Cell Aggregation Studies. Pages 211–215 in *Biologie des Spongiaires*, edited by C. Lévi and N. Boury-Esnault. Colloques internationaux du C.N.R.S. 291. Paris: Centre National de la Recherche Scientifique.
Fell, P. E., and K. B. Lewandrowski. 1981. Population Dynamics of the Estuarine Sponge, *Halichondria* sp., within a New England Eelgrass Community. *Journal of Experimental Marine Biology and Ecology*, 55:49–63.
Fell, P. E., K. Lewandrowski, and M. Lovice. 1979. Postlarval Reproduction and Reproductive Strategy in *Haliclona loosanoffi* and *Halichondria* sp. Pages 113–119 in *Biologie des Spongiaires*, edited by C. Lévi and N. Boury-Esnault. Colloques internationaux du C.N.R.S. 291. Paris: Centre National de la Recherche Scientifique.
Gerrodette, T., and A. O. Flechsig. 1979. Sediment–induced Reduction in the Pumping Rate of the Tropical Sponge *Verongia lacunosa*. *Marine Biology*, 55:103–110.
Harmelin, J. G. 1980. Etablissement des communautés de substrats dur en milieu obscur. Resultats préliminaires d'une experience á long terme en Meditérranée. *Memorie di Biologia Marina e di Oceanografia*, 10:29–52.
Hartman, W. D. 1958. Natural History of the Marine Sponges of Southern New England. *Peabody Museum of Natural History, Bulletin*, 12:1–155.
Hildemann, W. H., I. S. Johnson, and P. L. Jokiel. 1979. Immunocompetence in the Lowest Metazoan Phylum: Transplantation Immunity in Sponges. *Science*, 204:420–422.
Jackson, J. B. C. 1977. Habitat Area, Colonization and Development of Epibenthic Community Structure. Pages 349–358 in *Benthic Marine Organisms*, edited by B. F. Keegan, P. O'Ceidigh and P. J. Boaden. Oxford: Pergamon Press.

————. 1981. Interspecific Competition and Species Distribution: The Ghost of Theories and Data Past. *American Zoologist*, 21:889–901.

Jackson, J. B. C., and L. Buss. 1975. Allelopathy and Spatial Competition Among Coral Reef Invertebrates. *Proceedings of the National Academy of Sciences*, (U.S.A.), 72:5160–5163.

Johnson, M. F. 1979. Recruitment, Growth, Mortality and Seasonal Variations in the Calcareous Sponges *Clathrina coriacea* (Montagu) and *Clathrina blanca* (Miklucho-Maclay) from Santa Catalina Island, California. Pages 325–334 in *Biologie des Spongiaires*, edited by C. Lévi and N. Boury-Esnault. Colloques internationaux du C.N.R.S. 291. Paris: Centre National de la Recherche Scientifique.

Jokiel, P. L., W. H. Hildemann, and C. H. Bigger. 1982. Frequency of Intercolony Graft Acceptance or Rejection as a Measure of Population Structure in the Sponge *Callyspongia diffusa*. *Marine Biology*, 71:135–139.

Labate, M. 1965. Ecologia dei Poriferi della grotta della Regina (Adriatico meridionale). *Bollettino di Zoologia*, 32:541–553.

————. 1968. Variazioni temporali in un popolamento di Poriferi. *Archivio di Oceanogrografia e Limnologia*, 16:63–80.

Ott, J. A. 1981. Adaptive Strategies at the Ecosystem Level: Examples from Two Benthic Marine Systems. *Marine Ecology*, 2:97–180.

Pérès, J. M., and J. Picard. 1964. Nouveau manuel de bionomie benthique de la Mer Méditerranée. *Recueil de Travaux. Station Marine d'Endoume*, 31:1–137.

Reiswig, H. M. 1971. *In Situ* Pumping Activities of Tropical Demospongiae. *Marine Biology*, 9:38–50.

————. 1973. Population Dynamics of Three Jamaican Demospongiae. *Bulletin of Marine Sciences*, 23:191–226.

Rützler, K. 1970. Spatial Competition among Porifera: Solution by Epizoism. *Oecologia*, 5:85–95.

Sarà, M. 1966. Studio quantitativo della distribuzione dei Poriferi in ambienti superficiali della Riviera Ligure di Levante. *Archivio Oceanografia e Limnologia*, 14:365–386.

————. 1970. Competition and Cooperation in Sponge Populations. Pages 273–284 in *The Biology of the Porifera*, edited by W. G. Fry. London: Academic Press.

————. 1971. Ultrastructural Aspects of the Symbiosis Between Two Species of the Genus *Aphanocapsa* (Cyanophyceae) and *Ircinia variabilis* (Demospongiae). *Marine Biology*, 11:214–221.

————. 1985. Persistence and Changes in Marine Benthic Communities. *Nova Thalassia*, 7, Suppl. 3:7–30.

Sarà, M., and J. Vacelet. 1973. Ecologie des Démosponges. Pages 462–576 in *Traité de Zoologie, 3(1), Spongiaires*, edited by P.-P. Grassé. Paris: Masson.

Stone, A. R. 1970. Growth and Reproduction of *Hymeniacidon perleve* (Montagu) in Langstone Harbour, Hampshire. *Journal of Zoology*, 161:443–459.

Storr, J. F. 1964. Ecology of the Gulf of Mexico Commercial Sponges and Its Relation to the Fishery. *United States Fish and Wildlife Service, Special Scientific Report*, 466:1–73.

BENOIT VERDENAL
JEAN VACELET
Centre d'Océanologie de Marseille
Station Marine d'Endoume
13007 Marseille, France

Sponge Culture on Vertical Ropes in the Northwestern Mediterranean Sea

Abstract

A new attempt was made to breed commercial sponges in order to test the profitability of culturing sponges in view of the present state of knowledge in sponge ecology and the present market conditions. *Spongia officinalis* L., *S. agaricina* Pallas, and *S. nitens* (Schmidt) were bred using the cutting technique. The cultures were made on vertical ropes in a variety of environments, including one close to the Marseille sewage outlet and one in the pure waters of the Port-Cros National Park. Five hundred cuttings were fixed between depths of 12 and 92 m, and data on mortality and growth rates were collected for each over a period of 30 months. The results indicate that the highest growth rates are obtained in pure water and that *S. agaricina* has the greatest potential as a cultured species.

The drastic decline in the American sponge industry after the general blights of 1938 and 1947 (Smith, 1941; Storr, 1964) has led to overfishing on the Mediterranean sponge beds. In recent years, catches have slowly decreased, and there has been a sharp increase in price (Figures 1, 2). Sponge supply is now below the demand both in Europe and in America (Stevely et al., 1978), and the quality of the sponges available has also gone down. The most attractive species *(Spongia officinalis, S. agaricina)* have become sparse, and the present market relies mainly on the Mediterranean *Hippospongia communis,* which is less desirable as a bathing sponge.

In view of these circumstances and the fact that sponge dealers and governments in several countries have expressed some interest in sponge culture, we thought that it would be interesting to test once again the economic possibilities of sponge culture. Sponge breeding has been tried many times in the past (reviews in Crawshay, 1939; Vacelet, 1985) and although the cutting technique proved successful, its profitability was not conclusively demonstrated. Most of these attempts were conducted in the

416

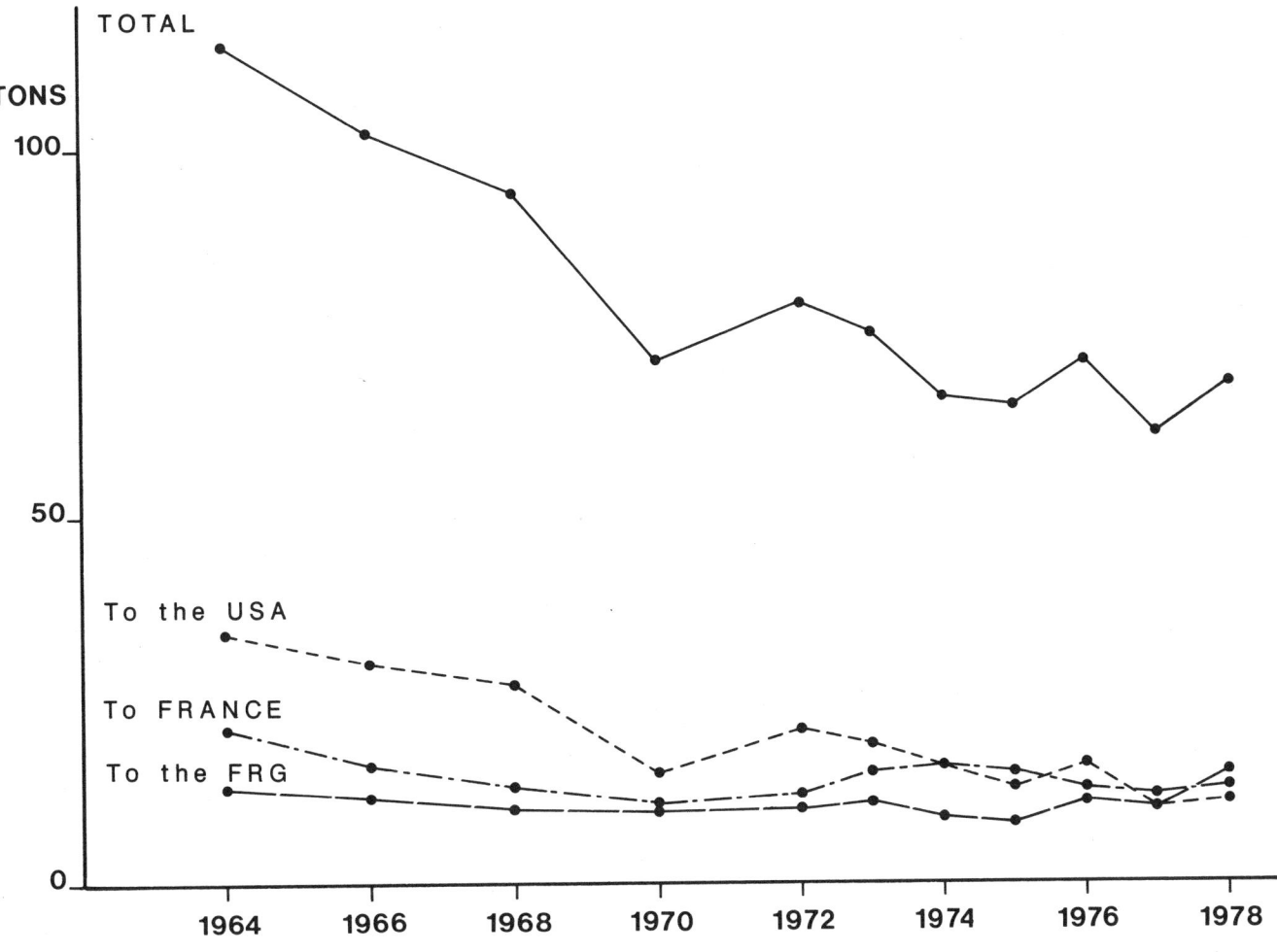

Figure 1. Decrease in commercial sponge catches, demonstrated by Greek exportations between 1964 and 1978. (Source: Greek custom house.)

Caribbean and in the Gulf of Mexico. Only a few experiments have been conducted in the Mediterranean, most of them in ancient times and with haphazard procedures, so that their results cannot be considered conclusive. In addition, the growth rate of noncommercial sponges in natural habitats or after transplantation is poorly documented (Sarà and Vacelet, 1973; Wilkinson and Vacelet, 1979).

In this work, we made a new attempt at sponge cultivation using the cutting technique, which differs in several respects from previous efforts: First, the experiments were conducted on the Mediterranean coasts of France, where there are no sponge fisheries but where natural populations of commercial sponges do occur. However, these populations prefer reduced illuminations and thrive mainly on walls or under ledges, in contrast to the sponge beds of the eastern Mediterranean (Vacelet, 1959). Second, the sponge species we chose are of high commercial value, but few have ever been bred before. Third, the cuttings were individually controlled in order to obtain precise data on growth rate and its variability. Fourth, in order to improve the growth rate, which seems to have

been the main problem in previous attempts, we selected an area with high nutrient level (in the proximity of the main sewer outlet of Marseilles, where natural populations occur under ledges and overhangs despite the high turbidity of the water), and we used the hanging wire culture method, which allows better utilization of the area of a "plantation" and represents an artificial situation in which a sponge could theoretically find many hydrodynamic advantages (Bidder, 1923). This method showed the most promise, to judge by ancient experiments (Moore, 1910), which, however, only had corrosive materials to work with.

More detailed results of this study were given by Verdenal (1986). For an economic appraisal of sponge culture based on these results, see Verdenal and Verdenal, 1987.

Material and Methods

Three species of the genus *Spongia* were studied: *S. officinalis* L. ("fine grecque"), *S. agaricina* Pallas ("elephant ear"), both with a high commercial value, and *S. nitens*

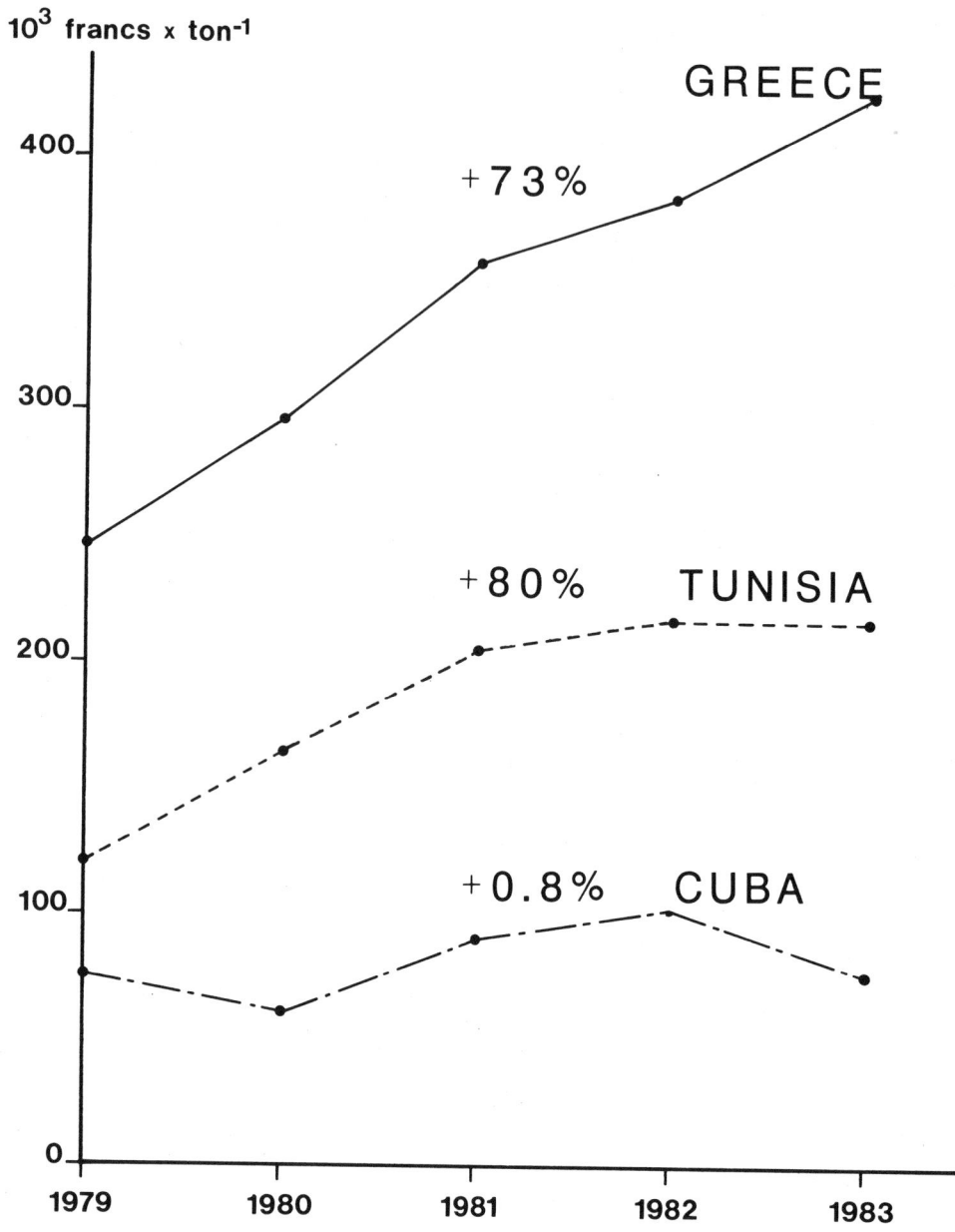

Figure 2. Increase in commercial sponge price, French imports between 1979 and 1983 (mean price according to origin). (Source: French custom house.)

(Schmidt), a very soft sponge that has not been exploited commercially, owing to its relative scarceness and to its small mean size but that has potentially good commercial value. *S. officinalis* has already been bred in the Adriatic sea in Schmidt's and Buccich's experiments (Marenzeller, 1879), but no attempts have yet been to culture the other two species. The "honey-comb" or "horse" sponge, *Hippospongia communis*, was not tried, as it is uncommon in this area. However, it is commercially important in Tunisia.

Six culture stations were established in different environmental conditions (Figure 3, Table 1), and 570 cuttings were studied—480 from *Spongia officinalis*, 60 from *S.*

agaricina, and 30 from *S. nitens*. The cuttings were obtained from freshly collected specimens and were fastened by plastic-coated metal wires on nylon ropes extending between a concrete base and a buoy (Figure 4a,b); the buoy was located a few meters under the surface for security reasons (Figure 5). The cuttings were placed at intervals of 0.5 m and were individually labeled. Their initial volume (about 30–60 ml) was estimated by measuring the three major axes. The ropes were raised briefly every 3 months or so and the cuttings were again measured. Growth was calculated as the percentage change in sponge volume.

3d Int. Sponge Conf. 1985

Table 1. Atomic absorption spectrophotometric quantification of metals

Metal	Wavelength (nm)	Bandwidth (nm)	Flame source
Cu	324.7	1	Air–acetylene
Pb	283.3	1	Air–acetylene
Zn	213.9	1	Air–acetylene
Mn	279.5	0.3	Air–acetylene
Ni	341.5	0.15	Air–acetylene
Cd	228.8	1	Air–acetylene
Cr	357.9	1	N_2O–acetylene
V	318.4	0.5	N_2O–acetylene

Results

MORTALITY RATE (FIGURE 6A,B)

At all stations, *Spongia nitens* experienced a high mortality rate: during the first month, up to 60% died at station 5, and almost all the cuttings died within a year. We therefore conclude that, although the number of cuttings was small, this species is not favorable for breeding by this technique.

Spongia agaricina and *S. officinalis* both showed a good recovery rate, with practically no mortality during the first month, regardless of the location. All cuttings of *S. agaricina* were alive after two years at station 2 (Figure 6a), and only one died in the other locations. *S. officinalis* showed a highly variable mortality rate, according to the location. On the open water ropes, the mortality rate increased with the pollution rate of the station. It reached 87% in 2 years at station 1. However, the mortality rate at the same station was lower when the cultures were protected from heavy sediment deposition by an overhang in a small cave. Therefore, the high mortality rate at this location seems to be due to the plugging of the canal system in water with a high particle content more than to chemical pollutants, which may accumulate considerably in the living tissue in these conditions (Verdenal et al., this volume). Apparently the good exposure to water currents in hanging cultures is insufficient to clean up the sponges

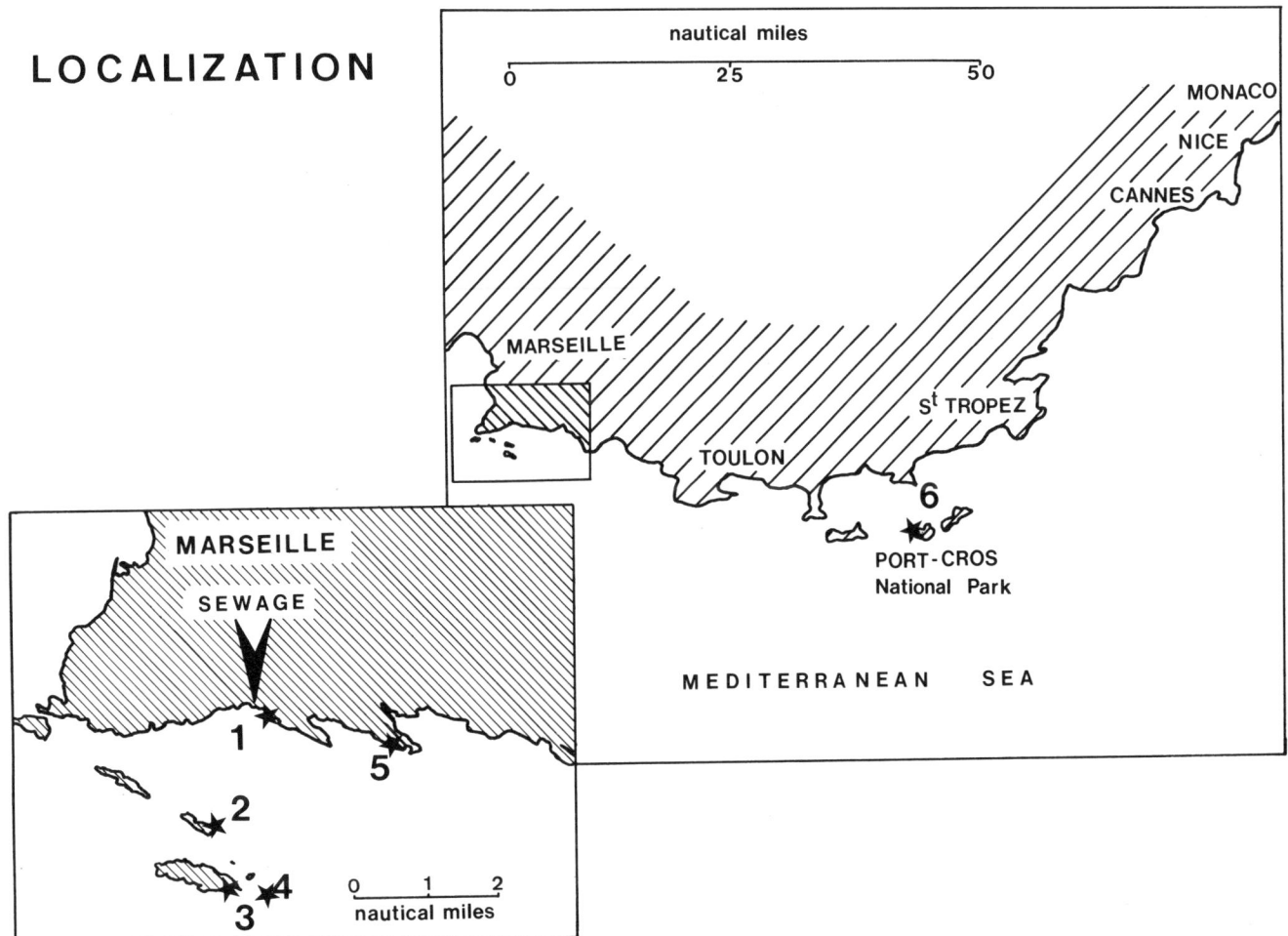

Figure 3. Geographic location of culture stations in southern France.

Figure 4. Sponge culture experiments: *a*, general view of culture station 2; *b*, newly made cutting from *Spongia officinalis*; *c*, healthy cutting, showing that after several months of culture the shape becomes round; *d*, unhealthy cutting with numerous epibionts.

in these turbid waters. The mortality rate is lower in pure water conditions (station 6), although for this species it is never less than 45% on open water ropes. In any case, the mortality rate is low in caves, where the sponges are protected from both sediment deposition and strong illumination, two factors that have an adverse effect on these sponges in the northwestern Mediterranean. How-

ever, there appears to be no correlation between mortality rate and depth of the cuttings on the same rope.

Some high mortality rates are clearly due to the physiological or pathological state of the mother sponge from which the cuttings were taken. In station 1b, for example, the cuttings that died within the first six months all came from the same mother sponge. The mortality rate does not seem to be related to the presence of reproduction stages in the mother sponge.

GROWTH RATE (FIGURE 7A,B)

High growth rates were observed in cuttings that adhered to their fastening wire or to the identification label (Figure 4c). Low growth rates seem to be associated with the presence of a cuticle (a thin, glossy pellicle) on the entire surface of some cuttings, which later was covered by numerous epizoans (Figure 4d). A sponge may survive a long time with this covering but does not grow.

As pointed out by Storr (1964) and Stevely et al. (1978), the growth rates of sponge cuttings may vary greatly, even when they are cultured in the same conditions and originate from the same mother sponge (Table 2).

For *Spongia officinalis* in open water cultures the highest growth rates were recorded in pure water conditions (station 6, Figure 7b), where the mean annual increase in volume was about 100% for both the first and second years. A low growth rate was recorded for cultures placed

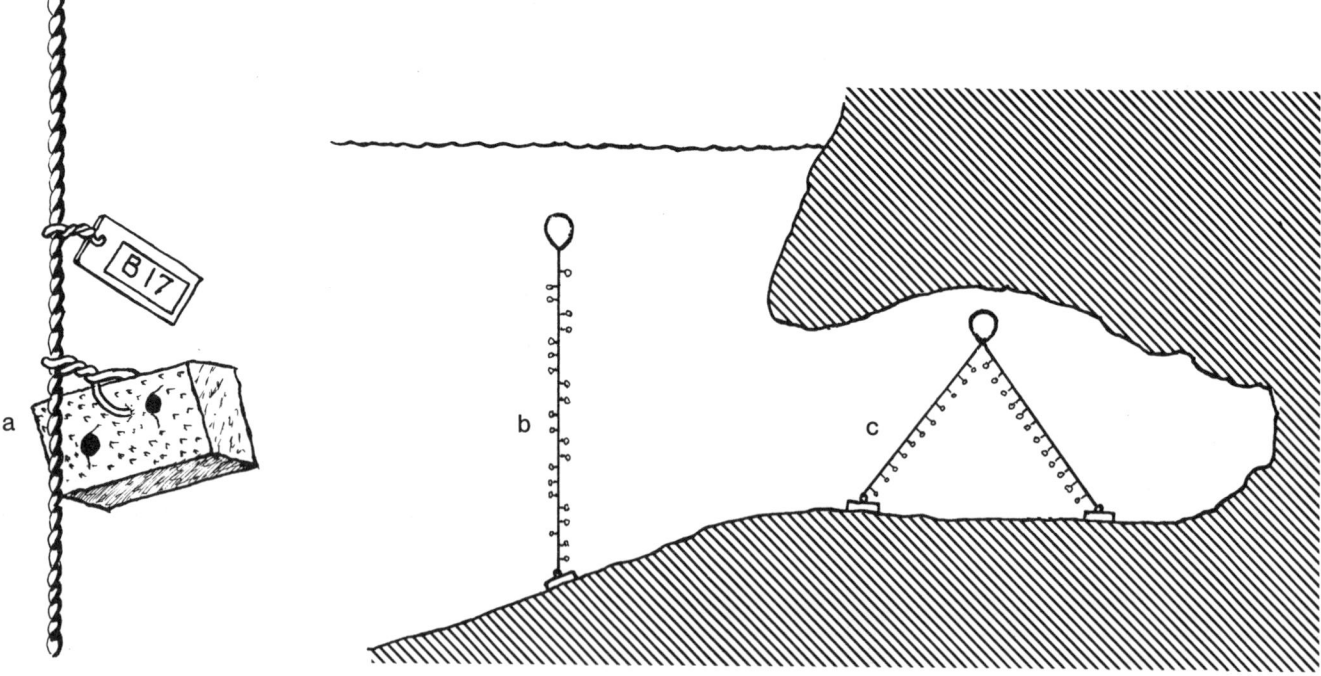

Figure 5. Culture on vertical ropes: *a*, cutting with its label; *b*, open water rope; *c*, underwater cave rope.

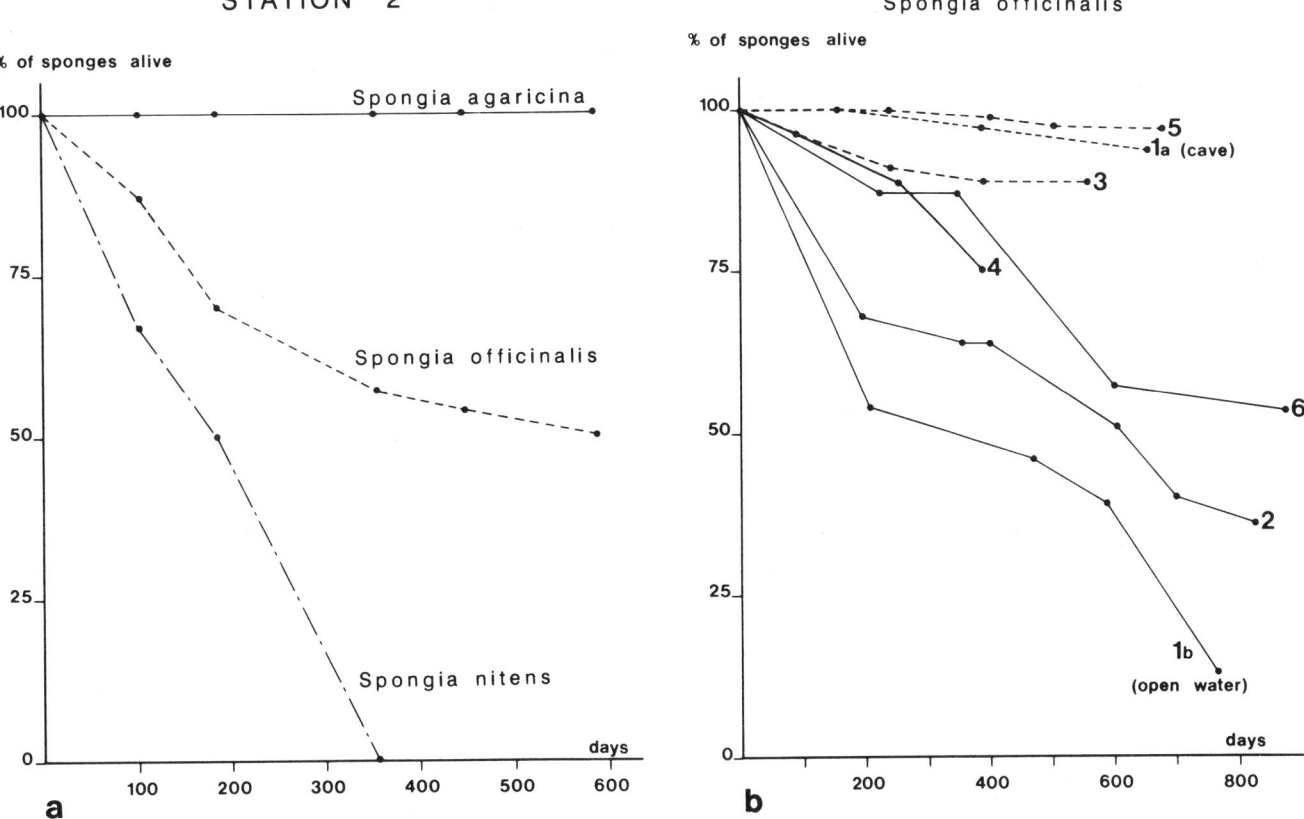

Figure 6. Sponge mortality in culture experiments: *a*, mortality according to species in open-water station 2; *b*, mortality of *Spongia officinalis* according to station; *solid line*, open water stations; *dotted line*, underwater cave stations.

in heavily polluted waters. There was no increase at station 1, whereas at station 2, growing periods alternated with periods of reduction in volume. This decrease occurred in winter; the spongin fibers became exposed at the periphery of the sponge fragment and the exposed skeleton was lost within 1 or 2 months. Similar processes were previously reported by Storr (1964) in the Gulf of Mexico. In the deep station (4), where temperatures rarely exceed 17–18°C in depths below 30 m, the sponges show almost no growth. In cave conditions, the growth rate is low, whatever the pollution rate.

Some cuttings displayed a sharp volume increase which reached 1,252% of the initial volume in 873 days (i.e., an annual growth rate of 523%) for one cutting. This was observed at station 6, where the best mean growth rates were found but where some cuttings showed low or even negative growth rates. These striking differences could not be correlated with the conditions observed, such as the features of the mother sponges, proportion of the surface or of the canal system preserved in the cutting, light, or current or depth conditions.

The two other species were studied less extensively. *Spongia nitens* cuttings showed no increase before their

death, whereas *S. agaricina* had very high growth rates, which, however, were lower in deep water and in highly polluted areas. In the same conditions (Station 2, Figure 7*b*), mean growth rates were significantly higher for this species than for *S. officinalis*. Unfortunately, we do not know the growth rate obtained in station 6, as the rope bearing the *S. agaricina* cuttings was accidentally lost. Since the highest values for *S. officinalis* were obtained at this station, this may also have been the case for *S. agaricina*. In any case, both species appear to have a highly variable growth rate.

Discussion

Of the species studied, *S. nitens* appears to be of little interest, at least in hanging cultures, because of its high mortality rate. The reasons for its susceptibility to transplantation are not evident.

Spongia agaricina, which unfortunately was studied less closely than *S. officinalis*, appears to be the most interesting species. Almost no mortality occurred among the cuttings and their growth rate was significantly higher than that of *S. officinalis* cultured under the same conditions. This may

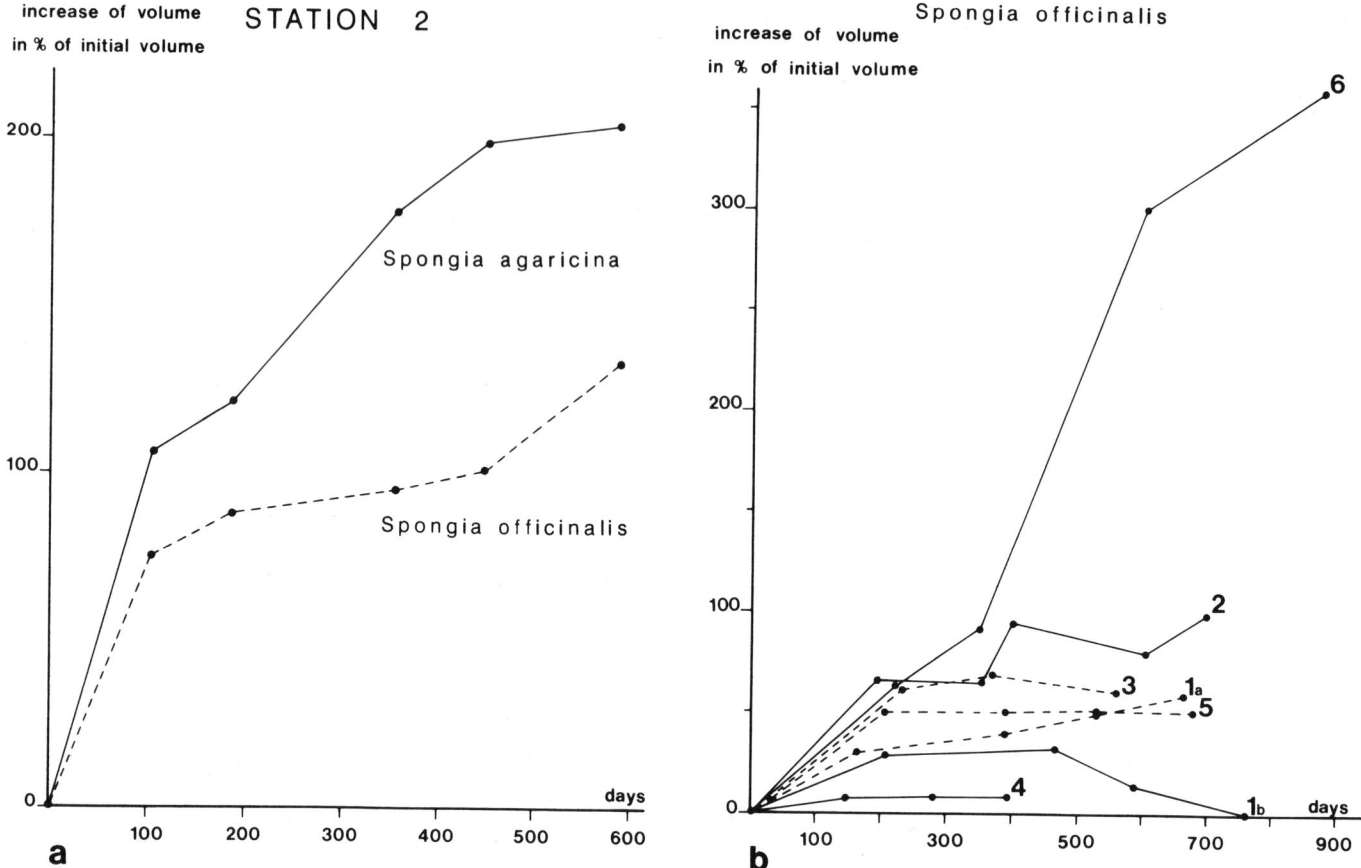

Figure 7. Sponge growth in culture experiments: *a*, growth according to species in open-water station 2; *b*, growth of *Spongia officinalis* according to station; *solid line*, open water stations; *dotted line*, underwater cave stations.

be explained by the need of a minor reorganization of the canal system in this lamellar sponge; cuttings consist of reduced but complete living units with an inhalant and an exhalant face, unlike the cuttings of the two other species, which are massive. Interestingly, *S. agaricina* maintained its lamellar shape although it was somewhat thicker in hanging cultures, where exposure to water flow differs from that on the rocky surfaces that the sponge normally lives on.

Spongia officinalis gives more inconclusive, irregular results, which vary with the environmental conditions, as discussed below. Clearly, this sponge should be studied further before any attempt is made to breed it for commercial purposes. The mean growth rate of *S. officinalis* is impressive in clear water conditions and is of the same order as the rate obtained in previous experiments with cultures on the bottom. In our study, some cuttings experienced a maximum increase of about 500% per year. It is interesting to know that such growth rates, which would be very profitable in a commercial culture, are possible. However, our results are quite erratic; the striking differ-

ences between similar cuttings cannot be explained and the mortality rate is too high. Although the survival is better in cave cultures, the growth rate in these conditions is very low where currents are reduced and irregular. The low growth rate seems to be related to the presence of a cuticle on the cutting. It would be useful to know more about this structure and the conditions of its formation.

The idea that hanging cultures would allow *Spongia officinalis* to live and to grow more rapidly if placed in waters with a high nutrient content must, according to our results, be dismissed. In our study, growth rates were low and many cuttings died in these conditions. Good growth rates were observed in areas with an average pollution level, but mortality rates were always high. Since natural, healthy populations of sponges live under ledges in highly polluted areas, it may be that sponges are able to withstand such an environment in natural conditions, but are unable to withstand the additional trauma of the cutting. It may also be that sponges occurring near the sewage outlet have very low growth rates, as they have to expend much energy on clearing the mineral particles from their

Table 2. Metal levels in *Spongia officinalis* (in ppm dry weight); for map of stations see Figure 1. (n = total number of sampling units in the sample, $\bar{x}$ = mean, s = standard deviation, – = no data)

Station		Cu	Pb	Zn	Mn	Cr	Fe	Ni	V	Hg	Cd
1	n	7	7	7	7	7	7	7	7	6	6
	$\bar{x}$	93.54	90.88	128.94	54.25	71.70	10754	44.92	56.85	0.45	1.18
	s	7.37	33.69	24.65	20.77	29.35	4189	4.03	16.22	0.08	0.30
2	n	6	6	6	6	6	6	6	6	–	–
	$\bar{x}$	82.08	17.56	101.13	40.48	17.01	2738	32.65	<10	–	–
	s	21.25	7.98	15.16	18.29	7.33	1054	3.69	–	–	–
3	n	3	3	3	3		3	3	2	4	4
	$\bar{x}$	92.10	15.13	66.63	51.16	66.76	1676	54.16	<10	0.47	1.75
	s	2.58	5.48	7.64	24.93	1.16	735	1.40	–	0.064	0.57
4	n	7	7	7	7	7	7	7	7	44	
	$\bar{x}$	92.68	17.38	81.17	41.54	21.84	1797	32.61	<10	0.927	0.220
	s	11.57	6.23	13.28	17.96	10.92	763	2.03	–	0.0095	
5	n	2	2	2	2	2	2	2	2	3	3
	$\bar{x}$	75.70	6.40	58.35	63.20	21.60	420	37.75	33.75	0.053	0.870
	s	11.17	0.84	0.92	17.25	7.92	57	2.33	3.74	0.006	0.135
6	n	6	6	6	6	6	6	6	6	–	–
	$\bar{x}$	84.73	29.98	93.40	62.90	11.46	3185	37.15	<10	–	–
	s	14.31	12.00	19.16	32.20	1.81	801	3.47	–	–	–
7	n	7	7	7	7	7	7	7	7	6	6
	$\bar{x}$	71.4	7.37	69.31	63.67	72.74	955	48.71	<10	0.041	1.35
	s	16.83	5.52	12.74	41.28	31.15	510	9.09	–	0.011	0.74

tissue, and are old. This hypothesis appears likely, since these sponges contain a considerable amount of iron particles (Lepidocrocite) on their skeleton (Verdenal et al., this volume; Vacelet et al., 1988) which tend to accumulate in the older parts of the skeleton of many keratose sponges.

It must be noted that cuttings died off throughout the period of the experiments, and not only at the beginning, when, in fact, the recovery rate was quite good. Therefore the traumatic damage of cutting and transplantation does not seem to be the main cause of death. Thus, practically speaking, there is no need to consider introducing a nursery period during which the dying cuttings would be discarded, as has sometimes been recommended.

Most estimates of the economic possibilities of sponge culture have been based on a doubling in volume in one year and a mortality rate of less than 10% a year. Our first data from the northwestern Mediterranean Sea show that using the hanging wire culture, which allows a better utilization of the water volume and would save time in a commercial plantation, can produce even better results, in terms of both growth and mortality rates, for *Spongia agaricina*. Interesting results also seem to be possible with *S. officinalis*, but they are too erratic at present to be considered reliable. Better control of both mortality and growth rate is needed for this species.

Acknowledgments

Financial support for this work was provided by the Agence Nationale pour la Valorisation de la Recherche (ANVAR), the Parc National de Port-Cros, the Région Provence-Alpes-Côte d'Azur (PACA), and the Institut Français de Recherche pour l'Exploitation de la mer (IFREMER). We thank Marie-Rose Causi for her excellent drawings and Jacques Millet for his field assistance.

Literature Cited

Bidder, G. P. 1923. The Relations of the Form of a Sponge to Its Currents. *Quarterly Journal of Microscopical Sciences*, 67:293–323.

Crawshay, L. R. 1939. Studies in the Market Sponges, 1: Growth from the Planted Cutting. *Journal, Marine Biological Association of the United Kingdom*, 23:553–574.

Marenzeller, E. von. 1879. Die Aufzucht des Badeschwammes aus Theilstücken. *Verhandlungen der Zoologisch-Botanischen Gesellschaft in Wien*, 28:687–694.

Moore, H. F. 1910. A Practical Method of Sponge Culture. *Bulletin of the U.S. Bureau of Fisheries*, 28(1):545–585.

Sarà, M., and J. Vacelet. 1973. Ecologie des Démosponges. Pages 462–576 in *Traité de Zoologie 3(1), Spongiaires*, edited by P.-P. Grassé. Paris: Masson.

Smith, F. G. W. 1941. Sponge Disease in British Honduras, and Its Transmission by Water Currents. *Ecology,* 22:415–421.

Stevely, J. M., J. C. Thompson, and R. E. Warner. 1978. The Biology and Utilization of Florida's Commercial Sponges. Florida Sea Grant College Program, Technical Report, 8. 45 pp.

Storr, J. F. 1964. Ecology of the Gulf of Mexico Commercial Sponges and Its Relation to the Fishery. *United States Department of the Interior, Fish and Wildlife Service, Special Scientific Report,* 466:1–73.

Vacelet, J. 1959. Répartition générale des éponges et systématique des éponges cornées de la région de Marseille et de quelques stations méditerranéennes. *Recueil des Travaux de la Station Marine d'Endoume,* 16(26):39–101.

———. 1985. Bases historiques et biologiques d'une éventuelle spongiculture. *Oceanis,* 11(6):551–584.

Vacelet, J., B. Verdenal, and G. Perinet. 1988. The Iron Mineralization of *Spongia officinalis* L. (Porifera, Dictyoceratida) and its relationships with the collagen skeleton. *Biology of the Cell,* 62:189–198.

Verdenal, B. 1986. Spongiculture en Méditerranée Nord-occidentale; aspects cultural, molysmologique et économique. Doctoral Dissertation, Université d'Aix-Marseille. 163 pp.

Verdenal, B., and M. Verdenal. 1987. Evaluation de l'intéret économique de la culture d'éponges commerciales sur les côtes méditerraneénnes françaises. *Aquaculture,* 64:9–29.

Wilkinson, C. R., and J. Vacelet. 1979. Transplantation of Marine Sponges to Different Conditions of Light and Current. *Journal of Experimental Marine Biology and Ecology,* 37:91–104.

JANIE L. WULFF*
Bingham Laboratories
Department of Biology
Yale University
New Haven, Connecticut 06511

Patterns and Processes of Size Change in Caribbean Demosponges of Branching Morphology

Abstract

The processes by which branching demosponges can change size and shape were studied in three species of different orders for almost two years. During this time, individuals of *Amphimedon rubens* (Haplosclerida), *Aplysina fulva* (Dictyoceratida), and *Iotrochota birotulata* (Poecilosclerida) were monitored undisturbed to determine the relative importance of gradual size change compared to growth and abrupt changes caused by fragmentation into smaller portions or fusion with other individuals. At the start of the study, detailed drawings and size measurements were made of 50 individuals of each species. The same data were collected after 9 months, and again after 22 months.

Analysis of the drawings and measurements indicates that these sponges grow by adding tissue exclusively to the tips of erect branches. A variety of circumstances favor initiation of new erect branches, but repent branches are formed only when erect branches become prone. Repent and basal portions of the sponges are disproportionately heavily affected by various agents of partial mortality, and a majority of fragments are generated by damage to basal and repent portions.

Patterns of addition and loss of tissues are very similar in these three species, reflecting their common branching morphology. Variation among these species in the relative importance of the different processes of site change may result from differences in skeletal materials and construction.

Demosponges can increase and decrease in size gradually, by growth and regression, and also abruptly, by fusion among conspecific individuals and by fragmentation into smaller portions. Unusually homogeneous construction, extreme morphological flexibility, and high regeneration capabilities allow some demosponges to add to or lose

---

*Present address: Bunting Institute, 34 Concorde Avenue, Cambridge, Massachusetts 02138.

from any portion of themselves without mortal loss of functional integrity. The processes and consequences of complex size change in sponges may therefore provide a challenging comparison to the biology of size in organisms that are only able to change size according to an inherent growth program.

The part of a sponge where material is lost or added may reflect an adaptive balance between avoiding the effects of various agents of mortality and reaping the benefits of particular positions on the substratum or in the water column. On the other hand, patterns of addition and loss of material may simply be the result of chance and the phylogenetic relationships and gross morphology that place bounds on the adaptive changes possible for a species.

Methods

Patterns of size change were examined in three species of branching Caribbean demosponges, each representing a different order: *Iotrochota birotulata* (Higgin), *Amphimedon rubens* (Pallas) (= *Amphimedon compressa* Duchassaing and Michelotti, sensu Wiedenmayer, 1977), and *Aplysina (= Verongia) fulva* (Pallas). For systematic discussion see de Laubenfels (1936) and van Soest (1978). The orders represented are Poecilosclerida, Haplosclerida, and Verongiida, respectively. Overall, the branching growth forms of these three species are very similar, but branch diameter and branching patterns vary both within and among species.

The three species are common on shallow-water Caribbean reefs and live attached to living or dead corals and to carbonate rubble. The populations studied live interspersed on a flat plane (2.1 m to 2.3 m below MLW) and on a slope of about 16 degrees (2.3 m to 5 m below MLW) off Guigala tupo, a small mangrove and coconut palm island near the San Blas Field Station of the Smithsonian Tropical Research Institute in Panama.

From the populations of each species I chose 50 healthy individuals of various sizes and tagged them with tiny plastic cable ties. I then made detailed drawings of these individuals on underwater slates and recorded the size of all branches and branch segments of each sponge (1) at the start of the study, (2) after 9 months, and (3) after 22 months (Figure 1). The sponges remained undisturbed in the field throughout the study.

Branch widths are sufficiently constant within individuals of these species so that total length of all branch segments is proportional to total volume. Total length is therefore a convenient measure of size. Comparisons of size among individuals must be made carefully, of course, and then only in terms of percentage increase or decrease. The cable ties and branching points provide reference points on the drawings so that comparisons of the drawings in time series for each individual indicate the locations in the sponges where material was added or lost.

Results

Net size changes in all individuals measured at 0, 9, and 22 months are plotted in Figure 2. These plots reflect three characteristics of net growth: (1) fragmentation into independent portions and fusion with other conspecifics in a large proportion of the sponges; (2) highly variable growth rates among individuals, but (3) relatively constant growth rates within individuals, even in some individuals that became fragmented or fused with other individuals.

The processes of growth, fusion, fragmentation, and partial mortality by which the sizes of sponges are changed differ in importance among species, both in terms of the numbers of individuals affected and the amount of biomass involved.

GROWTH

Growth was the only apparent process by which size changed for nearly half of the sponge individuals in the first 9 months (36.2%, 56.5%, 51.1% of the individuals of *Iotrochota birotulata*, *Amphimedon rubens*, and *Aplysina fulva*, respectively; see Figure 3). Wherever size measurements from both 0 and 9 months could be compared, new material was added only to the tips of erect branches (sample sizes were 27, 47, and 63 for *I. birotulata*, *A. rubens*, and *A. fulva*, respectively). No extension of the centers of branches of basal portions, or of branches lying on the substratum (repent branches) was observed.

The locations at which new branches were initiated could be determined from the drawings in time series (see Figure 1). New erect branches were initiated by bifurcation of tips or by sprouting from the sides of pre-existing erect branches, from the points at which other sponges had adhered to branches, from basal attachments or where repent branches had adhered to the substratum, and from the upward reorientation of the tips of repent branches. New erect branches were initiated in all of these ways in individuals of each species, but the relative proportions of each differed among the species (Figure 4). For example, nearly half (46.9%) of the new branches of *Iotrochota birotulata* sprouted from the sides of existing erect branches, whereas more than half (56.3%) of new branches in *Amphimedon rubens* resulted from the upward turning of the tips of repent branches (for typical examples, see Figure 1).

Repent branches, however, are initiated very differently. Repent branches do not appear to branch in the plane of the substratum. Thus, they do not arise through the branching of pre-existing repent branches but only from erect branches that bend over and become prone or

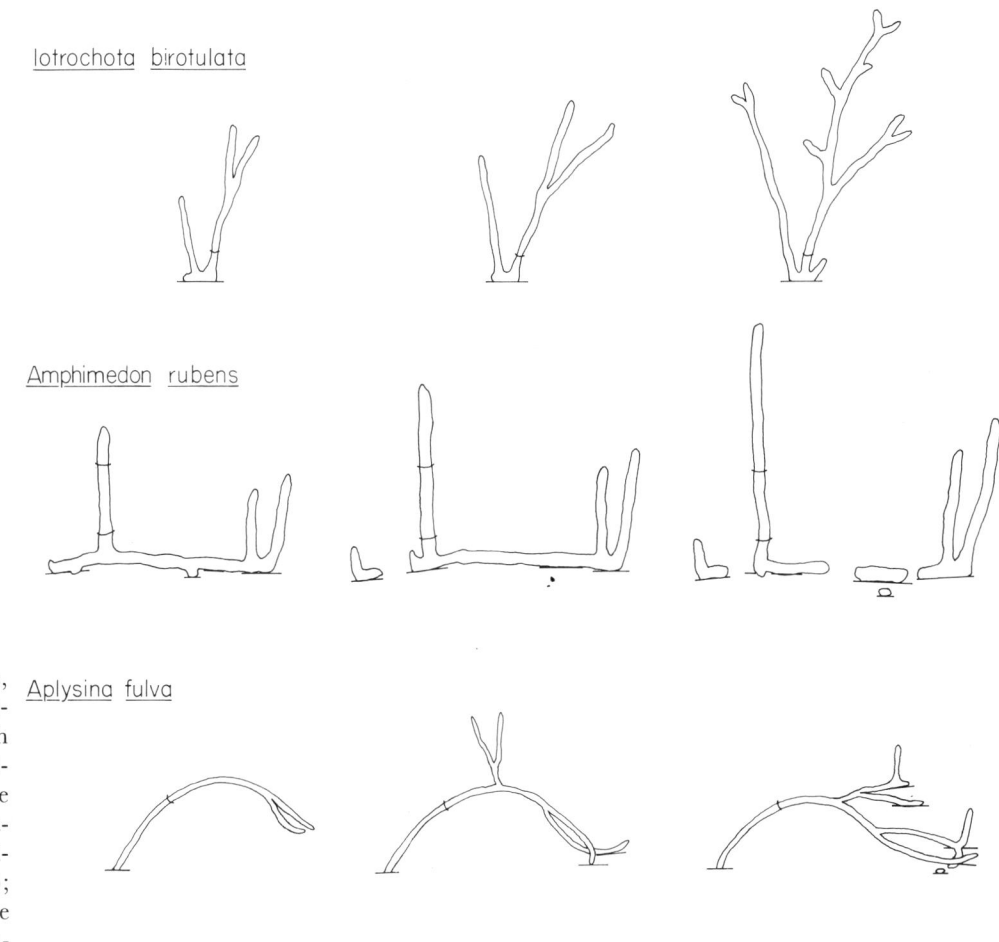

Figure 1. Schematic drawings, in time series, of one representative individual of each species studied. Straight horizontal lines indicate where sponges adhered to substratum (carbonate rubble, primarily from coral skeletons); lines crossing branches are growth reference markers (cable ties).

Iotrochota birotulata

Amphimedon rubens

Aplysina fulva

O months   9 months   22 months

from fragments of erect branches that come to rest in a prone position (Figure 5).

FUSION

Some monitored sponges that came into contact with neighboring conspecific individuals fused with them during the first 9 months (6.4% and 19.1% of the individuals of *Iotrochota birotulata* and *Aplysina fulva*, respectively; Figure 3) and between 9 and 22 months (7.7% and 13.3% in the same species). Fusion merges previously physiologically independent individuals into one confluent sponge. In the Guigala tupo populations of these species, fusion appears, in most cases, to indicate that the independent sponges are clone-mates—that is, that the sponges were derived from each other by an earlier fragmentation (Wulff, 1986a). With regard to change in the size of a physiologically confluent individual, the effects of fusion with another sponge can be striking. The total size of physiologically confluent sponges was increased by fusion by a mean of 177.5% for *I. birotulata* and of 486.6% for *A. fulva* (Figure 3). Obviously, fusion does not actually in-

crease the total amount of biomass. However, if the size of a physiologically continuous individual influences life history parameters, such as mortality and reproduction, fusion with a conspecific neighbor may change the subsequent life of the individual under consideration. The propensity for tissue-compatible branches to fuse is also illustrated by the many individuals in which branches became fused or that produced fragments that subsequently fused back onto the "parent" sponge during the first 9 months (19.2% 4.4%, 14.9% of the individuals of *I. birotulata*, A. *rubens*, and *A. fulva*, respectively) and between 9 and 22 months (38.5%, 23.8%, 60.0% for the same species).

FRAGMENTATION AND PARTIAL MORTALITY

Fragmentation and partial mortality are closely related in that some live tissue may be lost in the course of generating fragments, and some of the fragments produced may die. In addition, some sponges that appeared to have suffered partial mortality without producing fragments may have actually produced fragments that I was unable to

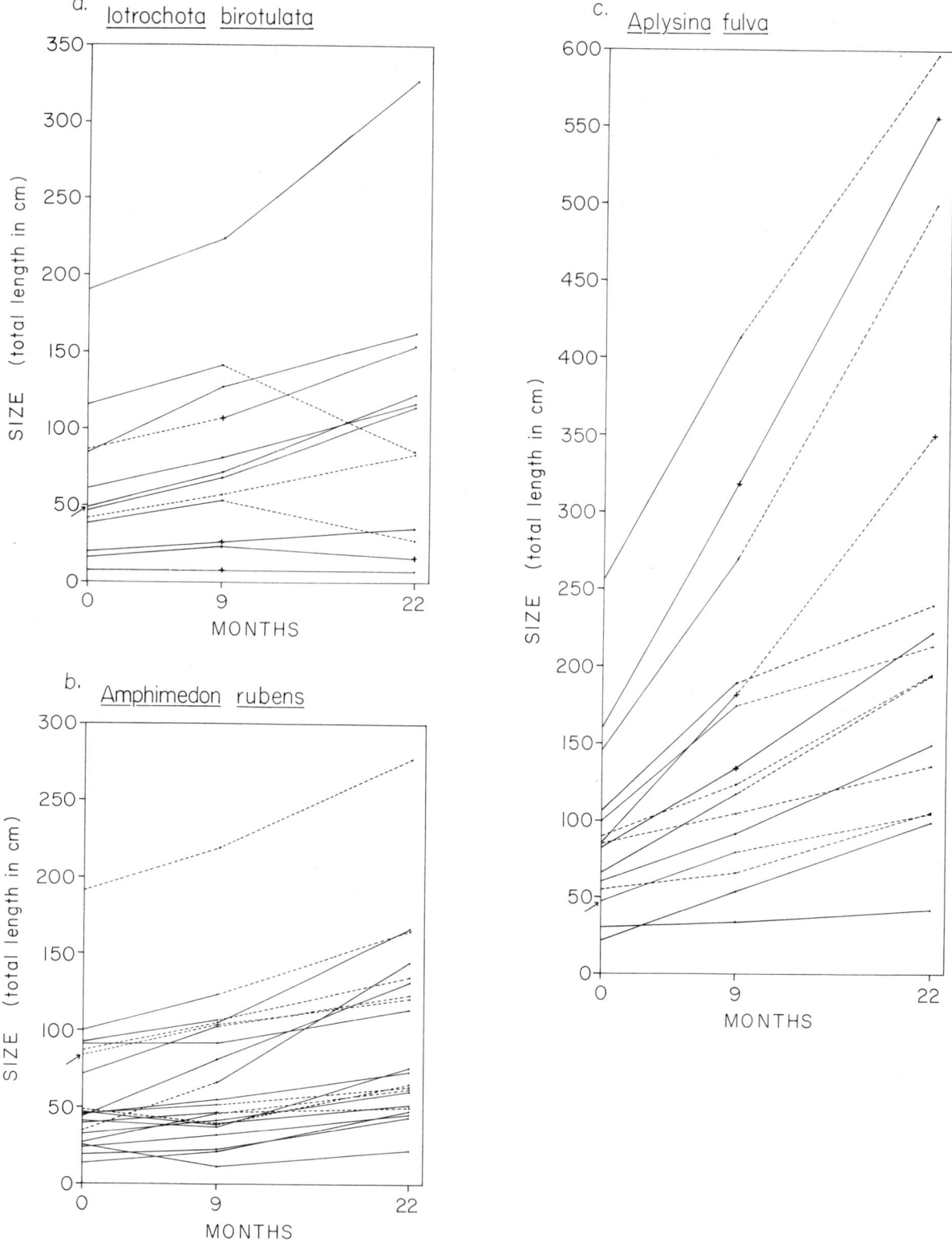

Figure 2. Change in size over time for all individuals for which measurements were made after 9 and 22 months: *a*, *Iotrochota birotulata*; *b*, *Amphimedon rubens*; *c*, *Aplysina fulva*. Fewer than 50 sponges of each species are represented because marker tags were lost from many individuals between observations at 9 and 22 months. Dotted lines indicate that sponge fragmented; all fragments found were combined for total size indicated. Each point at which a sponge was found to have fused with a previously independent conspecific individual marked with a plus *(+)*; additional individual not included in total size. Arrows indicate individuals illustrated in Figure 1.

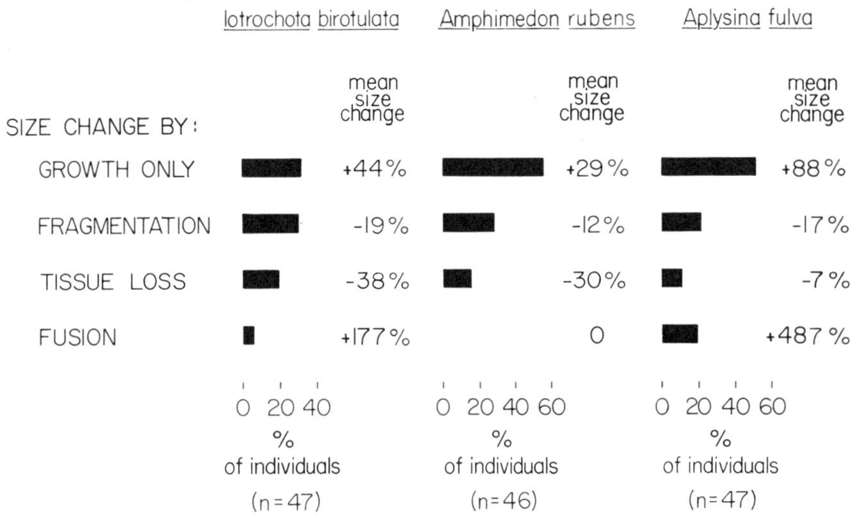

	Iotrochota birotulata		Amphimedon rubens		Aplysina fulva	
SIZE CHANGE BY:		mean size change		mean size change		mean size change
GROWTH ONLY	▬▬	+44%	▬▬▬	+29%	▬▬▬	+88%
FRAGMENTATION	▬▬	-19%	▬▬	-12%	▬	-17%
TISSUE LOSS	▬	-38%	▬	-30%	▪	-7%
FUSION	▪	+177%		0	▬	+487%
	0 20 40 % of individuals (n=47)		0 20 40 60 % of individuals (n=46)		0 20 40 60 % of individuals (n=47)	

Figure 3. Relative importance of different processes by which size can be changed. Bars represent proportions of individuals that changed size; percentages do not always add up to 100 because a few sponges both fragmented and fused with other conspecific individuals. Numbers indicate mean percent by which size was changed.

find because they had been dispersed. The importance of fragment production relative to partial mortality may therefore be underestimated by this tabulation. Together, fragmentation and partial mortality influenced net size change in 1/3 to 2/3 of the individuals in these populations (62%, 43%, 32% in *Iotrochota birotulata, Amphimedon rubens,* and *Aplysina fulva,* respectively) during the first 9-month period of observation (Figure 3) and also between 9 and 22 months (31%, 48%, 60% for these species). One way of assessing the importance of fragmenta-

tion in the dynamics of these sponge populations is to examine the rate of increase in numbers of physiologically independent individuals by asexual fragmentation. For example, of the 47 individuals of *I. birotulata* monitored during the first 9 months, 15 fragmented. These produced a total of 28 new individuals, 3 of which subsequently fused back onto their "parent" sponges. The net increase in numbers of individuals by asexual fragmentation was therefore 53.2% (= 25/47) of the original population size. The populations of *A. rubens* and *A. fulva* were similarly increased by additions of 41% and 23%, respectively, of the original number of individuals. Between 9 and 22 months, the populations of these three species were increased by additions of, respectively, 31%, 43%, and 60% of the number of individuals, solely by asexually produced fragments from monitored sponges.

Fragments can be produced by breakage of branches during storms, the bites of sponge-feeding fishes, localized infections by pathogens, mounds of sediment raised by burrowing shrimp or holothurians, grazing by starfish and foraging by eagle rays, and encroachment by other sessile organisms. Some of these agents of fragmentation or partial mortality tend to damage specific portions of the sponges. For example, violent currents may break erect branches, whereas mounds of sediment may sever repent branches by smothering portions of them. The majority of fragments that came from the monitored sponges resulted from damage to portions of sponges that were lying prone on the substratum, either discrete basal attachments or repent branches.. This, presumably, reflects the relative importance of the various agents of fragment generation. Fewer than 1/3 of the fragments produced (33%, 17%, 13% of fragments from *I. birotulata, A. rubens,* and *A. fulva,* respectively) resulted from damage to erect branches (Figure 6).

Patterns of the loss of material from these sponges, both in the course of fragment production and when fragments

GROWTH: INITIATION OF ERECT BRANCHES

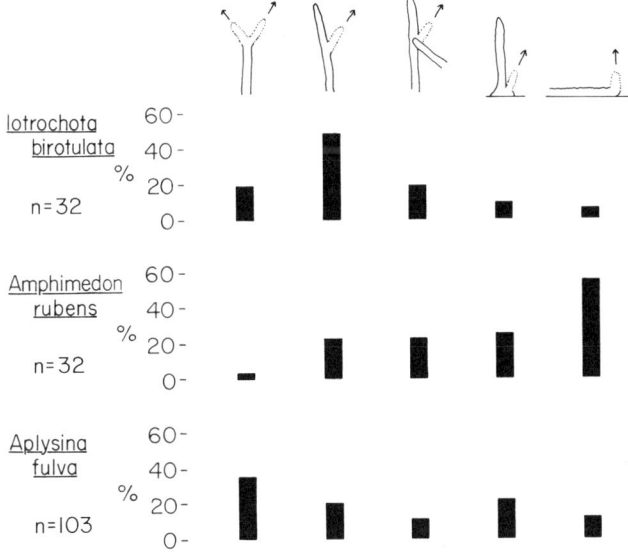

Figure 4. Positions from which new erect branches were initiated: bifurcation of erect branch, sprouting from side of erect branch, area where another sponge adhered to branch, basal attachment to substratum, upward reorientation of tip of repent branch. Bars represent percentage of erect branches initiated between 0 and 9 months.

3d Int. Sponge Conf. 1985

GROWTH: INITIATION OF REPENT BRANCHES

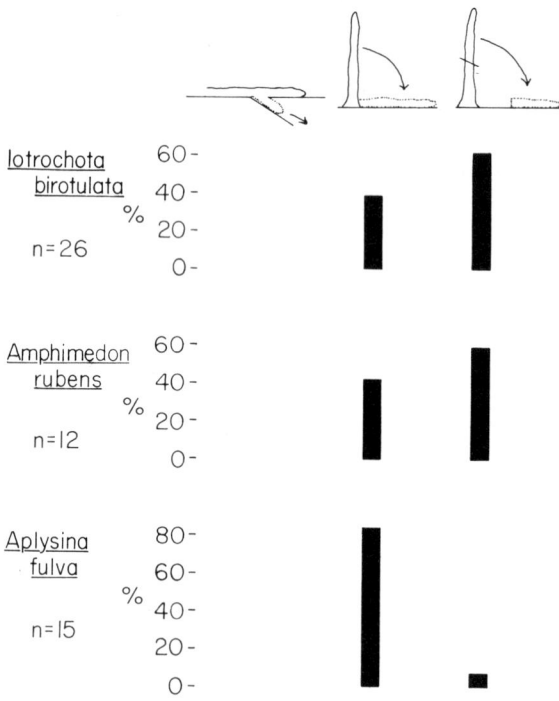

Figure 5. Initiation of new repent branches: branching of repent branch in plane of substratum, by bending down of erect branch to prone position, or by production of fragment from erect branch. Bars represent percentage of repent branches formed between 0 and 9 months.

were not found, also demonstrate disproportionately heavy partial mortality on basal and repent portions. Proportions of the total lengths of sponges that were in repent or basal portions at the initial measurement were less than one quarter (16.0%, 24.2%, 13.5% of the total length for *Iotrochota birotulata*, *Amphimedon rubens*, and *Aplysina fulva*, respectively). If partial mortality is random with respect to the part of the sponge affected, then less than one quarter of the material lost from these individuals would be expected to have come from basal and repent portions. However, about 1/2 of the material lost (48.4%, 45.3%, 58.3% for *I. birotulata*, *A. rubens*, and *A. fulva*, respectively) was from these portions that had been lying on the substratum (Figure 7).

SIGNIFICANCE OF INITIAL SIZE FOR CHANGE IN SIZE

Specific growth rates are compared directly in Figure 8, plotted as the percentage increase against initial size, for all individuals monitored during the first 9 months. Perhaps because of the substantial contributions of fragmentation, fusion, and partial mortality to net size change in

these sponges, initial size appears to be largely unrelated to subsequent size change. Although the few individuals that grew at the most rapid rates were all very small initially (Figure 8), growth appears, on the whole, not to be strongly influenced by initial size. In order to test for dependence of specific growth rate on size, for each species, all individuals that changed in size by growth only were divided evenly into three groups: the third with the largest, the third with the smallest, and the third with intermediate initial sizes (from the data plotted in Figure 8). These three groups did not differ significantly in specific growth rates for *Amphimedon rubens* and *Aplysina fulva* (no significant differences in pairwise comparisons within species by the Wilcoxon rank sum test; $p > 0.23$), but for *Iotrochota birotulata*, small sponges had significantly higher specific growth rates than sponges of intermediate or large initial sizes ($p = 0.03$ and $p = 0.05$ for small versus intermediate, and small versus large, respectively).

Smaller sponges were more likely to have changed size by addition of live tissue, that is, by growth or fusion, whereas larger sponges were more likely to have lost tissue (differences significant by the Wilcoxon rank sum test; $p = 0.015$, $p = 0.005$, $p = 0.025$ for, respectively, *Iotrochota birotulata*, *Amphimedon rubens*, and *Aplysina fulva*).

FRAGMENT GENERATION

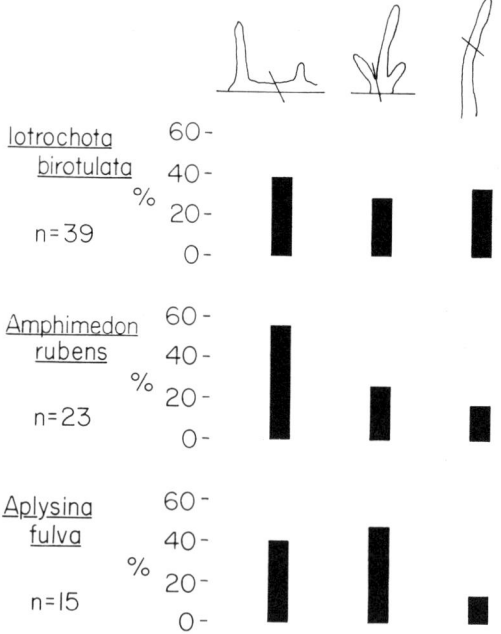

Figure 6. Production of fragments caused by damage to repent branches, to discrete basal attachments, or to erect branches. Bars represent percentage of fragments produced between 0 and 9 months.

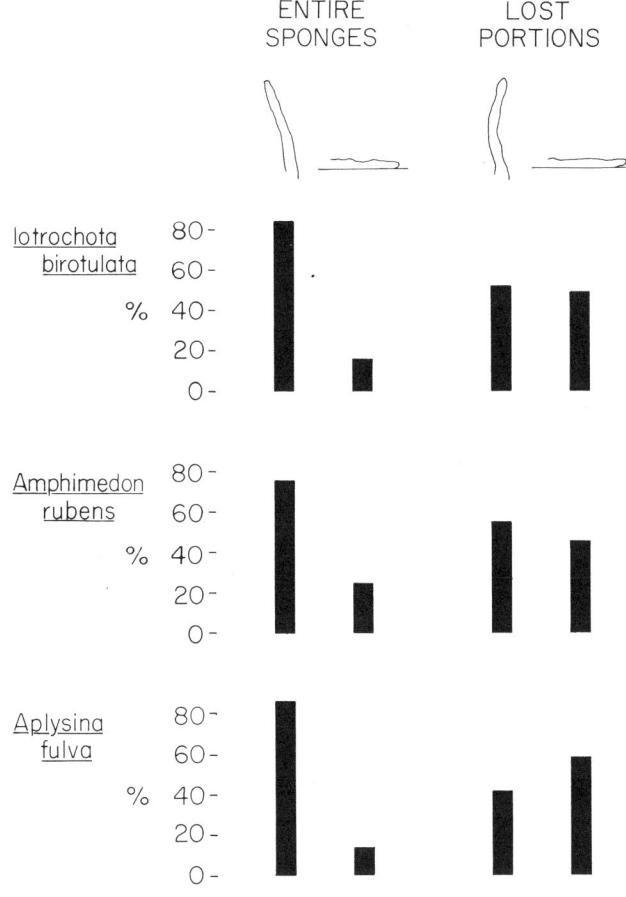

Figure 7. Effects of position relative to substrate on partial mortality. Pairs of bars on left of each set represent percentage of total branch lengths found in erect or repent portions of sponges at start of study. Pairs of bars on right represent percentage losses after 22 months. If partial mortality is random with respect to position of tissue in sponges, percentages lost from erect or repent portions should be the same as percentages of tissue found in erect or repent portions.

Discussion
PATTERNS OF SIZE CHANGE

Fewer than one half of the individuals in these sponge populations changed size by growth only, nearly one third became fragmented, and the others decreased in size by partial mortality or increased in size by fusion with other conspecific individuals during the first nine months.

Specific growth rates did not differ significantly between sponges of large, medium, or small initial size, except for *Iotrochota birotulata*. Susceptibility to partial mortality did, however, appear to be affected by size in these sponges. For each species, individuals that lost portions or became fragmented were significantly larger, initially, than those that only changed size by growth or fusion.

Increase in the size of individuals by growth occurs only at the tips of erect branches. The field observations on

which this conclusion is based did not provide information on where new cells originate or exactly where, on a microscopic scale, new material is added to the growing tip (see Simpson, 1984 for a review of cellular growth in sponges). New erect branches are formed in a variety of ways from previously existing erect branches and from basal attachments and repent branches. In contrast, repent branches are only formed by changes in the orientation of previously erect branches, or by fragmentation. The majority of fragments are produced by damage to repent or basal portions, and partial mortality is disproportionately heavy on basal attachments and repent branches.

Although casual observations suggest that these sponges appear to be growing as vines, extending along the substratum as they grow, this appearance is misleading. Here I continue to use the term "repent" to refer to the position of the branches that lie adherent to the substratum. However, if "repent" is used to imply active growth along the substratum, as is the case in many plants (e.g., the clover *Trifolium repens*), then this is a misnomer for these sponges. Branches seen in a repent position appear to result invariably from the repositioning of branches that achieved their length in an erect position.

Size change in these sponges can therefore be viewed as a dynamic cycle of addition of new material by growth at the tips, fragmentation and fusion, removal of material from the base by various agents of partial mortality, and the reorientation of toppled or severed branches to again grow upward.

INFLUENCE OF SIZE ON GROWTH AND SURVIVAL

Size has been demonstrated to influence growth, mortality, most physiological processes, and sexual reproduction in many organisms (e.g., Calder, 1984; Jackson, 1979). Effects of size may be of special interest when the organisms studied can both increase and decrease in size throughout their lives. For such organisms, primarily clonal invertebrates and plants, the importance of the decoupling of size from age has been recognized (e.g., Connell, 1973; Harper, 1977; Hughes and Jackson, 1980; Hughes, 1984, and references therein). In clonal organisms, such as bryozoans and corals, increased size may increase survival, increase sexual reproduction, and decrease growth rates (e.g., Connell, 1973; Loya, 1976; Jackson, 1979 and references therein).

For the three species of branching sponges considered in this study, the growth patterns of unmanipulated sponges that vary in size and occur in natural populations suggest that effects of size on growth may be nearly swamped by other influences. Although fusion with previously independent conspecific individuals can have a dramatic effect on the total size of a physiologically con-

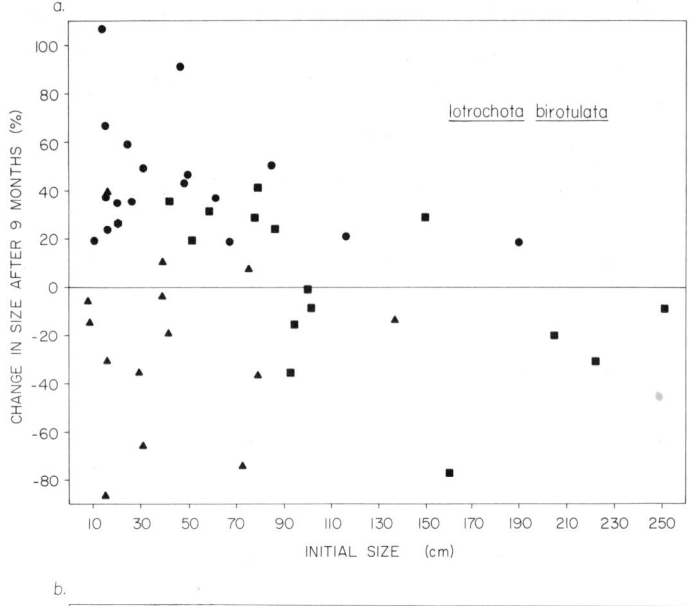

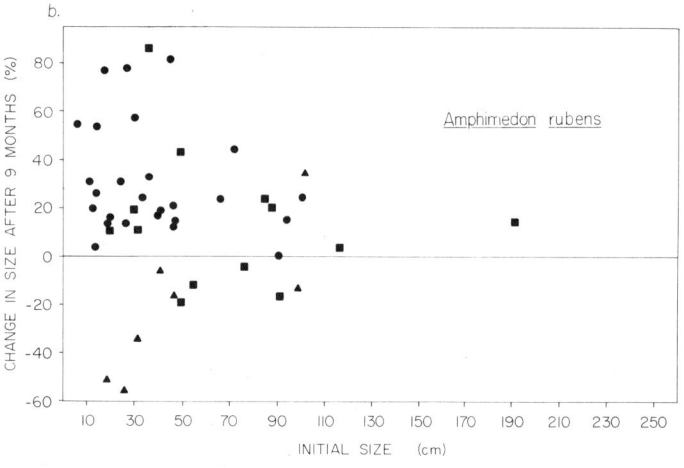

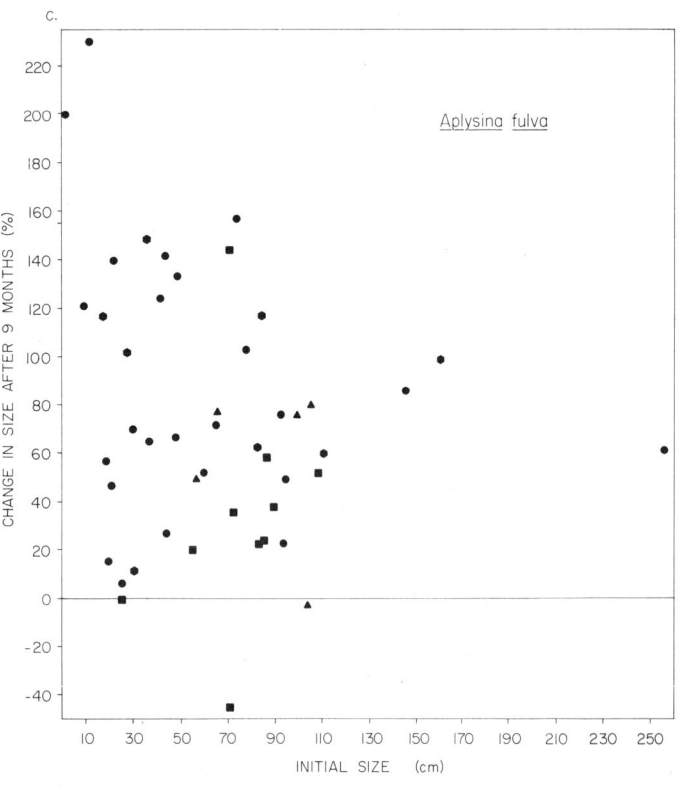

Figure 8. Specific growth rates expressed as percentage size change plotted against initial size for all individuals measured at 0 and at 9 months: *a, Iotrochota birotulata; b, Amphimedon rubens; c, Aplysina fulva.* Circles represent sponges that changed size only by growth; octagons represent sponges that fused with another conspecific individual and from which no material was lost; squares represent sponges that became fragmented (all fragments combined for total size); triangles represent sponges from which material was lost but no recognizable fragments were found.

fluent individual, growth rates of the originally monitored portions generally did not change when fusion brought about a sudden increase in size (see Figure 2). Similarly, the growth rates of many individuals that became fragmented did not change, even though two or more smaller individuals had taken the place of one larger sponge (Figure 2). On the other hand, continued linearity of these growth curves, plotted on arithmetic axes, demonstrates that the specific growth rate (growth standardized by initial size) of each individual does slow down as the organism increases in size. Enormous variation in growth rates among individuals, even among those that changed size only by growth, may make it impossible to determine the relationship between size and growth in these organisms without experiments in which genotype is controlled.

Slower growth rates have been demonstrated for larger individuals of some sponge species of massive or tubular morphology (e.g., Storr, 1964; Reiswig, 1973; Dayton, 1979). Differences in overall morphology may help to explain why the effects of size on growth rate appear to be more subtle in the three branching species in this study. The growth of massive or tube-shaped sponges is marked not only by changes in size, but also by changes in shape. In particular, ratios of surface area to volume tend to decrease with growth. This may have especially profound effects for filter-feeding organisms and therefore may be an important confounding factor of the interdependence of size and growth. By contrast, the surface-to-volume ratios of the branching species in this study do not change significantly with growth. If decrease in surface area relative to volume is a primary reason for the decreased rate of accumulation of biomass by larger massive sponges, then the apparently milder effect of overall size on growth rates of these branching species may be the expected result.

Partial mortality in these sponges disproportionately affected the larger individuals. Partial mortality has also been observed to disproportionately (with respect to numbers of physiologically independent individuals) affect larger corals in foliaceous species (Hughes and Jackson, 1980). Mortality of entire individuals has been demonstrated to decrease with increased size for scleractinian coral species (e.g., Connell, 1973; Highsmith et al., 1980; Hughes and Jackson, 1980; Hughes, 1984). It is not clear how partial mortality is related to the death of entire individuals in these sponges or perhaps in any clonal species. Fragmentation, which often involves partial mortality, may even increase the survival of the genotype in some cases. Asexual propagation might be predicted to increase with increased size simply because larger individuals have more material into which they can be fragmented. Sexual reproduction was not studied for the monitored individuals of these branching species, and it is possible that increased size increases the production of sexual propagules. However, virtually all successful recruitment into these populations is by asexual fragments (Wulff, 1986b) and so the overall importance of increased sexual reproduction to these sponges is not clear.

ADAPTIVE SIGNIFICANCE OF PATTERNS OF SIZE CHANGE

These three species exhibit strikingly similar patterns of size change. Although they share a common overall growth form (i.e., branching), their skeletal constructions are different to the extent that they are placed in different orders. These common patterns of size change may therefore be related to common branching morphology rather than taxonomic factors and thus may give clues to the advantages and disadvantages of this growth form for demosponges.

Disproportionately heavy mortality of basal and repent portions may give a selective advantage to sponges that actively add of material only to erect branches. Growth at the tips of erect branches also helps to raise more of the sponge into the current above the bottom, where there is, presumably, increased food and decreased resuspended sediment. If decreased partial mortality and access to better water quality accrue to the sponge that only builds upward, perhaps the high rates of fragmentation and production of repent branches seen in these sponges are incidental aspects of their life histories.

One often suggested advantage of repent branches is that they enable the "parent" organism to place genetically identical "offspring" in favorable habitat nearby (e.g., Leakey, 1981 and references therein; Lasker, 1983). This possibility has been explored in some detail for various plants. Extreme examples are plants that are known to "track" favorable microhabitats by the growth of runners or stolons (e.g., Kershaw, 1962; Turkington and Harper, 1979).

If repent branches of these sponges do not grow along the substratum, then it does not seem possible that they are tracking favorable microhabitats by growth. Depending on the relative importance of various sources of mortality, particular microsites may not be predictably good or bad, in any case, and distantly dispersed asexual fragments might provide better insurance against demise of the genotype (Wulff, 1986a). Unattached fragments of these species are able to disperse far, and many survive and become established as independent individuals (Wulff, 1985). Unattached fragments of *Aplysina fulva*, however, have significantly lower rates of survival than those of the other species, both after distant dispersal by a storm and under calm weather conditions (Wulff, 1985). This is also the species for which fewer fragments were formed from erect branches (Figure 6), a greater percentage of lost portions were from repent branches (Figure 7), a greater majority of repent branches were formed by bending over of erect branches rather than by fragmentation (Figure 5), a higher proportion of resettlement of cleared areas was by repent branches from neighboring individuals (Wulff, 1986b), and a higher proportion of monitored sponges fused with neighboring individuals (Figure 3). The relationship between low survival of unattached fragments and the traits that appear to emphasize fragmentation of repent branches is also evident from the clone structure of this population of *Aplysina fulva*. Those clones (recognized by both tissue compatibility and morphology) with especially large numbers of independent individuals are characterized by overall morphologies dominated by repent branches (Wulff, 1986a), as expected if repent branches play an important role in asexual propagation.

Comparisons among the syntopic populations of these three species might suggest, then, that *Iotrochota birotulata* and *Amphimedon rubens* are designed to propagate by dispersing fragments, whereas *Aplysina fulva* is designed to carefully place its clone members nearby because unattached fragments survive poorly. However, the discovery that repent branches do not grow in that orientation suggests a more cautious interpretation. The characteristics of *A. fulva* that may cause fragments to survive poorly (very narrow branches, elastic skeletal construction) may also be the very characteristics that cause branches to bend over and become prone. Dense tissue of sponges of this genus may also help to make repent branches especially susceptible to being severed by smothering under sediment mounds. Apparent adaptations for careful placement of asexual propagules may therefore be simple consequences of the construction of this species (that is, "exaptations", see Gould and Vrba, 1982). Comparisons of populations in habitats with different disturbance regimes and substratum availability are needed, along with more information on processes of growth on the cellular level, in order to fully understand the advantages and disadvantages, and the adaptive significance, if any, of these patterns of size change for demosponges of branching morphology.

Conclusions

Size and shape change in these three sponge species results from a combination of growth, fusion, fragmentation, and partial mortality. Fewer than half of the individuals changed size by growth only during the first nine months of the study. Growth adds tissue only to the tips of erect branches. Although repent branches may appear to be growing in the plane of the substratum, they must have achieved their length in an erect position. Specific growth rates of large sponges are either the same as or lower than those of smaller sponges. New erect branches are initiated in ways that vary in relative importance among the three species. New repent branches reult from toppling of erect branches and not from branching in the plane of the substratum. More than half of the sponges changed size abruptly, by fusion, fragmentation, or partial mortality. Large sponges were more likely to fragment or suffer partial mortality than were small sponges. Partial mortality was disproportionately heavy on basal and repent portions and, likewise, the majority of fragments resulted from damage to basal and repent portions. Size and shape change in these branching species is a dynamic cycle of addition of new material by growth at the erect tips, rearrangement by fragmentation and fusion, removal of material from the base by various agents of partial mortality, and the reorientation of toppled or severed branches to again grow upward.

Literature Cited

Calder, W. A. III. 1984. *Size, Function, and Life History.* Cambridge, Massachusetts: Harvard University Press.

Connell, J. H. 1973. Population Ecology of Reef Building Corals. Pages 205–245 in *Biology and Geology of Coral Reefs* edited by O. A. Jones and R. Endean. New York: Academic Press.

Dayton, P. K. 1979. Observations of Growth, Dispersal and Population Dynamics of Some Sponges in McMurdo Sound, Antarctica. Pages 271–282 in *Biologie des Spongiaires*, edited by C. Lévi and N. Boury-Esnault. Colloques Internationaux du C.N.R.S. 291. Paris: Centre National de la Recherche Scientifique.

Gould, S. J., and E. S. Vrba. 1982. Exaptation—A Missing Term in the Science of Form. *Paleobiology*, 8:4–15.

Harper, J. L. 1977. *Population Biology of Plants*. New York: Academic Press. 892 pp.

Highsmith, R. C., A. C. Riggs, and C. M. D'Antonio. 1980. Survival of Hurricane-Generated Coral Fragments and a Disturbance Model of Reef Calcification/Growth Rates. *Oecologia*, 46:322–329.

Hughes, T. P. 1984. Population Dynamics Based on Individual Size rather than Age: A General Model with a Reef Coral Example. *The American Naturalist*, 123:778–795.

Hughes, T. P., and J. B. C. Jackson. 1980. Do Corals Lie about Their Age? Some Demographic Consequences of Partial Mortality, Fission and Fusion. *Science*, 209:713–715.

Jackson, J. B. C. 1979. Morphological Strategies of Sessile Animals. Pages 499–555 in *Biology and Systematics of Colonial Animals*, edited by B. Rosen and G. Larwood. New York: Academic Press.

Kershaw, K. A. 1962. Quantitative Ecological Studies from Landmannahellir, Iceland. II. The Rhizome Behavior of *Carex bigelowii* and *Calamagrostis neglecta*. *Journal of Ecology*, 50:171–179.

Lasker, H. R. 1983. Vegetative Reproduction in the Octocoral *Briareum asbestinum* (Pallas). *Journal of Experimental Marine Biology and Ecology*, 72:157–169.

Laubenfels, M. W. de. 1936. A Discussion of the Sponge Fauna of the Dry Tortugas in Particular and the West Indies in General, with Material for a Revision of the Families and Orders of the Porifera. *Papers from Tortugas Laboratory*, 30:1–225.

Leakey, R. R. B. 1981. Adaptive Biology of Vegetatively Regenerating Weeds. *Advances in Applied Biology*, 6:57–90.

Loya, Y. 1976. The Red Sea Coral *Stylophora pistillata* is an r Strategist. *Nature*, 259:478–480.

Reiswig, H. M. 1973. Population Dynamics of Three Jamaican Demospongiae. *Bulletin of Marine Science*, 23:191–226.

Simpson, T. L. 1984. *The Cell Biology of Sponges*. New York: Springer. 662 pp.

Soest, R. W. M. van. 1978. Marine Sponges from Curaçao and other Caribbean Localities. Part 1. Keratosa. *Studies on the Fauna of Curaçao and Other Caribbean Islands*, 56:1–94.

Storr, J. F. 1964. The Ecology of the Gulf of Mexico Commercial Sponges and Its Relation to the Fishery. *U.S. Fish and Wildlife Service, Special Scientific Report*, 466:1–73.

Turkington, R., and J. L. Harper. 1979. The Growth, Distribution and Neighbour Relationships of *Trifolium repens* in a Permanent Pasture. IV. Fine-scale Biotic Differentiation. *Journal of Ecology*, 67:245–254.

Wiedenmayer, F. 1977. *Shallow-Water Sponges of the Western Bahamas*. Basel und Stuttgart: Birkhäuser Verlag. 287 pp.

Wulff, J. L. 1985. Dispersal and Survival of Fragments of Coral Reef Sponges. Pages 119–124 in *Proceedings of the Fifth International Coral Reef Congress, Tahiti*, 5. Moorea, French Polynesia: Antenne Museum-EPHE.

———. 1986a. Variation in Clone Structure of Fragmenting Coral Reef Sponges. *Biological Journal of the Linnean Society*, 27:311–330.

———. 1986b. Population Ecology of Caribbean Demosponges of Branching Morphology. Doctoral Dissertation, Yale University.

VANCE P. VICENTE[*]
Marine Ecology Division
Center for Energy and Environment Research
College Station, Mayaguez, Puerto Rico 00708

Overgrowth Activity by the Encrusting Sponge *Chondrilla nucula* on a Coral Reef in Puerto Rico

Abstract

The importance of overgrowth interactions was studied in the forereef of Cayo Enrique, Puerto Rico, where the encrusting sponge *Chondrilla nucula* was found to be a major aggressor. Overgrowths occurs more frequently (67%–80%) than standoff encounters (20–32%) in the 572 interspecific interactions observed. Significant differences were found between the mean number of overgrowths and standoffs ($p < 0.05$, t test) in the *Acropora palmata* zone, in the *A. cervicornis* zone, and in the base of the reef. The mean number of overgrowths also differed significantly between the three zones ($p < 0.05$, Kruskal-Wallis test), with a higher incidence of aggressions at the base of the reef. However, no linear correlation was found between frequency of overgrowths with depth. The variability in the frequency of aggressions within given depths also suggests that factors not necessarily related to depth may be important in determining frequency of overgrowths. On the other hand, microhabitat conditions appear to influence frequency of interactions. Only two organisms, the corallimorpharian *Ricordea florida* and the gorgonian *Erythropdium caribbaeorum*, were found to be successful competitors against *C. nucula*.

Although relict sclerosponges appear to have been displaced from shallow water habitats into deep or cryptic conditions by stronger competitors such as scleractinian corals (Hartman, 1980; Vacelet, 1983), many demosponges continue to be major constituents of coral reefs. Furthermore, various demosponges are encrusting and can readily overgrow corals and other sessile taxa in Atlantic (Vicente, 1978), and Pacific reefs (Bryan, 1973; Wilkinson, 1978).

The encrusting demosponge *Chondrilla nucula* has a Tethyan distribution and is ubiquitous in shallow-water marine environments (Hechtel, 1965; Vicente, 1975;

---

[*]Present address: U.S. Fish & Wildlife Service, SE/ES Caribbean Field Office, P.O. Box 491, Boqueron, Puerto Rico 00622.

Thorhaug and Roessler, 1977). This sponge is aggressive and can overgrow other sessile invertebrates such as corals (Glynn, 1973; Vicente, 1978; Suchanek et al., 1983). This study was conducted to determine if *Chondrilla nucula* is a dominant aggressor in a coral reef in Puerto Rico.

Material and Methods

The coral reef at Cayo Enrique lies on the southwestern shelf of Puerto Rico near the fishing village of La Parguera, approximately 1.6 km from the coast (Figure 1). This region is bordered by mangrove swamps, mud flats, and seagrass beds. Coral reef are particularly well developed in this area (Geonaga and Cintron, 1979). For this study, thirty one-square meter quadrats were permanently delineated with lines. They were located within the *Acropora palmata* zone (5–6 m in depth), *Acropora cervicornis* zone (5–6 m in depth), and at the base of Cayo Enrique reef (12 m in depth). All the interspecific interactions within each square meter were recorded in situ and classified according to the criteria utilized by Jackson (1979), except for case *d* (Figure 2).

Results

A total of 572 interspecific interactions distributed among 49 sessile species were found within the forereef. Overgrowths were found to occur more frequently (67%–80%) than standoff encounters (20%–32%). Significant differences were found between the frequency of overgrowths versus standoff situations ($p < 0.05$, t test), with a higher incidence of overgrowths in all zones (Figure 3). The mean number of overgrowths differed significantly between zones ($p < 0.05$, Kruskal-Wallis test), with a higher incidence of aggressions at the base of the reef. No significant linear correlation was found between the number of overgrowths with depth.

The sponge *Chondrilla nucula* accounted for about half

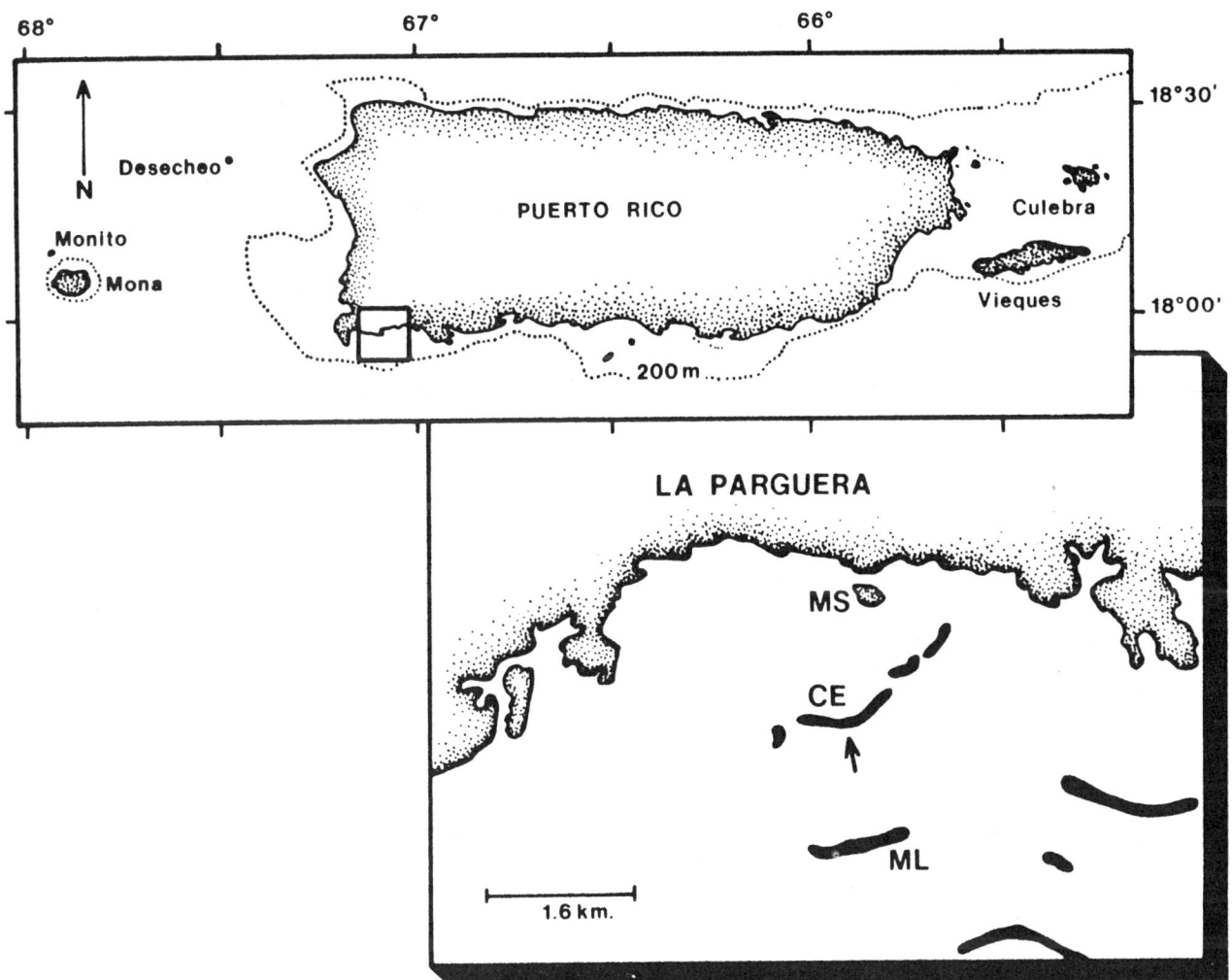

Figure 1. Coral reef at Cayo Enrique (CE) lies on southwestern shelf of Puerto Rico, 1.6 km from shore between reef Meida Luna (ML) and University of Puerto Rico marine station (MS). Arrow points to forereef studied.

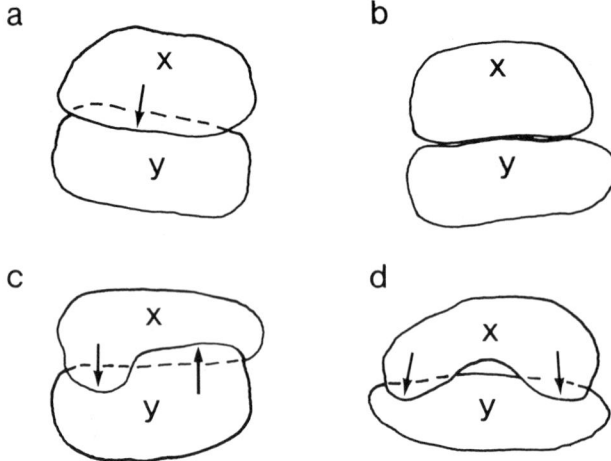

Figure 2. Method utilized for analysis of interspecific interactions according to Jackson (1979): *a*, one overgrowth encounter where species *x* overgrows species *y*; *b*, standoff encounter where neither species wins; *c*, this situation is recorded as two distinct overgrowths, one in which species *x* is overgrowing species *y* and in the other the outcome is reversed; *d*, in this study, contrary to Jackson (1979), this case is interpreted as one overgrowth encounter where species *x* is overgrowing species *y*. (Figure based on drawing in Jackson, 1979).

(43%–51%) of all the overgrowths. Furthermore, analysis by a contact data matrix (Buss, 1976) also indicates that *C. nucula* is the dominant aggressor of the forereef, followed by the encrusting gorgonian *Erythropodium caribbaeorum* (Figure 4).

Quantitative information on the interspecific interactions between *Chondrilla nucula* and scleractinian corals,

other sponges, and other taxa is given in Figure 5. Nine of the 13 species of corals interacting with *C. nucula* were found to be encrusted to some extent by the sponge. These include corals with irregular surfaces such as *Porites porites* (Figure 6). Even though some corals such as *Agaricia agaricites* may overgrow *C. nucula*, the sponge was invariably the net winner (Figure 5, set A). Interactions between *C. nucula* and other sponges occurred infrequently, except in the case of *Cliona aprica* which in 17 instances was found encrusted by *C. nucula* (Figure 5, set B). However, the level of aggression of *C. nucula* on the forereef may in part be balanced by the corallimorpharian *Ricordea florida*, which was observed overgrowing *C. nucula* on 15 occasions. Competitive reversals were found with *Erythropodium caribbaeorum*, which was also a net winner over *C. nucula* (Figure 5, set C).

Discussion

This study shows that overgrowth encounters occur more frequently than standoff situations among the sessile components of a forereef community. The variability in the frequency of interactions within specific zones and the lack of linear correlation between the number of interactions with increasing depth suggest that factors not necessarily related to depth are involved in determining the frequency of interactions. A potentially important factor may be substrate consolidation and its effect on the dispersion of sessile taxa of the reef. For example, the substrate at the *Acropora palmata* zone and at the base of the reef consisted of consolidated hard coral. Interactions, particularly overgrowths, occur more frequently in these two zones (Figure 3). On the other hand, few sessile inver-

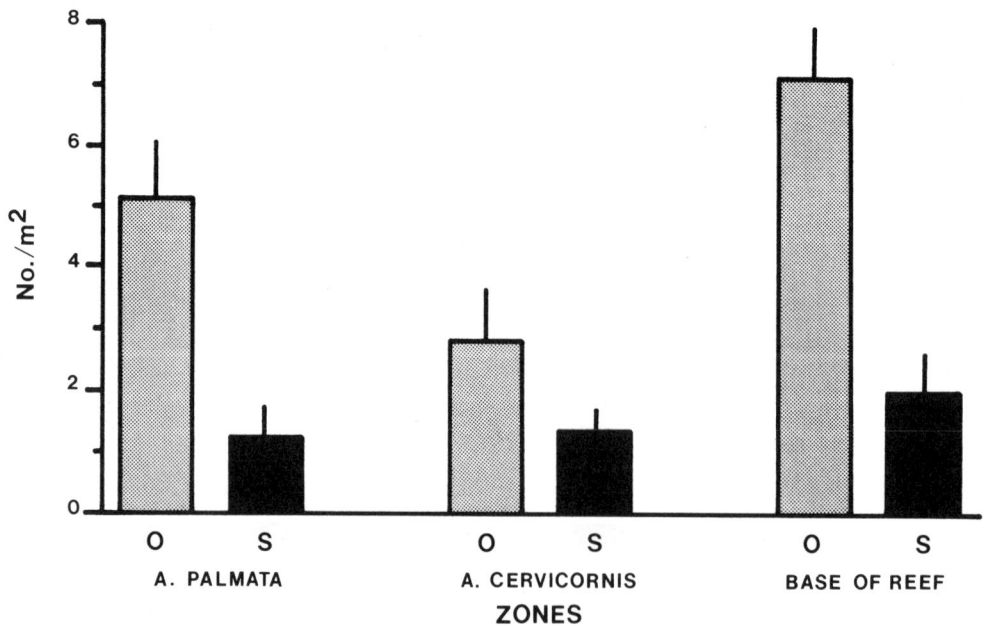

Figure 3. Mean number *(bars)* and standard error *(lines)* of overgrowth encounters and standoff situations on *Acropora palmata* zone, *A. cervicornis* zone, and on base of reef. *O*, overgrowth; *S*, standoff.

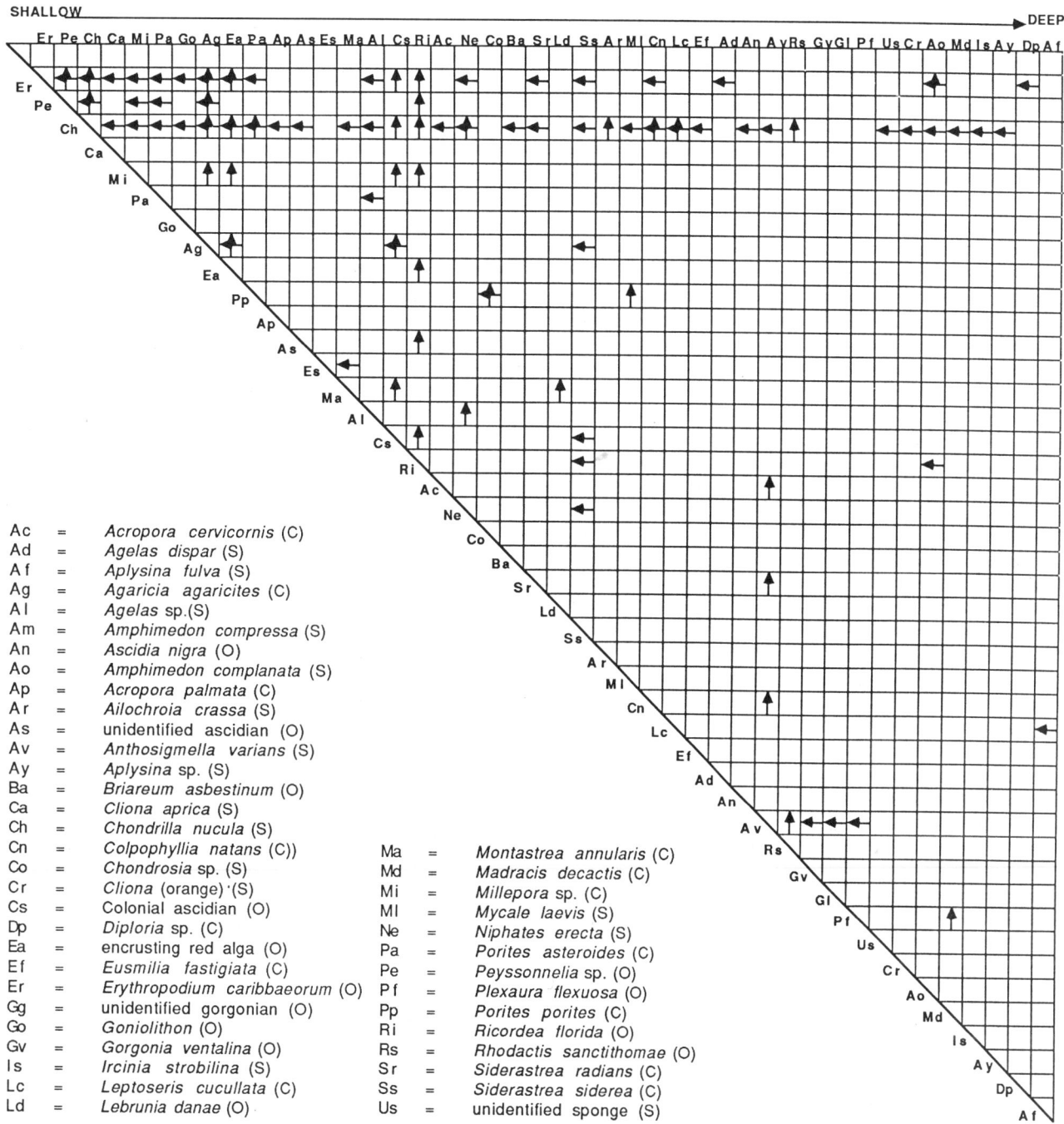

Figure 4. Contact data matrix of specific interactions found in front reef of Cayo Enrique. Arrows in each square point to winning species. Two arrows within a single square signify that either species may win.

Ac	=	*Acropora cervicornis* (C)
Ad	=	*Agelas dispar* (S)
Af	=	*Aplysina fulva* (S)
Ag	=	*Agaricia agaricites* (C)
Al	=	*Agelas* sp.(S)
Am	=	*Amphimedon compressa* (S)
An	=	*Ascidia nigra* (O)
Ao	=	*Amphimedon complanata* (S)
Ap	=	*Acropora palmata* (C)
Ar	=	*Ailochroia crassa* (S)
As	=	unidentified ascidian (O)
Av	=	*Anthosigmella varians* (S)
Ay	=	*Aplysina* sp. (S)
Ba	=	*Briareum asbestinum* (O)
Ca	=	*Cliona aprica* (S)
Ch	=	*Chondrilla nucula* (S)
Cn	=	*Colpophyllia natans* (C))
Co	=	*Chondrosia* sp. (S)
Cr	=	*Cliona* (orange) (S)
Cs	=	Colonial ascidian (O)
Dp	=	*Diploria* sp. (C)
Ea	=	encrusting red alga (O)
Ef	=	*Eusmilia fastigiata* (C)
Er	=	*Erythropodium caribbaeorum* (O)
Gg	=	unidentified gorgonian (O)
Go	=	*Goniolithon* (O)
Gv	=	*Gorgonia ventalina* (O)
Is	=	*Ircinia strobilina* (S)
Lc	=	*Leptoseris cucullata* (C)
Ld	=	*Lebrunia danae* (O)

Ma	=	*Montastrea annularis* (C)
Md	=	*Madracis decactis* (C)
Mi	=	*Millepora* sp. (C)
Ml	=	*Mycale laevis* (S)
Ne	=	*Niphates erecta* (S)
Pa	=	*Porites asteroides* (C)
Pe	=	*Peyssonnelia* sp. (O)
Pf	=	*Plexaura flexuosa* (O)
Pp	=	*Porites porites* (C)
Ri	=	*Ricordea florida* (O)
Rs	=	*Rhodactis sanctithomae* (O)
Sr	=	*Siderastrea radians* (C)
Ss	=	*Siderastrea siderea* (C)
Us	=	unidentified sponge (S)

tebrates occur in the *Acropora cervicornis* zone which consists mainly of *A. cervicornis* rubble, sand patches, and dead stands of *A. cervicornis* covered with thick algal tufts.

The data collected in this study leave little doubt that the encrusting sponge *Chondrilla nucula* is a dominant aggressor in the forereef of Cayo Enrique. This finding is consisted with the observations made by Suchanek et al. (1983) on a reef in St. Croix, where *C. nucula* and *Erythropodium caribbaeorum* are dominant aggressors. Personal ob-

servations on the forereef of Escollo Rodriguez on the west coast of Puerto Rico also indicate that *C. nucula* is a dominant aggressor. Furthermore, corals that were being overgrown by *C. nucula* over a year ago are still being displaced by the sponge (Figure 7). These observations suggest that overgrowth activity by *C. nucula*, may be important in modifying the biological structure and growth of some reefs in the Caribbean region. However, the effect of *C. nucula* on a coral reef may be counteracted by its suscep-

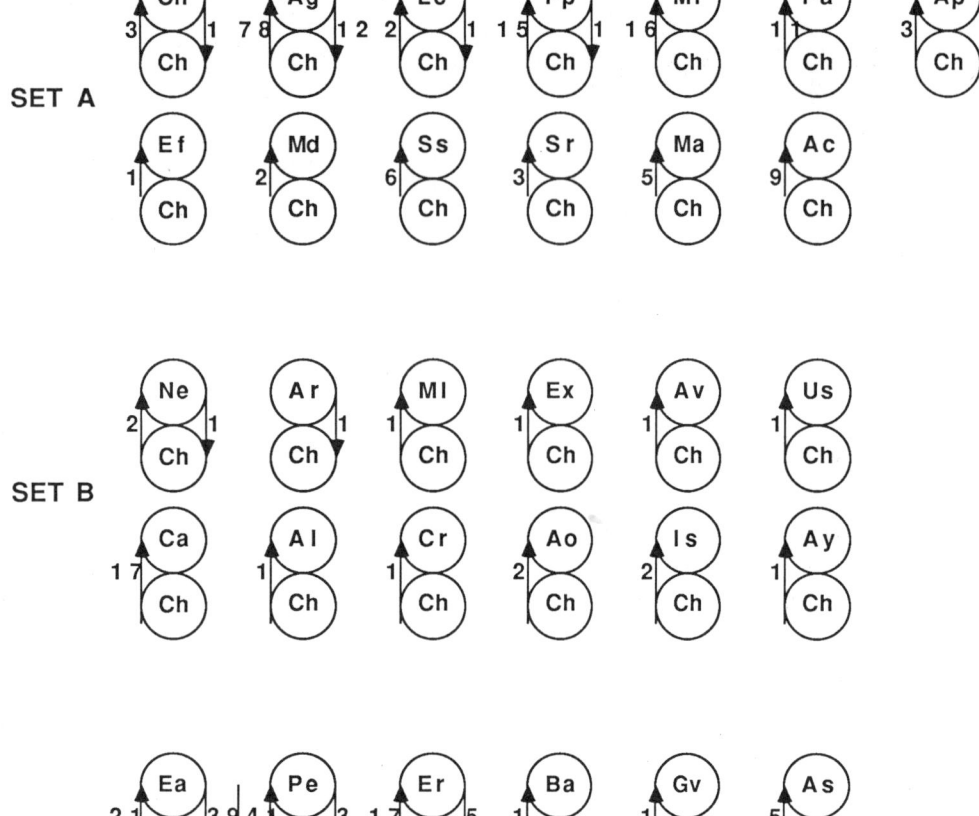

Figure 5. Frequency of specific interactions between encrusting sponge *Chondrilla nucula* and corals (Set A), other sponges (Set B), and other species (Set C). Abbreviations represent same species as in Figure 4.

tibility to become overgrown by *Ricordea florida* and by *E. caribbaeorum*. Other factors not yet identified may also limit the population of *C. nucula* in the coral reef.

The high competitive ability of *Chondrilla nucula* may be explained in part by the presence of cyanobacteria within the mesohyl of the sponge (Figure 8). These are very similar to the ones found in specimens from the Mediterranean sea (Gaino et al., 1976). The association between cyanobacteria and sponges is thought to be mutualistic (Rützler, this volume). The sponge *C. nucula* grows large in well lit environments perhaps as the result of this symbiotic relationship.

Conclusions

This study shows that overgrowth interactions may be an important factor in determining the distribution of corals and of other sessile invertebrates inhabiting coral reefs. Overgrowth encounters occur more frequently than standoff situations. Significant differences were also detected in the relative frequency between overgrowths versus standoffs at the three zones analyzed. Frequency of interactions appear to be related to microhabitat conditions such as substrate stability rather than to depth.

The encrusting sponge *Chondrilla nucula* is a dominant aggressor on the reef (it accounts for about half of all the aggression found). The fact that *C. nucula* is known to be a dominant aggressor in other Caribbean reefs implies that it may play an important role in modifying the community structure of some types of reefs. The particular susceptibility of corals to *C. nucula* also has implications for coral reef growth. The success of *Chondrilla nucula* in such a competitive environment may be explained in part by the presence of symbiotic cyanobacteria within the mesohyl of the sponge.

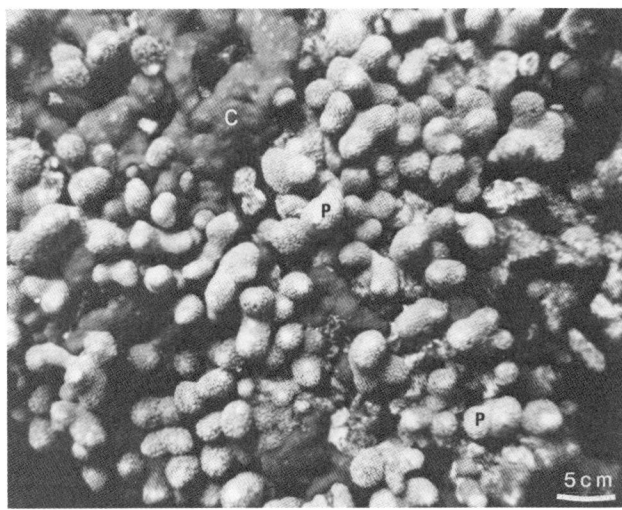

Figure 6. Encrusting sponge *Chondrilla nucula (C)* overgrowing coral *Porites porites (P)* in forereef of Cayo Enrique, Parguera, Puerto Rico.

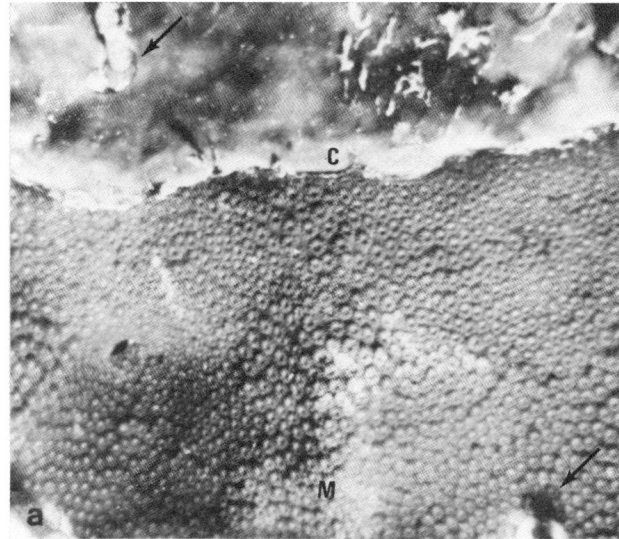

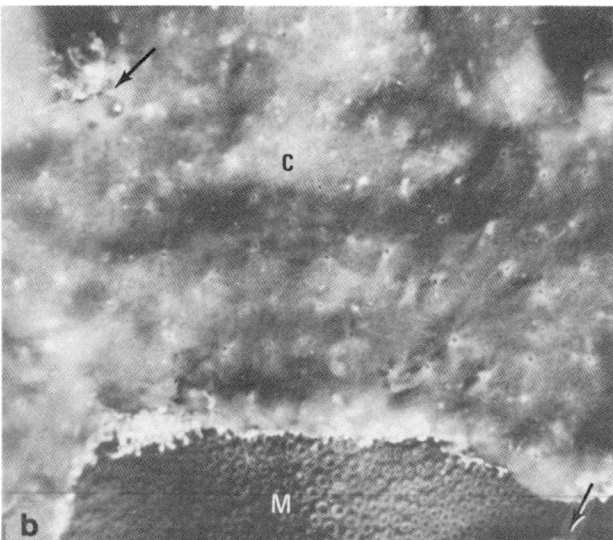

Figure 7. Sponge *Chondrilla nucula (C)* overgrowing scleractinian coral *Montastrea annularis (M)*, the major coral reef builder of Cayo Enrique in La Parguera, Puerto Rico: *a*, picture taken September 30, 1983; *b*, picture taken March 1, 1985. Picture area between nails *(arrows)* is 100 cm².

Acknowledgments

I am greatly indebted to the Department of Marine Sciences of the University of Puerto Rico for its continuous support. Economic assistance was received from the Laboratory Graduate Participation Program of the Oak Ridge Associated Universities. This fellowship was obtained through the Center for Energy and Environment Research of the University of Puerto Rico with the support of its Marine Ecology Division which provided facilities for this research. I also want to thank Gladys Collazo for preparing some of the figures on short notice.

I am grateful to Carlos Goenaga for his invaluable help, and to Patricia Hinds for spending so many cold hours with me underwater. Special thanks go to Curt Pueschel, the University Center at Binghamton, State University of New York for transmission electron microscopy. I also wish to thank my major professor Paul M. Yoshioka for his continuous advice during this study and for editing this manuscript, and Clive Wilkinson for discussing several important ideas with me.

Literature Cited

Bryan, P. G. 1973. Growth Rate, Toxicity, and Distribution of the Encrusting Sponge *Terpios* sp. (Hadromerida: Suberitidae) in Guam, Mariana Islands. *Micronesica*, 9:237–242.

Buss, L. W. 1976. Better Living Through Chemistry: The Relationship Between Allelochemical Interactions and Competitive Networks. Pages 315–327 in *Aspects of Sponge Biology*, edited by F. W. Harrison and R. R. Cowden. New York: Academic Press.

Gaino, E., M. Pansini, and R. Pronzato. 1976. Osservazioni sull'associazione tra una Cianoficea Croococcale e la Demospongia *Chondrilla nucula. Archivo di Oceanografia e Limnologia,* 18(supplement):545–552.

Glynn, P. W. 1973. Aspects of the Ecology of Coral Reefs in the Western Atlantic Region. Pages 271–319 in *Biology and Geology of Coral Reefs*, edited by O. A. Jones and R. Endean. New York: Academic Press.

Goenaga, C., and G. Cintron. 1979. Inventory of Puerto Rican Coral Reefs. Report submitted to the Coastal Management Zone of the Department of Natural Resources, Commonwealth of Puerto Rico. 66 pp.

Hartman, W. D. 1980. Ecology of Recent Sclerosponges. Pages 253–255 in *Living and Fossil Sponges*. Notes for a short course, edited by W. D. Hartman, J. W. Wendt, and F. Wiedenmayer. Miami: University of Miami.

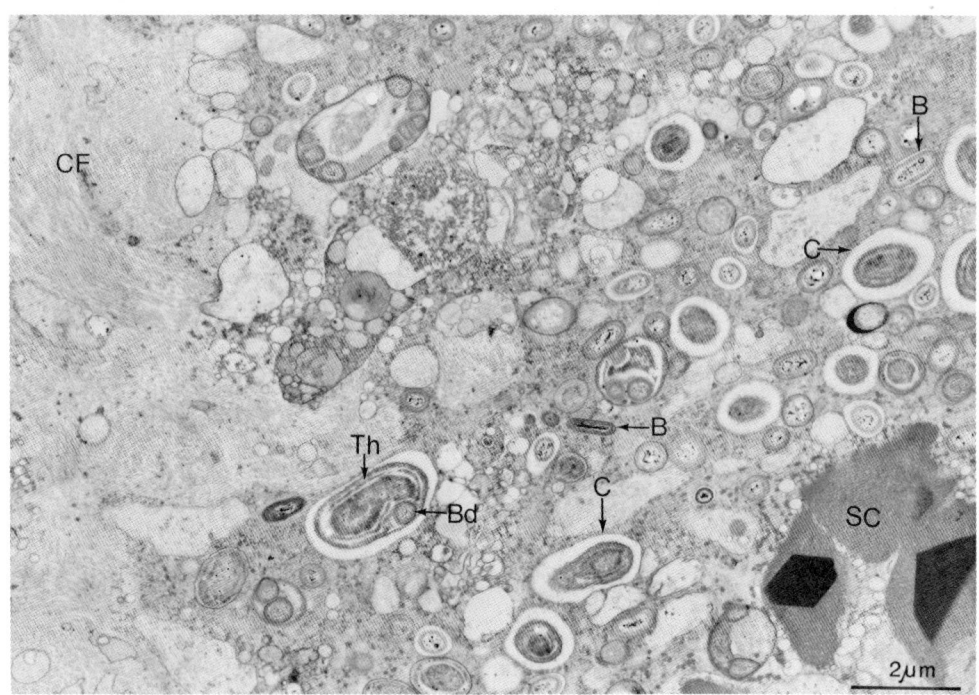

Figure 8. Transmission electron micrograph of tangential section of sponge *Chondrilla nucula*. *B*, bacteria; *Bd*, Bdello-vibrio-like bacterium; *C*, cyanobacteria; *CF*, collagen fibrills; *SC*, spherulous cell fragments; *Th*, thylakoids.

Hechtel, G. 1965. A Systematic Study of the Demospongiae of Port Royal, Jamaica. *Bulletin, Peabody Museum of Natural History,* 20:1–94.

Jackson, J. B. C. 1979. Overgrowth Competition Between Encrusting Cheilostome Ectoprocts in a Jamaican Cryptic Reef Environment. *Journal of Animal Ecology,* 48:805–823.

Suchanek, T. H., R. C. Carpenter, J. D. Witman, and C. D. Harvell. 1983. Sponges as Important Space Competitors in Deep Caribbean Coral Reef Communities. Pages 55–60 in *The Ecology of Deep and Shallow Coral Reefs,* edited by M. L. Reaka. Symposia Series for Undersea Research 1(1). Rockwille, Maryland: NOAA Undersea Research Program.

Thorhaug, A., and M. A. Roessler. 1977. Seagrass Community Dynamics in a Subtropic Estuarine Lagoon. *Aquaculture,* 12:253–277.

Vacelet, J. 1983. Les Eponges hypercalcifiees, reliques des organismes constructeurs de recifs du paleozoique et du Mesozoique. *Bulletin, Societé de Zoologie de France,* 108:547–557.

Vicente, V. P. 1975. The Seagrass Beds of Jobos Bay. *Puerto Rico Nuclear Center Jobos Bay, Annual Report,* 1975:119–153.

———. 1978. An Ecological Evaluation of the West Indian Demosponge *Anthosigmella varians* (Hadromerida: Spirastrellidae). *Bulletin of Marine Sciences,* 28:771–777.

Wilkinson, C. R. 1978. Microbial Associations in Sponges. 1. Ecology, Physiology, and Microbial Populations of Coral Reef Sponges. *Marine Biology,* 49:161–167.

JOSEPH P. SCHUBAUER*
THOMAS P. BURNS
THELMA H. RICHARDSON
Department of Zoology and Institute of Ecology
University of Georgia
Athens, Georgia 30602

Population Dynamics of Five Demospongiae in Jamaica: Variation in Time and Space

Abstract

The relationship between substrate type and the abundances and distributions of five species of Demospongiae was investigated at two sites on the west forereef slope in Discovery Bay, Jamaica, in 1984. Four years earlier, this reef had been heavily damaged by hurricane Allen and only one year earlier had experienced a massive die-off of an important grazer, *Diadema antillarum*. Observations of *Mycale laxissima* at one of the sites (Mooring I), were compared with the distribution and abundance data compiled by Reiswig (1973) at this same site in 1969. Of the five species examined at both sites in the present study, *Ircinia strobilina* and *Mycale laxissima* were the most abundant. Depth did not appear to influence the way species utilized different substrate types. In general, species densities were highest on isolated patches of reef, even though this substrate constituted no more than 2% of the surface area of the transects at either site. Sponge densities were observed to vary with depth and site. Overall, more specimen of *M. laxissima* were found in the transect at Mooring I in 1969 than in 1984, with abundance on natural substrate being comparable in both cases. A substantially higher percentage of the specimen of *M. laxissima* found at Mooring I were found in the sand channels in 1984 than in 1969. Our observations at Mooring I support Reiswig's (1973) hypothesis that *M. laxissima* is space limited on this reef.

Sponges are a major component of many benthic marine communities (Dayton et al., 1974; Hartman, 1977; Colin, 1978; Wenner et al., 1983). Due to the algal and bacterial symbionts many species host (Wilkinson, 1978, 1981; Rützler, 1981, this volume; Vacelet, 1981) many species are capable of fixing nitrogen or carbon (Wilkinson and Fay, 1979; Wilkinson, 1983) and of utilizing dissolved organic carbon as a nutritional source (Reiswig, 1974, 1981). Further, since they can contribute a significant

---

*Present address: Marine Science Research Center, SUNY, Stony Brook, New York 11794.

amount of nitrogen through excretion to the surrounding waters (Schubauer, 1985, 1988), they can be an important source of nutrients to primary producers in nutrient-impoverished waters. Thus, knowledge about the dynamics of these populations in time and space is important to our understanding of the synecology of benthic communities (particularly reefs) as well as the autecology of the individual species. This information is also needed to construct dynamic models of communities where sponges are major features.

In April 1984, the distribution and abundance of five Demospongiae were examined on the west forereef adjacent to the Discovery Bay Marine Laboratory of the University of the West Indies at Jamaica. The laboratory is on the north coast of Jamaica and is less than 500 m from the reef crest (Figure 1). In 1980, areas of the west forereef sustained heavy damage from Hurricane Allen (Woodley et al., 1981). In 1983, mass mortality of an important grazer, the long-spined urchin *Diadema antillarum*, occurred throughout the Caribbean. For detailed descriptions of these reefs prior to these disturbances, see Goreau (1959) and Goreau and Goreau (1973).

Materials and Methods

Locations of specimen of *Aplysina (= Verongia) fistularis, Ircinia strobilina, Neofibularia nolitangere, Mycale laxissima (= Mycale* sp. in Reiswig, 1973), and *Agelas* sp. (species B in Colin, 1978) were mapped by divers using scuba and compass-and-line methods at two locations, Mooring I, and Arena Mooring. The area surveyed at Arena Mooring drops off much more gradually and appears to have less relief than the area at Mooring I. The total area surveyed at Mooring I was 3625 m² and at Arena Mooring, 2500 m². The reef areas examined at Mooring I and Arena Mooring ranged in depth from 13 to 30 m and 12 to 23 m, respectively. Mooring I had already been surveyed in 1969 for *Mycale laxissima* ("*Mycale* sp.") and *Verongula reiswigi* ("*Verongia gigantia*") (Reiswig, 1973; prior to the hurricane and long-spined urchin die-off). Copies of Reiswig's original maps were used to compare the results of his census of this area with ours. The census at Arena Mooring was performed to provide insight into the amount of spacial variability expressed by the five species across the forereef. To our knowledge, our census at Arena Mooring is the first of its kind to be performed at this site.

Field maps were electronically digitized and substrate surface areas analyzed using a Tektronics 4051 microcomputer. Two-dimensional benthic surface areas were determined for three categories of substrate, which together equaled the total area of each study site. These substrate categories are defined as follows: (1) large continuous reefs refers to reef substrates larger than or equal to 10 m²; (2) miscellaneous patch reefs refers to reef substrates greater than 1 and less than 10 m² that are isolated from

other pieces of reef by sand; and (3) sand or sand channels are reef-rock structures less than 1 m² that are surrounded by sand or that occur in the sand channels. The number of each species attached to each type of substrate was then determined by re-examining the field maps. Finally, the site at Mooring I was subdivided into a shallow (13–20 m) and deep (21–30 m) zone and reanalyzed by zone. This division made it possible to compare the shallow zone at Mooring I with the Arena study site (12–23 m) and to compare the shallow and deep zones at Mooring I.

Results

SUBSTRATE AVAILABILITY

Over all, reef substrate constituted 41% and sand 59% of the substrate area available at the Mooring I site (Figure 2). In contrast, these two substrate types were almost equally available at the Arena Mooring site. Miscellaneous isolated patches of reef represented only 1% of the total substrate area available at both sites. The overall surface area in the shallow zone at the Mooring I site was greater than that available in the deep zone at this site (Figure 2). The major substrate type available in the shallow zone was large continuous reef substrate, and sand was the major type of substrate available in the deep zone. Isolated patches of reef made up 2% of the total substrate area available in the deep zone at Mooring I, in comparison with 1% in the shallow zone at the same site. The shallow zone at Mooring I (13–20 m), which approximately corresponds to the depth range of the entire site at Arena Mooring (12–23 m), was slightly smaller overall, had less sandy and isolated reef substrates, and had more large continuous reef substrate than the Arena Mooring site.

SPECIES DISTRIBUTIONS

At both study sites, the five species (each and taken together) occurred more frequently on large continuous reef substrate (Figure 3a,b) than in sand channels or on miscellaneous reef patches, and almost exclusively on a large continuous reef at the Arena Mooring site. The only exception was *Mycale laxissima* at Mooring I where, unlike *M. laxissima* at Arena Mooring, it was more frequently observed anchored in the sand. Further, a larger percentage of all five species of sponges observed at Mooring I were anchored in sand or attached to isolated patches of reef than at the Arena Mooring site. Neither *Neofibularia nolitangere* nor *Agelas* sp. were observed on sand or on isolated patches of reef at the Arena Mooring site but were found exclusively attached to reef substrate. At Mooring I all five species were found attached to all three substrate categories.

From species distribution patterns among substrate types in the two depth zones at Mooring I, and the overall

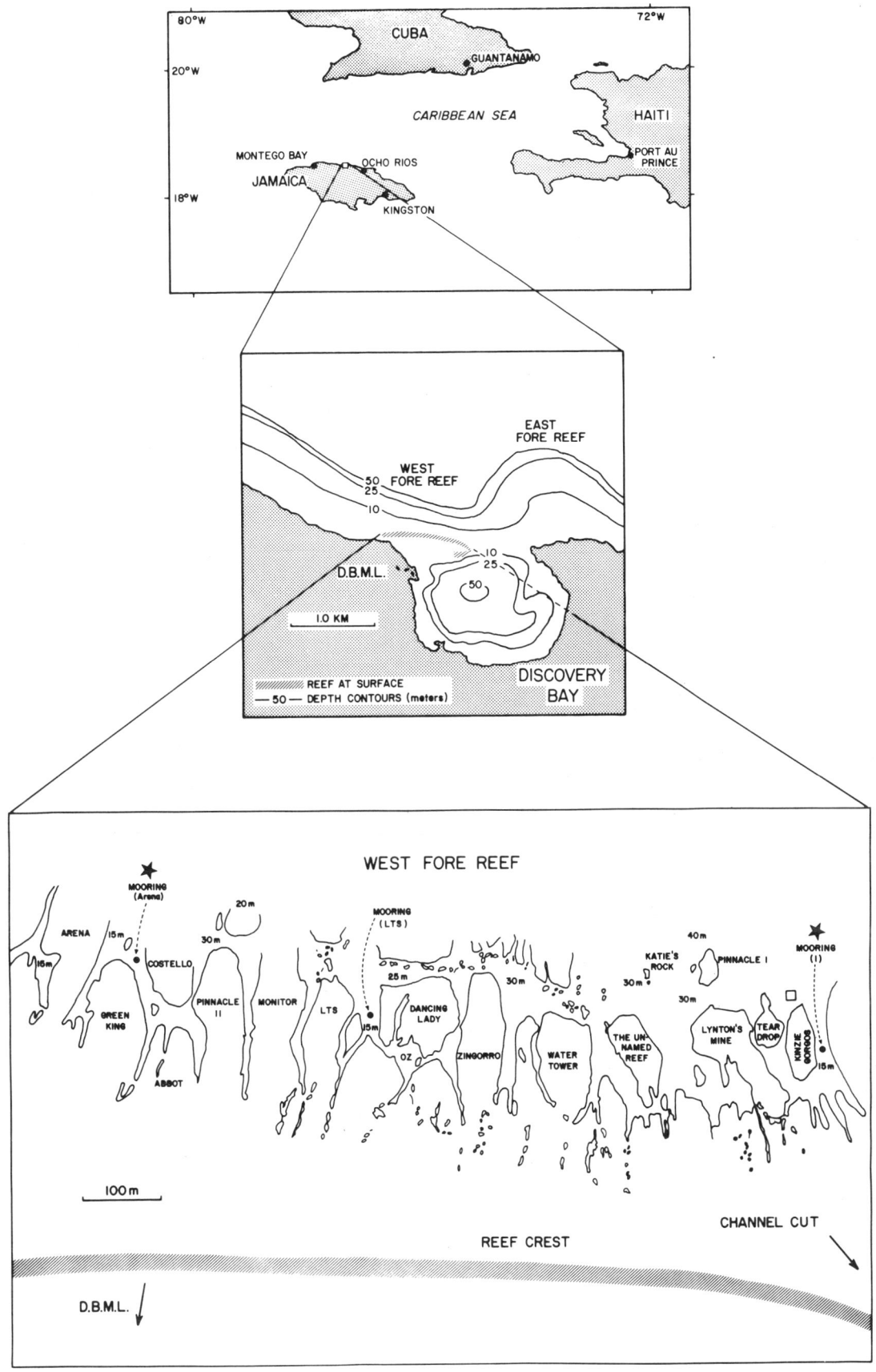

Figure 1. Map showing location of study site in Jamaica.

AVAILABLE BENTHIC SUBSTRATE

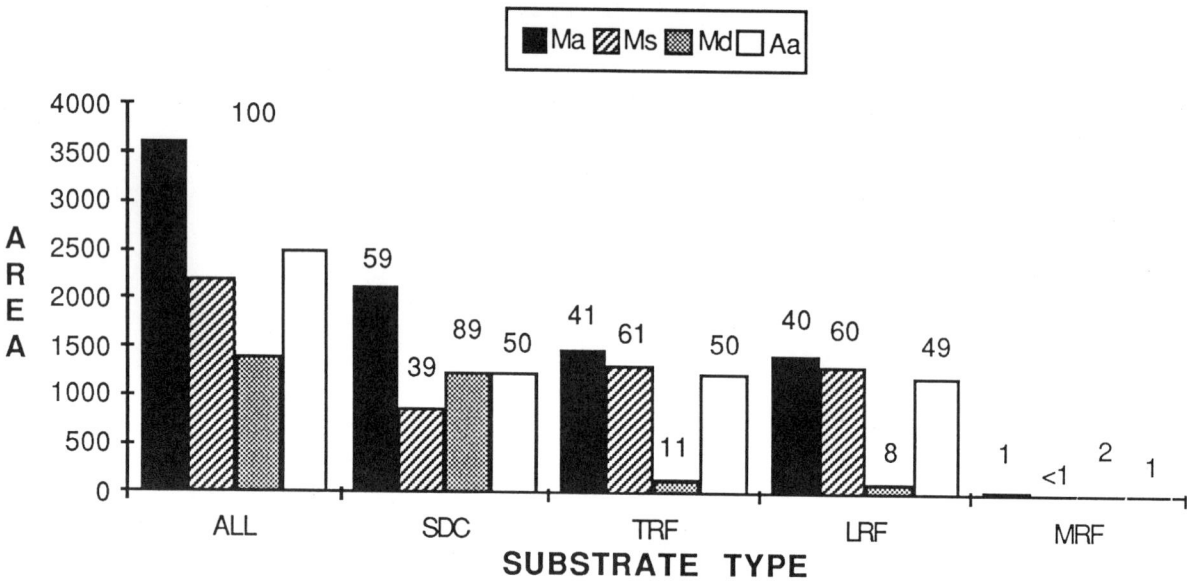

Figure 2. Histogram showing amount of substrate (area in m²) available in overall transect at Mooring I *(Ma)*, in shallow (13–20 m) zone at Mooring I *(Ms)*, in deep zone at Mooring I *(Md)*, and overall area at Arena Mooring *(Aa)*. Substrate categories indicated are: total transect area *(ALL)*, sand *(SDC)*, total reef area *(TRF)*, large continuous reef *(LRF)*, and isolated patch reef *(MRF)*. See text for definition of substrate types. Percentage each substrate category comprises of total area given above appropriate group.

species distributions at this site, it appears that the distribution patterns are comparable (Figures 3*b,c,d*), except in a few cases. Most notably, a number of sponge species observed on isolated patches of reef *(Mycale laxissima, Aplysina fistularis,* and *Neofibularia nolitangere)* and in the sand *Aplysina fistularis)* in the deep zone were not found on these substrates in the shallow zone. Note that two of the these species *(A. fistularis* and *N. nolitangere)* were also absent from these substrate types at the Arena Mooring site (which covers approximately the same depth range as the shallow zone at Mooring I). In addition, a larger percentage of all the sponges observed in the deep zone were found on isolated patches of reef or anchored in the sand as compared to the shallow zone.

SPECIES DENSITIES

Species occurrences were weighted by substrate surface areas available at the two sites, so that substrate specific densities could be compared between the sites and substrate utilization patterns could be compared between species. In general, the results from this analysis support the general trends indicated by the species distributions discussed above. The overall density of the five sponge species taken together at Arena Mooring (0.092/m²) was almost double that at Mooring I (0.052/m²) (Figure 4*a,b*). Densities of *Mycale laxissima* at Mooring I were higher on sand than on reef substrate, whereas the situation is re-

versed at Arena Mooring; moreover, the overall density of this species was slightly higher at Mooring I (Figure 4*a,b*). Of all species examined, *Ircinia strobilina* had the highest densities on all substrates at both sites, with the only exception *M. laxissima* in sand at Mooring I (Figure 4*b*).

Although isolated patches of reef accounted for only 2% or less of the substrate available at either site, sponge density was highest on this substrate type at both sites (Figure 5). At Arena Mooring, the cumulative density was only slightly higher (0.186/m²) on patch reefs, compared to 0.169/m² on large continuous reefs), but at Mooring I the cumulative sponge density on patch reefs was 15 times greater than that found on the other two substrate categories (0.861/m² on patch reefs versus 0.057/m² on large continuous reefs). Further, whenever one of the five sponge species was found on patch reef substrate at either site, it was found in higher densities than on any of the other substrate types. The only exception to this trend was *Ircinia strobilina* at Arena Mooring. The highest densities of this species were observed on large continuous reef substrate (Figure 4*a*).

At Mooring I, the cumulative sponge density in the deep zone was approximately 1.5 times that of the shallow zone at this site. However, sponge densities of most species attached to large continuous reef in the deep zone increased dramatically (Figure 4*c,d*) over those in the shallow zone. The overall density of sponges attached to patch reef substrates in the deep zone was 28% higher than in

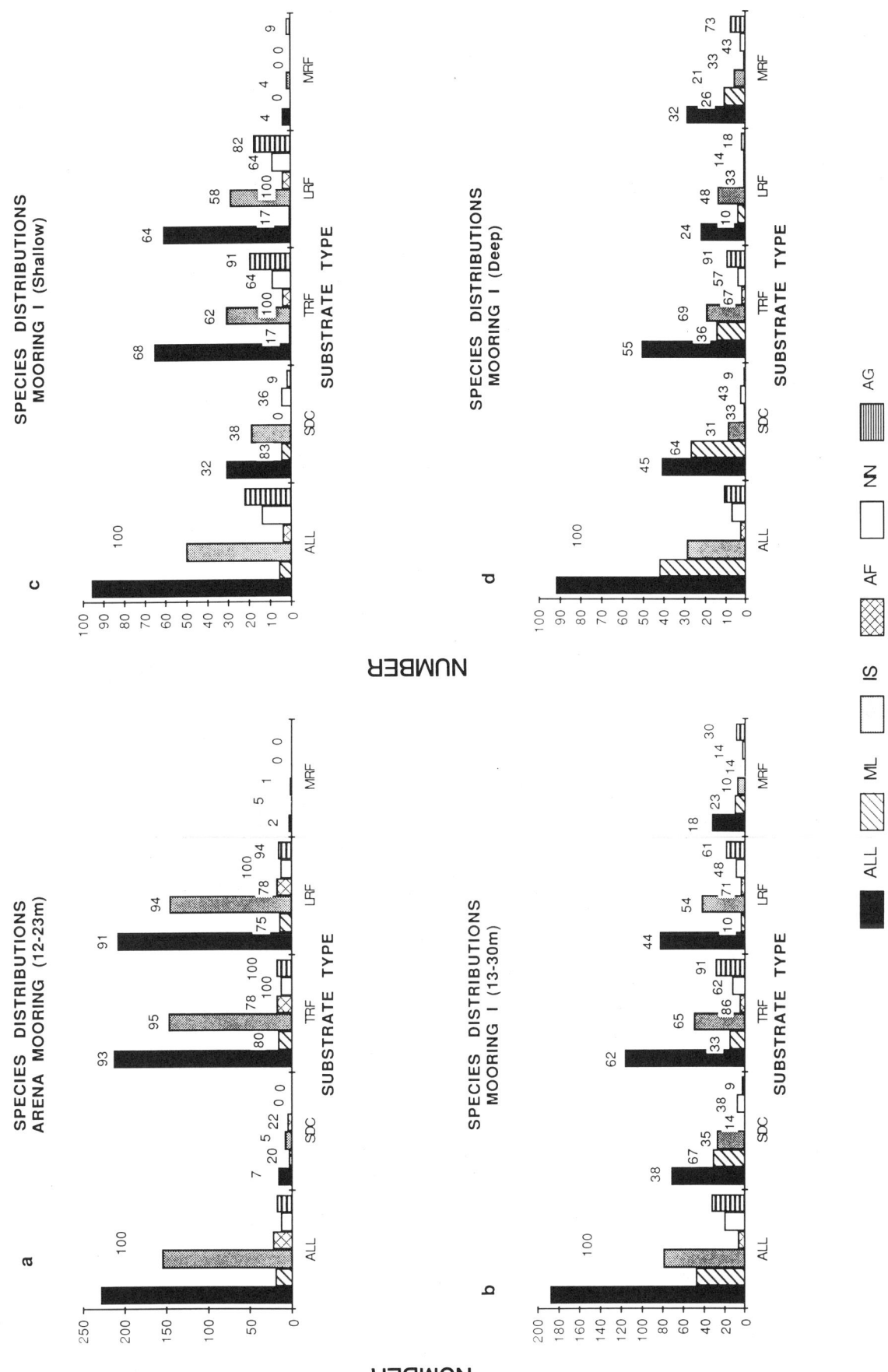

Figure 3. Histograms showing species distributions (number of individuals) on different substrate categories (compare Figure 2): *a*, Arena Mooring; *b*, Mooring I; *c*, shallow zone at Mooring I; *d*, deep zone at Mooring I. Species shown are as follows: total all 5 species (*ALL*), *Mycale laxissima* (*ML*), *Ircinia strobilina* (*IS*), *Aplysina fistularis* (*AF*), *Neofibularia nolitagere* (*NN*) and *Agelas* sp. (*AG*). Substrate types are as in Figure 2; percentages of individuals given above species bars.

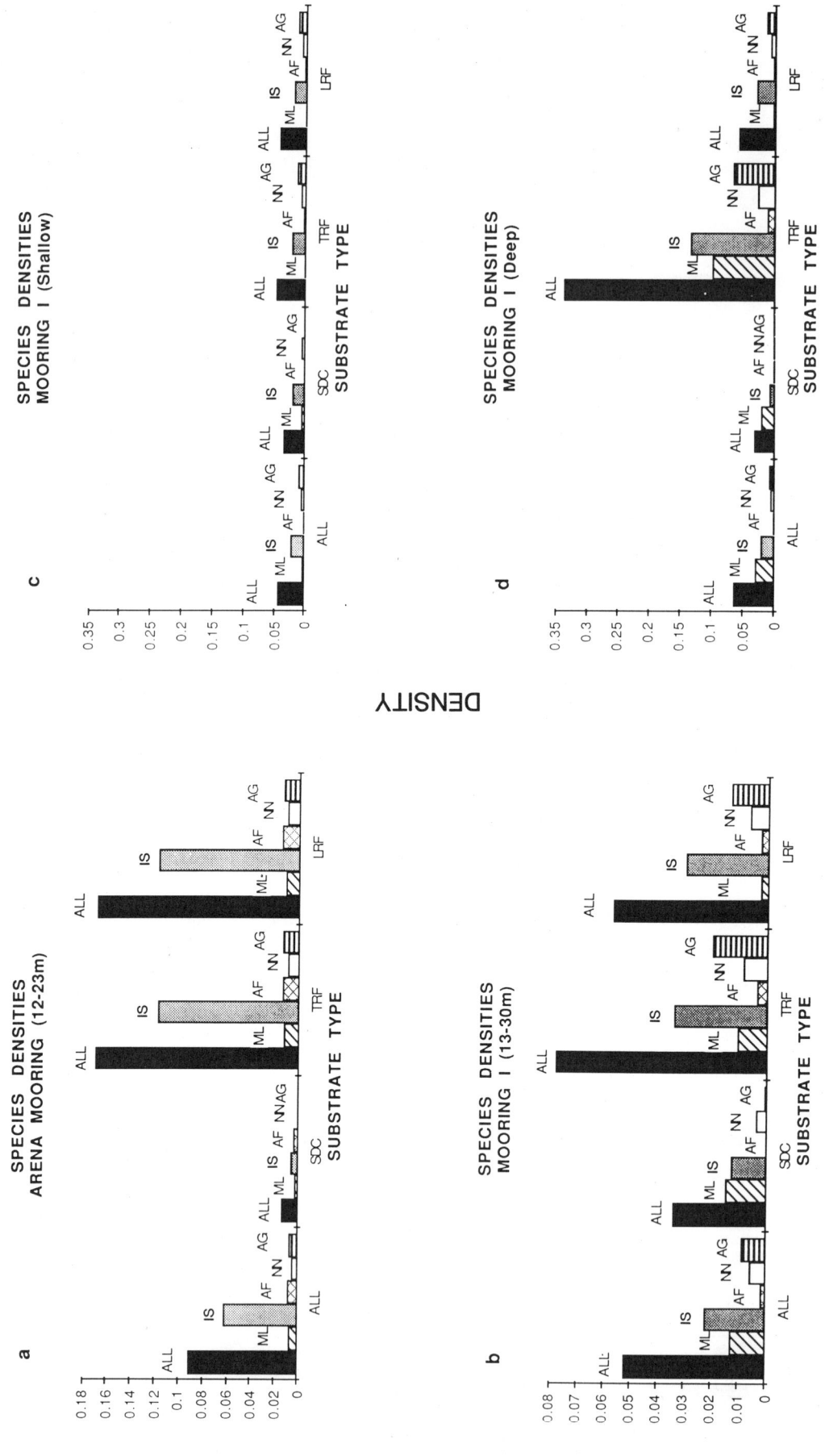

Figure 4. Histograms of species densities (specimens/m²) on different substrate types: *a*, Arena Mooring; *b*, Mooring I; *c*, shallow zone at Mooring I; *d*, deep zone at Mooring I. Abbreviations and symbols as in Figures 2 and 3.

the shallow zone (Figure 5*b*). *Mycale laxissima* was not found on this substrate in the shallow zone but was the dominant species on this substrate in the deep zone. Two species, *Ircinia strobilina* and *Agelas* sp., had lower densities on patch reef substrate in the deep zone than in the shallow zone (Figure 5*b*). Finally, sponge densities of the five species at Arena Mooring were generally lower or equal to those in the shallow zone at Mooring I (*M. laxissima* was higher) even though these sites are very similar in depth.

CHANGES IN DISTRIBUTION OF *Mycale laxissima*

Abundance of *Mycale laxissima* on natural substrate at the Mooring I site in 1969 and 1984 was very similar (39 and 42, respectively). However, distribution of *M. laxissima* among substrate types and the cumulative abundances varied considerably between these dates. Although most individuals were found attached to natural as opposed to artificial substrate in both years, the proportion of the total number of individuals observed on artificial substrate in 1984 had decreased by 31% from the sample in 1969 (Figure 6). From the distribution of *M. laxissima* on natural substrate in 1969 and 1984, it appears that a greater proportion of the individuals observed were attached to reef substrate (reef-rock structures greater than 1 m²) in 1969 than in 1984, when most were found in the sand channels between the reefs (Figure 6).

Discussion

In general, the sponge distributions on different substrates in this study are similar to those reported by others. However, earlier studies were restricted to shallow water. One notable exception was the study by Reiswig (1973), who observed *Mycale laxissima* from 13–62 m at Mooring I. Reiswig found this species to be primarily distributed at sand-substrate interfaces, partly because the individuals he observed were attached only to gorgonian stalks and loose coral rubble. Among the shallow-water sponges of the western Bahamas, *Aplysina fistularis*, *Ircinia strobilina*, and *Neofibularia nolitangere* are reported to occur on rock pavement, rocky substrate, and reef habitat, respectively (Wiedenmayer, 1977). In Port Royal, Jamaica, *Ircinia strobilina* has been found mainly on pilings and corals offshore, and *Aplysina fistularis* primarily on coral heads at depths of 10–15 ft (3–5 m). (Hechtel, 1965). Furthermore, in the Dry Tortugas, *A. fistularis* is commonly found on reefs from 5–40 m, and throughout the Caribbean, *N. nolitangere* is common on level bottoms among corals (from 3–46 m) (Colin, 1978). *I. strobilina* has also been reported to be the most abundant sponge in the Dry Tortugas between 0 and 17 m. Although *I. strobilina* was also the most abundant species in the study area, a considerable number of specimens were seen along the

deepest sections of our transects at Discovery Bay (down to 30 m).

Depth appears to have no obvious effect on the distribution of sponges among the substrate categories and over the range studied for three of the five species examined. It is not clear at this time why the other two species, *Aplysina fistularis* and *Neofibularia nolitangere*, were not found on isolated patches of reef or in the sand above 20 m in either transect. The reason does not appear to be simply the area of these substrates that is available.

The densities of all five species of sponges on all three substrate categories at Mooring I increase with depth. Similarly, Reiswig (1973) found an increase in sponge volume of *Mycale laxissima* and *Verongula reiswigi* as well as total sponge volume from approximately 20 to 50 m at Mooring I and cited wave stress and sedimentation as two possible factors limiting the upper distribution of many forereef sponge species. We concur with this evaluation.

Surprisingly, although isolated patches of reef represent no more than 2% of the substrate area available, densities of sponges on this substrate category are higher than on any other substrate at Mooring I, and are only slightly so at Arena Mooring. Again, the reasons for this are unclear. One possible explanation is that large continuous reef substrate may provide more refuges among its interstices and thus may harbor higher densities of sponge predators. Alternately, competition for space between sponges, other benthic invertebrates, and algae might be more intense on a large continuous reef substrate. Although the two study areas are clearly different topologically, there is no clear reason why the overall density of sponges is higher at Arena Mooring than at Mooring I.

Reiswig (1973) hypothesized that specimens of *Mycale laxissima* on rigid substrates would be dislodged more easily than those growing on more flexible substrates such as small loose coral rubble or gorgonian stalks, owing to their relatively narrow attachment stalks. Our observations appear to support this hypothesis in that although the overall abundance of sponges on natural substrate in 1969 (before Hurricane Allen) and in 1984 (after the hurricane) is relatively constant, the proportion of the specimens found on coral rubble in the sand has increased dramatically. Thus, the hurricane may well have scoured specimens on less suitable substrate (less flexible) and at the same time either created more suitable substrate (coral rubble), which ultimately has found its way into sand areas between the large sections of reef, or scoured other animals off suitable substrate. Further, the release of fast-growing fleshy benthic algae from heavy grazing by *Diadema* following die-off in 1983 (Morrison, 1986) may have prevented *M. laxissima* from recolonizing some of the areas it inhabited before the storm. Regardless of the mechanisms of these observed changes, the results support Reiswig's contention that *Mycale* is space limited on this reef.

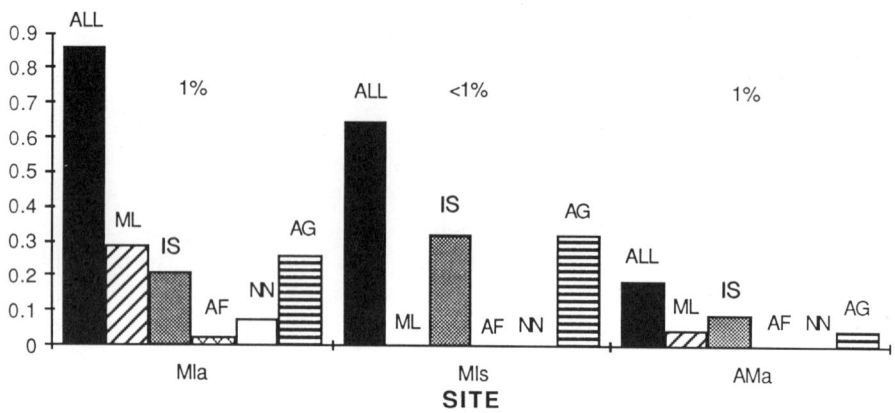

a   SPECIES DENSITIES ON ISOLATED PATCHES
     OF REEF AT BOTH SITES

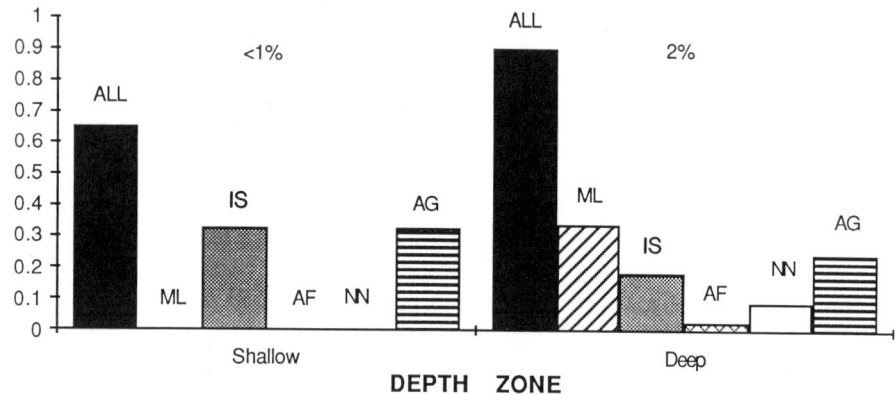

b   SPECIES DENSITIES ON ISOLATED PATCHES
     OF REEF AT MOORING I

DENSITY

Figure 5. Histograms of species densities (specimens/m²) on isolated patches of reef substrate: *a*, at both sites; *b*, Mooring I in two depth zones. Site abbreviations are: Mooring I all *(MIa)*, Mooring I shallow *(MIs)*, and Arena Mooring all *(AMa)*. Other abbreviations and symbols as in Figures 2 and 3.

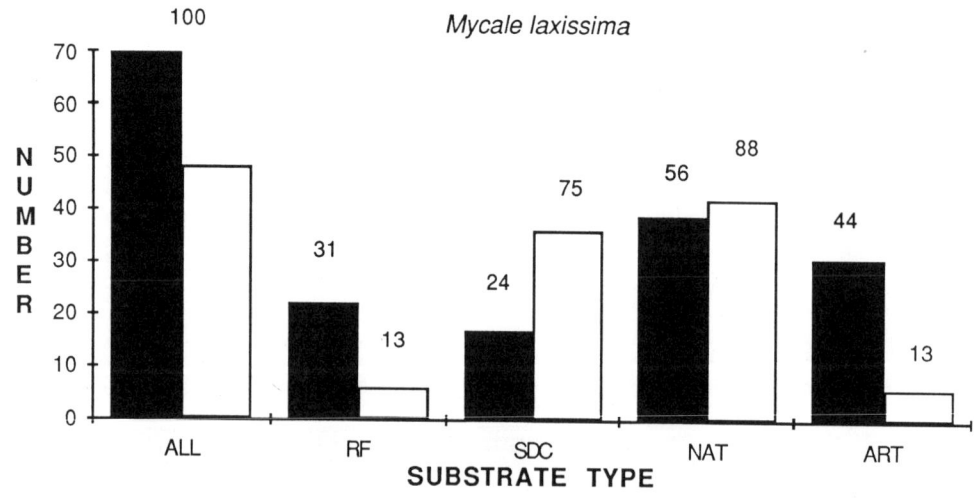

CHANGE IN DISTRIBUTION OF
Mycale laxissima

Figure 6. Histogram showing change in distribution of *Mycale laxissima* (number of individuals) on different substrate types at Mooring I in 1969 *(solid bars)* and 1984 *(open bars)*. Substrate categories as follows: total *(ALL)*, reef *(RF)*, sand *(SDC)*, natural substrate *(NAT)*, and artificial substrate *(ART)*. Percentages given above bars.

Conclusions

Ircinia strobilina and *Mycale laxissima* were the most abundant of the five species studied at both sites. Depth did not appear to influence the way species utilized different substrates. Densities of *M. laxissima* were dramatically higher in the deep part of the transect at Mooring I than in the shallow part. In general, sponge densities were highest on isolated patches of reef, even though this substrate constituted no more than 2% of the surface area of either transect. Overall, sponge densities were higher at the Arena Mooring site than at the Mooring I site. The abundance of *Mycale* at Mooring I was lower in 1969 as compared with 1984, mainly as a result of fewer individuals growing on artificial substrate. The abundance of *M. laxissima* on natural substrate was surprisingly similar in 1969 and 1984, only the location of the individuals differed. Although physical factors such as wave stress and sedimentation probably limit the upper distribution of many forereef sponges, the interaction of biological and physical factors determine their abundance and distribution below this limit. Our observations support Reiswig's contention that *Mycale* is space limited on the Discovery Bay reef.

Acknowledgments

We thank C. McDermott and B. Patten for their help in the field. We would especially like to thank H. Reiswig for allowing us access to his original field notes. This research was supported by a research grant from the University of Georgia and a grant-in-aid from the Society of the Sigma Xi. This is contribution no. 363 from the Discovery Bay Marine Laboratory, University of the West Indies, Discovery Bay, Jamaica.

Literature Cited

Colin, P. I. 1978. *Caribbean Reef Invertebrates and Plants*. Neptune City, New Jersey: T. F. H. Publications. 512 pp.

Dayton, P. K., G. A. Robilliard, R. T. Paine, and L. B. Dayton. 1974. Biological Accomodation in the Benthic Community at McMurdo Sound, Antarctica. *Ecological Monographs*, 44:105–128.

Goreau, T. F. 1959. The Ecology of Jamaican Coral Reefs I. Species Composition and Zonation. *Ecology*, 40:67–90.

Goreau, T. F., and N. I. Goreau. 1973. The Ecology of Jamaican Coral Reefs II. Geomorphology, Zonation, and Sedimentary Phases. *Bulletin of Marine Science*, 23:399–464.

Hartman, W. D. 1977. Sponges as Reef Builders and Shapers.

Pages 127–134 in *Reefs and Related Carbonates—Ecology and Sedimentology*, edited by S. H. Frost, M. P. Weiss, and J. B. Saunders. Tulsa, Oklahoma: American Association of Petroleum Geologists.

Hehtel, J. G. 1965. A Systematic Study of the Demospongiae of Port Royal, Jamaica. *Peabody Museum of Natural History Bulletin*, 20:1–103.

Morrison, D. 1986. Algal-Herbivore Interactions on a Jamaican Coral Reef. Doctoral Dissertation. University of Georgia, Athens, Georgia. 135 pp.

Reiswig, H. M. 1973. Population Dynamics of Three Jamaican Demospongiae. *Bulletin of Marine Science*, 23:191–226.

———. 1974. Water Transport, Respiration, and Energetics of Three Tropical Marine Sponges. *Journal of Experimental Marine Biology and Ecology*, 14:231–249.

———. 1981. Partial Carbon and Energy Budgets of the Bacteriosponge *Verongia fistularis* (Porifera: Demospongiae) in Barbados. *Marine Ecology*, 2:273–293.

Rützler, K. 1981. An Unusual Bluegreen Alga Symbiotic with Two New Species of *Ulosa* (Porifera: Hymeniacidonidae) from Carrie Bow Cay, Belize. *Marine Ecology*, 2:35–50.

Schubauer, J. P. 1985. Cycling of Nitrogen and Phosphorus by Marine Sponges. Abstract. American Society of Limnology and Oceanography Meeting, Minnesota.

———. 1988. Metabolism and Nutrient Cycling by Marine Sponges. Doctoral Dissertation. University of Georgia, Athens, Georgia. 119 pp.

Vacelet, J. 1981. Algal-Sponge Symbioses in the Coral Reefs of New Caledonia: A Morphological Study. Pages 713–719 in *The Reef and Man*, edited by E. D. Gomez et al. Proceedings, Fourth International Coral Reef Symposium, Manila, 2. Quezon City, Philippines: University of the Philippines.

Wenner, E. L., D. M. Knott, R. F. Van Dolah, and V. G. Burrell, Jr. 1983. Invertebrate Communities Associated with Hard Bottom Habitats in the South Atlantic Bight. *Estuarine, Coastal, and Shelf Science*, 17:143–158.

Wiedenmayer, F. 1977. *Shallow-Water Sponges of the Western Bahamas*. Basel und Stuttgart: Birkhäuser. 287 pp.

Wilkinson, C. R. 1978. Microbial Associations in Sponges. I. Ecology, Physiology and Microbial Pop[ulations of Coral Reef Sponges. *Marine Biology*, 49:161–167.

———. 1981. Significance of Sponges with Cyanobacteria on Davies Reef, Great Barrier Reef. Pages 705–712 in *The Reef and Man*, edited by E. D. Gomez et al. Proceedings, Fourth International Coral Reef Symposium, Manila, 2. Quezon City, Philippines: University of the Philippines.

———. 1983. Net Primary Productivity in Coral Reef Sponges. *Science*, 219:410–411.

Wilkinson, C. R., and P. Fay. 1979. Nitrogen Fixation in Coral Reef Sponges with Symbiotic Cyanobacteria. *Nature*, 279:527–529.

Woodley, W. D., et al. 1981. Hurricane Allen's Impact on Jamaican Coral Reefs. *Science*, 214:749–755.

Species Interaction
and Ecophysiology

KLAUS RÜTZLER
Department of Invertebrate Zoology
National Museum of Natural History
Smithsonian Institution
Washington, D.C. 20560

Associations between Caribbean Sponges and Photosynthetic Organisms

Abstract

Five types of associations between sponges and endobiotic autotrophs were discovered in Belize and Bermuda and are currently under study. At least 21 species of common and large shallow-water sponges (genera *Ircinia*, *Aplysina*, *Verongula*, *Cribochalina*, *Xestospongia*, *Neofibularia*, *Ulosa*, *Geodia*, *Chondrilla*) contain *Aphanocapsa*-type unicellular blue-green algae (cyanobacteria). One spongiid, *Oligoceras violacea*, is permeated by filaments of a *Phormidium* species, an oscillatorian cyanobacteria. Two species of clionid sponges, *Cliona caribbaea* and *C. varians*, contain dinophycean zooxanthellae. *Mycale laxissima* (Mycalidae), *Igernella notabilis* (Darwinellidae), and at least one unidentified dendroceratid were found to harbor two species of filamentous algae embedded in the spongin fibers of their skeletons; one, *Ostreobium*, is a chlorophyte, the other, *Acrochaetium*, a rhodophyte. Two sponges substitute their skeletons with live calcified red algae. *Dysidea janiae* (Dysideidae) is supported by *Jania adherens*, *Xytopsues osbornensis* (Myxillidae) by *J. capillaria*. All reported cases have parallels in the Mediterranean Sea or in the Indo-Pacific region but are difficult to compare because important new data, on ultrastructure and pigment properties are subject to variables that are not yet fully understood. Observations and experiments provide evidence that both partners generally profit from the symbiosis—the sponges by obtaining food supplements or body support, the plants by gaining nutrients and protection from grazers.

For more than a century biologists have been aware of the association between sponges and photosynthetic organisms (cyanobacteria or Cyanophyta, several groups of eukaryotic algae) (see Feldmann, 1933, and the literature cited therein), but research on the types of relationships between them did not begin in earnest until the 1970s.

Pioneers in the Mediterranean were the Italian and French groups gravitating around M. Sarà and J. Vacelet (see citations in Sarà and Vacelet, 1973:532–537). Subsequently, researchers became interested in the coral reef environments of the Australian Great Barrier Reef (Wilkinson, 1978, 1980, 1983; Larkum et al., 1987), the Red Sea (Wilkinson and Fay, 1979), and New Caledonia (Vacelet, 1981). Until recently, however, few spent much time on the photosynthetic sponge symbionts of the subtropical and tropical Atlantic ocean, except to comment on them in passing in studies on sponge taxonomy (de Laubenfels, 1950:174–175).

This report represents an attempt to fill some of the gaps in our knowledge about sponge symbionts. In it I summarize the results of extensive work in Bermuda and Belize that began in the early 1970s. I hope that the inventory and electron microscope survey of cyanobacterial and algal sponge associates discussed here will provide a data base for ongoing and future ecological and physiological investigations into complex host-symbiont relationships.

Material and Methods

The sponges examined in this study were collected in three types of habitat: shallow coral reef (2–20 m), lagoon (0.5–8.0 m), and mangrove (0–2 m). The localities surveyed were Bermuda, Bimini (Bahamas), St. John (U.S. Virgin Islands), Dominica, Jamaica, and Carrie Bow Cay (Belize), but electron microscope studies were restricted to specimens from Bermuda and Belize. The presence of photosynthetic organisms in the tissues was ascertained in the field by examining fresh microscope preparations and alcohol-soluble (chlorophylls, and others) and water-extractable (phycobilin) pigments.

Material for electron microscopy was fixed in 1.5% glutaraldehyde in 0.2 M cacodylate buffer (with 0.1 M sodium chloride and 0.35 M sucrose), pH 7.2 (2–4 h, 29°C). Postfixation in 1% osmic acid in the same buffer solution (1 h, 4°C) and dehydration to 95% ethanol (5 steps) followed immediately. Where necessary, desilicification was accomplished by adding 5% hydrofluoric acid to the washing solution (buffer) before the final rinse and dehydration. The material was stored in the last (95%) ethanol stop (1–3 weeks) until 100% dehydration and resin embedding (Spurr low viscosity embedding media Polysciences, Inc., Warrington, Pennsylvania) for transmission electron microscopy (TEM). Sections (TEM) were stained with uranyl acetate and photographed by Philips 200 or Zeiss EM9 S–2 microscopes at 1,600–45,000 times primary magnification.

Estimates of tissue proportions (sponge:symbiont) are based on point counts using a Weigel graticule inside the ocular of a light microscope with 100× oil immersion objective and stained sections.

Results

ASSOCIATIONS WITH UNICELLULAR CYANOBACTERIA OF THE *Aphanocapsa feldmanni* TYPE

At least 19 species of the large and common shallow-water Demospongea of the West Indies contain small unicellular cyanobacteria of the *Aphanocapsa feldmanni* type (Feldmann, 1933; Sará and Liaci, 1964). This number constitutes 45% of the 42 conspicuous sponge species examined in Bermuda and Belize. Sponges hosting this type of microorganism belong to the orders Astrophorida (*Geodia*, 2 species), Hadromerida (*Spheciospongia*, 1 species; *Chondrilla*, 1 species), Dictyoceratida (*Ircinia*, 2 species), Verongiida (*Aplysina*, 4 species; *Verongula*, 3 species), Petrosiida (*Cribochalina*, 2 species; *Xestospongia*, 3 species), and Poecilosclerida (*Neofibularia*, 1 species). The cyanobacteria are concentrated in the peripheral regions of their sponge hosts, where they may reach a maximum density of 28% of the cellular tissue volume (*Chondrilla nucula* Schmidt). The average symbiont concentration in the species examined was 10%.

The fine structure of this symbiont (Figure 1) is close to that described for its Mediterranean relative (Sarà, 1971; Vacelet, 1971; Gaino et al., 1977). Most differences can be attributed to variations in fixation techniques or physiological state of the organism at the moment of fixation because they also show up in of our own material.

These cyanobacteria are ovoid and measure 1.1–2.0 μm × 0.6–1.0 μm in median TEM cross sections. Adult cells are always separate from each other. Dividing stages occur in only about 2–4% of the population. Division occurs by median constriction in a plane perpendicular to the longer axis, so that cells form a figure 8 just before the daughter cells separate (Figure 1a).

The cell wall exhibits the characteristic four zones of electron density. In many sections it is undulating and separated from the plasmalemma by a clear space of 20–50 nm; this condition is probably a fixation artifact here. A thin sheath (up to 40 μm thick) is present, although it is not clearly discernible in all specimens. The thylakoid consists of a simple flat lamella wound in a spiral around the nucleoplasm, oriented parallel to the longer axis of the cell. This arrangement is best seen in the three-dimensional reconstruction given by Vacelet (1971:fig.1). Each lamella consists of a 10-nm electron-transparent zone sandwiched between two dense layers of equal thickness (30 nm total thickness). There is no vacuolisation, branching, or stacking in the photosynthetic apparatus. The lamellar spiral takes one to two-and-a-half turns; the innermost edge often rejoins the preceding turn. The nucleoplasm occupies about 40–50% of the central cell and displays a characteristic reticulation of fine filaments.

The four most common cytoplasmic inclusions of cyanobacteria could be identified on the electron micrographs of this sponge symbiont. Cyanophycin granules,

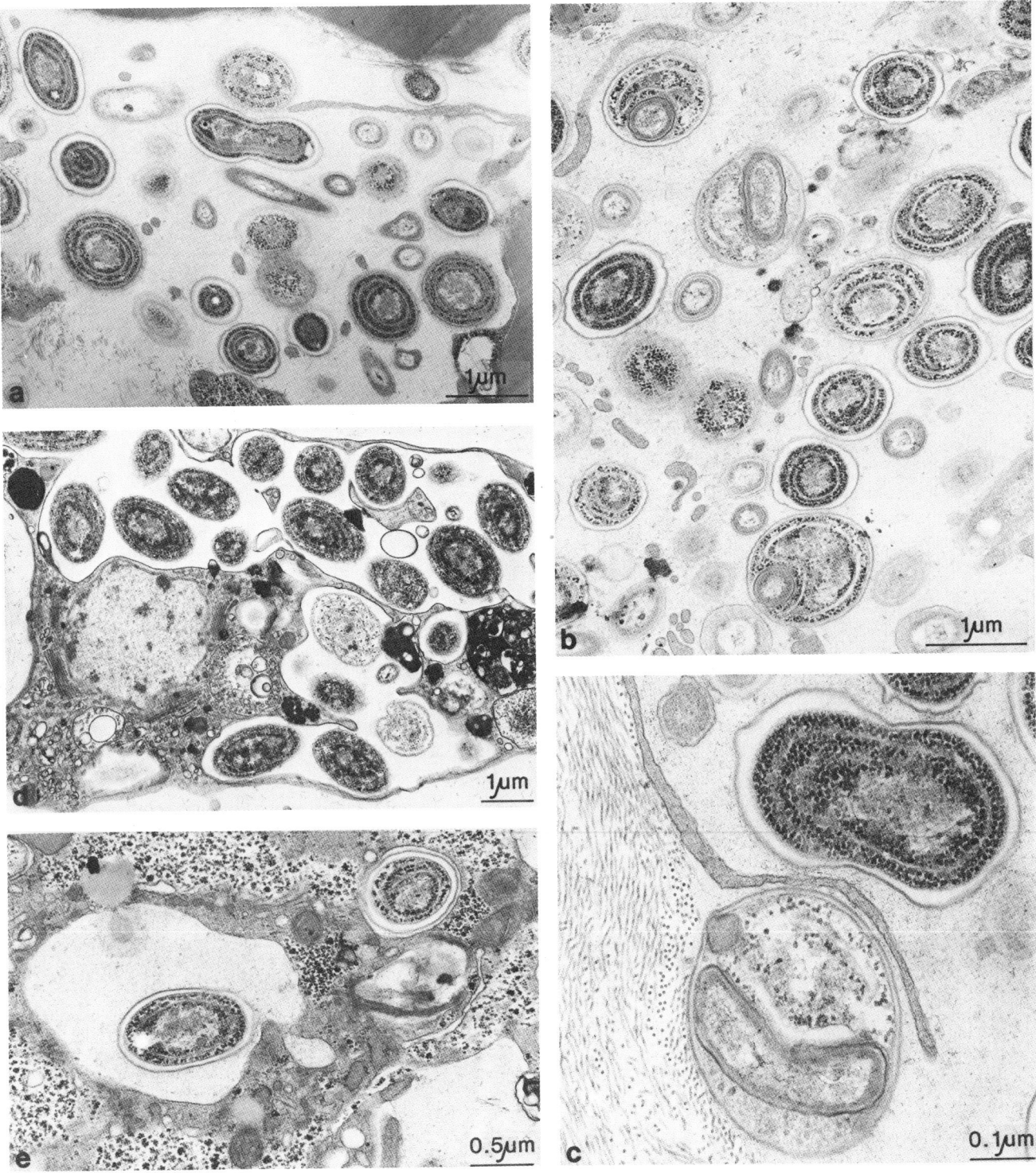

Figure 1. *Aphanocapsa feldmanni*-type cyanobacteria inside sponge hosts: *a–c*, in *Chondrilla* mesohyl; *d*, inside *Cribochalina* bacteriocyte; *e*, in stages of digestion inside *Chondrilla* archaeocyte. Note endoparasitic bacteria in *b* and *c*.

located primarily in the peripheral cell region, are electron dense (black) and measure 60–140 nm in diameter (cross sections). Polyglucan granules are also black, but only 15–20 nm in size. They occur between the lamellae of the thylacoids, often in great numbers. Clear areas 50–120 nm in diameter near the cell poles are presumed to be polyphosphate bodies. Polyhedral bodies are gray in outline, and measure 85–95 nm.

An interesting phenomenon is the endoparasitism of the symbiont by bacteria (Figure 1*b,c*) in associates of

Chondrilla nucula, both in the material of Belize as well as Bermuda. The afflicted cyanobacterial cell usually contains one rod-shaped bacterium, 0.6–0.8 μm × 0.3–0.4 μm in size; two bacteria were seen in only one specimen. These bacteria have a simple cell wall, a nucleoplasmic region filling 60–70% of the cell body, and no conspicuous inclusions. They look like many of the bacteria that occur freely in the mesohyle of the sponge. The pathogenic nature of the parasite is obvious from the progressive decay of the cyanobacterial host visible in the electron micrographs.

Cyanobacterial symbionts discussed in this section occur either free among the spongin fibrils of the sponge mesohyl or inside sponge cells. Most of the cyanobacteria are free. Intracellular symbionts are much rarer and occur either singly, inside small vacuoles of archaeocytes, or in groups, along with nonphotosynthetic bacteria, in bubble-shaped cells. In the former, the stages of disintegration indicate that the archaeocytes are digesting the symbiont. In the latter, all cells appear to be healthy and complete. They are inside a large (12–15 μm) bubble "vacuole" formed by a pinacocytelike cell (bacteriocyte). These cells have an ovoid (3.2 × 1.5 μm) anucleolate nucleus surrounded by the cytoplasm of a very small (5.0 × 1.8 μm) cell body that tapers to form a sheet 80–300 nm thick, the bubble wall. The lumen of the bacteriocyte contains not only populations of bacteria and cyanobacteria, but also cytoplasmic strands, apparently branched off the bubble wall, and fibrils of spongin identical to those surrounding the cell.

ASSOCIATIONS WITH UNICELLULAR CYANOBACTERIA OF THE *Aphanocapsa raspaigellae* TYPE

The relatively large blue-green symbiont *Aphanocapsa raspaigellae* has been found in association with sponges in only two cases in the Caribbean. The hosts are two closely related species of the Halichondriida genus *Ulosa.* The symbiont occurs thoughout the sponge body and makes up as much as 50% of the cellular tissue volume. It is never seen inside a sponge cell.

These cyanobacteria are conspicuous because of their bright green color and large (5–9 μm) size. They are spherical and divide by median constriction, and their thylacoids are unusual for their inflated sacs (Figure 2). The pigment composition of this organism is typical for cyanobacteria, although it has considerably less phycobiliprotein than other species in the group. This symbiont is described elsewhere in more detail (Rützler, 1981).

ASSOCIATION WITH A FILAMENTOUS CYANOBACTERIUM

The tissues of the dictyoceratid *Oligoceras violacea* (Duchassaing and Michelotti) are permeated by a filamentous

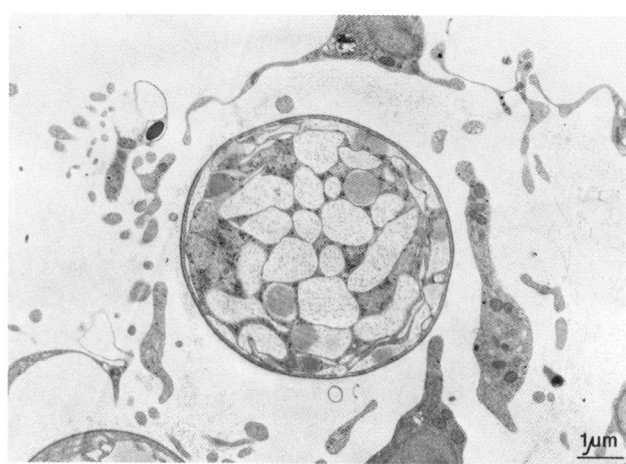

Figure 2. *Aphanocapsa raspaigellae*-type cyanobacterium inside sponge host *Ulosa.*

cyanobacterium (Figure 3). Although it contributes only about 13% to the total sponge volume , this procaryote may make up almost half of the cellular tissue volume in the host owing to the large quantities of sediment particles commonly incorporated in *Oligoceras.*

One-cell stages of the symbiont are spherical and 7–8 μm in diameter. After division the daughter cells remain rounded, except for some flattening at the cross walls, and the trichome has the appearance of a string of pearls (Figure 3a,b). The average trichome is 40 μm long and is made up of about 10 cells. A 16-cell filament of 63 μm was the longest observed. Cells in a trichome are, on average, 7 μm wide and 4 μm long. The largest cells are usually found in the center of the trichome.

The cell wall (Figure 3e) is composed of the usual four layers, which together are about 28 nm thick. The outermost layer (L IV) is slightly undulated and supports a 250–nm sheath displaying three structural zones. The innermost sheath is 70 nm thick and medium dense, with bundles of fibrils oriented perpendicular to the cell wall. The middle layer measures only 10 nm but is very dense, with fibrils intertwined and oriented parallel to the cell wall. The outer zone is less defined than the other two, is 170 nm thick, and is composed of a loose reticulation of fibrils that are more or less prependicular to the cell surface.

The thylakoids are fairly flat and radiate from the central zone of the nucleoplasm toward the cell surface, where they terminate near the plasmalemma (Figure 3c,d). The membranes are 8 nm thick and separated by intrathylacoid spaces 20–160 nm (rarely as much as 300 nm) wide. The nucleoplasm occupies nearly 50% of the cell.

Angular dark gray cyanophycin granules (Figure 3d), 400–700 nm in diameter, are common in the outer cell region and along the cross walls. Smaller light gray bodies (190 nm) of similar shape occur closer to the inner

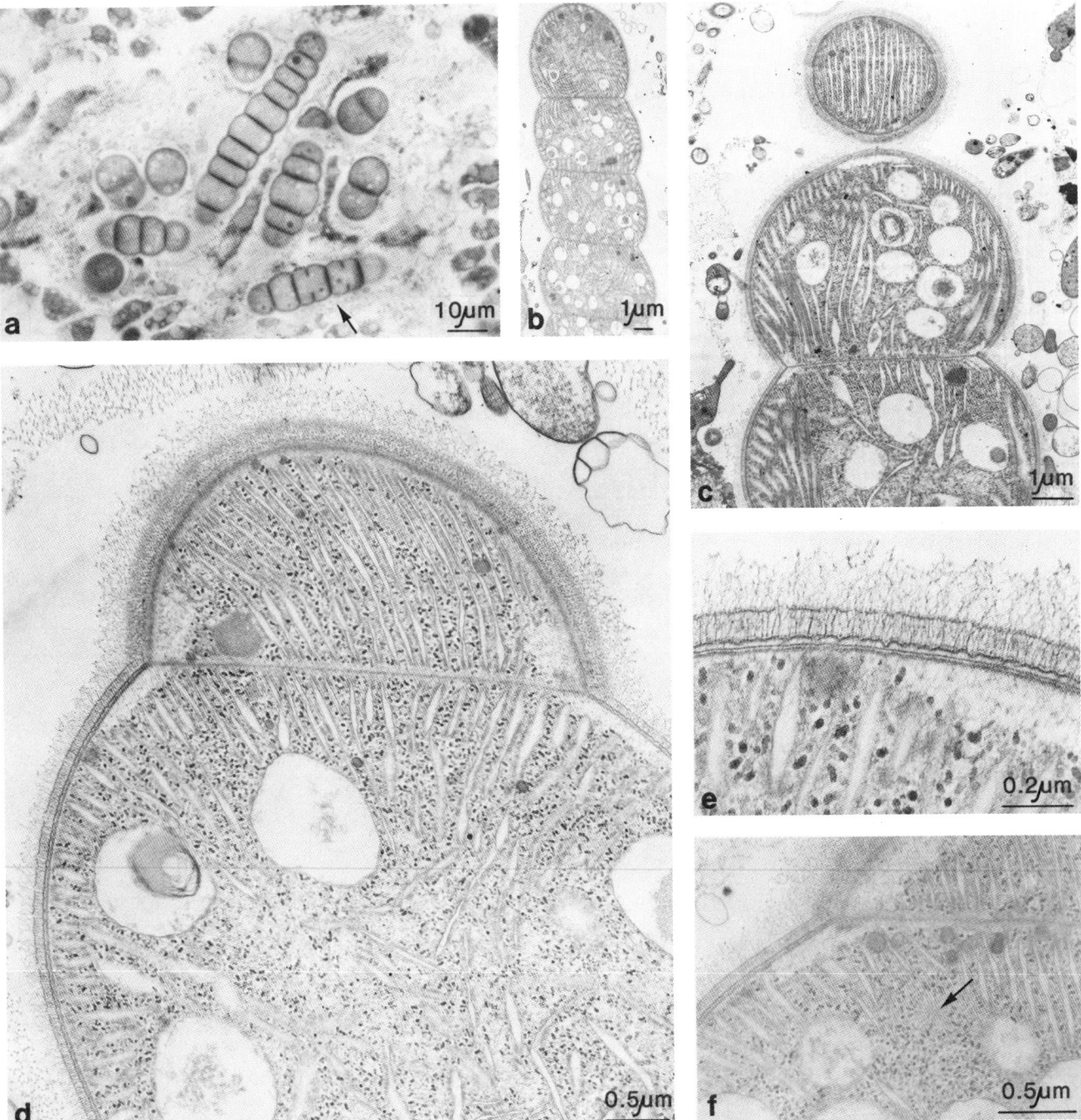

Figure 3. *Phormidium spongeliae*-type cyanobacterium inside sponge host *Oligoceras: a,* light micrograph of filaments in situ, some with dividing cells *(arrow); b,* TEM section of trichome; *c,* thylacoid arrangement in longitudinal section, one cell (top) cut tangentially; *d,* near-median longitudinal section showing cross wall and cell inclusions (note clear space separating symbiont from spongin fibrils, top left); *e,* cell wall and layered sheath; *f,* stellar body *(arrow).*

cytoplasm. They are interpreted as polyhedral bodies, but they are rare and not as distinctive as in other cyanobacteria examined. Rod-shaped polyglucan granules (50 × 15 nm) are abundant between the thylakoids.

Many large circular clear spaces (4–50 or more per cell section) are present in the cytoplasmic region of each cell

(Figure 3d). These spaces measure 300–800 nm (average 600 nm) in diameter, have the same electron transparency as the intrathylakoid spaces, and are bound by membranes strongly resembling thylacoid membranes. They were first interpreted as bubblelike inflations of the photosynthetic vesicles, but a connection with a thylacoid was established in only one of the hundreds of these structures examined. As a rule, thylacoids approaching the bubble vesicle narrow to a point and terminate right at the bubble membrane, just as they reach the plasmalemma near the cell or cross walls. Unlike the intrathylakoid spaces, the lumina of many bubble vesicles contain inclusions, some of which consist of medium gray granules that almost fill the lumen or are smaller, occasionally occur in pairs, and may be with or without fibrous or loosely flocculent material. On some images the granules seem to consist of interwoven fibrils, on others they appear homogenous (Figure 3d).

A small number of star-shaped inclusions (260–320 nm) were observed in the peripheral region of only three cells (Figure 3f). They consist of a bright central region and about 18 thin electron dense rays. These structures correspond to the stellar bodies described by Berthold et al. (1982).

This filamentous symbiont is always in extracellular position. Occasionally, sponge cells are grouped closely or extend pseudopodia around the symbionts, but no direct interaction with sponge cells was seen. Spongin fibrils of the mesohyle always maintain a some distance from the cyanobacterial filaments (Figure 3d).

ASSOCIATIONS WITH UNICELLULAR EUCARYOTIC ALGAE

Two species or groups of closely related species of greenish-brownish-blackish clionid excavating sponges (Hadromerida, genus *Cliona*, including *Anthosigmella*) are associated with the symbionts commonly referred to as zooxanthellae. The symbionts are concentrated in the outer sponge zones (in the papillae of alpha-stage forms) where they may attain 50% of the cell biomass. Algae from *C. caribbaea* Carter (Figure 4a) are morphologically the same as zooxanthellae from *Cliona* species (Vacelet, 1981) and from many other invertebrates, such as protozoans, cnidarians, and mollusks (Taylor, 1968, 1974; Bishop et al., 1976; Deane and O'Brien, 1978; Schoenberg and Trench, 1980; Tripodi and Santisi, 1982). They belong to the dinophycean species *Gymnodinium microadriaticum* (Freudenthal). Fully grown cells are spherical and measure 8–9 μm in diameter, which increases to about 11 × 9 μm just before division by (binary fission). Freshly separated daughter cells are oval, approximately 8 × 6 μm. Zooxanthellae from *Cliona* (= *Anthosigmella*) *varians* (Duchassaing and Michelotti) are similar in appearance but smaller, 3.5–4.0 × 4.0–5.0 μm (Figure 4b,c). All are intracellular, either fully embedded in a host archaeocyte vacuole or encircled by host cell filopodia.

The zooxanthellae of *Cliona caribbaea* from Bermuda and *C. varians* from Belize have many of the same structural features. Both have a two-membrane (new daughter cells) to five-membrane periplast and a single pyrenoid attached to the chloroplast by one or (rarely) two stems but free of thylakoids (Figure 4d). They also have an oval nucleolate nucleus, displaying chromosomes and a double nuclear envelope penetrated by pores, and a prominent branching chloroplast curved along the perimeter of the cell. Both have the usual inclusions, such as accumulation body, ovoid starch grains, and angular calcium oxalate crystals. The two algae differ primarily in the structural details of the nucleus and chloroplast. A section of the *C. caribbaea* symbiont shows nucleus to have 13–25 chromosome sections that are fine-fibrillar, tightly coiled structures, 0.3–0.6 μm in diameter (Figure 4a). In the *C. varians* zooxanthella, fewer than 10 chromosomes are visible per nuclear sections and the fibrils appear coarse and only loosely coiled (Figure 4b). The chloroplast in the *C. caribbaea* alga is traversed by parallel and closely spaced lamellae of three opposed thylakoids along its entire length and it lacks inclusions (Figure 4a,d). The same structure in the *C. varians* symbiont shows lamellar bands of various lengths (0.1–1.6 μm), which occur in loose arrangement and usually in various directions, and in places includes electron-dense granules of unknown origin (Figure 4b,c). It should be noted that mitochondria with tubular cristae, Golgi stacks, endoplasmatic reticulum, and some unidentified vesicles occur in both algal forms but are particularly prominent in the *C. varians* symbionts. No flagellae were found, nor evidence such as kinetosomes or centrioles, that would indicate a rudimental flagellar apparatus.

SPONGIN-PERMEATING FILAMENTOUS ALGAE

An unusual association was found in the spongin fiber skeleton of *Mycale laxissima* (Duchassaing and Michelotti) collected on shallow patch reefs in Belize. The rigid fibers here are 300 μm in diameter but fused in places to form 2–3 mm thick strands and are cored by staggered bundles of robust subtylostyle spicules, which occupy 30–50% of the fiber crossection. The fresh fibers are usually almost colorless, in contrast to the deep purple to blackish cellular tissue. At this location, however, most specimens have dark reddish to greenish fibers owing to the filamentous branching algae (at least two types) that are densely intertwined and fully embedded in the spongin strands (Figure 5a).

The algae have been identified as *Ostreobium* cf. *constrictum* Lucas (Chlorophyta) in the green fibers (Figure 5b,d) and *Acrochaetium spongicolum* Weber-van Bosse (Rhodophyta) in the red (M. J. Wynne, S. Fredricq; pers. comm.). The cells of *Ostreobium* measure 12–15 × 7–13 μm, those of *Acrochaetium* 8–21 × 3–9 μm; the latter are characterized by intercellular pit connections (Figure 5c). The filaments of both algae follow the direction of spongin layering of the host skeleton. On electron micrographs there is evidence of physical separation (tearing) of the

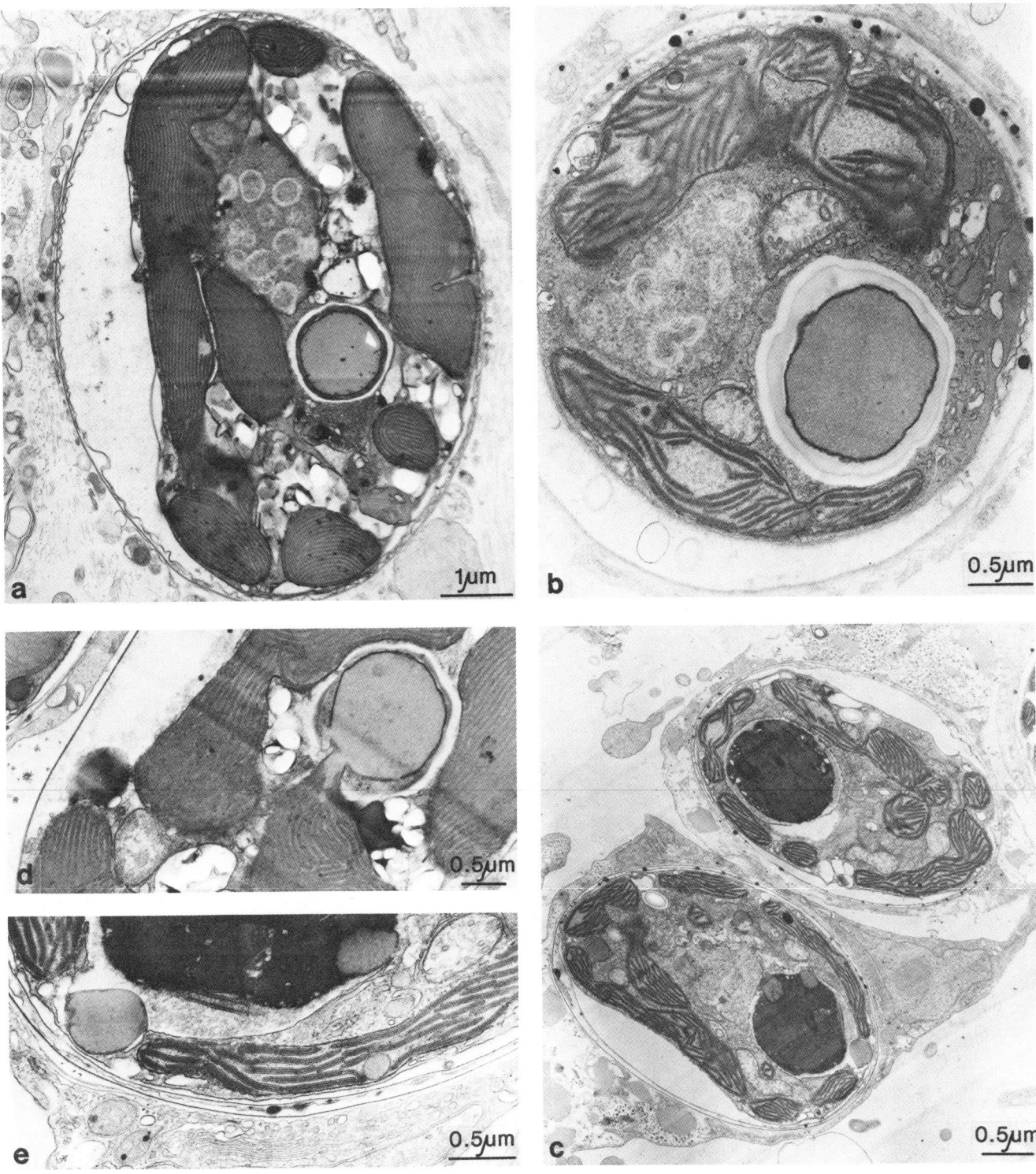

Figure 4. Zooxanthellae, *Gymnodinium*, inside sponge hosts: *a*, *G. microadriaticum* of *Cliona caribbaea*; *b,c*, *Gymnodinium* sp. of *C. varians* (note sponge cell enveloping recently separated daughter zooxanthellae in *c*); *d*, pyrenoid with stem attached to chloroplast, *G. microadriaticum*; *e*, chloroplast and accumulation body *(black)*, *Gymnodinium* sp.

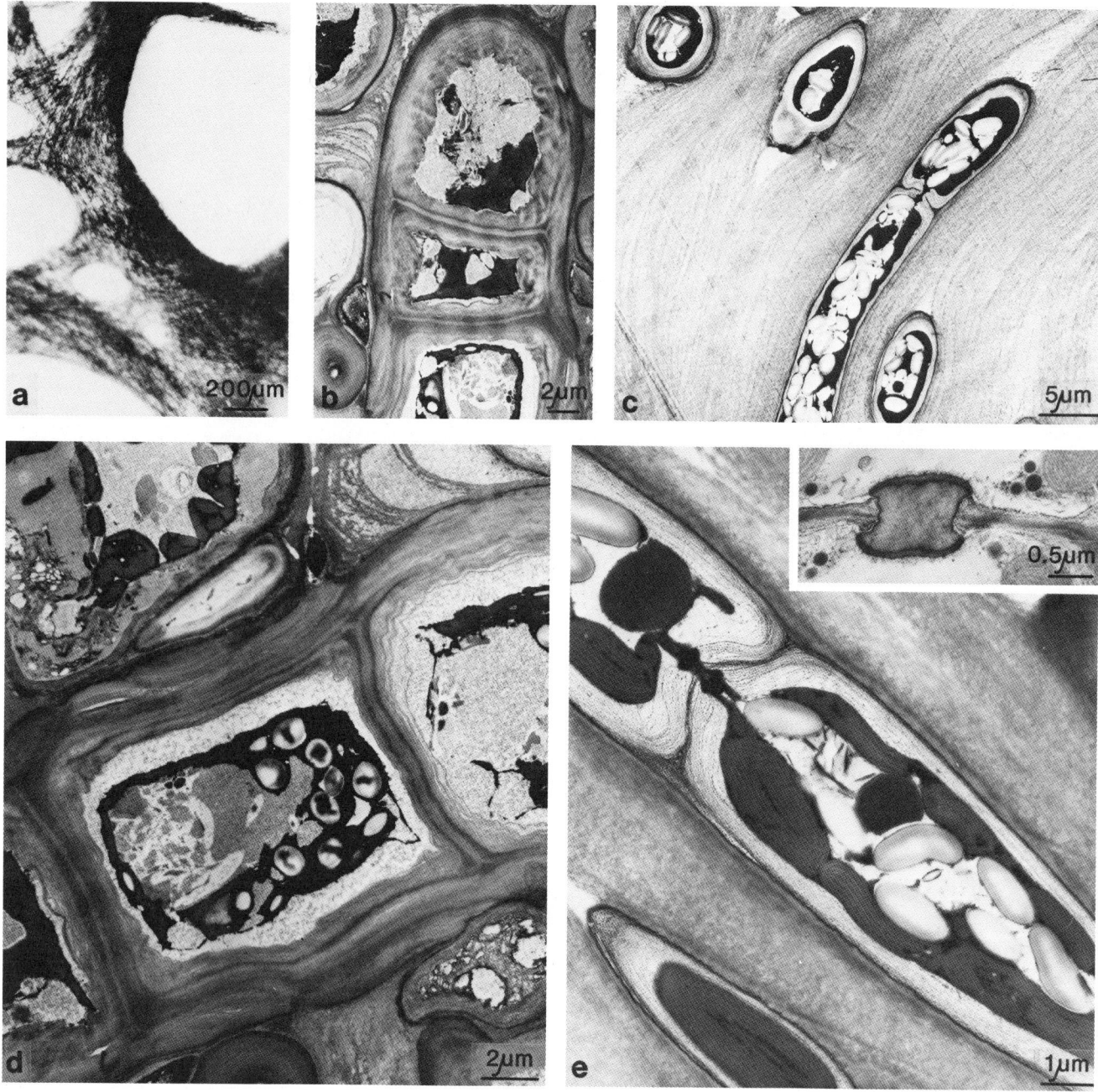

Figure 5. Spongin-permeating algae in *Mycale: a*, light micrograph of isolated spongin fibers riddled by algal filaments; *b*, longitudinal TEM section of filament tip of *Ostreobium* (note spongin layering left of top cell); *c*, sections of *Acrochaetium* embedded in spongin of *Mycale* fiber (note separation of spongin fibrils caused by algae); *d*, enlarged view of *Ostreobium* cell; *e*, enlarged view of *Acrochaetium* cell (insert: intercellular plug).

spongin layers by the growing algae (Figure 5*c*), but none of chemical dissolution (etching). On the other hand, spongin layering around some algal cells (Figure 5*b*) suggests that spongin was deposited onto the cell walls during simultaneous growth of sponge and alga, which is possible at the distal tips of the fibers that cause the conules on the sponge surface.

ALGAE AS SKELETAL SUPPORT

Two sponge species from the (sub-) tropical western Atlantic are known to substitute or reinforce their own skeleton by association with algae, the calcified Rhodophyta genus *Jania* (Figure 6). Both were found and studied in Bermuda. *Dysidea janiae* (Duchassaing and Michelotti) is

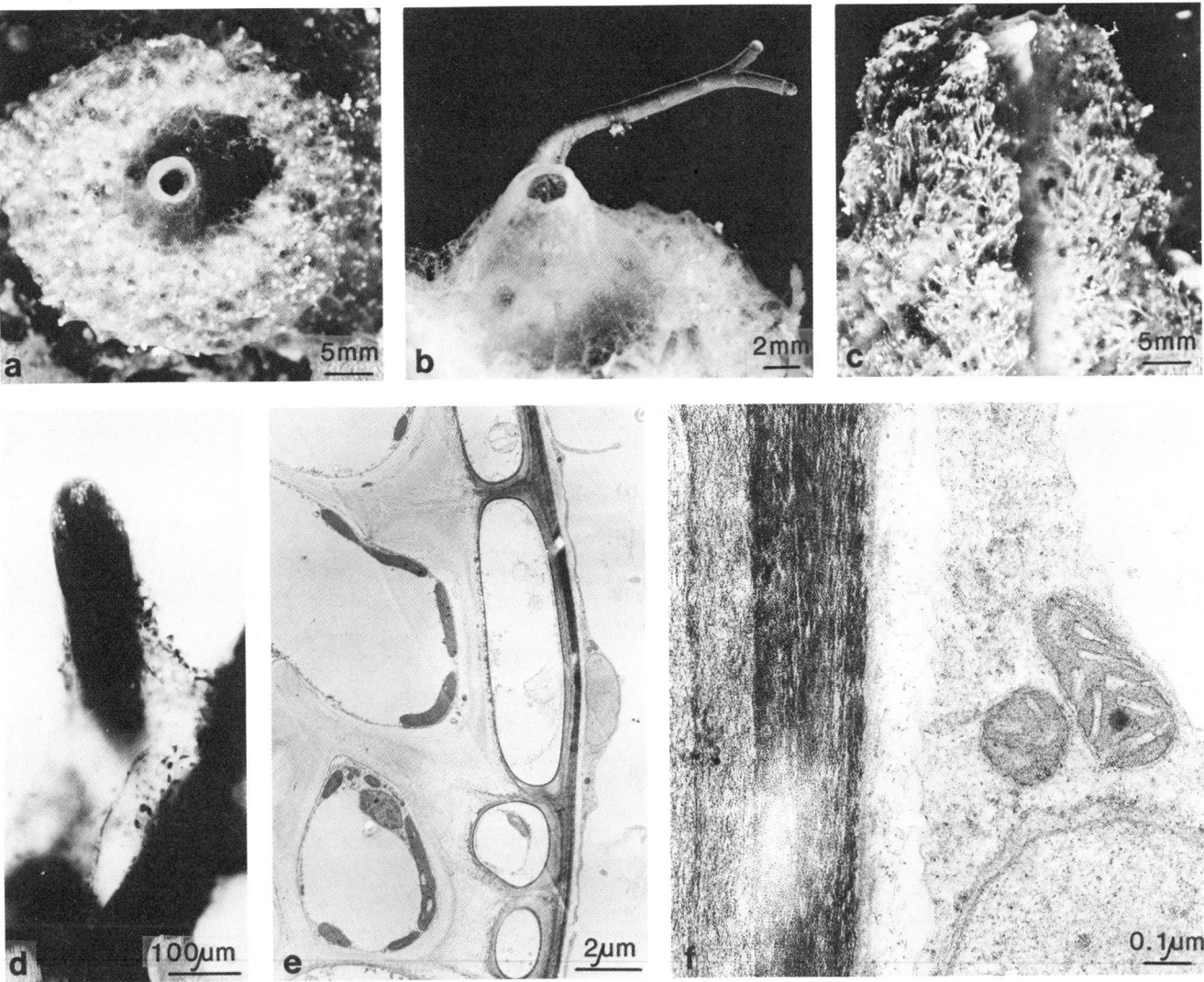

Figure 6. *Jania* as skeletal support: *a*, in *Dysidea janiae*, view of sponge chimney from above (osculum in center), algal growth tips appear as white dots; *b*, similar view as in *a* but one algal branch protruding from sponge; *c*, cutaway view of *Jania*-supported chimney of *Xystopsues* (principal exhalent canal in center leading to collapsed oscular cone on top); *d*, light micrograph of *Jania* branch enveloped by loosely organized cellular tissue of *Dysidea; e*, TEM cross section through *Jania* branch (chloroplasts and nucleus inside cells to left) coated by *Dysidea* pinacocyte *(right); f*, spongin-a fibrils in clear space between outer wall of *Jania* (blackish, striated), left, and *Dysidea* pinacocyte (with two mitochondria), right.

a keratose (Dictyoceratida) sponge that incorporates *J. adherens* Lamouroux instead of developing characteristic sediment-charged skeleton fibers. *Xytopsues osburnensis* (George and Wilson) is a myxillid (Poecilosclerida) full of the symbiont *J. capillacea* Harvey (W. Johansen, pers. comm.).

Dysidea janiae forms mounds or tubes 1–3 cm in diameter and 1–7 cm tall. The tissue is grayish tan but translucent, so the sponge takes the color of the underlying *Jania* algae, which varies from purplish red at the growth tips to greenish and whitish toward the base, which is often dead. The erect, branching algae have an average diameter of 100 μm. They constitute 60% or more of the sponge body and are fully contained by it. In other words, the sponge

determines the shape of this compound organism, and algal growth stops just below the level of the ecto-pinacoderm; very rarely does a single algal branch break this pattern and protrude beyond the sponge surface. Sponge cells attach directly to the surface of algal branches, except for a 200 nm space filled with collagen (spongin-a) fibrils. None of the typical sand-filled spongin-b fibers of *Dysidea* were seen in the sections. This sponge grows among *Jania* turfs and has not been found without the endozoic algae in Bermuda or elsewhere.

Xytopsues osburnensis forms irregular masses 5–10 cm in diameter. Color, anatomy, and histology are much the same as in *Dysidea*, except that the *Jania* branches are more brownish red and slightly thinner (80 μm in diameter), and the sponge develops its typical fiber strands with embedded spicules and sand grains. Judging from its pigmentation, *J. capillacea* seems healthier and more vigorous in its endozoic state than *J. adherens* of *Dysidea*, which is generally alive only in the distal part of its branches.

Discussion

Small (3 μm and less) ovoid chroococcoids are the most common photosynthetic associates of sponges in the Caribbean, just as they are in the Mediterranean Sea (Sará, 1966, 1971; Vacelet, 1971) and the Pacific Ocean (Wilkinson, 1978; Vacelet, 1981). Host sponges in the different oceans generally belong to the same families or even genera—for instance, *Chondrilla, Ircinia, Aplysina,* closely related petrosiids *(Petrosia, Calyx, Cribochalina, Xestospongia),* and *Neofibularia.* The systematic position of this cyanobacterium was discussed by Vacelet (1971) but is still not resolved because taxonomic revisions of the group have not taken into account the details of its fine structure. Sará and Liaci (1964) and several other authors assumed that the organism is identical with that from Mediterranean *Ircinia* and *Petrosia,* as described by Feldmann (1933) and therefore named it *Aphanocapsa feldmanni* Fremy. More recently, *Cyanothece* Komarek has been suggested as a more suitable generic allocation (Lafargue and Duclaux, 1979). Bacterial infection of symbiotic cyanobacteria has previously been observed in the host sponges *Neofibularia* and *Jaspis* from the Great Barrier Reef (Wilkinson, 1979a). This parasite is comparable in structure and size to the Australian organism classified as a bdellovibrio.

Large (5 μm or more), spherical, unicellular cyanobacteria from sponges are much less common than the *feldmanni* types, and published descriptions give few details. TEM micrographs were only taken on tissues of the Mediterranean keratose sponges *Ircinia* (Sarà, 1971) and *Aplysilla* (Duclaux, 1972). A Caribbean symbiont of this type described by Rützler (1981) is quite similar in size and structure but constitues as much as half of the biomass of the host sponge *Ulosa* (possibly better classified as genus *Dictyonella;* van Soest, pers. comm.). Another Caribbean cyanobacterial symbiont, described by Lafargue and Duclaux (1979) as *Synechocystis trididemni* from a didemnid ascidean, may be conspecific, but published TEM photomicrographs do not allow a positive identification. There may be more than one cyanobacterial species involved, although different fixation and processing techniques could account for the differences in structural details. On the other hand, we have evidence (Rützler and Muzik, unpublished) that crudely fixed and stored material may still retain taxonomically important details of the fine structure. Large (2-cm) chunks of coral rock coated by an undescribed encrusting sponge, *Terpios* sp., from Japan were fixed and stored in 2% cacodylate-buffered glutaraldehyde for more than three weeks before TEM

processing. Even so, the TEM images show cyanobacteria structurally undistinguishable from the *Ulosa* symbiont in the Caribbean. Similarly, Cox et al. (1985) show TEM images of a symbiont from Great Barrier ascidians and sponges that closely resemble those of the *Ulosa* associates and identify it as *S. trididemni.* Again, without modern revisions, one needs to be careful about the use of taxonomic designations. This kind of sponge symbiont was first described as *Aphanocapsa raspaigellae* (Hauck) by Feldmann (1933). Lafargue and Duclaux (1979) would use the genus *Synechocystis* Sauvageau for this species, and for their new *S. trididemni,* although they admit to never having observed subsequent cell division in perpendicular planes, the distinguishing feature of the genus.

Filamentous cyanobacteria have previously been found as sponge symbionts in the Mediterranean (Feldmann, 1933, Sarà, 1966), Red Sea (Wilkinson, 1979b), the Pacific (Vacelet, 1981; Berthold et al., 1982; Larkum et al., 1987), as well as the Caribbean (Wilson, 1902). Wilson's observations were made on his *Cacospongia (= Oligoceras) spongeliformis,* a species close to *Oligoceras violacea (= O. hemorrhages* de Laubenfels of authors). Only *Dysidea herbacea* Keller from the Pacific seems to harbor symbiont in close to the great numbers numbers found in *O. violacea* (30–40% of the sponge cell volume; Berthold et al., 1982). The systematic postion of the cyanobacterium is, as in the cases above, still open to question. *Phormidium* or *Oscillatoria spongeliae* (Schulze) have often been used in the literature because this organism was first observed in the 1870s in Mediterranian *Spongelia (= Dysidea).* The generic allocation varied, depending on whether the author detetected a sheath coating the trichome. However, other important taxonomic criteria—such as the number of cells per trichome, trichome branching, and presence of hormogonia and necridia—vary from one author to another, which suggests that different species may be involved. Morphologically, even with respect to the fine structure, the *O. violacea* symbiont compares best with recently described Pacific forms (Vacelet, 1981; Berthold, 1982; Larkum et al., 1987). The most significant difference seems to be that the Caribbean organism displays a distinctive, layered sheath.

It is curious that the classical "zooxanthellae" known to be common symbionts of many carbonate-secreting organisms such as corals occur only with limestone-excavating sponges, the clionids. This phenomenon was recently discussed by Vacelet (1981), who also noted that the zooxanthellae-bearing Pacific sponge *Spirastrella inconstans* should actually be classified as *Cliona.* This is a parallel case to our present observations on *Anthosigmella varians,* which has always been considered a spirastrellid. The delicate type and relative rarity of microscleres and the boring habit of early stages suggest that this sponge is a typical member of Clionidae. Observations on the fine structure of the *C. caribbaea* symbiont *Gymnodinium micro-*

adriaticum agree with Vacelet's (1981) findings for Pacific sponges. In particularly, the algae are alway located in intracellular postion, the chloroplast is denser than in cnidarians or in *Tridacna*, and flagellar structures are never seen. Structural differences in the nucleus and chloroplast of *G. microadriaticum* and the symbiont of *C. varians* are significant and indicate systematic differences on the species level or above. Some features of *G. simplex* (Lohmann) in TEM photomicrographs published by Dodge (1974) are similar to *C. varians* zooxanthellae, particularly the chloroplast structure, but the former is only known as a free-living form. The isolated alga will have to be cultured and more detailed taxomonic study before systematic relationships can be clarified.

It is remarkable to find algae flourishing inside the spongin skeleton of a densely pigmented, dark purple sponge such as *Mycale laxissima* at a water depth exceeding 5 meters. However, the body of this tubular species is fairly thin (10 to 20 mm), and light can reach the skeleton from both the outer and inner (atrial) surfaces. Both algae, as well as some close relatives, have been reported from similar cryptic habitats and even from sponge fibers. At least six *Ostreobium* species are known to occur as boring or endolithic algae in limestone substrates such as corals, mollusk shells, and calcified algae (Lukas, 1974). One member of the genus was once found with *Acrochaetium spongicolum* inside the skeleton of an unidentified keratose sponge in the Pacific (Weber-van Bosse, 1921). The photosynthetic efficiency of one of these algae (endolithic in the coral *Porites* in Hawai) was found to reach its peak at light levels between 10 and 100 Lux (Franzisket, 1968). Illumination at the main growth site of the alga, 10 mm below the coral surface, was measured at 15 Lux, which was only 0.03% of ambient light at the habitat of the host coral in a depth of 2 m. *Acrochaetium spongicolum* was first described from a keratose sponge fiber, as mentioned above. A similar form *(Rodochorton)* was subsequently reported from the skeletons of Mediterranean Keratosa, genera *Spongia, Cacospongia, Dysidea,* and *Aplysilla* (Sarà, 1966). Among our own samples that were not studied by TEM are *Acrochaetium*, such as *Igernella* from Belize, *Pleraplysilla* from Bermuda, *Niphates* from the U.S. Virgin Islands, and *Callyspongia* from the eastern Pacific (Chile; courtesy R. Desqueyroux). Some authors consider *Audouinella* to be the principal genus of this complex of filamentous Rhodophyta, but taxonomic studies are still inconclusive (Woelkerling, 1983).

Both cases of symbiosis with *Jania* algae were previously described from Bermuda (de Laubenfels, 1950: *Dysidea* as *D. fragilis* f. *algafera*, p. 22; *Xytopsues* as *X. griseus*, p. 75). *Dysidea janiae* is a species characterized by its obligatory relationship with *Jania*, and not just an algal-infested form, as de Laubenfels (1950) assumed. On the other hand, *Xytopsues osburnensis*, is known from its type locality off North Carolina as a thin encrustation without algal

symbionts and with proper skeleton fibers filled with sand grains (George and Wilson, 1919). Similar associations have been described from Pacific waters (e.g., *Gellius-Ceratodictyon*: Vacelet, 1981; Price et al., 1984), but there the algae determine the shape of the compound organism, whereas in our case the opposite occurs. It can be assumed that microclimatic conditions inside the sponge, a balance of growth-stimulating and impeding factors, control algal growth, but no details of such mechanisms are known. *Jania* is obviously capable of living without a host, unlike the Pacific *Ceratodictyon* symbiont, which is unable to survive by itself (Price et al., 1984).

Conclusions

Symbioses between shallow-water sponges and photosynthetic organisms are just as common and varied in the Caribbean as in other tropical and subtropical seas. More experimental work is needed to determine the benefits of the symbiotic life style for each partner, but numerous studies confirm that sponges receive phototrophically produced metabolic energy from associated cyanobacteria or algae (Vacelet, 1971; Wilkinson, 1979b; Wilkinson and Fay, 1979; Rützler, 1981; Wilkinson, 1983) and that phototrophic symbionts use animal waste for nutrients and a habitat protected from grazers (de Laubenfels, 1950; Price et al., 1984). An ecologicaly less significant but curious phenomenon is the skeleton-support function provided by certain algae. An interesting question is whether any of these associations could be harmful to one partner or could lead to parasitism. In one case under study, chroococcoid symbionts are able to damage the *Geodia* host tissue (Rützler, 1988). In addition, spongin-boring filamentous algae as seen in *Mycale laxissima* fibers could be considered parasitic, as they doubtlessly damage the skeleton needed to support this sponge.

Acknowledgments

Algal identifications and valuable suggestions were provided by M. J. Wynne, S. Fredricq, and W. Johansen. I thank P. Riordan and C. Rützler for TEM sectioning, R. Summers for TEM instructions, and K. Smith for field assistance and darkroom work. The Bermuda Biological Station and Philips Company provided invaluable assistance by allowing me to use their facilities. Financial support for this study was provided by the Smithsonian Fluid Research Fund, the Smithsonian Research Opportunities Funds, and a grant from the Exxon Corporation (for work in Belize). This is contribution no. 178, Caribbean Coral Reef Ecosystems (CCRE) program, National Museum of Natural History, Washington, D.C.

Literature Cited

Berthold, R. J., M. A. Borowitzka, and M. A. Mackay. 1982. The Ultrastructure of *Oscillatoria spongeliae*, the Blue-Green Algal

Symbiont of the Sponge *Dysidea herbacea*. *Phycologia*, 21:327–335.

Bishop, D. G., J. M. Bain, and W. J. S. Downton. 1976. Ultrastructure and Lipid Composition of Zooxanthellae from *Tridacna maxima*. *Australian Journal of Plant Physiology*, 3:33–40.

Cox, G. C., R. G. Hiller, and A. W. D. Larkum. 1985. An Unusual Cyanophyte, Containing Phycourobilin and Symbiotic with Ascidians and Sponges. *Marine Biology*, 89:149–163.

Deane, E. M., and R. W. O'Brien. 1978. Isolation and Axenic Culture of *Gymnodinium microadriaticum* from *Tridacna maxima*. *British Phycological Journal*, 13:189–195.

Dodge, J. D. 1974. A Rediscription of the Dinoflagellate *Gymnodinium simplex* with the Aid of Electron Microscopy. *Journal, Marine Biological Association of the United Kingdom*, 54:171–177.

Duclaux, G. 1972. Quelques données d'ultrastructure de zoocyanelles. I. Mise au point à propos de la paroi des Cyanophycées. *Botaniste*, 1972:251–271.

Feldmann, J. 1933. Sur quelques cyanophycées vivant dans le tissu des éponges de Banyuls. *Archives de Zoologie Expérimentale at Générale*, 75:331–404.

Franzisket, L. 1968. Zur Ökologie der Fadenalgen im Skelett lebender Riffkorallen. *Zoologische Jahrbücher, Physiologie*, 74:246–253.

Gaino, E., M. Pansini, and R. Pronzato. 1977. Aspetti dell' associazione tra *Chondrilla nucula* Schmidt (Demospongiae) e microorganismi simbionti (Batteri e Cianoficee) in condizioni naturali e sperimentali. *Cahiers de Biologie Marine*, 18:303–310.

George, W. C., and H. V. Wilson. 1919. Sponges of Beaufort (N.C.) Harbor and Vicinity. *Bulletin, Bureau of Fisheries*, 36:129–179.

Lafargue, F., and G. Duclaux 1979. Premiere exemple, en Atlantique tropical, d'une association symbiotique entre une ascidie Didemnidae et cyanophycée chroococcale: *Trididemnum cyanophorum* nov. sp. et *Synechocystis trididemni* nov. sp. *Annales Institut Oceanographique*, 55:163–184.

Larkum, A. W. D., G. C. Cox, R. G. Hiller, D. L. Parry, and T. P. Dibbayawan. 1987. Filamentous Cyanophytes Containing Phycourbilin and in Symbiosis with Sponges and an Ascidian of Coral Reef. *Marine Biology*, 95:1–13.

Laubenfels, M. W. de. 1950. The Porifera of the Bermuda Archipelago. *Transactions, Zoological Society of London*, 27:1–154.

Lukas, K. J. 1974. Two Species of the Chlorophyte Genus *Ostreobium* from Skeletons of Atlantic and Caribbean Reef Corals. *Journal of Phycology*, 10:331–335.

Price, I. R., R. L. Fricker, and C. R. Wilkinson. 1984. *Ceratodictyon spongiosum* (Rhodophyta), the Macroalgal Partner in an Alga-Sponge Symbiosis, Grown in Unialgal Culture. *Journal of Phycology*, 20:156–158.

Rützler, K. 1981. An Unusual Bluegreen Alga Symbiotic with Two New Species of *Ulosa* (Porifera: Hymeniacidonidae) from Carrie Bow Cay, Belize. *Marine Ecology*, 2:35–50.

———. 1988. Mangrove Sponge Disease Induced by Cyanobacteroa; Symbionts: Failure of a Primmitive Immune System? *Diseases of Aquatic Organism*, 5:143–149.

Sarà, M. 1966. Associazioni fra Poriferi e alghe in acque supericiali del litorale marino. *Ricerca Scientifica (Rome)*, 36:277–282.

———. 1971. Ultrastructural Aspects of the Symbiosis between Two Species of the Genus *Aphanocapsa* (Cyanophyceae) and *Ircinia variabilis* (Demospongiae). *Marine Biology*, 11:214–221.

Sarà, M., and L. Liaci. 1964. Associazione fra la Cianoficea *Aphanacapsa feldmanni* e alcune Demospongie marine. *Bollettino di Zoologia*, 31:55–65.

Sarà, M., and J. Vacelet. 1973. Ecologie des Demosponges. Pages 462–576 in *Traité de Zoologie 3(1), Spongiaires*, edited by P.-P. Grassé. Paris: Masson.

Schoenberg, D. A., and R. K. Trench. 1980. Genetic Variation in *Symbiodinium* (= *Gymnodinium*) *microadriaticum* Freudenthal, and Specificity in its Symbiosis with Marine Invertebrates. II. Morphological Variation in *Symbiodinium microadriaticum*. *Proceedings, Royal Society of London, B*, 207:429–444.

Taylor, D. L. 1968. In situ Studies on the Cytochemistry and Ultrastructure of a symbiotic Marine Dinoflagellate. *Journal, Marine Biological Association, United Kingdom*, 48:349–366.

———. 1974. Symbiotic Marine Algae: Taxonomy and Biological Fitness. Pages 245–262 in *Symbiosis in the Sea*, edited by W. B. Vernberg. Columbia, South Carolina: University of South Carolina Press.

Tripodi, G., and S. Santisi. 1982. A study on the Cell Covering of *Symbiodinium*, a Symbiote of the Octocoral *Eunicella*. *Journal of Submicroscopical Cytology*, 14:613–620.

Vacelet, J. 1971. Etude en microscopie électronique de l'association entre une cyanophycee chroococcale et une éponge du genre *Verongia*. *Journal de Microscopie*, 12:363–380.

———. 1981. Algal-Sponge Symbioses in the Coral Reefs of New Caledonia: A Morphological Study. Pages 713–719 in *The Reef and Man*, edited by E. D. Gomez et al. Proceedings, Fourth International Coral Reef Symposium, Manila, 2. Quezon City, Philippines: University of the Philippines.

Weber-van Bosse, A. 1921. Liste des algues du Siboga, II. Rhodophyceae. Pages 184–310, *Siboga Expedition, 59b*. E. J. Brill: Leiden.

Wilkinson, C. R. 1978. Microbial Associations in Sponges. III. Ultrastructure of the in situ Associations in Coral Reef Sponges. *Marine Biology*, 49:177–185.

———. 1979a. *Bdellovibrio*-Like Parasite of Cyanobacteria Symbiotic in Marine Sponges. *Archives of Microbiology*, 123:101–103.

———. 1979b. Nutrient Translocation from Symbiotic Cyanobacteria to Coral Reef Sponges. Page 373–380 in *Biologie des Spongiaires*, edited by C. Lévi and N. Boury-Esnault. Colloques Internationaux du C.N.R.S. 291. Paris: Centre National de Recherche Scientifique.

———. 1980. Cyanobacteria Symbiotic in Marine Sponges. Pages 553–563 in *Endocytobiology, Endosymbiosis and Cell Biology 1*, edited by W. Schwemmler and H. E. A. Schenk. Berlin and New York: Walter de Gruyter.

———. 1983. Net Primary Productivity in Coral Reef Sponges. *Science*, 219:410–411.

Wilkinson, C. R., and P. Fay. 1979. Nitrogen Fixation in Coral Reef Sponges with Symbiotic Cyanobacteria. *Nature*, 279:527–529.

Wilson, H. V. 1902. The Sponges Collected in Porto Rico in 1899 by the U.S. Fish Commission Steamer Fish Hawk. *Bulletin, U.S. Fish Commission*, 20:375–411.

Woelkerling, W. J. 1983. The *Audouinella (Acrochaetium-Rhodochorton)* Complex (Rhodophyta): Present Perspectives. *Phycologia*, 22:59–92.

YOSHIKI MASUDA
Department of Biology
Kawasaki Medical School
Kurashiki 701–01, Japan

Electron Microscopic Study on the Zoochlorellae of Some Freshwater Sponges

Abstract

Electron microscopy of some green freshwater sponges, *Spongilla lacustris*, *Radiospongilla cerebellata*, *R. sendai*, and *Heteromeyenia stepanowii* has revealed that zoochlorellae are intracellular and reside in membrane-bound host vacuoles. In the green adult (vegetative) tissue of all four species, many zoochlorellae were observed in the archeocytes and some of theses were in division. In the green gemmules of *S. lacustris*, many zoochlorellae were observed in the thesocytes and none were in division. In the brown gemmules of *R. cerebellata* and *R. sendai*, thesocytes contained considerably fewer zoochlorellae than archeocytes, and the zoochlorellae were not in division. Zoochlorellae in the archeocytes of *S. lacustris*, *R. sendai*, and *H. stepanowii* formed two autospores in division, whereas those of *R. cerebellata* formed four autospores. A pyrenoid was observed only in each zoochlorella of archeocytes, spongocytes, and young thesocytes of *R. cerebellata*. Zoochlorellae of the other species lacked pyrenoids at any stage. Zoochlorellae in the thesocytes of *R. cerebellata* looked much different from those in the archeocytes as they possessed degenerated chloroplasts that lacked arrayed stacks of thylakoids, starch grains, and pyrenoids. In the young archeocytes hatched from *R. cerebellata*, zoochlorellae possessed regenerated chloroplasts with arrayed stacks of thyrakoids, starch grains and pyrenoids. The zoochlorellae of the four species studied appear to fall into at least two large groups, which are defined by the presence of a pyrenoid and the mode of reproduction.

Zoochlorellae occur as intracellular symbionts in several invertebrate phyla, including the Protozoa, Porifera, Coelenterata, and Platyhelminthes. Some species of green freshwater sponges contain zoochlorellae (Van Tright, 1919; Gilbert and Allen, 1973; Wilkinson, 1980), but only a few studies have examined the ultrastructure of zoochlorellae, particularly of those in gemmules. Williamson (1979) has reported that zoochlorellae in *Spongilla lacustris* actively reproduce in adult green sponges but not in adult white sponges or gemmules and that pyrenoids are lacking in all zoochlorellae. In addition, Masuda (1985) has observed that the zoochlorellae in the archeocytes of *Radiospongilla cerebellata* resemble those of *Paramecium bursaria* and *Hydra viridis* in that both have a pyrenoid and their mode of reproduction is similar. In the gemmules of this species, the zoochlorellae looked much different from those in the archeocytes. Each zoochlorella had lost the normal chloroplast with arrayed stacks of thyrakoids, starch grains, and pyrenoid.

Materials and Methods

The population of *Spongilla lacustris* investigated was located in a pond near Lake Biwa in Shiga prefecture, Japan. Populations of *Radiospongilla cerebellata*, *R. sendai*, and *Heteromeyenia stepanowii* were located in ponds in Okayama prefecture. The green adult sponges and gemmules of four species were collected in summer and autumn, respectively. Young sponges of *R. cerebellata* were hatched from gemmules that had been stored in a refrigerator after collecting.

Before fixation, a fine slice was made through a portion of the gemmule coat with a double-edged razor blade to facilitate infiltration of fixatives and embedding media. All samples were first fixed in a 1:1 mixture of 2% glutaraldehyde in 0.1 M phosphate buffer (pH 7.3) and 2% paraformaldehyde in the same buffer and then in 1% osmium tetroxide in 0.1 M phosphate buffer. After the fixation, they were dehydrated in an ethanol series and embedded in epoxy resin. The sections were doubly stained with uranyl acetate and lead citrate solution and then examined under a Hitachi HS-9 electron microscope.

Results

Spongilla lacustris (Linnaeus).   Many zoochlorellae were found intracellularly within archeocytes (Figure 1). Each zoochlorella in the archeocyte was in an individual vacuole except during division. Some starch grains were found in the chloroplasts of the zoochlorellae in the archeocytes, but these were lacking in pyrenoids. They consistently formed two autospores in reproduction. Many zoochlorellae also occurred intracellularly within thesocytes (Figure 2). They contained some polyphosphate granules, lipid storage bodies, and chloroplasts which were lacking starch grains (Figure 3).

Radiospongilla sendai (Sasaki).   Many zoochlorellae were found intracellularly within archeocytes, as in *Spongilla lacustris* (Figure 4). The structures and dimensions of zoo-

chlorellae in the archeocytes closely resembled those of *S. lacustris*. They formed two autospores in reproduction. At early stages of gemmule formation, zoochlorellae in groups of several individuals were often located in one vacuole of each archeocyte (Figure 5). The number of zoochlorellae in one thesocyte was much lower than the number in one archeocyte. The structure and dimensions

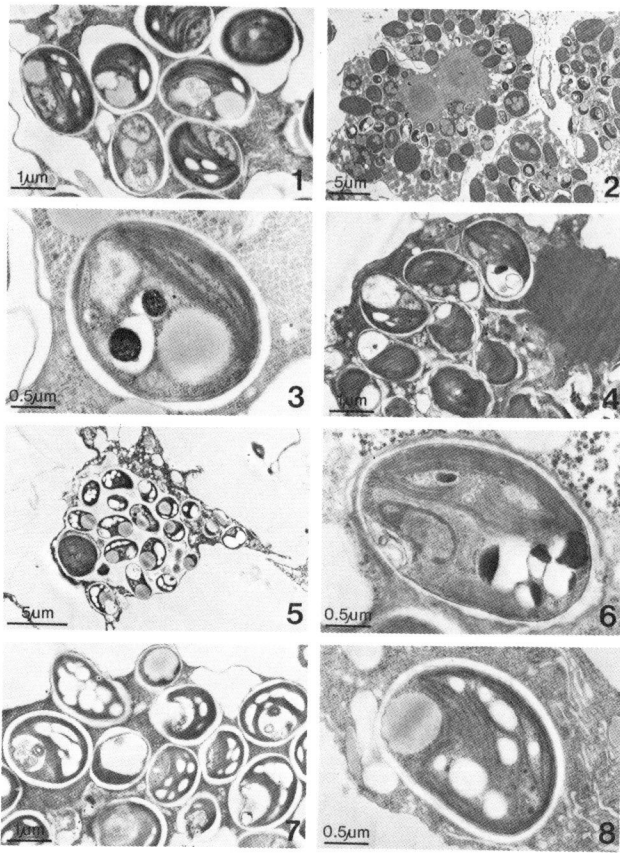

Figures 1–8. Electron micrographs of zoochlorellae in freshwater sponges: *1, Spongilla lacustris*, zoochlorellae in archeocyte of green adult sponge (vegetative tissue); each zoochlorella located in individual vacuole; *2, Spongilla lacustris*, thesocytes of green gemmule containing many zoochlorellae; each zoochlorella located in individual vacuole; *3, Spongilla lacustris*, a zoochlorella located in vacuole of thesocyte; chloroplast possesses closely arrayed stacks of thylakoids and lacks starch grains; *4, Radiospongilla sendai*, zoochlorellae in archeocyte of green adult sponge (vegetative tissue); each zoochlorella is located in individual vacuole; *5, Radiospongilla sendai*, zoochlorellae in groups of several individuals were often located in one vacuole on gemmulation; *6, Radiospongilla sendai*, zoochlorella located in vacuole of thesocyte; chloroplast possesses closely arrayed stacks of thylakoids and lacks starch grains; *7, Heteromeyenia stepanowii*, zoochlorellae in archeocyte of green adult sponge (vegetative tissue), each zoochlorella is located in individual vacuole; *8, Heteromeyenia stepanowii*, zoochlorella located in vacuole of archeocyte; several starch grains contained in chloroplast.

of zoochlorellae in the thesocytes closely resembled those of *S. lacustris* (Figure 6).

Heteromeyenia stepanowii (Dybowsky). Many zoochlorellae were found intracellularly within archeocytes (Figure 7), as in *Spongilla lacustris*. The structure and dimensions of zoochlorellae in the archeocytes closely resembled those in *S. lacustris* (Figure 8). They formed two autospores in reproduction. At the early stages of gemmule formation, zoochlorellae in groups of several individuals were often present in one vacuole of each archeocyte. Zoochlorellae could not be found in the thesocytes.

Radiospongilla cerebellata (Bowerbank). Many zoochlorellae were found intracellularly within archeocytes (Figure 9). Each zoochlorella was located in an individual vacuole except during division. The chloroplast was made up of arrayed stacks of thylakoids, starch grains, and a pyrenoid. The large pyrenoid, surrounded by a shell of starch grains, was found in the center of the chloroplast. Zoochlorellae in the archeocytes formed four autospores in reproduction (Figures 10, 11). At the early stages of gemmule formation, young thesocytes and spongocytes contained a few zoochlorellae. These zoochlorellae closely resembled those of adult green sponges (Figures 12, 13). When the gemmule coat was almost formed, each zoochlorella in the thesocytes contained a degenerated chloroplast that lacked the stacks of thylakoids and a few starch grains (Figures 14, 15). The number of lipid storage bodies had increased in the zoochlorellae (Figure 14). When the gemmule coat had formed, each zoochlorella in the thesocytes had a nucleus, many lipid storage bodies, some large electron dense polyphosphate granules, and a degenerated chloroplast (Figures 16, 17). In the young archeocytes in the gemmule close to hatching, each zoochlorella had a regenerated chloroplast with arrayed stacks of thyrakoids and starch grains (Figure 18). After hatching, the chloroplast eventually came to resemble those in the adult sponges (Figures 19, 20).

Discussion

The common algal symbionts of the freshwater sponges and other metazoans are unicellular algae of the order Chlorococcales, generally known as zoochlorellae. These symbionts have been found in four species of freshwater sponges from Japan. In *Spongilla lacustris*, *Radiospongilla sendai*, and *R. cerebellata*, they were observed in both archeocytes and thesocytes. In *Heteromeyenia stepanowii*, they could not be found in the thesocytes. Although further work is needed to follow up on this observation, if zoochlorellae are present in the thesocytes of *H. stepanowii*, the number will be small. Furthermore, in *R. sendai* and *R.*

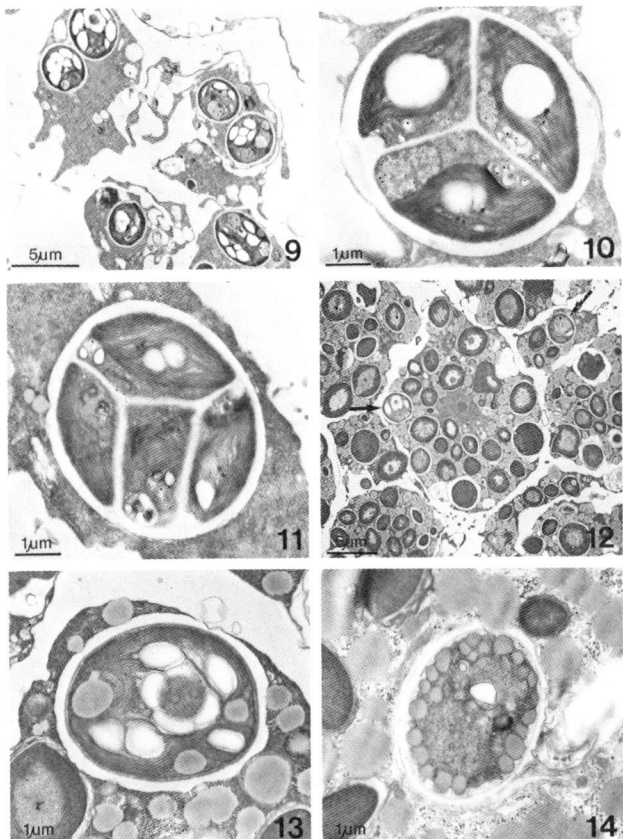

Figures 9–14. Electron micrographs of zoochlorellae in freshwater sponges: *9, Radiospongilla cerebellata*, zoochlorellae in archeocytes of green adult sponges (vegetative tissue); *10, Radiospongilla cerebellata*, reproducing zoochlorella in archeocyte of green adult sponge; oblique section shows three autospores; *11, Radiospongilla cerebellata*, oblique section through four autospores of reproducing zoochlorella; *12, Radiospongilla cerebellata*, two zoochlorellae *(arrows)* observed in young thesocyte containing many immature vitelline platelets; *13, Radiospongilla cerebellata*, zoochlorella in young thesocyte has pyrenoid in center of chloroplast; four starch grains surround pyrenoid; *14, Radiospongilla cerebellata*, zoochlorella in thesocyte as gemmule coat formation nears completion; degenerating chloroplast has small amount of starch grains.

cerebellata fewer zoochlorellae were present in the thesocytes than in the archeocytes. At an early stage of gemmule formation in *R. sendai* and *H. stepanowii*, groups of several zoochlorellae were often observed in one vacuole of some archeocytes. These zoochlorellae may be digested by the host or discharged from archeocytes to mesohyl.

Williamson (1979) reported that at no stage does the zoochlorella of *Spongilla lacustris* possess a pyrenoid. In this study, zoochlorellae of *S. lacustris*, *Radiospongilla sendai*, and *Heteromeyenia stepanowii*, also lacked a pyrenoid at each

470   YOSHIKI MASUDA

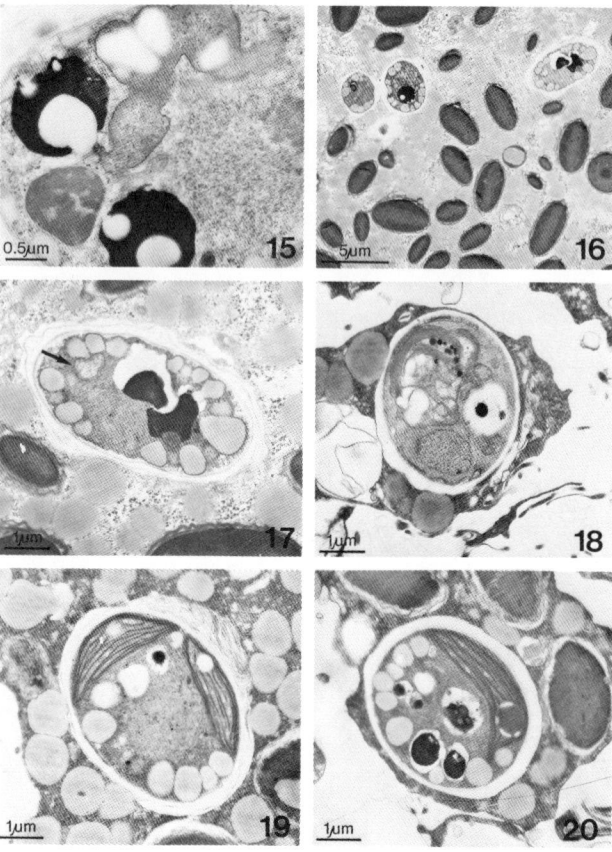

Figures 15–20. Electron micrographs of zoochlorellae in freshwater sponges: *15, Radiospongilla cerebellata*, part of zoochlorella in thesocyte at same stage as in Figure 14; degenerating chloroplast does not show arrayed stacks of thylakoids as seen in Figure 13; *16, Radiospongilla cerebellata*, thesocyte at finish of gemmule coat formation contains three zoochlorellae; zoochlorellae in which large polyphosphate granules are visible look much different from those of Figures 9, 12; *17, Radiospongilla cerebellata*, enlarged view of zoochlorella in Figure 16, it possesses many lipid storage bodies, two polyphosphate granules, and a small body seemingly a degenerating chloroplast *(arrow)*; *18, Radiospongilla cerebellata*, zoochlorella in young archeocyte of gemmule near hatching contains regenerate chloroplast possessing closely arrayed stacks of thyrakoids and starch grains; *19, Radiospongilla cerebellata*, zoochlorella in young archeocyte one day after hatching contains lens-shaped chloroplasts; *20, Radiospongilla cerebellata*, zoochlorella in young archeocyte one day after hatching contains chloroplast possessing pyrenoid; three vitelline platelets are being digested.

stage. However, each zoochlorella in the archeocytes, spongocytes, and young thesocytes of *R. cerebellata* possessed a chloroplast with a pyrenoid.

Williamson (1979) found that the zoochlorellae of *Spongilla lacustris* form only two autospores consistently. I also observed them in the archeocytes of *Spongilla lacustris*,

R. sendai, and *Heteromeyenia stepanowii.* But the mode of reproduction of zoochlorellae of *R. cerebellata* was to form four autospores (Figure 10, 11), as the zoochlorellae of *Hydra viridis* (Oshman, 1967) and *Paramecium bursaria* (Karakashian et al., 1968).

Williamson (1979) also reported that zoochlorellae in the thesocytes of *Spongilla lacustris* contain chloroplasts with a loosely packed thylakoid membrane. In this study, the zoochlorellae contain chloroplasts with a closely arrayed thylakoid membrane in the thesocytes of gemmules of *S. lacustris* (Figure 3) and brown gemmules of *Radiospongilla sendai* (Figure 6). However, starch grains were not observed in their chloroplasts. The zoochlorellae in the mature thesocytes of *R. cerebellata*, which looked different from those in the archeocytes, contained degenerated chloroplasts (Figure 17). The zoochlorellae in the young archeocytes contained regenerated chloroplasts (Figures 18, 19, 20). Osafune et al. (1975) have reported that the degeneration and regeneration of chloroplasts in *Chlorella protot|hecoides* are caused by nutritional factors as well as light conditions. Although there is no evidence that the zoochlorellae of *R. cerebellata* belong to the genus *Chlorella*, I suggest that the degeneration and regeneration of chloroplasts is caused by changes in environmental conditions, that is, differences of conditions inside archeocytes of vegetative tissue, thesocytes of gemmules, and developing young archeocytes.

The structure and dimensions of zoochlorellae of *Spongilla lacustris, Radiospongilla sendai* and *Heteromeyenia stepanowii* are similar and differ from those of *R. cerebellata*. Therefore, it appears that the zoochlorellae of these four species can be classified into at least two large groups, which are defined by the presence of a pyrenoid and their mode of reproduction.

Conclusions

Many actively reproducing zoochlorellae were found in the adult green sponges of *Spongilla lacustris, Radiospongilla cerebellata, R. sendai,* and *Heteromeyenia stepanowii.* Thesocytes in the green gemmules of *S. lacustris,* contained many zoochlorellae, whereas those in the brown gemmules of *R. cerebellata* and *R. sendai* contained considerably fewer zoochlorellae than archeocytes. A pyrenoid was lacking in the chloroplast of zoochlorellae of *S. lacustris, R. sendai,* and *H. stepanowii,* but was present in each chloroplast of zoochlorellae of *R. cerebellata,* except for the stages associated with mature thesocytes and young archeocytes. Zoochlorellac of *S. lacustris, R. sendai,* and *H. stepanowii* are similar in three respects: their dimensions, the lack of a pyrenoid, and their mode of reproduction. Zoochlorellae in the thesocytes of *R. cerebellata* looked much different from those in the archeocytes and possessed a degenerated chloroplast. Each zoochlorella in the young archeocytes that hatched from gemmules possessed a regenerated chloroplast.

Literature Cited

Gilbert, J. J., and H. L. Allen. 1973. Chlorophyll and Primary Productivity of Some Green Freshwater Sponges. *Internationale Revue der Gesamten Hydrobiologie*, 58:633–658.

Karakashian, S. J., M. W. Karakashian, and M. A. Rudzinska. 1968. Electron Microscopic Observation on the Symbiosis of *Paramecium bursaria* and Its Intracellular Algae. *Journal of Protozoology*, 15:113–128.

Masuda, Y. 1985. Electron Microscopic Study on the Zoochlorellae of Adult Green Sponges and Gemmules of *Radiospongilla cerebellata* (Bowerbank) (Porifera: Spongillidae). *Kawasaki Igakkai Shi Liberal Arts and Science Course*, 11:63–68.

Osafune, T., and E. Hase. 1975. Some Structural Characteristics of the Chloroplasts in the "Glucose-Bleaching" and Regreening Cells of *Chlorella protothecoides*. *Biochemie und Physiologie der Pflanzen*, 168:533–542.

Oshman, J. L. 1967. Structure and Reproduction of the Algal Symbionts of *Hydra viridis*. *Journal of Phycology*, 3:221–228.

Van Tright, H. 1919. Contribution to the Physiology of the Fresh-Water Sponges (Spongillidae). *Tijdschrift der Nederlandsche Dierkundige Vereeniging series 2*, 17:1–220.

Wilkinson, C. R. 1980. Nutrient Translocation from Green Algal Symbionts to the Freshwater Sponge *Ephydatia fluviatilis*. *Hydrobiologia*, 75:241–250.

Williamson, C. E. 1979. An Ultrastructural Investigation of Algal Symbiosis in White and Green *Spongilla lacustris* (L) (Porifera: Spongillidae). *Transactions of the American Microscopical Society*, 98:59–77.

ANNE MEYLAN*
Department of Zoology
University of Florida
Gainesville, Florida 32611

Nutritional Characteristics of Sponges in the Diet of the Hawksbill Turtle, *Eretmochelys imbricata*

Abstract

Organic content, energy content, and total nitrogen content were determined for ten species of sponges from the Florida Keys representing six orders of the Demospongiae. Five of these are genera or species that are preyed upon by the hawksbill turtle *(Eretmochelys imbricata),* a marine reptile that feeds almost exclusively on sponges. Its diet consists predominantly of astrophorid and hadromerid sponges that have a high silica (and thus low organic) content. The organic content of five prey taxa ranges from 35.5% to 74.9% of dry weight. Values for energy content of prey sponges range from 18.69 to 21.98 $kJ \cdot g^{-1}$ for ash-free tissue, which is lower than the expected range for most animal tissue. On a dry weight basis, energy values for prey sponges vary from 7.64 $kJ \cdot g^{-1}$ for *Geodia neptuni* to 15.66 $kJ \cdot g^{-1}$ for *Chondrilla nucula*. Determinations of total nitrogen content indicate that protein constitutes 25% to 59% of the dry weight of five prey sponges.

Marine sponges are not a usual food source for humans and have few natural predators. As a result, few data are available on their nutritional value. The only published records of human consumption of sponges to my knowledge document that *Chondrosia reniformis* is eaten by Dalmatian fishermen on the Adriatic coast of Yugoslavia (Steuer, 1904) and that *Chondrilla* is eaten in the Mediterranean (de Laubenfels, 1950). Heretofore, the only dedicated spongivores known among the vertebrates were a few species of highly evolved teleost fishes, including some angelfishes (Pomacanthidae), filefishes (Monacanthidae), siganids (Siganidae), and the moorish idol (Zanclidae) (Randall and Hartman, 1968; Hobsen, 1974; Sano et al., 1984).

*Present address: Department of Herpetology, American Museum of Natural History, New York, New York 10024.

My recent studies (Meylan, 1984, 1985) of the feeding ecology of a tropical, reef-dwelling marine turtle, the hawksbill *Eretmochelys imbricata* (L.), suggest that this species can be considered a dedicated spongivore. Sponges contributed 95.3% of the total dry weight of food items in the digestive tract contents of 61 turtles captured by fishermen in the Caribbean. The samples were drawn from 19 localities in seven countries. Evidence presented by Meylan (1984) has continued to accumulate and suggests that spongivory among hawksbills is widespread. Unpublished records now exist for widely separate localities in the Indian and Pacific oceans.

The discovery that such a large (up to 127 kg), mobile vertebrate feeds almost exclusively on sponges stimulated my interest in the nutritional value of this rarely used food resource. I was also intrigued by the fact that the sponges that were predominant in the samples contain extraordinary amounts of siliceous spicules—in some cases more than half the dry weight. More than 97% (dry-weight basis) of all sponges identified in the stomach contents of hawksbills are representatives of the Astrophorida and Hadromerida, two tetractinomorph orders that typically have well-developed siliceous skeletons. The highly siliceous astrophorid genera *Geodia*, *Ancorina*, *Ecionema*, and *Myriastra* were among the ten most important prey sponges, with *Ancorina* represented by three different species and *Myriastra* by no fewer than six. The order Hadromerida was also represented by some highly siliceous taxa such as *Placospongia*, which has a thick cortex of siliceous microscleres.

The silica in sponges is in an amorphous, hydrated form of glass similar to opal (Jones, 1979). Although the element silicon is not inert (there is some absorption by vertebrates for growth and bone development), siliceous sponge spicules appear to be unaffected by the digestive processes of hawksbills. The spicules are sharply pointed and many have multiple hooklike tips that would seem to pose a formidable mechanical problem for the digestive tract of any predator. Of greater relevance to the present paper, however, is the fact that these highly siliceous sponges contain little organic matter.

The implications for predators of a low-organic (high-ash) diet are many, but one of the most important concerns is the amount of food that must be consumed to meet nutritional requirements. Because the caloric value of most animal tissues falls within a fairly narrow range (Golley, 1961), the way a predator compensates for a diet with a large indigestible component is usually to eat larger amounts. This, in turn, has its consequences for many life history features, including daily activity patterns, competitive interactions, and range of movements.

The few data currently available on the nutritional value of marine sponges are widely scattered and in several instances have been compiled for other purposes. The nutritional parameter for which there is the most informa-

tion is organic content, or complementary ash content (Bergman, 1949; Vinogradov, 1953; Randall and Hartman, 1968; Reiswig, 1973; Dayton et al., 1974). Caloric values (energy content) appear to be available for only 23 species, none of which is included in the diet of the hawksbill turtle. Vinogradov (1953) summarized the few data that have been published on the nitrogen content of sponges. Because most of the studies he reviewed were carried out in the 1800s, he suggested that the analyses be repeated with modern techniques.

Because of this paucity of information, I carried out a limited number of nutritional analyses of prey sponges in order to gain some understanding of their potential food value. The scope of the work was restricted to prey sponges for which fresh material could be obtained. To allow at least limited comparison, I also analyzed five additional species that are common coral reef representatives of the major orders of non-prey sponges: Poecilosclerida, Haplosclerida, and Dictyoceratida. Only the following nutritional parameters were considered: organic content, energy content, and nitrogen content. It is important to point out that the data presented on nitrogen and energy content represent only potential values. Not all nutrients present in food are digestible, and, in any case, virtually nothing is known about the digestive physiology of the hawksbill (Bjorndal, 1985).

Methods

Live sponges were collected in the Florida Keys at Key Largo, Tavernier, and Big Pine Cay. Sediment adhering to the surface of the sponge or present in the aquiferous system was removed as thoroughly as possible with running water and a soft brush. All visible symbionts were removed with forceps. Large sponges were cut into blocks to facilitate drying. The samples were dried to a constant weight at 60°C in an oven with strong circulation and were stored in plastic bags. For analyses of nitrogen content, ash content, and energy content, dried sponges were ground in a Wiley mill (#20 screen). Several fragments taken from representative parts (ectosome, endosome) of each sponge were pooled. Because of the small size of some of the specimens of *Chondrilla nucula*, one of the samples is a composite of three individuals. Maximum storage time of all samples was five months.

One-gram samples of ground sponge were dried to a constant weight at 105°C and ashed in a muffle furnace for 3 h at 500°C (Allen, 1974). Each analysis was carried out in replicate; values for replicates were accepted within 2% error. Ash values were corrected for water of hydration of the silica in the spicules, in accordance with the findings of Vinogradov (1953) and Paine (1964). The correction factor was calculated from the weight loss observed upon ashing dry (105°C) cleaned spicules of *Geodia neptuni* for 3 h at 500°C. The spicules had been isolated by boiling

474

ANNE MEYLAN

pieces of sponge in concentrated nitric acid and collecting the spicules under vacuum on Whatman glass fiber filters (934AH Reeve Angel). Spicules were rinsed several times with distilled water to remove acid solids and were flushed with 95% ethanol into dry, weighed aluminum pans. They were then dried to a constant weight at 105°C. The average weight loss observed in ashing three samples was 3.95% (±0.16, $n = 3$).

Total nitrogen content was determined by a semimicro version of the Kjeldahl method, with the salicylic acid modification described by Nelson and Sommers (1972). Ground samples of approximately 0.5 g were used in the analyses. The amount of NH_3 in 10 ml aliquots of the digests was determined by steam distillation and hand titration. Replicates were accepted within 3% error, except in the case of one specimen of *Geodia neptuni* (3.6%) and one of *Spheciospongia vesparium* (4.8%). Values were corrected for percentage dry matter and percentage ash (corrected for water of hydration) on the basis of results of separate analyses using portions of the same powdered sample. Dry matter replicates were accepted within 1% error; ash replicates were within 2% error.

Energy content of sponges was determined by combustion of ground samples in a Parr oxygen bomb calorimeter (isothermal jacket). Corrections for percentage dry matter and percentage ash were obtained by separate analyses carried out on portions of the same samples. Replicate values were within 3% error, except for *Geodia neptuni* (4.1%).

My classification of sponges follows that of Lévi (1973) except where otherwise indicated. It was not possible to make determinations to the level of species for all sponges found in the stomachs of turtles. For example, it could not be determined whether *Geodia gibberosa*, *Geodia neptuni*, or both were represented. Conceivably, both could have been available to foraging hawksbills. The analysis was carried out with *Geodia neptuni* because of its availability at the collecting localities. These two species are highly similar in physical characteristics such as overall skeletal architecture and spiculation. *Cinachyra kuekenthali* was used in the nutritional analyses, although it is not known whether *C. kuekenthali* or *C. alloclada* was eaten by turtles.

Results and Discussion

Data on the organic content, energy content, and total nitrogen content of a representative series of demosponges are presented in Table 1. The analyses include both prey and non-prey species of *Eretmochelys*.

ORGANIC CONTENT

Organic content varies considerably both within and between orders. On a broadly comparative basis, one would expect a lower organic content (and hence higher ash content) in tetractinomorph sponges (such as astrophorids, spirophorids, and hadromerids) than in ceractinomorph sponges (poecilosclerids, haplosclerids, and dictyoceratids) because of the greater emphasis in the former on siliceous rather than organic fiber skeletons (Vinogradov, 1953). Fiber skeletons are completely lacking in the orders of sponges that were found to be important in the hawksbill's diet, a pattern that is discussed in detail elsewhere (Meylan, 1985). The data in Table 1 support the expected trend except in the case of *Chondrilla nucula*, which has the highest organic content of all sponges analyzed.

The placement of *Chondrilla* and the closely related genus *Chondrosia* in the Astrophorida is debatable (Lévi, 1973; Bergquist, 1978), but there can be little question that neither species is a typical astrophorid with respect to its organic content. The siliceous skeleton is greatly reduced in *Chondrilla nucula* and completely lacking in *Chondrosia*. Siliceous spicules make up only 5% of the dry weight of *Chondrilla nucula* (Meylan, 1984); total ash content is 25.1%. The remainder of ash can be partly attributed to foreign calcareous sediments, which cannot be completely eliminated from encrusting forms of *Chondrilla*.

Chondrilla nucula ranks highest in importance among the sponges consumed by *Eretmochelys*, calculated as the product of average percentage dry weight contribution and frequency of occurrence (Meylan, 1984). *Chondrilla nucula* and *Chondrosia* are among the few sponges known to be eaten by other species of marine turtles (de Laubenfels, 1950; Bjorndal, 1980; Balazs, 1980). They are frequently consumed by reef fishes (Randall and Hartman, 1968) and, as mentioned earlier, they are the only sponges reported to be eaten by humans.

Spheciospongia vesparium, a hadromerid consumed by *Eretmochelys*, has the lowest organic content of any of the sponges analyzed. Siliceous spicules account for 48.7% of the dry weight of the sponge (Meylan, 1984), so that the nonsiliceous ash component is significant (15.8%), as it is in *Chondrilla nucula*. Some previous studies (Dayton et al., 1974) have assumed that all ash in siliceous sponges is contributed by silica.

Dictyoceratid sponges such as *Ircinia stobilina* and *Spongia tubulifera* contain significant amounts of ash despite the fact that they do not have autochthonous spicules. Both have fiber skeletons in which foreign spicules, sand grains, and other debris are incorporated. Inorganic minerals contained in the spongin fibers undoubtedly contribute to the ash content. Junqua et al. (1974) have reported that iron can contribute up to 5.5% of the dry weight of spongin fibers of *Ircinia*.

ENERGY CONTENT

The energy content of most prey sponges is slightly lower than the expected range of 20.9–27.2 $kJ \cdot g^{-1}$ for ash-free

3d Int. Sponge Conf. 1985

Table 1. Organic matter, energy, and total nitrogen content of a representative series of demosponges (values are means (n = 1–3) ± SD when n = 3; – = no data)

Sponge	Organic matter (%, dry wt. basis)	Energy (kJ·g⁻¹)		Nitrogen (%)	
		Dry wt. basis	Ash–Free dry wt. basis	Dry Wt. basis	Ash–Free Dry wt. basis
Astrophorida					
+*Geodia neptuni* (Sollas)	41.5	7.64	18.69	4.98	11.99
Myriastra kallitetilla de Laubenfels	63.4	13.93	21.98	7.99	12.61
Chondrilla nucula Schmidt	74.9±3.21	5.66±0.61	21.06±0.19	9.44±0.24	12.70± 0.38
Spirophorida					
+*Cinachyra kuekenthali* Uliczka	47.9±3.9	9.99±0.90	20.84±0.39	5.54	11.50
Hadromerida					
Spheciospongia vesparium (Lamarck)	35.5±9.7	8.15	20.06	4.05±1.21	11.34± 0.32
Poecilosclerida					
Iotrochota birotulata (Higgin)	58.4 4.3	–	–	7.03	11.76
+*Agelas conifera* (Schmidt)	68.5	–	–	8.98	13.11
Haplosclerida					
Amphimedon rubens (Schmidt)	60.9	–	–	6.98	11.47
Dictyoceratida					
Ircinia strobilina (Lamarck)	62.8	–	–	8.31	13.24
Spongia tubulifera Lamarck	69.0±2.8	–	–	8.75±0.33	12.67± 0.29

* = species identified in the stomach contents of *Eretmochelys imbricata*.
+ = genus represented in the samples.
Iotrochota birotulata and *Agelas conifera* were found in trace amounts and are not considered important prey species for *Eretmochelys*.

animal tissue (Golley, 1961). Values for whole sponge (expressed here on a dry weight basis) are lowest for highly siliceous taxa, as one would expect. Energy content was not determined for non-prey sponges.

Paine (1964) suggested that some of the unusually low values recorded for the energy content of sponges were attributable to the noncombustibility of the water of hydration in the spicules. This potential source of error was corrected in the present study (see methods), and thus it is notable that values are still low. No data on these species of sponges appear to have been published elsewhere and thus it is not possible to compare results. Lower values of energy content have been published for other species, however. Reiswig (1973) reported a caloric value of only 15.82 kJ·g⁻¹ (ash-free) for *Mycale* sp. The most extensive data on the energy content of marine sponges are given by Dayton et al. (1974), who examined 19 species of sponges from Antarctica, including demosponges, calcareous sponges, and hexactinellids. Their values ranged 18.58–28.99 kJ·g⁻¹ for ash-free tissue.

A considerable part of the caloric content of sponges may be contained in spongin fibers (Reiswig, 1973). Spongin fibers of the poecilosclerid sponge *Mycale* sp., for example, account for 40% of the energy content (Reiswig,

1973). Because sponges with a fiber skeleton are rarely consumed by *Eretmochelys*, this information is of little consequence in the present context. However, it may be a significant limitation for sponge predators that do consume fibrous sponges. Some authors (Bloom, 1981) have assumed, possibly correctly, that predators do not digest spongin fibers at all. In the few cases where fibrous sponges were eaten by *Eretmochelys*, the spongin skeletal networks appeared to be unaffected by digestive processes. To my knowledge, however, the digestibility of spongin by any predator has never been determined.

Spongin fibers are known to be extremely resistant to enzymatic hydrolysis as well as to weak acid or alkaline hydrolysis (Garrone, 1978). Molecular interactions in the fibers are reported to be comparable to those binding polysaccharide chains in cellulose (Garrone, 1978). It is noteworthy that some predators appear to purposely avoid ingesting the fibers (Dayton et al., 1974).

Energy, as well as nutrients, may also be contained in collagen fibrils, which are present in all sponges but form an important part of the skeletal framework in certain species. Fibrils are particularly dense in several of the sponges consumed by *Eretmochelys* (e.g., *Chondrilla, Chondrosia, Tethya*) (Meylan, 1985). Their digestibility by predators is unknown. They are potentially an important source of nutrition because they are highly glycosylated, and they are associated with a wide variety of carbohydrate-rich compounds in the intercellular matrix (Garrone, 1978).

TOTAL NITROGEN CONTENT

The nitrogen content of the ten species analyzed varies widely when considered on a whole-(dry-) weight basis, as might be expected from the large variation in their ash contents. Values for total nitrogen are roughly equivalent on an ash-free basis.

There are few comparative data on this parameter. Vinogradov (1953) reviewed results reported by several investigators in the 19th and early 20th centuries. Nitrogen values for seven species of sponges, none of which are prey for *Eretmochelys*, ranged from 7.40% to 8.41% (dry-weight basis).

An estimate of protein content can be obtained by multiplying total nitrogen content by 6.25 (Lloyd et el., 1978). Values for prey sponges range from 25.3% to 59.0% on a whole-(dry-) weight basis. The average protein content of all sponges analyzed in the present study is 76.5% on an ash-free basis.

CONTRIBUTION OF SYMBIONTS

The nutritional value of sponges may be greatly enhanced by the presence of macrosymbionts, such as polychaetes, ophiuroids, amphipods, and shrimps. Pearse (1932) found more than 16,000 shrimps belonging to the genus *Synalpheus* in a single individual of *Spheciospongia vesparium*, which is a prey sponge of *Eretmochelys*. Macrosymbionts were poorly represented in the digestive tract contents of hawksbill turtles (Meylan, 1984). The extent to which they had been affected by digestive processes is not known. In the present study of nutritional characteristics of sponges, macrosymbionts were removed as thoroughly as possible.

Bacteria and cyanobacteria may also contribute to the nutritional value of sponges. Bacteria account for up to 38% of the mesohyl volume in *Verongia* (Vacelet, 1975). They are commonly found in astrophorid sponges (Vacelet, 1977). The abundance of bacteria and cyanobacteria in prey sponges and their nutritional contribution were not investigated in the present study.

Conclusions

Sponges that are important in the diet of the hawksbill turtle vary widely with respect to organic content, energy content, and protein content. Too few data are available from the present study and from the literature to allow any comparison of prey and non-prey sponges, even with respect to the few parameters considered here. The diet consists primarily of sponges that contain large amounts of indigestible silica (more than half the dry weight in several species) and thus could be described as low-quality. One species of sponge *(Chondrilla nucula)*, however, contains only 25% ash and is 59% protein on a dry-weight basis. Its close relative *Chondrosia* is also favored by hawksbills and could be expected to be comparable, if not higher, in protein content. It contains large quantities of densely packed collagen fibrils and lacks a siliceous or fiber skeleton. One of the important questions to be answered regarding the nutritional value of sponges to predators is whether the organic skeletal materials, spongin and collagen fibrils, are digestible. They constitute a large proportion of the organic matter of most sponges.

Acknowledgments

Funding for this study was provided by the World Wildlife Fund (Gland, Switzerland). I thank S. Pomponi, C. Curtis, B. Causey, J. Halas, and P. Meylan for assistance in collecting sponges. Collections made in Looe Key National Marine Sanctuary were carried out under permit KLNMS & LKNMS-04-83. K. Rützler and S. Pomponi kindly assisted with the identification of sponges. I thank J. Fiskell for the use of the IFAS Forest Soils Laboratory, University of Florida; and M. McLeod for teaching me the procedures for nitrogen analyses. I thank K. Bjorndal for many helpful discussions and technical advice. J. Ewel, F. Nordlie and F. Putz provided me with space and equipment in their laboratories. I thank P. Meylan, A. Carr, and R. Zweifel for comments on the manuscript.

Literature Cited

Allen, S. E. 1974. *Chemical Analyses of Ecological Materials*. Oxford: Blackwell Scientific Publications. 565 pp.

Balazs, G. H. 1980. Synopsis of Biological Data on the Green Turtle in the Hawaiian Islands. National Oceanic and Atmospheric Administration (USA), Technical Memorandum, National Marine Fisheries Service, Southwest Fisheries Center, 7. 141 pp.

Bergmann, W. 1949. Comparative Biochemical Studies on the Lipids of Marine Invertebrates, with Special Reference to the Sterols. *Journal for Marine Research*, 8:137–176.

Bergquist, P. R. 1978. *Sponges*. Berkeley: University of California Press. 268 pp.

Bjorndal, K. A. 1980. Nutrition and Grazing Behavior of the Green Turtle *Chelonia mydas*. *Marine Biology*, 56:147–154.

———. 1985. Nutritional Ecology of Sea Turtles. *Copeia*, 1985(3):736–751.

Bloom, S. A. 1981. Specialization and Noncompetitive Resource Partitioning among Sponge-Eating Dorid Nudibranchs. *Oecologia*, 49:305–315.

Dayton, P. K., G. A. Robilliard, R. T. Paine, and L. B. Dayton. 1974. Biological Accommodation in the Benthic Community at McMurdo Sound, Antarctica. *Ecological Monographs*, 44:105–128.

Garrone, R. 1978. *Phylogenesis of Connective Tissue*. Basel: S. Karger. 250 pp.

Golley, F. B. 1961. Energy Values of Ecological Materials. *Ecology*, 42:581–584.

Hobsen, E. S. 1974. Feeding Relationships of Teleostean Fishes on Coral Reefs in Kona, Hawaii. *Fisheries Bulletin*, 72:915–1031.

Jones, W. C. 1979. The Microstructure and Genesis of Sponge Biominerals. Pages 425–447 in *Biologie des Spongiaires*, edited by C. Lévi and N. Boury-Esnault. Colloques Internationaux du C.N.R.S. 291. Paris:Centre National de la Recherche Scientifique.

Junqua, S., L. Robert, R. Garrone, M. Pavans de Ceccatty, and J. Vacelet. 1974. Biochemical and Morphological Studies on Collagens of Horny Sponges. *Ircinia* Filaments Compared to Spongines. *Connective Tissue Research*, 2:193–203.

Laubenfels, M. W. de. 1950. An Ecological Discussion of the Sponges of Bermuda. *Transactions of the Zoological Society of London*, 27:155–201.

Lévi, C. 1973. Systematique de la Classe des Demospongiaria (Demosponges). Pages 577–631 in *Traité de Zoologie 3(1), Spongiaires*, edited by P.-P. Grassé. Paris: Masson.

Lloyd, L. E., B. E. McDonald, and E. W. Crampton. 1978. *Fundamentals of Nutrition 2d ed*. San Francisco: W. H. Freeman. 466 pp.

Meylan, A. 1984. Feeding Ecology of the Hawksbill Turtle *(Eretmochelys imbricata)*: Spongivory as a Feeding Niche in the Coral Reef Community. Ph.D. Dissertation, University of Florida. 118 pp.

———. 1985. The Role of Sponge Collagens in the Diet of the Hawksbill Turtle *(Eretmochelys imbricata)*. Pages 191–196 in *Biology of Invertebrate and Lower Vertebrate Collagens*, edited by A. Bairati and R. Garrone. New York: Plenum.

Nelson, D. W., and L. E. Sommers. 1972. A Simple Digestion Procedure for Estimation of Total Nitrogen in Soils and Sediments. *Journal of Environmental Quality*, 1:423–425.

Paine, R. T. 1964. Ash and Calorie Determinations of Sponge and Opisthobranch Tissues. *Ecology*, 45:384–387.

Pearse, A. S. 1932. Inhabitants of Certain Sponges at Dry Tortugas. *Papers, Tortugas Laboratory*, 28:117–124.

Randall, J. E., and W. D. Hartman. 1968. Sponge-Feeding Fishes of the West Indies. *Marine Biology*, 1:216–225.

Reiswig, H. M. 1973. Population Dynamics of Three Jamaican Demospongiae. *Bulletin of Marine Science*, 23(2):191–226.

Sano, M., M. Shimizu, and Y. Nose. 1984. Food Habits of Teleostean Reef Fishes in Okinawa Island, Southern Japan. *University Museum, University of Tokyo, Bulletin*, 25:1–128.

Steuer, A. 1904. Die Niedersten Seetiere als Nahrungsmittel des Menschen. *Österreichische Fischerei Zeitung*, 1:202–203.

Vacelet, J. 1975. Etude en microscopie électronique de l'association entre bacteries et spongiaires du genre *Verongia* (Dictyoceratida). *Journal de Microscopie et de Biologie Cellulaire*, 23:271–288.

———. 1977. Electron Microscope Study of the Association between some Sponges and Bacteria. *Journal of Experimental Marine Biology and Ecology*, 30:301–314.

Vinogradov. A. P. 1953. The Elementary Chemical Composition of Marine Organisms. *Memoir, Sears Foundation for Marine Research*, 2:1–647.

THOMAS M. FROST
JOAN E. ELIAS
Center for Limnology
University of Wisconsin
Madison, Wisconsin 53706

The Balance of Autotrophy and Heterotrophy in Three Freshwater Sponges with Algal Symbionts

Abstract

Magnitude and relative balance of autotrophy and heterotrophy were analyzed in three freshwater sponges with algal symbionts. The three species, *Corvomeyenia everetti, Ephydatia muelleri,* and *Spongilla lacustris* share a common habitat but exhibit distinct species-specific differences in their chlorophyll content. We tested three alternative models for the balance of autotrophy and heterotrophy: a null model in which heterotrophy and autotrophy are independent, a tradeoff model in which heterotrophy decreases as the potential return from autotrophy increases, and a mutualistic model in which heterotrophy is positively influenced by autotrophy. We found that *Spongilla lacustris*, with intermediate chlorophyll levels, consistently exhibited the highest levels of heterotrophy. Primary production per biomass of symbiotic association (sponge and algae) was also highest in *S. lacustris*. The efficiency of production (production/ chlorophyll) decreased among the three sponge species as chlorophyll increased. As a result we reject the null and tradeoff models and suggest a mutualistic interaction between sponges and their algal symbionts.

Algal-invertebrate symbioses are sustained by a combination of autotrophy and heterotrophy. Despite substantial research into the physiological aspects of such associations (e.g., Muscatine and Porter, 1977; Muscatine, 1980; Muscatine et al., 1981) the factors that control the relative balance of the two nutritional modes are poorly understood. Three contrasting models can be postulated for this balance. In the simplest case the components of the symbiosis may not affect each other: autotrophy and heterotrophy would be independent. Alternatively, if autotrophy and heterotrophy are considered fundamentally different processes, features that increase the efficiency of one process could affect the other adversely. For example, algal symbionts may be borne most efficiently in sponge cells that do not complement feeding activities. Likewise, a growth form that favors production may decrease efficiency in feeding. From this perspective, a balance or tradeoff would occur in algal invertebrate symbioses; that is, heterotrophy would be low when autotrophy was high, and vice versa. Porter (1976) has suggested that such a situation occurs in reef corals. Finally, mutualistic interactions between algae and invertebrates may yield increased efficiency for both means of procuring nutrients.

Here we examine the relative balance of autotrophy and heterotrophy in three freshwater sponge species that occur within the same habitat but that contain distinctly different concentrations of chlorophyll. We use this potential gradient in autotrophy to test the null, tradeoff, and mutualism hypotheses by measuring heterotrophy as the feeding rate of the sponges and autotrophy as the gross primary production of the sponge-algal association.

Two steps are necessary to evaluate the performance of algae and invertebrates in symbiotic associations. Feeding or production must be determined relative to (1) the biomass of the entire association and (2) the biomass of the component that is accomplishing the activity. For example, feeding by the sponge component may not be affected by the proportion of the association consisting of algae (Figure 1a, Model A). In this case the feeding rate per total biomass of algae and sponge will decline as the proportion of algae increases (Figure 1b, Model A). This simple situation represents a null model with no interaction between sponge and algae. As far as the association is concerned, however, it still leads to a decline in heterotrophy as autotrophy increases; provided that autotrophy is proportional to the amount of algal tissue in the association.

A more complex response could involve a decline in the feeding rate per sponge biomass alone as the proportion of algal tissue increased (Figure 1a, Model B). Here the presence of algae has a negative effect on feeding. Such a situation would yield a more precipitous drop in the feeding rate as the proportion of algal tissue increased (Figure 1b, Model B). Any such decline in the feeding rate per sponge biomass as the proportion of algal tissue increases would be consistent with the tradeoff hypothesis.

Alternatively, in a mutualistic situation, the feeding rate per sponge biomass would increase, up to a certain point as the proportion of algal tissue increased (Figure 1a, Model C). If this occurred, the feeding rate per biomass of sponge and algae would be highest in species that had intermediate concentrations of algae (Figure 1b, Model C). Even under these circumstances it seems unlikely that a high feeding efficiency of sponge tissue could occur when algae constituted a major portion of the biomass within the association. Thus the curves presented for this model predict maximum feeding rates at intermediate proportions of algal biomass.

Freshwater sponges are frequently green owing to the presence of algal symbionts (Penney and Racek, 1968). *Spongilla lacustris*, a common species, derives as much as 80% of its growth from autotrophy (Frost and Williamson, 1980). Techniques are available to measure both feeding activities by freshwater sponges (Frost, 1980a) and the primary production of algae within sponges (Gilbert and Allen, 1973). Thus it is possible to assess both autotrophy and heterotrophy for symbiotic associations that potentially depend on their algae for nutrition to a great extent.

Here we focus on three species: *Corvomeyenia everetti*, *Ephydatia muelleri*, and *Spongilla lacustris*, all of which exhibit distinct, species-specific differences in chlorophyll within a single habitat (Figure 2). Although the symbionts within *C. everetti* are yellow-green and differ markedly from the green algae in the other species, the differences in chlorophyll concentrations among these sponges represent a potentially major gradient in the contribution of autotrophy. We report (1) determinations of the proportion of total biomass consisting of sponge and algae in each association, (2) comparisons of the feeding activities of the

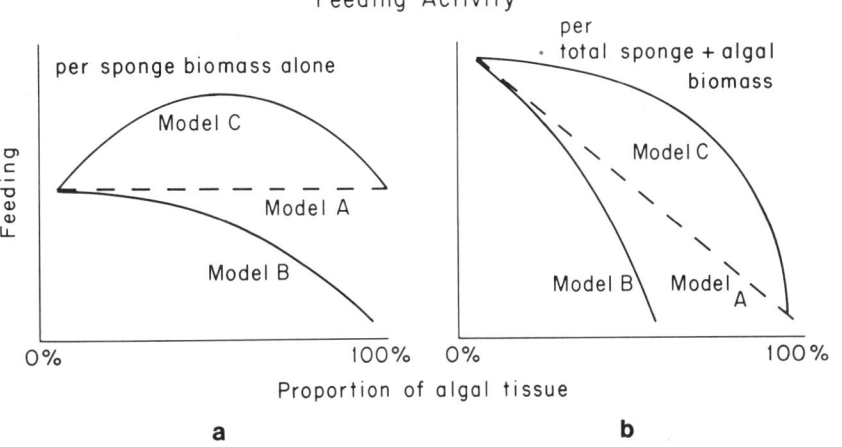

Figure 1. Three models of possible relationships between sponge feeding activity and proportion of algal tissue in symbiotic association of sponge and algae: a, plotted on basis of sponge biomass alone; b, on basis of entire association. A, null model; B, tradeoff model; C, mutualism model.

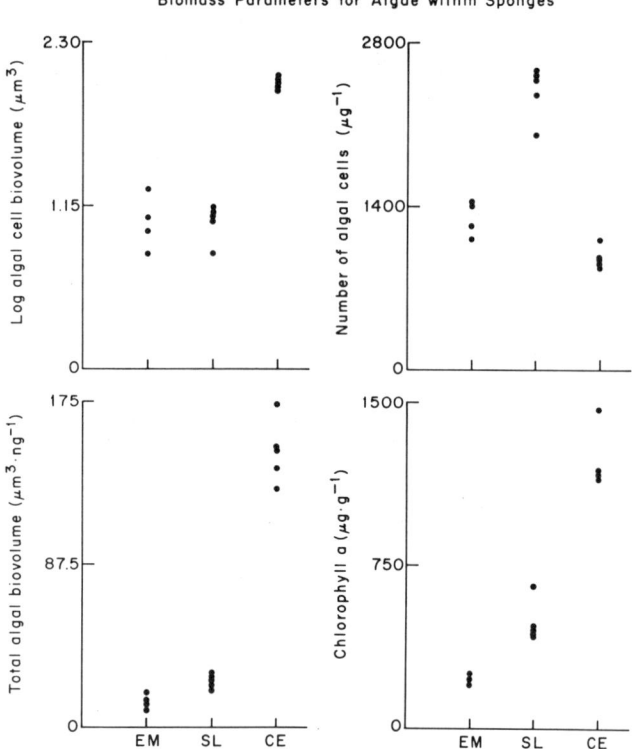

Figure 2. Biomass parameters for algae within *Ephydatia muelleri (EM)*, *Spongilla lacustris (SL)*, and *Corvomeyenia everetti (CE)*. Aside from algal cell volume, all data reported on basis of wet mass of symbiotic association (sponge and algae). Each point represents data from separate sponge specimen. For each parameter, one-way ANOVA on log-transformed data indicates a significant difference among sponge species ($p < 0.01$ in all cases). In addition, *T*-method analyses indicate that values for individual sponge species are significantly different ($p < 0.01$), except in case of EM and SL for algal cell volume.

three species under varied conditions throughout a summer, and (3) comparisons of primary production of the three species throughout one summer. We report feeding and production in terms of the total biomass of the association as well as the biomass of sponge or algae within each association. Thus we provide a direct assessment of the relative balance of autotrophy and heterotrophy that can be used to evaluate the null, tradeoff, and mutualistic models of algal-invertebrate symbioses.

Materials and Methods

We examined sponges in Little Rock Lake, Wisconsin (89°42′ W, 46°00′ N), a low ionic-strength, clear-water system (see Brezonik et al., 1986, for details on its limnology). Sponges occur extensively throughout the lake to depths greater than 4 m. Most *Ephydatia muelleri* and *Spongilla lacustris* grow either directly from the lake bottom

or attached to macrophytes (e.g., *Myriophyllum tennelum*). *Corvomeyenia everetti* appears to require macrophytes or other substrata for growth. We have observed no obvious patterns in the relative distribution of these sponges with depth; each species is as likely to occur near the surface as it is below 3 m.

The statistical techniques used in this study are described in Sokal and Rohlf (1981).

ANALYSES OF ALGAL PIGMENT CONCENTRATIONS AND VOLUMES

Chlorophyll analyses were conducted using a modification of a method introduced by Gilbert and Allen (1978). We crushed a weighed portion of wet sponge using a teflon pestle and extracted material using 90% methanol buffered with $MgCO_3$. The extracted material was boiled at 80°C for 30 seconds and then allowed to stand in the dark for 20 to 24 hours. We corrected for phaeopigments by acidifying to a pH of 2.7 (Moed and Hallegraeff, 1978). We confirmed that standard equations (Lorenzen, 1967) were appropriate for our spectrophotometer and extraction technique by calibrating with pure chlorophyll *a* (Sigma Chemical Company, St. Louis, Missouri).

The size and concentration of algal symbionts were measured on the same specimens analyzed for chlorophyll. To quantify algae, we crushed sponges as in the chlorophyll analyses, but suspended the material in 2% buffered formalin solution. Subsamples of this suspension were examined at 1000 × in a Palmer-Maloney Cell on a Zeiss inverted microscope using epifluorescent illumination appropriate for autofluorescence of chlorophyll. Diameters were measured for 15 cells per specimen and volume was calculated assuming that cells were spherical.

We tested for the destruction of algal cells in our grinding technique by comparing the distribution of chlorophyll in material that was removed from suspension by centrifugation with that remaining in the supernatant. Fluorometric determinations of chlorophyll (Turner Model 10–005 Fluorometer), which revealed a maximum of 3% of total sample chlorophyll in the supernatant of five replicated analyses, indicated that chlorophyll was associated with particles. Analyses of this material by fluorescence microscopy indicated a uniform distribution of chlorophyll in intact algal cells rather than in cell debris.

FEEDING RATES

Feeding was assessed as the rate of water processing based upon the removal of radioactively labeled yeast cells *(Rhodotorula glutinis)* from suspension. This method is a slightly modified version of a technique used in previous work (Frost, 1980a). Specimens of each species were placed in a plexiglass chamber submerged 40 cm below the water surface. Sponges were acclimated in 1 liter of water in the chamber for one half to one hour, and then 0.5 liter of a suspension of labeled cells was added to the chamber; thus the final concentration was between one and two labeled cells per ml. Actual cell concentration was determined directly in each feeding chamber two times during the feeding period. After 15 minutes, the sponges were removed from the feeding suspension and placed in cold, fresh lake water for return to the laboratory. Within less than an hour, two small portions of each sponge (approximately 10 mg dry mass) were placed in glass scintillation vials and dried for at least 24 hours at 60°C. Dried sponges were weighed and then digested as in Frost (1980b). Radioactivity was determined as disintegrations per minute (DPM) using a liquid scintillation counter (Beckman LS 1801). Samples of radioactive cells from the feeding chamber were filtered onto 0.45 μm cellulose membrane filters (Millipore), dried, and processed using the same techniques as for sponges.

Clearance rates were calculated as:

$$R = DPM_S \cdot DPM_F \cdot 15 \text{ min}^{-1},$$

where R = clearance rate in ml·min^{-1}, DPM_S = disintegrations per minute in a sponge sample, and DPM_F = disintegrations per minute in a ml of the yeast feeding suspension. These clearance rates are then reported on the basis of (1) the biomass of the entire symbiotic association (i.e., dry mass of sponge and algae) and (2) the approximate biomass of sponge tissue alone.

Feeding measurements were conducted on seven dates between June 25 and September 30, 1985. For each date, rates were determined for six specimens of each sponge species. These dates represent a broad range of environmental conditions that might influence feeding. To facilitate a comparison among species, our graphic presentations show rates that have been relativized to the mean clearance rate exhibited by *Spongilla lacustris* on that date. We tested for differences among species by blocking among dates in a two-way analysis of variance (ANOVA). All analyses were conducted on the nonrelativized, log-transformed mean clearance rates calculated from two replicate samples taken from each sponge specimen.

PRIMARY PRODUCTION

Gross primary production was measured for sponge-algal associations using a standard light-dark bottle oxygen difference measurement similar to that described by Gilbert and Allen (1973). Sponges were placed into 300-ml glass bottles that were either clear or made completely opaque with black vinyl tape and were incubated in the lake on a floating raft at a depth of 0.3 m for 3 hours. We measured oxygen using a self-stirring bottle probe and meter (YSI Model 58). After an incubation, sponges were dried and weighed as described above.

On each sampling date individual sponges were placed

into four light and four dark bottles. In addition, eight control bottles, four light and four dark, were filled with lake water only and processed along with the bottles containing sponges. All bottles were filled with water taken from the lake surface and mixed in a large sample vessel to ensure that initial conditions were homogeneous. Initial oxygen was also determined on samples from this vessel.

To determine gross primary production, we first calculated the amount of oxygen consumed per dry mass of sponge-algal association in each dark bottle. This provided an estimate of the average respiration rate during the experiment. We then estimated the expected respiration for each sponge in the light bottles on the basis of its dry mass. This respiration estimate was added on to the change in oxygen in each light bottle to yield gross primary production. The rate of production is based on (1) total dry biomass of sponge-algal association, and (2) chlorophyll content. We approximated the chlorophyll content on the basis of (1) the dry weight of the specimens in each experiment, (2) a separate wet mass-dry mass regression for each species based on samples collected between May 1 and September 30, 1985, and (3) the average chlorophyll content determined for each sponge species collected between 0.5 and 1 m in an extensive survey in mid-summer 1983.

Since it was possible to measure primary production for only one species on each sampling date, we used two sampling designs to facilitate comparisons among species. We conducted production measurements for all three species during eight separate periods between May 10 and September 12, 1985. Also, for each sampling date we measured the total amount of photosynthetically active radiation (PAR-moles [photons·m^{-2}]) that reached the lake surface during the incubation period by means of an integrating meter (Licor Instruments Model LI-1776 with a PAR sensor).

Although there was surprisingly little relationship between primary production and PAR for any of the sponge species, a seasonal trend was apparent. Thus, for comparisons among species we have blocked our data for the sampling period in a two-way analysis of variance.

BIOMASS COMPARISON BETWEEN SPONGES AND ALGAE

We estimated the portion of total biomass comprising algae and sponge within each species. We measured the wet mass in fresh sponge tissue, including algae, and the total biovolume for algae directly by assessing cell number and diameter in the sponge tissue. Both of these determinations were then converted to carbon. For sponges we determined wet mass:dry mass and dry mass:ash-free dry mass relationships for each species and assumed that carbon accounted for 50% of ash free dry mass (Reiswig, 1981). For algae we assumed that algal cells had approxi-

482   THOMAS M. FROST AND JOAN E. ELIAS

mately the same density as water and that carbon accounted for 10% of total wet mass (Vollenweider, 1974).

Results
ALGAL BIOMASS WITHIN SPONGES

Differences in the concentration of chlorophyll within sponge species are related to differences in both the number and the size of algal cells occurring within each species (Figure 2). *Corvomeyenia everetti*, which has the highest chlorophyll concentration, contains significantly fewer cells than the other species, but its algae are substantially larger. In contrast, the size of algal cells does not differ in *Ephydatia muelleri* and *Spongilla lacustris*. However, *S. lacustris* has a substantially higher chlorophyll content than *E. muelleri* because it has more algal cells.

Chlorophyll density is highly correlated with the total algal volume when all three species are taken together (r^2 = 0.89, $p < 0.01$). However, the ratio of chlorophyll to algal volume (fg Chl a:μm^3 of algae) in *Corvomeyenia everetti* (mean = 7.8) is significantly lower (one-way ANOVA, F = 15.7, $p < 0.001$) than in *Ephydatia muelleri* (17.8) and *Spongilla lacustris* (19.8), whose values are not significantly different (*T*-method). According to estimates of carbon in algae and in the total complex of sponge plus algae, algae account for 4% of biomass in *E. muelleri*, 9% in *S. lacustris*, and 67% in *C. everetti*.

ASSESSMENTS OF HETEROTROPHY

Distinct differences occurred in the feeding activities of the three sponge species, in measurements conducted over a broad range of conditions during the summer of 1985. Clearance rates per biomass of association (sponge and algae), for example, are significantly different among these species (Figure 3; two-way ANOVA, F = 31.8, $p < 0.01$). *Spongilla lacustris* exhibits the highest rate, followed by *Ephydatia muelleri* and then *Corvomeyenia everetti* (with mean rates of 21.7, 9.8, and 2.2 ml·min^{-1}·g dry mass^{-1}, respectively; all comparisons among species are significant at a $p < 0.01$ based on the *T*-method).

Clearance rates estimated per biomass of sponge alone are also significantly different (Figure 3, two-way ANOVA, F = 8.7, $p < 0.01$). The mean rate for *Spongilla lacustris* (25.4 ml·min^{-1}·g dry mass^{-1}) is substantially higher than either of the other species. However, for calculations on this basis, the mean rates for *Ephydatia muelleri* and *Corvomeyenia everetti* are not significantly different (10.7 and 8.55 ml·min^{-1}·g dry mass^{-1}, respectively) based on the *T*-method. Thus the sponge portions of *E. muelleri* and *C. everetti* feed at very similar rates while rates calculated per biomass of association for these species are substantial different.

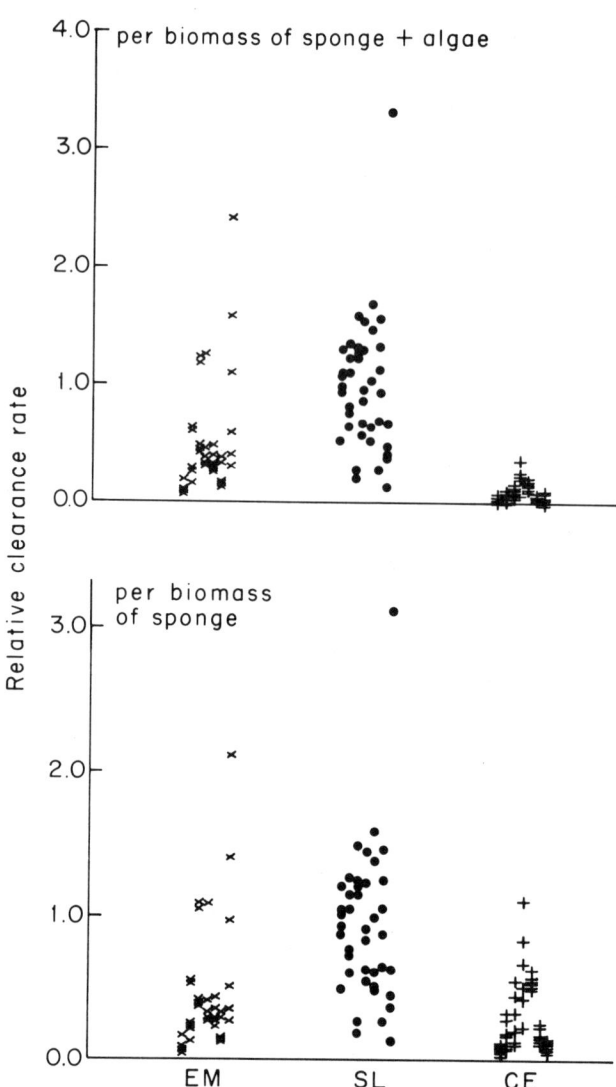

Figure 3. Clearance rates for *Ephydatia muelleri (EM)*, *Spongilla lacustris (SL)*, and *Corvomeyenia everetti (CE)*. Data represent experiments conducted throughout summer of 1985. Values recorded each day have been relativized to mean clearance rate exhibited by SL on that day. Each point represents mean of two determinations for a sponge specimen. Values reported on basis of total biomass of symbiotic association (sponge and algae, *top*) and on basis of estimated sponge biomass alone *(bottom)*.

ASSESSMENTS OF AUTOTROPHY

According to the rates based on the biomass of the association (sponge and algae), gross primary production among species was also significantly different during the summer of 1985 (Figure 4; two-way ANOVA; F = 56.5, p, < 0.01). Trends among species follow a similar pattern to that shown in feeding although the differences among species are less consistent. *Spongilla lacustris*, with intermediate

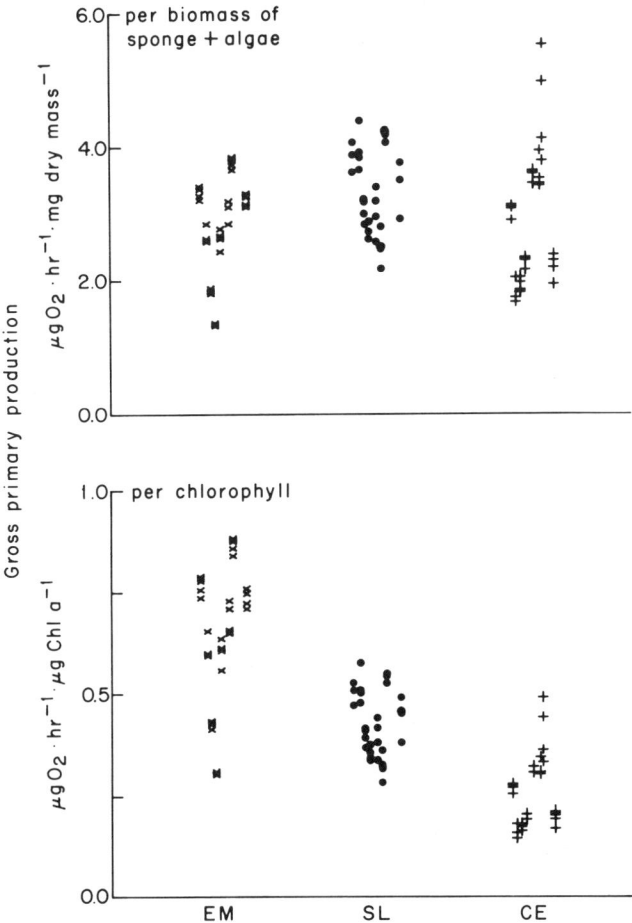

Figure 4. Rates of gross primary production for *Ephydatia muelleri (EM), Spongilla lacustris (SL)*, and *Corvomeyenia everetti (CE)*. Data represent experiments conducted throughout summer of 1985. Each point represents determination for a single sponge specimen. Values reported on basis of total biomass of symbiotic association (sponge and algae, *top*) and on basis of chlorophyll *(bottom)*.

levels of chlorophyll, exhibited the highest mean value for the year (Figure 2; mean value = 3.34 μg O₂·h⁻¹·mg dry mass⁻¹). We recorded the highest values on any sampling date for *Corvomeyenia everetti*, with the highest chlorophyll levels, but its mean rate for the year was not significantly different than that of *Ephydatia muelleri* (Figure 2; mean values = 2.87 and 2.73 μg O₂·h⁻¹·mg dry mass⁻¹, respectively), statistical inference based on the *T*-method.

Production rates per unit chlorophyll reveal a more conspicuous pattern. Significant differences are clearly apparent among the species (Figure 4; two-way ANOVA, $F = 1325$, $p < 0.01$) and indicate an inverse relationship between total sponge chlorophyll and production per unit pigment (mean values of 0.63, 0.44, and 0.26 μg O₂·h⁻¹·g chl *a*⁻¹ for *Ephydatia muelleri, Spongilla lacustris*, and *Corvomeyenia everetti*, respectively). Thus, as the proportion of

algae in sponge tissue increases, the efficiency of production decreases.

Discussion

Ephydatia muelleri, Spongilla lacustris, and *Corvomeyenia everetti* in Little Rock Lake represent a distinct gradient in the relative proportion of sponge and algal tissue in symbiotic associations. With 4% algal biomass, *E. muelleri* is predominantly sponge tissue, whereas *C. everetti*, with 67% algal biomass, is, in a sense, more plant than animal. The situation in *C. everetti* is similar to symbioses between marine sponges and heterotrophic bacteria in which bacterial biomass may exceed that of the host sponge (Reiswig, 1981). Our estimates of the absolute proportion of algal and sponge biomass must be interpreted with caution, however, since they are based on literature values for carbon in sponges and algae. Nonetheless, we are confident that the relative distribution of algal biomass among the three sponge species is accurate.

Our data for sponge feeding fail to support the null or tradeoff hypothesis regarding the balance of autotrophy and heterotrophy. Rather, our results suggest a mutualistic interaction between the presence of algae and the feeding performance of the sponges. *Spongilla lacustris*, with intermediate algal biomass, consistently exhibited feeding rates that were higher than either of the other two species. *Ephydatia muelleri*, with the lowest proportion of algae, fed at a higher rate than *Corvomeyenia everetti* when the total biomass of the association is taken into account, but the sponge-specific rates of these two species were fairly similar. Overall, these results are consistent with Model C in Figure 1 and suggest that a positive effect on sponge feeding was associated with intermediate levels of algal tissue within the symbiotic association.

The data on primary production are also consistent with a mutualistic model of interaction in sponge-algal associations. Production per biomass of association was generally highest in *Spongilla lacustris* with intermediate chlorophyll levels. Moreover, the efficiency of production, measured as production per unit chlorophyll decreases as absolute algal biomass increases. Thus the algae in the sponges exhibit higher efficiency when they are associated with higher proportions of sponge tissue.

Concomitant evaluations of feeding and production provide valuable insights into the function of autotrophy and heterotrophy in symbiotic associations. However the balance of these processes must also be evaluated in terms of their role in the overall energetic budget of symbiotic organisms. We are currently developing ways to evaluate the energetics of the sponge species in our study. Budgets are based on determinations of growth and respiration in addition to the feeding and production rates discussed here.

Although the three sponge species that we describe represent a straightforward gradient with respect to chlorophyll density, algal biomass, and potential return from autotrophy, on another level, taxonomic differences in the algal species in *Corvomeyenia everetti* confound our comparisons. It is true that comparisons solely between *Ephydatia muelleri* and *Spongilla lacustris*, which have apparently identical algal cells, reveal the same patterns as all three species do together. However, it is possible that the patterns reported here are species-specific rather than a general phenomenon in algal–sponge or algal–invertebrate symbioses. We are expanding our analyses by (1) adding other sponge species, (2) considering additional habitats in which *S. lacustris* exhibits distinctly different, habitat-specific chlorophyll concentrations, and (3) experimentally manipulating the energy available to sponges from autotrophy in situ to examine the response of feeding. This work will provide specific tests of a general, mutualistic interaction in algal-sponge associations.

Tradeoff concepts are sufficiently pervasive in ecological and evolutionary theory that they are often applied uncritically and without specific testing (Gould and Lewontin, 1979). Our results, which clearly warrant rejecting the tradeoff concept, emphasize the potential problems of such an approach. Although a mutualistic interaction between autotrophy and heterotrophy in symbiotic associations seems logical (energy available from algae could facilitate increased invertebrate feeding and nutrients available from an invertebrate host could facilitate increased algal production), such a relationship must be viewed against a null case in which alga and host do not significantly influence each other's nutritional modes. Thus specific testing is also required to demonstrate a mutualistic relationship. Our data for both feeding and primary production provide initial support for such a mutualism between freshwater sponges and their algal symbionts.

Acknowledgments

We thank Steve Hewett, Susan Knight, Daniel Schneider, and Michael Sierszen for reading and improving this manuscript. This work was supported by Grant BSR8315096 from the National Science Foundation.

Literature Cited

Brezonik, P. L., L. A. Baker, J. Eaton, T. Frost, P. Garrison, T. Kratz, J. Magnuson, J. Perry, W. Rose, B. Shepard, W. Swenson, C. Watras, and K. Webster. 1986. Experimental Acidification of Little Rock Lake, Wisconsin. *Water, Air, and Soil Pollution,* 31:115–121.

Frost, T. M. 1980a. Clearance Rate Determinations for the Freshwater Sponge *Spongilla lacustris*: Effect of Temperature, Particle Type and Concentration, and Sponge Size. *Archiv für Hydrobiologie,* 90:330–356.

———. 1980b. Selection in Sponge Feeding Processes. Pages 33–44 in *Nutrition in Lower Metazoa*, edited by D. C. Smith and Y. Tiffon. Oxford: Pergammon Press.

Frost, T. M., and C. E. Williamson. 1980. In Situ Determination of the Effect of Symbiotic Algae on the Growth of the Freshwater Sponge *Spongilla lacustris*. *Ecology,* 61:1361–1370.

Gilbert, J. J., and H. L. Allen. 1973. Chlorophyll and Primary Productivity of Some Green Freshwater Sponges. *Internationale Revue der gesamten Hydrobiologie,* 58:633–658.

Gould, S. J., and R. C. Lewontin. 1979. The Spandrels of San Marco and the Panglossian Paradigm: A Critique of the Adaptationist Programme. *Proceedings, Royal Society of London, B,* 205:581–598.

Lorenzen, C. V. 1967. Determination of Chlorophyll and Phaeo-Pigment: Spectrophotometric Equations. *Limnology and Oceanography,* 12:343–346.

Moed, J. R., and G. M. Hallengraeff. 1978. Some Problems in the Estimation of Chlorophyll *a* and Phaeopigment from Pre- and Post-Acidification Spectrophotometric Measurements. *Internationale Revue der gesamten Hydrobiologie,* 63:787–800.

Muscatine, L. 1980. Productvity of Zooxanthellae. Pages 381–402 in *Primary Productivity in the Sea* edited by Paul G. Falkowski. New York: Plenum Publishing Corporation.

Muscatine, L., L. R. McCloskey, and R. E. Marian. 1981. Estimating the Daily Contribution of Carbon from Zooxanthellae to Coral Animal Respiration. *Limnology and Oceanography,* 26:601–611.

Muscatine, L., and J. W. Porter. 1977. Reef Corals: Mutualistic Symbioses Adapted to Nutrient-Poor Environments. *Bioscience,* 27:464–460.

Penney, J. T., and A. A. Racek. 1968. Comprehensive Revision of a Worldwide Collection of Freshwater Sponges (Porifera: Spongillidae). *United States National Museum Bulletin,* 272:1–184.

Porter, J. W. 1976. Autotrophy, Heterotrophy, and Resource Partitioning in Caribbean Reef-Building Corals. *American Naturalist,* 110:731–742.

Reiswig, H. M. 1981. Partial Carbon and Energy Budgets of the Bacteriosponge *Verongia fistularis* (Porifera: Demospongiae) in Barbados. *Marine Ecology,* 2:273–293.

Sokal, R. R., and F. J. Rohlf. 1981. *Biometry* (2d edition). San Francisco: W. H. Freeman. 859 pp.

Vollenweider, R. A., editor. 1974. *A Manual on Methods for Measuring Primary Production in Aquatic Environments.* IBP Handbook No. 12. Oxford and Edinburgh: Blackwell Scientific Publication.

JOHN C. FRANCIS
LINDA BART
MICHAEL A. POIRRIER
Department of Biological Sciences
University of New Orleans
New Orleans, Louisiana 70148

Effect of Medium pH on the Growth Rate of *Ephydatia fluviatilis* in Laboratory Culture

Abstract

A laboratory culture system that has been developed for *Ephydatia fluviatilis* has several features of interest for experimental work: all components of the culture medium are defined, sponge growth rate can be controlled by providing growth-limiting quantities of live bacteria as food, and sponge cultures can be maintained indefinitely. When the effect of culture medium pH on sponge growth rate was measured and the results analyzed alongside field observations of the sponge, it was concluded that pH probably has a deterministic effect on the distribution of *E. fluviatilis* in nature.

A sponge culture system we have developed has some features that are especially attractive for experimental work. First, we are able to control the rate at which sponges grow and thus can obtain results beyond just growth versus no growth. Second, we have developed a completely defined culture medium in which the concentrations of all the chemical components are known. Third, we can maintain sponges in the culture system indefinitely. The experiments we conducted lasted several months, and the system itself was developed over several years (for the technical details see Poirrier et al., 1981; Francis et al., 1982; Harsha et al., 1983; Francis, 1984; Francis and Poirrier, 1986).

In the present study we use this culture system to determine the pH profile for *Ephydatia fluviatilis*. The ecological rationale for this work is derived from field studies (Cheatum and Harris, 1953; Jewell, 1935, 1939; Moore, 1953; Old, 1932; Penney, 1954; Poirrier, 1969; Racek, 1969; Wurtz, 1950) that indicated a correlation between the distribution of different species of freshwater sponges and chemical parameters of the habitat water. From Harrison's (1974) summary of chemical data on North American freshwater sponge species and discussion of the associations between pH and species distribution, it appears that different species of freshwater sponges are adapted to different pH conditions. Some species, such as *Spongilla mackayi*, are restricted to acidic waters and others, such as *Dosilia radiospiculata*, are restricted to alkaline waters. The ecological question we have addressed here is whether pH has a deterministic effect on the distribution of *E. fluviatilis* in nature or whether the sponge is just coincidentally associated with waters in the observed pH range. Poirrier (1974) reported growing colonies of *E. fluviatilis* in Louisiana in waters ranging in pH from 6.5 to 7.7. Our expectation in this study was that if pH does influence distribution in nature then we should be able to produce predictable and repeatable growth responses to different pH values in controlled culture.

Materials and Methods

SPECIMEN ORIGIN. Gemmules of *Ehpydatia fluviatilis* (Spongillidae) were collected from Lake Pontchartrain, Louisiana, and stored at 5°C until needed. Each sponge used for experimental purposes was grown from one gemmule.

CULTURE MEDIUM AND CONDITIONS. Concentrations of components were chosen to represent those common to Lake Pontchartrain, Louisiana, where *Ephydatia fluviatilis* is abundant. The amounts indicated are those required for preparation of 1 liter of culture medium: $NaHCO_3$, 77.2 mg; $CaCl_2 \cdot 2H_2O$, 114.1 mg; $Na_2SiO_3 \cdot 9H_2O$, 37.0 mg; $MgSO_4 \cdot 7H_2O$, 174.0 mg; KCl, 22.9 mg; $MgCl_2 \cdot 6H_2O$, 106.0 mg; NaCl, 718.0 mg; $NaNO_3$, 0.4 mg; $FeCl_3 \cdot 6H_2O$, 0.54 mg; $MnCl_2 \cdot 4H_2O$, 0.4 mg; H_3BO_3, 0.31 mg; $ZnCl_2$, 3.4 μg; $CoCl_2 \cdot 6H_2O$, 6.0 μg; $CuSO_4 \cdot 5H_2O$, 0.12 μg; ethylenediamine tetraacetic acid (EDTA), 6.0 mg. The pH of the bicarbonate-buffered medium was adjusted to the desired value with NaOH or HCl. Tetracycline, at initial concentrations of 50 μM, was used to control the growth of filamentous bacteria. One treatment per week was usually adequate.

The culture vessel has a working volume of approximately 100 ml, sufficant to cover a 25×75-mm glass slide leaning against the wall of the culture vessel at an angle of about 60°. Sponges were cultured on the underside of the slide. Medium was pumped through the culture vessel at a rate of approximately 100 ml/h. Medium in the culture vessel was stirred constantly by a magnetic stirbar. An air flow of 0.5 l/min was directed at the water surface. The laboratory was illuminated by fluorescent lighting for 12 h daily. Medium temperature in the culture vessel was maintained at 25°C by circulating water of that temperature through a water jacket surrounding the culture vessel.

FEEDING. Growth-limiting quantities of the bacterium *Escherichia coli* K-10 were pulsed into the sponge culture vessel once daily under general culture conditions. The bac-

teria were taken from glucose-limited chemostat cultures (Francis and Beary, 1983) and were added in the living state. (The bacterial culture medium contained a 0.06 M phosphate buffer; approximately 1 ml of this medium was delivered to the sponge culture vessel daily with the bacterial pulse, thereby providing phosphate in excess.) The rapid flow of culture medium through the culture vessel reduced the concentration of bacteria in the culture vessel to essentially zero by 6 h after the pulse. To provide perspective, a bacterial pulse of 10^7 cells (10^5 cells/ml) was sufficient to maintain a limited growth rate of 0.15 specific area/day for *Ephydatia fluviatilis*.

SPONGE AREA AND GROWTH RATE ESTIMATION. Sponge area was measured daily by projecting an image of the sponge on paper, tracing its outline and measuring the area of the tracing (in mm^2) with a planimeter (Dietzgen, D1802). The thin epithelial layer composed of pinacocytes, which forms the outer border of the sponge and includes the dermal membrane and oscula, was not included in these area measurements because the width of the subdermal cavity varies greatly under different conditions unrelated to growth.

Area estimates were converted to specific area by dividing each by the initial area of the sponge. Growth rate estimates in units of specific area per day were obtained through linear regression analysis. One would expect sponge area to increase exponentially with time. In this analysis, however, we assumed a linear relationship because the time of observation was relatively short. Standard error values reported in our earlier papers (Francis et al., 1982; Harsha et al., 1983; Poirrier et al., 1981) suggest that the assumption of linearity did not introduce much error. All sponges were trimmed to approximately the same initial area (about 2 mm^2) to minimize error due to size differences. Daily qualitative observations of sponge shape and density were made with an inverted microscope and with a dissecting microscope to ensure that increase in area was due to actual growth and was not just to an expansion related to a decrease in height. Height-width ratios remained constant and there was no evidence that an increase in area was due to decrease in height, increase in cell size, or expansion of the canal system.

Results

Quantities of the bacterium *Escherichia coli* were pulsed into the sponge culture vessel once daily to determine from the growth curve of *Ephydatia fluviatilis* (Figure 1). The results suggest that daily bacterial pulses of at least 5 × 10^7 bacterial cells establish maximum sponge growth rate. Experiments described below to monitor the effects of pH were conducted with bacterial pulses of 10^8

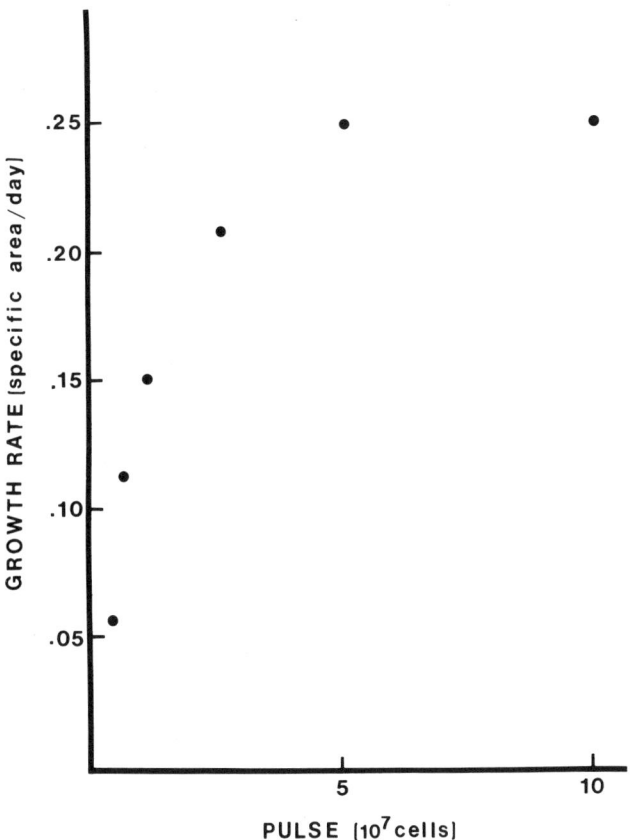

Figure 1. Growth rate of *Ephydatia fluviatilis* as a function of bacterial pulse delivered to sponge culture vessel.

Table 1. Growth rate of *Ephydatia fluviatilis* in defined medium at different concentrations of bicarbonate at pH 7.0

Bicarbonate (mg/l)	Growth Rate (specific area/d)
14.0	0.00
28.0	0.25
56.0	0.25
112.0	0.25

cells/day—a pulse that guarantees the maximum rate of growth.

Cultures of *Ephydatia fluviatilis* were maintained in culture media at pH 7.0 with bicarbonate concentrations ranging from 14 to 112 mg/l in order to measure the effect of bicarbonate concentration on sponge growth rate. This was a necessary control experiment because a bicarbonate buffer system was employed to manipulate pH in the experiments described below; furthermore, with a bicarbonate buffer the relative proportions of carbon dioxide, bicarbonate, and carbonic acid are different at different pH values. The bacterial pulse, 10^8 cells/day, was sufficient to produce maximum rate of growth at pH 7.0. The results presented in Table 1 suggest that bicarbonate concentration in the range 28 to 112 mg/l does not affect sponge growth rate at pH 7.0. Bicarbonate concentrations of 14 mg/l or lower, however, were unable to support viable sponge growth, probably because the medium lacked sufficient buffer strength to maintain pH at a level where growth could occur. Initial bicarbonate concentrations of 56 mg/l (prior to pH adjustment) were used in the pH experiments described below. That concentration was chosen because it provides good buffering capacity over a wide range of pH and because it is within the range of concentrations found in natural waters.

Cultures of *Ephydatia fluviatilis* were maintained in media buffered at different pH values to determine the effect of culture pH on sponge growth rate. The bacterial pulse, 10^8 cells/day, was sufficient to produce maximum growth rate at pH 7.0. The results presented in Table 2 suggest that culture pH over the range 6.7 to 7.4 does not affect sponge growth rate. At pH 6.6, sponge growth rate was reduced about 64%. At pH 6.5 and lower, no growth took place. At pH values of 7.5 and 7.6, sponge growth rates were reduced by 20% and 56% respectively. At pH 7.7 and higher, no growth was measurable.

Discussion

The effects of culture medium pH on the growth rate of *Ephydatia fluviatilis* in laboratory culture (Table 2) correlate well with field observations of *E. fluviatilis* in Louisiana waters ranging in pH from 6.5 to 7.7 (Poirrier, 1974). However, Moore (1953) reported *E. fluviatilis* from a drainage ditch in Louisiana with a lower pH range (5.9 to 6.8), and Old (1932) reported it in Michigan waters with a higher pH range (7.1 to 8.0). A precise correlation between laboratory and field data should not be expected because laboratory data pertain to sponge growth rate wheras field data pertain to the presence or absence of sponges. Sponges, may exist in nature outside the pH growth range for short periods of time during which no growth occurs and still be able to resume growth when favorable pH conditions return. Other explanations for

Table 2. Growth rate of *Ephydatia fluviatilis* in defined medium at different pH values

pH	Growth rate (specific area/d)	pH	Growth rate (specific area/d)
6.5	0.00	7.2	0.25
6.6	0.09	7.3	0.25
6.7	0.25	7.4	0.25
6.8	0.25	7.5	0.20
6.9	0.25	7.6	0.11
7.0	0.25	7.7	0.00
7.1	0.25		

differences between laboratory and field data may lie in genetic differences and different physiological adaptations among populations.

It is difficult to evaluate field data expressed as pH ranges when no information is provided on the number of measurements at particular pH values and when the condition of the sponge at the time of measurement is not indicated. The drainage ditch studied by Moore (1953) was a temporary habitat subject to fluctuating limnological conditions due to seasonal filling and drying. It is possible that the low-pH values recorded occurred during brief and unusual conditions. Poirrier (1974) and Old (1932) did not find *Ephydatia fluviatilis* in drainage basins with permanent low pH conditions. The absence of *E. fluviatilis* from these low pH waters supports laboratory findings. At the same time, Old (1932) reported *E. fluviatilis* occurring at pH values higher than those at which we could achieve growth in the laboratory, but it may be that those measurements were taken at marginal pH values where no growth or little growth occurred; genetic differences between Louisiana and Michigan populations is also a reasonable explanation. Although laboratory and field data are not in complete agreement, they provide good support for the conclusion that pH has a deterministic effect on the distribution of *E. fluviatilis* in nature.

One of the advantages of the culture system employed in the experiments reported here is the wide range of sponge growth rates available to the experimenter through manipulation of the bacterial pulse or the culture medium. We have exploited that feature in different ways in other experimental work. For example, we have examined the effects of calcium concentration and salinity (total dissolved salts) on the growth rate of *Ephydatia fluviatilis* in a 2^2 factorial experiment (Francis et al., 1982). An analysis of variance of the growth rate data indicated: (1) a significant calcium effect where growth rate increased with increasing calcium concentration; (2) a significant salinity effect such that growth rate decreased with increasing salinity; and (3) a significant interaction where the negative effect of higher salinity was overcome by increasing calcium concentration. The results agree with field observations of the sponge in Louisiana (Poirrier, 1974), where *E. fluviatilis* is most common in waters with relatively high calcium concentration and with relatively low salinity. The combined laboratory and field information supports the view that calcium concentration and salinity are important factors in determining the distribution of *E. fluviatilis* in Louisiana.

We have examined the effect of culture water temperature on growth rate of both *Spongilla alba* and *Ephydatia fluviatilis* (Harsha et al., 1983) to determine whether water temperature may have a deterministic effect on the seasonal life cycles of these species. (We have established a growth curve for *S. alba* (Harsha et al., 1983) similar to

that reported here for *E. fluviatilis*.) *E. fluviatilis* is active predominantly during cold months in Louisiana whereas *S. alba* is active predominantly during warm months. In laboratory culture of these two species one might therefore expect relatively faster growth rates for *E. fluviatilis* at lower temperatures than at higher temperatures and the opposite for *S. alba*—that is, relatively faster growth rates at higher temperatures than at lower temperatures. That expectation was realized. *S. alba* did not grow at 25°C or lower. The growth rate was slow at 28°C and then increased 400% as water temperature was raised to 30°C. At 32°C the growth rate was again very slow. *E. fluviatilis*, on the other hand, was capable of vigorous growth at water temperatures as low as 17°C. This result is especially interesting when compared with the observation that *S. alba* did not grow at all in experimental culture at 25°C or lower. The growth rate of *E. fluviatilis* increased with water temperature up to about 30°C, and then decreased rapidly at water temperatures above 30°C.

These culture results correlate well with field observations of *Spongilla alba* and *Ephydatia fluviatilis* in Louisiana. *S. alba* goes through its entire life cycle in water temperatures from 22°C to 37°C. It hatches from gemmules in the spring when water temperatures rise from 22°C to 25°C, it is active throughout the summer in water temperatures from 26°C to 31°C (which reach 37°C in shallow water on warm afternoons); and forms gemmules in mid-fall when water temperatures fall below 22°C (Poirrier, 1976). *E. fluviatilis*, on the other hand, hatches from gemmules in August and September in water temperatures decreasing from 29°C to 22°C. It is most active in water temperatures from 17°C to 30°C during fall and spring, and gemmulates in late spring when water temperatures rise above 30°C (Poirrier, 1974). The combined information from field observation and laboratory culture of *S. alba* and *E. fluviatilis* suggests that water temperature is a causal factor in the seasonal life cycles of these species.

A comparison of the growth curves for *Ephydatia fluviatilis* (Figure 1) and *Spongilla alba* (Harsha et al., 1983) reveals that *E. fluviatilis* utilizes bacterial cells *(Escherichia coli)* more efficiently to produce growth than does *S. alba*. In general, *S. alba* requires about ten times more bacterial cells than does *E. fluviatilis* to achieve a given limited growth rate, yet both species grow at the same maximum rate when bacterial cells are provided in excess. One explanation for this differential utilization of bacterial cells is differential uptake of bacterial cells by the two sponge species. Another explanation is that bacterial cells, although taken up at the same rate, are metabolized differently by the two sponge species; that is, differential digestion or egestion of bacterial cells is a contributing factor.

We tested the differential uptake hypothesis by measuring particle uptake rates with two sizes of latex beads and two sizes of bacterial cells in cultures of both sponge species under several different experimental conditions

(Francis and Poirrier, 1986). Our results warrant rejection of the uptake hypothesis—that is, that differential utilization of bacterial cells *(Escherichia coli)* by *Ephydatia fluviatilis* and *Spongilla alba* to produce growth is the result of the differential uptake of bacterial cells. The four particle types (two sizes of latex beads and two sizes of bacterial cells) were taken up at about the same rate by the two sponge species under several different experimental conditions. Also, in uptake experiments employing bacteria, a given growth-limiting bacterial pulse concentration established a much higher limited growth rate in *E. fluviatilis* than in *S. alba,* even though both sponge species took up the same number of bacterial cells per time. These observations, in conjunction with the additional observation that both sponge species grow at the same maximum rate when bacterial cells are provided in excess, are consistent with the hypothesis of differential metabolism (that is, differential digestion or egestion) of bacterial cells to produce growth.

Most of our applications of the culture system have addressed questions of ecological and physiological interest. In another application of the culture system, we have developed an experimental system for examining some aspects of sponge development. The system involves degeneration of a functional sponge into a reduction body in response to pH stress and then regeneration of a fully functional sponge from the same tissue when the pH stress is relieved. (The rationale for this developmental system is provided in two recent papers: Harrison and Davis, 1982, and Francis, 1984). Good reduction bodies form at pH 6.5 or lower. The lower the pH, the faster the response. When the pH stress is relieved—that is, pH returns to 7.0—regeneration of a fully functional sponge from the reduction body is realized in about four days under standard culture conditions. One feature of this system that is particularly useful for studying development is that degeneration (reduction) and regeneration can be produced in the same tissue and the rate at which both events occur can be controlled through manipulation of pH and bacterial pulse (i.e., feeding rate).

Finally, some aspects of the culture medium require further explanation. The medium was designed to represent the habitat of the sponge in southern Louisiana where it is abundant in estuarine waters with relatively high salinity. The concentrations of Na^+, Ca^{++}, Mg^{++} and Cl^- are higher than is characteristic of freshwater. Laboratory culture of *Ephydatia fluviatilis* should be possible in media with lower salinity, but we have not investigated that possibility. The ions Cu^{++}, Zn^{++} and Co^{++} are required at concentrations of 10^{-8} to 10^{-9} M for optimal sponge growth, that is, growth determined by the bacterial pulse. If any one of the ions is left out of the medium, there is no growth, just gradual deterioration of sponge tissue and eventual death. Reduction bodies do not form. At 10^{-6} M, roughly the concentration of Cu^{++} and Zn^{++}

in New Orleans tapwater, all three ions are toxic. They produce a relatively rapid deterioration of sponge tissue, but, again, reduction bodies do not form. The response is concentration dependent—the higher the concentration, the more rapid the deterioration. Co^{++} is required for B-12 synthesis. Cu^{++} and Zn^{++} are protein cofactors as well as chromatin components. A sponge culture system using the bacterium *Escherichia coli* as food source must, therefore, contain medium supplements of both Cu^{++} and Co^{++} since neither ion, in contrast to Zn^{++}, is a requirement for growth of the bacterium (Francis and Beary, 1983).

Conclusions

A laboratory culture system for *Ephydatia fluviatilis* was used to measure the effect of medium pH on sponge growth rate. Experimental results correlate well with field observations of the sponge, suggesting that pH probably has a deterministic effect on the distribution of *E. fluviatilis* in nature.

Literature Cited

Cheatum, E. P., and J. P. Harris. 1953. Ecological Observations Upon the Freshwater Sponges in Dallas County, Texas. *Field Laboratory,* 23:97–103.

Francis, J. C. 1984. Reduction Body Formation and Subsequent Regeneration of *Ephydatia fluviatilis* in Laboratory Culture. *Transactions, American Microscopical Society,* 103:347–352.

Francis, J. C., and D. A. Beary. 1983. Chemostat Competition of Alkaline Phosphatase Regulatory Mutants in *Escherichia coli. Canadian Journal of Microbiology,* 29:174–180.

Francis, J. C., and M. A Poirrier. 1986. Particle Uptake in Two Freshwater Sponge Species, *Ephydatia fluviatilis* and *Spongilla alba. Transactions, American Microscopical Society,* 105:11–20.

Francis, J. C., M. A. Poirrier, and R. A. LaBiche. 1982. Effects of Calcium and Salinity on the Growth Rate of *Ephydatia fluviatilis. Hydrobiologia,* 89:225–229.

Harrison, F. W. 1974. Sponges (Porifera: Spongillidae). Pages 29–63 in *Pollution Ecology of Freshwater Invertebrates,* edited by C. W. Hart and S. L. H. Fuller. New York: Academic Press.

Harrison, F. W., and D. W. Davis. 1982. Morphological and Cytochemical Patterns during Early Stages of Reduction Body Formation in *Spongilla lacustris. Transactions, American Microscopical Society,* 101:317–324.

Harsha, R. E., J. C. Francis, and M. A. Poirrier. 1983. Water Temperature: A Factor in the Seasonality of Two Freshwater Sponge Species, *Ephydatia fluviatilis* and *Spongilla alba. Hydrobiologia,* 102:145–150.

Jewell, M. E. 1935. An Ecological Study of the Freshwater Sponges of Northeastern Wisconsin. *Ecological Monographs,* 5:461–504.

———. 1939. An Ecological Study of the Freshwater Sponges of Wisconsin, II. The Influence of Calcium. *Ecology,* 20:11–28.

Moore, W. G. 1953. Louisiana Freshwater Sponges, With Ecological Observations on Certain Sponges of the New Orleans Area. *Transactions, American Microscopical Society,* 32:24–32.

Old, M. C. 1932. Environmental Selection of the Freshwater Sponges (Spongillidae) of Michigan. *Transactions, American Microscopical Society,* 51:129–136.

Penney, J. T. 1954. Ecological Observations on the Freshwater Sponges of the Savannah River Project Area. *University of South Carolina Publication Series, Biology,* 1:156–172.

Poirrier, M. A. 1969. Louisiana Freshwater Sponges: Ecology, Taxonomy and Distribution. Ph.D. Dissertation, Louisiana State University, Baton Rouge. 173 pp.

———. 1974. Ecomorphic Variation in Gemmoscleres of *Ephydatia fluviatilis* with Comments upon Its Systematics and Ecology. *Hydrobiologia,* 44:337–347.

———. 1976. A Taxonomic Study of the *Spongilla alba, S. cenota, S. wagneri* Species Group with Ecological Observations of *S. alba.* Pages 203–213 in *Aspects of Sponge Biology,* edited by F. W. Harrison and R. R. Cowden. New York: Academic Press.

Poirrier, M. A., J. C. Francis, and R. A. LaBiche. 1981. A Continuous–Flow System for Growing Freshwater Sponges in the Laboratory. *Hydrobiologia,* 79:255–259.

Racek, A. A. 1969. The Freshwater Sponges of Australia (Porifera: Spongillidae). *Australian Journal of Marine and Freshwater Resources,* 20:267–310.

Wurtz, C. B. 1950. Freshwater Sponges of Pennsylvania and Adjacent States. *Notulae Naturae, Academy of Natural Sciences, Philadelphia,* 228:1–10.

MICHAEL A. POIRRIER
MARGARET WINTER
Department of Biological Sciences
University of New Orleans
New Orleans, Louisiana 70148

Gemmule Pneumatic Layer Development in Louisiana Populations of *Spongilla lacustris*

Abstract

In field studies of *Spongilla lacustris* in Louisiana the structure of gemmules varied with the chemical content of the habitat water. The thickness of the pneumatic layer and the number of gemmoscleres appeared to increase with increasing pH and alkalinity, and decreasing free carbon dioxide. Experimental studies were conducted to test these field observations. Cuttings of *S. lacustris* were placed in chemically defined media until gemmules formed. Media with different alkalinities were prepared by adding different amounts of $CaCO_3$ (5–75 mg/l). The thickness of the pneumatic layer and the number of gemmoscleres increased with the amount of $CaCO_3$ added. Maximum gemmule development occurred at 25 mg/l $CaCO_3$. Unlike typical branched forms from northern United States and Europe, Louisiana specimens of *S. lacustris* were always encrusting and always had short microscleres (40–60 μm in length). *S. lacustris* was most common in acidic streams and was active throughout the year.

Spongilla lacustris L. (Spongillidae) is a common freshwater sponge that occurs throughout the northern hemisphere (Penney and Racek, 1968). It is a variable species with intraspecific differences in gemmule, spicule, and colonial morphology. This variation has caused considerable taxonomic confusion. Different forms were first described as distinct species but later regarded as varieties of *S. lacustris* (Potts, 1887).

The degree of development of the gemmule pneumatic layer in *Spongilla lacustris* is known to vary widely (Potts, 1887; Girod, 1899; Stephens, 1920; Sasaki, 1939; Jorgensen, 1946; Penney and Racek, 1968; and Gilbert and Simpson, 1976). Early workers attempted to associate different phenotypes with a particular habitat or region, but this notion was abandoned when Jorgensen (1946) reported gemmules with and without a pneumatic layer from the same locality. Penney and Racek (1968) suggested that differences in pneumatic layer development might be associated with speciation trends among populations. Gilbert and Simpson (1976) reported that gemmules with and without pneumatic layers are produced in *S. lacustris* during different seasons in the same habitat.

Our field studies of Louisiana Spongillidae indicated that the degree of development of the gemmule pneumatic layer and associated gemmoscleres in *Spongilla lacustris* may vary with the chemical composition of the habitat water. Laboratory studies were conducted to test this hypothesis and to determine whether other factors not monitored may also have been responsible. Laboratory studies focused on the effects of alkalinity (the total quantity of base that can be determined by titration with a strong acid expressed as milligrams per liter equivalents of $CaCO_3$) on pneumatic layer development, because it was thought that it might be the most important factor. Calcium is related to alkalinity in most natural waters because bicarbonate forms from the dissolution of $CaCO_3$. During the experimental studies, we added different amounts of $CaCO_3$ to synthetic habitat water and observed the thickness of the pneumatic layer of gemmules that formed in sponges exposed to the different media. We also took into account the ecology and taxonomy of Louisiana populations of *Spongilla lacustris*.

Material and Methods

FIELD STUDIES. Samples of *Spongilla lacustris* were obtained from diverse aquatic habitats throughout Louisiana. Temperature, free carbon dioxide, pH, and alkalinity, were measured in water where *S. lacustris* was found. Analytical methods for chemical parameters followed Standard Methods (American Public Health Association, 1976), and collecting methods and preparation of study material followed Pennak (1978).

EXPERIMENTAL STUDIES. Sponges were collected from Abita Creek at Louisiana Highway 435, St. Tammany Parish, Louisiana, during April and June of 1979. Green sponges, 2 mm thick and without gemmules, were removed from wood substrata and transported to the laboratory in plastic containers filled with habitat water. In the laboratory, sponges were cut into 2-cm^2 pieces while submerged in habitat water.

Water from Abita Creek was analyzed for major ionic components, and a synthetic medium was prepared to match this ionic composition; 5 mg KCl, 6.6 mg $MgSO_4 \cdot 7H_2O$ and 45.7 mg $Na_2SiO_3 \cdot 9H_2O$ were added to each liter of distilled water. Seven experimental media were prepared by adding the following amounts of $CaCO_3$ in mg to each liter of experimental solution: 0, 5, 10, 15, 20, 25, and 75. An additional experimental solution was prepared with the sodium metasilicate omitted and with 10 mg/l $CaCO_3$ added. The pH of each solution was adjusted to 6.2 (the pH of Abita Spring water) with 0.12 M HCl.

Sponge cuttings were placed in 24 plastic dishes measuring 6.4 × 6.4 × 8.9 cm containing 400 ml of the experimental solution. Three replicates of each experiment were performed. The plastic dishes were placed in a Percival Instruments (Boone, Iowa) environmental chamber and maintained at 17–21°C in total darkness for one month. Gemmules usually form in two weeks under these conditions, but sponges were held for one month to ensure that complete gemmulation had occurred.

After one month, sponges were removed and pH and alkalinity of the experimental media measured. Gemmules were removed from sponges and placed in 95% ethyl alcohol for 5 minutes, air-dried, soaked in xylene until it penetrated and cleared the pneumatic layer, and then mounted on microscope slides with balsam. The thickness of the pneumatic layer was measured using a compound microscope equipped with an ocular micrometer. A linear regression analysis (Remington and Schork, 1970) was used to determine whether a relationship existed between pneumatic layer thickness and the amount of calcium carbonate added.

Results

FIELD STUDIES. *Spongilla lacustris* was collected from 12 streams and 3 lakes in Louisiana. It was most abundant in the headwaters of sandy-bottomed streams less than 10 m wide. Other sponges found with it were *Eunapius mackayi*, *Trochospongilla pennsylvanica*, and *Anheteromeyenia ryderi*.

Spongilla lacustris colonies were usually green, but tan, gray, and black specimens were found in heavily shaded areas. They were always of the encrusting type ranging from 0.1 to 1 cm in thickness. They were found throughout the year but were most abundant during the winter.

The microscleres measured 40–60 μm in length and the megascleres 220–350 μm. The number of microscleres and the thickness of the pneumatic layer varied. Forms with few microscleres in the dermal membrane and with thin pneumatic layers in the gemmule were found in one lake and in the headwaters of streams during periods of high water. Forms with numerous microscleres and well-developed pneumatic layers were found in downstream areas, standing-water habitats, and the headwaters of streams during periods of low water.

Chemical data from sites where gemmules with and without pneumatic layers were collected are presented in Table 1. Overall, these data indicate that in Louisiana, *Spongilla lacustris* occurs in acidic waters which are high in free carbon dioxide and low in alkalinity. Sponges in which the gemmules had thin pneumatic layers and few gemmoscleres were from habitats with low pH and alkalinity (Table 1). Sponges with thick pneumatic layers

Table 1. Chemical data from sites where *Spongilla lacustris* gemmules with and without pneumatic layers were obtained

Chemistry	Pneumatic layer absent (10 sites)		Pneumatic layer present (4 sites)	
	Range	Mean	Range	Mean
pH	5.0–6.6	5.9	6.5–6.9	6.7
Free CO_2 (ppm)	5–24	15.6	5.5–11	8.8
Alkalinity (ppm)[a]	15–15	11.6	23–75	44.3

[a]Equivalent $CaCO_3$ in mg/l.

and numerous gemmoscleres in almost a radial arrangement were from habitats with high pH and alkalinity, and low free CO_2 values (Table 1).

EXPERIMENTAL STUDIES. The thickness of the pneumatic layer increased with increasing concentration of $CaCO_3$ in the experimental medium (Table 2). At 25 mg/l $CaCO_3$, the thickness of the pneumatic layer reached 43 to 50 μm; higher concentrations produced only slight increases in thickness. The thickness of the pneumatic layer as a function of $CaCO_3$ concentration is presented graphically in Figure 1. Linear regression analysis of these data indicated a high positive relationship ($R^2 = 0.97$) with the pneumatic layer increasing 1.76 μm per mg/l $CaCO_3$; the maximum layer thickness observed in nature in Louisiana was 50 μm. A comparison of these field and laboratory results suggests that maximum development of the pneumatic layer in nature probably occurs at $CaCO_3$ concentrations of about 25 mg/l.

Gemmules with a thin pneumatic layer (2.5–3.5 μm) and no gemmoscleres were produced in medium 9, which did not contain sodium metasilicate (Table 2). There was no correlation between $CaCO_3$ concentration and the number of gemmules produced. Fewer gemmules were formed in media with less than 15 mg/l $CaCO_3$ and also in the medium lacking silica, while the greatest number of gemmules formed in the medium with 15 mg/l $CaCO_3$.

The number and arrangement of gemmoscleres in laboratory culture varied with the thickness of the pneumatic layer. This result coincides with our field observations of *Spongilla lacustris*.

Alkalinity measurements taken at the end of our experiments were reasonably close to the levels expected on the basis of the amount of $CaCO_3$ added, but pH values increased from 6.2 at the beginning of the experiment to values near neutrality—6.7 to 7.2—at the end. The increase in pH was probably due to the release of carbon dioxide from the medium.

Discussion

Louisiana populations of *Spongilla lacustris* differ from the population described by Penney and Racek (1968) in that they are never branched and the microscleres are always small. The microscleres of Louisiana specimens range from 40 to 60 μm in length, whereas those reported by Penney and Racek (1968) typically range from 70 to 130 μm. These differences are not readily explained because encrusting forms with long microscleres and branching forms with short microscleres have been reported from several areas, notably the northern United States, Ireland, and Japan (Potts, 1887; Stephens, 1920; Sasaki, 1939;). At the same time, we have found this short-microsclere form elsewhere in in the southeastern United States (Alabama, Florida, Georgia, Mississippi, Texas, and South Carolina), but have never observed the branched morph in this area.

The two morphs also differ ecologically. *Spongilla lacustris* from Michigan waters is reported to have a broad pH range and a slight preference for alkaline waters (Old, 1932). In Wisconsin its distribution does not appear to be related to water calcium content (Jewell, 1935, 1939). In Louisiana, on the other hand, *S. lacustris* appears to be limited to acidic waters with high concentrations of free carbon dioxide and low alkalinities. Unlike northern U.S. populations, Louisiana specimens are active throughout the year with no seasonal period of synchronous gemmule

Table 2. Comparison of gemmule coat thickness and number of gemmules produced in seven experimental media with different amounts of $CaCO_3$ added, and one medium with 10 mg/l $CaCO_3$ and no Na_2SiO_3 added (three replicate experiments were performed)

Gemmules (experiment)		$CaCO_3$ addition to medium (mg/l)							
		0	5	10	15	20	25	75	10[a]
Coat thickness (μm)	(1)	7.3	12.4	19.3	18.8	37.8	43.1	56.8	3.5
	(2)	2.2	5.5	18.5	30.3	41.3	43.4	50.7	2.2
	(3)	3.6	13.5	15.4	27.2	42.8	49.5	50.9	2.8
Number produced	(1)	6	6	8	48	10	12	12	4
	(2)	5	8	7	8	4	9	9	5
	(3)	5	11	5	36	9	10	12	7

[a]No Na_2SiO_3 added

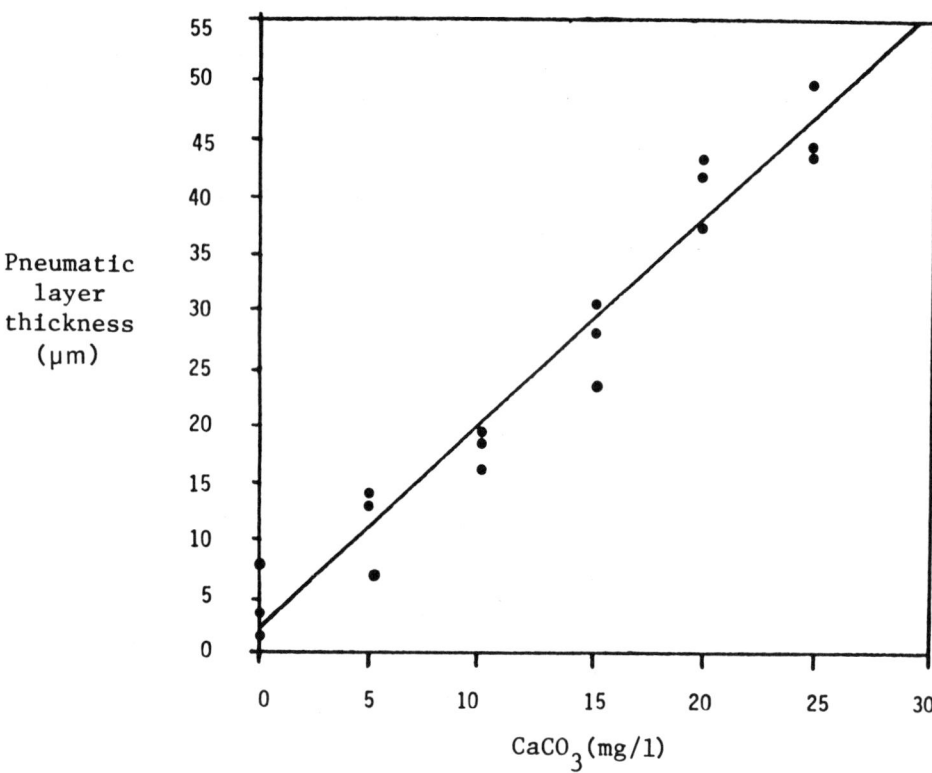

Figure 1. Graph showing relationship between mean thickness of gemmule pneumatic layers and amount of $CaCO_3$ added to six experimental media. Three replicate experiments were performed.

formation. There may indeed be genetic differences between the two forms, but current data are too limited to assign them to different species.

The high correlation between $CaCO_3$ concentration and the thickness of the pneumatic layer (Figure 1) suggests that gemmule development is affected by chemical conditions associated with alkalinity. Since calcium, bicarbonate, and free carbon dioxide increased with $CaCO_3$ concentration and pH varied slightly among experiments, it is not clear which of these factors directly affects pneumatic layer development. Since calcium has been shown to affect distribution (Jewell, 1939), growth (Francis et al., 1982) and gemmule hatching (Ostrom and Simpson, 1978) in freshwater sponges, it is probably the most important factor in this relationship.

Our experimental studies also demonstrated that gemmules with thin pneumatic layers and no gemmoscleres are produced in the absence of silica. The absence of gemmoscleres would be expected because of their siliceous composition. However, the thin pneumatic layer in gemmules without gemmoscleres might indicate that gemmoscleres are necessary for maximum pneumatic layer development.

How calcium might affect pneumatic layer development is unclear. The pneumatic layer is composed of collagen (Simpson, 1984), and calcium is not known to be a major component. However, calcium may affect the synthesis of pneumatic collagen or gemmoscleres. Thus thin layers would be expected in calcium-poor waters. The

precise function of the gemmular coat is also unknown, but a thick coat with numerous gemmoscleres probably provides a better environmental barrier to predation by small invertebrates, such as snails and insects, digestion by fish and waterfowl, and degradation by bacteria and fungi. Because acidic, calcium poor waters are generally lower in biological activity and diversity than waters with higher calcium concentrations, the production of thick gemmule coats may be an adaptation not needed in calcium-poor waters. The gemmular coat may also perform other functions; for example, it may help increase resistance to complete desiccation or provide buoyancy. The calcium content of temporary habitats increases through evaporation and, after drying, fluid in the pneumatic layer is replaced by air. Gemmules with thick coats float after drying. Floating gemmules would be transported by surface currents and perhaps exposed to surface-feeding fish and waterfowl. If the thick coat provides more resistance to vertebrate digestion, this might provide another means of dispersal.

Although the nonbranching, short-microsclere form of *Spongilla lacustris* found in the southeastern United States may be different from the typical branched form, gemmule variation in the branched form is probably also related to calcium. Girod (1899) reported *S. lacustris* gemmules without pneumatic layers from the mountain lakes of Auvergne, in France and found that the pneumatic layer and number of gemmoscleres increased downstream in the Allier River and its tributaries. The variation he

3d Int. Sponge Conf. 1985

observed was probably due to a downstream increase in minerals, including calcium usually found in streams. In reviewing the taxonomy of *S. lacustris*, Penney and Racek (1968) observed that the majority of cold-temperate and subarctic specimens they examined had gemmules with ill-defined or no pneumatic layers, whereas, specimens from central and southern Europe had gemmules with thick, pneumatic coats that contained gemmoscleres. They postulated that future studies might indicate a speciation trend in these populations. We suggest that these differences may be due to the generally lower calcium content in the waters of coniferous and tundra habitats in comparison with the waters of central and southern Europe.

Gilbert and Simpson (1976) found that, in New Hampshire populations, gemmules with thick pneumatic layers were produced in the summer and gemmules with thin pneumatic layers were produced in October. Thin gemmules were rare and were restricted to the bases of encrusting growths and along the axes of thick branches. They did not give water chemistries, but others studying primary production in *Spongilla lacustris* from the same habitat (Gilbert and Allen, 1973) have reported an alkalinity of 8.6 mg $CaCO_3$/l. The production of gemmules with thin pneumatic layers would be expected at this low alkalinity. The production of a few well-developed gemmules in areas of thick sponge growth during the summer is probably due to different conditions in the thicker, older regions of the sponge. Gemmules can generally be found at the base of large colonies of most freshwater sponges during the growing season. Louisiana specimens of *S. lacustris* with different degrees of gemmule development often had thick layered gemmules at the base and thin layered gemmules in other regions of the colony.

The coat of gemmules of *Ephydatia fluviatilis* and *Anheteromeyenia ryderi* also exhibits different degrees of development under different chemical conditions (Poirrier, 1974, 1977). In these species, the number, length, and spination of the birotulate gemmoscleres increase as the mineral content of water increases. In gemmules with maximum development, the rotules of the gemmoscleres overlap and the spined shafts fit close together in a thick pneumatic layer. As in *Spongilla lacustris*, the thicker coat with more gemmoscleres appears to be an adaptation to provide a superior environmental barrier for the protection of the gemmular contents. This morphological trend should not be confused with gemmosclere malformations that are caused by adverse environmental conditions. It is not clear how water chemistry affects this relationship and whether the calcium concentration affects species other than *S. lacustris*. Because gemmosclere morphology has been regarded as an important taxonomic character in freshwater sponges, the functional morphology of the gemmoscleres of variable species should be investigated to provide a better understanding of their taxonomy.

3d Int. Sponge Conf. 1985

Conclusions

Louisiana populations of *Spongilla lacustris* differ from the typical branched form reported from the northern United States and Europe. All the specimens we examined were encrusting, had short microscleres, occurred in acidic environments generally low in calcium, and did not exhibit synchronous gemmule formation during autumn. Our field and laboratory studies demonstrated a correlation between water alkalinity and the thickness of the pneumatic layer in the gemmule coat. We think that calcium is the most important factor in this relationship. Calcium concentration increases with alkalinity in most waters, and it has known effects on freshwater sponges. The thick coat appears to be an adaptation that provides better protection for the gemmule contents. How calcium affects pneumatic layer development is unknown. Since higher calcium concentrations are usually associated with drying and high biological productivity in habitats, the production of thick coats may be an adaptation to provide increased resistance to drying, predation, or digestion not needed in calcium-poor environments.

Acknowledgments

We thank John Francis, Roy Ary, and Robert Cashner for their suggestions and help in preparing the manuscript, and Marie LeBlanc for her good humor, patience, and typing skill.

Literature Cited

American Public Health Association. 1976. *Standard Methods for the Examination of Water and Wastewater* (14th ed.). New York: American Public Health Association. 1193 pp.

Francis, J. C., M. A. Poirrier and R. A. LaBiche. 1982. Effects of Calcium and Salinity on the Growth Rate of *Ephydatia fluviatilis* (Porifera: Spongillidae). *Hydrobiologia*, 89:225–229.

Gilbert, J. J., and H. L. Allen. 1973. Chlorophyll and Primary Productivity of Some Green Freshwater Sponges. *Internationale Revue Gesamten Hydrobiologie*, 58:633–658.

Gilbert, J. J., and T. L. Simpson. 1976. Gemmule Polymorphism in the Freshwater Sponge *Spongilla lacustris*. *Archiv für Hydrobiologie*, 78:268–277.

Girod, P. 1899. Les Eponges des eaux douces d'Europe. *Le Micrographie Preparateur*, 7:106–115.

Jewell, M. E. 1935. An Ecological Study of the Fresh-Water Sponges of Northeastern Wisconsin. *Ecological Monographs*, 5:461–504.

————. 1939. An Ecological Study of the Fresh-Water Sponges of Wisconsin. II. The Influence of Calcium. *Ecology*, 20:11–28.

Jorgensen, C. B. 1946. On the Gemmules of *Spongilla lacustris* auct. Together with some Remarks on the Taxonomy of the Species. *Videnskabelige Meddelelser Dansk Nafraurhistorisk Forening*, 109:69–79.

Old, M. C. 1932. Environmental Selection of the Fresh-Water Sponges (Spongillidae) of Michigan. *Transactions, American Microscopical Society*, 51:129–136.

Ostrom, K. M., and T. L. Simpson. 1978. Calcium and the Release from Dormancy of Freshwater Sponge Gemmules. *Developmental Biology*, 64:332–338.

Pennak, R. W. 1978. *Fresh-Water Invertebrates of the United States*. New York: Ronald Press. 803 pp.

Penney, J. T., and A. A. Racek. 1968. Comprehensive Revision of a World-wide Collection of Freshwater Sponges (Porifera: Spongillidae). *United States National Museum, Bulletin*, 272:1–184.

Poirrier, M. A. 1974. Ecomorphic Variation in Gemmoscleres of *Ephydatia fluviatilis* Linnaeus (Porifera: Spongillidae) with Comments Upon its Systematics and Ecology. *Hydrobiologia*, 44:337–347.

———. 1977. Systematic and Ecological Studies of *Anheteromeyenia ryderi* (Porifera: Spongillidae) in Louisiana. *Transactions of the American Microscopical Society*, 96:62–67.

Potts, E. 1887. Contribution towards a Synopsis of the American Forms of Fresh-Water sponges. *Proceedings, Academy of Natural Sciences of Philaldelphia*, 39:158–279.

Remington, R. D., and M. A. Schork. 1970. *Statistics with Applications to the Biological and Health Sciences*. Englewood Cliffs, New Jersey: Prentice–Hall. 418 pp.

Sasaki, N. 1939. Fresh-Water Sponges Obtained in South Saghalin. *Science Reports of the Tohoku Imperial University*, 9:219–247.

Simpson, T. L. 1984. *The Cell Biology of Sponges*. New York: Springer. 662 pp.

Stephens, J. 1920. The Fresh-Water Sponges of Ireland. *Proceedings, Royal Irish Academy*, 35:205–254.

PAUL E. FELL
Department of Zoology
Connecticut College
New London, Connecticut 06320

Tolerances of the Dormant Forms of Some Estuarine Sponges, Notably *Microciona prolifera*

Abstract

In laboratory experiments the dormant forms of several estuarine sponges showed different tolerances to three stresses that they may face during the winter: freezing of the habitat, exposure to low air temperatures, and low salinity. The dormant forms of *Microciona prolifera* and *Halichondria* sp., which lack choanocyte chambers, survived culture in seawater (25‰) supercooled to −5°C for 5 days, but freezing of the seawater resulted in the death of the sponges. In contrast, gemmules of *Haliclona loosanoffi* and *Prosuberites epiphytum* survived enclosure within ice at −5°C for 5 days. Dormant *Microciona* was killed by a 4-h exposure to air at −10°C, but the gemmules of *Haliclona* and *Prosuberites* were able to tolerate this stress. After a 4-h exposure to air at −15°C, the gemmules of *Haliclona* did not germinate and those of *Prosuberites* exhibited a low survival. Dormant *Microciona* survived a salinity of 10‰ at 5°C for up to 9 days and a salinity of 5‰ for up to 1 day when the sponge was gradually brought down to the test salinity and then was gradually returned to 20‰.

A number of estuarine sponges, as well as many freshwater sponges, become dormant during the winter months. Some sponges, including *Haliclona loosanoffi* and *Prosuberites epiphytum*, form gemmules (Hartman, 1958; Fell et al., 1984). Others, including *Microciona prolifera* and occasionally *Halichondria* sp., undergo tissue regression during which choanocyte chambers and frequently also canals disappear (Hartman, 1958; Simpson, 1968; Fell et al., 1984). Both overwintering states are similar in that functional tissue disappears during the dormant period and must be redeveloped in the spring when the animals become active (Fell, 1976; Simpson, in press). Sponge gemmules and other dormant forms of sponges are often assumed to be tolerant of various environmental stresses. However, this assumption is based primarily on field observations; and there is little experimental evidence to support or refute it (Fell, 1975). Furthermore, it seems probable that the tolerances of different species to specific stresses vary.

Gemmules can be stored for months or even years at low temperature and then they can be induced to develop into small sponges within a few days by raising the temperature (Rasmont, 1962; Fell, 1974). This report indicates that the dormant forms of two sponges that undergo tissue regression can be handled in a similar manner. Thus dormant sponges provide an ideal system for studying stress tolerances. The development of an active sponge from gemmules or from dormant tissue following exposure to stressful conditions provides a clear-cut criterion for survival.

Estuarine sponges of the northern temperate zone are often confronted with three potential stresses during the winter: (1) freezing of the habitat, (2) exposure to low air temperature at low tide, and (3) low salinity. This paper presents the results of a preliminary study of the tolerances of the dormant forms of several estuarine sponges to some or all of these conditions. Since almost nothing is known about the significance of tissue regression in sponges, the discussion concentrates on the tolerances of dormant *Microciona prolifera*.

Material and Methods

Erect, branching specimens of *Microciona* prolifera were collected in the lower Mystic Estuary in Connecticut during late November 1978 and early December 1983, and they were suspended in Instant Ocean (Aquarium Systems, Mentor, Ohio) or Dayno (Dayno Manufacturing Co., Lynn, Massachusetts) aquaria containing natural seawater (28–30‰) at about 3–4°C. After the sponges were maintained under these conditions for one month, they were fully dormant, lacking canals and choanocyte chambers. *Halichondria* sp. was also collected in the Mystic Estuary and handled in a similar manner. Gemmules of *Haliclona loosanoffi* (attached to eelgrass and algae) and those of *Prosuberites epiphytum* (on the shells of the gastropod *Urosalpinx cinerea*) were obtained in or near the Mystic and Thames estuaries during the late summer and fall. They were stored in the dark at 4–5°C in covered finger bowls filled with natural seawater which was changed at irregular intervals.

Small pieces of dormant sponge (ca. 5 mm in greatest dimension) and small groups of gemmules were stimulated to develop into functional sponges by culturing them under the following conditions. They were usually placed on sheets of lens paper (but occasionally on fine nylon mesh) in covered 4 inch (10 cm) finger bowls containing 100–150 ml of Instant Ocean seawater. Usually 5 pieces of sponge or clusters of gemmules were cultured in each bowl. The cultures were maintained in the dark at 20°C

(± 1°C) in B.O.D. incubators, and the seawater was changed at 5-day intervals. For most of the experiments, the salinity of the seawater, which was made with glass distilled water, was 25‰. In experiments in which the explants of dormant sponge were exposed to air or to changes in salinity, they were cultured in seawater (25‰) at 5°C for about 24 h to allow the cut surfaces to heal before the sponges were placed under the test conditions. For these short-term experiments, the sponge exlants were not fed.

The capacity of the sponges to withstand freezing of the habitat was examined by placing explants of the dormant sponge or clusters of gemmules in Instant Ocean seawater (25‰) at 5°C and then keeping them in an incubator at −5°C for 5 days. Some of the containers of seawater froze, while others supercooled. In some instances supercooled seawater was seeded with small ice crystals to bring about freezing. After 5 days at −5°C the cultures were placed at 20°C. Controls were maintained at 5°C for 5 days.

The effect of brief exposure to low air temperatures was studied by placing explants of dormant *Microciona* or clusters of gemmules on sheets of lens paper in dry finger bowls and incubating them at a low temperature for from 1 to 4 h. At the end of the exposure period, the cultures were flooded with Instant Ocean seawater (25‰) at 5°C and incubated at 20°C.

Low salinity tolerance of dormant *Microciona* was studied in two ways. In some experiments explants of dormant sponge were taken from Instant Ocean seawater at 25‰ salinity, lightly blotted on paper toweling, and placed in seawater of lower salinity (5, 10, 15, or 20‰) at 5°C for 15 days. On the 15th day the seawater was replaced with 25‰ seawater at 5°C, and the cultures were incubated at 20°C. In other experiments the dormant sponges were gradually (see below) brought down to the test salinity and then were gradually returned to a salinity of 20‰ before they were cultured at 20°C. The explants were kept one day each at 25‰ and 20‰ and then they were maintained at 15‰ for 5 days. Some explants were subsequently exposed to a salinity of 10‰ for periods ranging from one to 15 days. They were then transferred back to 15‰ seawater for 5 days before they were put in 20‰ seawater at 5°C and incubated at 20°C. Other explants were kept at 10‰ for 2 or 2.5 days and then were placed in 5‰ sea water for from 3 h to one day. After culture in 5‰ seawater, the explants were kept at 10‰ for 2 or 2.5 days and then at 15‰ for 5 days before they were transferred to 20‰ seawater at 5°C and incubated at 20°C. The explants were gently blotted on paper toweling when they were transferred from one salinity to another.

Water temperature, ice cover and salinity (surface and bottom) at the northern inlet between Bebee cove and the lower Mystic estuary were recorded at regular intervals from December 1983 through November 1984. Water samples were filtered through Whatman No. 1 filter paper, and salinities were measured with a Goldberg Refractometer (American Optical Instrument Co., Buffalo, New York). Minimum air temperature data were obtained from the water filtration plant in Groton, Connecticut.

Results

The population of *Microciona prolifera* studied in this report inhabited an area under a causeway across an inlet connecting Beebe Cove with the lower Mystic Estuary. At extreme high tide the depth of the water under the causeway was about 1.5 m, and at extreme low tide there were only isolated pools a few centimeters deep. Consequently, most of the erect, branching specimens of *Microciona* were partly or completely exposed during extreme low water of spring tides. Water temperatures dropped to below 0°C during the middle of the winter, and ice formed on both sides of the inlet. There was a rapid flow of water through the inlet and the water beneath the causeway rarely froze. Air temperatures dropped to −18°C. The bottom salinity ranged from 19‰ to 31‰ and was lowest during the winter and spring. Surface salinities sometimes fell to 16‰ (Figure 1).

In the dormant form of *Microciona* there was no evidence of subdermal spaces or of incurrent or excurrent canals. Occasionally, spicules penetrated through the surface of the sponge, especially at the tips of branches. The surface of the sponge was covered by a thin layer of material rich in diatoms. Histological study revealed a compact mass of cells in which spicules and bundles of collagen were embedded. There were no choanocyte chambers or canals (also see Simpson, 1968).

When explants of dormant *Microciona* were cultured at 20°C and at salinities ranging from 20 to 30‰, all of them ($n = 190$) developed into active sponges with canals and subdermal spaces within 5 to 8 days. A small number (5 out of 80) of the explants cultured in 25‰ seawater developed prominent oscular tubes. The activated explants had a normal tissue organization with many choanocyte chambers. Dormant *Microciona* could be stored in an Instant Ocean aquarium at 3–4°C for up to six months before being activated at 20°C.

Activated explants of *Microciona*, transferred to aerated seawater and fed Liquefy No. 1, a nutrient suspension for small fish (Interpet Ltd., Surrey, England), formed prominent oscular tubes and exhibited pumping activity. Some of them also developed extensive outgrowth zones that covered up to several times the area of the original explant. One such explant remained healthy for 88 days, at which time the experiment was terminated. Although fewer cultures of *Halichondria* were studied, they were similar to those of *Microciona*, but they typically formed prominent oscular tubes without feeding and aeration.

When explants of dormant *Microciona* or of *Halichondria* were cultured in Instant Ocean seawater (25‰) super-

Table 1. Activation of dormant explants of *Microciona prolifera* and *Halichondria* sp. and gemmules of *Haliclona loosanoffi* and *Prosuberites epiphytum* at 20° C following exposure to low water temperature or freezing for 5 days; seawater was 25‰ (− = no data)

Species	Conditions prior to attempted activation		
	+5° C	−5° C	−5° C (frozen)
M. prolifera	100% ($n = 30$)	100% ($n = 25$)	0 ($n = 40$)
Halichondria sp.	87% ($n = 63$)	100% ($n = 20$)	0 ($n = 65$)
H. loosanoffi	100% ($n = 27$)	−	100% ($n = 27$)
P. epiphytum	−		100% ($n = 2$)

cooled to −5°C for 5 days and then were cultured at 20°C, they developed into active sponges and were indistinguishable from controls maintained at 5°C for 5 days. However, if the seawater froze at −5°C, the sponges were killed (Table 1). In contrast, the gemmules of *Haliclona* and *Prosuberites* were able to survive enclosure within ice at −5°C for 5 days. However, the development of gemmules of *Haliclona*, which had been enclosed in ice, was delayed by about 2 days in comparison with controls.

Explants of dormant *Microciona* exposed to air at −5°C for 4 h exhibited 100% survival, but those exposed to air at −10°C for the same period were killed (Table 2). When explants were placed at −10°C for 1 h, most of them died; and of those that survived, many possessed only small regions (Haliclona and *Prosuberites* are more tolerant of exposure to low air temperatures than is the dormant form of *Microciona* (Table 2). Gemmules of both species exhibited good survival following exposure to air at −10°C for 4 h and those of *Prosuberites* showed a limited survival after being kept at −15°C for 4 h.

When explants of dormant *Microciona* were transferred directly from 25‰ Instant Ocean seawater at 5°C to lower salinities at the same temperature for 15 days, those maintained at 20‰ underwent no obvious change, some of the explants at 15‰ became slightly lobular, and those at 5‰ or 10‰ exhibited numerous lobulations (Figure 2, inset). Following transfer of the explants back to 25‰ seawater at 5°C and incubation at 20°C, those that had been at salinities of 15‰ or more developed into active sponges within 5–8 days (Figure 2). Histological examination of such explants revealed numerous choanocyte chambers, as well as canals and subdermal spaces. On the other hand, explants that had been maintained at 5‰ or 10‰ became moribund.

Explants of dormant *Microciona* were able to tolerate short exposures to very low salinities at 5°C when they were gradually brought down to the test salinity and then gradually returned to 20‰ Instant Ocean seawater before they were cultured at 20°C (Figure 3). Although they became very lobular, the explants exhibited good survival at 10‰ for up to 9 days and at 5‰ for up to one day. When four of the explants which had been cultured at 5‰ (three for 12 h and one for 24 h) were placed in aerated natural seawater and fed Liquify No. 1, they developed prominent

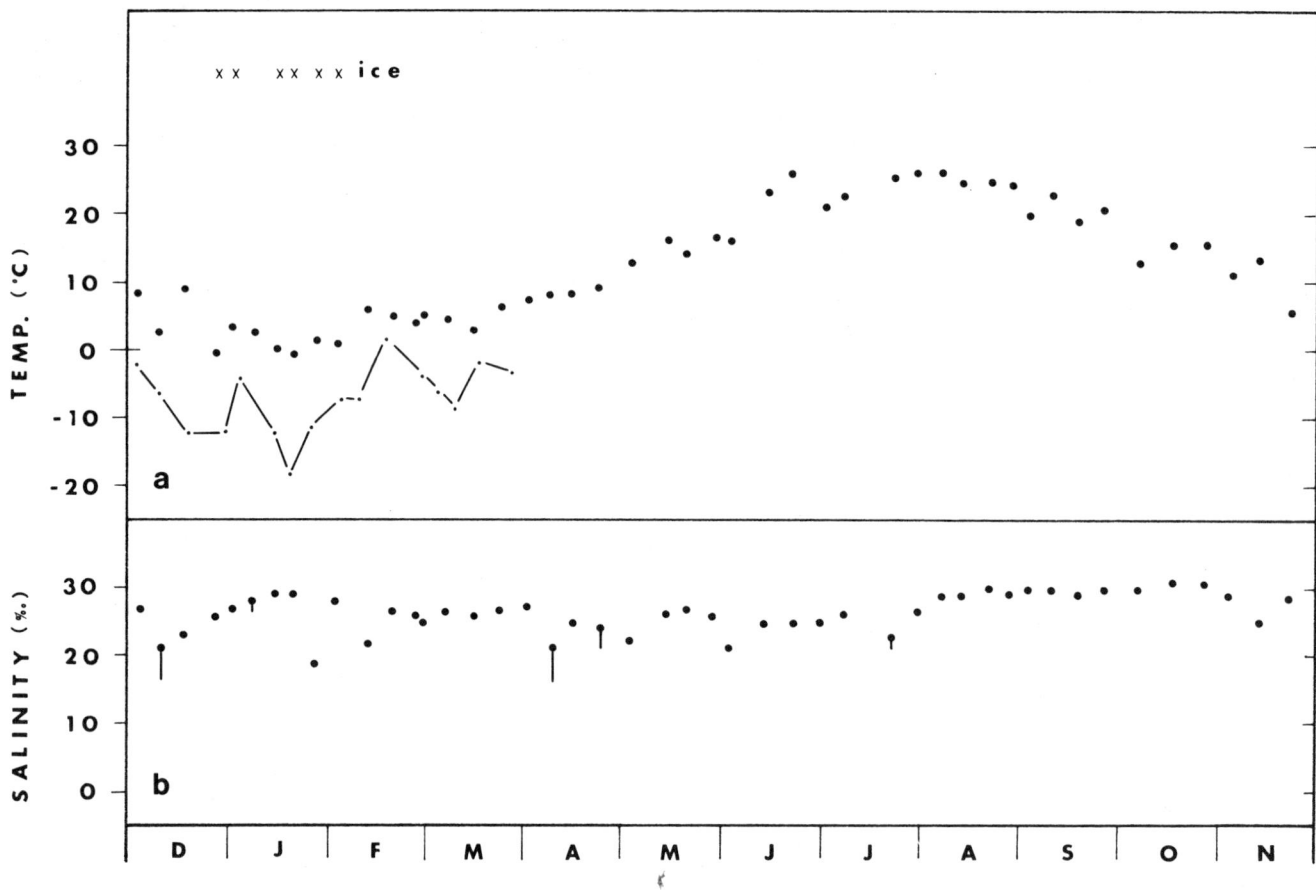

Figure 1. Environmental conditions in lower Mystic Estuary at inlet into Bebee Cove over one-year cycle (1983–1984): *a*, surface water temperatures indicated by large dots, minimum air temperatures shown by small dots connected by lines, and ice formation in estuary near study site indicated by + at top of graph; *b*, bottom salinities represented by large dots. Vertical line indicates where surface salinity was lower than that at bottom.

oscular tubes and exhibited pumping activity. Three of the four explants also formed extensive outgrowth zones. The cultures were maintained for 38–54 days, and all of the explants appeared healthy when the experiments were terminated.

Discussion

This study has shown that gemmules of *Haliclona loosanoffi* and *Prosuberites epiphytum* can tolerate enclosure in ice at

−5°C for 5 days, but that the dormant forms of *Microciona prolifera* and *Halichondria* sp. cannot. Many specimens of *Halichondria* and gemmules of *Haliclona* are sometimes found embedded in ice in the Mystic Estuary during the winter. When the ice melts, moribund specimens of *Halichondria* may be abundant. In the study area where *Microciona* is abundant, the water usually does not freeze, presumably because of its rapid flow. However, ice forms a short distance away where currents are slower. Although *Microciona* and *Halichondria* are apparently eliminated

Table 2. Activation of dormant explants of *Microciona prolifera* and gemmules of *Haliclona loosanoffi* and *Prosuberites epiphytum* at 20° C following exposure to low air temperatures (− = no data)

Species	Time (h)	Temperature				
		+5° C (water)	+5° C (air)	−5 ° C (air)	−10 ° C (air)	−15 ° C (air)
M. prolifera	1	—	95% (*n* =20)	—	30% (*n* =40)[a]	—
	4	98% (*n* =52)	100% (*n* = 20)	100% (*n* =20)	0 (*n* =48)	—
H. loosanoffi	4	—	100% (*n* = 6)	—	100% (*n* =3)	0 (*n* =3)
P. epiphytum	4	—	—	—	100% (*n* =2)	20% (*n* =5)[b]

[a]Seven of 12 cultures that showed at least some development had only small regions (10% their mass) exhibiting active tissue.
[b]Only a few of the gemmules germinated.

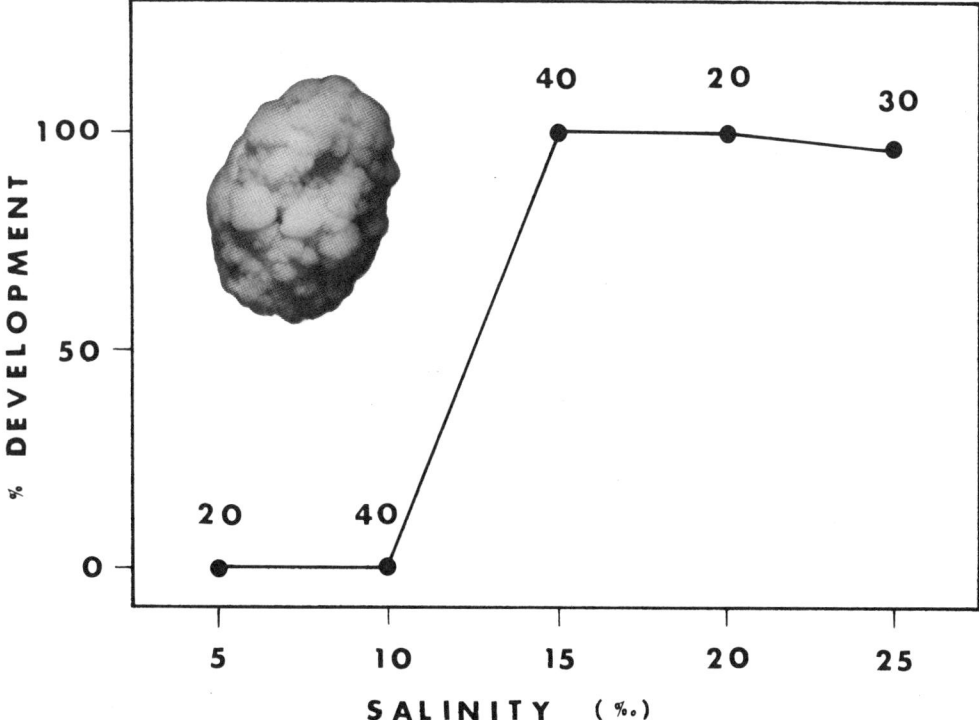

Figure 2. Development of dormant explants of *Microciona prolifera* transferred directly from 25‰ Instant Ocean seawater to lower salinities of 5°C for 15 days and then returned to 25‰ seawater at 5°C before they were cultured at 20°C. Number of explants tested shown above each point. Inset shows explant which had been at 5‰ and 5°C for 10 days. Note extensive lobulations.

from certain shallow areas with relatively slow currents when seawater freezes, gemmules of at least two estuarine sponges seem to be able to tolerate this stress. As might be expected, the gemmules of a number of freshwater sponges—*Anheteromeyena ryderi*, *Spongilla lacustris*, *Heteromeyenia tubisperma*, and *Eunapius fragilis*—can also tolerate

freezing of their habitat (Fell and Bazer, unpublished observations).

It was found that dormant *Microciona* is killed by a 4-h exposure to air at −10°C but not by a 4-h exposure to air at −5°C. In Connecticut, air temperatures fall to as low as −19°C during the winter. In midwinter, dead specimens of

Figure 3. Development of dormant explants of *Microciona prolifera* following culture in either 10‰ or 5‰ Instant Ocean seawater at 5°C for different periods of time. Explants were gradually brought down to test salinity and then gradually returned to a salinity of 20‰ before they were cultured at 20°C (see material and methods for details). Solid circles represent explants placed at 10‰ with 5 days culture at 15‰ after they were exposed to lower salinity. Triangle indicates explants kept at 15‰ for only one day following exposure to 10‰. Open circles represent explants cultured at 5‰. Number of explants tested shown above each point.

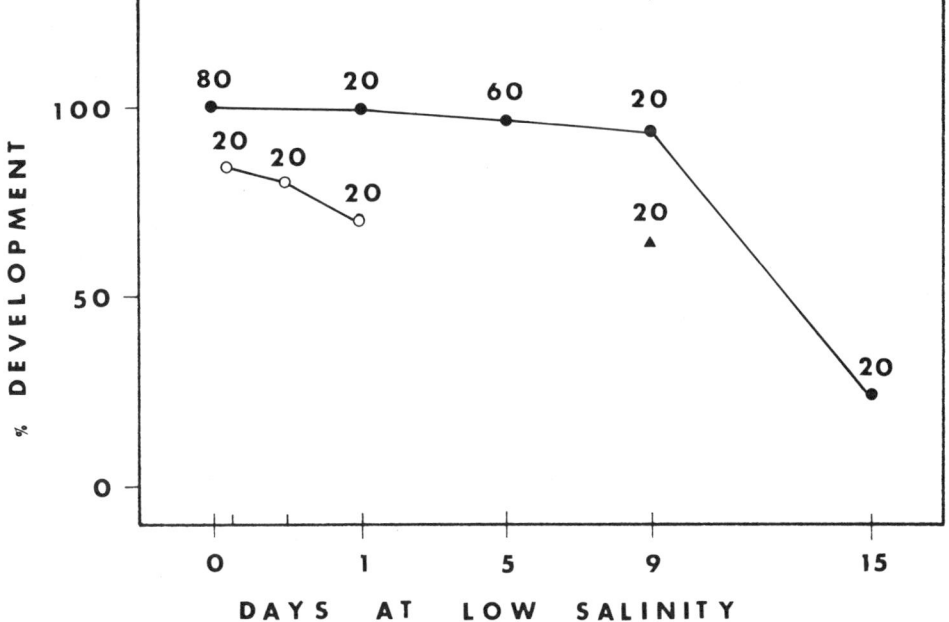

Microciona and specimens with only the tips of their branches dead are observed in the Mystic Estuary at extreme low tide. Apparently a low freezing tolerance is among the factors that limit *Microciona* to the lower intertidal (from mean low water) and subtidal zones in northern temperate regions (Hartman, 1958). Juniper and Steele (1969) reported the temporary elimination of several intertidal sponges by severe winter conditions at Portsmouth, England.

Although only a few clusters of gemmules were tested, those of *Haliclona* and *Prosuberites* appear to be somewhat more tolerant of low air temperatures than the dormant form of *Microciona*. They survive at least 4 h at −10°C, but at −15°C the gemmules of *Haliclona* are killed, and those of *Prosuberites* exhibit only a low level of survival. *Haliclona loosanoffi* generally does not extend up the shore above the mean low water level (Hartman, 1958). By contrast, the gemmules of the freshwater sponges *Anheteromeyenia ryderi* and *Eunapius fragilis* survive a 24-h exposure to air at −20°C (Bazer, Harvey, and Fell, unpublished observations).

Microciona prolifera generally exhibits a polyhaline to upper mesohaline distribution (Galtsoff, 1925; Wells et al., 1964; Marsh, 1970; Andrews, 1973; Sutherland and Karlson, 1977) and may experience a range in salinity from 16‰ to 27‰ during a single tidal cycle (Galtsoff, 1925). In Galtsoff's (1925) study of the effect of salinity on the reaggregation of dissociated cells of *Microciona*, good aggregate formation occurred at salinities ranging from 15.5‰ to 31.0‰ (at 18°C), and some aggregation took place at salinities as low as 12.4‰ and as high as 43.4‰. Furthermore, it has been found that when explants of dormant *Microciona* were transferred directly from 25‰ Instant Ocean seawater at 5°C to lower salinities and incubated at 20°C, development into active sponges occurred at all salinities down through 12.5‰ but not at 10‰ (Fell, unpublished observations). This sponge was noticeably damaged in some Virginia estuaries and temporarily eliminated from others when run-off resulting from a tropical storm (in June) lowered salinities to below 10‰ for many days (Andrews, 1973).

In Virginia estuaries, *Microciona* has been found where summer salinities are generally about 15‰ or higher but winter and spring salinities are somewhat lower (Andrews, 1973). This observation suggests that: (1) the sponge may be more tolerant of lower salinities at lower temperatures; or (2) the winter (dormant) form of *Microciona* may be somewhat more tolerant of lower salinities than the summer (active) form. The present study has shown that dormant *Microciona* can tolerate a salinity of 10‰ for up to 9 days and a salinity of 5‰ for up to 24 h when changes in salinity are gradual. Therefore *Microciona* (at least in the dormant state) appears to be able to withstand brief exposure to low salinities, provided the

changes are not too abrupt. Experiments in our laboratory indicate that when active explants of *Microciona* (cultured at 20°C) are gradually brought down to a salinity of 10‰, they undergo reversible tissue regression similar to that occurring during the winter (Knight and Fell, 1987).

Although the dormant form of *Microciona* appears to tolerate a wide range of salinity conditions, it is not as tolerant of low salinity as the gemmules of *Haliclona loosanoffi*. Gemmules of *Haliclona* can tolerate direct transfer from 25‰ to 5‰ seawater at 20°C and can survive exposure to the lower salinity for at least 2 weeks. Germination of the gemmules occurs at salinities ranging from 20‰ to 40‰; and at lower salinities, it is reversibly inhibited (Fell, 1975).

Conclusions

Under the test conditions in this study, the dormant forms of *Microciona prolifera* and *Halichondria* sp. appeared to be less tolerant of environmental stresses than are the gemmules of *Haliclona loosanoffi* and *Prosuberites epiphytum*. Although the gemmules of *Haliclona* and *Prosuberites* can withstand freezing of their habitat, the dormant forms of *Microciona* and *Halichondria* are unable to survive this stress. The gemmules of *Haliclona* and *Prosuberites* are more tolerant of exposure to low air temperatures than is dormant *Microciona*, but all of these dormant sponges have a low cold tolerance. Dormant *Microciona* is able to survive brief exposure to low salinities at 5°C; however, it appears to be more readily killed by low salinity than the gemmules of *Haliclona*.

Acknowledgments

Sincere thanks are due to Ruth Fell and Patricia Knight for their assistance in parts of this study. This work was supported by grants from the Research Corporation (Cottrell College Science Grants).

Literature Cited

Andrews, J. D. 1973. Effects of Tropical Storm Agnes on Epifaunal Invertebrates in Virginia Estuaries. *Chesapeake Science*, 14:223–234.

Fell, P. E. 1974. Diapause in the Gemmules of the Marine Sponge, *Haliclona loosanoffi*, With a Note on the Gemmules of *Haliclona oculata*. *Biological Bulletin*, 147:333–351.

———. 1975. Salinity Tolerance and Desiccation Resistance of the Gemmules of the Brackish-Water Sponge, *Haliclona loosanoffi*. *Journal of Experimental Zoology*, 194:409–412.

———. 1976. Analysis of Reproduction in Sponge Populations: An Overview with Specific Information on the Reproduction of *Haliclona loosanoffi*. Pages 51–67 in *Aspects of Sponge Biology*, edited by F. W. Harrison and R. R. Cowden. New York: Academic Press.

Fell, P. E., E. H. Parry, and A. M. Balsamo. 1984. The Life Histories of Sponges in the Mystic and Thames Estuaries (Connecticut), With Emphasis on Larval Settlement and Postlarval Reproduction. *Journal of Experimental Marine Biology and Ecology*, 78:127–141.

Galtsoff, P. S. 1925. Regeneration after Dissociation (An Experimental Study on Sponges) I. Behavior of Dissociated Cells of *Microciona prolifera* Under Normal and Altered Conditions. *Journal of Experimental Zoology*, 42:183–221.

Hartman, W. D. 1958. Natural History of the Marine Sponges of Southern New England. *Bulletin, Peabody Museum of Natural History*, 12:1–155.

Juniper, A. J., and R. D. Steele. 1969. Intertidal Sponges of the Portsmouth Area. *Journal of Natural History*, 3:153–163.

Knight, P., and P. E. Fell. 1987. Low Salinity Induces Reversible Tissue Regression in the Estuarine Sponge *Microciona prolifera* (Ellis & Solander). *Journal of Experimental Marine Biology and Ecology*, 107:253–261.

Marsh, G. A. 1970. A Seasonal Study of *Zostera epibiota* in the York River, Virginia. Ph.D. Thesis, College of William and Mary, Williamsburg, Va. 156 pp.

Rasmont, R. 1962. The Physiology of Gemmulation in Freshwater Sponges. Pages 3–25 in *Regeneration*, edited by D. Rudnick. New York: Ronald Press.

Simpson, T. L. 1968. The Biology of the Marine Sponge *Microciona prolifera* (Ellis and Solander) II. Temperature–Related, Annual Changes in Functional and Reproductive Elements with a Description of Larval Metamorphosis. *Journal of Experimental Marine Biology and Ecology*, 2:252–277.

―――. In press. Porifera. In *Reproductive Biology of Invertebrates 6, Asexual Reproduction, Life Cycles, and Reproductive Strategies*, edited by K. G. Adiyodi and R. G. Adiyodi. New Delhi: Oxford and IBH Publishing Company.

Sutherland, J. P., and R. H. Karlson. 1977. Development and Stability of the Fouling Community at Beaufort, North Carolina. *Ecological Monographs*, 47:425–446.

Wells, H. W., M. J. Wells, and I. E. Gray. 1964. Ecology of Sponges in Hatteras Harbor, North Carolina. *Ecology*, 45:752–767.

HENRY M. REISWIG
Redpath Museum
McGill University
Montreal, Quebec, Canada H3A 2K6

In Situ Feeding in Two Shallow-Water Hexactinellid Sponges

Abstract

The feeding activity of two species of hexactinellid sponges, *Rhabdocalyptus dawsoni* and *Aphrocallistes vastus*, was investigated to determine if a pattern exists that might relate to, and help explain, the unique syncytial organization of these organisms. Water samples were collected from ambient water and from atrial openings of specimens at depths of 20–37 m in Saanich Inlet and Barclay Sound, British Columbia. The samples were analyzed for total, dissolved, and particulate carbon and culturable bacteria. The results indicate that patterns of retention (use) are distinctive for the two species. *R. dawsoni* retains only dissolved organic carbon and shows no tendency to capture bacteria, whereas *A. vastus* retains particulate material as its primary organic carbon source and captures bacteria at moderate rates. Because of high variance in the data, interpretation of these patterns is tentative. The data, as interpreted here, are taken to indicate that hexactinellid sponges feed principally upon colloidal organic matter. The apparent use of dissolved material by *R. dawsoni* may be attributed to the activity of the veil-supported biotic jungle of that species.

The Hexactinellida have been recognized as a unique group of sponges since the earliest work on their tissue organization (Schulze, 1880, 1886). Their syncytial organization—long suspected and recently demonstrated (Mackie and Singla, 1983)—contrasts strikingly with the cellular nature of all other members of the phylum. This organizational pattern appears to serve as a mechanism for propagating stimuli over a continuous membrane system (Mackie et al., 1983), but otherwise its functional significance remains unknown. In the cellular Porifera, a reasonably consistent relationship appears to exist between microscopic anatomy and particulate food retention (Reiswig, 1971). Particles in the upper nanoplankton (5–20 μm) and microplankton (20–200 μm) size range appear to be ingested by cells lining inhalant canals, while smaller nanoplankton (2–5 μm), picoplankton (0.2–2 μm), and colloidal materials appear to be captured pri-

504

marily on the specialized reticulum of the choanocyte collar and, less abundantly, on other exposed cell surfaces.

The collar of the Hexactinellida is similar in general features and dimensions to that of the cellular sponges. The microvilli are spaced about the same distance and bridging synapticula of presumed glycoprotein substances form a fine-mesh reticulum, as in other Porifera. Besides the syncytial versus cellular organization, the most obvious differences between the groups are the open trabecular inhalent system of the hexactinellids in place of the inhalent canal system of cellular sponges, and the presence of a secondary reticulum lying within the choanocyte chambers at the midlevel of the more widely spaced collars. In order to determine whether these structural differences might be related to differences in the pattern of particle capture, I investigated organic carbon and bacteria retention in natural populations of two hexactinellid sponges. The working hypothesis to be tested was: Hexactinellida are optimally designed to capture and utilize dissolved organic matter.

Materials and Methods

The study populations comprised *Rhabdocalyptus dawsoni* (Lambe) and *Aphrocallistes vastus* Schulze at depths of 20–37 m on the granitic western shelf of Senanus Island, Saanich Inlet, Vancouver Island, British Columbia, Canada (48°35.6' N, 123°29.3' W). The two species present striking contrasts within the Hexactinellida, not only in their supporting skeletal systems (lyssacine vs. dictyonine), but more obviously in the general nature of their outer surfaces. The external surface of *R. dawsoni* is covered by a dense veil of prostal spicules that supports a thick layer of accumulated sediment and a variety of epifauna and epiflora—a so-called biotic jungle (Mackie and Singla, 1983). In contrast, the outer surface of *A. vastus* is obviously "naked", lacking a veil and accumulated sediment. The Senanus Island populations were repeatedly visited by use of scuba between November 1982 and August 1983 for a variety of purposes, one of which was to investigate feeding. A secondary population of *R. dawsoni* located at depths of 30 to 32 m off San Jose Island, Barclay Sound (48°54.5' N, 125°02' W) was also investigated for a short period, 2 to 6 April, 1983.

Water samples were collected in situ in 50 ml glass syringes cleaned by acid-dichromate washing and muffle-furnace oxidation (500°C). Samples were collected in pairs consisting of (a) ambient water alongside a specimen approximately 10 to 30 cm from both the specimen surface and the substrate, and (b) water from the distal portion of the atrial cavity (obtained by holding the syringe in the large, > 2 cm diameter, oscular aperture during slow filling). Samples were collected during slack tide to minimize contamination with resuspended flocculent sediment. Capped sample syringes were trans-

ported to the laboratory in a water-filled ice chest (0°C) for processing within 2 h of collection.

At the laboratory each sample syringe was vigorously mixed, rinsed in distilled water, and supported vertically in a clamping holder for removal of subsamples. The first 5 ml were discarded to avoid possible cap contamination. The next 5 ml were ejected directly into a filter-holder loaded with a 13 mm diameter 0.2 μm discrete-pore filter for scanning electron microscopy (analysis still in progress). Then three subsamples of approximately 5 ml (4–6.5 ml) were ejected directly into precombusted 10 ml glass ampules that had been preloaded with 0.5 g potassium persulfate and preweighed. These constitute the samples used to measure total organic carbon (TOC).

Next, a precombusted 25 mm diameter filter holder loaded with a Whatman GF/C glass-fiber filter (effective pore size 0.5 μm) was inserted into the sample syringe fitting. The first 5 ml driven through the filter were discarded; three subsamples of approximately 5 ml were then slowly delivered through the filter directly into three ampules prepared, as above, for measuring dissolved organic matter (DOC). Finally, the filter unit was removed from the sample syringe, the remainder of the sample was ejected into a sterile vial, and 0.1 to 0.7 ml were transferred by sterile micropipette into 5 ml of sterile seawater (from the collection location). This diluted sample was then filtered through a sterile 47 mm diameter 0.2 μm pore-size Millepore filter for enumeration of culturable bacteria. The filter was transferred to a nutrient pad saturated with standard marine nutrient medium, Oppenheimer and Zobell 2216 E (Gunkel and Rheinheimer, 1972) in a locking culture dish. Most bacterial subsamples were single, but duplicate and triplicate subsamples were occasionally prepared to assess variability of the procedure.

Each set of ampules was processed completely before work began on the next sample so as to minimize contamination of standing subsamples by airborne materials. The ampules were reweighed to permit accurate calculation of the subsample added, were dosed with 0.2 ml of 10% phosphoric acid to oxidize inorganic carbonates, and were finally purged for 5 min in CO_2-free oxygen and flame-sealed with an Oceanography International Carbon Analyzer Purge-Seal Unit. At this point, processing of the next sample was begun, using separate sterile sets of apparatus where required (filter-holders, pipettes, and so on). A random sequence of sample processing (ambient or atrial) was used to avoid systematic effects due to time of ice-water storage.

At the termination of all field collections, the ampule subsamples for TOC and DOC measurement were autoclaved for 4 h before being processed in Oceanography International Total Carbon Ampule Analyzer, Model 0524B. The procedure included wet digestion in the ampule and CO_2 detection by nondispersive infrared analyzer. A reagent blank correction and a sucrose reference curve were produced following the manufacturer's recommendations. A volume correction curve was produced with filtered bottom water from the Senanus Island sampling site. The mode of measurement selected was "curve height" as monitored by chart recorder, since this produced the best correlation to reference solutions. Values for TOC and DOC are given in sucrose-equivalent organic carbon, weight per volume, with particulate organic carbon (POC) being calculated by difference of the means of TOC and DOC.

Bacterial sample filters and control blanks were cultured at room temperature (20–25°C), in darkness and high humidity. After 5.7 to 6.7 days (determined by trial as optimum), filters were fixed in formalin fumes and photographed in grazing incident illumination. Colony numbers, representing original colony forming units (cfu's), were enumerated in photonegatives at high magnification. Although it is widely recognized that culturable bacteria represent a subsample of only 1 to 3% of total bacteria in water samples, they provide a relatively simple method of indexing the population. Fluorescence microscopy was not readily available for direct counting of stained bacterial cells at the laboratories utilized.

Twenty-eight sample pairs were collected, processed, and analyzed; 20 from the Senanus Island populations and 8 from Barclay Sound. Owing to contamination, breakage, and accidental dessication of cultures, only 18 pairs were suitable for organic carbon tests and 13 for bacterial analysis. In computing means and standard deviation of percentage change between samples (proportions), values were arcsine-transformed; where proportion differences exceeded 100%, inverse hyperbolic transformations were employed.

Results

The TOC content in the 18 sets of ambient samples averaged 2.42 ± 0.23 mg C·l⁻¹. This was divided between between dissolved and particulate fractions in the common 90:10 proportions: 2.18 ± 0.17 mg DOC·l⁻¹ and 0.24 ± 0.14 mg POC·l⁻¹ There were no statistical differences between ambient carbon values of the sample sets at the two locations (Saanich Inlet vs. Barclay Sound), nor between those of the two species. Both sponges exhibit high variability in the retention (utilization) of TOC and each of the two organic carbon fractions (Figure 1).

The veiled species *Rabdocalyptus dawsoni* retains a mean of only 56 μg TOC·l⁻¹, with a very high standard deviation of ± 250 μg·l⁻¹ ($n = 12$). In other words, only 2.3% of the available TOC is used in these sample sets. DOC accounts for virtually all of this retention (mean retention is 71 ± 135 μg·l⁻¹, while a slight production of 15 ± 181 μg·l⁻¹ of POC is indicated. The clearly uneven distribution of DOC retention suggests that some consideration should

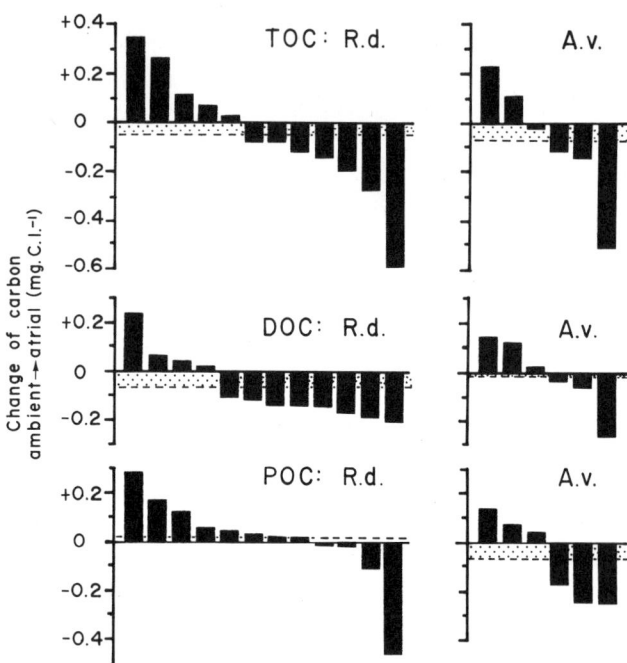

Figure 1. Change in total *(TOC)*, dissolved *(DOC)*, and particulate *(POC)* organic carbon content of water samples, atrial and ambient, from two hexactinellid sponges, *Rhabdocalyptus dawsoni (left)* and *Aphrocallistes vastus (right)*. Positive values indicate net production; negative values indicate net removal of that fraction. Sample sequence from greatest positive to greatest negative change within a set. Means indicated by dashed lines with stippling.

be given to the use of the median (removal of ca. 130 μg C·l⁻¹) or mode (removal of ca. 150 μg C·l⁻¹) as an estimate of activity of this species. Without additional information on the condition of the specimens involved in the "producing" samples, the arithmetic mean is accepted here as the best estimator. In terms of available (ambient) levels of these fractions, removal of DOC amounts to only 3.2% of that large pool, while the apparent slight production of POC consitutes 6.8% of that relatively small fraction. In terms of organic carbon, the diet of *R. dawsoni* appears to consist entirely (100%) of DOC.

In contrast, the nonveiled species *Aphrocallistes vastus* retains a mean of 77 μg TOC·l⁻¹ with a similar high standard deviation of ± 256 μg·l⁻¹ ($n = 6$). In this case, utilization appears to be only 3.2% of the TOC available in the appropriate ambient samples. For this species, TOC use is mainly attributed to the particulate fraction, 68 ± 175 μg·l⁻¹, with a negligible use of 8 ± 150 μg·l⁻¹ as DOC. That is to say 25% of the available POC and only 0.4% of available DOC are retained. Thus the organic carbon diet of *A. vastus* consists of about 89% POC and 11% DOC.

The absolute values for removal of carbon fractions presented here must be considered tentative in view of the high variance and small sample sizes. Much of this variance is attributable to the single-sample resolution limit of the analyzer, which the manufacturer considers to be ± 2% in the range of these samples, or ± 45 µg C·l⁻¹. Tests of reagent blanks, sucrose standards, and volume standards confirmed this level of resolution during sample processing. Thus approximately 50% of the variance can be attributed to analytical variance; a large portion of the residual variance remains inherent in the samples and must be attributed to the activities of the sponges themselves. Significance tests (paired T-test) indicate that only one set of results is near rejection of a null hypothesis: the removal of DOC by *R. dawsoni* ($0.05 < p < 0.10$).

Enumeration of cultured bacterial colonies also showed high variation in the differences between ambient (inhalant) and atrial (exhalant) samples (Figure 2). The mean density of culturable bacteria in the 13 ambient samples acceptable for analysis was $5.46 ± 3.48·10^3$ cfu per ml. Although the patterns of bacterial interaction with the two sponges appear to differ, these cannot be statistically distinguished from each other or from zero owing to the very high variance in the sample sets. *Rhabdocalyptus dawsoni* did not exhibit a net capture (and use) of bacteria, since mean retention was only $0.3 ± 49.2%$ ($n = 6$). Samples from *Aphrocallistes vastus* had a mean moderate net retention of $18 ± 63%$ ($n = 7$). Here, again, the uneven distribution of the small sample set may suggest that the median (65%) or mode (nearly 80%) might better provide an estimate of bacterial retention for healthy, active individuals, but since information on pumping rate and tissue biopsy is unavailable, the arithmetic mean is said to reflect overall activity of the population.

The high variance in the difference between samples (atrial and ambient) is unlikely to be related to procedure or the chance selection of strikingly different bacteria by restricted culture conditions. The mean difference be-

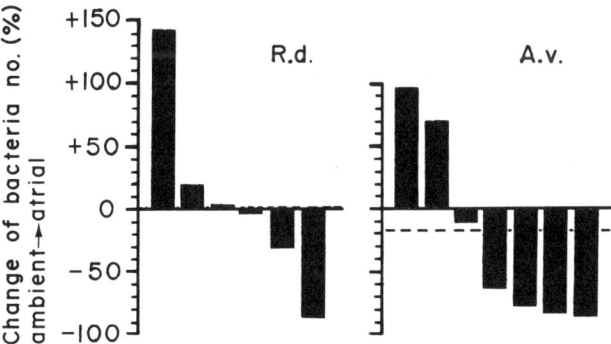

Figure 2. Percentage change in bacteria (cfu) numbers on cultured membrane filters from water samples, atrial and ambient, from two hexactinellid sponges, *Rhabdocalyptus dawsoni (left)* and *Aphrocallistes vastus (right)*. Positive values indicate net production; negative values indicate net removal of bacteria. Means represented by dashed lines.

tween replicate samples was $21 ± 9%$ ($n = 13$) and controls were consistently free of cfu's.

Normally, bacteria numbers (cfu's) can be expected to correlate positively with organic carbon values, especially for POC, since a significant portion of that fraction consists of bacterioplankton. Bacteria counts of ambient water of the 12 sample sets acceptable for both analysis procedures correlated most highly with TOC ($r = 0.88$), less strongly with DOC ($r = 0.76$), and weakest with POC ($r = 0.61$); all are highly significant ($p < 0.025$). In atrial (exhalant) samples, the bacteria counts did not correlate significantly with TOC but correlated positively with DOC ($r = 0.57$; $p = 0.025$). Surprisingly, there was a slight negative overall correlation with POC, although it was not statistically significant. When the samples of the two species were analyzed separately, there appeared to be no correlation between bacteria counts and POC in *Rhabdocalyptus dawsoni* samples ($r = 0.08$; $n = 6$; $p > 0.25$) and a moderate, but statistically insignificant negative correlation in *Aphrocallistes vastus* samples ($r = -0.51$; $n = 7$; $0.10 < p < 0.25$).

Discussion

The great variation in the data from the study can be attributed to several possible sources. About half of the variance in the carbon values is due to analytical procedure, but the remainder is apparently related to the samples themselves and to their handling prior to analysis.

SAMPLING ERROR. Ambient samples were necessarily collected 1 to 2 min before atrial samples to avoid resuspension of light sediments by contact by the diver with the substrate (required during atrial sampling). Two types of change could have taken place between sampling points: horizontal exchange of water micropatch, and physical disruption of fine vertical stratification by intrusion of the diver. These disruptions in the water around the specimens are the direct result of collecting on-site samples may be the cause of significant variation between samples (e.g., uneven distribution of microaggregates in micropatches; Goldman, 1984), but their effects cannot be evaluated with the data available. Such problems can be avoided either by using a remote-operated sampling system or by only working with specimens in the artificial conditions of laboratory aquaria.

PUMPING ACTIVITY. *Rhabdocalyptus dawsoni* is known to spontaneously cease pumping current in nature (G. Silver, pers. comm.), and its response to stimulation in the laboratory by arrest of pumping has been described in detail (Mackie et al., 1983). If some of the sponges sampled in the field were in full or partial arrest during sample collection, either because of spontaneous behavior or as a response to inadvertant stimulation by the diver, a significant shift in the "removal" pattern would be expected—most likely a loss of change between the samples as a result of intrusion of ambient water into the atrium. Under the restricted field conditions of this study it was impractical to assess pumping rate for each specimen, before or during sample collection without compromising the quality of the samples. Assessments of pumping activity made shortly (a few minutes) after sample collection did not correlate with any pattern of carbon or bacteria removal in the respective samples. In the future, activity should be measured with a low-velocity current meter while the atrial sample is being collected.

FILTRATION AND RETENTION. Variations of net filtration-retention activity of the sponges may also be an important source of variance in the results. Irregular emission of amorphous aggregates from the oscula of actively pumping *Rhabdocalyptus dawsoni* in laboratory aquaria is clear evidence of lack of uniformity in time. The aggregated vary from 200 μm to a few μm in diameter and probably extend to submicroscopic sizes. Many are visible to the naked eye in oblique illumination, but they are best seen under a dissecting microscope. They are eliminated from specimens of all sizes of this species (3–30 cm long), and nothing suggests that they represent artifacts resulting from artificial conditions. If even one of the rare larger aggregates was included in a 50-ml sample and a 5 ml ampule subsample, it would greatly increase the variance of the entire sample set.

The mechanism of aggregate formation in *Rhabdocalyptus dawsoni* (and the similar aggregates formed by probably all members of the phylum Porifera) remains unknown. Although trabecular tissues and chamber walls are unlikely to be "contractile" in the normal sense of the term, it is not impossible that such structureless tissues may flow and coalesce over short distances. Trapped accumulations of small particles could simply be moved "through" an imperforate membrane by slow migration of the syncytium around the aggregate, bringing it to the exhalant side. This notion is consistent with the size of the aggregates that are far too large to have passed through any known pores in the membrana reticularis. Some aggregates inspected under high magnification were found to consist of a mixture of typical suspended particulate material bound together by a diaphanous network of thin transparent threads. That similar "packages" of detrital material can be found in histological sections of both species studied here supports the interpretation that these accumulate in the inhalant system and pass through the syncytium wall by tissue migration.

HISTOLOGICAL EVIDENCES. There is as yet only scanty histological evidence concerning particle capture and ingestion in hexactinellid sponges. In the electron microscopy study of *Rhabdocalyptus dawsoni* by Mackie and Singla (1983), phagosomes with cellular objects (bacterial and algal particles) of 0.3 to 1.0 μm diameter are only occasionally seen, and these are invariably in the syncytium of the flagellated chamber wall. Smaller noncellular particles of about 0.05 μm in diameter are seen in phagosomes of the collar bodies. These observations suggest that fine colloidal particles are captured in the synapticular network of the supposed glycoprotein filaments bridging the collar microvilli, either through chance interception or actual sieving on the 0.05×0.20 μm mesh, and are ingested by the collar bodies. Material in this size range bridges the division between dissolved and particulate fractions. The cellular nano- and picoplankton may be ingested not only by the syncytial network of flagellated chamber walls, but and probably also by the trabecular network; little information is yet available on these regions. Similar electron microscopy studies of *Aphrocallistes vastus* are in preparation, but are not yet available for comparison.

ECTOSYMBIOSIS. The histological evidence presented by Mackie and Single (1983) and their conclusions appear to conflict with the results of the organic carbon and bacteria analysis for *Rhabdicalyptus dawsoni* obtained here. This species seems to derive all of its net carbon retention from dissloved sources, with net production of particulate matter and little or no evidence of bacterial capture. The suggestion offered by Mackie and Singla that the community of organisms inhabiting the "spicule jungle" may be an important food source provides a resolution of the apparent contradiction. If the hyperdermal veil community includes a metabolically active microbial component that can effectively capture both DOC and POC, and if the community produces a downstream particulate effluent in the colloidal and bacterial sizr ranges towards the sponge surface, this may be the major source of particulate food for the sponge component. The net exchange of the "jungle" and hexactinellid combination could easily account for DOC use and lack of bacteria cropping as indicated in the analysis of water samples. This association, which may be referred to as an "ectosymbiosis", occurs in every *Rhabdocalyptus dawsoni* longer than 1 to 2 cm in length and may indeed be a nutritional strategy common to all heavily-veiled Hexactinellida.

Aphrocallistes vastus, which lacks a spicule veil and is thus more likely to be representative of the majority of the class Hexactinellida, does not appear to utilize DOC at all, but feeds solely upon particulate materials. Mean retention efficiency of POC is moderate (25%) and comparable to mean retention for bacterial particles (18%). The pattern of retention by *A. vastus* may be typical of hexactinellids, including *R. dawsoni* minus its veil community, but more extensive investigation and innovative approaches will be needed to separate the dynamic exchanges of the sponge and veil community for testing of this assumption.

COMPARISON WITH DEMOSPONGES. Care should be taken in comparing these hexactinellids with demosponges since similar data on demosponges (in terms of sample sizes, analytical methods, and habitat) are not yet available. However, a general pattern has been established for demosponges from a wide variety of habitats that suggests high efficiency of net bacterial removal, 70 to 100% (Reiswig, 1971; Wilkinson, 1978; Gili et al., 1984; Stuart and Klumpp, 1984). Analyses of large water samples have shown that POC retention in demosponges is generally moderate (25–45%), and that colloidal material (microscopically invisible material retained by filters) is probably retained at nearly the efficiency of bacterial particles (close to 100%). According to the mean of the small data set in this study, *Aphrocallistes vastus* and probably other hexactinellids are comparatively poor at retaining and using bacteria, but are relatively efficient at retaining and using colloidal fractions.

The mean net retention of TOC by these two hexactinellids is considerably higher than that determined for POC retention for tropical demosponges (around 17–40 μg C·l⁻¹; Reiswig, 1971, 1981) in relatively low-nutrient waters, the only region for which quantitative data are available at present. Demosponges in temperate coastal waters probably retain and utilize greater net quantities of organic matter since a larger pool is available. Hexactinellids in their typical deep, low-nutrient waters would be expected to exhibit much lower levels of net retention, but possibly greater retention efficiencies of the limited resources available. The hexactinellids and demosponges cannot be properly compared without some measurements of water transport, respiration, growth, and reproductive effort in *Rhabdocalyptus dawsoni* and *Aphrocallistes vastus*. Ideally, a similar set of data should be collected for demosponges from the same, or at least an equivalent, habitat.

Conclusions

The retention patterns of the two hexactinellids *Rhabdocalyptus dawsoni* and *Aphrocallistes vastus* differ from each other and from that of typical demosponges. The veiled solitary lyssacine sponge *R. dawsoni* retains a mean of 71 μg C·l⁻¹ as dissolved organic matter and generates an excess of 15 μg C·l⁻¹ as particulate matter. This species exhibits no net retention of bacteria. In contrast, *A. vastus* retains a mean of 68 μg C·l⁻¹ as particulate material and 8 μg C·l⁻¹ as dissolved material. It retains bacteria at a mean efficiency of only 18% of the cells available, al-

though higher retention (60–80%) may occur in individual samples. No variations in internal structure have been detected between these species to account for these differences in feeding pattern. The presence of a dense microfauna and microflora associated with the superficial veil of *R. dawsoni* may be responsible for the apparent use of dissolved matter and the presumed production of particulate matter before the water reaches the sponge surface proper. The original working hypothesis that hexactinellid sponges are likely users of dissolved organic resources is not supported.

The overall retention pattern of particulate matter by *Aphrocallistes vastus* differs from that of demosponges in the relatively low efficiency of mean bacterial retention, although the data suggest that individual specimens or perhaps the entire population at specific times may approach that of demosponges. The mechanism by which most bacteria traverse the filtration system of the flagellated chambers and collar surfaces of *A. vastus* (and presumably *Rhabdocalyptus dawsoni*) is as yet unknown. The lack of discrete canals in the noncellular hexactinellids may restrict the group to a diet composed mainly of colloidal material and prevent them from effectively capturing and using cellular components of the suspended organic spectrum.

Acknowledgments

I thank Elizabeth Day, Maria Byrne, Sabina Leader, and Catherine von Carolsfeld for assistance in diving; George O. Mackie, Ralph O. Brinkhurst, C. S. Wong, and R. E. Foreman for providing laboratory facilities; Daniel Lachance for help in counting bacteria; and Sam Salley for help with statistical tests. This work was supported by a research grant by Fisheries and Oceans, Canada. Institutions that kindly provided facilities include the Biology Department of the University of Victoria, the Institute of Ocean Sciences, Sidney, B.C., and the Bamfield Marine Station, Bamfield, B.C.

Literature Cited

Gili, J. M., M. A. Bibiloni, and A. Montserrat. 1984. Tasas de filtracion y retencion de bacterias "in situ" de tres especies de esponjas litorales, estudio preliminar. *Miscellanea Zoologica* (Barcelona), 8:13–21.

Goldman, J. C. 1984. Conceptual Role for Microaggregates in Pelagic Waters. *Bulletin of Marine Science*, 35:462–476.

Gunkel, W., and G. Rheinheimer. 1972. Bacteria. Pages 156–175 in *Research Methods in Marine Biology*, edited by C. Schlieper. Seattle: University of Washington Press.

Mackie, G. O., I. D. Lawn, and M. Pavans de Ceccatty. 1983. Studies on Hexactinellid Sponges. II. Excitability, Conduction and Coordination of Responses in *Rhabdocalyptus dawsoni*

(Lambe, 1873). *Philosphical Transactions, Royal Society of London, B, Biological Sciences*, 301:401–418.

Mackie, G. O., and C. L. Singla. 1983. Studies on Hexactinellid Sponges. I. Histology of *Rhabdocalyptus dawsoni* (Lambe, 1873). *Philosophical Transactions, Royal Society of London, B, Biological Sciences*, 301:365–400.

Reiswig, H. M. 1971. Particle Feeding in Natural Populations of Three Marine Demospongiae. *Biological Bulletin*, 141:568–591.

———. 1981. Partial Carbon and Energy Budgets of the Bacteriosponge *Verongia fistularis* (Porifera:Demospongiae) in Barbados. *Marine Ecology*, 2:273–293.

Schulze, F. E. 1880. On the Structure and Arrangement of the Soft Parts in *Euplectella aspergillum*. *Transactions, Royal Society of Edinburg*, 29:661–673.

———. 1886. Über den Bau und das System der Hexactinelliden. *Abhandlungen der Deutschen Akademie der Wissenschaften, Berlin, Klasse Physik und Mathematik*, 1886:1–97.

Stuart, V., and D. W. Klumpp. 1984. Evidence for Food-Resource Partitioning by Kelp-Bed Filter Feeders. *Marine Ecology Progress Series*, 16:27–37.

Wilkinson, C. R. 1978. Microbial Associations in Sponges. I. Ecology, Physiology and Microbial Populations of Coral Reef Sponges. *Marine Biology*, 49:161–167.

GYSELE VAN DE VYVER
Laboratoire de Biologie Animale et Cellulaire
Université Libre de Bruxelles
50 av. F. D. Roosevelt
1050 Bruxelles, Belgium

BERNARD VRAY
Laboratoire de Parasitologie
Université Libre de Bruxelles
50 av. F. D. Roosevelt
1050 Bruxelles, Belgium

SAMIA BELAOUANE
DOMINIQUE TOUSSAINT
Laboratoire de Biologie Animale et Cellulaire
Université Libre de Bruxelles
50 av. F. D. Roosevelt
1050 Bruxelles, Belgium

Efficiency and Selectivity of Microorganism Retention by *Ephydatia fluviatilis*

Abstract

The epuration capacity of the freshwater sponge *Ephydatia fluviatilis* was studied by exposing in vitro specimens to four species of bacteria and one yeast known to cause infectious diseases. The kinetics and selectivity of retention were measured using either ^{3}H thymidine or iodine 125 labeled microorganisms, In addition, TEM observations were carried out to determine whether the captured microorganisms were phagocytosed and digested. The retention ratios were close to 92% for Escherichia *coli*, 85% for *Staphylococcus aureus*, 53% for *Klebsiella pneumoniae*, 46% for *Pseudomonas aeruginosa*, and 14% for *Candida albicans*. The ultrastructural observations clearly showed that, whereas *E. coli* and *P. aeruginosa* were phagocytosed and digested, only a few *C. albicans* were phagocytosed and none were digested.

Our quantitative and qualitative data clearly demonstrate the selectivity of capture, phagocytosis, and digestion of microorganisms in the diet of the freshwater sponge *Ephydatia fluviatilis*. The results also point to a new aspect of the ecological role of sponges in reducing microbial contamination. These filter feeders appear to be very effective against some types of microorganisms, but almost completely ineffective against others.

An important aspect of sponge feeding ecology is the nature of the food particles taken up. Earlier investigations have shown that, sponges, depending on their anatomy, can retain a wide range of particles, including dissolved organic molecules, colloid particles, free bacteria, and even unicellular algae (e.g., Madri et al., 1967, 1971; Reiswig, 1971, 1975; Schmidt, 1970; Vacelet, 1975, 1979).

The importance of free bacteria in the diet of sponges (Rasmont, 1961, 1975; Mank and Kilian, 1979; Francis, 1984) and their ability to concentrate large numbers of microorganisms suggest that these filter feeders could be effective in reducing the microbial pollution caused by fecal contamination (Madri et al., 1967). However, before this hypothesis can be assessed, one must know if sponges capture, phagocytose, and digest different types of microorganisms with the same efficiency.

In the present study, we present a quantitative and qualitative evaluation of the efficiency of capture, phagocytosis, and degradation of the freshwater sponge *Ephydatia fluviatilis* exposed in laboratory conditions to five pathogen microorganisms.

Material and Methods

SPONGES. Ten-day old *Ephydatia fluviatilis* (Spongillidae) sponges of the strain δ (Van de Vyver, 1970, 1975) raised from single gemmules were grown on cover glasses placed in the wells of microtest plates at 20°C. Each well received a batch of 50 gemmules and 2 ml of M medium (Rasmont, 1961) for the quantitative experiments and 1 gemmule for the ultrastructural studies.

MICROORGANISMS. Four bacteria species *(Escherichia coli, Klebsiella pneumoniae, Pseudomonas aeruginosa,* and *Staphylococcus aureus)* and one yeast species *(Candida albicans)* were used. These microorganisms were chosen because they are the most common pathogen microorganisms in natural freshwater.

A strain of *Escherichia coli* was isolated from a pond containing sponges (Willenz and Van de Vyver, 1984); *Klebsiella pneumoniae, Pseudomonas aeruginosa, Staphylococcus aureus,* and *Candida albicans* were clinically isolated from human pathological material. All the strains were stored at 4°C on agar. Bacteria were grown overnight at 37°C in yeast extract (oxoid) 0.25%, Brain Heart Infusion (Difco) 1% and dextrose (Difco) 0.1%. *C. albicans* were grown overnight at 37°C in Sabouraud medium (2% glucose, 2% neopeptone, and 2% agar).

RADIOACTIVE LABELING OF MICROORGANISMS. Two milliliters of an overnight culture were added to 18 ml of medium containing 15 μl of tritiated thymidine (New England Nuclear, 20 Ci/mmol). The culture was incubated at 37°C with constant stirring for 3 h, until a maximum population was reached. The optical density was measured every 30 min in a spectrophotometer (Jobin Yvon 201) at 620 nm. Bacteria counts were obtained by reference to the McFarland standard. After incubation, the bacteria were washed by three successive centrifugations (5,000 g at 4°C in a Jovan CR 4 11 centrifuge) in M medium. A sample of this preparation was monitored for effectiveness of radioactive labeling before any further handling.

In some cases *(Pseudomonas aeruginosa* and *Candida albicans),* thymidine was not sufficiently incorporated by bacteria; the labeling was performed with ^{125}I as follows: 2 ml of a bacteria suspension in PBS (Phosphate Buffer Saline pH 7.2) were added to 1 ml of Na^{125}I carrier free (Amersham, United Kingdom). Iodination was initiated by adding 3 or 4 iodo-beads (Pierce, Rockford, Illinois).

The suspension was stirred 10 min at room temperature. The iodo-beads were removed and bacteria were washed three times by centrifugation in PBS.

FEEDING. The sponges used were each raised from single gemmules. Experiments were started by substituting of the culture medium with 1 ml of labeled suspensions of microorganism. Fifty 1-gml sponges were exposed to initial bacterial concentrations of 2×10^7/ml water. The culture plates were stirred continuously to prevent microorganisms from settling (rotary shaker, 80 RPM). At different intervals—10, 30, and 40 min and 1,.2, 5, 10, and 24 h—the excess microorganisms were removed by three successive rinsings and were saved for counting. Both the excess microorganisms and those retained by the sponges were dissolved in NP 40 1% (Nonidet P 40, Sigma) and in Insta-gel scintillation fluid (Packard), and counted in a Beckman LS 7500 liquid scintillation analyzer. All experiments were performed in duplicate.

TRANSMISSION ELECTRON MICROSCOPY. After different incubation times, sponges were fixed for 3 h at 20°C in 0.35% glutaraldehyde–0.025 M cacodylate buffer, pH 7.4. They were rinsed three times for 10 min in the buffer, postfixed for 1 h in 1% osmium tetroxide in the same buffer, and dehydrated through a graded ethanol series. Specimens were embedded in ERL 4206 according to Spurr (1969). Before they were sectioned, the siliceous spicules of the sponges were dissolved in hydrofluoric acid. Thin sections were stained with uranyl acetate and lead citrate. The procedure introduced by Willenz and Van de Vyver (1984) was used to display acid phophatases.

Results
KINETICS OF MICROORGANISM RETENTION

Incubation was carried out with an initial concentration of 2×10^7 microorganisms per milliliter. For each microorganism tested, the amount of sponge-associated radioactivity, determined in disintegration per minute (DPM), and the corresponding number of bacteria were measured in the course of time (Figure 1). The correlation with the number of bacteria was made by referring to a microtitre assay adapted from Phillips et al. (1979). However, in order to more readily compare the retention efficiency of *Ephydatia fluviatilis* for the five microorganisms tested, we converted all the results into percentage of retention expressed as a function of time (Figure 2). This graph clearly indicates, the obvious differences in retention capacity of the sponges for the five microorganisms. The maximum capacity of retention is close to 92% of the initial concentration for *Escherichia coli,* 87% for *Staphylococcus aureus,* 53% for *Klebsiella pneumoniae,* 46% for *Pseudomonas aeruginosa,* and only 14% for *Candida albicans.*

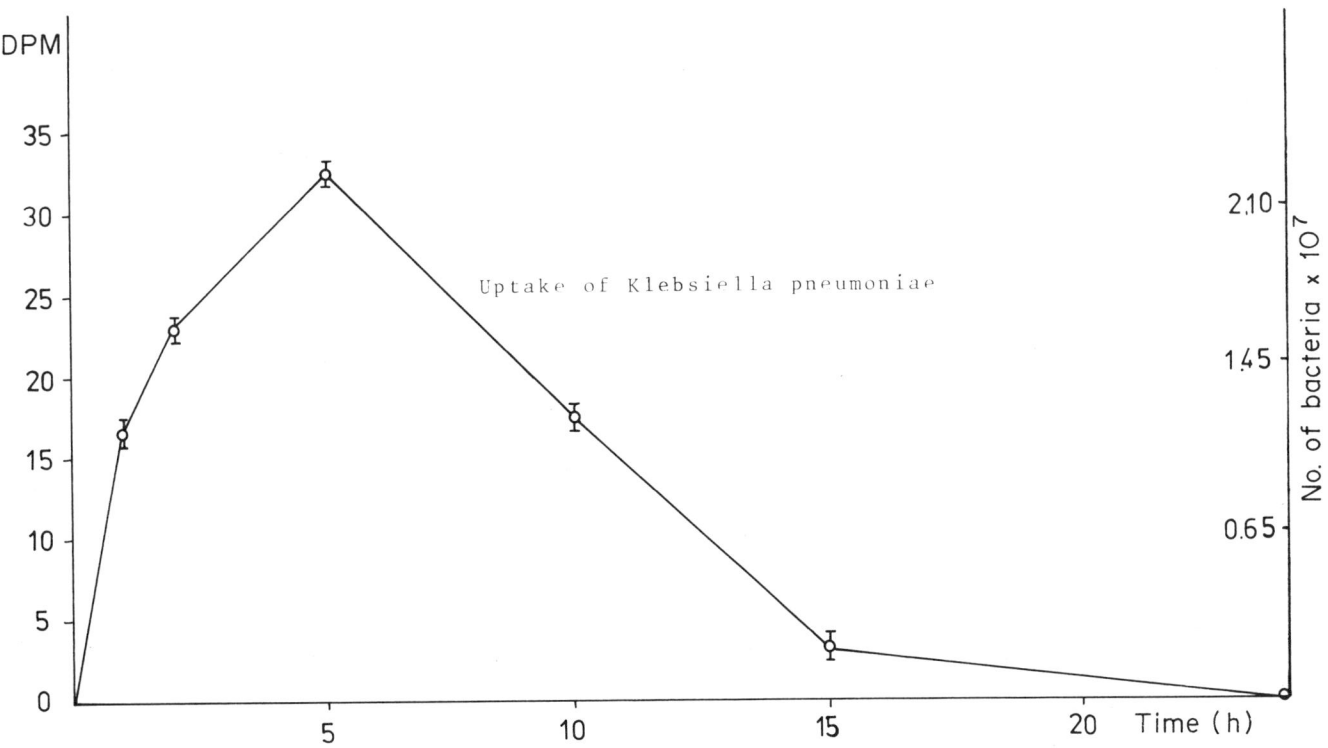

Figure 1. Retention by *Ephydatia fluviatilis* (means and *S.D.*) of ³H labeled *Klebsiella pneumoniae* over time.

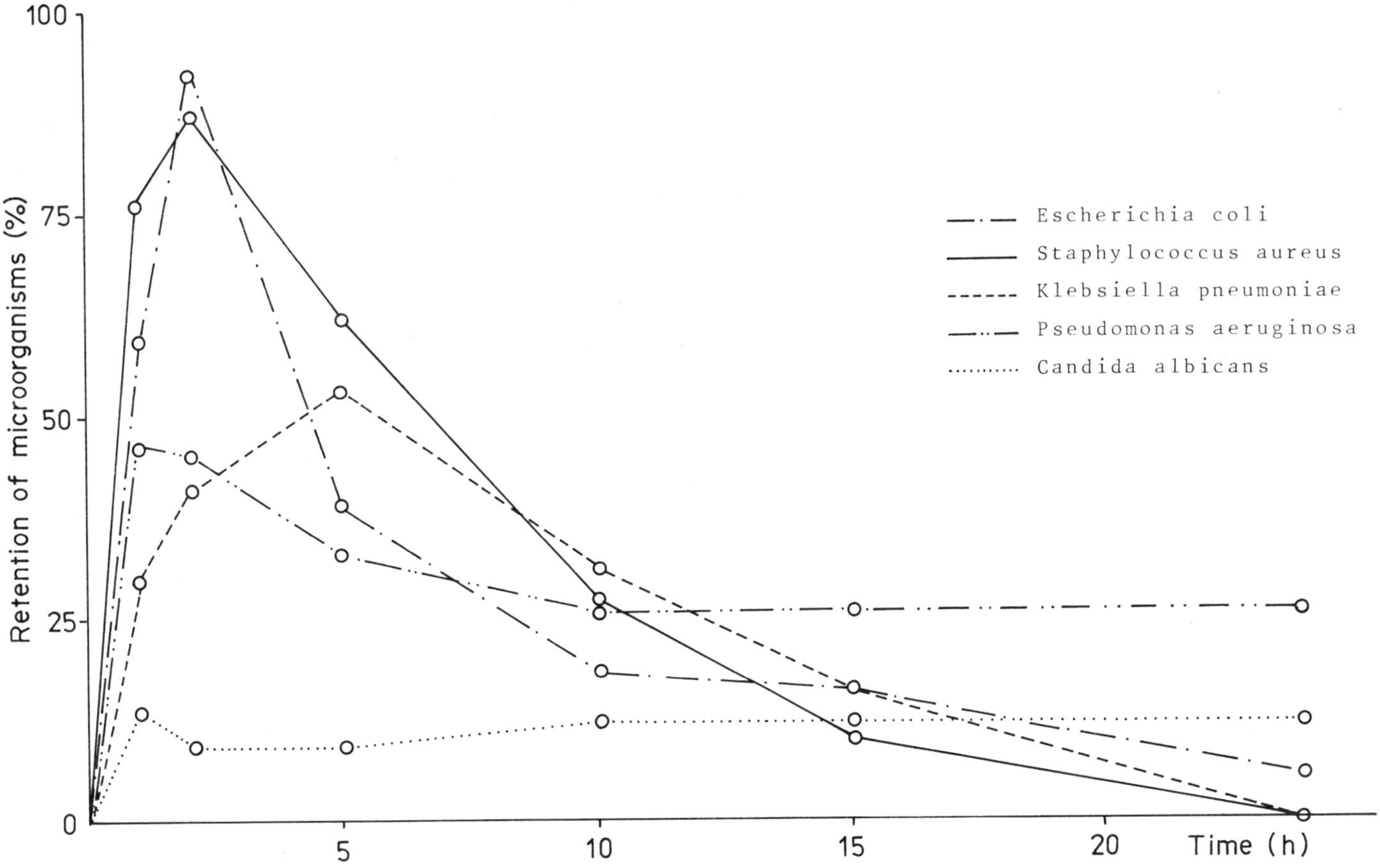

Figure 2. Percentage of retention by *Ephydatia fluviatilis* of five microorganisms.

3d Int. Sponge Conf. 1985

For *E. coli, P. aeruginosa, S. aureus,* and *C. albicans,* the initial retention rate was high during the first two hours of incubation, and then continued at a lower rate for 24 h or more. For *K. pneumoniae,* the maximum rate of retention was only reached after 5 h.

These quantitative results confirm that *Ephydatia fluviatilis* is a selective particle feeder. They led us to investigate further the relationship between the percentage of microorganisms captured and the related phagocytic and digestive capacities of the sponge cells.

ELECTRON MICROSCOPIC STUDIES OF PHAGOCYTOSIS AND DIGESTION

In the present work, electron microscopic observations were made for *Pseudomonas aeruginosa* and *Candida albicans.* They will be compared with the exhaustive study of capture, endocytosis and excocytosis realized with *Escherichia coli* (Willenz and Van de Vyver, 1984; Willenz et al., 1986).

Pseudomonas aeruginosa. These microorganisms entered the sponges following the incurrent canals only. No bacteria were seen to be phagocytosed by the pinacoderm. Forty minutes after the beginning of incubation, bacteria were seen to be phagocytosed by choanocytes and even archeocytes. At that time, one bacterium was found per phagosome, but after 2 h, 10 to 12 bacteria were present in large vesicles. This change suggested multiple fusions of individual phagosomes (Figure 3). Five hours after the bacteria were added, acid phosphatases were detected in the phagosomes (Figure 4). Their presence attested to fusion between lysosomes and phagosomes and provided evidence that the bacteria were digested. At no time during the experiment, were any bacteria—undigested or partially digested—seen in the excurrent canals.

Candida albicans. One hour after the beginning of the experiments, some pinacocytes were found to contain one phagosome with one yeast particle. However, most of the yeasts were captured and carried by the incurrent canals (Figure 5). After being taken up, many *C. albicans* were not phagocytosed by the choanocytes and the archeocytes, but were immediately expelled in the excurrent canals. This is indicated by the fact that after 40 min of incubation, yeast particles were already accumulated in the excurrent canals and that after 5 h, most of them were expelled undigested in the excurrent system (Figure 6). If a few *Candida* were phagocytosed by the choanocytes and the archeocytes, none of them were digested. Indeed, at the end of the experiment (24 h), no acid phosphatases could be detected in the phagosomes containing yeast (Figure 7).

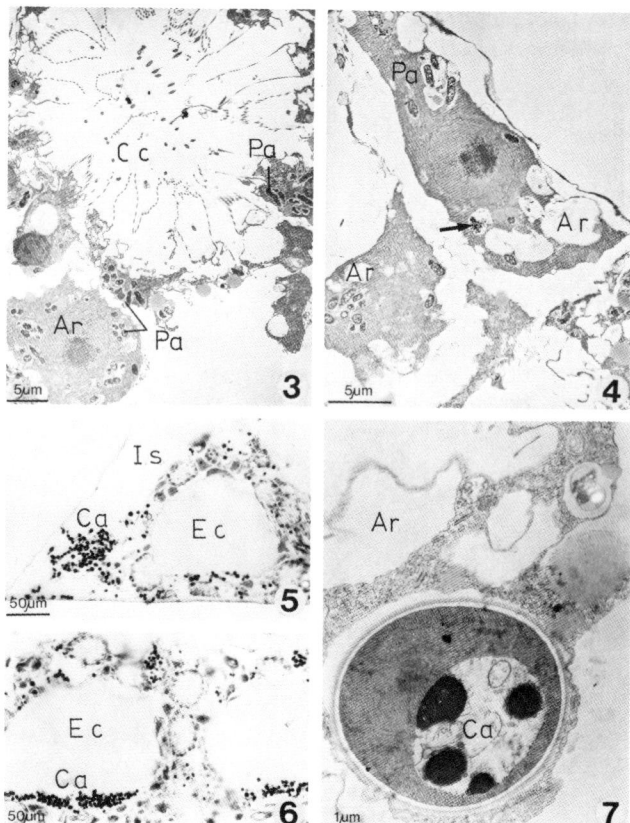

Figures 3–7. *Ephydatia fluviatilis* (1-gml sponges), TEM and light micrographs of tissue after exposure to two microorganisms: *3,* choanocytes and one archeocyte of sponge provided with *Pseudomonas aeruginosa* for 5 h; several bacteria enclosed in large phagosomes; *4,* archeocyte of sponge provided with *Pseudomonas aeruginosa* for 5 h; arrowheads indicate presence of acid phosphatase in phagosomes; *5,* sponge provided with *Candida albicans* for 40 min; most yeast particles are concentrated in incurrent spaces; *6,* sponge provided with *Candida albicans* for 5 h; most yeast particles are expelled in excurrent canals; *7,* archeocyte of *Candida albicans*-exposed sponge containing large phagosome provided with yeast particle; no deposit of acid phosphatase was observed. *Ar,* archeocyte; *Ca, Candida albicans; Cc,* choanocyte chamber; *Ec,* excurrent canal; *Is,* incurrent space; *Pa, Pseudomonas aeruginosa.*

Discussion and Conclusions

The importance of bacteria and other microorganisms in the diet of sponges has been reported many times, but little work has yet been done on the selectivity of capture, phagocytosis, and digestion of these filter feeders. Our quantitative results show obvious selectivity in the efficiency of microorganism uptake by *Ephydatia fluviatilis.* Of the five particles tested, *Escherichia coli* (a gram-negative bacillus) and *Staphylococcus aureus* (a gram-positive coccus) were highly taken up, whereas two other gram-negative bacilli—*Klebsiella pneumoniae* and *Pseudomonas aeruginosa*—were only captured now and then and the yeast *Candida albicans* hardly at all.

Such striking differences are obviously related to many factors. The process by which sponges retain microorganisms can be divided into three steps: (1) the particles are taken up by the incurrent system; (2) they adhere to the sponge cells, followed by phagocytosis; (3) lysosomial digestion takes place. The adherence properties obviously play a key role in the selective retention rate observed for the five microorganisms tested. Adherence depends on the biochemical composition of the cell wall, the antiphagocytic properties of the polysaccharide capsule, the presence or absence of pili and flagellae, and electric charges and hydrophobicity (see review in Willenz and Van de Vyver, 1984). Such characteristics vary not only from species to species, but also from strain to strain. In turn, it is clear that the biochemical composition of the glycocalyx and its negative charges and the types of sponge cell interactions with the microorganisms contribute to the adherence and finally to the retention rate.

The low efficiency of uptake measured for *Pseudomonas aeruginosa* can be explained by the fact that, despite their morphological similarities, *Ephydatia fluviatilis* captures great amounts of *Escherichia coli* by its pinacoderm (Willenz et al., 1986), whereas *P. aeruginosa* enters the sponges following the incurrent canals only. Note, however, that if *P. aeruginosa* are never seen in the pinacocytes, the bacteria carried by the incurrent system are readily phagocytosed and digested by the choanocytes and the archeocytes, as occurs for *E. coli*.

The slightly low uptake of *Klebsiella pneumoniae* and the 3-h delay of its maximal retention rate may be correlated to the mucous capsule that characterizes the strain tested. Indeed, such polysaccharide capsules are known for their antiphagocytic properties.

The very low retention ratio of *Candida albicans* is certainly due to the fact that great amounts of yeast carried by the incurrent canals are rarely phagocytosed by the sponge cells, but are readily expelled from the incurrent spaces to the excurrent canals. Moreover, the few yeast particles phagocytosed by pinacocytes, choanocytes, or archeocytes were never seen to be digested.

The present results demonstrate the selectivity of uptake of five microorganisms and provide a new perspective on the ecological role of sponges in the reduction of microbial contamination. Indeed, while sponges appear to act efficiently against some type of microorganism pollution, they are inefficient with others.

Literature Cited

Francis, J. C. 1984. Reduction Body Formation and Subsequent Regeneration of *Ephydatia fluviatilis* in Laboratory Culture. *Transactions of the American Microscopical Society*, 103:347–352.

Madri, P. P., G. Claus, S. M. Kunen, and E. E. Moss. 1967. Preliminary Studies on the *Escherichia coli* Uptake of the Red-beard Sponge *Microciona prolifera* (Verrill). *Life Science*, 6:889–894.

Madri, P. P., M. Hermel, and G. Claus. 1971. The Microbial Flora of the Sponge *Microciona prolifera* and Its Ecological Implications. *Botanica Marina*, 14:1–5.

Mank, A., and E. Kilian. 1979. The Ingestion and Digestion of Food of the Freshwater Sponge *Spongilla lacustris*. Pages 353–360 *in Biologie des Spongiaires*, edited by C. Lévi and N. Boury-Esnault. Colloques Internationaux du C.N.R.S. 291. Paris: Centre National de la Recherche Scientifique.

Phillips. W. A., M. J. Shelton, and C. S. Hosking. 1979. A Simple Micro-Assay for Neutrophil Bactericidal Activity. *Journal of Immunological Methods*, 26:187–191.

Rasmont, R. 1961. Une technique de culture des éponges d'eau douce en milieu contrôlé. *Annales, Societé Royale Zoologique de Belgique*, 91:147–156.

———. 1975. Freshwater Sponges as a Material for the Study of Cell Differentiation. *Current Topics in Developmental Biology*, 10:141–159.

Reiswig, H. M. 1971. Particle Feeding in Natural Populations of Three Marine Demospongiae. *Biological Bulletin*, 141:568–591.

———. 1975. Bacteria as Food for Temperate-water Marine Sponges. *Canadian Journal of Zoology*, 53:582–589.

Schmidt, I. 1970. Phagocytose et pinocytose chez les Spongillidae. *Zeitschrift der Vergleichenden Physiologie*, 66:398–420.

Spurr, A. R. 1969. A Low Viscosity Epoxy Resin Embedding Medium for Electron Microscopy. *Journal of Ultrastructure Research*, 26:31–43.

Vacelet, J. 1975. Etude en microscopie électronique de l'association entre bactéries et spongiaires du genre *Verongia*. *Journal de Microscopie et de la Biologie Cellulaire*, 23:271–288.

———. 1979. La place des spongiaires dans les systèmes trophiques marins. Pages 259–270 *in Biologie des Spongiaires*, edited by C. Lévi and N. Boury-Esnault. Colloques Internationaux du C.N.R.S. 291. Paris: Centre National de la Recherche Scientifique.

Van de Vyver, G. 1970. La non-confluence intraspécifique chez les spongiaires et la notion d'individu. *Annales d'Embryologie et de Morphogènese*, 3:251–262.

———. 1975. Phenomena of Cellular Recognition in Sponges. *Current Topics in Developmental Biology*, 10:123–140.

Willenz, P., and G. van de Vyver. 1984. Ultrastructural Localization of Lysomial Digestion in the Freshwater Sponge *Ephydatia fluviatilis*. *Journal of Ulstrastructure Research*, 87:13–22.

Willenz, P., G. van de Vyver, and B. Vray. 1986. A Quantitative Study of the Retention of Radioactively Labeled *E. coli* by the Freshwater Sponge *Ephydatia fluviatilis*. *Physiological Zoology*, 59:495–504.

BENOIT VERDENAL
Centre d'Océanologie de Marseille
Station Marine d'Endoume
13007 Marseille, France

CATHERINE DIANA
ANDRE ARNOUX
Laboratoire d'Hydrobiologie et de Molysmologie
Aquatique, Faculté de Pharmacie
13385 Marseille, France

JEAN VACELET
Centre d'Océanologie de Marseille
Station Marine d'Endoume
13007 Marseille, France

Pollutant Levels in Mediterranean Commercial Sponges

Abstract

Several heavy metals and pesticides were measured by atomic absorption (V, Cr, Mn, Fe, Ni, Cu, Zn, Cd, Hg, and Pb) or by chromatography (PCB, DDT and its by-products) in three commercial sponges from more or less polluted areas of the northwest Mediterranean sea: *Spongia officinalis*, *S. agaricina*, and *S. nitens*. Levels of metals were similar in all three species. Iron accumulates heavily on the skeleton in the form of lepidocrocite granules and in polluted stations may represent up to 7.5% of the dry weight. V (max 366 ppm of the dry weight), Pb, Cr, and Zn also accumulate in the skeleton. Mn, Ni, Cd, and Hg are more abundant in the living tissue than on the skeleton. Cu seems to be concentrated equally in the skeleton and in the living tissue. Concentrations of Cu, Mn, and Cd are constant, whatever the pollution level of the station, with high values for Cu (100 ppm) and low values for Cd (1.2 ppm). Pesticides are heavily concentrated (up to 25 ppm for PCB and 3.5 ppm for DDT). They seem to be more concentrated by *S. officinalis* than by the other two species. This fixation occurs mainly in the living tissue; the levels on the skeleton are 50 to 200 times lower. The concentrations of these pollutants on the skeleton offer no toxicity problem for commercial use, even in sponges living in heavily polluted areas, such as those influenced by the sewage of Marseille. However, the market quality may be reduced by the concentration of lepidocrocite granules, which diminishes the mechanical resistance of the skeleton. The possible use of these sponges as biological pollution indicators is emphasized.

With the overfishing of commercial sponges on Mediterranean beds, an effort has been launched to grow these sponges industrially. In order to improve their growth rate, we plan to take advantage of the organic and bacterial enrichment provided by a sewage outlet near Marseille, which seems to favor the development of such filter feeders. However, this type of effluent also carries many chemical pollutants. Therefore, it is essential to learn how sponges respond to such pollutants.

We decided to study the accumulation of some of these

516

elements both in the skeleton and in the living tissue of natural populations of sponges living in diverse conditions of pollution.

More detailed results of this study were given by Verdenal (1986) and, for iron, by Vacelet et al. (1988).

Materials and Methods

The metals analyzed were chosen for either their high toxicity in animals (Hg, Cd, Pb), or their potential toxicity when they occur in great quantities although they may play a metabolic role (Cu, Zn, Mn, Ni Cr, V). Fe was also quantified because it may cause the sponge skeleton to deteriorate. We also searched for organochlorinated compounds (PCB, DDT and its by-products) because they are often mixed with domestic effluents and they may make sponges unsuitable for commercial use.

SPECIES AND STATIONS. The analyses were carried out on three of the most common commercial species on the French Mediterranean coast: *Spongia officinalis* L., *Spongia*

agaricina Pallas, and *Spongia nitens* (Schmidt). The sponges were sampled in 1983 and 1984, that is, before the starting of a large water-treatment plant which has been operating since 1987. Seven sampling stations (Figure 1) were selected for their degree of fertility, as determined by the proximity to the sewer of Marseille and the prevailing currents. The "A" current that heads out to sea, under west and northwest winds, represents 45% of the local system and creates an upward circulation of clean bottom waters at station 3. The "B" current appears under south and southeast winds and represents 30% of the local system. This current is most important with respect to pollution of the zone. The clean reference station is located in the Port-Cros National Park.

EXTRACTION AND QUANTITATIVE ANALYSIS. Immediately after collection, samples were frozen at −20°C. They were then lyophilized, pulverized, and homogenized. Heavy metal analysis was carried out on the digestion product of 1 g of dry sponge powder by 10 ml of nitric acid (Suprapur Merck) at 150°C, during 2 h in a steel-sheathed teflon

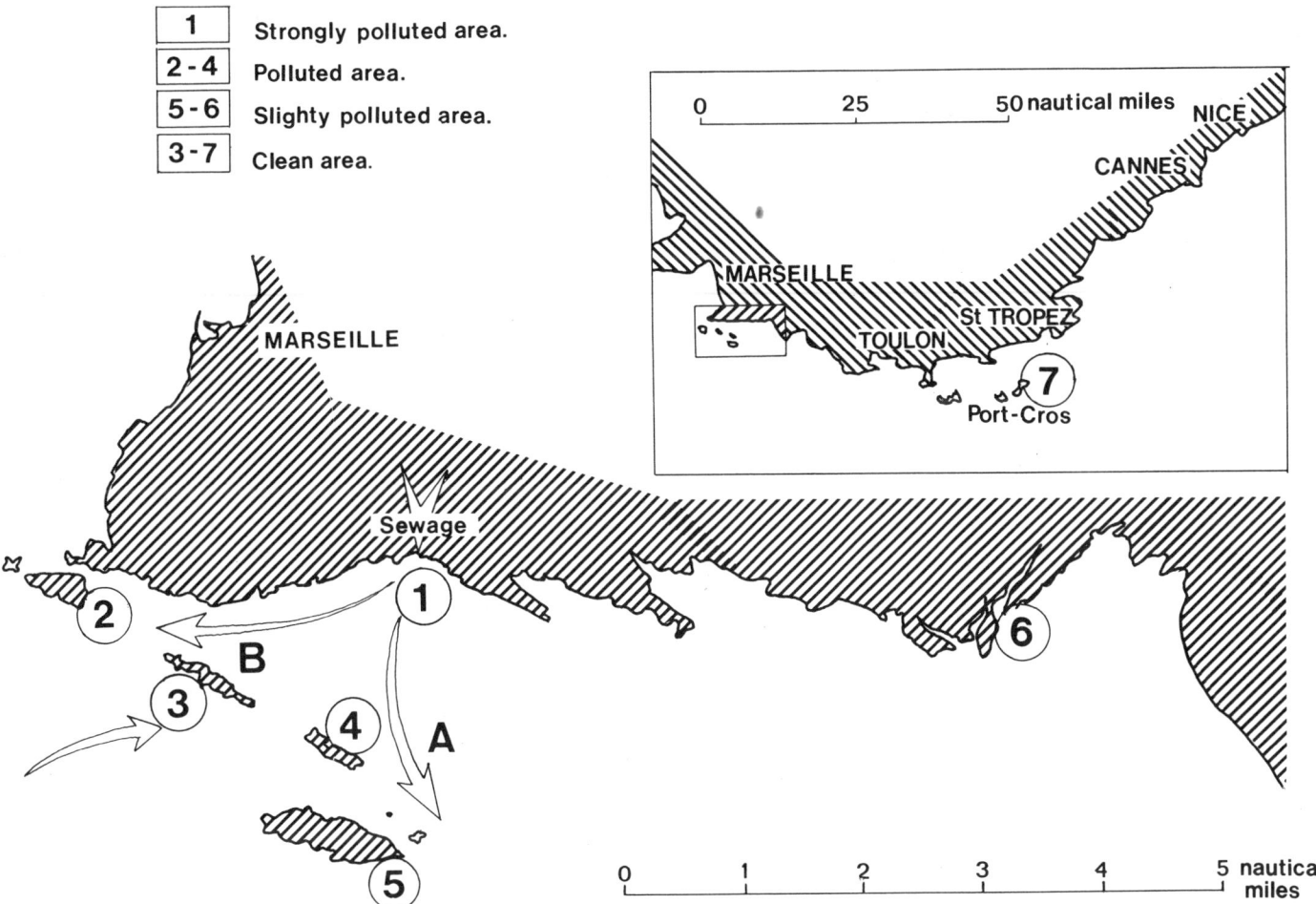

Figure 1. Map of study location in southern France, indicating sampling stations and current flow.

Table 1. Sponge culture station characteristics (for locations see Figure 3)

Station	Distance from sewage outlet	Quality of the water	Other culture conditions	Depth (m)
1a	0.3 km east	Very strongly polluted	One rope in a cave	5–15
1b	0.3 km east	Very strongly polluted	One rope in open water	5–20
2	3.0 km south	Strongly polluted	Three ropes in open water	15–30
3	5.0 km south	Strongly polluted	Two ropes in open water	15–30
4	4.5 km south	Open sea conditions with intermittent supplies from the sewer	The entire rope is below 30 m depth in open water	32–92
5	3.0 km east	Slightly polluted	Four ropes in a cave	15–25
6	Port–Cros National Park	Clean	One rope in open water	15–30

container. Metals were quantified by atomic absorption spectrophotometry (spectrophotometer I.L. 251), in the conditions given in Table 1. Hg was quantified by vapor generation atomic absorption spectrophotometry in the following conditions: reagents, 2 ml $SnCl_2$ (200 g·l^{-1}), 40 ml HCl 1N + 10 ml of the sample solution; wavelength, 253.7 nm; bandwidth, 1 nm; and gas vector, N_2. Fe was quantified by molecular absorption spectrophotometry at 505 nm after combination of Fe^{++} with orthophenantroline at pH 4.8.

The organochlorinated compounds were extracted by Hexane in a Soxhlet apparatus. After purification, the PCB and the derivatives of the DDT family were quantified by gas chromatography (chromatograph Tracor 560) according to the recommendations of Marchand (1983).

The results are given in ppm of dry weight and statistically analyzed through nonparametric techniques. All the series have been compared with the Kruskall-Wallis test (Elliot, 1977).

TRANSMISSION ELECTRON MICROSCOPY (TEM) AND ENERGY DISPERSION SPECTROMETRY (EDS). Pieces of *Spongia officinalis* were fixed in 2.5% glutaraldehyde in seawater and 0.4 M cacodylate buffer 1/1, postfixed in Osmium tetroxide 2% in sea water and embedded in Araldite. Ultrathin sections were cut using a diamond knife and were contrasted with uranyl acetate and lead citrate. EDS analysis was performed in a scanning electron microscope on the sectioned surface of the embedded samples used for TEM studies.

Results and Discussion
METALS

The observed differences in the three sponge species seem to be due only to intrinsic variability. Therefore, the three sponges are thought to accumulate the metals in question in the same way, and we may extend to *Spongia agaricina* and to *S. nitens* the results for *S. officinalis,* which are far more detailed.

Cu, Mn, and Cd are accumulated by *Spongia officinalis* at the same rate at every location, with high values for Cu and Mn and low ones for Cd, relative to the rates documented for other filter feeders of the same zone, such as *Mytilus* (Table 2). Means (with confidence limit for $\alpha <$ 5%) were 85 ± 5 ppm for Cu, 53 ± 8 ppm for Mn, and 1.24 ± 0.22 ppm for Cd (Figure 2) in *Spongia officinalis*. Other heavy metal accumulation in *S. officinalis* varied with the location in which the sponge was growing (Kruskall-Wallis test, null hypothesis rejected for $\alpha = 0.5\%$; Figures 3, 4).

A cluster analysis of the metal content of the sponges by stations provides a classification that agrees quite well with the theoretical classification of the pollution level of the stations (Figure 5). The "metallic charge" of the sponges clearly depends on the metal content of its environment, except for Cu, Mn, and Cd. However, we must note that at the cleanest stations (stations 3 and 6) sponges have the lowest accumulation of heavy metals, but the highest Ni and Cr content (Figure 3). This phenomenon has not yet been explained.

Analyses of similar samples of cleaned skeletons confirm these results and invalidate the hypothesis of a preferential accumulation by the macroinhabitants of the sponge. In order to compare the accumulation of the metals in the entire sponge and in the skeleton, metal concentrations were determined before and after the skeleton of the same specimen was cleaned.

Cu was found in the same concentration both in the skeleton and in the living tissue. Ni, Cd, Hg, Mn, had accumulated in the living tissue, while the other metals—Fe, Cr, Zn, Pb, V—were fixed in the skeleton (K.W. test, null hypothesis rejected for $\alpha < 0.5\%$).

Among the metals fixed on the skeleton, Fe can reach up to 7.5% of the skeletal dry weight (Robin, 1935), giving the skeleton a dark rust color. More detailed results on this metal have been published elsewhere (Vacelet et al., 1988). EDS analysis shows that Fe is situated essentially around the fibers (Figure 6c). In transmission electron microscopy, these granules appear as clusters of thin crystals, sometimes covered with a layer of spongin, but set out

Table 2. Sponge growth rates at the culture stations for *Spongia officinalis* and *S. agaracina* (in parentheses)(– = not determined)

Station	Time (days)	Number of cuttings	Death rate (%)	Growth rate (%)				
				Minimum	Maximum	Mean	Standard deviation	Coefficient of variation
1a	647	30 (10)	14 (0)	–25 (27)	538 (134)	77 (67)	103 (45)	133 (67)
1b	764	30	87	–39	20	0	25	–
2	825	90 (20)	64 (0)	–23.5 (42)	195 (452)	87 (203)	66 (124)	76 (61)
3	538 (20)	60 (0)	14 (37)	–2 (247)	232 (119)	66 (58)	43 (49)	65
4	375	120	25	–45	83	7.7	27	350
5	530	120 (10)	11 (0)	–7.5 (–1)	277 (100)	59 (45)	58 (42)	98 (93)
6	873	30	46	–37	1252	360	357	99

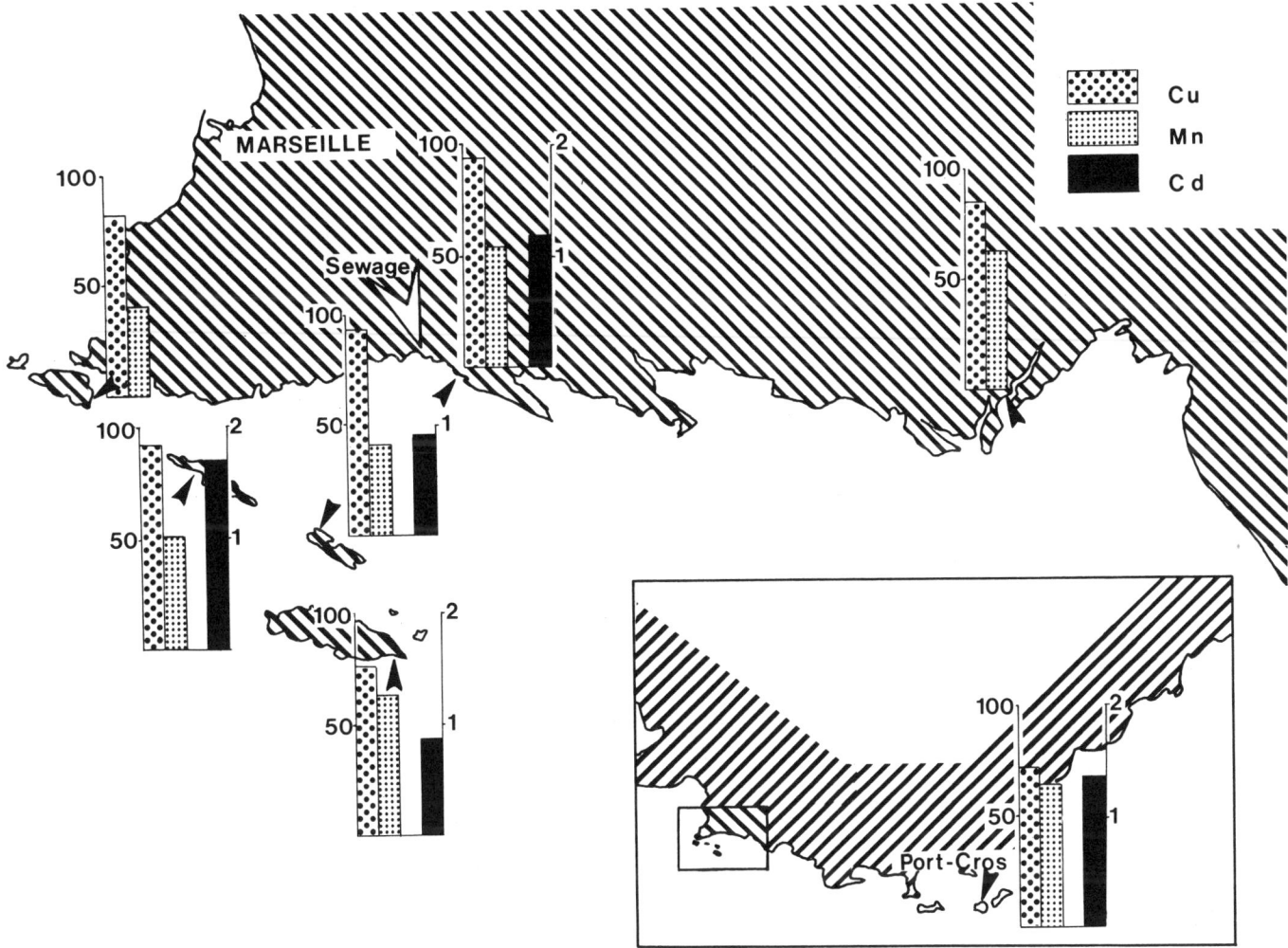

Figure 2. Distribution of copper, manganese, and cadmium in *Spongia officinalis* (in ppm dry weight) in study area.

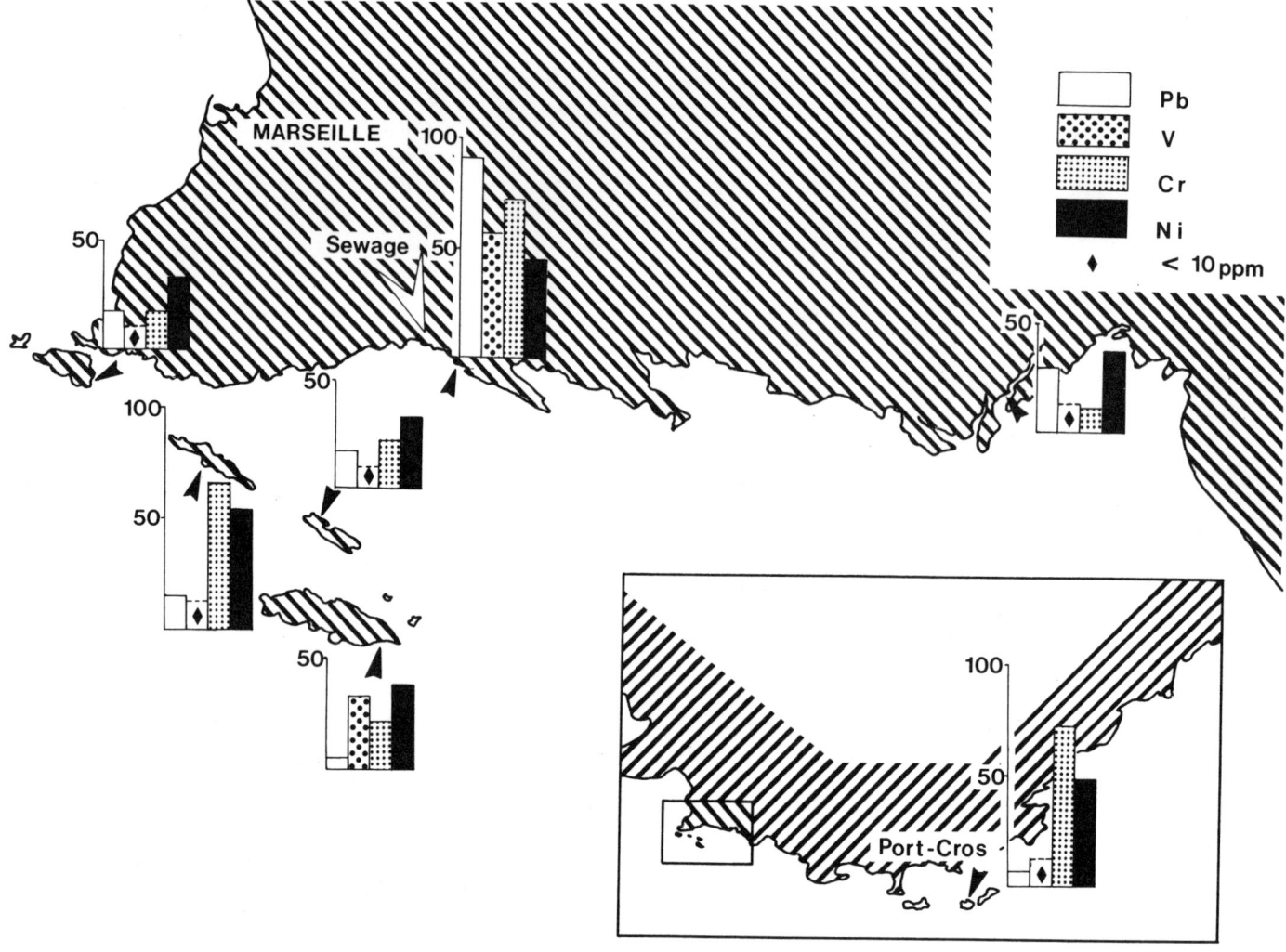

Figure 3. Distribution of lead, vanadium, chromium, and nickel in *Spongia officinalis* (in ppm dry weight) in study area.

in concentric circles along the discontinuities of the outer spongin layers (Figure 6a,b).

Through X-ray diffraction, Towe and Rützler (1968) have shown that the precipitate is a hydrated ferric oxyd called lepidocrocite. Under the physiological pH and redox conditions, these crystals are extremely insoluble; one can observe phantom matrices of Fe in the living tissue after some fibers have disappeared (Figure 6d,e). This confirms that the sponge may reshape its skeleton through digestion of the fibers in the living tissue although sponge collagen is highly resistant to commercial collagenase (Garrone, 1978).

The Pb, Cr, Zn, and V that are also found in great quantities in the skeleton are possibly coprecipitated with Fe and trapped in the crystals.

Among the highly toxic metals, Pb is precipitated in an insoluble form; Cd is kept at a modest accumulation rate. Only Hg, which is accumulated in the living tissue, might be toxic for the sponge, although it never reaches high concentrations, with a maximum of 0.52 ppm in the sta-

tion nearest to the sewer outlet. In the same area, *Mytilus* sp. contains twice as much Hg (Arnoux et al., 1980). The constant uptake rate of Cu and Mn is perhaps an indication of their role in sponge physiology.

Among the accumulated metals that are very likely coprecipitated with Fe on the surface of fibers, V is particularly interesting. Its accumulation reaches a maximum of 366 ppm on the skeleton, which far surpasses the values commonly quoted for the animal kingdom (0.2–2.0 ppm). The V concentrations of the *Spongia* skeleton do not reach the extremes found in the blood cells of some ascidians (6500 ppm) but we can no longer consider these ascidians to be the only concentrative agents of the animal kingdom (Monniot, 1978). The physiological role of V is still unknown.

ORGANOCHLORINATED BIOCIDES

Before we could analyze the organochlorinated biocides, we had to pool the powder of several sponges in order to

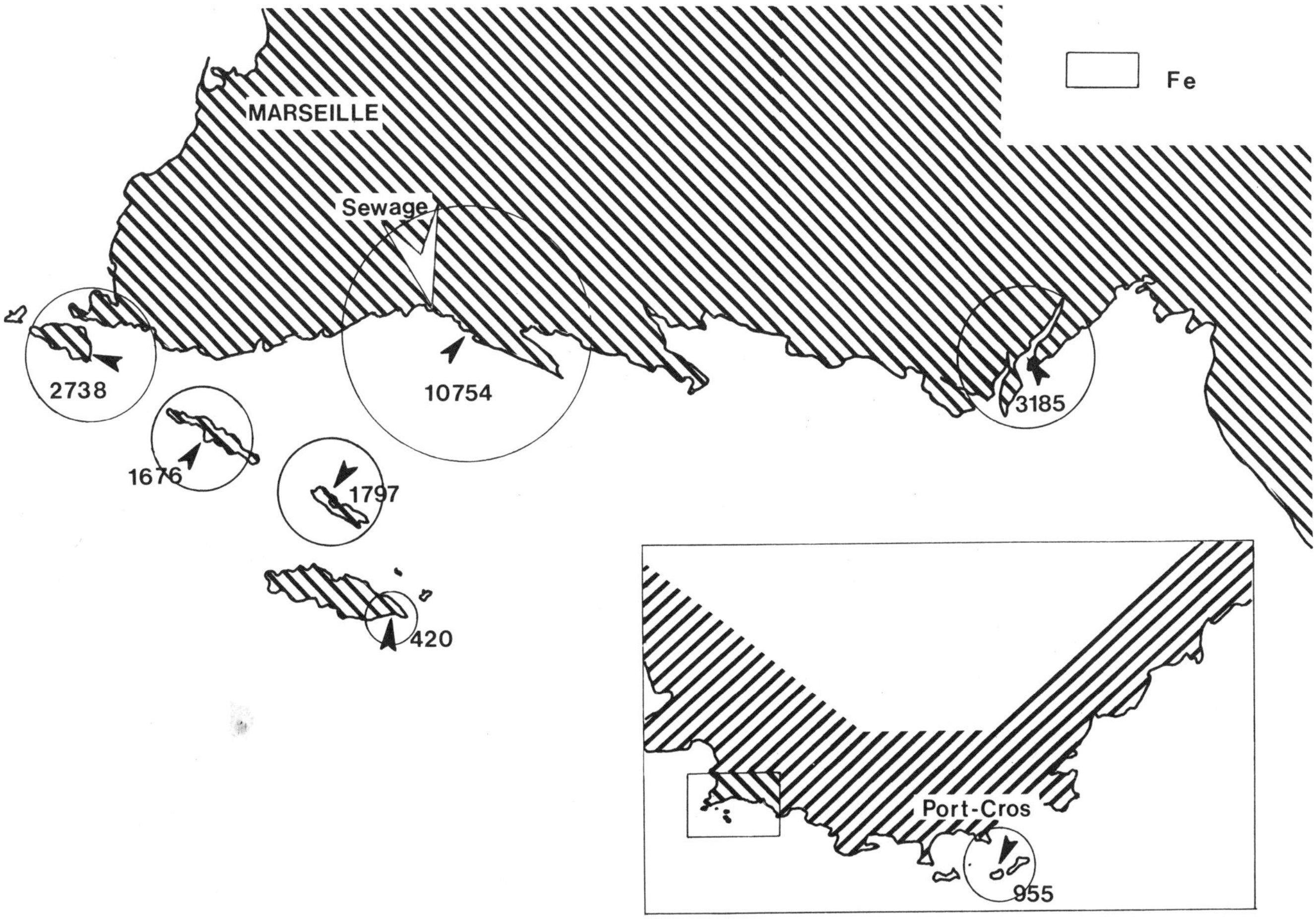

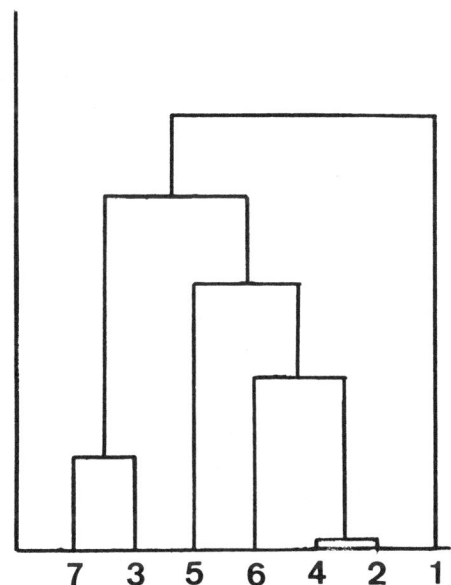

Figure 4. Distribution of iron in *Spongia officinalis* (in ppm dry weight) in study area. Areas of disks are proportional to quantity of iron.

obtain an adequate supply of material. Consequently, the results cannot be statistically treated. Nevertheless the data are of some interest, as long as the possibility of strong intraspecific or intrastation variability is kept in mind.

Spongia officinalis seems to accumulate the organochlorinated biocides much more efficiently than the other two species (Figure 7). The sponges at Cortiou contain more PCB than the ones at the other stations (Figure 8). The PCB is strongly accumulated in polluted areas. The great disproportion of its distribution shows the massive outflow of PCB through the sewer. Note that the level of this contamination remains significant despite the legal measures that have been taken to stop this type of pollution.

This is less evident for DDT and its metabolites. Their distribution in the littoral waters is more homogeneous than that of PCB, and they are accumulated by *Spongia* only under special conditions. DDT is not an exclusive industrial pollutant, but also appears in agricultural zones. On the other hand, analysis of DDT content in organisms 20 years ago showed a heterogeneity and a preferential supply by the sewer of Marseille that was as

Figure 5. Results of cluster analysis of stations according to metal content of sponges (barycenter criterium).

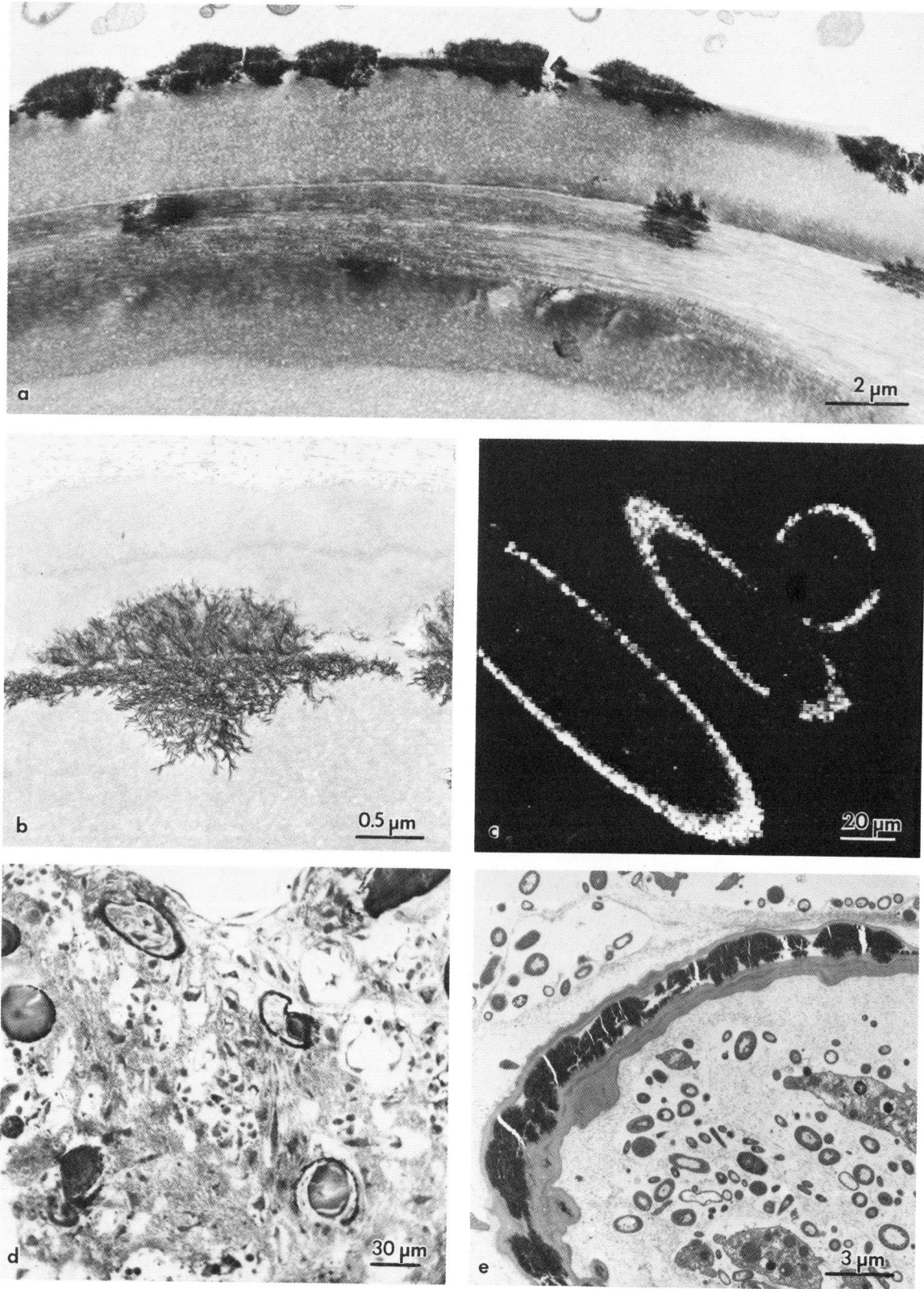

Figure 6. Iron in *Spongia officinalis:* *a,* TEM picture of iron granules on discontinuities of fiber; *b,* TEM picture of cluster of lepidocrocite crystals; *c,* distribution map showing that iron granules are exclusively situated on periphery of fibers; *d,* fibers with iron matrix and one iron phantom matrix (semithin section); *e,* TEM picture of an iron phantom matrix.

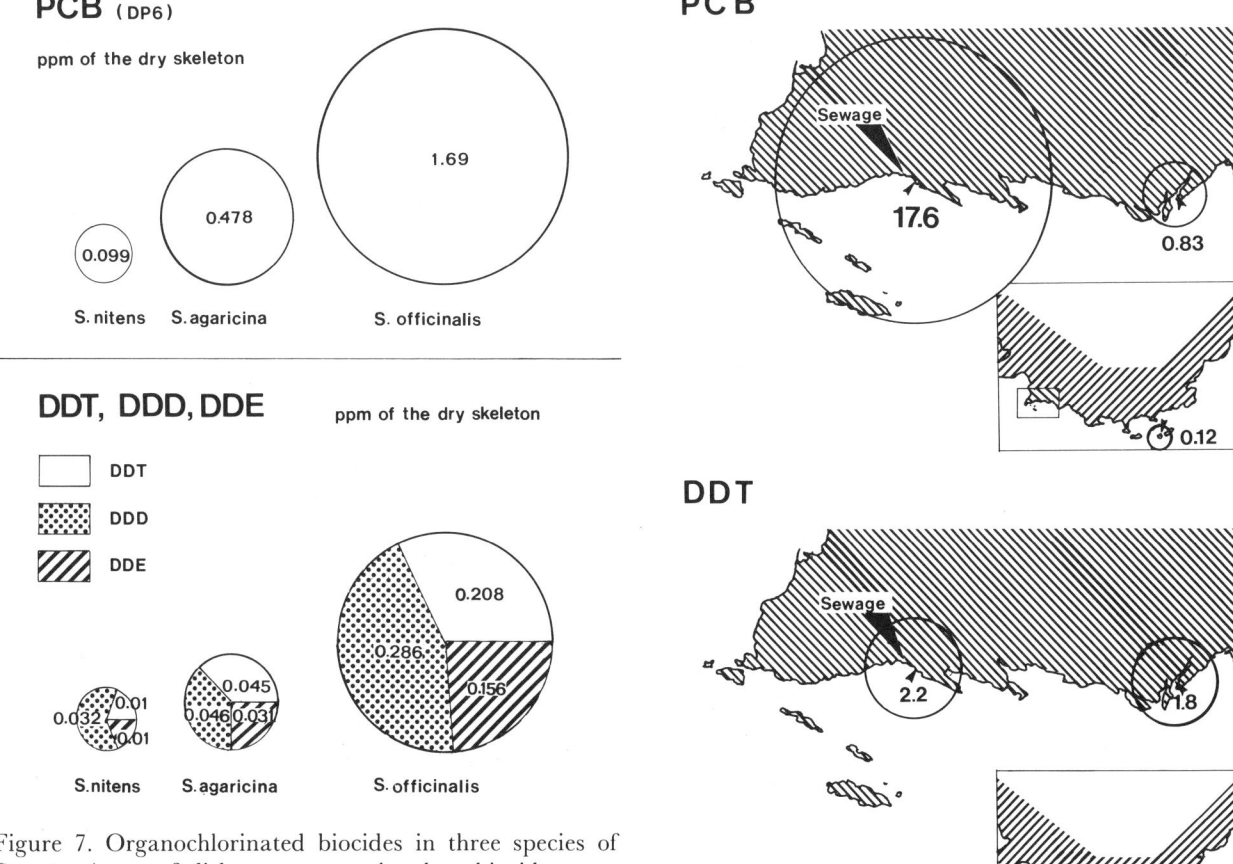

Figure 7. Organochlorinated biocides in three species of *Spongia*. Areas of disks are proportional to biocides concentrations.

Figure 8. Organochlorinated biocides in *Spongia officinalis* from different localities. Areas of disks are proportional to biocide concentration.

spectacular as the situation is today for PCB. This clearly demonstrates the efficiency of years of well-observed prohibition concerning the use of DDT.

The concentration ratios of the entire sponge to the skeleton for PCB and DDT are 50 and 250 respectively. The organochlorinated biocides are accumulated in the living tissue of the sponge and their fixation by the skeleton represents only 2% to 0.4% of the total concentration. That seems to be logical because these hydrophobic and lipophile molecules tend to accumulate in the fat of living beings. The sponges of the *Spongia* genus contain up to 30% of bacteria in volume (Vacelet and Donadey, 1977). These bacteria are gram negative and often have very thick and very lipid-rich walls. It is likely that those bacteria walls contribute significantly to the accumulation of these organochlorinated compounds.

Most of these compounds are eliminated during commercial preparation. These sponges are very heavy concentrators of organochlorinated biocides, especially *Spongia officinalis*, in which we find 30 times more PCB and 100 times more DDT than in *Mytilus sp.* samples from the same area (Arnoux et al., 1980). We do not know what the

consequences may be on the physiology of the sponges, but this seems to inhibit the regeneration and the growth of the cuttings.

Conclusions

From a commercial point of view, pollutant concentrations do not create a toxicity problem that would prohibit the use of the skeleton. However, the market quality of sponges may be reduced somewhat because of the great quantity of lepidocrocite granules, which seem to affect the mechanical properties of the skeleton.

For a long time mussels have been considered a good biological pollution indicator. Sponges of the genus *Spongia* also accumulate metals, such as Fe, Pb, V, Zn, Cr, Ni, Hg, as well as organochlorinated compounds. Metal uptake has been reported in some tropical demosponges and may be quite high for some metals: for example, Ni in *Dysidea crawshayi* (= *Ulosa ruetzleri*) (Bowen, 1951) and in *Spirastrella cuspidifera* (Patel et al., 1985). However, it is difficult to use these sponges as indicators of pollution because we cannot estimate their age. *Spongia* species, for example, live longer than 10 years and accumulate the pollutants during this period of time. Furthermore, the modalities and the kinetic and physiological role of these accumulations remain unknown.

It would be interesting to examine the pollutants in mussels and in sponges in order to compare the short term effects of the accumulation in the mussels with the long-term effects in sponges. Sponges could be used to detect even small amounts of lasting pollution. Before they could be used in this way, however, it would be necessary to establish small sponge cultures in the zones influenced by pollution, so that sponges can be sampled and analysed from areas in which age and initial pollutant levels were known.

Literature Cited

Arnoux, A., J. Tatossian, J., L. Monod, and A. Blanc. 1980. Etude des teneurs en métaux lourds et composés organochlorés dans les organismes marins prélevés dans le secteur de Cortiou (Marseille). Pages 471–482 in *V^emes Journées d'Etudes sur les Pollutions marines en Méditerranée*. Cagliari: Commission internationale pour l'exploitation scientifique de la Méditerranée.

Bowen, V. T. 1951. Comparative Studies of Mineral Constituents of Marine Sponges. *Journal of Marine Research*, 10:153–167.

Elliott, J. M. 1971. *Some Methods for the Statistical Analysis of Samples of Benthic Invertebrates*. Ambleside, United Kingdom: Freshwater Biological Association Scientific Publication 25. 144 pp.

Garrone, R. 1978. *Phylogenesis of Connective Tissue, Morphological Aspects and Biosynthesis of Sponge Intercellular Matrix*. Basel: S. Karger. 250 pp.

Marchand, M. 1983. Dosages des pesticides et polychlorobipheniles dans l'eau, les sédiments et les organismes marins par chromatographie gazeuse. Pages 367–380 in *Manuel des analyses chimiques en milieu marin*, edited by A. Aminot and M. Chaussepied. Brest: Centre National de l'Exploitation des Océans.

Monniot, F. 1978. Connaissances actuelles sur les ions métalliques chez les ascidies. Pages 185–194 in *Les métaux en milieu marin, phosphore et dérivés phosphorés*, 2. Paris: Centre National de la Recherche Scientifique.

Patel, B., M. C. Balani, and S. Patel. 1985. Sponge "Sentinel" of Heavy Metals. *Sciences of the Total Environment*, 41:143–152.

Robin, M. P. 1935. Sur la présence du fer dans les éponges communes. *Journal de Pharmacie et de Chimie (serie 8)*, 21:600–604.

Towe, K. M., and K. Rützler. 1968. Lepidocrocite Iron Mineralization in Keratose Sponge Granules. *Science*, 162:268–269.

Vacelet, J., and C. Donadey. 1977. Electron Microscope Study of the Association between some Sponges and Bacteria. *Journal of Experimental Marine Biology and Ecology*, 30:301–314.

Vacelet, J., B. Verdenal, and G. Perinet. 1988. The Iron Mineralization of *Spongia officinalis* L. (Porifera, Dictyoceratida) and Its Relationships with the Collagen Skeleton. *Biology of the Cell*, 62:189–198.

Verdenal, B. 1986. Spongiculture en Méditerranée Nord-occidentale; aspects cultural, molysmologique et économique. Doctoral Dissertation, Université Aix-Marseille. 163 pp.

Participants

Conference Participants

NOTE: Names and affiliations are those valid at the time of the conference, November 1985.

Pedro Manuel Alcolado, Instituto de Oceanologia, Cuidad de la Habana, Cuba [represented by K. Rützler]

Belinda Alvarez, Fundación Científica de los Roques, Caracas, Venezuela

William Austin, Khoyatan Marine Labortory, Cowichan Bay, British Columbia, Canada

Gerald Bakus, Allan Hancock Foundation, University of Southern California, Los Angeles, California, USA

Christopher Battershill, The University of Auckland, Marine Laboratory, Leigh, New Zealand

Patricia Bergquist, Zoology Department, University of Auckland, Auckland, New Zealand

Charles H. Bigger, Department of Biological Sciences, Florida International University, Miami, Florida, USA

John W. Bisbee, Department of Biology, Western Carolina University, Cullowhee, North Carolina, USA

Calhoun Bond, Jr., Department of Biology, University of North Carolina, Chapel Hill, North Carolina, USA

Nicole Boury-Esnault, Station Marine d'Endoume, Marseille, France

Max M. Burger, Department of Biochemistry, Biocenter of the University of Basel, Basel, Switzerland

Bruno Burlando, Istituto di Zoologia, Universitá di Genova, Genova, Italy

Rodolfo A. Caberoy, National Museum of the Philippines, Manila, Philippines

Adam S. G. Curtis, Department of Cellular Biology, University of Glasgow, Glasgow, Scotland, United Kingdom

Louis De Vos, Laboratoire de Biologie Animale et Cellulaire, Université Libre de Bruxelles, Bruxelles, Belgium

Ruth Desqueyroux-Faundez, Muséum d'Histoire Naturelle, Genève, Switzerland

Maria Cristina Diaz, Fundación Científica de los Roques, Caracas, Venezuela

Claude Donadey, Station Marine d'Endoume, Marseille, France

David W. Elvin, Shelburne, Vermont, USA

D. John Faulkner, Scripps Institute of Oceanography, La Jolla, California, USA

Paul Fell, Department of Zoology, Connecticut College, New London, Connecticut, USA

Robert Finks, Department of Geology, Queens College, Flushing, New York, USA

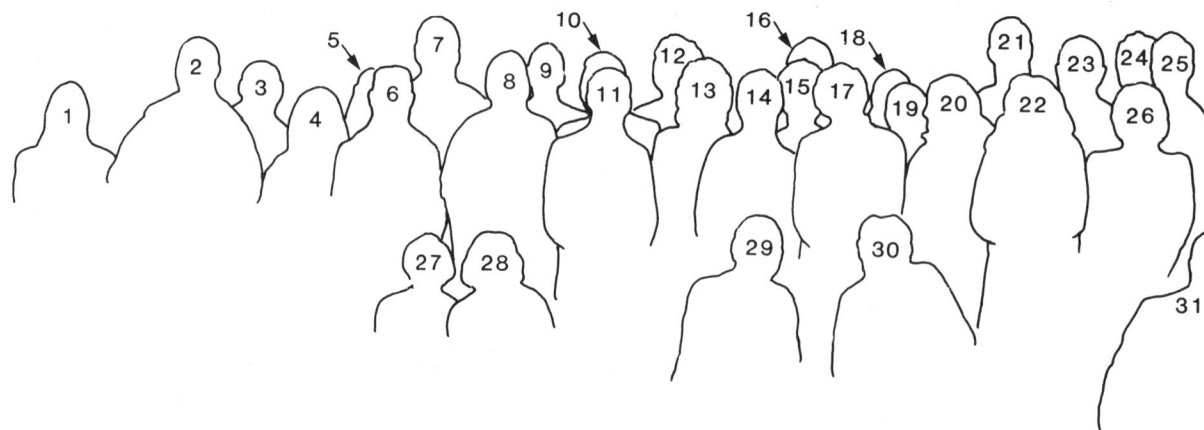

1, Janie Wulff; **2**, John Stevely; **3**, Charles Bigger; **4**, Carolyn Teragawa; **5**, Mishelle Lawson; **6**, Ma Yun Tao; **7**, Felix Wiedenmayer; **8**, Heather Kaye; **9**, Peter Wainwright; **10**, Norbert Weissenfels; **11**, Clifford Jones; **12**, Paul-Friedrich Langenbruch; **13**, Valerie Jones; **14**, Keiko Tanaka-Ichihara; **15**, Cecilia Volkmer-Ribeiro; **16**, Antonio Solé-Cava; **17**, Patricia Bergquist; **18**, Eric Horgan; **19**, María Uriz; **20**, Anna Traveset; **21**, Paul Fell; **22**, Jane Fromont; **23**, John Francis; **24**, John Hooper; **25**, Wilfred Hoppe; **26**, Shirley Pomponi; **27**, Chung Ja Sim; **28**, Yoko Watanabe; **29**, Christopher Battershill; **30**, Theo van Kempen; **31**, Clive Wilkinson.

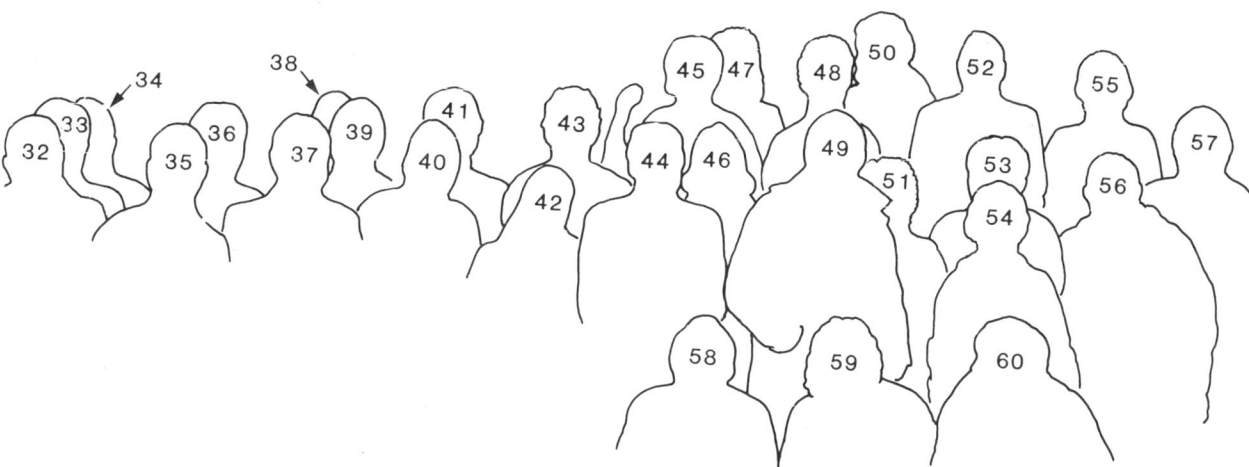

32, Tracy Simpson; **33**, John Faulkner; **34**, Jean Vacelet; **35**, Hazime Mizoguchi; **36**, Bruno Burlando; **37**, Claude Donadey; **38**, Rob van Soest; **39**, Robert Garrone; **40**, Joachim Reitner; **41**, David Elvin; **42**, Rachel Wood; **43**, Calhoun Bond; **44**, Robert Finks; **45**, Nicole Boury-Esnault; **46**, Sarah Klontz; **47**, Philippe Willenz; **48**, Gysèle Van de Vyver; **49**, Frederick Harrison; **50**, Louis De Vos; **51**, Ruth Desqueyroux-Faundez; **52**, Welton Lee; **53**, Benoît Verdenal; **54**, Nancy Kaye; **55**, Ruth Williams; **56**, Gordon Kaye; **57**, William Austin; **58**, Vance Vicente; **59**, Sven Zea; **60**, Joseph Schubauer.

TOP ROW, LEFT TO RIGHT: Willard Hartman, Yoko Watanabe, Chung Ja Sim; Yoshiki Masuda, Carolyn Teragawa; Tracy Simpson.
CENTER ROW, LEFT TO RIGHT: Rodolfo Caberoy, Tom Humphreys; Keiko Tanaka-Ichihara, Michele Sará.
BOTTOM ROW, LEFT TO RIGHT: Klaus Rützler; Claude Donadey, Benoît Verdenal; Maurizio Pansini.

TOP ROW, LEFT TO RIGHT: Ma Yun Tao, Nancy Kaye, Frederick Harrison, Gordon Kaye, Anna Traveset; Yoko Watanabe, Yoshiki Kurahashi, Cecilia Wolkmer-Ribeiro, Hazime Mizoguchi, Takaharu Hoshino.
CENTER ROW, LEFT TO RIGHT: Louis De Vos, Gysèle Van de Vyver; Jean Vacelet, Robert Garrone.
BOTTOM ROW, LEFT TO RIGHT: Kate Smith with symposium proceedings manuscript; Bruno Burlando; Patricia Bergquist.

John Francis, Department of Biological Sciences, University of New Orleans, New Orleans, Louisiana, USA

Jane Fromont, Department of Zoology, University of Auckland, Auckland, New Zealand

Thomas M. Frost, Center for Limnology, University of Wisconsin, Madison, Wisconsin, USA

Patricia D. Fry, Department of Science, Luton College of Higher Education, Luton, Bedfordshire, England, United Kingdom

Marie-France Gallissian, Station Marine d'Endoume, Marseille, France

Robert Garrone, Laboratoire d'Histologie et de Biologie Tissulaire, Université Claude-Bernard, Villeurbanne, France

Frederick W. Harrison, Department of Biology, Western Carolina University, Cullowhee, North Carolina, USA

Willard D. Hartman, Peabody Museum of Natural History, Yale University, New Haven, Connecticut, USA

Sandrino Holvoet, Laboratoire de Biologie Animale et Cellulaire, Université Libre de Bruxelles, Bruxelles, Belgium

John Hooper, Division of Natural Science, Northern Territory Museum of Arts and Sciences, Darwin, Australia

Wilfred F. Hoppe, Carmabi Institute, Willemstad, Curaçao

Takaharu Hoshino [deceased], Mukaishima Marine Biological Station, Hiroshima University, Hiroshima, Japan

Tom Humphreys, University of Hawaii, Kewalo Marine Laboratory, Honolulu, Hawaii, USA

Ian S. Johnston, Department of Biology, Northwestern College, Orange City, Iowa, USA

W. Clifford Jones, School of Animal Biology, University College of North Wales, Bangor, United Kingdom

Peter Karuso, Department of Chemistry, University of Hawaii at Manoa, Hawaii, USA

Gordon Kaye, Department of Anatomy, Albany Medical School, Albany, New York, USA

Heather Kaye, Redpath Museum, McGill University, Montreal, Quebec, Canada

Nancy Kaye, Department of Anatomy, Albany Medical School, Albany, New York, USA

Theo van Kempen, Geological Institute, University of Amsterdam, Amsterdam, The Netherlands

Sarah Ward Klontz, Department of Invertebrate Zoology and Geology, California Academy of Sciences, San Francisco, California, USA

Paul-Friedrich Langenbruch, Zoologisches Institut der Universität Bonn, Bonn, Federal Republic of Germany

Mishelle P. Lawson, Stanford University, Stanford, California, USA

Welton Lee, Department of Invertebrate Zoology, California Academy of Sciences, San Francisco, California, USA

Yoshiki Masuda, Department of Biology, Kawasaki Medical School, Kurashiki, Japan

Henry Mautner, Department of Biochemistry, Tufts University School of Medicine, Boston, Massachusetts, USA

Anne Meylan, Department of Herpetology, American Museum of Natural History, New York, New York, USA

Gradimir N. Misevic, Department of Biochemistry, Biocenter of the University of Basel, Basel, Switzerland

Hazime Mizoguchi, Division of Biology, Junior College of Rissho University, Kumagaya City, Saitama, Japan

Maurizio Pansini, Istituto di Zoologia dell'Università di Genova, Genova, Italy

Sheila M. Pauls, Museu de Ciencias Naturais, Porto Alegre, Rio Grande do Sul, Brazil

Michael Poirrier, Department of Biological Science, University of New Orleans, New Orleans, Louisiana, USA

Shirley Pomponi, State College, Pennsylvania, USA

Henry Reiswig, Redpath Museum, McGill University, Montreal, Quebec, Canada

Joachim Reitner, Institute of Paleontology, Berlin, Federal Republic of Germany

J. Keith Rigby, Department of Geology, Brigham Young University, Provo, Utah, USA

Manuel Rubió, Instituto de Investigaciones Pesqueras di Barcelona, Blanes, Gerona, Spain

Klaus Rützler, Department of Invertebrate Zoology, National Museum of Natural History, Smithsonian Institution, Washington, D.C., USA

Michele Sarg, Istituto di Zoologia dell'Universitg di Genova, Genova, Italy

Paul Scheuer, Department of Chemistry, University of Hawaii at Manoa, Honolulu, Hawaii, USA

George Schmahl, Summerland Key, Florida, USA

Joseph Schubauer, Institute of Ecology, The University of Georgia, Athens, Georgia, USA

Chung Ja Sim, Department of Biology, Han Nam University, Daejeon, Korea

Tracy Simpson, Department of Biology and Health Sciences, University of Hartford, West Hartford, Connecticut, USA

L. Courtney Smith, Division of Biology, California Institute of Technology, Pasadensa, California, USA

Kathleen Smith, Department of Invertebrate Zoology, National Museum of Natural History, Smithsonian Institution, Washington, D.C., USA

Rob W. M. van Soest, Institute of Taxonomic Zoology, University of Amsterdam, Amsterdam, The Netherlands

Antonio M. Solé-Cava, Marine Biological Station, Port Erin, Isle of Man, United Kingdom

John Steveley, Bradenton, Florida, USA

Shirley Stone, Department of Zoology, British Museum (Natural History), London, England, United Kingdom

Keiko Tanaka, Shimoda Kita High School, Shimoda, Shizuoka 415, Japan

Carolyn Teragawa, Department of Molecular Genetics and Cell Biology, University of Chicago, Chicago, Illinois, USA

Janice Thompson, Department of Chemistry, Stanford University, Stanford, California, USA

Anna Traveset, Department of Biology, University of Pennsylvania, Philadelphia, Pennsylvania, USA

María-Jesús Uriz, Instituto de Investigaciones Pesqueras di Barcelona, Blanes, Gerona, Spain

Jean Vacelet, Station Marine d'Endoume, Marseille, France

Gysèle Van de Vyver, Laboratoire de Biologie Animale et Cellulaire, Université Libre de Bruxelles, Bruxelles, Belgium

Benoît Verdenal, Station Marine d'Endoume, Marseille, France

Vance Vicente, Department of Marine Science, University of Puerto Rico, Center for Energy and Environmental Research, Mayaguez, Puerto Rico, USA

Cecilia Volkmer-Ribeiro, Museu de Ciencias Naturais, Porto Alegre, Rio Grande do Sul, Brazil

Yoko Watanabe, Department of Biology, Ochanomizu University, Tokyo, Japan

Norbert Weissenfels, Zoologisches Institut der Universität Bonn, Bonn, Federal Republic of Germany

Felix Wiedenmayer, Frauenfeld, Switzerland

Clive Wilkinson, Australian Institute of Marine Science, Townsville, Queensland, Australia

Philippe Willenz, Peabody Museum of Natural History, Yale University, New Haven, Connecticut, USA

Ruth Williams, University of Wisconsin, Milwaukee, Wisconsin, USA

Jon D. Witman, Zoology Department, University of New Hampshire, Durham, New Hampshire, USA

Rachel Wood, Department of Earth Sciences, The Open University, Bucks, England, United Kingdom

Janie Wulff, Bingham Laboratories, Yale University, New Haven, Connecticut, USA

Sven Zea, Division of Biological Science, The University of Texas at Austin, Texas, USA